W9-DHH-566

ARTIFICIAL LIFE II

ARTIFICIAL LIFE II

PROCEEDINGS OF THE WORKSHOP ON ARTIFICIAL LIFE HELD FEBRUARY, 1990 IN SANTA FE, NEW MEXICO

Editors

Christopher G. Langton
Los Alamos National Laboratory

Charles Taylor
University of California

J. Doyne Farmer
Los Alamos National Laboratory

Steen Rasmussen
Los Alamos National Laboratory

Proceedings Volume X

SANTA FE INSTITUTE
STUDIES IN THE SCIENCES OF COMPLEXITY

Addison-Wesley Publishing Company
The Advanced Book Program
Redwood City, California • Menlo Park, California • Reading, Massachusetts
New York • Don Mills, Ontario • Wokingham, United Kingdom • Amsterdam
Bonn • Sydney • Singapore • Tokyo • Madrid • San Juan

Publisher: *Allan M. Wylde*
Marketing Manager: *Laura Likely*
Production Manager: *Pam Suwinsky*

Director of Publications, *Santa Fe Institute: Ronda K. Butler-Villa*
Technical Assistant, Santa Fe Institute: *Della L. Ulibarri*

This volume was typeset using TEXtures on a Macintosh II computer. Camera-ready output from an Apple LaserWriter Plus Printer.

Library of Congress Cataloging-in-Publication Data

Artificial life II: the proceedings of an interdisciplinary workshop on
the synthesis and simulation of living systems held 1990 in
Los Alamos, New Mexico/edited by Christopher G. Langton...[et al.],
p. cm.—(Santa Fe Institute studies in the sciences of complexity.
Proceedings: v. 10)
Proceedings of the Second Artificial Life Workshop.
Includes bibliographical references.
1. Biological systems—Computer simulation—Congresses.
2. Biological systems—Simulation methods—Congresses.
I. Langton, Christopher G. II. Artificial Life Workshop (2nd: 1990:
Los Alamos, N.M.) III. Title: Artificial Life two.
IV. Title: Artificial life 2. V. Series: Proceedings volume in the
Santa Fe Institute studies in the sciences of complexity; v. 10.
QH324.2.A75 1991 577—dc20 91-14184
ISBN 0-201-52570-4.—ISBN 0-201-52571-2 (pbk.)

1 2 3 4 5 6 7 8 9 10-MA-95 94 93 92 91

About the Santa Fe Institute

The *Santa Fe Institute* (SFI) is a multidisciplinary graduate research and teaching institution formed to nurture research on complex systems and their simpler elements. A private, independent institution, SFI was founded in 1984. Its primary concern is to focus the tools of traditional scientific disciplines and emerging new computer resources on the problems and opportunities that are involved in the multidisciplinary study of complex systems—those fundamental processes that shape almost every aspect of human life. Understanding complex systems is critical to realizing the full potential of science, and may be expected to yield enormous intellectual and practical benefits.

All titles from the *Santa Fe Institute Studies in the Sciences of Complexity* series will carry this imprint which is based on a Mimbres pottery design (circa A.D. 950–1150), drawn by Betsy Jones.

Santa Fe Institute Studies in the Sciences of Complexity

PROCEEDINGS VOLUMES

Volume	Editor	Title
I	David Pines	Emerging Syntheses in Science, 1987
II	Alan S. Perelson	Theoretical Immunology, Part One, 1988
III	Alan S. Perelson	Theoretical Immunology, Part Two, 1988
IV	Gary D. Doolen et al.	Lattice Gas Methods for Partial Differential Equations, 1989
V	Philip W. Anderson et al.	The Economy as an Evolving Complex System, 1988
VI	Christopher G. Langton	Artificial Life: Proceedings of an Interdisciplinary Workshop on the Synthesis and Simulation of Living Systems, 1988
VII	George I. Bell & Thomas G. Marr	Computers and DNA, 1989
VIII	Wojciech H. Zurek	Complexity, Entropy, and the Physics of Information, 1990
IX	Alan S. Perelson & Stuart A. Kauffman	Molecular Evolution on Rugged Landscapes: Proteins, RNA and the Immune System, 1990
X	Christopher G. Langton et al.	Artificial Life II: Proceedings of the Second Interdisciplinary Workshop on the Synthesis and Simulation of Living Systems, 1991
XI	John A. Hawkins & Murray Gell-Mann	Evolution of Human Languages, 1991

LECTURES VOLUMES

Volume	Editor	Title
I	Daniel L. Stein	Lectures in the Sciences of Complexity, 1988
II	Erica Jen	1989 Lectures in Complex Systems
III	Daniel L. Stein & Lynn Nadel	1990 Lectures in Complex Systems

LECTURE NOTES VOLUMES

Volume	Author	Title
I	John Hertz, Anders Krogh, & Richard G. Palmer	Introduction to the Theory of Neural Computation, 1990
II	Gérard Weisbuch	Complex Systems Dynamics

Contributors to This Volume

David Ackley, Cognitive Science Research Group, Bellcore

Richard J. Bagley, Complex Systems Group, Theoretical Division, and Center for Nonlinear Studies, Los Alamos National Laboratory

Louis Bec, Institut Scientifique de Recherche Paranaturaliste

Mark A. Bedau, Philosophy Department, Dartmouth College

Richard K. Belew, Cognitive Computer Science Research Group, Computer Science & Engineering Department, University of California at San Diego

Alletta d'A. Belin, Shute, Mihaly, and Weinberger

Maarten Boerlijst, Bioinformatica

Peter Cariani, Department of Systems Science, State University of Binghamton, and Eaton-Peabode Labs, Massachusetts Eye and Ear Infirmary

Robert Collins, University of California, Los Angeles

Claus Cooper, University of California, Los Angeles

Martin J. M. de Boer, Theoretical Biology Group, University of Utrecht

Michael G. Dyer, Artificial Intelligence Laboratory, Computer Science Department, University of California, Los Angeles

J. Doyne Farmer, Complex Systems Group, Theoretical Division, Los Alamos National Laboratory

Rasmus Feldberg, Center for Modelling, Nonlienar Dynamics, and Irreversible Thermodynamics, The Technical University of Denmark

Margot Flowers, University of California, Los Angeles

Walter Fontana, Complex Systems Group, Theoretical Division, and Center for Nonlinear Studies, Los Alamos National Laboratory

F. David Fracchia, Department of Computer Science, University of Regina

Brian Hazelhurst, Department of Anthropology, University of California at San Diego

W. Daniel Hills, Thinking Machines Corporation

Pauline Hogeweg, Bioinformatica

Edwin Hutchins, Department of Cognitive Science, University of California at San Diego

Bernie Jackson, Department of Computer Science, Stanford University

David Jefferson, Department of Computer Science, University of California, Los Angeles

Sonke Johnson, Department of Biochemistry and Biophysics, School of Medicine, University of Pennsylvania

Stuart A. Kauffman, Department of Biochemistry and Biophysics, School of Medicine, University of Pennsylvania, and the Santa Fe Institute

Carsten Knudsen, Center for Modelling, Nonlienar Dynamics, and Irreversible Thermodynamics, The Technical University of Denmark

Richard Korf, University of California, Los Angeles

John R. Koza, Computer Science Department, Stanford University

Christopher G. Langton, Complex Systems Group, Theoretical Division, Los Alamos National Laboratory

Kristian Lindgren, Nordita and the Institute for Physical Reseource Theory, Chalmers University of Technology

Michael Littman, Cognitive Science Research Group, Bellcore

Bruce MacLennan, Computer Science Department, University of Tennessee

John McInerney, Cognitive Computer Science Research Group, Computer Science & Engineering Department, University of California at San Diego

Norman Packard, Physics Department, and Center for Complex Systems Research–Beckman Institute, University of Illinois at Urbana-Champaign

Przemyslaw Prusinkiewicz, Department of Computer Science, University of Regina

Steen Rasmusseen, Complex Systems Group, Theoretical Division, and Center for Nonlinear Studies, Los Alamos National Laboratory

Thomas S. Ray, School of Life and Health Sciences, University of Delaware

Nicol N. Schraudolph, Cognitive Computer Science Research Group, Computer Science & Engineering Department, University of California at San Diego

Peter Schuster, Institut für Theoretische Chemie der Universität Wien

Alvy Ray Smith

Elliott Sober, Philosophy Department, University of Wisconsin

Eugene H. Spafford, Software Engingeering Research Center, Department of Computer Sciences, Purdue University

David G. Stork, Department of Electrical Engineering, Stanford University, and Ricoh California Research Center

Charles E. Taylor, Department of Biology, University of California, Los Angeles

Scott Walker, Deparment of Computer Science, Stanford University

Alan Wang, University of California, Los Angeles

Gregory M. Werner, Artificial Intelligence Laboratory, Computer Science Department, University of California, Los Angeles

Dedication

This volume is dedicated to the memory of Aristid Lindenmayer (1925–1989). He is shown here at the first Artificial Life workshop held September, 1987 in Santa Fe, New Mexico. Photo by Kevin Kelly.

Christopher G. Langton
Complex Systems Group, MS-B213, Theoretical Division, Los Alamos National Laboratory, Los Alamos, New Mexico, 87545 and Santa Fe Institute, 1120 Canyon Road, Santa Fe, NM 87501; e-mail: cgl@t13.lanl.gov

Preface

THE WORKSHOPS

In September 1987, the first workshop on Artificial Life was held at the Los Alamos National Laboratory. That workshop brought together 150 researchers from a wide variety of disciplines—from anthropology to zoology—interested in the simulation and synthesis of biological phenomena.

That seminal workshop gave birth to the growing field of Artificial Life. Since that first workshop, a large number of people have become interested in the field and its methodological approaches and have initiated new research projects.

Many of these new research projects were reported at the Second Artificial Life Workshop, held in February 1990, in Santa Fe, New Mexico. This volume constitutes the proceedings of that second workshop.

It is difficult to compare the two workshops. We will never recapture the novelty and excitement of the first workshop. It was the first time many of the participants became aware of the true depth and breadth of the questions and techniques they had been toying with until then. For many of us, it was the first confirmation that we were not crazy—that there was a solid basis for the line of inquiry we had all been pursuing in isolation: studying biological phenomena without studying biological things.

The primary purpose of the first workshop was to find out what kinds of approaches researchers were pursuing in their attempts to simulate or even synthesize life, evolution, ecological dynamics, and so forth. Another goal was to find out what kinds of fundamental biological questions were most appropriately addressed by such techniques.

The proceedings from that first workshop constituted an overview of the possibilities. Although there were no major theoretical breakthroughs reported in any of the papers, and many of the research efforts were clearly in the very earliest stages of development, it was easy to see the power implicit in these early tentative experiments. However, it was not easy to predict the direction in which the field would progress from such a start.

These proceedings of the 1990 Artificial Life Workshop provide a second data point on the direction in which the field of Artificial Life is progressing, and I am very pleased with the trajectory that it appears to be on. Overall, there is more good science and less "gee-whiz" than in the first proceedings. This is not to put down those first proceedings—they were exactly what they had to be. However, it is clear that the field has matured a good deal in the intervening two years, and this is a sign of more good things to come.

In additional to this volume, a video proceedings is available from Addison-Wesley, either separately or as a specially priced package including the videotape, a paperback version of the proceedings volume, and a limited edited poster. This videotape includes some of the presentations and systems discussed at the workshop and some historical footage.

The next data point on the trajectory of Artificial Life will be provided by the third Artificial Life workshop, which is tentatively scheduled to be held in June of 1992, in Santa Fe, New Mexico.

WHAT IS ARTIFICIAL LIFE?

Artificial Life is a field of study devoted to understanding life by attempting to abstract the fundamental dynamical principles underlying biological phenomena, and recreating these dynamics in other physical media—such as computers—making them accessible to new kinds of experimental manipulation and testing.

The essence of the Artificial Life approach was captured by A.W. Burks, in a description of von Neumann's investigations on the phenomenon of self-reproduction:

> What kind of logical organization is sufficient for an automaton to reproduce itself? This question is not precise and admits to trivial versions as well as interesting ones. Von Neumann had the familiar natural phenomenon

> of self-reproduction in mind when he posed it, but he was not trying to simulate the self-reproduction of a natural system at the level of genetics and biochemistry. *He wished to abstract from the natural self-reproduction problem its logical form.*[1] [emphasis added]

To understand the breadth and scope of the field of Artificial Life, it is only necessary to note that we can replace the references to "self-reproduction" above with references to many other biological phenomena, including: the origin of life, molecular self-assembly, embryogenesis (including growth, development, and differentiation), animal behavior (ethology), insect-colony dynamics, evolution, speciation, ecological dynamics, and even linguistic and socio-cultural evolution.

In addition to providing new ways to study the biological phenomena associated with life here on Earth, *life-as-we-know-it*, Artificial Life allows us to extend our studies to the larger domain of the "bio-logic" of possible life, *life-as-it-could-be*, whatever it might be made of and wherever it might be found in the universe.

Thus, Artificial Life is not only about studying existing life, but also about the possibility of synthesizing *new* life, within computers or other "artificial" media. The life that is realized in these alternative media will force us to broaden our understanding of the proper domain of biology to include self-organizing, evolving, and even "living" machines, regardless of the specific physical stuff of which they are constituted, or whether or not they are based upon the same chemical and physical principles as the life that has evolved here on Earth.

ACKNOWLEDGMENTS

Neither the workshop itself nor these proceedings could have succeeded without the efforts of a great many people. As I will inevitably forget to mention some set of absolutely vital helpers, let me just begin by thanking everybody who had anything to do with this whole process: I couldn't have done it without you all!

THE WORKSHOP

First of all, I owe a great debt to the two people who really pulled this thing off: Ginger Richardson and Andi Sutherland, both of the Santa Fe Institute. They showed amazing resilience, fortitude, and vision as the organizing committee fumbled and stumbled blindly towards the approaching conference.

I would also like to thank the other members of the workshop "self-organizing" committee: Charles Taylor, Doyne Farmer, and Steen Rasmussen. I can't say that a great deal of organization emerged spontaneously in this group; in fact, there was significantly more chaos than order! However, we all worked hard, made a lot of phone calls, and carried tons of equipment. And, of course, we had Ginger and Andi to back us up and make us look good! Doyne Farmer also did a great job organizing the events on the "Artificial Night."

Thanks are also due to George Cowan and Mike Simmons, President and Vice-President, respectively, of the Santa Fe Institute. They provided much of the money and put the personnel and resources of the Institute at our disposal. Congratulations to George on being awarded the Fermi Prize! We all wish him the best of luck in his retirement—the entire SFI family will miss his gentle, steady hand at the helm.

Robin Justice of the SFI was primarily responsible for assembling and maintaining the flock of computers, video-projectors, and other electronic paraphernalia required to run a workshop of this nature. Everything seemed to work splendidly.

Andre Longtin did an admirable job translating for zoosystémician Louis Bec, not only between French and English, but also between two quite different philosophical traditions. I'm not sure that the latter wasn't the more challenging task.

Marcella Austin, comptroller at the SFI, balanced the budget and paid all the bills—thanks, Marcella!

Thanks also to all of the SFI people who helped out at the registration desk during the workshop.

David Campbell and Gary Doolen of the Center for Nonlinear Studies at Los Alamos National Laboratory put up the rest of the money and provided personal and logistic support. David gave a fine welcoming speech at the workshop, and has generally been a strong voice in support of Artificial Life at both LANL and the SFI. I am personaly deeply indebted to him for his continued faith in me and in the validity of this field.

Marian Martinez of the CNLS coordinated the mountain of LANL and DOE forms and dealt with the LANL bureaucracy, usually in triplicate. She is a real trooper and a true friend.

Chris Shaw created the wonderful image which became the icon for the workshop. Helen Eubank embellished Chris Shaw's artwork and created a truly beautiful and captivating poster. She is also responsible for the graphic design of the official workshop T-shirt and the limited edition poster accompanying the Video Proceedings and the paperback version of these proceedings in a special sales package.

Thanks also to Mark Pauline, Raymond Drewery, and Jonathan Levine of the Survival Research Laboratory (SRL) for coming to the workshop and bringing one of the tempermental actors (a sonic cannon) in their ongoing "sensitive machine-age social allegory."

The staff of the Sweeny Convention Center in Santa Fe, where the workshop was held, has probably not yet recovered from the sudden and deafening resurrection of SRL's sonic cannon. I owe them an apology along with my thanks—the blasts were felt blocks away.

Kenny Tennyson, the Sweeny Center building manager, remained surprisingly affable and calm throughout all kinds of minor and major crises. He also played a major role in the resurrection of the sonic cannon when he pulled a replacement relay out of an old dishwasher. The Survival Research Lab owes its survival to this sort of selfless, mechanical altruism. Kenny was assisted by Mitchell Schmitt and William Trujillo.

The workshop was supported in part by the U.S. Department of Energy under contract # W-7405-ENG-36. Research and workshops cosponsored by the Santa Fe Institute are supported in part by grants from the John D. and Catherine T. MacArthur Foundation, the National Science Foundation (PHY-8714918), and the U.S. Department of Energy (ER-FG05-88ER25054).

THE PROCEEDINGS

For the proceedings, I have to thank first and foremost the contributors who provided the intellectual content. Also, a great deal of credit is due to the army of reviewers who helped fine tune the contributors' contributions. Over 60 papers were contributed for inclusion in these proceedings, of which we have selected a little over 25 for publication. Again, my apologies to those whose papers did not end up in these proceedings. The overall quality of the submissions was very high, but we simply could not include all of the papers we would like to have printed.

Thanks especially to Charles Taylor, who took on a good portion of the editing load and the task of locating and harassing reviewers, and to my other co-editors, Doyne Farmer and Steen Rasmussen.

Thanks again to George Cowan and Mike Simmons of the Santa Fe Institute for including these proceedings in their series: *Santa Fe Institute Studies in the Sciences of Complexity*.

The actual work of bringing these proceedings into physical reality fell to Ronda Butler-Villa, Director of Publications at SFI, and her able assistant Della Ulibarri. Once again, I cannot thank Ronda enough. She has been through this whole process with me once before and I am quite grateful that she agreed to do it again nonetheless. Third time's the charm, Ronda!

Allan Wylde, the director of Addison Wesley's Advanced Book Program, has been a solid backer of the entire SFI series, and an enthusiastic supporter of the Artificial Life volumes. He's flexible, personable, and manages an excellent outfit. Laura Likely, Allan's chief assistant, has probably contributed several meters of fingernails to this project, but has borne up admirably and is always pleasant to deal with. We have also benefited from the assistance of Kathleen Palmer and Karl Matsumoto.

Jolene Manning of Southwest Color Separations oversaw the production of the negatives for the color plates.

ONCE AGAIN LAST BUT STILL NOT LEAST...

My wife Elvira and my sons Gabriel and Colin have once again provided aid and comfort throughout this whole project. I am eternally grateful for their love and understanding, despite the large amount of time spent with my other, "artificial" family.

Now, kids, about those 10,000 lollypops I promised you....

REFERENCES

1. Burks, A. W. "Introduction." *Essays on Cellular Automata.* Urbana: University of Illinois Press, 1970.

Contents

Overview

Christopher G. Langton
Complex Systems Group, MS-B213, Theoretical Division, Los Alamos National Laboratory, Los Alamos, New Mexico, 87545 and Santa Fe Institute, 1120 Canyon Road, Santa Fe, NM 87501; e-mail: cgl@t13.lanl.gov

Introduction

THE PAPERS

Of the more than 60 papers submitted for consideration, we selected a little over 25 for publication in these proceedings. This necessarily means that some of the work that was presented at the workshop is not represented here explicitly. However, these papers, taken together with those included in the proceedings of the first workshop, provide nearly complete coverage of the methods that are being employed and the types of problems to which they are being applied.

The talks at the second workshop were arranged to reflect the chronological and hierarchical ordering evident in the natural world. Thus, talks treating the origin of life came first, followed by talks on the evolutionary process, on development and differentiation, learning, computer life, ecological dynamics, and finally to discussions on the future evolution of life. Progressing through the material in this natural sequence continuously reinforces the "emergent" nature of biological structures and functions, as the phenomena under discussion on one day can be seen to have emerged "on the shoulders," so to speak, of the phenomena under discussion the day before.

It also reflects the bottom-up methodology employed by most AL investigations themselves, in which rules of behavior are specified for the individual parts of a system, which are then "turned loose," and allowed to interact with one another. With the correct rules governing the individual parts, and the correct arrangement of the parts themselves, the phenomenon of interest should emerge spontaneously out of all of the interactions between parts, in much the same way that it emerges in the natural system under study.

In these proceedings, we follow the natural ordering of the topics employed during the workshop, from studies of the origin of life using "artificial chemistries" to a discussion of the future of life in the solar system.

OVERVIEW OF THE SECOND ARTIFICIAL LIFE WORKSHOP

In the first paper, Charles Taylor (one of the workshop organizers) provides a comprehensive overview of the Second Alife Workshop, and offers his personal perspective on the current state of the "art" of Artificial Life, why it should be pursued, and—a topic of great import—why it should be funded.

ARTIFICIAL CHEMISTRIES, SELF-ORGANIZATION, AND THE ORIGIN OF LIFE

The next five papers describe research with various "artificial chemistries," addressing questions about functional self-organization and the origin of life. In artificial chemistries, a network of chemical reactions among molecules in, say, a chemostat is approximated by a network of logical "reactions" among functions in a computer.

For example, reactions in which enzymes cleave or join substrate molecules, resulting in the production of new molecules with new functional properties, can be approximated in a computer by reactions in which "function-strings" cleave or join other function-strings, resulting in the production of new strings with new functional properties. Such approximations allow investigators to explore the conditions under which complex, self-organized dynamics can emerge spontaneously in systems which are like chemistries, which capture certain specific properties of "real" chemistries, but which are much simpler and easier to experiment with.

The paper by Chris Langton describes research on the conditions under which complex dynamics can be expected to emerge in spatially extended systems obeying "local" update rules (cellular automata). The results of this research indicate that a complex dynamics involving information processing emerges spontaneously when such systems are in the vicinity of a second-order—or "critical"—phase-transition between ordered and disordered behavior. This suggests a fundamental connection between information processing, or computation, and critical dynamics. Langton suggests that life may have originated near a critical phase transition, and that evolution naturally brings systems to, and keeps them at, this phase transition at "the edge of chaos."

The paper by Richard Bagley and Doyne Farmer addresses the following question. Contemporary organisms are built out of large, complicated molecules that depend on each other for their own synthesis. Proteins and RNA are synthesized from DNA templates, while DNA is synthesized from replication reactions that require proteins and RNA. In even the simplest organisms, the chemical machinery required to maintain the process of replication is highly complex. There seems to be a minimum level of complexity below which a replicating machine based on the same principles as those of a contemporary organism simply cannot function. How then did life ever begin?

Bagley and Farmer are exploring the possibility that prior to living organisms there were other, simpler processes that set the stage for the emergence of life. They demonstrate that under appropriate conditions, a simple metabolism of polymers can spontaneously emerge, without the need for information storage in templates. This metabolism consists of a specific collection of polymers, called an *autocatalytic set*, that help each other by catalyzing each others' formation. Under appropriate conditions autocatalytic sets can evolve out of a simple, undifferentiated initial state, generating a sequence of complex, highly differentiated, final states. These final states consist of specific sets of long polymers, similar to those of contemporary organisms. While the autocatalytic set does not replicate itself in the usual sense, it propagates itself by taking over any medium with suitable properties. It sustains itself as long as the appropriate conditions are met. Furthermore, it is capable of generating a lineage of related autocatalytic sets.

One autocatalytic set can seed the formation of another, similar set. The autocatalytic sets of this model can be viewed as "proto-life forms." They share many of the properties of living systems: They have a metabolism, as well as the ability to evolve and store information. They also reproduce, although they do so more continuously and with less fidelity than contemporary organisms.

The papers by Walter Fontana and Steen Rasmussen et al. describe research employing more abstract artificial chemistries. In these efforts, the goal is not so much to determine what actually happened within the specific pre-biotic soup that existed on Earth, but to come to a better theoretical understanding of the way in which lifelike dynamics could emerge in principle within a wide class of possible pre-biotic soups. Thus, they abstract certain functional properties of known chemical reaction networks, and concentrate on delineating the spectrum of dynamical behaviors such abstract functional "soups" can exhibit in principle. Once the full spectrum is known, they can concentrate on locating the specific behavioral regimes most clearly associated with the spontaneous emergence of lifelike dynamics.

Fontana builds his artificial chemistry upon the well-known lambda-calculus, a formal symbol manipulation system which formed the basis for the Lisp programming language. In Fontana's system, functions expressed in the lambda-calculus are represented as character strings that "react" with other such functional character-strings via function composition, producing new function-strings in the process. These new function-strings then enter into reactions with existing function-strings, and the process continues. Fontana finds that the natural dynamics of this "Algorithmic Chemistry" ("Alchemy," for short) gives rise to the spontaneous emergence of many cooperative reactions between function-strings, including self-replicators, self-replicating sets, autocatalytic cycles, symbiotic and parasitic sets of functions, and so forth.

The paper by Steen Rasmussen, Carsten Knudson, and Rasmus Feldberg describes another such "self-programming" system based on a parallel von Neumann machine. This system is even further abstracted from "real chemistry," yet gives rise to similar results. In their "Venus" and "Luna" simulation systems, an area of computer memory is initially randomly filled with instructions and execution is initiated at many places simultaneously. As these points of execution wander through memory, instructions are executed, memory is altered, new execution pointers are created and old execution pointers are deleted. In the process, the memory becomes self-organized: programs emerge that copy themselves and create execution pointers to their copies or to other memory areas; programs overwrite other programs or capture their execution pointers; different programs cooperate with each other or parasitize other programs; and so forth. The evolutionary dynamics in this system passes through several successive "epochs," each characterized by a different dominant functional dynamics. Although these emergent functional structures are still far simpler than, and probably also different from, any contemporary biological organisms, the structures are able to channel and focus the available computational resources very effectively. They are stable to perturbations, they are able to propagate themselves (e.g., reproduce), and they can undergo evolution. Functional cooperation, as seen in life, is an inherent property of the dynamics of such self-programmable systems.

The paper by Martin Boerlijst and Pauline Hogeweg illustrates very nicely the vital role played by spatial dynamics in distributed systems. In their research, they treat the classic hypercycle model for the origin of life, originally due to Eigen[5] and treated analytically via a system of Ordinary Differential Equations (ODE's) by Eigen and Schuster.[6] In this model, Eigen and Schuster proposed that multiple autocatalytic cycles could themselves get caught up in higher-order cycles, which they called "hypercycles." They hypothesized that life originated with the emergence of these hierarchical hypercycles.

However, an analysis of the ODE hypercycle model showed that hypercycles are unstable.[13] They are subject to disruption by parasites, who draw catalytic support from the cycle but do not give catalytic support back to the cycle. Such parasites will inevitably emerge and kill any hypercycle by draining off all of its resources.

Boerlijst and Hogeweg, however, have performed a PDE-like analysis of the Hypercycle model using a cellular automaton approach that allows for spatial inhomogeneity, and hence for the emergence of spatial structure and dynamics. The results are significantly different. In their spatially extended version, they observe that the autocatalytic cycles are realized in space as multi-armed spiral waves of catalytic activity, much like the spiral waves that appear in reaction-diffusion systems like the classic Belousov-Zhabotinsky reaction. Furthermore, these spiral waves of catalytic activity are stable against invasion by parasites, because they push the parasites out to the periphery of the spirals, into the boundary region between spiral waves. There, if they are very lucky, the parasites can eke out just enough catalytic support to stay alive, but they cannot invade the central regions of the spirals. Thus, the emergent spatial dynamics drastically alters the ultimate behavior of the hypercycle system, and this model provides renewed support for the viability of the hypercycle as a model for the origin of life.

The paper by Schuster is a review of the work that his group in Vienna has been doing on the fitness surface of hypothetical RNA molecules. The prediction of the tertiary structure of polynucleotides is still a computationally intractable problem. In contrast, the prediction of secondary structure has been solved for small polynucleotides. Applying these tools to simplified RNA molecules, Schuster is able to compute various phenotypic properties, such as rate constants for replication or degradation, and the free energy of conformation (in two-dimensions).

His system is useful for several reasons. First, it characterizes the adaptive surface in a more biologically plausible system than anyone has done before. Second, it makes progress on the study of the general genotype/phenotype problem by concentrating on the simplest relation of this kind found in nature. Third, he is studying the issue of evolution in an artificial system that has many of the properties assumed for early life, and which serves as an Artificial Life model for how natural life could have evolved in the RNA-ribozyme world that is currently thought to be an early step in life on Earth.

In summary, this collection of papers treating artificial chemistries provides insight into the conditions under which we might expect a wide variety of "real" chemical systems to become self-organized and to exhibit the emergence of the kinds of complex structural and functional dynamics that we would expect to characterize the origin of life.

EVOLUTIONARY DYNAMICS

Modern organisms owe their structure to the complex process of biological evolution, and it is very difficult to discern which of their properties are due to chance, and which to necessity. If biologists could "re-wind the tape" of evolution and start it over, again and again, from different initial conditions, or under different regimes of external perturbations along the way, they would have a full *ensemble* of evolutionary pathways to generalize over. Such an ensemble would allow them to distinguish universal, necessary properties (those which were observed in all the pathways in the ensemble) from accidental, chance properties (those which were unique to individual pathways). However, biologists cannot rewind the tape of evolution, and are stuck with a single, actual evolutionary trace out of a vast, intuited ensemble of possible traces.

Although studying computer models of evolution is not the same as studying the "real thing," the ability to freely manipulate computer experiments, to "rewind the tape," perturb the initial conditions, and so forth, can more than make up for their "lack" of reality.

It has been known for some time that one can evolve computer programs by the process of natural selection among a population of variant programs. Each individual program in a population of programs is evaluated for its performance on some task. The programs that perform best are allowed to "breed" with one another via *genetic algorithms*.[8,10] The offspring of these better-performing parent programs replace the worst-performing programs in the population, and the cycle is iterated. Such evolutionary approaches to program improvement have been applied primarily to the tasks of function optimization and machine learning.

However, such evolutionary models have rarely been used to study evolution itself.[16] Researchers have primarily concentrated on the *results*, rather than on the *process*, of evolution. In the spirit of von Neumann's research on self-reproduction via the study of self-reproducing *automata*, the following papers study the process of evolution by studying evolving populations of "automata."

The paper by Lindgren is a clear example of the insights that can be gained from studying simple abstractions of complex biological phenomena. In this paper, Lindgren studies evolutionary dynamics within the context of a well-known game-theoretic problem: the *Iterated Prisoners' Dilemma* model (IPD). This model has been used effectively by Axelrod and Hamilton in their studies of the evolution of cooperation.[1,2]

In the Prisoners' Dilemma model, the payoff matrix (the fitness function) is constructed such that individuals will garner the most payoff collectively in the long run if they "cooperate" with one another by *avoiding* the behaviors that would garner them the most payoff individually in the short run. If individuals only play the game once, they will do best by not cooperating ("defecting.") However, if they play the game repeatedly with one another (the "iterated" version of the game), they will do best in the long run by cooperating with one another.

In Lindgren's version of this game, strategies can evolve in an open-ended fashion by evolving larger memories, allowing them to base their decisions on whether to cooperate or defect upon longer histories of previous interactions.

Many complicated and interesting strategies evolve during the evolutionary development of this system. More importantly, however, are the various phenomenological features exhibited by the dynamics of the evolutionary process. First of all, the system exhibits a behavior that is remarkably suggestive of the *punctuated equilibria* proposed by Elderidge and Gould.[7] After an initial irregular transient, the system settles down to relatively long periods of stasis "punctuated" irregularly by brief periods of rapid evolutionary change.

Second, the diversity of strategies builds up during the long periods of stasis, but often collapses drastically during the short, chaotic episodes of rapid evolutionary succession. These "crashes" in the diversity of species constitute "extinction events." In this model, these extinction events are observed to be a natural consequence of the dynamics of the evolutionary process alone, without invoking any catastrophic, external perturbations (there are no comet impacts or "nemesis" stars in this model!). Furthermore, these extinction events happen on multiple scales: there are lots of little ones and fewer large ones.

This provides a very nice basis for comparison, as in order to understand the dynamics of a system that is subjected to constant perturbations, one needs to understand the dynamics of the *un*perturbed system first. We do not have access to an unperturbed version of the evolution of life on Earth; consequently, we could not have said definitively that extinction events on many size scales would be a natural consequence of the process of evolution itself. By comparing the perturbed and unperturbed versions of model systems like Lindgren's, we may very well be able to derive a universal scaling relationship for "natural" extinction events, and therefore be able to explain deviations from this relationship in the fossil record as due to external perturbations.

Third, the phenomenon of *co-evolution* is also illustrated nicely by Lindgren's model. It is usually the case that a mix of several different strategies dominates the system during the long periods of stasis. In order for a strategy to do well, it must do well by cooperating with other strategies. These mixes may involve three or more strategies whose collective activity produces a stable interaction pattern that benefits all of the strategies in the mix. Together, they constitute a more complex, "higher-order" strategy, which can behave as a group in ways impossible for any individual strategy.

It is important to note that, in many cases, the "environment" that acts on an organism, and in the context of which an organism acts, is primarily constituted of the other organisms in the population and their interactions with each other and the physical environment. There is tremendous opportunity here for evolution to discover that certain collections of individuals exhibit emergent, collective behaviors that reap benefits for the whole collection. Thus, evolution can produce major leaps in biological complexity, without having to produce more complex individuals, by discovering the many ways in which collections of individuals at one level can work together to form aggregate individuals at the next higher level of organization.

This is thought to be the case for the origin of eukaryotic cells, which are viewed as originating from cooperative collections of simpler, prokaryotic cells.[12] It is also the process involved in the origin of multicellular organisms, which lead to the Cambrian Explosion of diversity some 700 million years ago. It was probably a significant factor in the origin of the prokaryotes themselves, and it has been discovered at least seven times independently by the various social insects (including species of wasps, bees, ants, and termites).

Another study involving co-evolution is reported in the paper by Hillis. In this paper, Hillis exploits host-parasite interactions among a population consisting of sorting *programs* (the hosts) and sorting *problems* (the parasites). The programs are trying to do a better job of sorting the problems, while the problems are trying to make the programs perform badly.

This results in an evolutionary "arms race," in which which programs that reach local optima become targets for the population of problems, which reduces the fitness of the programs, and diminishes the height of their local fitness peak, leaving them free to climb another peak. Thus, the interactions between the two populations changes the shape of the fitness landscape dynamically. When the population of programs reach the top of a fitness "hill," the population of problems turns the hill into a fitness "valley," which the population of programs can climb out of, and can then continue to climb to the top of another (hopefully higher) hill. When the population of programs reaches a hill that the population of problems cannot turn into a valley, the population of programs has reached a global optimum. Thus, the joint, coupled, host-parasite evolutionary dynamics finds better solutions to the sorting problem in less time than the evolutionary dynamics of a population consisting of sorting programs alone.

The paper by Stuart Kauffman and Sonke Johnsen reports on theoretical and experimental approaches to understanding the dynamics of co-evolving organisms via a study of their coupled fitness landscapes. In their words:

> We consider co-evolution occurring among species each of which is itself adapting on a rugged, multipeaked fitness landscape. But the fitness of each genotype, via the phenotype, of each species is affected by the genotypes via the phenotypes of the species with which it is coupled in the ecosystem. Adaptive moves by one co-evolutionary partner, therefore, may change the fitness and the *fitness landscapes* of the co-evolutionary partners. Anecdotally, development of a sticky tongue by the frog alters the fitness of the fly, and what it ought to do: it should develop slippery feet. In this framework, adaptive moves by any partner may *deform* the fitness landscapes of other partners.

Their research is carried out in the context of Kauffman's NK family of rugged, multipeaked fitness landscapes. In the NK family, N is a parameter governing the number of genes in a genotype, while K is a parameter governing the fitness contribution of each gene. By tuning K, the "ruggedness" of the fitness landscape can be adjusted. To study co-evolution, they couple the individual NK fitness landscapes via another parameter, C, which determines the number of genes in each of the *other* species that contribute to the fitness of an individual. Via this coupling, adaptive changes by one species can alter the fitness landscapes of the other species.

One interesting result is that co-evolutionary systems seem to achieve a "poised" state "at the edge of chaos," lending support to an idea originally due to Norman Packard and Chris Langton,[15,11] that evolutionary dynamics will bring populations to the vicinity of a phase transition between ordered and disordered behaviors (see the description of Langton's paper, above).

Tom Ray reports on the application of his "Tierra" simulation system to the study of the origin of biological diversity. In this paper, he describes the emergence of an incredibly diverse array of co-evolutionary phenomena. In his simulation system, programs reproduce themselves and compete for residency in the memory (the "soup.") In the resulting dynamics, Ray, a tropical ecologist by training, notes the emergence of whole "ecologies" of interacting species of computer programs. Furthermore, he is able to identify many phenomena familiar to him from his studies of real ecological communities, such as competitive exclusion, the emergence of keystone predators, parasites, hyper-parasites, symbiotic relationships, and so forth.

Thus, in co-evolving systems collections of individuals can effectively pursue strategies that are unachievable on the part of individuals. Given a robust enough system with enough potential diversity, evolution apparently quickly discovers such collections.

Evolution does not usually get a chance to solve every new problem from scratch. Rather, it takes whatever structures or processes it has at hand and applies them to new tasks—tasks which may be quite different from those for which the newly recruited structures were developed in response to in the past. Thus, many of the structures and processes in organisms appear somewhat sub-optimal. From an engineering point of view, the task could easily have been performed much more simply and elegantly.

The paper by David Stork, Bernie Jackson, and Scott Walker illustrates how a certain non-optimal neural structure in crayfish can be explained on the basis of selection for different behaviors in the past. Specifically, there is a synapse in the neuronal circuitry used by crayfish for tail-flipping which appears to be useless. Its presence can be explained, however, as a vestige of a previous adaptation when the same neural circuitry was used for swimming, and only later adapted to tail-flipping.

The paper by Bedau and Packard provides a clean treatment of a very messy problem: the quantification of such notions as "evolutionary activity" and "evolutionary progression." In their research, they define evolutionary activity as "the rate at which useful genetic innovations are absorbed into the population." They are able to provide a quantitative measure of "useful genetic innovations," as well as a measure of the rate at which such innovations are absorbed into the population, in a way that would be impossible, or at least extremely difficult, to do for "real" organisms.

By keeping a count of how many times each "gene" in their population of organisms is invoked, they can view the increase of use of a set of genes in a population as the propagation of a "lump" of higher usage counts through a global histogram of overall gene usage. They refer to such a lump as a "wave of evolutionary activity." When the global histogram of overall gene usage is plotted as a function of time, each activity wave traces out a line with some characteristic slope. If time is plotted on the vertical axis, the shallower the slope, the faster the activity wave is spreading through the population. Steep slopes indicate that the activity wave is not spreading very rapidly.

These activity waves measure more than just changes in gene frequency in the gene pool; they also measures changes in actual gene usage. Thus, this measure would indicate evolutionary change even if genes never changed their representation in the gene pool, but were merely getting triggered more and more frequently, indicating an evolution in the overall global dynamics of genetic activity. This would be extremely hard to measure in a biological population, but is probably an important factor in the evolutionary process.

It is also interesting to note that the activity is "bursty." Long periods of relatively low genetic activity are punctuated by bursts of high genetic activity, which start many activity waves propagating through the count histogram. This lends further support to the "genericity" of the punctuated equilibrium model.

Finally, they use their quantitative measure of evolutionary utility to shed some light on the problem of teleology in biology. Philosophers and biologists have wrestled for centuries with the difficult question of "purpose" in the natural world. Is the evolutionary process "striving" in some way towards some ultimate goal? Are organisms themselves striving purposefully towards their own individual ends? How much of these purposes, goals, strivings, ends, and etc., in terms of which we typically *describe* the actions and structures of organisms, can be attributed to the organisms themselves? Or to the evolutionary process that produced them? Bedau and Packard provide a fresh, remarkably clean treatment of this often muddled topic.

In summary, the study of Evolution with a capital "E" via the study of evolving populations of processes within computers is one of the most promising research areas within the field of Artificial Life. After hearing Kristian Lindgren's talk at the workshop, many researchers went back to their models, took another look at the ongoing evolutionary dynamics, and discovered a similar pattern of long periods of stasis punctuated by brief periods of rapid evolutionary change. I think a lot of these researchers kicked themselves for not having noted this feature in their data earlier, a consequence of concentrating on the products of the process, rather than on the process itself.

The occurence of such a characteristic feature of the evolutionary process in a number of different computer implementations of evolution is highly suggestive that it is, in fact, a *generic* property of evolution, and would be observed in any evolutionary process anywhere in the universe. However, although the data are highly suggestive, even persuasive, they are hardly definitive. This early work needs to be built upon, extended to include spatial dynamics, a broad range of implicit fitness landscapes, and so forth.

Furthermore, our experience with the evolutionary dynamics of systems embedded in computers has convinced many of us that, although correct in the large, our theories about Darwinian evolution must be extended to include the effects of nonlinear interactions within large populations of individuals. It is now apparent that such systems can give rise to emergent, collective behaviors that evolution could easily discover and reinforce.

DEVELOPMENT

Multicellular organisms start life as a single cell. Through repeated cell divisions, cell movements, and cellular differentiation, the mature organism develops via the complex process known as *morphogenesis*. The mechanics of the process by which the growing body of cells shapes itself and produces all the right kinds of cells in all the right places is still largely a mystery. All the cells contain the same genetic program, and it is not well understood how the different types of cells diverge into different domains of genetic program space.

Lindenmayer systems (L-systems) have found broad application within the Artificial Life community for analyzing the complex process of morphogenesis. The use of L-systems is in keeping with the overall spirit of the Artificial Life approach. L-systems abstract the "logical form" of the natural problem of morphogenesis. They allow researchers to work and experiment with systems that grow, develop, and differentiate *like* real organisms, but which are not, themselves, real organisms. They are another instance of the general class of growing, developing, and differentiating things. Therefore, by studying them, and by comparing and contrasting them with the growth, development, and differentiation of living systems, we can learn more about the "real" system by understanding the general class of systems it belongs to.

Sadly, Aristid Lindenmayer, the inventor of L-systems, recently passed away after a battle with cancer. However, he has inspired many researchers to carry on his work. From their original applications to the study of branching, one-dimensional filamentary structures, L-systems have been extended to be able to treat growing two-dimensional "tissues" of cells. The paper by Martin DeBoer, David Fracchia, and Przemyslaw Prusinkiewicz employs context-free, map L-systems to describe the growth and development of cellular layers, including the development of tissue layers that form three-dimensional surfaces, like those found in young embryos. They demonstrate that several of the complex cell division patterns observed in nature can be captured by relatively simple L-systems. They are not claiming that L-systems capture cellular growth and division in the same way that real cells do it. However, as they state, "spatial-temporal regularities of cell divisions found with the help of map L-systems may reveal properties of cell division control."

EVOLUTION, LEARNING, AND COMMUNICATION

Evolutionary adaptation can be viewed as a kind of "learning" that takes place on time scales much longer than the lifetimes of individual organisms. Traditional computational learning methods, such as back-propagation applied to neural nets, often require that each action of the learning system be evaluated by a "teacher" or "critic," who "knows" what the correct action should have been, and provides the learning system with an evaluation its performance, ranging from a simple "yes" or "no" to providing it with the "correct" answer. However, in many real learning situations, there is no "correct" answer, or even if there is, it may be computationally intractable to determine what it is, so there can be no teacher that provides the system with the correct feedback. In such learning situations, neural networks perform rather poorly.

Genetic algorithms (GA), on the other hand, seem to work well in this kind of learning situation, whereas they fare poorly in situations requiring a teacher, as there are no mechanisms for incorporating the feedback about the "correct answer" into the system so that it performs better the next time.

Thus, there are different kinds of learning situations, and organisms must develop different kinds of learning techniques to deal with them.

The next two papers treat the relative advantages and disadvantages of evolutionary "learning" versus the more traditional kinds of learning that occur during an organism's lifetime. The relative effects of these two different forms of learning can be studied by constructing systems that are capable of both kinds of learning and comparing the results using both learning strategies with the results when one or the other kind of learning is turned off.

It is generally found in such studies that a combination of evolutionary and more traditional learning works best. For instance, one might study evolving populations of organisms whose behavior is controlled by neural nets whose specific wiring diagram and initial weights are specified by a genetic strings, but which can change these weights during their lifetimes via a standard learning method such as back-propagation. Genetic algorithms can tune the networks to do better on the tasks for which the "world" provides no immediate feedback about performance, while back-propagation can take advantage of whatever kinds of immediate feedback the "world" provides.

The paper by David Ackley and Michael Littman employs an adaptation strategy they call *Evolutionary Reinforcement Learning* (ERL). ERL "combines genetic evolution with neural network learning, and an artificial life 'ecosystem' called AL, within which populations of ERL-driven adaptive 'agents' struggle for survival." During the course of their simulations, they observe that the best performance occurs when both evolutionary learning and short-term learning are employed. More importantly, however, they observe the occurrence of a number of subtle effects at the level of population genetics that appear to enhance overall learning, such as the classic "Baldwin effect," a mechanism by which changes in behavior learned over the short term can eventually be transferred to the "genetics" of long-term learning.

The paper by Richard K. Belew, John McInerney, and Nicol N. Schraudolph specifically explores the tradeoffs involved in hybrids of evolutionary and neural-network learning schemes. In particular, they explore the set of mutual constraints imposed by each learning scheme on the genetic representation of the neural-networks. They find that the optimal genetic encodings of networks are quite different from the most obvious representation consisting of a simple concatenation of weights. They also find that hybrids are successful largely because they combine a *global* search of the weight space, mediated by the genetic algorithm, with a *local search*, mediated by traditional gradient descent methods.

The behavior of social insects has proved particularly intriguing to Artificial Life researchers. This is because colonies of social insects—ants, bees termites, wasps, etc.—provide the most accessible natural examples of the manner in which large aggregates of fairly simple organisms can give rise to emergent patterns of global dynamical activity. These global dynamical structures, in turn, organize the behavior of the collection of individuals by determining the local context within which individuals decide what to do. This leads to the formation of a new, higher level individual—the colony itself—which can pursue its own "colonial" ends. The highly purposeful behavior exhibited by insect colonies provides an excellent metaphor for the way in which purposeful behavior might emerge in the large aggregate of molecules making up a living cell, or in the large aggregate of neurons constituting the brains of higher organisms.[9] The properties and possibilities inherent in these sorts of emergent, collective dynamics are of primary import to the study of life and intelligence.

The paper by Jefferson et al. describes a study of the evolution of trail-following by ants. The study has several motivations. First, the study is an attempt to understand the complexity of the trail-following problem: How hard is it to learn to follow "noisy" trails? Second, they address the question of representation: What effect do different representation schemes have on the complexity of the problem and the nature of the solutions. Third, they are interested in the question of knowledge transfer: How much general trail-following ability is gained from learning to follow a single trail? After a population has learned to follow one specific trail, is it harder or easier to learn to follow a new, different trail?

They investigate two different representation schemes: Finite State Machines (FSM) and Artificial Neural Nets (ANN). They implement both so that they are roughly equivalent in computational power (i.e., an equivalent number of internal memory states, etc.) As a result of this comparison, they propose a number of properties that should be shared by representation schemes for organisms in biologically motivated studies, including *computational completeness, syntactic closure, scalability,* and so forth.

The research of Robert Collins is aimed at understanding the evolution of collective behavior. In his paper in this volume, he reports on current work with his "AntFarm" simulation system implemented on the massively parallel Connection Machine. Using this system, Collins has explored the evolution of cooperative foraging behavior in colonies of artificial organisms that resemble ants. This study has a number of implications for evolution in complex environments, the evolution of cooperation among closely related individuals, and the evolution of communication.

It should be clear by this point that *representation* is a major issue in Artificial Life research. A representation scheme for implementing algorithms that is understandable by human beings (such as a familiar programming language like Pascal or Fortran) may not be amenable to manipulation by genetic algorithms. Thus, programmable representations may not be evolvable, and evolvable representations may not be programmable.[4] Many of the research efforts reported in these proceedings employ representations that were chosen for their evolvability, not for their programmability.

John Koza attempts to effect a compromise between programmability and evolvability. By applying a genetic algorithm to the *parse tree* of a program (rather than to its linear representation as a bit-string,) and by respecting syntactic categories, Koza is able to achieve evolution within the context of a programmable language. In the context of the tree representation of a program, the crossover operator of the genetic algorithm is allowed to swap subtrees between two parent programs when the roots of those subtrees belong to the same syntactic category. In Koza's system, *all* subtrees are S-expressions in Lisp. Since all S-expressions belong to the same syntactic category, any subtree can be interchanged with any other. Koza has applied this system to evolve programs to solve a number of hard problems, including a number of Artificial Life problems.

Ethology is the study of animal behavior, and hence constitutes a prime domain for the implementation of Artificial Life models. A growing number of ethologists are turning to Artificial Life techniques for testing hypotheses about the mechanisms underlying different classes of animal behaviors. As mentioned above, behaviors of individuals are often shaped by co-evolutionary processes, and, consequently, it is difficult to explain many behaviors without an understanding of this co-evolutionary development. Therefore, a number of ethological investigations using Artificial Life models involve more than just attempts to generate known behaviors: they involve attempts to recreate the evolutionary development of known behaviors.

One aspect of animal behavior (or human behavior for that matter) that clearly involves a good deal of historical depth is *communication*. The explanation for many of the specific features of any communication scheme—from a repertoire of squeaks and grunts to fully developed natural languages—necessarily rests on historical processes. Communication schemes and natural languages are excellent examples of the kinds of unique historical processes that make theory construction so difficult. Many linguists and ethologists believe that there are universal principles afoot in the development of languages and other animal behaviors, but teasing them out from all of the unique historical events is an extremely difficult task. As discussed above, this sort of situation is one of those which can be remedied in part by Artificial Life models, which allow the construction of ensembles of closely related historical traces, leading (hopefully!) to an ability to distinguish between universal and accidental features.

The following two papers investigate the development and evolution of communication schemes within populations of synthetic organisms. The paper by Bruce MacLennan employs Artificial Life methods to study issues in the development of communication. In his words:

> If we want to understand what makes symbols meaningful (and related phenomena such as intentionality), then AI—at least as currently pursued—will not do. If we want genuine meaning and original intentionality, then communication must have real relevance to the communicators. Furthermore, if we are to understand the pragmatic context of the communication and preserve ecological validity, then it must occur in the communicators' natural environment, that to which they have become coupled through natural selection. Unfortunately, the natural environments of biological organisms are too complicated for carefully controlled experiments.

MacLennan's answer to this dilemma is to introduce the field of *Synthetic Ethology: the study of synthetic life forms in synthetic worlds to which they have become coupled through evolution.* By studying the development of communication behavior in simplified, controllable worlds, he is able to draw conclusions about the development of communication behavior in the more messy, real world.

The paper by Greg Werner and Michael Dyer investigates the emergence of a simple communications protocol in a population of artificial organisms. As the communications protocol is used to guide potential mates to each other, mating success is directly related to communicative success. Since their world evolves within the context of a two-dimensional lattice, spatial dynamics play a significant role in the development of the protocols. In particular, they observe the formation of different "dialect" domains, which restrict mating to occur within distinct dialect lineages. This promises to be an excellent vehicle for the study of speciation, including the development of mechanisms for restricting mating that are thought to precede some speciation events.

The paper by Edward Hutchins and Brian Hazlehurst discusses learning in the cultural process. Ultimately, linguistic evolution has allowed the encoding of culture, providing the capacity to transmit learned knowledge across generations extra-genetically. Any communication scheme requires the production of an *artifact* in the physical world that mediates the communication among organisms embedded in that world. "Barring mental telepathy, one mind can only influence another by putting some kind of structure in the environment of the other mind." Thus, the authors argue, the evolution of culture must involve the establishment of correlations between three different types of structures in the world: natural structures in the environment, structures in the heads of organisms (internal representations), and *artifacts* (structures in the environment that are put there by organisms). Artifacts mediate the "heavy traffic in representations of the world" that "move within and among individuals." Spoken or written languages provide a basis for such artifacts, but many other "physical" entities or processes can bear the weight of mediating between internal representations.

Hutchins and Hazlehurst address the cultural process through a study of the development of such artifactual structures in an artificial world. In their words:

> It is important to approach this subject with the understanding that culture is not a thing or any collection of things; it is a process. In the human sphere, myths, tools, understandings, beliefs, practices, artifacts, architectures, classification schemes, etc. alone or in combination do not in themselves constitute culture. Each of these structures, whether internal or external, is a *residue* of the cultural process. The residues are, of course, indispensable to the process, but taking them to be culture itself diverts our attention from the nature of the cultural process.

COMPUTER LIFE

One of the goals of Artificial Life is to expand the empirical data base upon which we rely for theory construction. One way we can do this is to create processes within computers and other "artificial" media that exhibit lifelike behaviors but which are not explicit simulations or models of any particular known biological organisms.

In this section, we review several papers treating computer based "life-forms" as objects worthy of studying on their own rights.

Von Neumann's studies of self-reproduction involved the construction of structures which were not at all like any known living things. Yet, through their study, he was able to determine principles which are likely to be true of *any* self-reproducing system, including known biological organisms. For instance, he found that the genetic description of an organism must be used in two different ways: 1) *interpreted* as instructions to construct its offspring, and 2) *uninterpreted* as data to be copied to create duplicates of the description to be passed on to its offspring. This was found to be the case for biological organisms once Watson and Crick revealed the workings of the DNA molecule.

Von Neumann's proof of the possibility of machine self-reproduction is achieved via a book-length constructive proof. The paper by Alvy Ray Smith provides a *one-page* proof of the possibility of machine self-reproduction. Whereas von Neumann required both computation universality and construction universality of his self-reproducing machines, Smith shows that computational universality alone suffices. Smith's proof relies on an elegant application of the famous Recursion Theorem of recursive function theory.

Another class of computer life-forms is already in existence, and is, in fact, becoming quite a problem. *Computer viruses* have been in existence for approximately ten years now, and since the introduction of the first viruses, not a single species has been eradicated. In fact, more sophisticated computer viruses are being introduced more and more rapidly, and the problem promises to become epidemic in the near future.

Biologists argue over whether or not biological viruses can be considered to be alive. Although they reproduce, they have no metabolism of their own to produce the required parts, and must take over the metabolic and reproductive machinery of a host cell in order to produce offspring. Allowing for the different medium, computer viruses appear to be every bit as much alive as biological viruses. The paper by Eugene Spafford presents an overview of the workings of different classes of computer viruses and argues that the release of computer viruses onto a network is as morally reprehensible as would be the release of biological viruses into a public reservoir.

There is a caution here that we all must attend to. Attempts to create Artificial Life may be pursued for the highest scientific and intellectual goals, but they may have very devastating consequences in the real world, if researchers do not take care to insure that the products of their research cannot "escape," either into computer networks or into the biosphere itself. There is little danger of that at this stage in the field, but the danger is increasing rapidly as we understand more and more about the synthesis of biological phenomena. What will the world be like when the means to produce self-reproducing robots is as widely available as the means to produce self-reproducing computer programs?

PHILOSOPHY AND EMERGENCE

Artificial Life presents a new forum within which to address a large number of philosophical questions, both classical and novel, about the nature of life, intelligence, and existence. There is a good deal of philosophical debate surrounding the field of Artificial Intelligence (AI), which all too often gets hopelessly mired down within the seemingly bottomless pit of subjective concepts like "mind" and "consciousness."

The debate about the possibility of machine life will have something in common with the debate about the possibility of machine intelligence. Both AI and AL study their respective subject matter by attempting to realize it within computers. Although AI has not yet achieved anything that even its most ardent supporters would call genuine machine intelligence, AI has completely changed the way in which scientists think about what it is to be "intelligent," and has, therefore, made a major scientific contribution, even though it hasn't achieved its overall goal.

Similarly, Artificial Life will force us to rethink what it is to be "alive." The fact is, we have no commonly agreed upon definition of "the living state." When asked for a definition, biologists will often point to a long list of characteristic behaviors and features shared by most living things (such as the list collated by Mayr[14]) which includes things like self-reproduction, metabolic activity, mortality, complex organization and behavior, etc. However, as most such lists are constituted of strictly behavioral criteria, it is quite possible that we will soon be able to exhibit computer processes that exhibit all of the behaviors on such a list. When that happens, we will have two choices, either 1) admit that the computer process is alive, or (more likely!) 2) change the list so as to exclude the computer process from being considered alive. Part (but only part) of the failure of AI to achieve "intelligence" is due to the fact that it is chasing a moving target. It used to be thought that playing chess required true intelligence. Now that computers can beat most humans at it, the ability to play chess is no longer considered an indicator of intelligence.

As is the case for AI, there are two claims that can be made with respect to the ontological status of artificial life models. The *Weak Claim* holds that these computer processes are nothing more than *simulations* of life, which are clearly useful for addressing certain scientific questions about "real" life, but which could never be considered to be *instances* of life themselves. The *Strong Claim*, on the other hand, holds that any definition or list of criteria broad enough to include all known biological life will also include certain classes of computer processes, which, therefore, will have to be considered to be "actually" alive.

The position that biological function can be abstracted from biological "stuff" is a "functionalist" position. Functionalism has been a major school of thought in the philosophical debate on the ontological status of intelligence, and many of the issues discussed in that debate have direct analogs in the debate on the ontological status of life. The paper by Elliott Sober compares and contrasts the functionalist approaches to Artificial Intelligence and Artificial Life. In doing so, Sober reflects on the successes and failures of the functionalist position with respect to AI, and, by extrapolation, predicts a similar suite of successes and failures for the functionalist position with respect to Artificial Life.

Just as many researchers in the field of Artificial Intelligence believe in the Strong Claim with respect to the possibility of synthesizing genuinely *intelligent* processes within computers, many researchers in the field of Artificial Life believe in the Strong Claim with respect to the possibility of synthesizing genuinely *living* processes within computers. The paper by Steen Rasmussen attempts to uncover the logical structure that must underlie such beliefs. This is a valuable effort, primarily because it attempts to reduce grandiose but untestable claims to a set of (hopefully more accessible and testable) assumptions that must be true if the larger claims are to be true.

The paper by Peter Cariani addresses the difficult concept of "emergence," a central, but poorly defined, concept in many fields, including Artificial Life. Cariani identifies the different senses in which the term "emergence" is typically used, and argues that computers cannot exhibit "genuinely" emergent processes. Rather, Cariani argues, only nature can support emergence in the most important sense of the term. Although his conclusions are in opposition to some of the beliefs held by many Artificial Life researchers, he is clearly correct that our usage of the term "emergence" must be more carefully qualified, if it is to have any real meaning or utility in the theories that we ultimately construct. There is some debate on whether or not "emergence" can be given a useful working definition at the present time. Like "life," it is one of those terms that we all use because the concept it refers to is a useful one, but we often use it without being able to say precisely what it means. Cariani's paper is a welcome opening position statement in what I hope will be a long and fruitful debate.

The posing of questions about the "essence" of life has never been an activity restricted to scientists alone, but has occupied the minds of philosophers, poets, writers, artists, and, indeed, every human being that has ever lived. In fact, it is impossible for any living, thinking being to think about living and being from a strictly objective viewpoint. We cannot ask about the nature of life with the same objective dispassion with which we ask about the nature of stars. We are not stars, but we are alive. Life pervades us and everything we do. Every living thing is an "expert" on life, and there are as many legitimate ways to express this expertise as there are living things. In short, the term "life" in common human usage covers more than is captured by its scientifically accessible core, and scientists are not the only ones we should listen to in our attempts to understand "life" in this broader sense.

Furthermore, the *practice* of science involves more than science. Although scientists often work within a world of abstraction and mathematics, the results of their abstract mathematical musings often have very tangible effects on the real world. The mastery of any new technology changes the world, and the mastery of a fundamental technology like the technology of life will necessarily change the world fundamentally. Because of the potentially enormous impact that it will have on the future of humanity, on Earth and beyond, it is extremely important that we involve the entire human community in the pursuit of Artificial Life.

It is a bit of a challenge to characterize the contribution by Louis Bec—zoosystémician, artist, poet, and philosopher in the grandest French tradition. Perhaps the best thing to do is to quote from Andre Longtin's referee report on Louis' paper (with Andreś permission):

> In this paper, the author describes the approach of the zoosystémician to the study of artificial life. After an introduction in which three "operators" defining the zoosystémician's intervention in the realm of Artificial Life are outlined, the author proceeds to explain how the archaic caesura between the arts, on one side, and science and technology, on the other side, has been displaced. With the advent of the sciences of the "artificial" and of communication, as well as the explosion of the technosciences and the sciences of the living, *a lieu* has emerged in which total integration of arts, sciences, and technology can be achieved. There are now two different "epistemological poles" that encompass this integration. The first strives to link "poetic," "symbolic" descriptions of nature's mechanisms to scientific ones, producing "metaphorical expressions." The second involves activities (cybernetics, artificial intelligence,...) which, among other ends, ultimately aim to simulate and act on the world, to better understand it by transforming it. In order to discover the places where thought originates and develops into the acts of creation and research, the author argues that a "fabulatory epistemology" is necessary, an epistemology enlarged to ethics, esthetics, ideology, and technology, and perfused by an activity of the imaginary.
>
> I recommend the paper for publication under the condition that it be reproduced in French. In fact, Louis' "fabulatory epistemology" is inextricably bound to its linguistic living substrate; it is at once an exercise in zoological creation and a fabulation on the French (and also Greek) languages. Any translation is sure to degrade this "symbiosis," unless done by someone who has precisely Louis's synthetic view of epistemology, philosophy, linguistics, arts, science, technology and culture.

THE FUTURE

Finally, the article by Doyne Farmer and Alletta Belin, is forward looking and intentionally provocative. This paper is the written version of an invited talk presented at a symposium to celebrate Murray Gell-Mann's 60th birthday. For this symposium, Gell-Mann posed a series of questions concerning the major social and technical challenges facing mankind in the next century. In his talk, Doyne suggested that the emergence of Artificial Life will present mankind with many such challenges. In his words:

> With the advent of artificial life, *we may be the first species to create its own successors.* What will these successors be like? If we fail in our task as creators, they may indeed be cold and malevolent. However, if we succeed,

> they may be glorious, enlightened creatures that far surpass us in their intelligence and wisdom. It is quite possible that, when the conscious beings of the future look back on this era, we will be most noteworthy not in and of ourselves but rather for what we gave rise to. Artificial life is potentially the most beautiful creation of humanity. To shun artificial life without deeper consideration reflects a shallow anthropocentrism.

To this I would only add that to *woo* Artificial Life without deeper consideration would be equally shortsighted. The creation of life is not an act to be undertaken lightly. We must do what we can to ensure that the future is equally bright for both our technological and our biological offspring.

REFERENCES

1. Axelrod, R., and W. D. Hamilton. "The Evolution of Cooperation." *Science* **211** (1981): 1390–1396.
2. Axelrod, R. *The Evolution of Cooperation.* New York: Basic Books, 1984.
3. Burks, A. W. Introduction to his collection: "Essays on Cellular Automata," University of Illinois Press, 1970 (emphasis added).
4. Conrad, M. "The Price of Programmability." In *The Universal Turing Machine: A Half-Century Survey*, edited by R. Herken. Oxford: Oxford University Press, 1988.
5. Eigen, M. "Self-Organization of Matter and the Evolution of Biological Macromolecules." *Naturwissenschaften* **10** (1971).
6. Eigen, M., and P. Schuster. *The Hypercycle: A Principle of Natural Self-Organization.* New York: Springer-Verlag, 1979.
7. Eldredge, N., and S. J. Gould. "Punctuated Equilibria: An Alternative to Phyletic Gradualism." In *Models in Paleobiology*, edited by T. J. M. Schopf, 82–115. Berlin: Freeman Cooper, 1972
8. Goldberg, D. E. *Genetic Algorithms in Search, Optimization, and Machine Learning.* Reading, MA: Addison-Wesley, 1989.
9. Hofstadter, D. R. *Gödel, Escher, Bach: An Eternal Golden Braid.* New York: Basic Books, 1979.
10. Holland, J. H. *Adaptation in Natural and Artificial Systems.* Ann Arbor: University of Michigan Press, 1975.
11. Langton, C. G. "Computation at the Edge of Chaos: Phase Transitions and Emergent Computation." *Physica D* **42** (1990): 12–37.
12. Margulis, L. *Origin of Eukaryotic Cells.* New Haven: Yale University Press, 1970.
13. Maynard-Smith, J. "Hypercycles and The Origin of Life." *Nature* **280** (1979): 445.

14. Mayr, E. *The Growth of Biological Thought.* Cambridge, MA: Harvard University Press, 1982.
15. Packard, N. "Adaptation Toward the Edge of Chaos." Center for Complex Systems Research Technical Report, CCSR-88-5, University of Illinois, 1988.
16. Wilson, S. W. "The Genetic Algorithm and Simulated Evolution." In *Artificial Life*, edited by C.G. Langton. Santa Fe Institute Studies in the Sciences of Complexity, Proc. Vol. VI, 157–166. Redwood City, CA: Addison-Wesley, 1989.

Charles E. Taylor
Department of Biology, University of California, Los Angeles, CA 90024, Internet address: taylor@cs.ucla.edu

"Fleshing Out" Artificial Life II

The field of study called Artificial Life (Alife) is concerned with human-made systems that exhibit behaviors which are more or less characteristic of natural living systems. While there are no artificial systems that would generally be regarded as alive at this time, a number of systems possess subsets of those properties and are being actively developed and expanded. Many, but not all, of those systems have been discussed in this volume and in the volume preceding it.[40] Others were described at the Artificial Life II meeting, in Santa Fe, New Mexico, February 5-9, 1990. Here I shall describe, however briefly, some of those systems which could not be included here, but which are nonetheless part of the Alife endeavor, and attempt to put the diverse approaches into some sort of order.

INTRODUCTION

I must first distinguish between those qualities which *define* life, and those which are simply *properties* of life. Many concepts are useful for thinking and for everday discourse but are difficult, or even impossible, to define precisely. Familiar examples include " machine," "game," and "love."[32,46] "Living" is another one. The theory of vitalism, the belief that living systems are fundamentally different from non-living ones, was popular for many years and directed biologists toward identifying just what delineates the living from the non-living. Few biologists today think it is worthwhile to pay much attention to that distinction.

It seems that no satisfactory formal *definition* of life has been proposed, though a number of *properties* that nearly all natural living systems possess can be identified. Mayr[45] provides a thoughtful review of this issue. Among those properties of living systems he feels to be the most important are: their complexity and organization; the chemical uniqueness of the molecules making them up; the uniquess and variability of individual living beings; their possession of a genetic program, which directs the unfolding of the ultimate phenotype; a history shaped by natural selection; and apparent indeterminacy of their actions. Each property by itself, even when considered with others, is unable to clearly delineate the living from the non-living, but together they do help to characterize what makes living things unique.

The problem is illustrated by self-replication, or reproduction. Reproduction is unquestionably an important property of most naturally occurring organisms, and is possibly the single most important one for distinguishing life from non-life. But the ability to reproduce is not posessed by *all* naturally living systems. At the borderline of life, but generally regarded as still alive, are such things as prions, which consist only of protein and contain no nucleic acids, and AIDS retroviruses, which contain RNA, but no DNA and are unable to reproduce on their own; they can do so only after they have directed host organisms to make DNA for them, and then "taken over" the host's metabolic system. More clearly demonstrating that reproduction is not essential for life are mules, or castratos and post-reproductive individuals who walk, talk, think, and are by all other criteria quite alive, yet are unable to produce offspring. It is apparent that reproduction, by itself, is not a defining feature of life, but is only one feature that most, though not all, living systems possess. A similar discussion could be directed at any of the other properties of life which are listed above.

MODELS OF ALIFE

There are many parallels between the early history of Artificial Intelligence and the current state of Artificial Life (Alife). The futility of precisely defining the subject matter is one of these. Another is the diversity of approaches that have been taken. During its early years Artificial Intelligence seemed to be best described by a simple listing of related research programs.[3,72] The same seems true of Alife today. My own classification of Alife research programs, made largely along methodological lines, with examples and a few remarks about each form, is as follows:

1. *Computer Viruses (and Worms, Bacteria, Rabbits, etc.)*: Computer viruses offer a graphic example of artificial life and illustrate many of the important

properties that living systems typically posess—reproduction, integration of parts, unpredictability, etc. They are discussed in this volume by several authors, in most detail by Spafford. While the potential that these systems hold for perniciousness is quite evident, they may also contain possibilities for good. Harold Thimbleby, in his talk at the meeting, suggested that such virus-like programs might be helpful if they were disbursed by individuals on a network, and directed to "report back" interesting or useful information to their "owners." For better or for worse, the number of such entities has increased explosively during recent years, and their multiplication, it seems, will continue in that phase for some time to come.

2. *Evolving Computer Processes*: Similar to computer viruses are a host of systems that embody the metaphor of living organism as processes or patterns in time-space that are capable of reproducing, absorbing information from their environment, modifying their surroundings, etc. Several examples were reported at the first AL conference.[50,66] It is important to recognize that it is not the *computer* that is said to be alive, but the *processes* themselves, interacting with the material substrate that supports them (processors and memory, including disks when there is virtual memory) that are properly regarded to have the properties of life.

   It may appear that computer processes are fundamentally different from natural life processes because the association between a process itself and the material supporting it is less intimate. For example, a computer process which is occupying the CPU can be interrupted and sent out to memory, perhaps to disk, while some other process gets to be executed on the CPU. The difference is only superficial. Some seeds can remain dormant for thousands of years, being neither metabolizing nor irritable during their dormancy, but they are unquestionably alive then and are capable of germinating when they encounter the right conditions.[26] Likewise, the computer processes can "survive" out in memory someplace, just waiting for the right conditions to reappear so that they may resume their active state.

   While such artificial life forms have much in common with computer viruses, they are typically not rogues, but are designed with the purpose of controlling and observing their behavior. Such programs have been around for many years,[20] and are beginning to see practical application in simulations of biological systems for research,[24,66] for integration into pest management,[23,22] and in education, to illustrate to young people the emergent properties that collections of interacting life forms may produce.[56]

3. *"Biomorphs" and Ontogenetically Realistic Processes*: The ability of living systems to evolve and their possession of a full morphogenesis that generates complex patterns of integrated parts, seem almost unique in the natural world. Complex and beautiful patterns, often with fractal geometry, may emerge from the repeated and collective result of simple actions by living cells. Examples in this category include the "biomorph" programs of Dawkins[12,13] and L-systems of Lindenmeyer and Prusinkiewicz,[42,53] which were developed in greater biological detail by Rob DeBoer[14] at the meeting.

At present, the emphasis and techniques of the biomorph-like systems and the evolving, self-reproducing ones distinguished above do seem different. But as our understanding and our ability to generate complex phenotypes by evolving bit strings develops, the line between those approaches will disappear. Computer graphics and animation are becoming more widely available, so we shall see more groups developing the potential for generating beautiful, life-like, and often unpredictable patterns and behaviors. For example, Larry Yeager[74] has programmed evolving neural nets that behave and move about, signaling to one another, with fleshed-out morphologies in a 3-dimensional environment on a Silicon Graphics workstation. An overview of the environment can be seen in one window, with the perspectives of each individual in "subwindows."

There will also be more emphasis on studying the chain of events between genes and the morphology of a full organism. In most systems that exist now, the genotypes are just bit strings which code parameters like weights or thresholds in an artificial neural net, and display none of the richness of form that would come were they to code for processes which can generate the nets recursively or in other, more biologically realistic ways.[17] The issues of how best to generate modularity, growth, and encoding of neural networks or other features of morphologically complex organisms by evolvable "genetic" representations are in need of study.

4. *Robotics*: Robots present a quite different approach to life like systems. Like Biomorphs, these typically lack the ability for self-reproduction, though some computer companies (e.g., NeXT), are not far from having microcomputers controlling robots which generate more of themselves.[44] Whether in their full literary and imaginative potential or as they actually exist now (e.g., the Collection Machine and the Confection Machine, described at the meeting by Rod Brooks), or as they will be in the reasonably forseeable future,[15,16,47,48] robots clearly possess many of the properties that are normally associated only with life—complexity, integration of parts, irritability, movement, etc.
5. *Autocatalytic Networks*: The antipode of high-technology robots, which are designed and constructed by humans to the highest levels of technology, are simple systems that spontaneously generate life-like, self-reproducing (or self-promoting) processes in a simulated chemical soup. Two types of these systems are under active investigation—autocatalytic networks and cellular automata. While fundamentally alike, these approaches differ in the assumptions they make and in the mathematical techniques that they employ.

There are many variants of autocatalytic networks; see, for example, those developed by Farmer,[2] Stuart Kauffman,[35,36] Rasmussen,[54] and others reviewed by Schuster.[61] An especially well-developed example of this approach is the theory of hypercycles, pioneered by Manfred Eigen and Peter Schuster.[18] Loosely speaking, hypercycles are connected networks of reacting elements (e.g., RNAs) that allow coherent evolution of functionally coupled, self-replicating entities. These entities are typically coupled networks of chemical reactants.

All naturally living systems of which I am aware exemplify hypercycles. The class of hypercycles is theoretically very much larger, however, and includes most serious candidates for the entities bridging the gap between unquestionably non-living and unquestionably living intermediates in the history of life on earth. Eigen and Schuster[18] have developed a reasonable theory about when hypercycles will spontaneously increase their complexity, and when they will collapse. The mathematical treatment of these and other autocatalytic networks is based on continuous reaction kinetics and has typically assumed complete mixing of elements, with no spatial structure. This encourages images of interaction and flux, which help to fill in the steps that must be imagined to have one day existed in the primordial soup. It has been less successful in dealing with spatial heterogeneity; it is here that models of cellular automata excel.

6. *Cellular Automata*: Cellular automata are arrays of cells, each of which assumes a discrete state. These states may change through (discrete) time according to well-defined rules. The transition rules are typically taken to be the same throughout the array, and take into account the current state of a cell together with the states of its immediate neighbors. All cells are assumed to be updated simultaneously. Both autocatalytic networks and cellular automata generate self-organizing, self-replicating patterns which persist through time. The mathematical techniques for studying them are, however, quite different. The cellular automata studies of Langton,[41] and those of Wolfram,[73] are examples of this approach.

The history of cellular automata is unusually distinguished. They were invented in the 1940s by John von Neumann, who "envisaged a systematic theory which would be mathematical and logical in form and which would contribute in an essential way to our understanding of natural systems (natural automata) as well as to our understanding of both analog and digital computers (artificial automata).[9]" Von Neumann developed this parallel with cellular automata over a series of years, until his death.[9,69] With the development of massively parallel SIMD computers[27,43,68] and the wider availability of low-cost color graphics, these systems are becoming much easier to study. The work of Langton,[41] directed at discerning which sets of cellular automata rules will support self-organizing processes and which will not, and then relating these findings to phase transitions in natural systems, is a step toward the fulfillment of von Neumann's vision.

Coming from another direction, Vladimir Kuz'min and his colleagues in the USSR have been using cellular automata to simulate the events that accompanied the early origin of life, especially the events surrounding the spontaneous origin of chirality which preceeded the first self-reproducing molecules.[25] Because there is such a close tie between cellular automata and the theory of computation, this approach has also been providing a helpful bridge for investigating the apparent Turing-machine equivalence of certain chemical systems and living cells. Ron Fox,[21] in a paper presented at the meeting, described both the insights that these mathematical approaches have provided, and the remaining gaps that need to be bridged.

In this volume, the paper by Boerlijst and Hogeweg[7] combines the two approaches toward generating self-reproducing processes, hypercycles and cellular automata, in a very satisfying solution to problems posed by hypercycle theory alone, and illustrates how natural systems may have become organized and spontaneously increased in complexity.

7. *Artificial Nucleotides*: It is important to recognize that artificial life is not confined to computers. While there could be many substrates which *might* support life, we know only one that unquestionably *does*, and the variety of life like systems that these chemical systems will support is only beginning to be explored.

   In the 1960s Sol Spiegelman and colleagues[63] combined the minimal set of molecules known at that time to permit RNA reproduction in a test tube—nucleotide precursors, inorganic molecules, energy source, replicase enzyme, and RNA primer from the $Q\beta$ bacteriophage. With all their needs supplied, the bacteriophage RNA molecules no longer needed to infect bacterial hosts, but did have to reproduce quickly in order to remain at an appreciable frequency. Through a sequence of serial transfers the RNA molecules increased rapidly in number, but at the same time became smaller, until some sort of minimum size was reached. The population of molecules evolved from a large, infectious form to smaller, non-infectious forms, presumably because it was favorable to shed the nucleotides which coded for properties no longer needed. These replicating and evolving RNA molecules were clearly akin to primitive, artificial life forms.

   Since then, new developments in molecular biology have permitted ever more interesting and artificial molecules to evolve. A key factor was the discovery by Cech and Altman that certain RNAs posess both enzymatic and replication abilities. This has permitted the evolution of single RNA molecules that carry out prescribed new reactions *in vitro*.[49,58] Another significant development has been the polymerase chain reaction for amplifying DNA.[4]

   Many societal issues will need to be identified and solved before this can be encouraged on a large scale.[64] But with sufficient care, many of the dangers from genetically engineering new life forms can be expected to be ameliorated, if not altogether eliminated, perhaps by using alternative genetic codes and other self-enforced methods.

8. *Cultural Evolution*: Ideas may come into being, perhaps originating from others, may be around for a while, may change the cultural milleau in ways favorable for their survival, may spread to others, and then "die out." Collections of ideas often form mutually supporting environments for their propagation. Ersatz religions and fads like the Teenage Mutant Ninja Turtles are only two examples. The relation to computer processes that have been the object of Alife research is more than superficial. It may be extreme to think of mutually supporting groups of ideas as actually posessing life, hypercycle-like, but the notion is not just fanciful. Richard Dawkins,[12] in particular, has stressed the similarity of real viruses, computer viruses, and groups of ideas in their life-like properties. The foundations for a serious, formal, theory of cultural evolution are being

laid.[10] It seems quite possible that this will someday become an exciting and useful application from studies in Alife.

This list, while far from comprehensive, illustrates the wide variety of research programs that Alife embraces. It also demonstrates that what joins researchers in this endeavor is not just their use of a common set of techniques, nor a common background in traditional areas of science. Rather, the element most in common to this research is the shared perspective (conviction if you like) that the signal feature of life is not the carbon-based substrate that supports the naturally occuring forms which we see on earth, and which consitute living systems as traditionally viewed. The important thing about life is that *"the local dynamics of a set of interacting entities (e.g., molecules, cells, etc.) supports an emergent set of global dynamical structures which stabilize themselves by setting the boundary conditions within which the local dynamics operates.* That is, these global structures can "reach down" to their own, physical bases of support and fine tune them in the furtherance of their own, global ends. Such LOCAL to GLOBAL back to LOCAL, inter-level feedback loops are essential to life, and are the key to understanding its origin, evolution, and diversity."[41]

Most researchers at the Alife meeting would agree that the projects described above represent only the most accessible and exploitable openings to a large set of systems that could possess life—the opening wedge into a new field that is likely to provide a satisfying synthesis of the worlds of Nature and of computation.

REASONS TO STUDY ALIFE

Why study Alife, and why should research in this area be supported? There are many reasons for doing so; these range from engineering new applications for control over our environment to providing a new, better perspective on our place in the Natural world.

Though still nascent, the field of artificial life has begun to provide the means for better understanding *emergent properties.* Collections of units at a lower level of organization, through their interaction, often give rise to properties that are not the mere superposition of their individual contributions, but gives the ensemble substantially new, emergent, properties. Such phenomena are found at all scales throughout nature, but seem especially apparent in systems that are alive. Indeed, life, itself, is an emergent property. When such ensembles are disassembled into their constituents, the interactions which give rise to the emergent properties are necessarily lost. Reductionist science, the method which characterizes most serious academic research today, is largely *analytic*, breaking things down, and so misses these emergent properties. Reductionist science has been enormously successful in many respects, but many features of Nature are being ignored. This occurs not because those features are uninteresting or unimportant. To the contrary, one would love to study them, but the appropriate tools and effective methods for researching them have been lacking. The field of Alife, because it is essentially *synthetic*—putting elements together to create life forms rather than dissecting them—offers the possibility to examine and study emergent properties in more detail than was previously possible.

There are many ways in which new properties can emerge, though none is so accessible or as easy to manipulate as that of increasing complexity by evolution through natural selection. Hence, there has been an emphasis on reproduction and evolvability in this volume[31,51] and in the previous one.[13] It may also happen that the same tools and methods found useful in Alife will be useful for problems on other scales as well. I suspect that the opportunity for studying emergent properties contributes to the unusual number of physicists that have become interested in Alife. Be that as it may, Alife represents a nontraditional approach to the study of emergent properties in nature, and permits the study of issues that have otherwise been very difficult to address.

I believe that Alife will prove to be especially useful as a tool for studying problems in biology. This is true for studying abstract issues like what sorts of entropy conditions will support life, but also for the specific and concrete problems which arise in the laboratory every day. Biology has long benefited from the use of simple model organisms to study phenomena which, more often than not, generalize to more complex ones, including humans. The laboratory rat, fruit fly (*Drosophila*), and colon bacteria (*Escherichia coli*) are just a few examples of such model organisms. Not only does Alife offer systems that have shorter life spans for experimentation (20 years for humans, 10 days for *Drosophila*, 1 second for Hillis'[28] sorting algorithms describe in this volume), but Alife systems offer unparalleled opportunities for control and reproducibility. They also offer the opportunity to see when interesting but unpredicted features emerge, then "roll back" time and study their genesis.[1] Further, because attention is necessarily directed toward systems that have some, but not all, of the properties of life, Alife systems help provide ideas of intermediate states that undoubtedly occurred on the path to life as we know it today. At the meeting there was much attention to ethology,[6,7,11] to growth and development[14] to ecology,[55] to the origin of life on earth,[21,39,60] and the possibility of life on other planets.[75] As a tool for basic research in biology, the ability to have organisms and communities of arbitrary simplicity, even if artificial, is extremely helpful.

Alife systems hold potential for developing new technologies and for otherwise enlarging our control over nature. This is evident with robotics, discussed during the meeting by Rod Brooks.[8] And there are many other applications as well. One that has recently begun to receive interest is evolving software. Today there are several broad classes of techniques for engineering a program for a specific task: explicitly writing the code from specifications of the task; automatic programming, where the formal specifications for the task are written in some specification language and then refining the resulting program to improve its performance; and machine learning, where the computer is "trained" by providing desired input and output, as by back propogation over neural nets.[59]

Work in Alife suggests that a fourth means of developing programs is to first choose a computational architecture and then provide a sequence of environments to evolve programs that will accomplish the specified tasks. For some time now the genetic algorithm community has been evolving parameter sets that will provide good, if not necessarily the best, solutions for a variety of tasks. At the meeting Stephanie Forrest described how this could be used to arrive at solutions to problems in game theory, based on early work by John Holland.[29]

That the programs themselves, and not simply their parameter sets, will evolve is more novel, but there are many such cases described in this volume.[11,28,31,70] The best ways to engineer new abilities will undoubtedly depend on both the task and on the computer architectures that are available. The acquisition of new abilities through learning, as by Artificial Neural Nets, is inherently a serial process, while that acquisition through evolution is inherently parallel. Massively parallel computers like Connection Machines were essential for several of the papers given at the meeting[11,28,41]; even greater parallelism would find immediate use.[43,68]

Another obvious application of Alife is through genetic engineering. The ability to evolve new life forms, capable of living in new places or carrying out reactions that traditional life cannot, through molecular biology offers many exciting possiblities. Work described by Gerald Joyce[33] at the meeting is a step consciously directed toward that goal.

Less appreciated in the past, but now receiving more of the attention it deserves, is the need for developing new educational tools. It is apparent that our country is not doing the job it would like to do, and needs to do, in teaching science and mathematics to the next generation. By synthesizing new life forms, it may be that students will be better able to understand and learn about emergent properties—especially in biology, where they are so evident, but in other areas like social behavior as well. Efforts in this direction by several groups were described at the meeting by the Apple Vivarium project (Alan Kay,[37] Larry Yaeger[74]) and the MIT Media Lab (Mitchel Resnick[56,57]); these represent applications of Alife technology. One of these software packages, SimCity, which was demonstrated by James Kalin[34] at the meeting, is already being marketed to aid in urban planning.

The effectiveness of Alife models for simulations in ecology and evolution has been mentioned above. Each year millions of children die from insect-borne disease, and a substantial part of the world's crops are lost to insects. The toll would be much greater if chemical pesticides were not available. At the same time, overuse of pesticides has led to very serious problems for the health of the world's ecosystems. There is an urgent need to discover ways for using pesticides more effectively and more wisely. How to do so will require better simulations than have been possible in the past.

Our research group has been developing Alife programs for simulating populations of insects for mosquito control and for agriculture. Some of this work has been described in Taylor et al.,[66] Fry, Taylor, and Devgan,[22] and Fry and Taylor.[23] By viewing groups of insects in separate breeding sites as separate subpopulations that can reproduce, interact with other subpopulations through migration, and disappear for a variety of reasons (drying up, scouring out by rains, etc.), we have been able to reproduce past population cycles of these insects and to examine alternate methods of controlling them. These simulations capture the relevant ecological interactions and their emergent properties more accurately and more robustly than have past simulations, and are beginning to find use for guiding mosquito-control programs.

In view of the many profound consequences from the developing field of artificial life, it is clear that social, ethical, and philosophical issues must be addressed now or in the immediate future. There is a widespread conviction among those involved in this research that the association and support among the "two cultures" in our society must be more than token. That point is stressed in papers by scientists Farmer and Belin,[19] Moravec[48] and Drexler,[16] and by artists Bec[5] and Pauline[52] (who, with the Survival Research Laboratory, gave a performance at the meeting). They are exploring the relation between humans and machines, and are helping researchers to develop a wider awareness of what may result.

Attention from the popular press can sometimes be a mixed blessing for scientific research, but the many journalists present at the meeting are certainly contributing to the appreciation of what an altered viewpoint on the meaning of life may hold for the problems our society is having with borderline cases of life. It is especially gratifying to see that the artists and the press were as concerned as those who are involved with the Alife research endeavor itself about learning and appreciating the larger issues that accompany an expanded view of life.

The Alife II meeting, and this volume that comes from it, are part of an attempt by people coming from many different directions and backgrounds, not just scientists and engineers, to develop a new view of life, and to understand the consequences that accompany that altered perspective. That viewpoint is different from the prevailing one because it regards the life we see in the plant and animal kingdoms as only one kind of many possible instantiations of living things. Other processes might properly be regarded as alive, and still other forms of life might well be created. Seen this way, humans become less separate from the natural world around us, and more continuous, more unified, with it. The consequences for science, engineering, and our view of ourselves are so great that it seems the Alife endeavor will not only continue to be exciting intellectually, but will be important in many other ways, as well. The challenge is to be aware of those consequences, and to direct them in ways that are beneficial.

ACKNOWLEDGEMENTS

I thank the UCLA Artificial Life Group for their comments and criticisims of the ideas expressed here. This work was supported by the W. M. Keck Foundation.

REFERENCES

1. Ackley. D. H., and M. L. Littman. "Interaction Between Learning and Evolution." Artificial Life II. This volume.
2. Bagley, R., and J. D. Farmer. "Emergence of Robust Autocatalytic Networks." Artificial Life II. This volume.
3. Barr, A., and E. A. Feigenbaum. *The Handbook of Artificial Intelligence*, Vol I. Los Altos, CA: Kauffman, 1981.
4. Beardsley, T. "New Order: Artificial Evolution Creates Proteins Nature Missed." *Sci. Am.* (1990): 18–19.
5. Bec, L. "Elements D'Epistemologie Fabulatoire." Artificial Life II. This volume.
6. Belew, R. K., J. McInerney, and N. N. Schraudolph. "Evolving Networks: Using the Genetic Algorithm with Connectionist Learning." Artificial Life II. This volume.
7. Boerlijst, M. C., and P. Hogeweg. "Self-Structuring and Selection: Spiral Waves as a Substrate for Prebiotic Evolution." Artificial Life II. This volume.
8. Brooks, R. "Real Artificial Life." Paper presented at the 1990 Workshop on Artificial Life held in Santa Fe, NM, February, 1990.
9. Burks, A. W. "Preface and Editor's Introduction." In *Theory of Self-Reproducing Automata*. Urbana, IL: University Illinois Press, 1966.
10. Cavalli-Sforza, L. L., and M. W. Feldman. *Cultural Transmission and Evolution: A Quantitative Approach*. Princeton, NJ: Princeton University Press, 1981.
11. Collins, R. J., and D. R. Jefferson. "AntFarm: Towards Simulated Evolution." Artificial Life II. This volume.
12. Dawkins, R. *The Blind Watchmaker*. New York: Norton, 1986.
13. Dawkins, R. "The Evolution of Evolvability." *Artificial Life*, edited by C. G. Langton. Studies in the Sciences of Complexity, Proc. Vol. VI, 201–220. Reading, MA: Addison-Wesley, 1989.
14. DeBoer, M., R. D. Fracchia, and P. Prusinkiewicz. "Analysis and Simulation of the Development of Cellular Layers." This volume.
15. Drexler, K. E. *Engines of Creation*. New York: Doubleday, 1986.
16. Drexler, K. E. "Biological and Nanomechanical Systems: Contrasts in Evolutionary Capacity." *Artificial Life*, edited by C. G. Langton. Studies in the Sciences of Complexity, Proc. Vol. VI, 501–520. Reading, MA: Addison-Wesley, 1989. .
17. Edelman, G. M. *Neural Darwinism: The Theory of Neuronal Group Selection*. New York: Basic Books, 1987.
18. Eigen, M., and P. Schuster. *The Hypercycle: A Principle of Natural Self-Organization*. New York: Springer Verlag, 1979.
19. Farmer, J. D., and A. Belin. "Artificial Life: The Coming Evolution" This volume.

20. Fogel, L. J., A. J. Owens, and M. J. Walsh. *Artificial Intelligence Through Simulated Evolution.* New York: Wiley, 1966.
21. Fox, R. "Synthesis of Artificial Life." Paper presented at the 1990 Workshop on Artificial Life held in Santa Fe, NM, February, 1990.
22. Fry, J., C. E. Taylor, and U. Devgan. "An Expert System for Mosquito Control in Orange County, California." *Bull. Soc. Vector Ecol.* **2** (1989): 237–246.
23. Fry, J., and C. E. Taylor. "Mosquito Control Simulation on the Connection Machine." *Bull. Calif. Mosquito and Vector Control Association.* In press.
24. Gibson, R. M., C. E. Taylor, and D. R. Jefferson. "Lek Formation by Female Choice: A Simulation Study." *Behav. Ecol.* **1** (1990): 36–42.
25. Goldanskii, V. I., and V. V. Kuz'min. "Spontaneous Mirror Symmetry Breaking in Nature and the Origin of Life." *Z. Phys. Chemie.* **269** (1988): 216–274.
26. Harrington, J. F. "Seed and Pollen Storage for Conservation of Plant Gene Resources." In *Genetic Resources in Plants–Their Exploration and Conservation*, edited by O. H. Frankel and E. Bennett, 501–517. Philadelphia, PA: Davis, 1970.
27. Hillis, W. D. *The Connection Machine.* Cambridge, MA: MIT Press, 1985.
28. Hillis, W. D. "Co-Evolving Parasites Improve Simulated Evolution as an Optimization Procedure." This volume.
29. Holland, J. H. *Adaptation in Natural and Artificial Systems.* Ann Arbor: University of Michigan Press, 1975.
30. Holland, J. H. "Echo: Explorations of Evolution in a Minature World." Paper presented at the 1990 Workshop on Artificial Life held in Santa Fe, NM, February, 1990.
31. Jefferson, D., R. Collins, C. Cooper, M. Dyer, M. Flowers, R. Korf, C. Taylor, and A. Wang. "Evolution as a Theme in Artificial Life: The Genesys/Tracker System." This volume.
32. Johnson-Laird, P. N. *The Computer and the Mind: An Introduction to Cognitive Science.* Cambridge, MA: Harvard University Press, 1988.
33. Joyce, G. "RNA Evolution." Paper presented at the 1990 Workshop on Artificial Life held in Santa Fe, NM, February, 1990.
34. Kalin, J. "Sim City." Demonstration given at the 1990 Workshop on Artificial Life held in Santa Fe, NM, February, 1990.
35. Kauffman, S. A. "Autocatalytic Replication of Polymers." *J. Theor. Bio.* **119** (1986): 1–24.
36. Kauffman, S. A., and S. Johnson. "Coevolution to the Edge of Chaos: Coupled Fitness Landscapes, Poised States, and Coevolutionary Avalanches." This volume.
37. Kay, A. "Point of View is Worth 80 IQ Points." Paper presented at the 1990 Workshop on Artificial Life held in Santa Fe, NM, February, 1990.
38. Koza, J. R. "Co-Evolution of Populations of Computer Programs." This volume.
39. Kuz'min, V. "Symmetry Breaking as a Possible Conducting Principle in Evolution: Example of Chirality." Paper presented at the 1990 Workshop on Artificial Life held in Santa Fe, NM, February, 1990.

40. Langton, C. G., ed. *Artificial Life*. Santa Fe Institute Studies in the Sciences of Complexity, Proc. Vol. VI. Reading, MA: Addison-Wesley, 1989.
41. Langton, C. G. "Life at the Edge of Chaos." This volume.
42. Lindenmeyer, A., and P. Prusinkiewicz. "Developmental Models of Multicelluar Organisms: A Computer Graphics Perspective." In *Artificial Life*, edited by C. G. Langton. Santa Fe Institute Studies in the Sciences of Complexity, Proc. Vol. VI, 221–250. Reading, MA: Addison-Wesley, 1989.
43. Margolus, N., and T. Toffoli. "Cellular Automata Machines." *Lattice Gas Methods for Partial Differential Equations*, edited by G. Doolan. Santa Fe Institute Studies in the Sciences of Complexity, Proc. Vol. IV, 219–249. Reading, MA: Addison-Wesley, 1990.
44. Markoff, J. "All NeXT's Factory Lacks is Orders: Computers Operate Robot Assemblers." *New York Times*, Monday, December 24, 1990, p. 23.
45. Mayr, E. *The Growth of Biological Thought.* Cambridge, MA: Harvard University Press, 1982.
46. Minsky, M. L. *Computation, Finite and Infinite Machines.* Englewood Cliffs, NJ: Prentice Hall, 1967.
47. Moravec, H. *Mind Children.* Cambridge, MA: Harvard University Press, 1988.
48. Moravec, H. "Human Culture: A Genetic Takeover Underway." Artificial Life, Studies in the Sciences of Complexity, Proc. Vol. VI, edited by C. G. Langton,167–200. Reading, MA: Addison-Wesley, 1989.
49. North, G. "Expanding the RNA Repertoire." *Nature* **345** (1990): 576–578.
50. Packard, N. H. "Evolving Bugs in a Simulated Ecosystem." In *Artificial Life*, edited by C. G. Langton. Studies in the Sciences of Complexity, Proc. Vol. VI, 141–156. Reading, MA: Addison-Wesley, 1989.
51. Packard, N. H. "Measurement of Evolutionary Activity and Teleology." Paper presented at the 1990 Workshop on Artificial Life held in Santa Fe, NM, February, 1990.
52. Pauline, M. Performance by the Survival Research Laboratory presented at the 1990 Workshop on Artificial Life held in Santa Fe, NM, February, 1990.
53. Prusinkiewicz, P., and A. Lindenmeyer. *The Algorithmic Beauty of Plants.* New York: Springer Verlag,1990.
54. Rasmussen, S. "Toward a Quantitative Theory of Life. In *Artificial Life*, edited by C. G. Langton. Studies in the Sciences of Complexity, Proc. Vol. VI, 79–104. Reading, MA: Addison-Wesley, 1989.
55. Ray, T. S. "An Approach to the Synthesis of Artificial Life." This volume.
56. Resnick, M. "Overcoming the Centralized Mindset: Toward an Understanding of Emergent Phenomena." Epistemology and Learning Memo Number 11, MIT Media Laboratory Technical Reports, 1990.
57. Resnick, M. "Logo: A Children's Environment for Exploring Self-Organizing Behavior." Paper presented at the 1990 Workshop on Artificial Life held in Santa Fe, NM, February, 1990.
58. Robertson, D. L., and G. F. Joyce, "Selection *in vitro* of an RNA Enzyme that Specifically Cleaves Single-Stranded DNA." *Nature* **344** (1990): 467–468.

59. Rumelhart, D. E., J. L. McClelland, and the PDP Research Group. *Parallel Distributed Processing: Explorations in the Microstructure of Cognition.* Cambridge, MA: MIT Press, 1986.
60. Schopf, J. W. "What Must be True for Life to Have Evolved on this Planet." Paper presented at the 1990 Workshop on Artificial Life held in Santa Fe, NM, February, 1990.
61. Schuster, P. "Dynamics of Autocatalytic Reaction Networks." *Molecular Evolution on Rugged Landscapes*, edited by A. Perelson and S. Kauffman. Studies in the Sciences of Complexity, Proc. Vol. IX, 281–306. Redwood City, CA: Addison-Wesley, 1990.
62. Spafford, E. H. "Computer Viruses: A Form of Artificial Life." This volume.
63. Spiegelman, S., N. R. Pace, D. R. Mills, R. Livisohn, T. S. Eikhorn, M. M. Taylor, R. L. Peterson, and D. H. L. Bishop. "Chemical and Mutational Studies of a Replicating RNA Molecule." *Proc. XII. Int. Cong. Genetics* **3** (1969): 127–154.
64. Suzuki, D., and P. Knudson. *Genethics: The Ethics of Engineering Life.* Toronto: Stoddart, 1988.
65. Taylor, C. E., D. R. Jefferson, S. R. Turner, and S. R. Goldman. "RAM: Artificial Life for the Exploration of Complex Biological Systems." In *Artificial Life*, edited by C. G. Langton. Studies in the Sciences of Complexity, Proc. Vol. VI, 275–295. Reading, MA: Addison-Wesley, 1989.
66. Taylor, C. E., L. Muscatine, and D. R. Jefferson. "Maintenance and Breakdown of the Hydra-Chlorella Symbiosis: A Computer Model." *Proc. Roy. Soc. Lond. B* **238** (1989): 277–289.
67. Thimbleby, H., I. Witten, and D. Pullinger. "Laws for Cooperative Artificial Life." Paper presented at the 1990 Workshop on Artificial Life held in Santa Fe, NM, February, 1990.
68. Toffoli, T. "Programmable Matter." Paper presented at the 1990 Workshop on Artificial Life held in Santa Fe, NM, February, 1990.
69. von Neumann, J. *Theory of Self-Reproducing Automata.* Urbana, IL: University Illinois Press, 1966.
70. Werner, G. M., and M. G. Dyer. "Evolution of Communication in Artificial Organisms." This volume.
71. Wilson, S. Paper presented at the 1990 Workshop on Artificial Life held in Santa Fe, NM, February, 1990.
72. Winston, P. H., and R. H. Brown. *Artificial Intelligence: An MIT Perspective.* Cambridge, MA: MIT Press, 1979.
73. Wolfram, S. *Theory and Applications of Cellular Automata.* Singapore: World Scientific, 1986.
74. Yeager, L. Personal communication at the 1990 Workshop on Artificial Life held in Santa Fe, NM, February, 1990.
75. Zuckerman, B. "Constraints on and Prospects for Live Elsewhere in the Universe." Paper presented at the 1990 Workshop on Artificial Life held in Santa Fe, NM, February, 1990.

Origin/Self-Organization

Christopher G. Langton
Complex Systems Group, MS-B213, Theoretical Division, Los Alamos National Laboratory, Los Alamos, New Mexico, 87545, and the Santa Fe Institute, 1120 Canyon Road, Santa Fe, NM 87501

Life at the Edge of Chaos

In living systems, a dynamics of information has gained control over the dynamics of energy, which determines the behavior of most non-living systems. How has this domestication of the brawn of energy to the will of information come to pass?

By studying the conditions under which a complex dynamics of information can emerge spontaneously in a class of formal systems known as Cellular Automata (CA), we suggest that information can come to dominate the dynamics of physical systems in the vicinity of a second-order (or *critical*) phase transition.

We discuss the implications of this finding for our understanding of the origin and evolution of life and intelligence.

1. INTRODUCTION

What is Life? What is meant by the notion of "the living state?" What properties or characteristics distinguish the living from the non-living?

There have been many criteria proposed to identify the "living state." Biologists today will often point to a long list of properties that are shared by *most* living things. Such lists often include capacities for self-reproduction and metabolism, the exhibition of complex structure and interdependence among parts, reliance on a genetic code, a genotype/phenotype relation, and so forth.[1]

However, the most salient feature that distinguishes living organisms is that their behavior is clearly based on a complex dynamics of information. In living systems, information processing has somehow gained the upper hand over the dynamics of energy that dominates the behavior of most non-living systems.

Therefore, the fundamental question motivating the research described here is:

Under what conditions can we expect a dynamics of information to emerge spontaneously and come to dominate the behavior of a physical system?

If we can answer this question we will have gone a long way towards understanding how life could emerge from non-life, not only here on Earth but elsewhere in the Universe as well.

1.1 THE DYNAMICS OF CELLULAR AUTOMATA

In this paper, we will address this question via a study of the conditions under which a complex dynamics of information can emerge spontaneously within a class of discrete approximations to physical systems known as cellular automata (CA).

CA are appropriate formalisms within which to pursue this study for a number of reasons.

- CA are spatially extended, nonlinear dynamical systems.
- As nonlinear dynamical systems, CA exhibit the entire spectrum of dynamical behaviors, from fixed-points, through limit cycles, to fully developed chaos.
- CA are capable of supporting universal computation. Thus, they are capable of supporting the most complex known class of information dynamics.
- There is a very general and universal representation scheme for all possible CA: a look-up table. This form of representation allows us to parameterize the space of possible CA, and to search this space in a canonical fashion.
- CA are very physical, a kind of "programmable matter."[45] Thus, what we learn about information dynamics in CA is likely to tell us something about information dynamics in the physical world.

[1]Mayr[35] has collated a representative list of such properties.

So, we can address the fundamental question above in the context of CA by asking:

> *Under what conditions can we expect a complex dynamics of information to emerge spontaneously and come to dominate the behavior of a CA?*

PREVIEW

The investigation of this particular question will take up much of the rest of this paper. First, we will introduce cellular automata more formally, describe methods for sampling the space of possible CA dynamics, and review statistical measures used to characterize these dynamics. Then, we will turn to qualitative and quantitative overviews of CA dynamics, which demonstrate the existence of a *phase transition* in the space of CA. Next, we observe that we can locate the most complex information dynamics in the vicinity of this phase transition, and show how this can serve to explain a great deal about the structure of the space of computations in general. Finally, we return to the main theme of this paper and discuss what all of this has to tell us about life, its origin, and evolution.

2. CELLULAR AUTOMATA

Formally, a cellular automaton is a D-dimensional lattice with a finite-state automaton (FSA) residing at each lattice site. Each automaton takes as input the states of the automata within some *finite, local* region of the lattice, defined by a neighborhood template $\mathcal{N}$, where the dimension of $\mathcal{N} \leq D$. The size of the neighborhood template, $|\mathcal{N}|$, is just the number of lattice points covered by $\mathcal{N}$. By convention, an automaton is considered to be a member of its own neighborhood.

Each FSA consists of a finite set of *cell states* Σ, a finite *input alphabet* α, and a *transition function* Δ, which is a mapping from the set of neighborhood states to the set of cell states. Letting $N = |\mathcal{N}|$:

$$\Delta : \Sigma^N \rightarrow \Sigma.$$

The *state* of a neighborhood is the cross product of the states of the FSA covered by the neighborhood template. Thus, the input alphabet α for each automaton consists of the set of possible neighborhood states: $\alpha = \Sigma^N$. Letting $K = |\Sigma|$ (the number of cell states) the size of α is equal to the number of possible neighborhood states

$$|\alpha| = |\Delta| = |\Sigma^N| = K^N.$$

To define a transition function Δ, one must associate a unique next state in Σ with each possible neighborhood state. Since there are $K = |\Sigma|$ choices of state to assign as the next state for each of the $|\Sigma^N|$ possible neighborhood states, there are $K^{(K^N)}$ possible transition functions Δ that can be defined. We use the notation $\mathcal{D}_N^K$ to refer to the set of all possible transition functions Δ which can be defined using N neighbors and K states.

For example, consider a two-dimensional cellular automaton using eight-states per cell, a rectangular lattice, and the five-cell neighborhood template shown above. Here $K = 8$ and $N = 5$, so $|\Delta| = K^N = 8^5$ so there are 32,768 possible neighborhood states. For each of these, there is a choice among eight states for the next cell state under Δ, so there are $|\mathcal{D}_N^K| = K^{(K^N)} = 8^{(8^5)} \approx 10^{30,000}$ possible transition functions using the five-cell neighborhood template with eight-states per cell.

2.1 PARAMETERIZING THE SPACE OF CA RULES

$\mathcal{D}_N^K$, the set of possible Δ functions for a CA of K states and N neighbors, is fixed once we have chosen the number of states per cell and the neighborhood template. However, there is no intrinsic order within $\mathcal{D}_N^K$; it is a large, undifferentiated *space* of CA rules.

Imposing a *parameterization scheme* on this undifferentiated space of CA rule allows us to define a natural ordering on the rules. The ideal ordering scheme would partition the space of CA rules in such a manner that rules from the same partition would support similar dynamics. Such an ordering on $\mathcal{D}_N^K$ would allow us to observe the way in which the dynamical behaviors of CA vary from partition to partition.

The location in this space of the partitions supporting universal computation, relative to the location of partitions supporting *other* possible dynamical behaviors, would then provide us with insights into the conditions under which we should expect a complex dynamics of information to emerge in CA.

2.1.1 THE λ PARAMETER We will consider a subspace of $\mathcal{D}_N^K$, characterized by the parameter λ.[26,27,28]

The λ parameter is defined as follows. We pick an arbitrary state $s \in \Sigma$, and call it the *quiescent* state s_q. Let there be n_q transitions to this special quiescent state in a transition function Δ. Let the remaining $K^N - n_q$ transitions in Δ be filled by picking randomly and uniformly over the other $K - 1$ states in $\Sigma - s_q$. Then

$$\lambda = \frac{K^N - n_q}{K^N}. \tag{1}$$

If $n_q = K^N$, then *all* of the transitions in the Δ function will be to the quiescent state and $\lambda = 0.0$. If $n_q = 0$, then there will be *no* transitions to s_q and $\lambda = 1.0$. When all states are represented equally in the Δ function, then $\lambda = 1.0 - 1/K$.

The parameter values $\lambda = 0$ and $\lambda = 1 - 1/K$ represent the most homogeneous and the most heterogeneous Δ functions, respectively. The behavior in which we will be interested is captured between these two parameter values. Therefore, we experiment primarily with λ in the range $0 \leq \lambda \leq 1 - 1/K$.

λ discriminates well between dynamical regimes for "large" values of K and N, whereas λ discriminates poorly for small values of K and N. For example, for the "elementary" one-dimensional CA with $K = 2$ and $N = 3$, λ is only very roughly correlated with dynamical behavior, which probably explains why the relationships reported here were not observed in earlier work on classifying CA dynamics,[49,51] as these investigations were carried out using CA with minimal values of K and N.

For these reasons, we employ CA for which $K \geq 4$ and $N \geq 5$, which results in Δ functions of size $4^5 = 1024$ or larger.

2.1.2 SEARCHING CA SPACE WITH THE λ PARAMETER In the following, we use the λ parameter as a means of sampling $\mathcal{D}_N^K$ in an ordered manner. We do this by stepping through the range $0.0 \leq \lambda \leq 1.0 - 1/K$ in discrete steps, randomly constructing Δ functions for each λ point. Then we run CA under these randomly constructed Δ functions, collecting data on various measures of their dynamical behavior. Finally, we examine the behavior of these measures as a function of λ.

Δ functions are constructed in two ways using λ: the *random-table method* and the *table-walk-through method.* In the random-table method, λ is interpreted as a bias on the selection of states from Σ as we sequentially fill in the transitions that make up a Δ function. To do this, we step through the table, flipping a λ-biased coin for each neighborhood state. If the coin comes up tails (with probability $1.0 - \lambda$), we assign the state s_q as the next cell state for that neighborhood state. If the coin comes up heads (with probability λ), we pick one of the $K - 1$ states in $\Sigma - s_q$ at uniform random as the next cell state.

In the table-walk-through method, we start with a Δ function consisting entirely of transitions to s_q, so that $\lambda = 0.0$ (but note restrictions below). New transition tables with higher λ values are generated by randomly replacing a few of the transitions to s_q in the current Δ function with transitions to other states, selected randomly from $\Sigma - s_q$. Tables with *lower* λ values are generated by randomly replacing a few transitions that are *not* to s_q in the current Δ function by transitions to s_q.

Thus, under the table-walk-through method, we progressively perturb "the same table," whereas under the random-table method, each table is generated anew, independently of the others.

2.1.3 FURTHER RESTRICTIONS ON CA In order to make our studies more tractable, we impose two further conditions on the rule spaces:

1. *Strong Quiescence:* all neighborhood states uniform in cell state s_i will map to state s_i under Δ.
2. *Spatial Isotropy:* all planar rotations of a neighborhood state will map to the same cell state under Δ.

The first restriction implies that uniform regions of the array remain uniform under the action of a rule. The second restriction implies that the local physics cannot tell which way is up, so to speak. That is, local rules cannot make use of the global property of absolute orientation with respect to the lattice, although they can distinguish between left and right handedness, which is a local property.

2.2 CLASSIFYING CA BEHAVIOR

The most widely known scheme for classifying cellular automata on the basis of their dynamical behaviors is due to Wolfram,[49] who proposed the following four qualitative classes of CA behaviors:

- **Class I CA** evolve to a fixed, homogeneous state.
- **Class II CA** evolve to simple separated periodic structures.
- **Class III CA** yield chaotic aperiodic patterns.
- **Class IV CA** yield complex patterns of localized structures.

Wolfram finds the following analogs for his four classes of cellular automaton behaviors in the field of dynamical systems.[49]

- **Class I CA** evolve to *limit points.*
- **Class II CA** evolve to *limit cycles.*
- **Class III CA** evolve to *chaotic* behavior of the kind associated with *strange attractors.*
- **Class IV CA** have very long *transients*, and "no direct analog for them has been identified among continuous dynamical systems."

Various suites of statistical measures have been employed in the attempt to quantify these qualitative differences, with somewhat limited success.[12]

However, classification alone is not enough. What is needed is a deeper understanding of the structure of cellular automata rule spaces, one which provides an explanation for the existence of the observed classes and for their relationships to one another. Choosing an appropriate parameterization of the space of cellular automata rules, such as λ, allows direct observation of the way in which different statistical measures are related as a function of the parameter(s). These relationships in turn provide an explanation for the existence and ordering of the various qualitatively distinguishable classes of cellular automata dynamics. From the vantage point of the resulting understanding of the "deep structure" of CA rule spaces, the various qualitative classes proposed by Wolfram and others make sense, and are even to be expected. However, it is also obvious that such classification schemes can only serve as rough approximations to the more subtle, underlying structure.

2.2.1 CLASS IV CELLULAR AUTOMATA Of the four Wolfram classes, Class IV is both the most interesting and the least well characterized. It is the most interesting class because it contains the CA rules exhibiting the most "complex" behaviors. Conway's game of LIFE[16] is a well-known example of a Class IV CA. Class I CA exhibit the maximal possible order: a uniform, homogeneous structure, like a crystal. This order exists on both *local* and *global* scales. Class II CA also exhibit both local and global order, although not maximal. Class III CA can exhibit maximal *disorder*, and this disorder exists on both local and global scales. Class IV CA exhibit a great deal of local order, but little apparent global order, although with time, global order may emerge.

Wolfram suggests that Class IV CA are capable of supporting computation, even universal computation, and that it is this capacity that makes their behavior so complex. The game of LIFE has been shown to be capable of universal computation.[6] In this paper, we provide support for this hypothesis and offer an explanation for why we should expect to see a capacity for information processing in certain regions of CA rule space. The association of Class IV CA with "very long transients" will figure "critically" in the explanation, and helps to explain why Class IV is hard to characterize.

3. QUALITATIVE OVERVIEW OF CA DYNAMICS

In this section, we present a *qualitative* overview of the change in dynamical behavior of a typical one-dimensional CA as we pass through the most interesting region of CA rule space by varying λ via the table-walk-through method.[2] In the next section, we will present a *quantitative* overview.

3.1 A SURVEY OF ONE-DIMENSIONAL CA DYNAMICS USING λ

For these one-dimensional CA, $K = 4$ and $N = 5$ (i.e., *two* cells on the left and *two* cells on the right are included in the neighborhood template). The arrays consist of 128 sites connected in a circle, providing periodic boundary conditions. Each array is started from a random initial configuration on the top line, and successive lines show successive time steps in the evolution of the dynamical behavior.

For each value of λ, we show two evolutions. The arrays in Figure 1 are started from a uniform random initial configuration over all 128 sites. This series illustrates the kinds of structures that develop, as well as the typical transient times before these structures are achieved.

[2] It is not always the case that the table-walk-through method passes through the most interesting region of CA rule space. This topic will be taken up in the next section.

The arrays in Figure 2 are initialized with a patch of 20 randomized sites, with the remaining sites set to state 0 (the "quiescent" state). This series illustrates the relative rates of the spread (or collapse) of the area of dynamical activity.

For those values of λ exhibiting long transients, we have reduced the scale of the arrays in order to display longer evolutions.

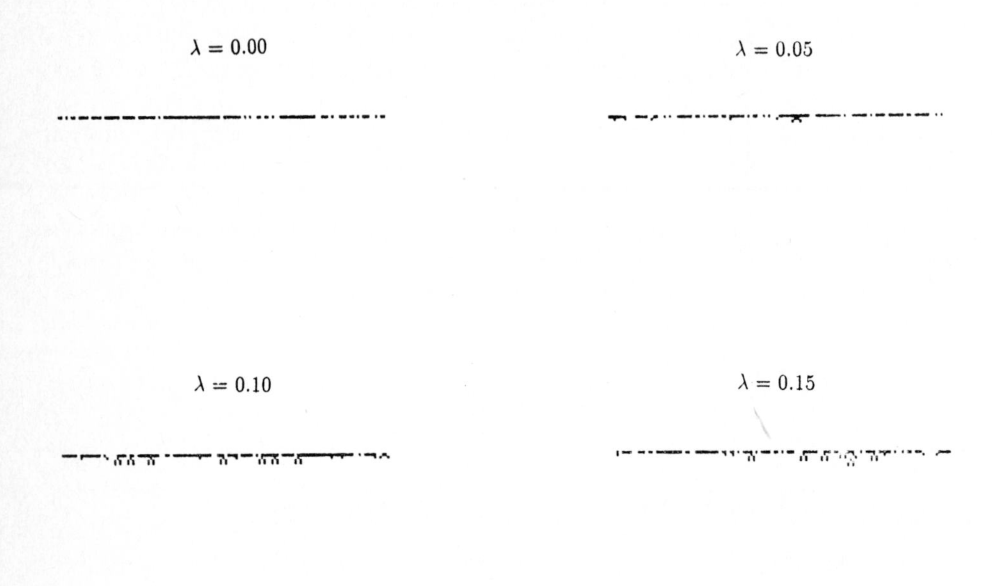

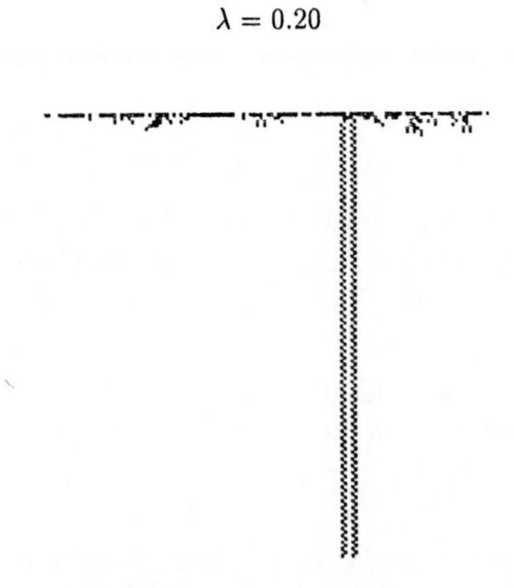

FIGURE 1 Evolution of one-dimensional CA over $0 < \lambda \leq 0.75$ from fully random initial conditions.

$\lambda = 0.25$

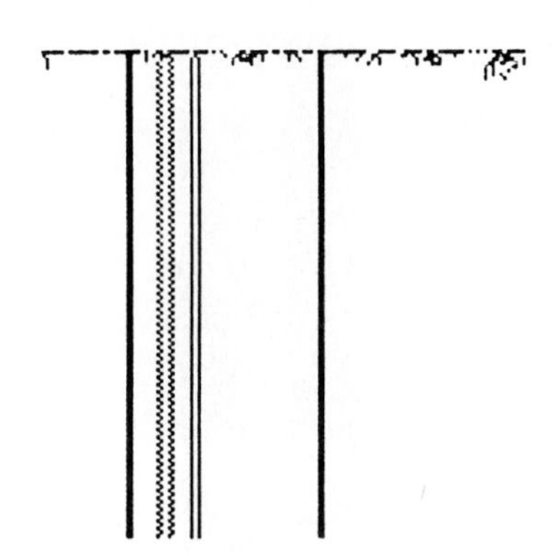

$\lambda = 0.30$

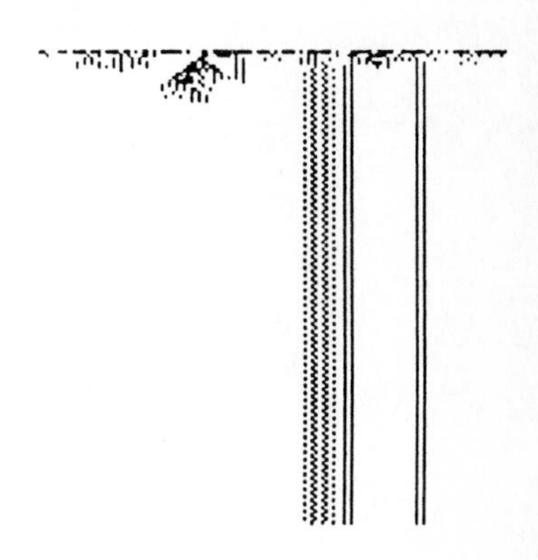

$\lambda = 0.35$

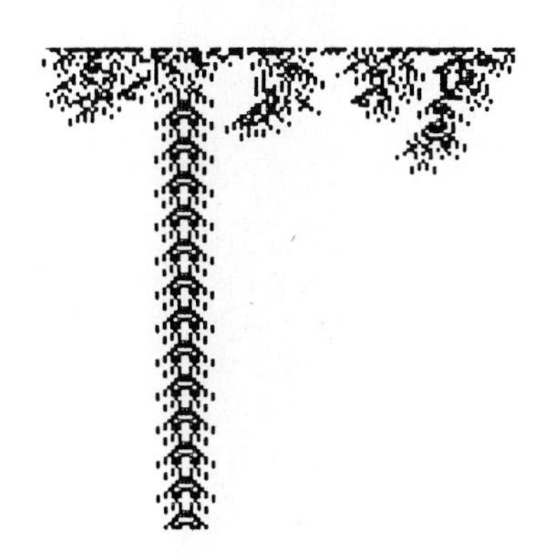

$\lambda = 0.40$

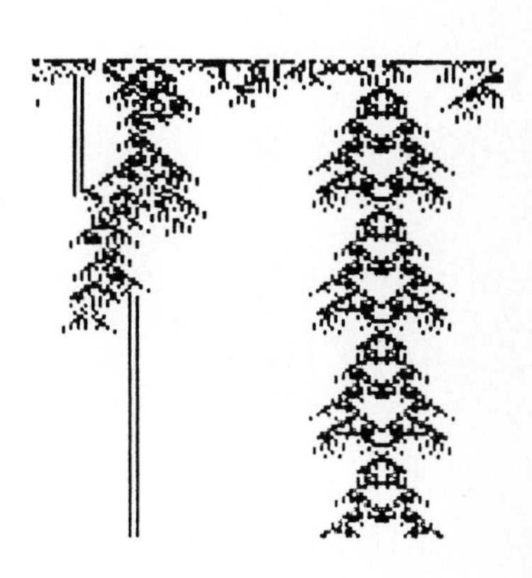

FIGURE 1 (continued)

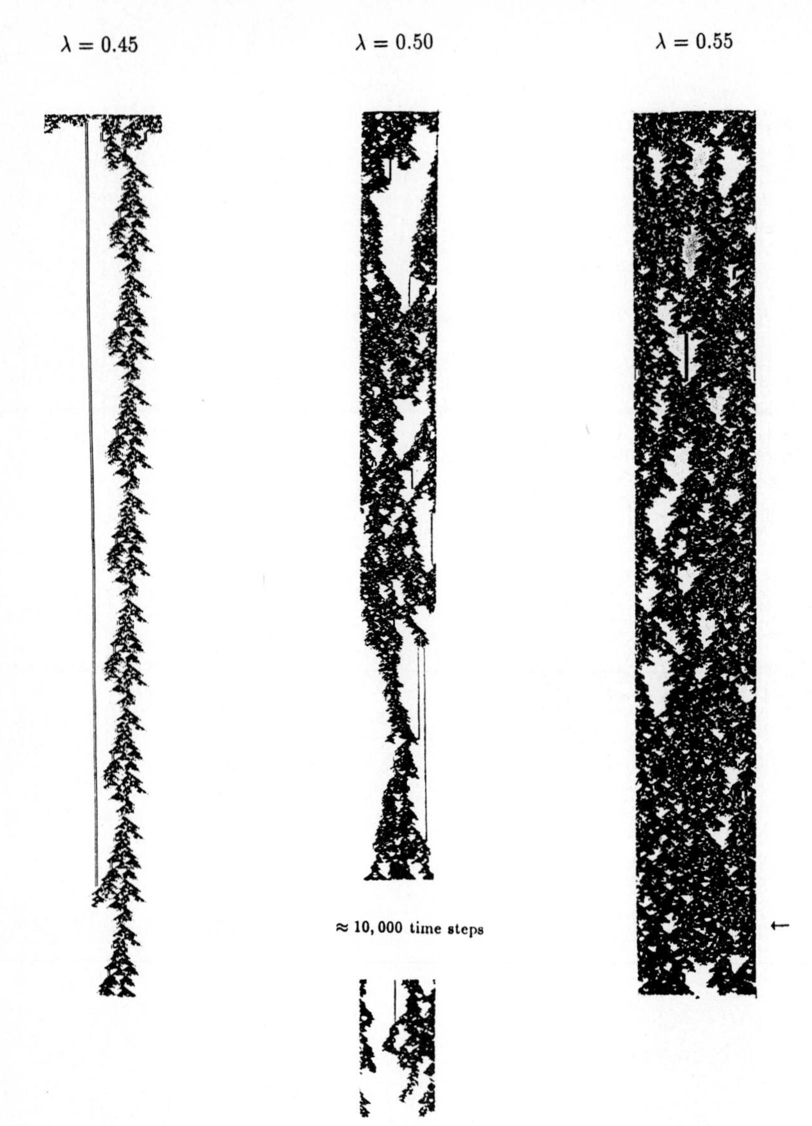

FIGURE 1 (continued)

$\lambda = 0.60$

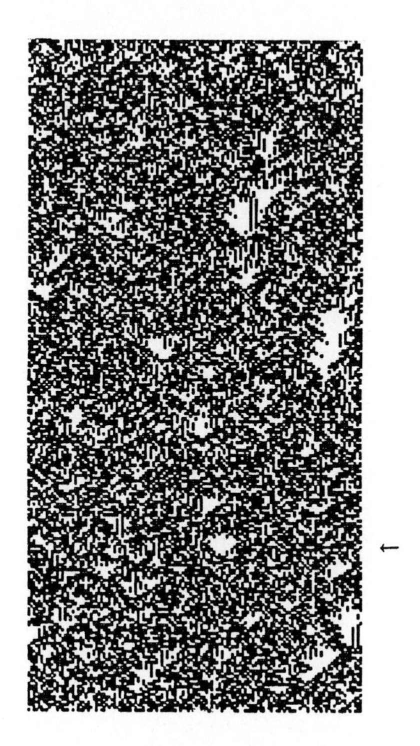

$\lambda = 0.65$

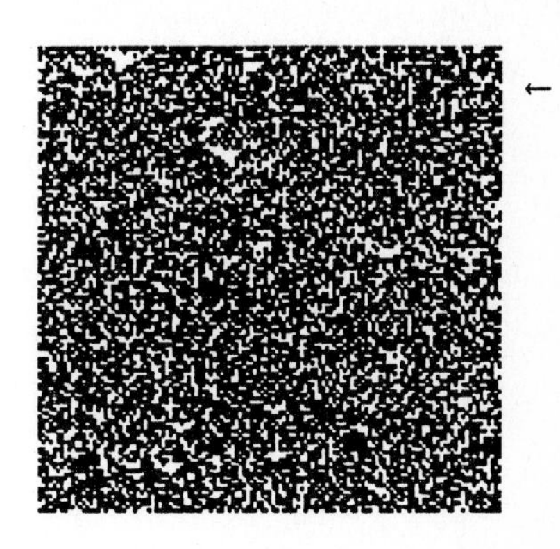

$\lambda = 0.70$

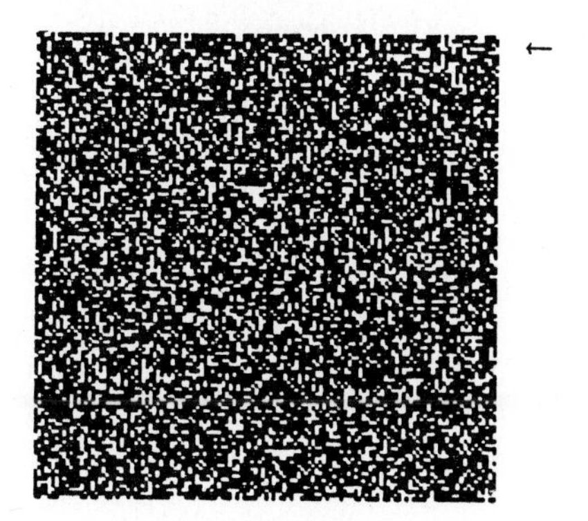

$\lambda = 0.75$

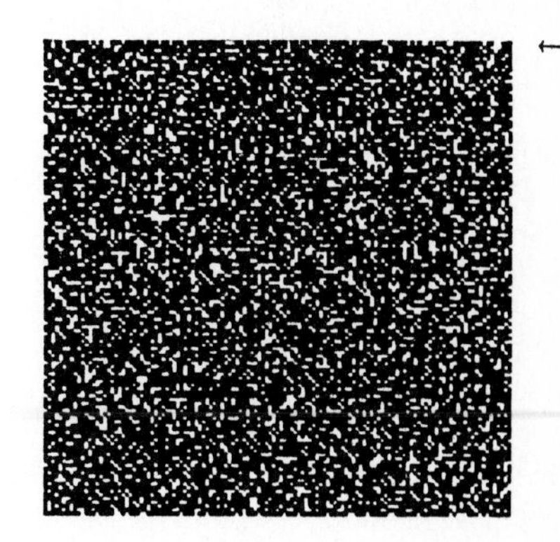

FIGURE 1 (continued)

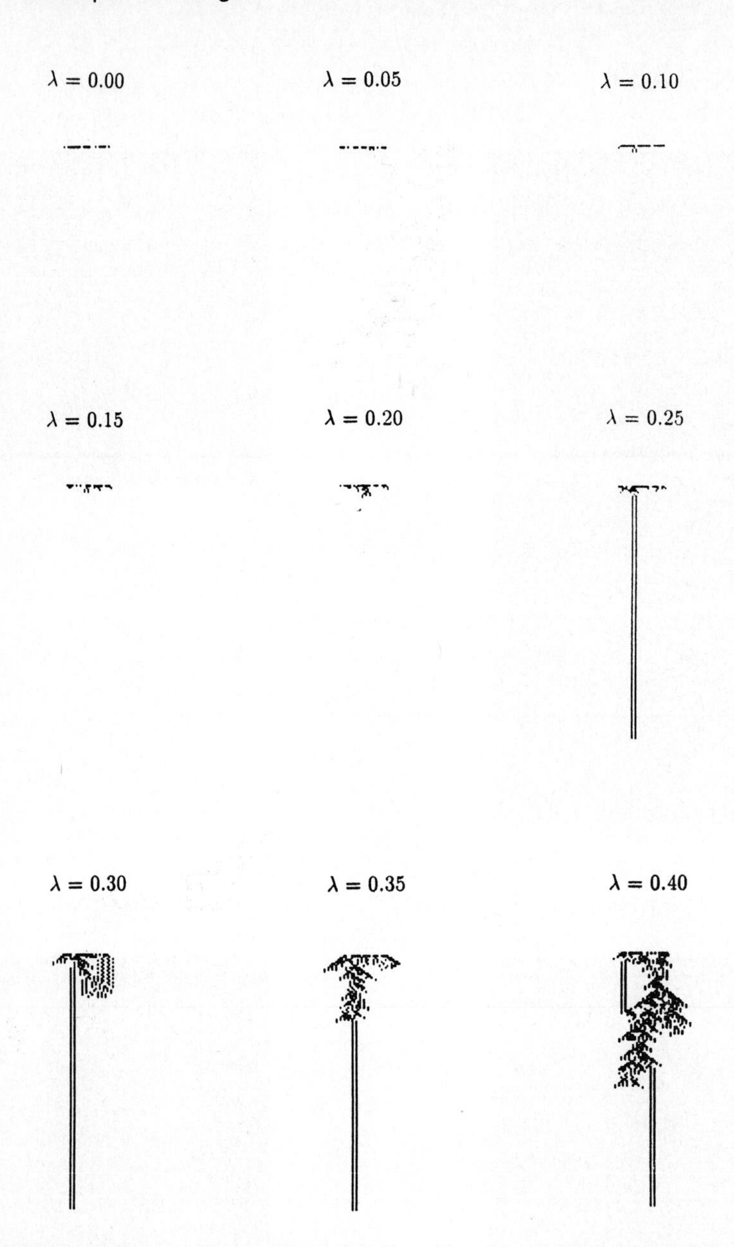

FIGURE 2 Evolution of one-dimensional CA over $0 < \lambda \leq 0.75$ from partially random initial conditions.

FIGURE 2 (continued)

$\lambda = 0.60$

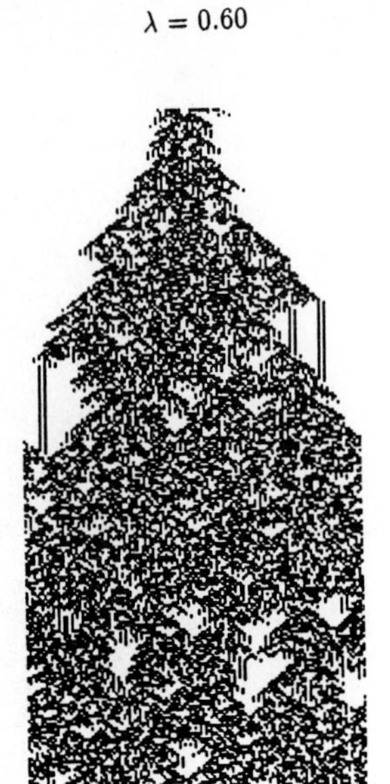

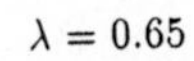

$\lambda = 0.70$

$\lambda = 0.75$

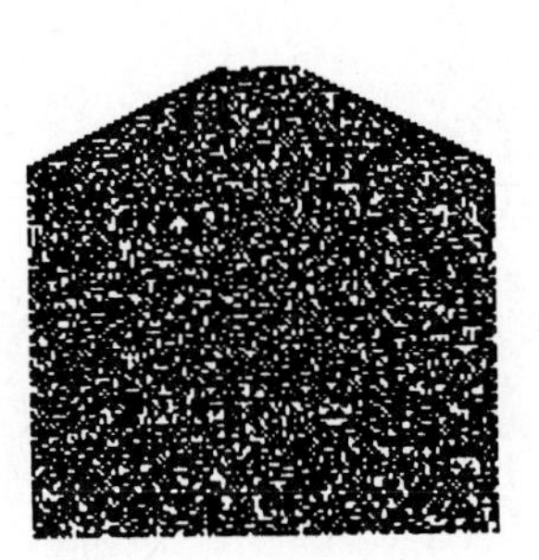

FIGURE 2 (continued)

We start with $\lambda \approx 0$. Note that under our strong quiescence condition, we cannot have λ be identical to 0. The salient features observed as we vary λ away from 0 are itemized below. Note that these examples are the result of a *single* table walk-through, and therefore the association of specific dynamical behaviors with specific values of λ is not particularly significant.

$\lambda \approx 0.00$: All dynamical activity dies out after a single time step, leaving the arrays uniform in state s_q. The area of dynamical activity has collapsed to nothing.

$\lambda \approx 0.05$: The dynamics reaches the uniform s_q fixed-point after approximately 2 time steps.

$\lambda \approx 0.10$: The homogeneous fixed point is reached after 3 or 4 time steps.

$\lambda \approx 0.15$: The homogeneous fixed point is reached after 4 or 5 time steps.

$\lambda \approx 0.20$: The dynamics reaches a periodic structure, which will persist forever (Figure 1.20). Transients have increased to 7–10 time steps as well. Note that the evolution does not necessarily lead to periodic dynamics (Figure 2.20).

$\lambda \approx 0.25$: Structures of period 1 appear. There are now three different possible outcomes for the ultimate dynamics of the system, depending on the initial state. The dynamics may reach a *homogeneous* fixed point (consisting entirely of state s_q,) or it may reach a *heterogeneous* fixed point (consisting mostly of cells in state s_q, with a sprinkling of cells stuck in the other states,) or it may settle down to periodic behavior. Notice that the transients have lengthened even more.

$\lambda \approx 0.30$: Transients have lengthened again.

$\lambda \approx 0.35$: Transient length has grown significantly, and a new kind of periodic structure with a longer period has appeared (Figure 1.35). Most of the previous structures are still possible; hence, the spectrum of dynamical possibilities is broadening.

$\lambda \approx 0.40$: Transient length has increased to about 60 time steps, and a structure has appeared with a period of about 40 time steps. Areas of dynamical activity are still collapsing down onto isolated periodic configurations.

$\lambda \approx 0.45$: Transient length has increased to almost 1,000 time steps. (Figure 1.45). Here, the structure on the right appears to be periodic, with a period of about 100 time steps. However, after viewing several cycles of its period, it is apparent that the whole structure is moving to the left, and so this pattern will not recur precisely in the same position until it has cycled at least once around the array. Furthermore, as it propagates to the left, this structure eventually annihilates a period one structure after about 800 time steps. Thus, the transient time before a cycle is reached has grown enormously. It turns out that even after one orbit around the array, the periodic structure does not return exactly to its previous position. It must orbit the array 3 times before it repeats itself exactly. As it has shifted over only 3 sites after its quasi-period of 116 time steps, the true period of this structure is 14,848 time steps. Here, the area of dynamical activity is at a balance point between collapse and expansion, as illustrated in Figure 2.45.

$\lambda \approx 0.50$: Typical transient length is on the order of 12,000 time steps. After the transient, the dynamical activity settles down to periodic behavior, possibly of period one as shown in the figure. Although the dynamics eventually becomes simple, the transient time has increased dramatically. Note in Figure 2.50 that the general tendency now is that the area of dynamical activity *expands* rather than contracts with time. There are, however, large fluctuations in the area covered by dynamical activity, and it is these fluctuations that lead to the eventual collapse to periodic dynamics in a finite array.

$\lambda \approx 0.55$: Whereas before, the dynamics *eventually* settled down to periodic behavior, we are now in a regime in which the dynamics settles down to effectively *chaotic* behavior. Furthermore, the previous trend of transient length *increasing* with increasing λ is reversed. The arrow to the right of the evolutions in Figures 1.55–1.75 indicates the approximate time by which the average site occupation density has settled down to within 1% of its long-time average. Note that the area of dynamical activity expands more rapidly with time.

$\lambda \approx 0.60$: The dynamics are quite chaotic, and the transient length to "typical" chaotic behavior has decreased significantly. The area of dynamical activity expands more rapidly with time.

$\lambda \approx 0.65$: Typical chaotic behavior is achieved in only ten time steps or so. The area of dynamical activity is expanding at about one cell per time step in each direction, approximately half of the maximum possible rate for this neighborhood template.

$\lambda \approx 0.70$: Fully developed chaotic behavior is reached in only two time steps. The area of dynamical activity is expanding even more rapidly.

$\lambda \approx 0.75$: After only a single time step, the array is essentially random and remains so thereafter. The area of dynamical activity spreads *at* the maximum possible rate.

Therefore, by varying the λ parameter throughout $0.0 < \lambda \leq 0.75$ over the space of possible $K = 4$, $N = 5$, one-dimensional cellular automata, we progress from CA exhibiting the maximal possible order to CA exhibiting the maximal possible disorder. At intermediate values of λ, we encounter a *phase transition* between periodic and chaotic dynamics, and while the behavior at either end of the λ spectrum seems "simple" and easily predictable, the behavior in the vicinity of this phase transition seems "complex" and unpredictable.

3.2 COMMENTS ON QUALITATIVE DYNAMICS

A few comments on these examples are in order.

First, the progression through the spectrum of dynamical behaviors as a function of λ is clearly:

$$\text{fixed-point} \rightarrow \text{periodic} \rightarrow \text{“complex”} \rightarrow \text{chaotic}\,.$$

In terms of the Wolfram classes, the sequence is:

$$I \rightarrow II \rightarrow IV \rightarrow III\,. \tag{2}$$

That is, the complex rules are located *inbetween* the periodic and the chaotic rules.

Second, there is clearly a phase transition between periodic and chaotic behavior, with $\lambda_c \approx 0.50$ (in these examples).

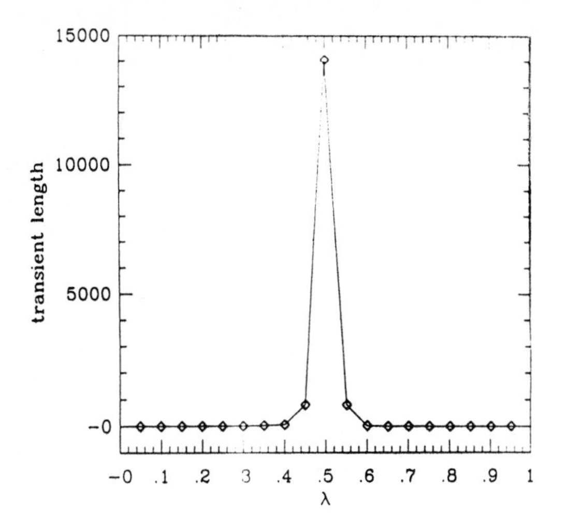

FIGURE 3 Average transient length T versus λ. Transient length apparently diverges rapidly in the vicinity of the transition.

Third, transients are short at either end of the range of λ and get longer as we approach the middle of the range.[3]

Transients clearly diverge in the vicinity of λ_c, evidence of the second-order phenomenon of *critical slowing down.* As we continue to raise λ beyond λ_c, although the dynamics are now settling down to effectively chaotic behavior instead of periodic behavior, the transient lengths are getting *shorter* with increasing λ, rather than longer. The relationship between transient length and λ for these examples is plotted in Figure 3.

Fourth, clearly, in CA for which $\lambda \ll \lambda_c$ or $\lambda \gg \lambda_c$, we do not have to wait very long before we can be almost certain about the ultimate outcome of the dynamics. For CA with $\lambda \approx \lambda_c$, however, the dynamics will look very much the same, whether it will ultimately result in a periodic state, or will remain non-periodic "forever." In general, the outcome is undecidable. We will discuss this further in relation to Turing's famous Halting problem.

Fifth, the size of the array has an effect on the dynamics only for intermediate values of λ. For low values of λ, array size has *no* discernible effect on transient length. Not until, $\lambda \approx 0.45$, do we begin to see a small difference in the transient length as the size of the array is increased. For $\lambda = 0.50$, however, array size has a significant effect on transient length. The growth of transient length as a function of array size for $\lambda = 0.50$ is plotted in Figure 4. The essentially linear relationship on this log-normal plot suggests that transient length depends *exponentially* on array size for $\lambda \approx 0.50$.

[3]Note that "transients" for chaotic CA are defined differently than for periodic CA. Transients for periodic CA are defined in terms of the number of iterations before a cycle is reached, while transients for chaotic CA are defined in terms of iterations before statistical convergence is achieved. This differs from the traditional CA literature, but is more in line with the literature on dynamical systems and chaos theory.

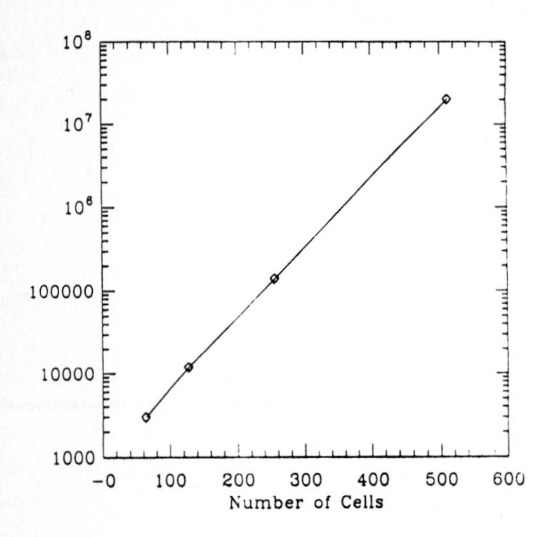

FIGURE 4 Average transient length T versus array size N for $\lambda = 0.50$. This plot suggests that transient length is growing exponentially with N in the vicinity of λ_c.

Furthermore, transient times exhibit *decreasing* dependence on array size as λ is increased past the transition point. For the highest value of λ ($= 0.75$ in this case) when all states are represented uniformly in the transition table, the transient lengths exhibit *no* dependence on array size, as was the case for the lowest values of λ. We will discuss this further when we use this fact to help explain the existence of computational complexity classes.

Sixth, the behavior of the CA dynamics is most complicated in the vicinity of the transition. Compare the length of the description of the dynamics at each λ value in the previous section. It takes longer to describe what is going on near the transition than it does to describe what is going on far from the transition. The dynamics becomes simply periodic for low λ, whereas for high λ the randomness simply spreads outwards in a uniformly expanding "circle" at the maximum possible rate. The mutual information and entropy data presented in the next section will quantify the important distinction between complexity and randomness.

Finally, it is important to note that the transition region supports both static and propagating structures (Figure 1.45). The propagating structures are essentially *solitary waves*, quasi-periodic patterns of state change, which—like the "glider" in Conway's game of LIFE[16]—propagate through the array, constantly moving with respect to the fixed background of the lattice. Figure 5 traces the time evolution of an array of 512 sites, and shows that the rule governing the behavior of Figure 1.45 supports several different kinds of "particles," which interact with each other and with the static periodic structures in complicated ways (e.g., note the manner in which the collision of a propagating particle with a static periodic structure produces a particle traveling in the opposite direction). Such propagating and static structures can form the basis for signals and storage, and interactions between them can modify either stored or transmitted information in the support of an overall computational process.

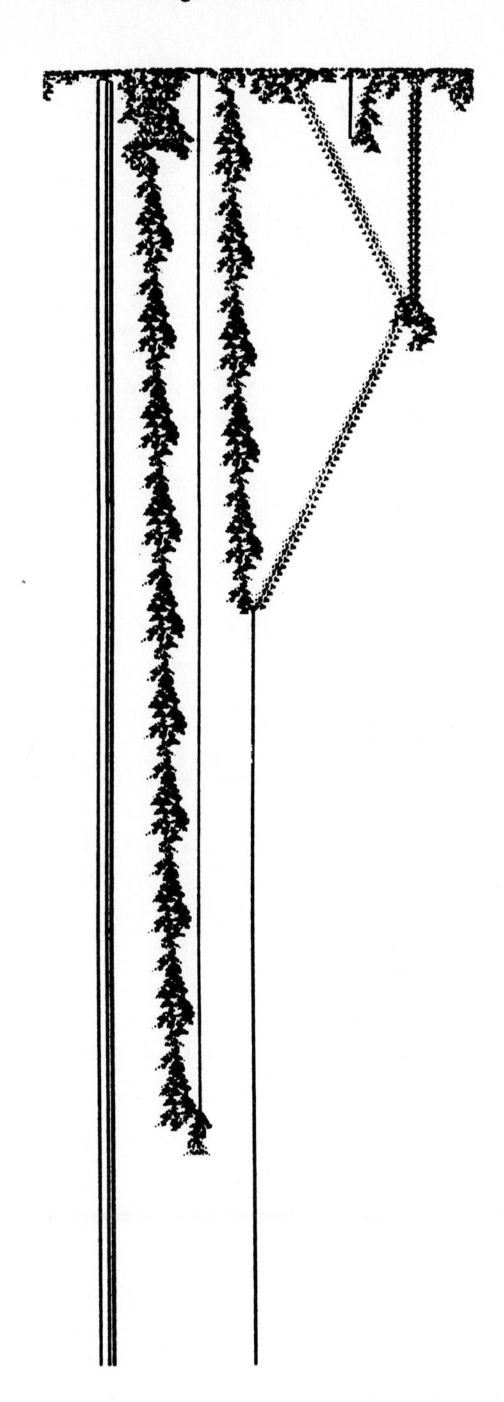

FIGURE 5 Evolution of a one-dimensional array consisting of 512 sites under the rule governing the evolution illustrated in Figure 1, $\lambda = .45$.

3.3 QUALIFICATIONS

It must be pointed out that although the examples presented here illustrate the most interesting change in dynamics as a function of λ, the story is not quite as simple as we have presented it here.

The situation is complicated by two factors. First, different traversals of λ space using the table-walk-through method make the transition to chaotic behavior at different λ values, although there is a well-defined distribution around a mean value.

Second, one does not always pass through the "complex" regime as neatly as in this example. More often, the dynamics jumps directly from fairly ordered to fairly disordered behavior. We will discuss both of these complications in the next section.

Despite these qualifications, the overall picture is clear: as we survey CA rule-spaces using the λ parameter, we encounter a phase transition between periodic and chaotic behavior, and the most complex behavior is found in the vicinity of this transition.

4. QUANTITATIVE OVERVIEW OF CA DYNAMICS

In this section, we present a brief quantitative overview of the structural relations among the dynamical regimes in CA rule spaces as revealed by the λ parameter.[4]

The results of this section are based on experiments using two-dimensional CAs with $K = 8$ and $N = 5$. Arrays are typically of size 64 x 64, and again, periodic boundary conditions are employed.

4.1 MEASURES OF COMPLEXITY

The measures employed were chosen for their collective ability to reveal the presence of information in its various forms within CA dynamics.

4.1.1 SHANNON ENTROPY We use Shannon's Entropy H to measure basic information capacity. For a discrete process A of K states[5]:

$$H(A) = -\sum_{i=1}^{K} p_i \log p_i \,. \tag{3}$$

[4]For a more detailed review, see Langton.[28]

[5]Throughout, log is taken to the base 2; thus, the units are bits.

Figure 6 shows the average entropy per cell, $\overline{H}$, as a function of λ for approximately 10,000 CA runs. The random-table method was employed, so each point represents a distinct random transition table.

First, note the overall envelope of the data and the large variance at most λ points. Second, note the sparsely populated gap over $0.0 \leq \lambda \lesssim 0.6$ and between $0.0 \leq \overline{H} \lesssim 0.84$. This distribution appears to be bimodal, suggesting the presence of a phase transition. Third, note the rapid decrease in variability as λ is raised from ~ 0.6 to its maximum value of 0.875.

Two other features of this plot deserve special mention. First, the abrupt cutoff of low $\overline{H}$ values at $\lambda \approx 0.6$ corresponds to the *site-percolation* threshold $P_c \approx 0.59$ for this neighborhood template. Thus, we may suppose that, since λ is a dynamical analog of the site occupation probability P, the *dynamical* percolation threshold for a particular neighborhood template is bounded above by the *static* percolation threshold P_c. This is borne out by experiments with other neighborhood templates. For instance, the nine-neighbor template exhibits a sharp cutoff at $\lambda \approx 0.4$, which corresponds well with the site percolation threshold $P_c \approx 0.402$ for this lattice.

The second feature is the "ceiling" of the gap at $\overline{H} \approx 0.84$. This turns out to be the average entropy value for one of the most commonly occurring chaotic rules. In such rules the dynamics has collapsed onto only two states—s_q and one other—and the rule is such that a mostly quiescent neighborhood containing one non-quiescent state maps to that non-quiescent state. In one-dimensional cellular automata, such rules gives rise to the familiar triangular fractal pattern known as the Serpinski Gasket. There are many ways to achieve such rules, and they can be achieved at very low λ values. Most of the low-λ chaotic rules are of this type.

The entropy data of Figure 6 suggest an anomaly at intermediate parameter values, possibly a phase transition between two kinds of dynamics. Since there seems to be a discrete jump between low and high entropy values, the evidence points to

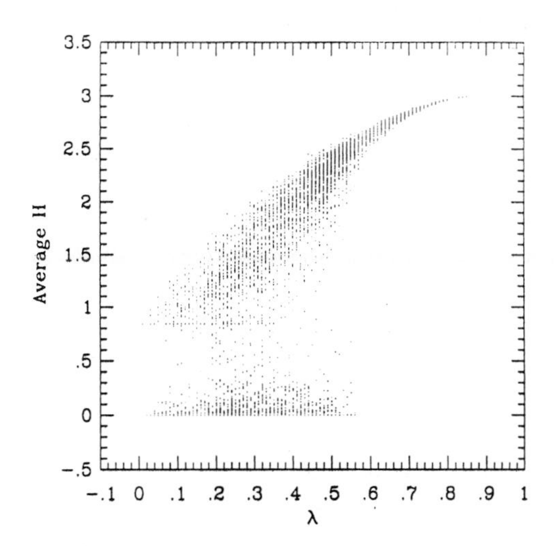

FIGURE 6 Average single cell entropy $\overline{H}$ over λ space.

a first-order transition, similar to that observed between the solid and fluid phases of matter. However, the fact that the gap is not completely empty suggests the possibility of second-order transitions as well.

The table-walk-through method of varying λ reveals more details of the structure of the entropy data. Figure 7 shows four superimposed examples of the change in the average cell entropy as we vary the λ value of a table. Notice that in each of the four cases, the entropy remains fairly close to zero until—at some critical λ value—the entropy jumps to a higher value, and proceeds fairly smoothly towards its maximum possible value as λ is increased further. Such a discontinuity is a classic signature of a first-order phase transition. Most of our complexity measures exhibit similar discontinuities at the same λ value *within a particular table*.

Notice also that the λ value at which the transition occurs is different for each of the four examples. Obviously, the same thing—a jump—is happening as we vary λ in each of these examples, but it happens at different values of λ. When we superimpose 50 runs, as in Figure 8, we see the internal structure of the entropy data envelope plotted in Figure 6.

Since we have located the transition events, we may line up these plots by the events themselves, rather than by λ, in order to get a clearer picture of what is going on before, during, and after the transition. This is illustrated in Figure 9. The abcissa is now measured in terms of $\Delta\lambda$: the distance from the transition event. Figure 10 shows the same data as Figure 8 but lined up by $\Delta\lambda$.

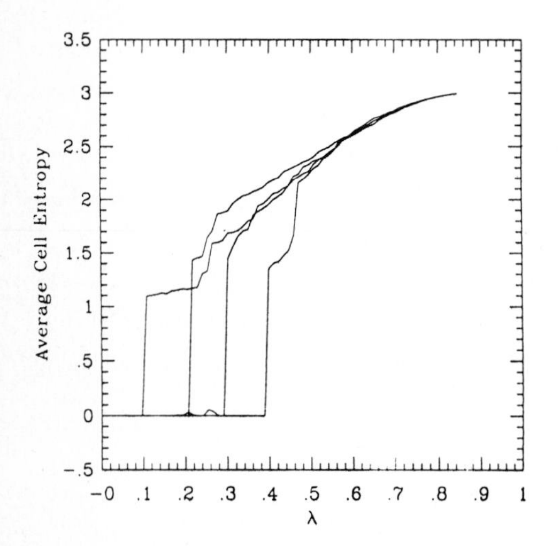

FIGURE 7 Superposition of entropy versus λ curves for four separate runs using the table-walk-through method.

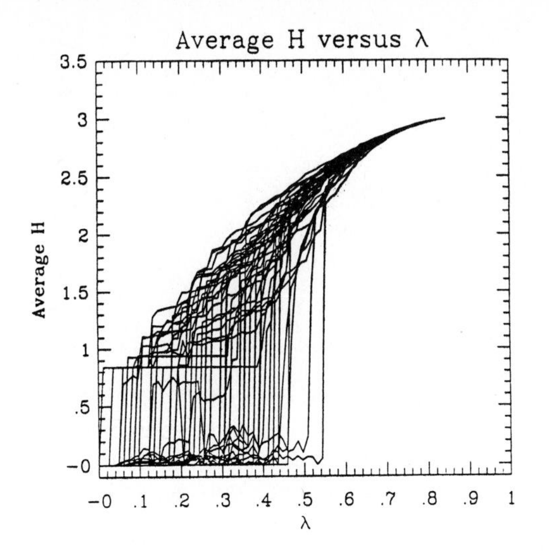

FIGURE 8 Superposition of entropy versus λ curves for 50 separate runs using the table-walk-through method.

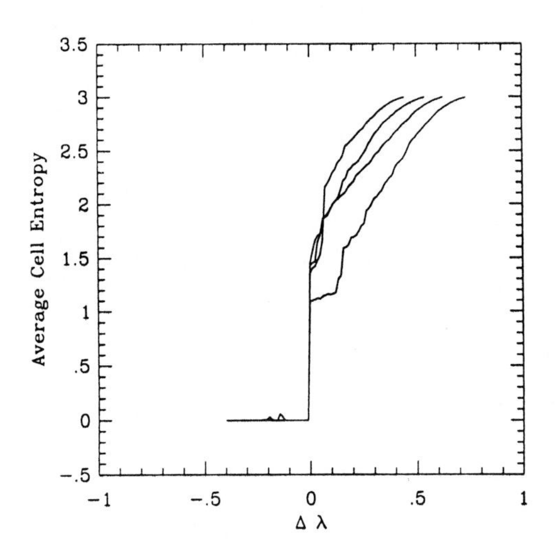

FIGURE 9 Superposition of entropy versus λ curves for the four runs plotted in Figure 7, lined up by $\Delta\lambda$.

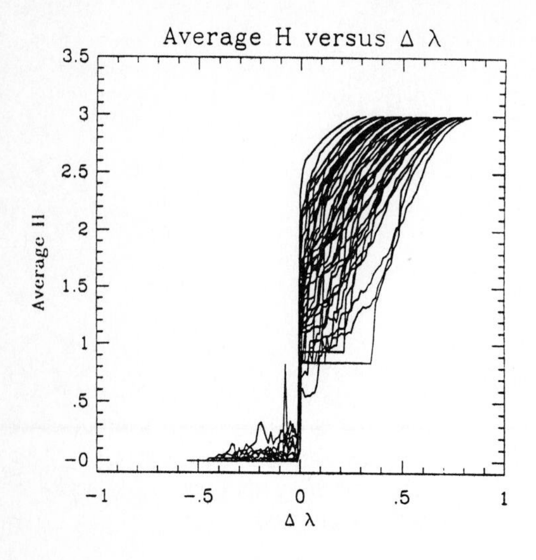

FIGURE 10 Superposition of entropy versus λ curves for the 50 runs plotted in Figure 8, but lined up by $\Delta\lambda$.

4.1.2 MUTUAL INFORMATION In order for two distinct cells to cooperate in the support of a computation, they must be able to affect one another's behavior. Therefore, we should be able to find correlations between events taking place at the two cells.

The mutual information $I(A;B)$ between two cells A and B can be used to study correlations in systems when the values at the sites to be measured cannot be ordered, as is the case for the states of the cells in cellular automata.[30]

The mutual information is a simple function of the individual cell entropies, $H(A)$ and $H(B)$, and the entropy of the two cells considered as a joint process, $H(A,B)$:

$$I(A;B) = H(A) + H(B) - H(A,B) \tag{4}$$

This is a measure of the degree to which the state of cell A is correlated with the state of cell B, and *vice versa.*

Figure 11 shows the average mutual information between a cell and itself at the next time step. Note the tight convergence to low values of the mutual information for high λ and the location of the highest values.

The increase of the mutual information in a particular region is evidence that the correlation length is growing in that region, further evidence for a phase transition.

Figure 12 shows the behavior of the average mutual information as λ is varied, both against λ and $\Delta\lambda$. The average mutual information is essentially zero below the transition point, it jumps to a moderate value at the transition, and then decays slowly with increasing λ. The jump in the mutual information clearly indicates the onset of the chaotic regime, and the decaying tail indicates the approach to effectively random dynamics. The lack of correlation between even adjacent cells at high λ values means that cells are *acting* as if they were independent of each other, even though they are causally connected. The resulting global dynamics is the same as if each cell picked its next state at uniform random from among the K states, with no consideration of the states of its neighbors. This kind of global dynamics is predictable in the same statistical sense that an ideal gas is globally predictable. In fact, it is appropriate to view this dynamical regime as a hot gas of randomly flipping cells.

Figure 13 shows the average mutual information curves for several different temporal and spatial separations. Note that the decay in both time and space is slowest in the middle region.

At intermediate λ values, the dynamics support the preservation of information locally, as indicated in the peak in correlations between distinct cells. If cells are cooperatively engaged in the support of a computation, they must exhibit some—but not *too* much— correlation in their behaviors. If the correlations are too strong, then the cells are overly dependent, with one mimicking the other—not a cooperative computational enterprise.

On the other hand, if the correlations are too small, then the cells are overly *independent*, and again, they cannot cooperate in a computational enterprise, as each cell does something totally unpredictable in response to the state of the other. Correlations in behavior imply a kind of common code, or protocol, by which changes of state in one cell can be recognized and understood by the other as a *meaningful signal.* With no correlations in behavior, there can be no common code with which to communicate information.

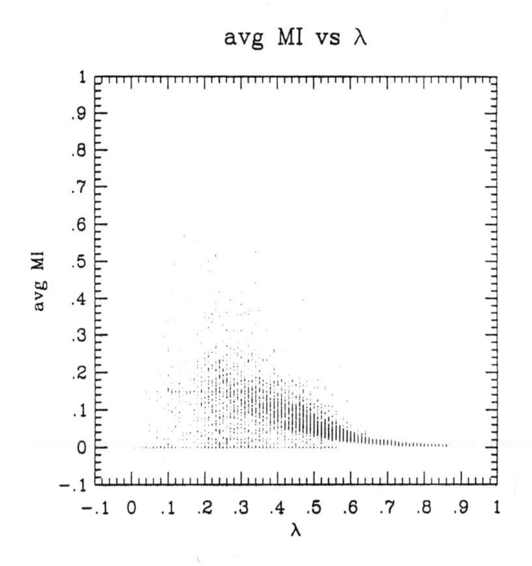

FIGURE 11 Raw data on mutual information between a cell and itself at the next time step plotted over λ.

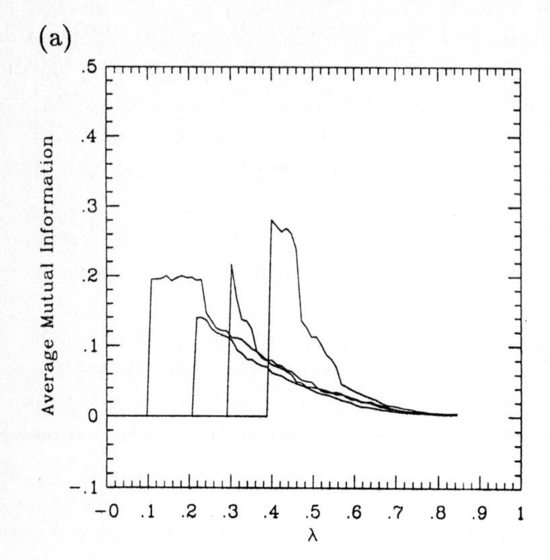

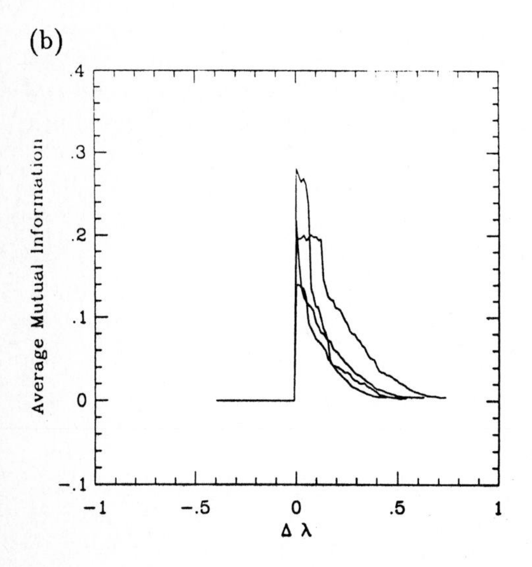

FIGURE 12 Plots of mutual information versus a) λ and b) $\Delta\lambda$ for four runs using the table-walk-through method. The steep rise indicates the transition from periodic to chaotic behavior.

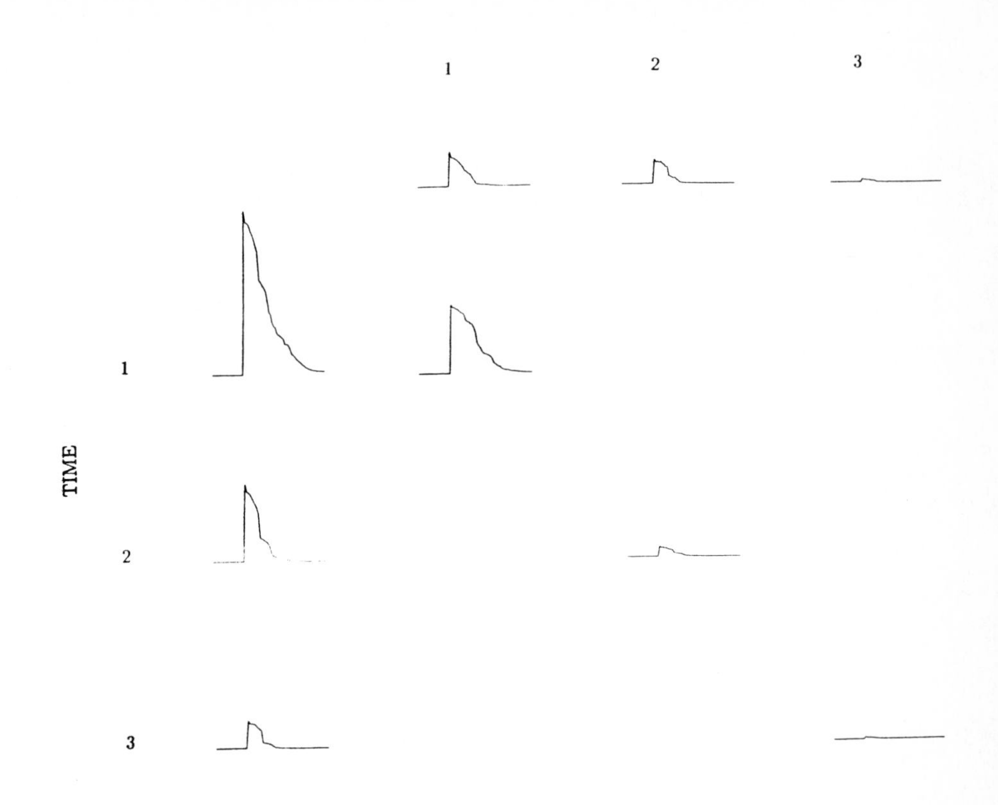

FIGURE 13 Decay of mutual information in space and time.

4.1.3 MUTUAL INFORMATION AND ENTROPY It is often useful to examine the way in which observed measures behave when plotted against one another, effectively removing the (possibly unnatural) ordering imposed by the control parameter.

Of the measures we have looked at, the most informative pair when plotted against each other are the mutual information and the average single cell entropy. The relationship between these two measures is plotted in Figure 14. Again, we see clear evidence of a phase transition.

The envelope of the relationship is bounded below the transition by the linear bound that H places on the mutual information. All of the points on this line are for periodic CAs. This line intersects the curve bounding the envelope *above* the transition at an entropy value $H_c \approx 0.32$ on the normalized entropy scale.

This is a *very* informative plot. There is a clear, sharply defined maximum value of mutual information at a specific value of the entropy, and the mutual information falls off rapidly on either side. This seems to imply that there is an *optimal working entropy* at which CAs exhibit large spatial and temporal correlations. Why should this be the case?

Briefly, information storage involves *lowering* entropy while information transmission involves *raising entropy.*[18] In order to compute, a system must do both, and therefore must effect a tradeoff between high and low operating entropy. It would seem from the work reported here that this tradeoff is optimized in the vicinity of a phase transition.

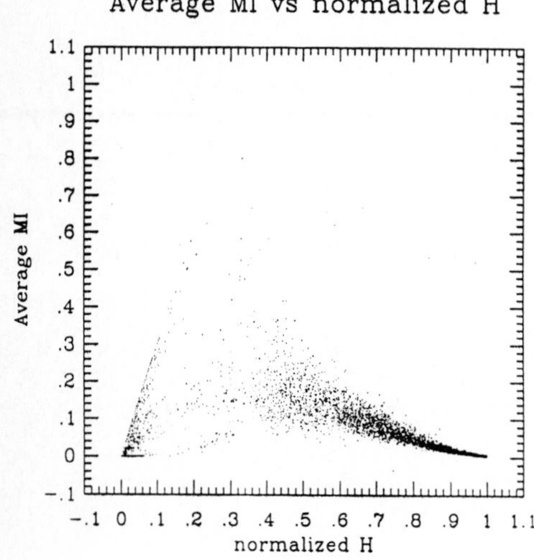

FIGURE 14 Mutual information versus normalized entropy.

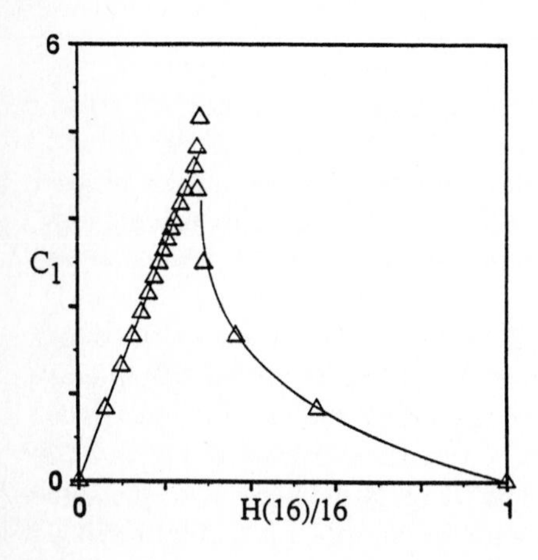

FIGURE 15 Crutchfield's plot of complexity versus normalized entropy for the logistic map (from Crutchfield.[11])

A similar relationship has been observed by Jim Crutchfield at Berkeley in his work on the transition to chaos in continuous dynamical systems.[11] This relationship is illustrated in Figure 15. Briefly, the ordinate of this plot—C—is a measure of the size of the minimal finite state machine required to recognize strings of 1's and 0's generated by a dynamical system (the logistic map, in this case) when these strings are characterized by the normalized per-symbol entropy listed on the abscissa. The observance of this same fundamental entropy/complexity relationship in these different classes of dynamical systems is very exciting.

These relationships support the view that, rather than increasing monotonically with randomness—as is the case for the usual measures of complexity, such as that of Chaitin and Kolmogorov[9,25]—complexity increases with randomness only up to a point—*a phase transition*—after which complexity *decreases* with further increases in randomness, so that total disorder is just as "simple," in a sense, as total order. Complex behavior involves a mix of order and disorder.

5. COMPLEX CA AND COMPUTATION

Now that we have demonstrated the existence of a "complex" domain within the space of CA rules—a domain that typically supports complex interactions between propagating and static structures—what reasons do we have to believe that such interactions could support complex information processing, even universal computation?

This is the question of *constructability:* Can these complex interactions be pressed into useful service as logical building blocks in the construction of a universal computing device? In this section, we provide evidence that the dynamics typically exhibited by complex CA rules can be applied to the construction of embedded general-purpose computers.

5.1 PROVING UNIVERSALITY OF CA RULES

There can be no universal algorithm for deciding whether an arbitrary CA rule will support universal computation. The only way to prove such a capacity for a particular rule is by construction. If one can construct a general-purpose computer under a particular CA rule—or show that all of the necessary components can be constructed and that they could be wired together in principle—then that is a sufficient proof. Failure to find such a construction does *not* constitute a proof that a particular rule will *not* support universal computation; it merely leaves the question open.

The thing that makes this problem hard is that the dynamics of interactions between simple elements supported by complex rules are often more reminiscent of chemistry than of logic. Figuring out how to go about constructing logical elements from an arbitrary given rule is like figuring out how to build logical elements out of interactions between nucleotides and proteins. It is clearly possible that one could build a biological universal computer,[6] however, designing computers based on an arbitrary "chemical" logic is much more difficult than designing computers based on more familiar logical elements.

In short, logical functions such as **AND**, **OR**, and **NOT** are typically not primitives under an arbitrary rule, and it may be a very difficult task to find how to construct them out of the functions that ARE primitive under a particular rule.

Constructive proofs of this type often have two stages. In the first stage, one attempts to incorporate the set of low-level interactions supported by a complex rule into a basic set of more familiar logical switches and elements, such as **AND**, **OR**, and **NOT** gates, clocks, wire-crossings, and etc. In the second stage, one works with these more familiar logical switches to construct an embedded computer, which might be a general-purpose RAM computer, a universal Turing Machine, or an instance of some other class of universal computational devices.

This is not to say that computation could not be taking place under a particular rule unless its dynamics support **AND**, **OR**, and **NOT** gates. Computation might be taking place based on a very different set of logical primitives naturally supported by a particular rule. It might be very difficult to figure out exactly what those primitives are. However, most constructive proofs first show that things like **AND**, **OR**, and **NOT** gates can be constructed because those are the logical primitives that we are familiar with and know how to assemble into universal computers. Thus, such proofs simply demonstrate a *capacity* for universal computation, which might be realized much more "naturally" in a very different manner under a particular rule, but which we might not recognize as constituting universal computation.

In the following sections, we will demonstrate the "constructability" of a complex rule by showing how some of the logical switches involved in the first stage discussed above may be built. A complete construction involving the second stage is beyond the scope of this paper. Here, I merely want to provide convincing evidence that such a construction could be carried through.

5.1.1 CATALOGUING THE BASIC INTERACTIONS The first stage involves performing a large set of experiments to determine all of the most basic interactions between the simple structures supported by a rule.

Sometimes these interactions have simple results (e.g., a "glider" is reflected or a "blinker" is displaced). At other times, these interactions may have more complicated results (such as the production of more gliders or blinkers), and the site of the interactions may exhibit complex activity for some time, generating lots of by-products in the process.

Once these interactions and their products have been carefully catalogued, the experimenter has a set of reactions to work with, and can try to arrange configurations in a CA so that the products of one reaction become the inputs to other reactions, on just the right trajectories, and with just the right timing. In other

[6]A human brain with the ability to mark and read the environment constitutes an existence proof.

words, one has to work with the "material" at hand, and try to mold its natural dynamics and properties into the kind of computational building blocks that are to be used in the construction of a general-purpose computational device.

5.2 THE PROOF THAT LIFE IS UNIVERSAL

The game of LIFE is a complex rule that has been proven capable of supporting universal computation via a constructive proof of the kind discussed above.[6] $\lambda_{LIFE} \approx 0.273$, which locates it near the critical regime for $K = 2$, $N = 9$ CA rules.

In this section, we will review this constructive proof. In the next section, we will show that the same kind of construction can be carried out for rules in the complex regime, using one of the complex rules generated by the table-walk-through method as an example.

In the proof of the universality of the game of LIFE, a stream of regularly spaced, propagating structures called "gliders" represents a string of bits. In such a stream, the presence of a glider at a potential bit-position represents a "1" while the absence of a glider at a bit-position (a "hole") represents a "0." The **AND**, **OR**, and **NOT** gates are implemented by colliding together such streams of gliders and holes.

An important configuration in the game of LIFE is the *Glider Gun:* a periodic configuration that produces a steady stream of regularly spaced gliders, i.e., a steady

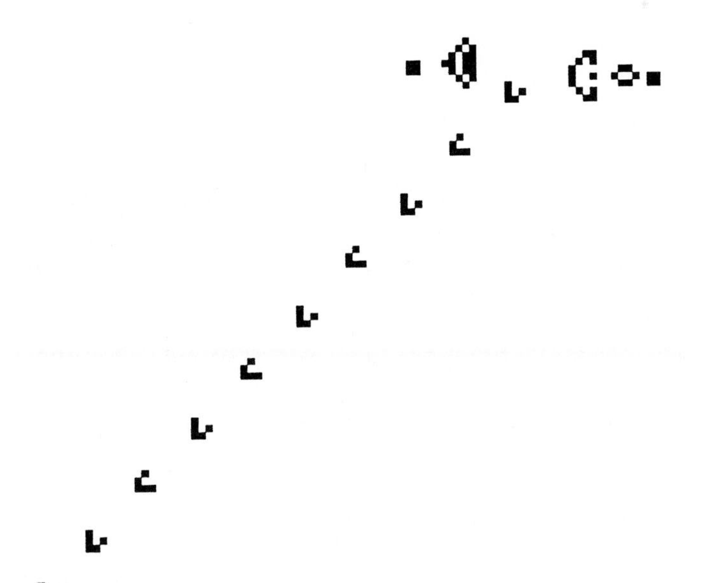

FIGURE 16 A Glider Gun, a periodic LIFE configuration which produces a steady stream of gliders.

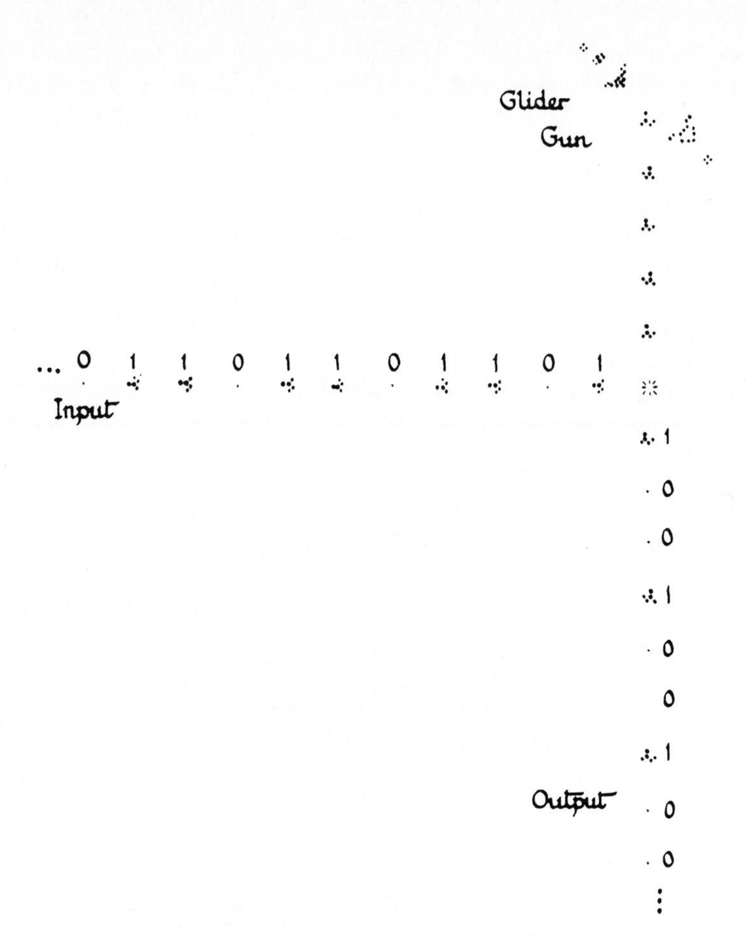

FIGURE 17 A NOT gate in the game of LIFE (from Berlekamp[6]).

stream of "1's." A glider gun producing a stream of gliders is shown in Figure 16. Such periodic "all-1" bit streams are useful for timing and for providing fixed inputs to logical elements.

Figure 17 shows the construction of a **NOT** gate in the game of LIFE. The output of a Glider Gun is collided at right angles with a stream of gliders encoding the input to the **NOT** gate. The two streams are timed so that when two gliders collide, they annihilate each other. Consequently, wherever there is a glider (a 1) in the input stream, it collides with a glider from the Glider Gun and they annihilate, leaving a hole (a 0) in the output stream. Wherever there is a hole in the input stream, the associated glider in the stream from the Glider Gun passes through untouched into the output stream. Therefore, the output stream contains a glider at every position where there was a hole in the input stream, and contains a hole at every position where there was a glider in the input stream. That is, the output stream contains a 1 for every 0 in the input stream, and a 0 for every 1 in the input stream, thus implementing the **NOT** function.

The **AND** gate is illustrated in Figure 18. **AND** builds on the **NOT** function by using the output of **NOT** to gate a second input stream. $A \wedge B$ is implemented as follows. Input stream B is fed into the **NOT** gate. What comes out is $\neg B$. If we now collide this $\neg B$ stream with the stream representing A, holes in the $\neg B$ stream will allow passage of gliders in the A stream, and gliders in the $\neg B$ stream will annihilate gliders in the A stream, leaving holes. Thus, the stream A gated by stream $\neg B$ contains a glider only at positions where stream A had a glider and stream $\neg B$ had a hole, which was wherever stream B had a glider. That is, the output stream contains a 1 at every location where both inputs were 1, and has 0's everywhere else, thus implementing the **AND** function.

The **OR** gate is illustrated in Figure 18. **OR** combines the **AND** gate with another **NOT** gate as follows. The *other* stream coming out of the **AND** gate described above is just stream $\neg B$ gated by stream A. It has a hole wherever $\neg B$ had a hole, and in addition it has a hole wherever $\neg B$ had a glider *and* A had a glider. It has a glider wherever stream $\neg B$ had a glider and stream A had a hole. Thus, it implements the logical function $(\neg A \wedge \neg B)$, which simplifies to $\neg(A \vee B)$ by DeMorgan's law. The output from this is then inverted by a **NOT** gate, which yields $(A \vee B)$.

There are further elements required to construct a complete working computer, such as elements that will turn a signal stream by right angles, introduce delays for timing, implement an extendable memory, and so forth. We will not go into the details of these here. The interested reader can consult Berlekamp.[6]

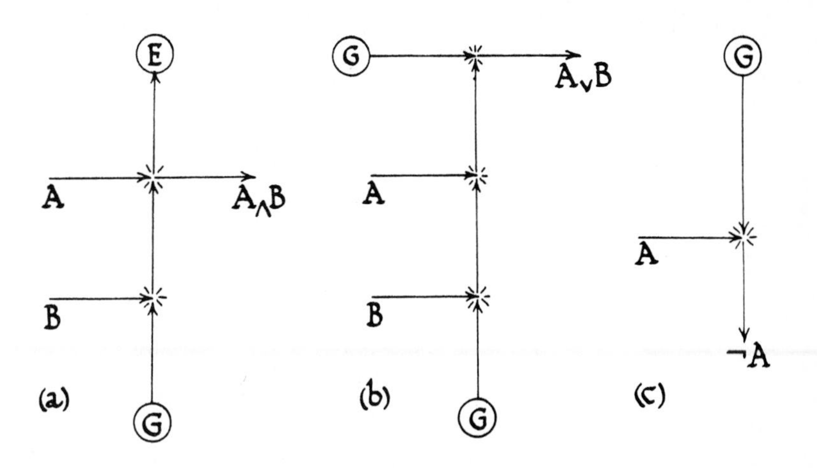

FIGURE 18 AND, OR, and NOT gates in the game of LIFE (from Berlekamp[6]).

5.3 CONSTRUCTABILITY FOR OTHER COMPLEX RULES

Constructions like those for **NOT**, **AND**, and **OR** above clearly demonstrate "constructability" for the game of LIFE. In this section, we show that similar constructive proofs can be provided for other complex rules by demonstrating the construction of **NOT**, **AND**, and **OR** gates within a particular $K = 2$, $N = 9$ complex rule generated via our table-walk-through method. Like the game of LIFE, this rule supports a rich variety of propagating gliders, glider guns, periodic blinkers, and a complex "chemistry" of interactions between such structures.

Figure 19 shows a Glider Gun under this rule producing a stream of gliders.[7] By colliding together streams of gliders and holes representing inputs with streams of gliders produced by glider guns and with each other, we can construct the same **AND**, **OR**, and **NOT** gates exhibited above for the game of LIFE. The **NOT** gate is shown in Figure 20. The other two gates are built up from this gate in exactly the same manner as they were in the game of LIFE.

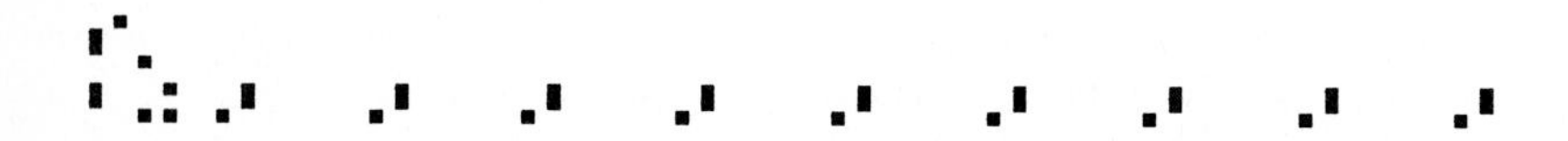

FIGURE 19 Glider gun under a randomly generated "complex" rule.

Glider Gun

Input

Output

FIGURE 20 NOT gate under a "complex" rule.

[7]This particular Glider Gun propagates through space, i.e., it is a "Gliding" Glider Gun.

Again, other components will be needed to construct a complete working computer, and a proof of universality will require assembling these components together with the proper signal routing and delay elements, but the existence of these gates demonstrates "constructability" for this sample complex rule.

Similar constructions can be made for other complex rules, but not all complex rules tried have yielded to such simple constructions. In these latter cases, some aspect of the necessary componentry or interactions proved elusive. For instance, for some complex rules, none of the possible glider collisions resulted in the annihilation of both gliders. In other rules, Glider Guns were not found. These negative results do not prove that these rules are not capable of supporting universal computation—they merely suggest that this particular construction technique may not go through for these rules, whereas it is entirely possible that some other construction would go through.

6. PHASE TRANSITIONS AND COMPUTATION

In this section, we summarize and discuss the main points that can be derived from the evidence provided in the last several sections.

First, we review the phase-transition structure of CA rule space implied by the evidence from the previous sections, and show how this "deep structure" explains the surface-level phenomenology of CA dynamics, which includes the qualitative Wolfram classes, the existence of complexity classes, the capacity for universal computation, undecidability, and so forth.

Second, we note that the surface-level phenomenology of CA systems is remarkably similar to the surface-level phenomenology of computational systems, which includes the existence of complexity classes, computability classes, the capacity for universal computation, undecidability, etc.

Third, we conclude that the structure of the space of computations is dominated by a second-order phase transition, in terms of which the existence of—and relationships between—the surface-level features of computational systems find a relatively straightforward explanation.

6.1 THE FUNDAMENTAL STRUCTURE OF CA RULE SPACE

Piecing together the results of the previous sections results in a clear picture of the fundamental structure of cellular automata rule space. This structure is illustrated schematically in Figure 21.

There are two primary regimes of rules—*periodic* and *chaotic*— separated by a *transition regime.* This transition regime is *not* simply a smooth surface between the other two domains, but itself has a complicated structure. From the perspective provided by the λ parameter, it seemed that much of the transition regime is

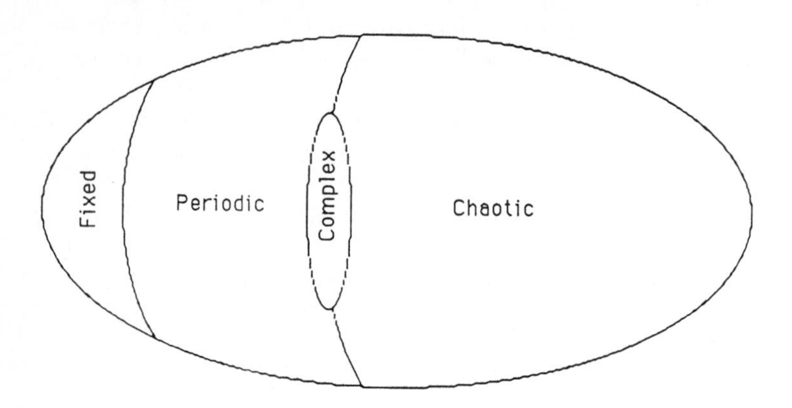

FIGURE 21 Schematic drawing of CA rule space indicating relative location of periodic, chaotic, and complex regimes.

simply a boundary between periodic and chaotic rules, containing no rules within it. Crossing the transition regime at such a boundary appeared to give rise to a discrete jump from "simple" periodic dynamics to "simple" chaotic dynamics, accompanied by a discrete jump in statistical measures of the kind usually associated with first-order transitions.

However, from a more detailed investigation[8] it is apparent that the phase transition is primarily a second-order, or *critical*, transition. Crossing this critical transition region gives rise to "complex" dynamics, accompanied by relatively "smooth" changes in statistical measures.

6.2 PHASE TRANSITIONS AND CA PHENOMENOLOGY

The phase-transition structure underlying CA rule spaces provides a coherent framework for organizing and explaining much of the documented phenomenology of CA dynamics.

First, this picture of a phase transition separating a domain of ordered dynamics from a domain of disordered dynamics (each of which might be further subdivided), provides a simple explanation for the existence of—and the relationships between—the four qualitative classes of CA behavior identified by Wolfram. Class IV rules are found in the transition regime separating the periodic rules (Classes I & II) from the chaotic rules (Class III). Figure 21 illustrates the way in which the Wolfram classes fit into the phase-transition picture of CA rule space.

[8] See Langton.[28]

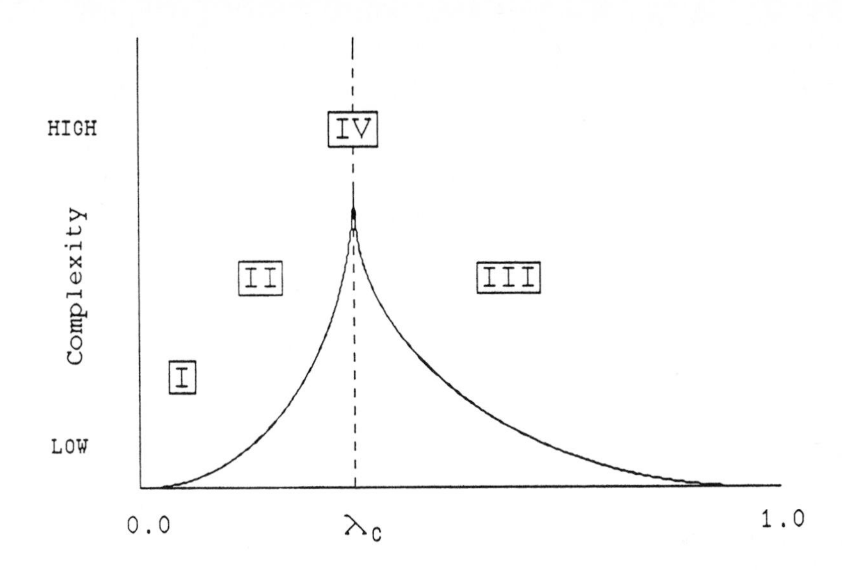

FIGURE 22 Schematic drawing of complexity versus λ over CA rule space, showing the relationship between the Wolfram classes and the underlying phase-transition structure.

However, it is clear that the Wolfram classes, or any other scheme which attempts to partition CA rule space into a small number of "classes" or "categories," can only constitute coarse-grained approximations to the real situation. The phase portrait of CA rule space is a much more "continuous" structure with few, if any, well-defined boundaries besides the transition itself.

As one varies λ over its full range, one sees the full spectrum of CA dynamical behaviors. Some are simple, some are complex. Varying λ throughout its range produces a reasonably smooth progression through what appear to be two, mirror-image complexity hierarchies, one on either side of the transition regime. Approaching the transition from "below," one progresses from simple fixed-point to simple periodic behavior, and then on to longer and longer period behavior, accompanied by longer and longer transients showing more and more sensitivity to array size, until the transition regime is reached. This regime exhibits very complex behavior accompanied by extremely long transients, which show roughly exponential dependence on array size. Furthermore, finite initial configurations started in small portions of very large arrays *may or may not* exhibit transients leading to periodic behavior. The slow growth, and occasional local collapse of complex dynamics to periodic behavior, makes the ultimate outcome of a particular rule operating on a particular initial configuration impossible to predict in the general case.

As one proceeds past the transition regime into the chaotic regime, one traverses another complexity hierarchy, but in reverse order. Even just slightly past the transition regime, ultimate collapse to periodic behavior becomes extremely rare in finite arrays, and would never occur in infinite arrays. Also, transient times to typical chaotic behavior, and their dependence on array size, are shrinking rapidly with movement away from the transition regime. As one approaches the upper limit of the range for λ, CA become completely randomized after minimal transients with *no* dependence on array size. Although maximally chaotic, these dynamics are easily predictable in a statistical sense, and are therefore much simpler than the dynamics observed in the transition regime.

Thus, by varying λ, one proceeds *up* through one complexity hierarchy, from simple to complex dynamics, and then proceeds *down* through another such hierarchy, from complex to simple dynamics. At either end, behavior is simple and predictable, while in the middle, behavior can become arbitrarily complex and highly unpredictable. The entire spectrum of CA dynamical behaviors can be ordered with respect to "distance" from a critical transition point in CA rule space.

6.3 THE PHENOMENOLOGIES OF CA AND COMPUTATIONS

From the discussions above, it is apparent that an underlying phase-transition structure is responsible for the existence of most of the surface-level features of the diverse phenomenology of CA behaviors. Furthermore, the relationship between many of these features is explained coherently by the existence of this underlying phase-transition structure.

When we look to the diverse phenomenology of computations, we see a remarkably similar set of surface-level features as those documented above for CA, including complexity hierarchies, computability classes, arbitrarily long transients leading to unpredictability, terminating and non-terminating dynamics, universal computational capacity, and so forth.

In this section, we briefly review the phenomenology of computations and show how similar it is to the phenomenology of CA described above. In the next section, we see that the phenomenology of computation can be equally well explained by assuming a phase transition underlying the space of computations.

6.3.1 COMPUTABILITY CLASSES As we have seen, there are three possibilities for the ultimate outcome of the evolutions of CA. Some CA rapidly "freeze-up" into short-period behavior from any possible initial configuration. On the other hand, most CA will never become periodic, rapidly settling down to chaotic behavior instead. We can predict the ultimate dynamics of many CA with a high degree of certainty.

However, there exist some CA for which both of these ultimate dynamical outcomes are possible, and because they are typically associated with extended transients, it is effectively undecidable whether a particular CA rule operating on a particular initial configuration will ultimately lead to a periodic state or not.

Similarly, some computations halt, some do not. For most computations, we can decide whether or not they will halt. However, Turing demonstrated that this "Halting problem" is, in general, undecidable. There exist computations for which it is *not* possible to decide whether or not they will halt when started on arbitrary inputs.

Thus, with respect to our ability to decide the ultimate behavior of both CA and computations, there are essentially three possibilities: we can determine that they will halt, we can determine that they will not halt, or we *cannot* determine whether or not they will halt.

6.3.2 COMPLEXITY CLASSES As we saw for CA, some CA relax to their ultimate dynamics after transients whose lengths are independent of the size of the array, while for other CA this transient time can depend exponentially on array size. This is true for CA both in the periodic and in the chaotic regime.

Similarly, for computations that *do* halt, some halt in an amount of time which is only a linear—or even a constant—function of the "size" of the input, while other computations exhibit polynomial, or even exponential, dependence on input size.[17] These different classes of functions describing the relationship between input size and "transient time" to halting constitute a complexity hierarchy, with the simplest computations at one end, and the most powerful computations at the other end.

Furthermore, by resorting to the device of "oracles,"[21] it has been determined that there is a similar complexity hierarchy for *non-halting* computations.

6.3.3 UNIVERSALITY We have seen that the most complex CA have the interesting property that they are arbitrarily "programmable." This means that by merely manipulating details of the initial configuration, without changing the rules of the CA itself, any computable function can be implemented

Similarly, it is known that there exist Turing Machines—representatives of the maximal class of computing devices—that can imitate any other Turing Machine merely through manipulations of the initial state of the input tape.

Thus, both CA and the more familiar realizations of computing devices can be constructed so that their global dynamics is arbitrarily sensitive to subtle details of their initial state.

6.4 THE "DEEP STRUCTURE" OF COMPUTATION

As is clear from the section on CA and phase transitions above, these familiar features of the phenomenology of computation have a clear and simple explanation if we assume that a critical phase transition dominates the structure of the space of computations. This structure is illustrated in Figure 23, which should be compared to Figure 22.

First, the existence of computability classes is explained by the fact that the phase transition separates the space of computations into an ordered and a disordered regime, which we refer to as the halting and the non-halting computations, respectively.

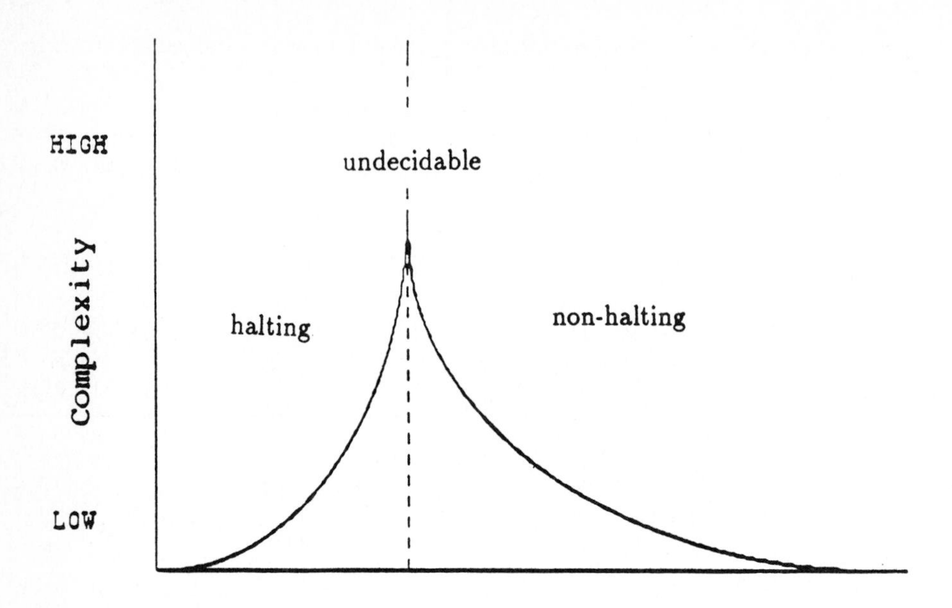

FIGURE 23 Schematic drawing of computation space, indicating relative location of halting, non-halting, and undecidable regimes.

Second, the dynamics in the vicinity of a second-order transition gives rise to the phenomenon of *critical slowing down*, and therefore the transients along the way to ultimately periodic (halting) behavior can diverge to infinity, which explains the existence of undecidability, as characterized by Turing's famous Halting Problem.

Third, the existence of complexity hierarchies is explained by the functional relationship between expected transient time, system size, and distance (with respect to some parameter) from the actual transition point observed for systems exhibiting second-order phase transitions. This functional relationship in the divergence of transient times occurs whether approaching the transition from above or below, and hence, we find that the most complex computations are to be found at a phase transition, between the halting and the non-halting domains, sandwiched in between two complexity hierarchies that diverge as one approaches the phase transition from either the halting regime or from the non-halting regime. Away from the transition regime, transient times are independent of system size, while close to the transition itself, transients can grow exponentially with system size.

Fourth, the existence of universal computation is explained by the fact that the dynamics of physical systems in the vicinity of critical transitions exhibit a divergence in their "susceptibility," that is, in their sensitivity to minute details of their internal structure and to external perturbations. This self-sensitivity in the vicinity of a critical transition is manifested in universal computers as their "programmability."

Thus, if we assume that a critical phase transition dominates the structure of the space of computations, *we expect to see exactly what we do see* when we review the phenomenology of computation.

We *expect* to see a divergence in transient lengths as one approaches the computations in the vicinity of the transition, and we expect to see this increase whether we approach these computations from either the ordered (halting) or the disordered (non-halting) regimes. We *expect* that the functional relationship between transient length, system size, and "distance" from the transition should be governed by different classes of bounding functions, increasing in complexity from constant to exponential as one approaches the computations in the vicinity of the phase transition. We *expect* to see a divergence in susceptibility (programmability) for the computations in the vicinity of the transition. We *expect* to see critical slowing down (the Halting Problem) for the computations in the vicinity of the phase transition.

Thus, the assumption that a critical transition underlies the space of computations provides a simple and straightforward explanation for the existence of, and the relationship between, many significant features of the phenomenology of computation, features which are currently known only through a loose collection of theorems, hypotheses, lemmas, corollaries, and observations.

7. IMPLICATIONS AND QUESTIONS

Now, returning to the main theme of this paper, what does all of this have to do with life? Simply stated, it means that we now have evidence for a *natural domain of information* in the physical world. We can now provide a tentative answer to the question posed at the beginning of this paper:

> *We expect that information processing can emerge spontaneously and come to dominate the dynamics of a physical system in the vicinity of a critical phase transition.*

As the origin of life is intimately associated with the spontaneous emergence of a dynamics of information in the physical world, we suggest that the origin of life occurred when some physico-chemical process underwent a critical phase transition in the early history of the Earth.

In the following sections, we will review what I believe to be some of the implications of the existence of a fundamental connection between computation and phase transitions. What follows is, of necessity, more speculative than the previous sections, as it anticipates future work, rather than describing work already accomplished. However, a reasonable case can be made for everything that follows.

7.1 PHYSICS AND COMPUTATION

The results of this research point to a fundamental equivalence between the dynamics of phase transitions and the dynamics of information processing. Such a connection is a two-edged sword. With one edge, we can apply what we know about the physics of phase transitions to further our understanding of computation. With the other edge, we can apply what we know about computation to further our understanding of phase transitions.

Thus, the establishment of such a connection will bring us fundamentally new insights into the nature of computation. Some of the phase-transition phenomena mentioned above as analogs of computational phenomena are well understood in the context of such bodies of theory as thermodynamics, statistical mechanics, and renormalization group theory. By understanding computation as a special case of phase-transition phenomena, much of the apparatus of the above bodies of theory can be brought to bear on problems in the theory of computation.

Likewise, the apparatus of the theory of computation can be brought to bear on problems in the theory of phase transitions. In particular, it may become clear that certain aspects of phase transitions are not treatable by *any* theory in principle, because they effectively involve undecidable problems. This connection already tells us something about the phenomenon of critical slowing down: it is the physical manifestation of the fact that the system is engaged in "solving" an intractable problem. A material near its critical transition point between the liquid and gas states must, in effect, come to a global decision about whether it will settle down to a liquid or to a gas. This sounds almost anthropomorphic, but the results reported here suggest that we must think of such systems as effectively *computing* their way to a minimum energy state.

It also lends support to the Church-Turing hypothesis that *no* system will be found that computes a wider class of functions than a Turing machine. The fact that physical systems having to "decide" between two qualitatively different physical states can take arbitrarily long to make the "decision"—the phenomenon of critical slowing down—suggests that physical systems in general are bound by the same *in principle* limitations as computing devices.

In fact, it is perhaps misleading to claim that the phenomenology of phase transitions explains the phenomenology of computation when, in fact, one could equally well claim that the phenomenology of computation explains the phenomenology of phase transitions. On the basis of the evidence presented here, we could equally well say that the phenomenon of critical slowing down is simply a physical manifestation of the halting problem.

What all this is really telling us is that these two phenomenologies are not "two" at all but rather "one." We are observing one and the same phenomenon reflected in two very different classes of systems and their associated bodies of theory.

7.2 SOLIDS, FLUIDS, AND DYNAMICS

With CA, and other spatially extended dynamical systems, we are experimenting with "artificial matter" (or "programmable matter" in Toffoli's words[45]). That is, we are experimenting with "material" for which we have precise control over the behaviors of the individual "atoms" or "molecules" of which it is constituted.

It is somewhat surprising that, despite the many different varieties of atoms and molecules that constitute "real" materials, almost every known substance comes in one of three flavors: solid, liquid, or gas. As it is possible to continuously transform liquids into gases and *vice versa* without passing through a phase transition, they are taken to constitute a single, more general phase of matter: *fluids*. Therefore, there are really just *two* fundamental phases of matter—solids and fluids—and so we should not be surprised to find two similar fundamental phases in our "artificial" materials, despite the large number of ways that we can put them together. The important point here is that solids and fluids are *dynamical*, rather than merely *material*, categories.

We know solids and fluids primarily as states of matter, rather than as universal classes of dynamical behavior, because up until quite recently, everything that exhibited dynamical behavior was fundamentally material in constitution. Now, however, with the availability of computers, we are able to experiment with dynamical behaviors *per se*, abstracted from any particular material substrate. What we find is that, despite having abandoned the material basis of solids and fluids, *we are nonetheless left in possession of solid and fluid dynamics!* Thus, we are safe in assuming that these fundamental classes of dynamical behavior do not inhere in material *per se*, but rather in the way in which the material is organized.

The most important point, however, is that these two universality classes of dynamical behavior are separated by a *phase transition*. As we have seen, the dynamics of systems operating near this phase transition provides the basis of support for embedded computation. Thus, a third category of dynamical behaviors exists in which materials—or more broadly, dynamical systems in general—can make use of an intrinsic computational capacity to avoid either of the two primary categories of dynamical behavior by maintaining themselves on indefinitely extended transients.

Since computers and computations are specific instances of material and dynamical systems respectively, they are also ultimately bound by these same universality classes. Therefore, computer "hardware" can behave like a solid, like a fluid, or like something in between.

An interesting open question to pursue here is whether one can resolve the fluid phase of CA (and other dynamical systems) even further into liquid and vapor phases. The careful reader may have noticed that I have consistently used the phrase "in the vicinity of a second-order phase transition" rather than the phrase "*at* a second-order phase transition." This is in part due to the fact that it is hard to determine whether or not one is precisely "at" a critical point when working with finite systems. But it is also due to the fact that in most of the experiments on CA leading to these results, the regime exhibiting the most complex dynamics appears to be just slightly below (to the ordered side) of the critical transition itself. Most physical systems exhibit a liquid phase or the "ordered side" of their critical point, and liquids exhibit extremely complex dynamics.

It is somewhat surprising that liquids are *very* poorly understood. Although the structures of both solids and gases are well studied, it has proven difficult in practice to characterize the structure of liquids. As might be expected from their location between the solid and the gas phases, liquids exhibit much more complicated behaviors than either of the other two phases. Recent work by Stanley[43] suggests that the hydrogen-bond network of liquid water exhibits very complex dynamics, and super-cooled water can apparently exhibit critical dynamics.

It may be appropriate to view liquids as constituting a broadened phase transition in the phase portrait of a material. Liquids occupy just the right spot "at the edge of chaos" in the phase portrait of many materials to be candidates for further investigation along the lines of the results reported in this paper. The complex transition regime might be co-extensive with the liquid regime for many real and "artificial" materials.

7.3 LIFE, INTELLIGENCE, AND EVOLUTION

This generalization of the solid, liquid, and vapor phases of matter to universal classes of behavior for dynamical systems in general, has important consequences for our understanding of the kinds of behaviors achievable by computer "hardware." It has generally been thought that computer hardware and biological "wetware" are fundamentally different kinds of "stuff," and that, because of this fundamental difference, computer hardware could never achieve some of the more exotic dynamical behaviors exhibited by wetware, such as life and intelligence. However, if it is properly understood that hardness, wetness, or gaseousness are properties of the organization of matter, rather than properties of the matter itself, then it is only a matter of organization to turn "hardware" into "wetware" and, ultimately, for "hardware" to achieve everything that has been achieved by wetware, and more.

In addition, this perspective implies that a material system operating in the vicinity of a phase transition can behave like a computer. This has important consequences for our understanding of the kinds of behaviors achievable by "wetware."

Life is clearly founded on a capacity to sense, process, and act on information. The primary macromolecules of life are "informational" molecules, and these molecules are engaged in a finely tuned dynamics of information processing which we have only just begun to understand.

It has long been an open question how such a dynamics of information could have gotten started within the pre-biotic soup, and how it could have refined itself and evolved through time to produce such magnificent information processors as human brains. The results of this research suggest a new set of answers to these questions.

We now know that a dynamics of information can spontaneously emerge in physical systems near a critical phase transition. The results of this paper suggest the possibility that the information dynamics which gave rise to life came into existence when global or local conditions brought some medium—perhaps H_2O, perhaps some other material—through a critical phase transition. As we have seen in CA, the dynamics in the vicinity of a critical phase transition support a complex dynamics of metastable structures, which is very much of a kind with the complex dynamics of metastable structures that characterizes life. Furthermore, we also know that, in CA at least, this dynamical regime can support arbitrarily complicated information processing. Thus, this regime is a good candidate for the origin of the kind of information dynamics that we would consider to be synonymous with the origin of life.

Now, Nature could not have been so beneficent as to have maintained this medium near a critical phase transition for very long. This means that the nascent information dynamics must have gained control over some parameters that allowed them to maintain *local* conditions near the phase transition while *global* conditions drifted away from the phase transition. Of course, there must be many parameters that could push such systems away from their vital transition point, and many of these probably varied widely, destroying a large proportion of these early information-processing systems. Evolution can be viewed as the process of gaining control over more and more "parameters" affecting a system's relationship to the vital phase transition.

Living systems can perhaps be characterized as systems that dynamically *avoid* attractors. The periodic regime is characterized by limit-cycle or fixed-point attractors, while the chaotic regime is characterized by strange attractors, typically of very high dimension. Living systems need to avoid either of these ultimate outcomes, and must have learned to steer a delicate course between too much order and too much chaos—the *Scylla* and *Charybdis* of dynamical systems.

They apparently have done so by learning to maintain themselves on extended transients—i.e., *by learning to maintain themselves near a "critical" transition.* Once such systems emerged near a critical transition, evolution seems to have discovered the natural information-processing capacity inherent in near-critical dynamics, and to have taken advantage of it to further the ability of such systems to maintain themselves on essentially open-ended transients.

Of course, climbing out of one attractor just pushes the problem back to a higher-dimensional phase space, in which the system is again in the basin of some attractor. It is therefore possible to view evolution as a repeated iteration of the process whereby a system climbs out of one attractor into a higher-dimensional phase space, only to find itself in the basin of a higher-dimensional attractor, a process that gives evolutionary significance to the phrase "out of the frying pan, into the fire!"

In the context of the work of Stanley, described in the previous section, it is interesting to consider the possibility that the dynamics of information which eventually lead to the origin of life may have emerged as the structural dynamics of liquid water. Theories of the origin of life have assumed that life originated in the dynamics of molecules *embedded* in liquid water. However, rather than liquid water having merely provided the "nursery" for the origin of life among molecules embedded within it, life may have originated in the dynamics of water itself. The dynamics may then have eventually spread to the embedded molecules, setting the stage for a "genetic takeover" of the kind proposed by Cairns-Smith,[8] in which the induced dynamics of the embedded molecules took over from the original dynamics of liquid water. Or, perhaps the natural dynamics of liquid water still plays a "vital" role in modern life.

There is ample evidence in living cells to support an intimate connection between phase transitions and life. Many of the processes and structures found in living cells are being maintained at or near phase transitions. Examples include the lipid membrane, which is kept in the vicinity of a sol-gel transition; the cytoskeleton, in which the ends of the microtubules are held at the point between growth and dissolution; and the naturation and de-naturation (zipping and unzipping) of the complementary strands of DNA.

In the case of intelligence, there is also qualitative evidence for this phase transition view in the dynamics of the brain. It is vital that the brain be kept very near to 98.6°F in order to work properly. We've all experienced the chaotic nature of our thinking processes when we have a fever. Some have experienced the seizures (periodic dynamics) that accompany hypothermia, when the brain gets too cold. On the temperature scale, clearly, the brain operates in a very narrow regime between periodic and chaotic dynamics, and a great amount of physiological machinery has evolved to keep it at this critical point. Our mental capabilities are apparently only possible in the vicinity of this phase transition between periodic and chaotic neural dynamics.

There is also evidence that evolutionary dynamics brings populations to "the edge of chaos." Eldredge and Gould claim that the fossil record exhibits *punctuated equilibria*.[14] That is, the fossil record seems to indicate that long periods of evolutionary stasis are irregularly interrupted by periods of chaotic and rapid evolutionary change. Evidence from recent computer experiments indicates that this phenomenon may be generic for evolutionary processes.[33] Many nonlinear dynamical systems exhibit a remarkably similar phenomenon known as "intermittency," in which long periods of regular periodic behavior are interrupted irregularly by bursts of rapidly fluctuating chaotic behavior. Most importantly, intermittency is generically observed in these systems when they are in the vicinity of a transition from periodic to chaotic behavior. Thus, punctuated equilibria in the fossil record can be viewed as evidence for the proposition that biological evolution has maintained populations in a state of near-critical dynamics.

7.4 RELATED WORK

There are other lines of research that have intimate connections with this work.

As mentioned earlier, Jim Crutchfield has observed a similar increase in effective computational complexity in the vicinity of phase transitions in continuous dynamical systems.[11]

Vichniac, Tamayo, and Hartman[46] discovered that the Wolfram classes could be recovered by varying the frequency of two simple rules in an inhomogeneous cellular automaton. They also suggested a relation between critical slowing down and the halting problem.

Norman Packard and Wentian Li have mapped out the space of "elementary" $K = 2$, $N = 3$, one-dimensional CA fairly completely, using a parameterization scheme similar to λ.[31]

Packard has also performed an interesting series of experiments in which he "adapts" CA rules by selecting for certain properties of the global dynamics.[39] He finds that an initially random population of rules will migrate towards the phase-transition region. His interpretation of this phenomenon is that it is easier to find rules that will *compute* the desired behavior—by making use of a general computational capacity—than it is to find rules that are "hard-wired" to produce *only* the desired behavior.

Stuart Kauffman has investigated a class of dynamical systems known as *boolean nets*,[22,23] in which he finds a phase transition between ordered and disordered dynamics as a function of the *connectivity* of the network. These nets also exhibit phase transitions as a function of an "internal homogeneity parameter," a parameter very much like λ, which controls the number of 1's in the boolean functions found at each node in the network.[47] Kauffman is currently working on a set of experiments to see if evolving populations of these boolean nets will converge on the transition regime in their parameter space (see the contribution by Kauffman and Johnsen in these proceedings.)

Harold McIntosh[36] has applied the mean-field approach of Gutowitz[19,20] and suggests that the Wolfram classes can be distinguished on the basis of simple features of the mean field theory curves. These simple features clearly locate Class IV (complex rules) at the transition between Class II (periodic rules) and Class III (chaotic rules.)

Bill Wootters has applied mean-field theory to explain the λ parameter results, and has been able to reproduce many of the features of Figure 6.[52]

The computer-optimization procedure known as *simulated annealing*[24] has a strong connection with this work. Annealing schedules call for extended stays in the vicinity of the freezing point. This is partially due to the phenomenon of critical slowing down: it takes longer for the systems to relax near phase transitions. However, we have seen that slowing down is intimately associated with complex computational dynamics, and it is interesting that the transition is the very point at which we expect information processing to emerge spontaneously within the system being annealed. This suggests that the real reason for hovering in the vicinity of the freezing point—and the reason that it takes longer to relax there—is that even the simple act of relaxation in this regime is computationally complex: it cannot simply relax, it must *compute* its way to a minimum energy state. The system at the freezing point is effectively caught up in running an embedded computation.

The concepts reported here clearly have some relation to the notion of self-organized criticality.[3] Bak et al. have proposed that a number of systems, including some CA, exhibit a magical capacity to organize themselves towards a critical state, *in the absence of any parameter tuning.* Bak has suggested that Conway's game of LIFE is a self-organized critical system, although he does not bring LIFE's computational capacity into the discussion.[4] Bennett[5] has recently demonstrated that LIFE is generically *sub-critical*, but suggests that under a small set of initial conditions, LIFE might be *super-critical.*

Finally, Andy Wuensche and Mike Lesser have recently produced an atlas of basin of attraction fields for simple one-dimensional CA.[53] These basin fields are constituted of directed graphs on the state-spaces of CA. When organized according to the λ parameter, or according to a related parameter Z which they define, one can observe that the phase transition in CA dynamics is associated with a phase transition in the associated state-space graphs—a transition analogous to the emergence of a "giant component" in random graphs. Fixed-point and periodic CA are associated with a high degree of convergence in the state space, while chaotic CA are associated with very low convergence. Complex CA have irregular and extremely long transient trees, exhibiting a low but positive degree of convergence. Their parameter Z is a global measure of the overall degree of convergence in state-space, and is probably a more accurate predictor of the phase transition than λ.

Thus, there is a growing body of experimental and theoretical evidence pointing to a fundamental association between complex dynamics, computational capacity, and phase transitions in both abstract-formal and concrete-physical systems, organic as well as inorganic.

8. CONCLUSION

In this paper, we have attempted to uncover the conditions under which a complex dynamics of information processing can emerge spontaneously and come to dominate the dynamics of a physical system. This has lead us to the observation that a second-order, or *critical*, phase transition underlies the space of a number of classes of dynamical systems, including cellular automata (CA) and computations, and that sytems in the vicinity of such a phase transition can support arbitrary information processing.

We have concluded that phase transitions figure "critically" in the origin and evolution of life and intelligence.

ACKNOWLEDGMENTS

I would like to thank the following friends and colleagues for stimulating discussions on the topics discussed here: Richard Bagley, Jim Crutchfield, Doyne Farmer, Stephanie Forrest, Walter Fontana, Howard Gutowitz, Hyman Hartman, Stuart Kauffman, Wentian Li, Norman Packard, Steen Rasmussen, Rob Shaw, Creighton Walker, and Bill Wootters.

REFERENCES

1. Atkins, P. W. *Physical Chemistry.* San Francisco: W. H. Freeman, 1978.
2. Atkins, P. W. *The Second Law.* New York: W. H. Freeman, 1984.
3. Bak, P., C. Tang, C., and K. Wiesenfeld. "Self-Organized Criticality." *Phys. Rev. A* **38(1)** (1988): 364–374.
4. Bak, P., K. Chen, and M. Creutz. *Nature* **342** (1989): 780–782.
5. Bennett, C., and M. S. Bourzutschky. "Non-Criticality of the Cellular Automaton 'Life.'" *Nature* (1990).
6. Berlekamp, E., and J. H. Conway, and R. Guy. *Winning Ways for Your Mathematical Plays.* New York: Academic Press, 1982.
7. Burks, A. W. *Essays on Cellular Automata.* Urbana: University of Illinois Press, 1970.
8. Cairns-Smith, A. G. *Genetic Takeover and the Mineral Origins of Life.* Cambridge: Cambridge University Press, 1982.
9. Chaitin, G. *J. Assoc. Comput. Mach.* **13** (1966): 145.
10. Codd, E. F. *Cellular Automata.* New York: Academic Press, 1968.
11. Crutchfield, J. P., and K. Young. "Computation at the Onset of Chaos." *Complexity, Entropy, and Physics of Information*, edited by W. H. Zurek, 223–270. Santa Fe Institute Studies in the Sciences of Complexity, Proc. Vol. VIII. Redwood City, CA: Addison-Wesley, 1990.
12. Culik, K., and S. Yu. "Undecidability of CA Classification Schemes." *Complex Systems* **2** (1988): 177–190.
13. Derrida, B., and Y. Pomeau. "Random Networks of Automata: A Simple Annealed Approximation." *Europhys. Lett.* **1(2)** (1986): 45–49.
14. Eldredge, N., and S. J. Gould. "Punctuated Equilibria: An Alternative to Phyletic Gradualism." *Models in Paleobiology*, edited by T. J. M. Schopf, 82–115. Freeman Cooper, 1972.
15. Fredkin, E., and T. Toffoli. "Conservative Logic." *Int'l J. Theor. Phys.* **21** (1982): 219–253.
16. Gardner, M. "Mathematical Games: The Fantastic Combinations of John Conway's New Solitaire Game 'Life.'" *Sci. Am.* **223(4)** (1970): 120–123.
17. Garey, M. R., and D. S. Johnson. *Computers and Intractability.* San Francisco: W. H. Freeman, 1979.
18. Gatlin, L. L. *Information Theory and the Living System.* New York: Columbia University Press, 1972.
19. Gutowitz, H. A., B. K. Knight, and J. D. Victor. "Local Structure Theory for Cellular Automata." *Physica D* **48** (1987): 18.
20. Gutowitz, H. A. "A Hierarchical Classification of Cellular Automata." *Proceedings of the 1989 Cellular Automata Workshop*, edited by H. A. Gutowitz. Amsterdam: North Holland, 1990.
21. Hopcroft, J. E., and J. D. Ullman. *Introduction to Automata Theory, Languages, and Computation.* Menlo Park, CA: Addison-Wesley, 1979.

22. Kauffman, S. A. "Metabolic Stability and Epigenesis in Randomly Constructed Genetic Nets." *J. Theor. Biol.* **22** (1969): 437–467.
23. Kauffman, S. A. "Emergent Properties in Random Complex Automata." *Physica D* **10** (1984).
24. Kirkpatrick, S., C. D. Gelatt, and M. P. Vecchi. "Optimization by Simulated Annealing." *Science* **220(4598)** (1983): 671–680.
25. Kolmogorov, A. N. *Prob. Inf. Transm.* **1(1)** (1965).
26. Langton, C. G. "Studying Artificial Life with Cellular Automata." *Physica D* **22** (1986): 120–149.
27. Langton, C. G. "Computation at the Edge of Chaos." *Physica D* **42** (1990): 12–37.
28. Langton, C. G. "Computation at the Edge of Chaos: Phase-Transitions and Emergent Computation." Ph.D. thesis, University of Michigan, 1991.
29. Langton, C. G., C. Taylor, J. D. Farmer, and S. Rasmussen, eds. This volume.
30. Li, W. "Analyzing Complex Systems." Columbia University, 1989.
31. Li, W., and N. Packard. "Structure of Elementary Cellular Automata Rule-Space." *Complex Systems* (1990).
32. Lindgren, K., and M. G. Nordahl. "Universal Computation in Simple One-Dimensional Cellular Automata." *Complex Systems* **4** (1990): 299–318.
33. Lindgren, K. "Evolutionary Phenomena in Simple Dynamics." This volume.
34. Mandelbrot, B. B. *Fractals: Form, Chance, and Dimension.* San Francisco: W.H. Freeman, 1977.
35. Mayr, E. *The Growth of Biological Thought.* Cambridge: Harvard University Press, 1982.
36. McIntosh, H. V. "Class IV Cellular Automata and a Good LIFE." *Proceedings of the 1989 Cellular Automata Workshop*, edited by H. A. Gutowitz. Amsterdam: North Holland, 1990.
37. von Neumann, J. *Theory of Self-Reproducing Automata*, edited and completed by A. W. Burks. Urbana: University of Illinois Press, 1966.
38. Packard, N. H., and S. Wolfram. "Two-Dimensional Cellular Automata." *J. Stat. Phys.* **38** (1985): 901.
39. Packard, N. H. "Adaptation Toward the Edge of Chaos." *Complexity in Biological Modelling*, edited by S. Kelso and M. Shlesinger. 1988.
40. Reichl, L. E. *A Modern Course in Statistical Physics.* Austin: University of Texas Press, 1980.
41. Smith III, A. R. "Simple Computation-Universal Cellular Spaces." *JACM* **18(3)** (1971): 339–353.
42. Stanley, H. E. *Introduction to Phase Transitions and Critical Phenomena.* Oxford: Oxford University Press, 1971.
43. Sciortino, P. H. P., H. E. Stanley, and S. Havlin. "Lifetime of the Bond Network and Gel-Like Anomalies in Supercooled Water." *Phys. Rev. Lett.* **64(14)** (1990): 1686–1689.
44. Stauffer, D. *Introduction to Percolation Theory.* Philadelphia, PA: Taylor and Francis, 1985.

45. Toffoli, T., and N. Margolus. *Cellular Automata Machines.* Cambridge MA: MIT Press, 1987.
46. Vichniac, G. Y., P. Tamayo, and H. Hartman. "Annealed and Quenched Inhomogeneous Cellular Automata." *J. Stat. Phys.* **45** (1986): 875–883.
47. Walker, C. Personal communication.
48. Wolfram, S. "Stastical Mechanics of Cellular Automata." *Rev. Mod. Phys.* **55** (1983): 601–644.
49. Wolfram, S. "Universality and Complexity in Cellular Automata." *Physica D* **10** (1984): 1–35.
50. Wolfram, S. "Twenty Problems in the Theory of Cellular Automata." *Physica Scripta* (1985).
51. Wolfram, S. *Theory and Applications of Cellular Automata.* Singapore: World Scientific, 1986.
52. Wootters, W. T., and C. G. Langton. "Is There a Sharp Phase Transition for Deterministic Cellular Automata?" *Physica D* **45** (1990): 95–104.
53. Wuensche, A., and M. Lesser. *An Atlas of Basin of Attraction Fields of One-Dimensional Cellular Automata* (working title). Santa Fe Institute Studies in the Sciences of Complexity, Reference Volume I. Redwood City, CA: Addison-Wesley, 1992.

Richard J. Bagley and J. Doyne Farmer
Complex Systems Group, Theoretical Division, and Center for Nonlinear Studies, Los Alamos National Laboratory, Los Alamos, NM 87545 and Santa Fe Institute, 1120 Canyon Road, Santa Fe, NM 87501

Spontaneous Emergence of a Metabolism

Networks of catalyzed reactions with nonlinear feedback have been proposed to play an important role in the origin of life. We investigate this possibility in a polymer chemistry with catalyzed cleavage and condensation reactions, studying the properties of a well-stirred reactor driven away from equilibrium by the flow of mass. Near equilibrium the distribution of material is uninteresting; it favors short polymers but is otherwise homogeneous. However, under appropriate non-equilibrium conditions, the situation changes radically: The nonlinear feedback of the reaction network focuses the material of the system into a few specific polymer species, whose concentrations can be orders of magnitude above the background. Like a metabolism, the network of catalytic reactions "digests" the material of its environment, incorporating it into its own form. For this reason we call it an *autocatalytic metabolism*. We vary the diet of an autocatalytic metabolism, and demonstrate that under some variations it persists almost unchanged, while in other cases it dies. We argue that the dynamical stability of autocatalytic metabolisms gives them regenerative properties that allow them to repair themselves and to propagate through time.

CONTENTS

1. MOTIVATION

1.1 SETTING THE STAGE FOR AN ORIGIN OF LIFE

When Miller and Urey discovered that amino acids could be formed under conditions that might be similar to those of the prebiotic earth,[28] the spontaneous synthesis of proteins seemed just around the corner. However, this turns out to be much more difficult than the spontaneous synthesis of individual amino acids. Equilibrium conditions tend to favor dissociation, and generate a concentration profile that is fairly uniform. Except for occasional fluctuations, for long polymers the population of any given molecular species is typically zero. The population distribution of polymers is homogeneous, nonspecific, and uninteresting. This is in contrast to living organisms, which have high concentrations of a few *specific* polymer species.[1]

Contemporary organisms achieve specificity through a codependent relationship between templates and enzymes. Proteins and nucleic acids synthesize each other through a replication mechanism in which none of the components synthesizes itself. Even for the simplest organisms, this process is highly complex. There seems to be a minimum level of complexity below which a replicating machine based on proteins and nucleic acids simply cannot function. While it is easy to understand how such a replicating machine perpetuates itself, it is difficult to understand how the necessary initial conditions ever arose on their own. The probability that both enzymes and templates could be created through a statistical fluctuation is effectively nil. This suggests that other processes preceded contemporary life.

The idea that enzymatic activity might have set the stage for the origin of life was developed by Oparin, who suggested that coacervates may have played a major role.[33] Early experiments unsuccessfully attempted to use clays and other materials as nonspecific catalysts for polymerization.[34] Calvin studied several different scenarios through which catalytic activity could provide a selection mechanism, even without self-replication.[5,6] In 1971 Rössler,[37,38] Eigen,[10] and Kauffman[23] developed this idea further. In particular, Rössler envisioned a form of chemical evolution similar to that studied here. He emphasized the importance of specific catalysts which catalyze only a small fraction of all possible reactions. Along these same lines, Kauffman[24] later modeled the problem in terms of random graphs, and showed that under reasonable assumptions the probability of catalytic closure is quite high.[2] The random graph model was developed into a kinetic model that could be simulated on a computer by Farmer et al.[11] This line of investigation, which attempts to find possible precursors facilitating the emergence of life should be contrasted with

[1] An exception is provided by the experiments of Sidney Fox, who by heating a mixture of amino acides demonstrated the formation of polypeptides, called proteinoids.[20,21] The structure of proteinoids is not random; some subsequences, such as certain hexapeptides, occur much more frequently than others. In contrast, our goal is to increase the concentration of entire molecules, so that it is several orders of magnitude above equilibrium.

[2] Another toy model investigating the possibility that a metabolism might have spontaneously emerged without a replicator is due to Dyson.[8]

other work that addresses the (also very interesting) question of the early evolution of life once replication has already begun.[1,9,10,39]

In this paper we study the behavior of a network of catalyzed chemical reactions, along the lines laid out by Farmer et al.[11] We make several enhancements of the model and analytically study a few simple cases to gain better intuition about the dynamics. We also improve the simulation so that it is several orders of magnitude faster. This allows us to simulate the kinetics of a complicated reaction network in a matter of seconds. As a result, we are able to widely explore the parameter space and answer many of the questions originally raised in earlier papers.

Our main result is that, under appropriate conditions, a catalytic reaction network can focus most the material of its environment into a few chemical species. For this to happen the system must be driven the appropriate distance from equilibrium, polymerization must be favored, and it must have diverse kinetic parameters. Favoring polymerization may require the addition of energy, for example, through pyrophosphates energized by light. In spite of these restrictions, there is a wide range of parameters in which the material of the system is focused into only a few species, which dominate over the background.

Focusing radically alters the material composition of the environment. The species that emerge reinforce each others' production and largely take over the reaction vessel, excluding other possibilities. Since this behavior is analogous to that of a metabolism, we call the resulting set of species and reactions an *autocatalytic metabolism*. Under appropriate conditions autocatalytic metabolisms can evolve out of a simple, undifferentiated initial state, generating a sequence of complex, highly differentiated, final states. Like contemporary organisms, these final states are composed of a highly focused, specific set of long polymers. While the autocatalytic metabolism does not replicate itself in the usual sense, it propagates itself by taking over any medium with suitable properties, sustaining itself as long as the appropriate conditions are met. Furthermore, it may generate a lineage of related autocatalytic metabolisms.

The model that we study here applies to any system in which polymers can catalyze the formation of other polymers through cleavage and condensation reactions. If the basic building blocks are amino acids, then the polymers are called polypeptides or proteins. Such reactions are common among proteins, forming the basis for many of the functions of living organisms. If the basic building blocks are nucleotides, then the polymers are called nucleic acids (RNA or DNA). It is well known that polypeptides possess a large repertoire of catalytic activities of this type; the recent discovery of specific catalysis reactions in RNA suggests that nucleic acids may also possess the necessary properties.[7] Whether the polymers are polypeptides or nucleic acids changes the parameters but not the basic form of the model.

Even if the model we discuss here has nothing to do with the actual origin of life on earth, it might provide a *possible* origin of life in the laboratory. Although at present we cannot predict the outcome of experiments in detail, we can make qualitative predictions that provide broader experimental guidelines. The accumulation of more experimental knowledge can be used to determine the unknown parameters of our model, which in turn should sharpen its predictive value for experiments. Although many important experimental details are still unknown, and some important questions await further study, our numerical simulations suggest that it may be possible to synthesize autocatalytic metabolisms in the laboratory.

1.2 NONTRIVIAL DISSIPATIVE STRUCTURES

The problem of the emergence of life is embedded in the broader problem of understanding self-organizing phenomena, which from the point of view of a physicist may be more interesting anyway. Many non-living systems exhibit self-organizing properties, albeit much weaker than those of living systems. Is there a sharp distinction between living and nonliving systems? Or can there exist levels of organization that are between those of present living and non-living systems? Can evolution and other self-organizing properties of living systems be viewed as manifestations of a general law that describes the tendencies of matter to organize itself?

There are many simple examples that have been cited as instances of self-organizing phenomena in nature. For example, when a fluid is heated from below, under appropriate conditions, patterns of convection cells form. The macroscopic structure of these cells is internally generated by the system itself, and is not apparent in its initial conditions. Such patterns are often called *dissipative structures*, because they occur when energy flows into a system and then is dissipated.[32]

Several researchers have asserted that life is a non-equilibrium phenomenon, associated with dissipative structures.[32] This is certainly true, but it is a very weak statement. While deviation from equilibrium is a necessary condition for life, it is far from sufficient. Driving a system away from equilibrium does not necessarily cause the emergence of order—in fact, it often has precisely the opposite effect. A central question that must be addressed in a theory of self-organizing phenomena is: Why do some non-equilibrium situations foster the spontaneous emergence of organization, while others do not?

There is a big gap between the dissipative structures of simple non-living systems, such as patterns in fluid convection, and the much richer dissipative structures associated with living systems. The model discussed here is intended to bridge this gap, at least to some extent, by showing the possibility for dissipative structures that are intermediate in complexity between living and non-living systems. Autocatalytic metabolisms are more complex than convection patterns, in that they propagate specific information through time. One autocatalytic metabolism can seed the formation of another, similar metabolism. The autocatalytic metabolisms of this model can be viewed as proto-life forms, since they have a metabolism, they evolve and store information, and they reproduce (although more continuously and with less fidelity than contemporary organisms). They are also dynamically stable, and so capable of self-repair. They, thus, have many of the essential properties of living systems, albeit in a much less sophisticated form.

Besides demonstrating the possibility for the spontaneous generation of autocatalytic metabolisms, one of our main purposes in this paper is to discover under what conditions they can be expected to form. How does their formation depend on the parameters of the system, such as the flow of energy, or the inherent diversity of the underlying dynamics? Although our results are specific to this model, they nonetheless suggest several rules that may pertain to the more general problem of self-organization.

1.3 A SIMPLE MODEL FOR STUDYING EVOLUTION WITH AN EMERGENT NOTION OF FITNESS

In principle, it is possible to describe biological systems at a fundamental level in terms of their dynamics. At this level of description, "selection" is an emergent property of the dynamics. In practice, however, for most systems this is hopelessly intractable. As a result, studies of evolution are typically couched in terms of the fitness function, which is an empirical construct, disconnected from the laws of physics. Even so, in most systems the fitness function is known only in very special circumstances where all but a few relevant factors are neglected. In general, fitness is a complicated function of the external environment, which includes other organisms. As a result, most theoretical models for evolution make many *ad hoc* assumptions, postulating fitness functions that may be qualitatively different from those in the real world.

As pointed out by Eigen,[10] Rössler,[37] and others, chemistry provides an excellent forum for studying evolution. The laws of chemical kinetics are well understood, and make it possible to model the behavior of the system at a fundamental level. These laws determine population levels and therefore determine fitness. As in biological systems, fluctuations are always present, generating random variation. Thus, for chemical networks we can describe the fitness at the fundamental level of dynamics.

Even though autocatalytic metabolisms do not have templates or a genetic code, because of their specificity they are nonetheless capable of evolution. This is discussed in a companion paper.[4] Autocatalytic metabolisms, therefore, provide an interesting alternative for studying evolution in a chemical setting. It is also interesting to note that autocatalytic structures analogous to those we study here occur spontaneously in more abstract environments, as observed by Fontana[16,17] and Rasmussen et al.[36]

2. BACKGROUND

In this section we discuss some of the properties of catalyzed reaction networks, providing a background for the development of the simulation in Section 3. We discuss the reactions we are going to consider, and show why they are uninteresting at equilibrium. We then explain how the situation is altered as we move away from equilibrium, and how catalysis can play an important role in focusing the material of the system into just a few chemical species. We define autocatalytic sets and the related notion of autocatalytic metabolisms.

2.1 SPONTANEOUS REACTIONS

We are interested in reversible polymerization reactions, in which two polymers either *condense* to form a single longer polymer, or a single polymer *cleaves* into two shorter polymers. Cleavage and condensation can be considered together as a single reversible reaction. The reaction in which polymers A and B join together to form C, giving off water, or equivalently, in which C hydrolyzes into A and B, can be written

$$A + B \rightleftharpoons C + H, \tag{1}$$

where H represents water.

Providing the concentrations are sufficiently high and the solution is well stirred, the law of mass action provides a good approximation of the kinetics. Let k_f be the rate constant for the forward reaction, $A+B \rightarrow C+H$, and k_r be the rate constant for the backward reaction $C + H \rightarrow A + B$. The rate equation for C is then

$$\dot{C} = \frac{dC}{dt} = k_f AB - k_r HC. \tag{2}$$

For convenience, whenever the meaning is unambiguous, we use the same symbol to represent both a polymer and its concentration. Similar equations apply for $\dot{A}$ and $\dot{B}$.

2.2 EQUILIBRIUM DISTRIBUTION OF POLYMERS

At equilibrium the concentrations of the polymers of a given length can be computed analytically using the classical theory of polycondensation reactions developed by Flory and Stockmayer.[15,40,44] For simplicity we assume that all the reactions have the same forward and backward rate constants, and that the reaction vessel is well stirred. Furthermore, we assume that the monomers are oriented so that each monomer has two sites, which we arbitrarily designate as the "+" site and "–" site.

We follow the treatment of Macken and Perelson.[27] Rather than solving for the concentrations of each polymer, it is more convenient to use an aggregate variable y, which is the concentration of free sites of a given kind (either "+" or "–"). We assume that the polymers are unbranched, and that they cannot form rings. For reactions of the form of Eq. (2),

$$\dot{y} = -k_f y^2 + k_r(m_0 - y)H, \tag{3}$$

where m_0 is the total concentration of monomers, which is equal to the concentration of free sites if nothing is bound. At steady state $\dot{y} = 0$ and the concentration of free sites is

$$y = \frac{(1 + 4\kappa y_0)^{1/2} - 1}{2\kappa}, \tag{4}$$

where $\kappa = k_f / H k_r$ is the equilibrium constant.

We now compute the concentration of polymers of length n. At equilibrium, let ρ be the probability for the formation of a bond. This is the ratio of bound sites to the total number of sites, i.e.,

$$\rho = \frac{m_0 - y}{y_0}. \tag{5}$$

Assume that each binding event is independent. For a given free site, the probability that it is attached to $n - 1$ bonds followed by another free site is $\rho^{n-1}(1 - \rho)$. Solving Eq. (5) for y shows that the concentration of free sites for a given value of ρ is $y = m_0(1 - \rho)$. Thus, the concentration of polymers of length n is

$$x_n = m_0(1 - \rho)^2 \rho^{n-1}. \tag{6}$$

At equilibrium, inserting Eq. (4) into Eq. (5) gives

$$\rho = 1 - \frac{(1 + 4\kappa m_0)^{1/2} - 1}{2\kappa m_0}. \tag{7}$$

Note that $\rho < 1$. Thus, Eq. (6) implies that the concentration of polymers of length n decreases exponentially with n, at a rate that depends only on the product of the equilibrium constant and the concentration of monomers. In a system with m distinct monomers, present initially at equal concentrations, the concentration of any particular polymer species of length n is further decreased by a factor of m^{-n}. For $m > 1$, even if $\rho \approx 1$, so that polymerization is favored, for large n the concentration of any particular species is quite small. For example, for polypeptides $m = 20$; the concentration of a polypeptide of length $n = 30$ is roughly 20^{-30} less than that of a monomer. For a container of finite size, this implies that, except for occasional fluctuations, most longer species are not present.

2.3 CATALYZED REACTIONS

The presence of a catalyst (enzyme) E can accelerate a reaction.

$$A + B \overset{E}{\rightleftharpoons} C + H. \tag{8}$$

At equilibrium the rate of the forward reaction equals that of the backward reaction, so that $\dot{A} = \dot{B} = \dot{C} = 0$. Catalysis speeds up the rate at which the system approaches equilibrium, but does not change the concentrations at equilibrium.

However, when the reaction is driven away from equilibrium, for example by externally supplying one of the participants in the reaction, catalysis can shift the steady state. This is the basis for the effect we study here.

Catalysis increases both the forward and backward rate constants by the same amount. This can be taken into account by defining a quantity ν that we call the *catalytic efficiency.* For fixed concentration of the reactants, the increase in the velocity of the reaction is proportional to the product of ν and the concentration E of the catalyst. The kinetic equation for C can be crudely approximated as

$$\dot{C} = (1 + \nu E)(k_f AB - k_r HC). \tag{9}$$

Similar equations apply to $\dot{A}$ and $\dot{B}$. When the catalytic efficiency $\nu = 0$, this reduces to the kinetic equation for a spontaneous reaction.

Note that, for a population of polymers, this reaction is just one reaction in a network of many. A given polymer may play the role of A in some reactions, and the role of C in others. To compute the rate of production of any given species, it is necessary to sum all the relevant reaction terms.

The approximation made in Eq. (9) neglects the effect of saturation, which comes about because the enzyme and the reactants are bound together for a finite time. During this time they cannot participate in new reactions, which lowers the effective reaction rates. If this is a dominant effect, so that most of the enzyme or product is bound at any given time, the reaction is *saturated.* To take this into account we do not use Eq. (9), but rather use a more accurate approximation. We keep track of the concentration of any given species x_i that is bound into complexes through an auxiliary variable $\overline{x}_i$, which is equal to the sum of the concentrations of all the complexes in which x_i is bound. To keep the simulation tractable, we assume that all complexes unbind at the same rate k_u. This approximation is described in more detail in the Appendix.

2.4 DRIVING FROM EQUILIBRIUM

To make anything interesting happen in a reaction network it must be driven away from equilibrium. In this model we investigate two different mechanisms. The first involves a flux of mass, and the second involves the formation of energetic pyrophosphate molecules, driven by light.

2.4.1 MASS FLOW We model a reaction vessel with a steady input flux of monomers or short polymers, and an output flux due either to diffusion or overflow of the reaction vessel. This might correspond to a prebiotic environment, or it might correspond to a chemostat in a laboratory experiment. The chemical species that are input are collectively called the *food set.* For simplicity, we assume an inflow rate δ of concentration per unit time, and an outflow that is proportional to concentration, with rate constant K.

For an element of the food set, the kinetic equations are of the form

$$\frac{dx_i}{dt} = k_u\overline{x}_i + \sum(\textit{reaction terms}) + \delta - Kx_i, \tag{10}$$

$$\frac{d\overline{x}_i}{dt} = -k_u\overline{x}_i + \sum(\textit{reaction terms}) - K\overline{x}_i. \tag{11}$$

The (*reaction terms*) are defined in the appendix in the discussion following Eq. (25). For a species outside the food set, the kinetic equations are of the same form, except that $\delta = 0$. There is a net flow of mass from the food set to the other elements of the system, which drives it away from equilibrium.

Because the reaction terms conserve mass, the total mass in the reaction vessel always goes to a fixed point, independent of initial concentrations. To see this, note that only the last two terms in Eq. (10) and the last term in Eq. (11) change the total mass. The total mass concentration is proportional to[3] $m = \sum n(i)x_i$, where $n(i)$ is the length of the ith species. Letting N_f be the number of elements in the food set, the rate of change of the total mass concentration is given by a simple differential equation,

$$\frac{dm}{dt} = N_f\delta - Km, \tag{12}$$

which has a global fixed point $m_0 = N_f/K\delta$. This means the initial mass is irrelevant anyway, and so for convenience in our simulations, we choose $m(0) = m_0$. Thus, there are effectively only two parameters relating to the flow of mass through the system, which can be δ and K, or equivalently δ and m_0.

Since δ and K are not intuitively easy to interpret, it is sometimes useful to quote results in terms of the *mean reaction number* r. This is defined as the mean number of times a given monomer participates in a reaction, on average, from the time it enters the vessel until the time it is flushed out. At equilibrium r is infinite, and when the other parameters are fixed, it decreases monotonically as δ increases.

2.4.2 PYROPHOSPHATES As we demonstrate in subsection 2.6, catalytic focusing requires conditions that favor polymerization. The tendency to polymerize can be enhanced by an appropriate input of energy. The mechanism that we investigate here involves pyrophosphate molecules, which play a role analogous to that of *ATP* in contemporary organisms. This mechanism is supported by early experiments.[35] The detailed sequence of reactions was suggested by Ron Fox,[4] and is illustrated in Table 1.

It proceeds as follows: Light causes the formation of pyrophosphate molecules (p_2), which is balanced by hydrolysis. When a pyrophosphate molecule binds to polymer A, it creates the energized form A^* and releases a phosphate atom in the

[3] To convert this to units of mass/volume there is a constant of proportionality, that depends on the mass of a monomer.

[4] Private communication.

TABLE 1 Pyrophosphate energizes and enhances polymerization. The first column lists the reactions and the second column the reaction rate. γ represents a photon (and in the column on the right represents the intensity of light), p a phosphate atom, and H water. A and B represent the polymers that condense to form C. E is the catalyst. A bar indicates a complex that is bound to the enzyme E. k_f is the rate constant for polymerization, k_r for hydrolysis, k_u for the dissociation of the bound complex, k_e for the polymerization of phosphate, and k_a for the activation of a polymer. Activated polymers are indicated with a "*" superscript. k_f^* is the rate constant for the condensation of an activated polymer with another polymer.

Reaction			Rate
$2p$	$\xrightarrow{\gamma}$	$p_2 + H$	$k_e \gamma p^2$
$p_2 + HC$	$\rightarrow$	$2p$	$k_r p_2 H$
$A + p_2$	$\rightarrow$	$A^* + p$	$k_a A p_2$
$A^* + H$	$\rightarrow$	$A + p$	$k_r A^* H$
$A^* + B + E$	$\rightarrow$	$\overline{CE} + p$	$k_f^* \nu E A^* B$

process. A^* may hydrolyze, releasing the other phosphate atom, or it can bind to another polymer B (in the presence of the catalyst E). This occurs with a rate constant k_f^*, which is greater than the unenergized rate constant k_f. Thus, the addition of energy favors polymerization.

Based on simulations involving the full reaction scheme shown in Table 1, we found that when the concentration of pyrophosphate and the input of light are sufficiently high, the behavior is roughly equivalent to that obtained by simply using the equations given in the Appendix, with the effective forward rate constant equal to k_f^*. For convenience, in the numerical experiments described here we simply assume that the parameters quoted correspond to k_f^*, and use the simpler equations which do not involve pyrophosphate.

2.5 REACTION GRAPH

Each distinct monomer can be assigned a character from a fixed alphabet, $a, b, c, \ldots$. A polymer can then be represented as a character string, for example $(acabbacbc\ldots)$. We assume that the polymers are oriented, so that abc and cba are different strings. The topological structure of a network of reactions, each of the form of Eq. (8), can be represented as a polygraph with two types of nodes and two types of connections,[11] as illustrated in Figures 1 and 2. One type of node represents the polymer species and is labeled by the corresponding string. The other type of node represents the catalyzed reaction and is labeled by a black dot. The polymers that participate in a reaction are connected to the corresponding reaction node by reaction links (black arrows), which point in the direction of condensation. Each polymer is connected to the reactions it catalyzes by a catalytic link (dotted line). Each reaction has at least four links: three reaction links, and one or more catalytic links.

2.6 CATALYTIC FOCUSING

Under appropriate conditions catalysis can focus most of the material of a reaction network into only a few species. The basic idea can be grasped by considering the simple reaction network

$$ba + H \rightleftharpoons a + b \overset{E}{\rightleftharpoons} ab + H,$$

as shown in Figure 1. Assume a and b are supplied at rate δ, and diffuse out of the container with rate constant K, as described in subsection 2.4. For simplicity, assume the concentrations of E and H are maintained at fixed values. Neglecting saturation, according to the approximation of Eq. (9), the rates of change of ab and ba are

$$\dot{[ab]} = \gamma(k_f[a][b] - k_r H[ab]) - K[ab], \tag{13}$$

$$\dot{[ba]} = k_f[a][b] - k_r H[ba] - K[ba], \tag{14}$$

where $[ab]$ is the concentration of polymer ab. Setting the derivatives to zero and using the mass conservation condition of Eq. (12) gives

$$\frac{[ab]}{[ba]} = \frac{1+\beta}{1+\frac{\beta}{\gamma}}. \tag{15}$$

$\beta = \delta/m_0 k_r H$ is a dimensionless parameter related to the deviation from equilibrium, where $m_0 = a(0) + b(0)$ is the total concentration of monomers. Note that $\beta \geq 0$. $\gamma = 1 + \nu E$ is a dimensionless parameter that characterizes the strength of catalysis. $\gamma \geq 1$, and $\gamma = 1$ corresponds to an uncatalyzed reaction.

Under what circumstances is the concentration of ab much greater than that of ba? At equilibrium $\beta = 0$ and the ratio of $[ab]$ to $[ba]$ is one. This ratio can become large only when $\beta \gg 0$, i.e., only when the system is driven well away from equilibrium. When $\beta \gg \gamma$ this ratio approaches γ; when $\beta \ll \gamma$ it approaches β. Thus, by varying γ and β the concentration of ab relative to ba can be made arbitrarily large.

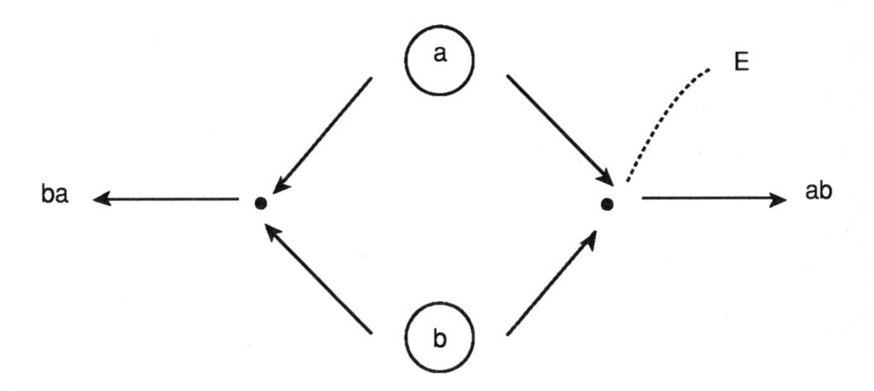

FIGURE 1 A simple network illustrating how steady-state concentration can be boosted by catalytic activity. a and b are driven at a fixed rate δ, and the enzyme E is maintained at a fixed concentration.

Note that the ability to focus comes about because the formation of one species is catalyzed, while that of the other is not. If all reactions were catalyzed equally, with equal kinetic parameters, there would be no focusing; the concentration of ab would equal that of ba. Focusing thus requires *specific catalysis*, in which some reactions are catalyzed more strongly than others.

2.7 AUTOCATALYTIC SETS AND METABOLISMS

To achieve catalytic focusing the enzyme E must be maintained at high concentration. One way for the system to accomplish this by itself is through an autocatalytic reaction, in which one of the products catalyzes its own formation. A simple example is

$$A + B \overset{C}{\rightleftharpoons} C + H. \tag{16}$$

If we set $C = ab = E$ in reaction (13), then the enzyme is produced automatically, and the focusing maintains itself.

Simple autocatalytic reactions such as reaction (16) are obviously very special. A more common situation occurs when autocatalysis involves a cooperation between reactions, in which one species catalyzes the formation of another. An *autocatalytic set* is defined as a set of chemical species such that each member of the set is produced by at least one catalyzed reaction involving only members of the set. This notion was introduced by Calvin,[6] Eigen,[10] Kauffman,[23] and Rössler.[37] Since the reactions we are considering are reversible, a species can be produced either by cleavage or condensation. Thus reaction (16) is an autocatalytic set, and so is

$$A + B \overset{A}{\rightleftharpoons} C + H. \tag{17}$$

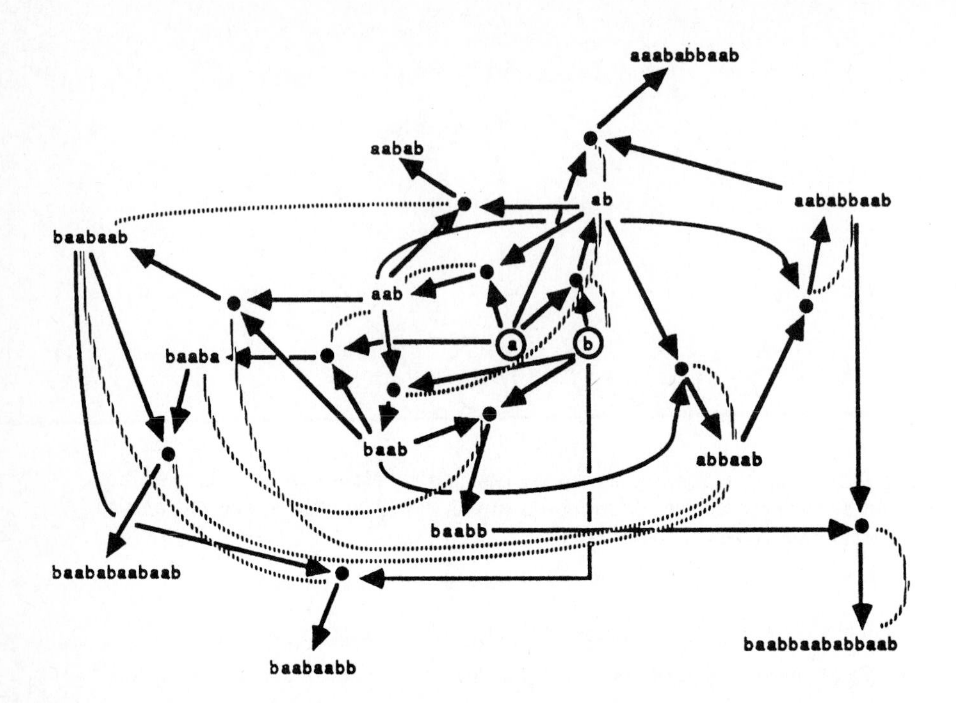

FIGURE 2 An autocatalytic network consisting of 15 species. The monomers a and b, which are circled, comprise the food set. Character strings represent the polymers. The fat dots represent reaction nodes. The arrows connecting the reactants to the nodes are reaction links and point in the direction of condensation. The broken lines connecting polymer species to reactions are catalytic links, indicating which species catalyze which reactions.

A more complicated (and more typical) autocatalytic set is shown in Figure 2. This network happens to have one catalytic link per reaction, a rather special property. We will use this reaction network, or variants with more catalytic links, for many of our numerical experiments. Note that autocatalytic sets may contain other autocatalytic sets as subsets.

Using the random graph model described in the next section, Kauffman[24] showed that for any given probability of catalysis, if the food set is sufficiently large, the resulting graph will almost always have an autocatalytic set. This result is encouraging, since it suggests that autocatalytic sets exist under fairly reasonable conditions. We must emphasize, however, that the presence of an autocatalytic set in a reaction network, in and of itself, does not imply that there will be any interesting departures from equilibrium behavior. To achieve catalytic focusing it is critical that the kinetic parameters are favorable. Thus the graph-theoretic notion of an autocatalytic set is a necessary but not a sufficient condition.

To make this distinction, we define an *autocatalytic metabolism* as an autocatalytic set whose concentrations make significant departures from the values they would have if none of the reactions were catalyzed. The phrase "significant departures" is subjective, and is admittedly rather vague. However, from an operational point of view, in our simulations we often see a clear distinction between autocatalytic sets that can function as metabolisms and those that cannot, as shown in Section 6.

3. SIMULATION

In principle the kinetic equations are all we need to know in order to simulate the behavior of a reaction network. In practice, however, there are two major problems: The first is that the kinetic parameters cannot be determined from first principles. To deal with this we construct an artificial chemistry, as discussed in subsection 3.1. The second problem is that there are an infinite number of possible reactions, and it is intractable to solve all of them; we must focus our computational resources on only the most relevant ones. Our method for doing this is discussed in subsection 3.2.

3.1 ARTIFICIAL CHEMISTRY

In a real chemical system the efficiencies and rate constants of the reactions depend on detailed properties of chemical composition, as well as on thermodynamic parameters such as temperature and pressure. While a computation of these constants from quantum mechanics and statistical mechanics is possible in principle, from a practical point of view, at this point in time it is hopelessly intractable.

To circumvent this problem, the approach introduced by Kauffman,[24] Farmer et al.,[11,12] and Bagley et al.[2] is to invent an artificial chemistry, a set of rules stating which catalyzed reactions occur, and with what strength. An artificial chemistry cannot reproduce the behavior of a real chemistry in detail, but it may reproduce many of the correct qualitative properties. An artificial chemistry can produce complex behavior, even though it is simple from a calculational point of view. By exploring different artificial chemistries, we can discover which properties cause significant changes in behavior, and which do not. We can begin with simple chemistries and move toward more complex chemistries, adding layers of realism as needed. The knowledge gained in this way can be useful in guiding experimental investigations of real systems, by pinpointing the essential quantities that need to be measured in experiments in order to make the model more realistic.

Since our primary interest is in understanding the effect of catalysis, we first address the problem of assigning a catalytic efficiency to each reaction. We do this using two different methods. In the first, we construct a completely disordered artificial chemistry, by assigning catalytic efficiencies at random, and in the second, we construct a highly ordered artificial chemistry, assigning them with a string matching algorithm. The random method is more disordered than real chemistry, and the string matching method is more regular than real chemistry. From a qualitative point of view, we hope that real chemistry lies somewhere between these two extremes.

To be strictly correct, every possible reaction should be included in the reaction graph. However, in practice the reaction graph must be trimmed, so that computational resources are used only for the most essential reactions. We take advantage of the fact that the vast majority of reactions are catalyzed only weakly and can be treated essentially as spontaneous reactions. The graph represents only those reactions with sufficient catalytic efficiency to make them significantly different from the corresponding spontaneous reactions. Thus, when we refer to a "catalyzed reaction," we mean a "strongly catalyzed reaction," and when we refer to a spontaneous reaction, we mean a "weakly catalyzed reaction."

3.1.1 RANDOM ASSIGNMENT OF REACTIONS In some cases changing a single monomer can have a dramatic effect on the chemical properties of a polymer, either because it causes a drastic change in the configuration of the polymer or because it alters the properties of a critical site. If this were always the case, then chemistry would be random. For a random chemistry there is no correlation between chemical formulas and chemical properties. This is unrealistic. However, it does have the advantage of being easy to implement, and lies at one extreme in the space of all possible chemistries.

Following Kauffman,[24] we assume that out of all possible spontaneous reactions, only a fraction p are catalyzed with sufficient strength to be significantly different from spontaneous reactions. The set of reactions that is catalyzed is chosen at random. To see the basic idea, imagine creating a list of all possible catalyzed reactions. For m distinct monomers the number of species of length n is m^n, the number of possible spontaneous reactions is the order of m^{2n}, and the number of possible catalyzed reactions is the order of m^{3n}. The reactions that are strongly catalyzed can be determined by flipping a biased coin that returns heads with probability p. Reactions that receive heads are assigned a non-zero value of ν, and all others are assigned $\nu = 0$. The random rule generates an ensemble of possible chemistries, corresponding to all possible sequences of random choices.

Operationally the procedure described above would be very time consuming, since m^{3n} can be a very large number. It can be made much more efficient by decomposing the problem properly, focusing attention only on reactions involving species that are already present in the reaction vessel, and taking care to avoid double counting, as described by Farmer et al.[11]

To determine catalytic efficiencies, one simple possibility is to set $\nu =$ constant, so that all the reactions on the graph are catalyzed with the same efficiency. Another natural possibility is to choose the catalytic efficiencies at random according to a given probability distribution, for example by making the probability of a given efficiency uniform within given maximum and minimum values. We employ both of these in our simulations.

3.1.2 ASSIGNMENT OF REACTIONS BY STRING MATCHING The match rule provides an alternative artificial chemistry that is probably closer to real chemistry than the random rule discussed above. It lies at the opposite extreme—while the random rule is too disordered, the match rule is probably too ordered. For the match rule, changing a single monomer in a given polymer only causes a small change in its chemical properties. Two similar polymer strings always have similar chemical properties. The match rule assumes that the information contained in the string of a given polymer contains all the information needed to specify its chemical properties.

We roughly follow the approach used to model the immune system by Farmer et al.[12] For convenience, in this discussion we assume a two-letter alphabet consisting of a and b, although the rule is easily generalized to a larger alphabet. The two reactants A and B in Eq. (8) join together to form C. The character string corresponding to C is matched against that of enzyme E. There are several possible alignments; we require that the string E span the binding site between A and B. Each allowed alignment of E against C is given a match score according to the number of complementary matches, i.e., the number of cases where an "a" is paired against a "b." We then compute the probability P that a score as good or better would be obtained if the strings E and C were generated at random. We use this to define a quantity we call the *specificity* $s = 1/P$. The catalytic efficiency depends on the specificity through a function $\nu(s)$. We typically assume that high specificity corresponds to higher catalytic efficiency, and choose $\nu(s)$ to be linear. For a given choice of A, B, and E, the total catalytic efficiency is the sum of the efficiencies computed for each of the allowed alignments. For a more detailed description, see Bagley.[3]

The match rule assigns a catalytic efficiency to every possible catalyzed reaction. The reactions with catalytic efficiencies above a given threshold ν_0 are installed in the reaction graph. The match rule is completely deterministic, and generates a unique chemistry. The requirement that the specificity exceed a fixed threshold implies that very short polymers cannot participate in catalyzed reactions. Thus, the properties of the match rule are quite different from those of the random rule.

At this point we have studied the random rule more thoroughly than the match rule. We intend to present results using the match rule in a future paper.

3.1.3 OTHER KINETIC PARAMETERS The other relevant kinetic parameters that may vary from reaction to reaction are the forward reaction rate k_f, the backward reaction rate k_r, and the unbinding constant k_u. In order to use the approximation for the saturation problem described in the Appendix, it is necessary that k_u be the same for all reactions. Since we can vary the catalytic efficiency for each individual reaction, this is not a serious problem.

The rate constants k_f and k_r play an important role. At equilibrium diverse rate constants cause the polymers of a given length to have nonuniform concentrations. However, since to first approximation the rate constants only depend on the two monomers at the binding site, the resulting nonuniformities in the concentration profile are much more regular and less pronounced than those resulting from catalytic focusing. For convenience we typically assume that k_f and k_r are the same for all reactions, although in some cases we also vary them randomly.

3.2 METADYNAMICS

The number of possible reactions is infinite. Of course, in reality only a finite number are important. Unfortunately, it is usually impossible to state in advance which reactions can be neglected. A metadynamical simulation attempts to solve this problem by restricting attention to a variable set of reactions and chemical species, and adding or deleting as needed.

The problem of determining the essential reactions is particularly severe for polymer chemistry, where the number of possible reactions grows exponentially with the length of the polymers. For $m = 20$ and $n = 10$, for example, there are already more than Avogadro's number of possible catalyzed reactions. Even if we knew the rate constants for every reaction, we could never hope to keep track of all of them in a computer simulation.

A real chemostat only contains a finite number of molecules and hence a finite number of species, with a finite number of possible reactions between them. Continuous differential equations fail to exploit this. Suppose, for example, that at $t = 0$ all the concentration is placed in a few species, and the concentrations of the rest are set to zero. According to the laws of continuous kinetics, for $t > 0$ there are generically *an infinite number of species* with non-zero concentrations. This unrealistic result comes from the approximation made in treating a discrete population of molecules as though it were continuous. This is equivalent to the assumption of a well-stirred reaction vessel with infinite volume. For the problem that we address here, the fact that the reaction vessel is finite is of critical importance.

One solution to this problem is to treat the kinetics as a random process, using integer populations and simulating molecular collisions through Monte Carlo techniques. This method is efficient when the populations are low, but is computationally inefficient when the populations are high. If the population is 10^6, for example, a differential equation solver might change the concentration by as much as one percent in a single step and still retain reasonable accuracy, while the same change with a stochastic simulator requires 10^4 steps.

The metadynamics approach offers an alternative that is computationally more efficient when there is a wide disparity in population size.[2,11,12] At any given time the reaction network is modeled by a finite set of continuous differential equations, representing the dominant reactions. The topological structure of this set of differential equations is represented as a graph containing only species that are either present in the reactor, or that can be produced by other species present in the reactor. As the concentrations change the dominant species and reactions may also change. The graph is changed to reflect this, which in turn changes the differential equations. The dynamics occur on two time scales, the faster time scale of the differential equations and the slower time scale for changing the graph. A typical example of a metadynamics simulation is shown in Figure 3.

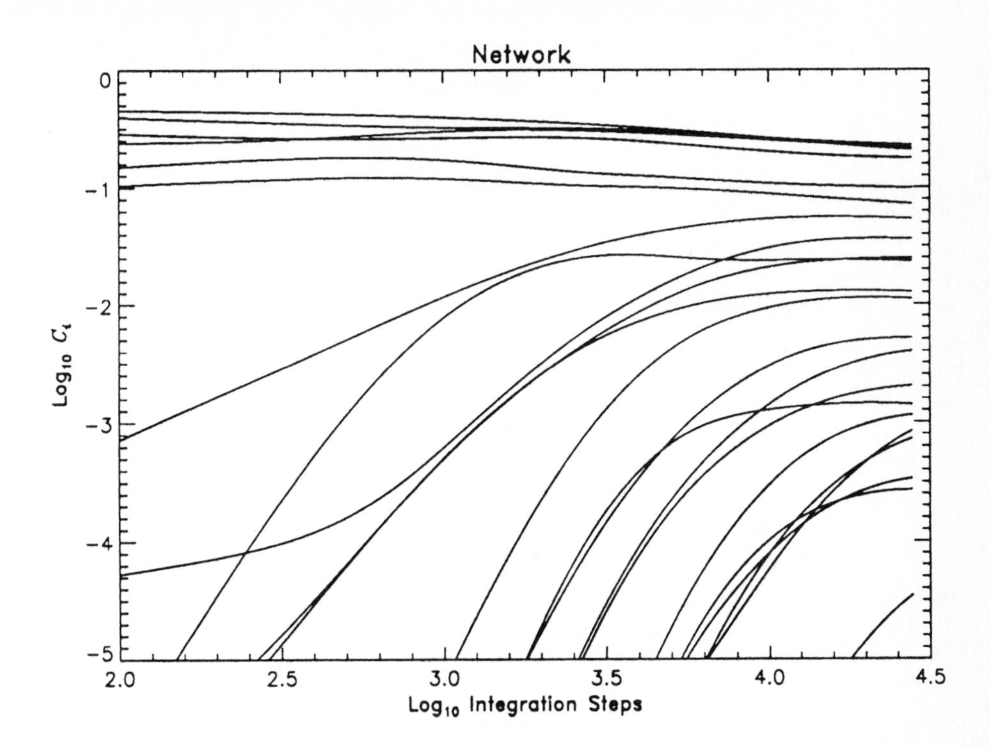

FIGURE 3 A typical metadynamics simulation. The logarithm of the concentration of each species is plotted against the logarithm of time. Initially only the four food set species a, b, ab and ba have non-zero concentration. Catalyzed reactions within the food set produce new species, which in turn have more catalyzed reactions. At the point where we begin the graph at $t = 100$, four new species have appeared, so there are a total of eight polymer species. Each new species can be seen appearing at the bottom of the graph as it crosses above the threshold. The system eventually approaches a steady state solution with 22 species above the threshold. The parameters are the same for all reactions, and are $k_f = 10^2$, $k_r = 10$, $\nu = 10^4$, $k_u = 10^4$, $H = 1$, $\delta = 10^2$, and $m_0 = 3$.

To take into account the fact that the container is finite, we impose a concentration threshold corresponding to the presence of a single molecule. If the concentration of a given species is significantly below this threshold, then that species is unlikely to be present. Therefore it cannot participate in reactions that produce other new species, and these reactions can be safely ignored. We only include reactions between species that are above threshold. They may produce new species, which rise above threshold; when this happens the new species are installed in the graph and allowed to catalyze new reactions. Similarly, species that were formerly above threshold may fall below it and be removed.

Since the number of species with concentrations above threshold at any given time is finite, the graph is also finite. Adjusting the threshold makes it possible to keep the graph from becoming unmanageably large—smaller reaction vessels have higher concentration thresholds, corresponding to smaller graphs. By making the reaction vessel sufficiently small, we can insure that the graph has less than a given number of elements. We typically strive to simulate a container that is roughly the size of a bacterium, although because of limitations in computer resources we are often forced to use smaller containers.

In our simulations the system always approaches a unique fixed point. This appears to be true independent of the way we model the reactions, i.e., whether or not we account for saturation or whether or not we include pyrophosphates. This suggests that there is a Lyapunov function for catalyzed kinetic equations in this class.[5]

We are able to speed up our simulations by several orders of magnitude by assuming the existence of a unique fixed point, which implies that the steady-state deterministic equations can be solved algebraically. We call the fixed point for any given set of differential equations a *dynamical fixed point.* The metadynamical simulation is simplified as follows: We find the dynamical fixed point of the current set of differential equations. We then examine it and check to see whether any species have moved above or below threshold in comparison to the previous dynamical fixed point. If so, we change the differential equations accordingly and find a new dynamical fixed point, and repeat the process. Eventually there are no changes compared to threshold and this procedure stops. We call the final dynamical fixed point the *metadynamical fixed point.* By reducing the metadynamics to a sequence of algebraic operations, on a typical workstation a simulation such as that of Figure 3 is compressed into a few seconds.

Note that although the deterministic simulation described above always reaches a metadynamical fixed point, once we reincorporate the stochastic effects of spontaneous reactions, the system may hop between many different metadynamical fixed points and the long-term dynamics become quite interesting. This is the basis for our claim that autocatalytic metabolisms can evolve, and is discussed in more detail in a companion paper.[4]

3.3 THE BACKGROUND OF UNCATALYZED REACTIONS

Restricting attention solely to the reaction graph is a good assumption as long as the reactions on the graph dominate over everything else. This is not always the case. For example, spontaneous reactions always dominate near equilibrium. To study the competition between catalyzed and spontaneous reactions, the spontaneous reactions must be taken into account, at least as an aggregate.

[5]These equations are reversible, which makes them different from many other autocatalytic equations that are known to display limit cycles, chaos, and hysteresis.

For the purposes of this discussion it is convenient to divide the chemical species into those of the catalytic network and those of the background, which are produced only by spontaneous reactions. This is illustrated in Figure 4. The *shadow* is a special subset of the background, consisting of species that can be directly produced by reactions involving only members of the reaction network. In a situation where the

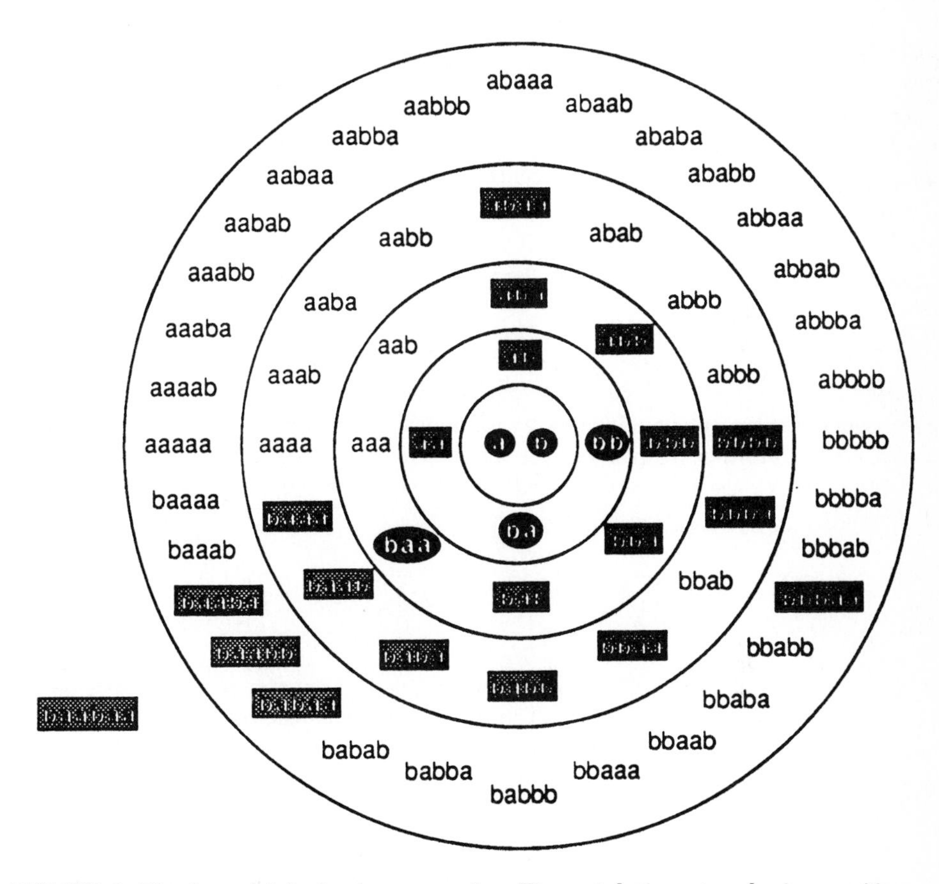

FIGURE 4 The "mandala" of polymer species. The *catalytic network*, shown with white letters on black ovals, includes the food set, but also consists of other species produced by catalytic reactions. There must be a continuous path in the corresponding graph of catalyzed reactions from members of the food set to each member of the catalyzed network; furthermore, the concentration must be above a concentration threshold corresponding to the presence of a single molecule. The *background* consists of everything that is not in the catalytic network. The *shadow*, shown by white letters on a grey background, is a special subset of the background, consisting of species that can be produced by a spontaneous reaction involving only themselves and members of the catalytic network.

concentrations of the reaction network are high, the shadow plays a special role because it is maintained at concentrations above the rest of the background. The shadow is in this sense similar to Eigen's quasi-species.[10]

For clarity we will first discuss the problem of modeling spontaneous reactions alone, without catalyzed reactions, and then return to discuss the case when spontaneous reactions are in competition with catalyzed reactions.

The difficulty of modeling spontaneous reactions comes about because there are so many of them. To make the problem tractable, we assume that all reactions have the same forward and backward rate constants k_f and k_r. The problem is then equivalent to that of modeling the spontaneous reactions in a system with only one distinct type of monomer ($m = 1$). We can lump together all polymers of a given length n, adding together their concentrations to get a combined concentration s_n. The allowed reactions are of the form

$$s_i + s_j \rightleftharpoons s_k + H. \tag{18}$$

A reaction is only possible if $i + j = k$. The contributions to the kinetics are now described in the order that the corresponding terms appear in Eq. (19): s_k receives contributions from condensation reactions between smaller species and from cleavages of longer species. It loses to condensations between polymers of length k and polymers of other lengths, as well as to cleavages of itself. If there are polymers of length k that are part of the food set, s_k will also gain from external driving. The resulting rate equation for s_k is

$$\frac{ds_k}{dt} = k_f \sum_{i+j=k} s_i s_j - k_r \sum_{n>k} (n-k+1) s_n - k_f s_k \sum_{j=1} m^j s_j - k_r (k-1) s_k + n_k \delta s_k - K s_k, \tag{19}$$

where $\delta = 0$ if k is outside the food set, and n_k is the number of elements of length k in the food set. For an alphabet of m monomers, with the assumption of uniform reaction rates, the concentration computed above is divided evenly among all the species present. It is

$$a_k = \frac{s_k}{m^k}. \tag{20}$$

We now outline our approach for treating the competition between spontaneous and catalyzed reactions: Let s_k now represent the sum of the concentrations of the polymer species of length k, excluding the catalytic network. The dynamics of the background due to its interactions with itself can still be taken into account by equations of the form of Eq. (19). However, the concentration for a single species is now defined to be

$$a_k = \frac{s_k}{(m^k - l_k)}, \tag{21}$$

where l_k is the number of elements of the catalytic network of length k. The coupling to the autocatalytic set can be taken into account by carefully counting all of the interactions. The members of the catalytic network contribute to the spontaneous background through their condensation and cleavage reactions. Similarly, they receive concentration from the spontaneous background. The only approximation necessary comes from the assumption that all the elements of the background have the same concentration. This is not strictly true. In particular, the concentrations in the shadow are typically above those of the rest of the background. However, this is a second-order effect, and we feel that in most cases our simulations are approximately correct. The problem of coupling the catalytic network to the spontaneous background is discussed in more detail by Bagley.[3]

4. EMERGENCE

In this section we present simulations of catalytic reaction networks under several different conditions. We begin by demonstrating that, under appropriate circumstances, autocatalytic metabolisms emerge at concentrations that are several orders of magnitude higher than those of the background. We explore the parameter dependence and show that there is a range of parameter values where autocatalytic metabolisms thrive.

4.1 DEPARTURE FROM EQUILIBRIUM

In Figure 5 we show a sequence of three simulations in which we drive the system further away from equilibrium by increasing the parameter δ. In each case we let the reaction vessel approach its steady-state behavior, and then plot the concentration of all the polymers in the vessel of length less than twelve. In Figure 5(a) $\delta = 0.01$, the mean reaction number $r \approx 210,000$, and the system is nearly at equilibrium. The observed behavior is in quantitative agreement with the predictions of subsection 2.2. The concentration falls off exponentially with length, all species of a given length have the same concentration, and there is no interesting structure in the concentration profile.

In Figure 5(b) $\delta = 10^5$, which corresponds to a mean reaction number of approximately 33. The concentrations of the catalyzed reaction network are orders of magnitude above those of the background, and most of the mass of the system is concentrated in the autocatalytic metabolism.

Finally, in Figure 5(c) we show the case when $\delta = 10^{7.5}$, corresponding to a mean reaction number of roughly 0.5. The dominance of the autocatalytic metabolism is less evident—the flow of matter through the system is so high that the reaction network has a much smaller effect on the composition of the system. This is because the flow of mass through the system is so large that there is no time for a given species to react before it is flushed from the system.

These simulations demonstrate the ambiguity of the word "non-equilibrium" in this context. On one hand, the parameter δ is a control parameter that can be used to drive the system away from equilibrium. On the other hand, as demonstrated in Figure 5, increasing δ does not necessarily make the *physical properties* of the system deviate further from those at equilibrium. In terms of the non-uniformity of the concentration profile, $\delta = 10^5$ produces a larger deviation from equilibrium properties than $\delta = 10^{7.5}$. The deviation from equilibrium properties is at a maximum when the driving away from equilibrium is finite. A flow of energy is needed to move the system away from the structureless equilibrium state, but too large a flow of energy again results in a structureless state.

We have explored several quantitative measures of the deviation of the physical properties of the system from those at equilibrium. One of these is the steady-

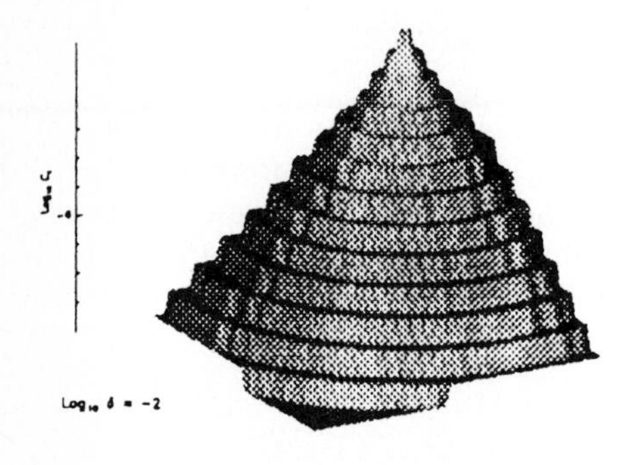

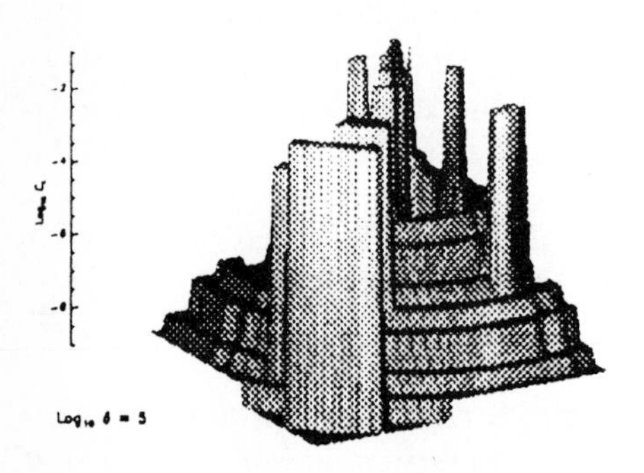

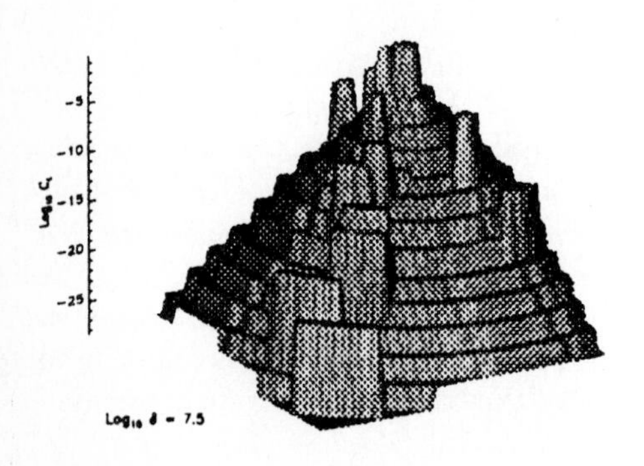

FIGURE 5 Concentration "landscapes" at different displacements from equilibrium. The two horizontal axes correspond to the polymer species, arranged concentrically with the shortest polymers in the center, as in Figure 4. The vertical axis corresponds to the logarithm (base 10) of the concentration of each species. The food set consists of the polymers a and b, using a variation of the network of Figure 2 with 118 catalytic links. In (a) $\delta = .01$, and the system is near equilibrium; the concentration falls off exponentially with length but is otherwise featureless. In (b) $\delta = 10^5$, which gives the behavior that is most distinct from that at equilibrium; in this case the concentration of the autocatalytic set is many orders of magnitude above the corresponding equilibrium values. Finally, in (c) $\delta = 10^{7.5}$, and the system is so far from equilibrium that it loses most of its interesting structure. (Note the change of scale in comparison with (b).) The other parameters are $k_f = 6.49 \times 10^2$, $k_r = 2.50$, $\nu = 8.97 \times 10^5$, $k_u = 5.00 \times 10^4$, and $m_0 = 2.0$.

state slope Λ of the concentration profile. In Figure 6 we plot the concentrations of the species in the network as a function of their length and compare them with the background, for a favorable set of parameters where the system exhibits an autocatalytic metabolism.[6] The logarithm of concentration as a function of length in Figure 6 gives roughly lines of slope Λ, indicating that the concentrations decrease roughly exponentially with length. To measure Λ we use a least-mean-squares fit. The longer polymers of the network are at much higher concentrations than those of the background. Consequently, for the network Λ is much larger (less negative) than that it is for the background. Λ thus provides a measure of the deviation from the equilibrium profile.

In Figure 7 we plot Λ as a function of the mass flow δ. For small values of δ the system is near equilibrium, and Λ for the network is nearly equal to Λ for the background. As δ increases Λ increases, reaches a maximum, and then decreases. Near the maximum, Λ approaches zero, indicating the concentration profile is almost flat.

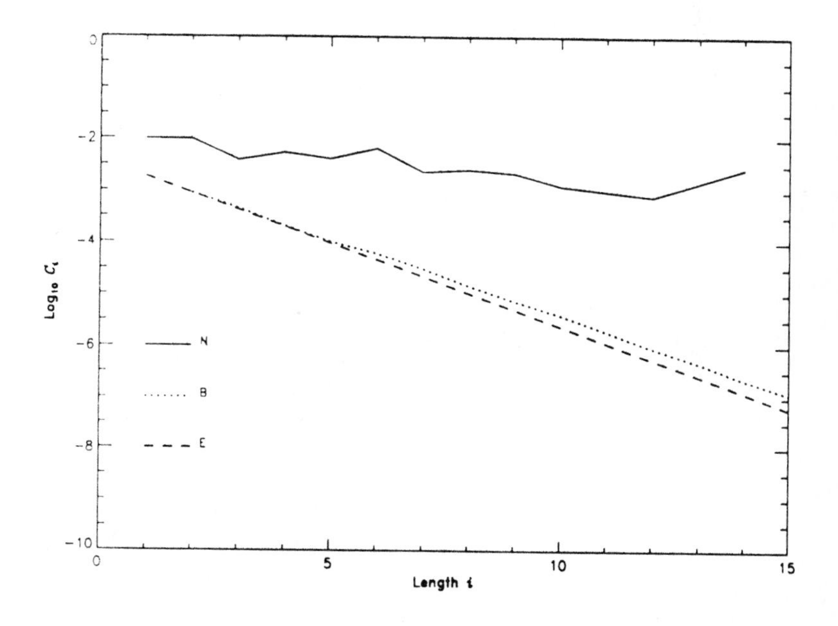

FIGURE 6 Mean concentration as a function of polymer length at steady state. The solid line (N) corresponds to the network, the dotted line (B) corresponds to the background, and the dashed line (E) corresponds to the equilibrium concentrations. The simulation is performed for the network of Figure 2 with parameters $k_f = 6.49 \times 10^2$, $k_r = 2.50$, $\nu = 8.97 \times 10^5$, $k_u = 5.00 \times 10^4$, $\delta = 1.79 \times 10^1$, and $m_0 = 2.0$.

[6]Note that this network has different parameters than those of Figure 5.

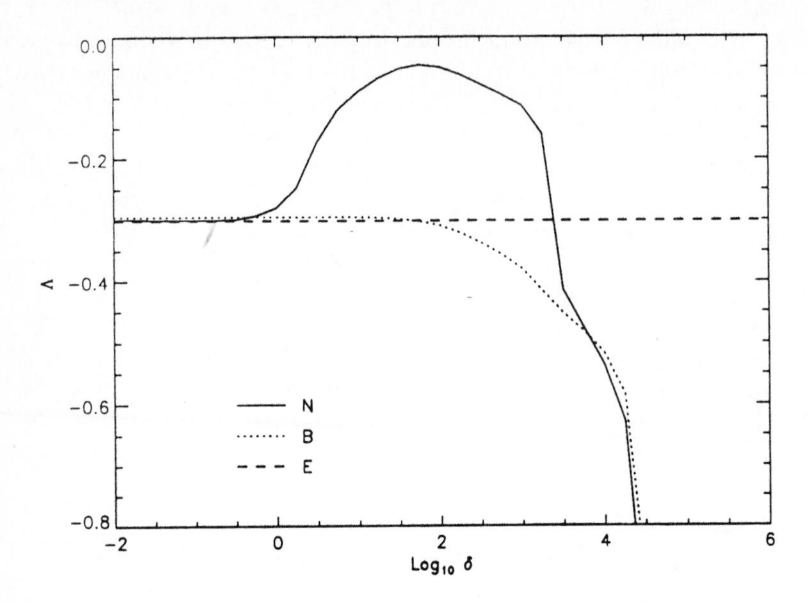

FIGURE 7 The slope of the concentration profile vs. the mass flux δ. (See Figure 6 for an example of a concentration profile.) The network (N) corresponds to the solid curve, the background (B) to the dotted curve, and equilibrium (E) to the dashed curve. Parameters are the same as those for Figure 6, except that δ is varied.

The mass concentrated in the metabolism provides another natural measure of the deviation of its properties relative to those at equilibrium. For example, in Figure 8 we plot the fraction of the mass in the background, the food set, and the catalytic network (subtracting the food set), as a function of δ. There is a central regime where the majority of the mass of the system is concentrated in the autocatalytic metabolism. Note that this regime overlaps with the regime where Λ is large. However, the two are somewhat skewed; Λ peaks at roughly $\delta = 10^2$, while the mass peaks when $\delta > 10^3$.

We feel that the need for a balanced energy flow for "interesting behavior" reflects a general principle. Another possible example is the fact that life evolved on Earth and not on Mercury or Pluto.

4.2 DEPENDENCE ON PARAMETERS

How special are the parameters for which autocatalytic metabolisms occur? To answer this question as quantitatively as possible, we have systematically explored the parameter space, testing for the presence of autocatalytic metabolisms. We used two measures of the dominance of the autocatalytic set. One of these is Λ, and

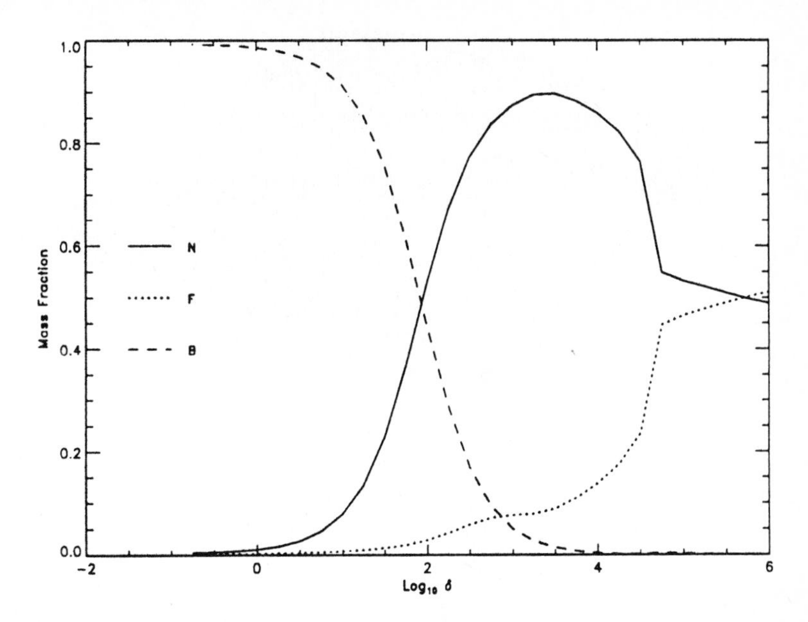

FIGURE 8 The fraction of the total mass for the network (N), the background (B), and the food set (F). (The fraction for the network excludes the food set.) The simulation is the same as that of Figure 7.

the other is the mass ratio $R = N/(B + F)$, where N is the mass of the network (neglecting the food set), B is the mass of the spontaneous background, and F is the mass of the food set.

In Figure 9 we show the behavior of Λ under variations of δ and ν. Note that Λ remains near zero for a broad range of parameter values. For comparison, in Figure 10 we plot the mass ratio R as a function of the same parameters, but for a network with fewer catalytic links. The behavior is more sharply peaked, but there is a broad range in which the autocatalytic set contains the majority of mass in the reaction vessel.

In Figure 11 we show the behavior of Λ under variations of ν and the unbinding rate constant k_u. This figure illustrates how distinct the behavior of the autocatalytic metabolism is from that of the background. In one regime, roughly corresponding to lower values of ν and lower values of k_u, Λ behaves just as it does for the spontaneous background. It is more negative and forms a relatively flat surface as a function of parameters. The other regime, which corresponds to larger values of the two parameters, has higher values of Λ. The transition from one regime to the other is quite sharp.

Finally, to illustrate the effect of varying the forward rate constant k_f, in Figure 12 we show the effect of varying k_f and δ. Once again we see a broad parameter

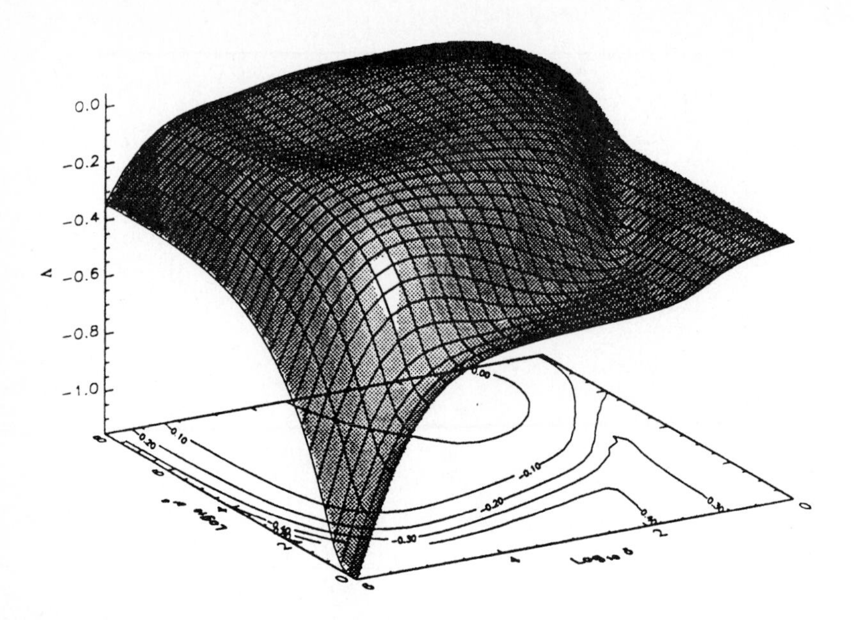

FIGURE 9 Λ vs. $\log_{10}\delta$ and $\log_{10}\nu$ for the variation of the network of Figure 2 with 118 catalytic links. The parameters are $k_f = 3.02\times10^4$, $k_r = 2.70$, $k_u = 7.11\times10^4$, and $m_0 = 2.0$.

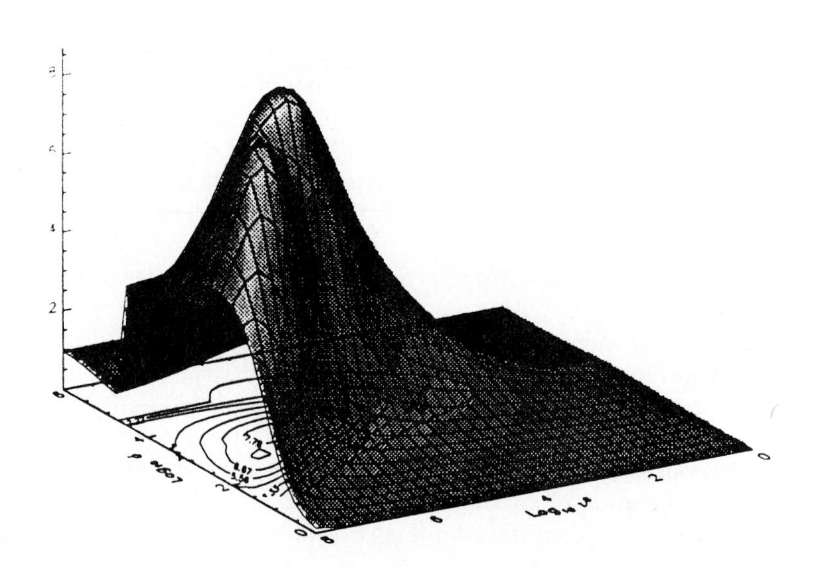

FIGURE 10 The mass ratio $R = N/(B+F)$ vs. $\log_{10}\delta$ and $\log_{10}\nu$ for the network of Figure 2. $k_f = 6.49 \times 10^2$, $k_r = 2.50$, $k_u = 5.00 \times 10^4$, $m_0 = 2.0$.

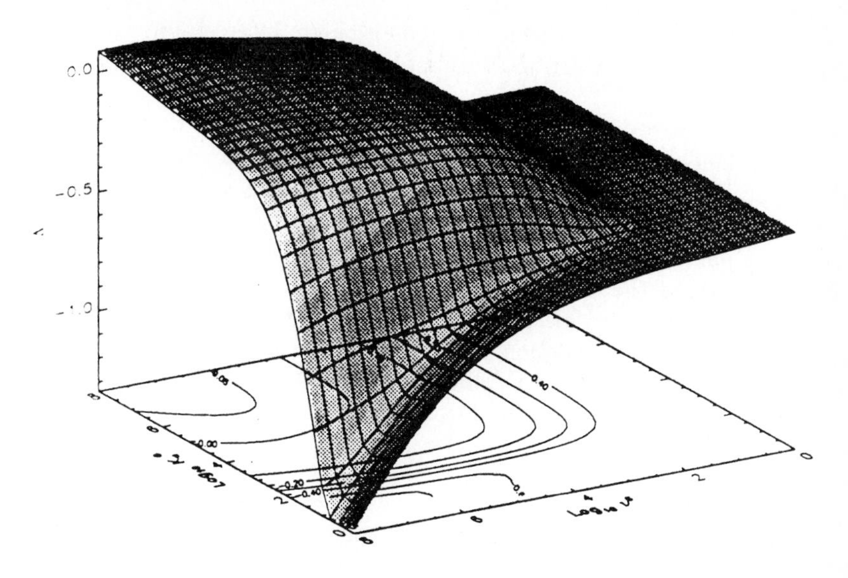

FIGURE 11 Λ vs. $\log_{10}\nu$ and $\log_{10}k_u$, for the variation of the network of Figure 2 with 118 catalytic links. $k_f = 3.02 \times 10^4$, $k_r = 2.70$, $\delta = 1.41 \times 10^2$, and $m_0 = 2.0$.

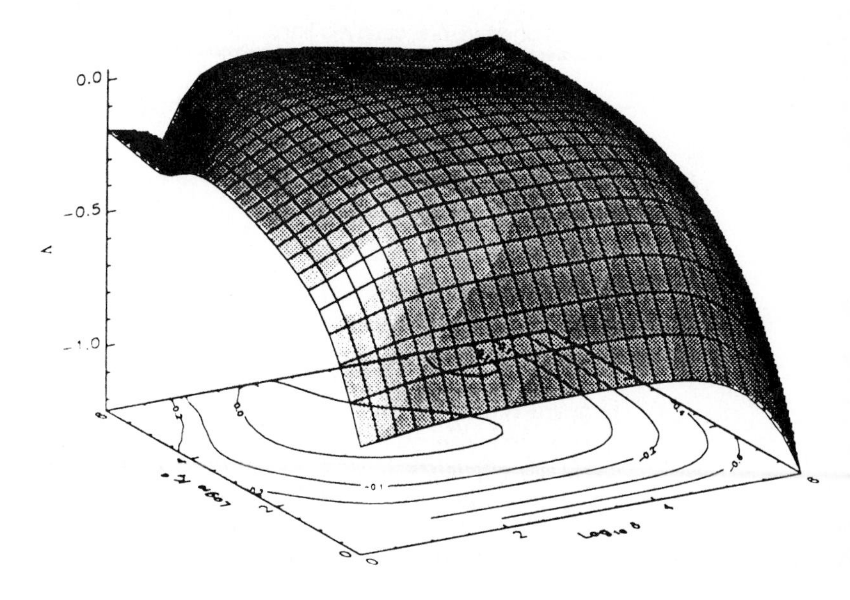

FIGURE 12 Λ vs. $\log_{10}\delta$ and $\log_{10}k_f$ for the variation of the network of Figure 2 with 118 catalytic links. $k_r = 2.70$, $\nu = 5.26 \times 10^5$, $k_u = 7.11 \times 10^4$, and $m_0 = 2.0$.

regime in which the values of the slope are quite high. Interestingly, for small values of δ there is a regime in which increasing k_f beyond roughly 10^4 causes a decrease in Λ.

It is evident from Figures 6–12 that for reaction networks with fixed topologies and uniform kinetic parameters, the dependence on parameters is fairly smooth. The parameters that approximately give the maximum value of Λ are shown in Table 2. To measure the size of the regime where autocatalytic sets are dominant, we swept the parameters of the system one at a time, holding the other parameters constant at those of the maximum. We somewhat arbitrarily say that the system supports an autocatalytic metabolism if the mass ratio is greater than one. For convenience we held the reaction graph fixed. In one case we used the graph of Figure 2 which has a minimal number of catalytic links; in the other we used a graph with the same reactions, but 118 catalytic links. The results are summarized in Table 2.

TABLE 2 Parameter regime that supports autocatalytic metabolisms. The kinetic parameter values roughly corresponding to the maximum value of Λ are shown in the third column. To measure the size of the parameter regime that supports autocatalytic metabolisms, we varied each parameter one at a time, holding the others constant at their optimum values. We say that the system supports an autocatalytic metabolism when the mass ratio $R>1$. For the experiment shown in (a) we kept the reaction graph fixed at that of Figure 2, which has 15 reactions and 15 catalytic links. In (b) we used a variation with the same reactions but 118 catalytic links.

Global parameters	Description	Optimum	Range
(a)			
k_f	Forward reaction rate	6.49×10^2	$[< 1, 10^{4.5}]$
k_r	Reverse reaction rate	2.50	Not computed
ν	Catalytic efficiency	8.97×10^5	$[10^{1.6}, > 10^{10}]$
k_u	Unbinding rate	5.00×10^4	$[10^1, 10^{3.5}]$
δ	Mass flux	1.79×10^1	$[10^1, > 10^6]$
(b)			
k_f	Forward reaction rate	3.02×10^4	$[< 1, 10^5]$
k_r	Reverse reaction rate	2.70	$[10^3, 10^{3.2}]$
ν	Catalytic efficiency	5.26×10^5	$[< 1, 10^4]$
k_u	Unbinding rate	7.11×10^4	$[10^7, > 10^{10}]$
δ	Mass flux	1.41×10^2	$[10^2, > 10^6]$

The overall conclusion that we draw from these studies is that in our artificial chemistry there is a broad regime in which autocatalytic metabolisms dominate. At this point, since the relevant kinetic parameters for real polymers are unknown, we cannot say whether they lie within the favorable regime. In particular, it seems that real catalytic efficiencies and forward rate constants are certainly on the low end of the favorable regime. Determining whether or not real systems lie within the favorable regime requires further investigation.

5. SENSITIVITY

The experiments of the previous section demonstrate that as long as the topology is fixed and the parameters are uniform, the behavior of autocatalytic metabolisms depends smoothly on parameters. In this section we investigate the sensitivity of catalytic networks to variations in topology and to non-uniform variations of the parameters of each individual reaction.

5.1 SENSITIVITY TO TOPOLOGY

The topology of a reaction network is clearly important. In Table 2, for example, we have already seen that a reaction network with more catalytic links produces autocatalytic metabolisms across a wider parameter regime. To test the dependence on topology more directly, we simply generate many different reaction networks at random and simulate their dynamics. For example, in Figure 13 we fix the reactions and the kinetic parameters, but randomly vary the catalytic links. When there are a small number of catalytic links, most reaction networks have highly negative values of Λ near the equilibrium value. When there is an intermediate number of catalytic links, Λ is highly sensitive to the topology—some networks depart significantly from equilibrium, while others do not. Finally, when there is a large number of catalytic links, almost all networks have roughly the same value of Λ, making a significant departure from equilibrium.

This result is intuitively reasonable. When there are only a small number of catalytic links, the configuration has to be rather special to generate an autocatalytic set. As the number of links increases the existence of an autocatalytic set becomes more and more likely. Note, however, that the existence of an autocatalytic set is by no means a sufficient condition for the system to generate a metabolism; there are many graphs that have autocatalytic sets in them, yet show no significant departure from equilibrium. Thus, there is also an additional dynamic effect—adding more and more catalytic links enhances the nonlinear feedback necessary to deviate from equilibrium properties.

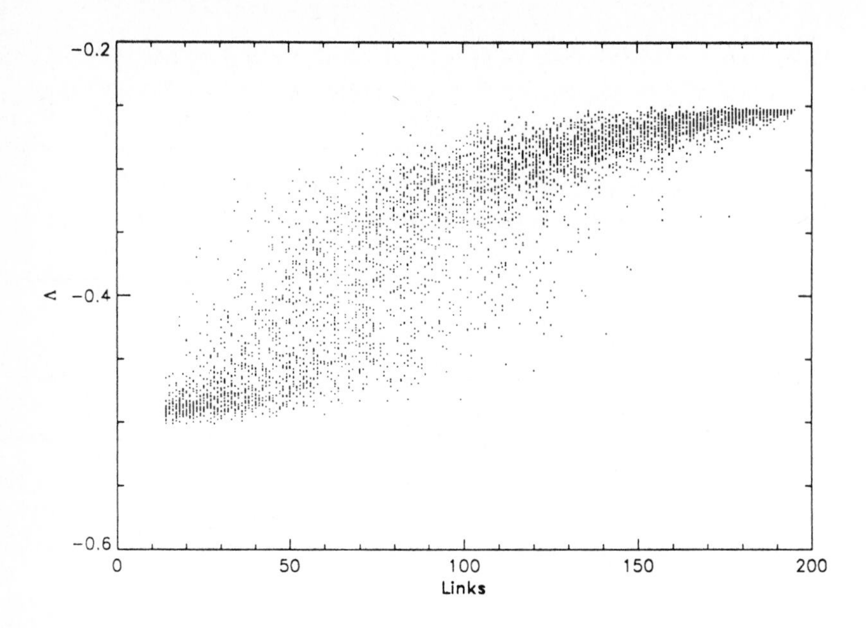

FIGURE 13 Dependence of non-equilibrium properties on catalytic connectivity. The steady state slope Λ is shown for 5000 catalytic networks with different connectivity. The reactions are fixed as in Figure 2, but the catalytic links are assigned randomly. The total number of catalytic links is varied, as plotted on the horizontal axis. The kinetic parameters are $k_f = 10^2$, $k_r = 10$, $\nu = 10^2$, $k_u = 10^4$, $\delta = 10^2$, and $m_0 = 2.0$.

5.2 SENSITIVITY TO PARAMETERS

For simplicity, in the results presented so far we assumed the kinetic parameters were the same for all the reactions in the network. This is obviously unrealistic; in any real chemical network, the parameters of each reaction are different. To test this sensitivity, in Figure 14 we vary the catalytic efficiency of each individual reaction while we hold the efficiency of the other reactions constant. Varying the efficiency of some reactions has a large effect on the steady-state slope, while varying others does not.

The autocatalytic set of Figure 14 has only 15 links, one per reaction, so perhaps it is not surprising that it is highly sensitive to the value of each catalytic efficiency. If a similar experiment is performed using the same reaction network, but with 118 catalytic links, the opposite occurs. The properties are relatively insensitive to the catalytic efficiency, as shown by Bagley.[3] This robustness comes about because there are typically many catalysts for each reaction. This result is consistent with that of Figure 13.

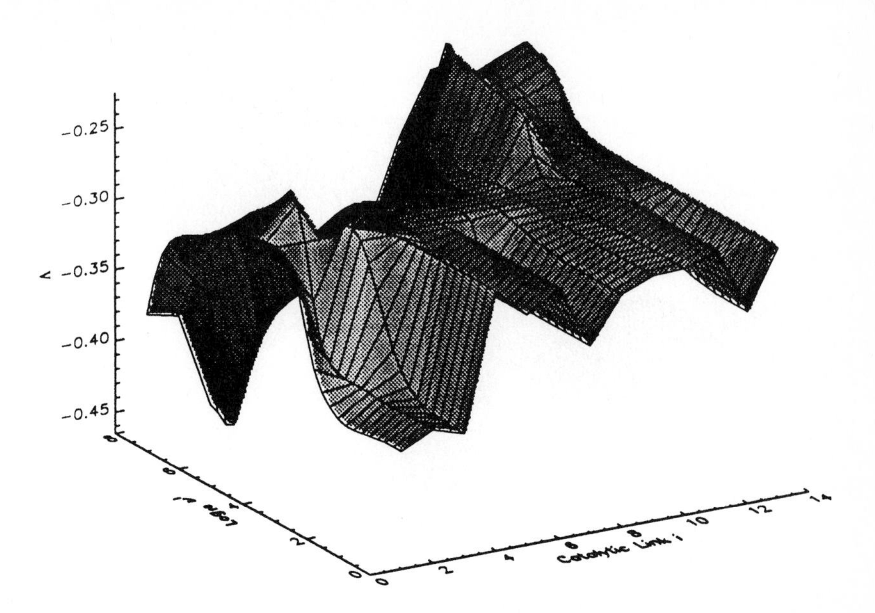

FIGURE 14 Sensitivity to variation of catalytic efficiency. The slope of the network at steady state is plotted as a function of the catalytic efficiency of each reaction, using the graph of Figure 2 which conveniently has only one catalytic link per reaction. The index i labels the catalysts, which are ordered according to length. The catalytic efficiency of each individual reaction is varied while that of other reactions is held constant. The parameters are $k_f = 6.49 \times 10^2$, $k_r = 2.50$, $\nu = 8.97 \times 10^5$, $k_u = 5.00 \times 10^4$, and $m_0 = 2.0$.

Figure 15 shows the results of varying the the forward rate constant k_f for the catalytic network of Figure 2. As we see, there is also considerable sensitivity to variations of k_f.

The experiments of this section demonstrate that the sensitivity of autocatalytic metabolisms to variations of parameters depends on the density of the reaction graph. In some parameter regimes the properties of autocatalytic metabolisms are quite sensitive to parameters. This is not surprising—the kinetic equations are highly nonlinear, and such sensitive behavior is common in highly nonlinear systems. This enhances the perception of autocatalytic metabolisms as "emergent phenomena." Without actually performing a simulation, in these parameter regimes the behavior of an autocatalytic metabolism is difficult to predict in advance. In other parameter regimes, however, almost all parameter settings yield similar behavior.

FIGURE 15 Sensitivity to variations of the forward rate constant. Using the reaction network of Figure 2, the forward reaction rate k_f of each reaction is varied in turn, while k_f for all the other reactions is held constant. The steady-state slope Λ of the network is plotted on the vertical axis. The parameters are $k_f = 6.49 \times 10^2$, $k_r = 2.50$, $\nu = 8.97 \times 10^5$, $k_u = 5.00 \times 10^4$, and $m_0 = 2.0$.

5.3 ROLE OF DIVERSITY

In order for catalysis to be able to concentrate the material of the system into only a few species, it is necessary for some catalytic efficiencies to be high, while others are low. Thus, it is clear that diversity of parameters is a necessary condition of catalytic focusing. When we set $\nu > 0$ for some reactions and $\nu = 0$ for the remainder, we are automatically introducing diversity. It is our impression that additional diversity in the kinetic parameters, such as a distribution of values for ν, k_f, and k_r, tends to increase the ability of the system to focus, but at present we have not demonstrated this quantitatively.

6. METABOLIC ROBUSTNESS

In this section we make more precise the sense in which the autocatalytic activity we have observed is similar to that of a metabolism, and investigate robustness to changes in the food set.

6.1 WHAT IS A METABOLISM?

A metabolism takes in material from its environment and reassembles it in order to propagate the form of the host organism. Autocatalytic metabolisms take in material from the food set, and through a chain of reactions whose rates are accelerated by catalysis, boost the concentration of the members of the autocatalytic set. They differ from contemporary metabolisms in that they do not require a host organism.

Another function of a metabolism is to extract free energy from the environment and make it available for the functions of the organism. Through the pyrophosphate mechanism discussed in subsection 2.4.2, autocatalytic sets can extract energy from light and use it to energize polymers. This in turn alters the effective forward and backward rate constants, enhancing the tendency for polymerization. As we see in Figure 15, for an autocatalytic set to take over the medium in which it resides, it is critical that the rate constants be in the proper regime. For real chemistry the unenergized values of k_f are probably in the unfavorable regime. This mechanism for extracting energy is critical to the existence and therefore the "functioning" of an autocatalytic metabolism.

In conclusion, although autocatalytic metabolisms are certainly much less sophisticated than the metabolisms of contemporary organisms, they exhibit the same rudimentary properties.

6.2 ROBUSTNESS

The fact that their underlying deterministic dynamical equations appear to have a unique stable fixed point endows autocatalytic metabolism with a considerable degree of robustness to trauma. A perturbation in the concentration of one of its elements, for example, quickly dies out.[7] Thus, "self-repair" of autocatalytic metabolisms is built into their chemical kinetics.

Another notion of robustness concerns variations in diet. Some metabolisms can digest only one kind of food, and are, therefore, fragile to changes in their environment. An example is the Panda, which can only digest bamboo. Others can digest many different types of material, and are, therefore, quite robust to changes in their environment. An example is the cockroach, which to the average urban dweller seems capable of digesting anything.

In this section we investigate the robustness of autocatalytic metabolisms by simply varying their food set. For example, in Figure 16 we simulate the steady-state concentration profile for five different food sets in conditions where everything else is held constant. In the first case, called the "default," the food set consists of the four species $\{a, b, ab, bb\}$, which are all input at the same rate. This food set gives rise to clear dominance of the autocatalytic metabolism over the background. In the other cases the supply of at least one of these four species is eliminated, and

[7]However, a perturbation that adds *new* elements to the autocatalytic metabolism may be strongly amplified. This possibility is discussed in detail in the companion paper.[3]

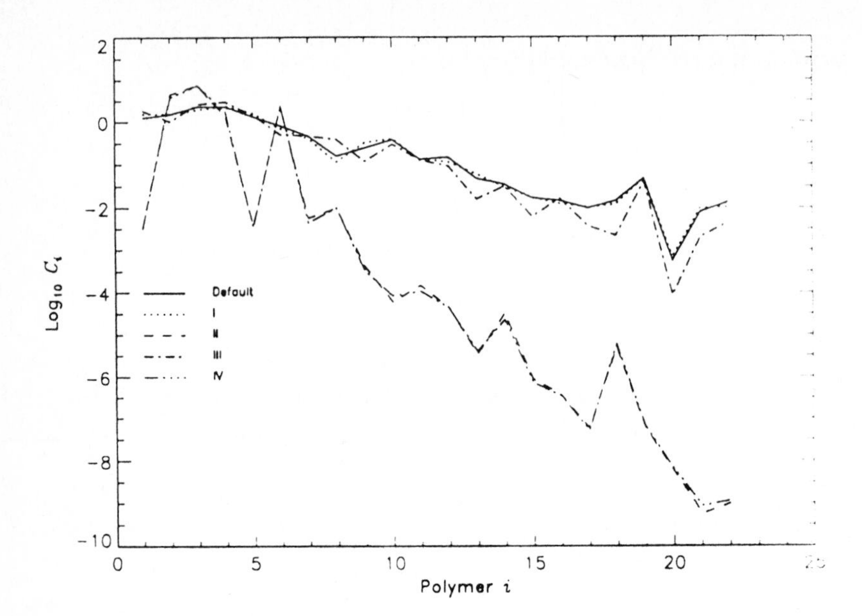

FIGURE 16 A test of the robustness of an autocatalytic metabolism. A reaction network with 22 polymers, corresponding to the endpoint in the simulation of Figure 3, is used to test four different variations of the food set. The table shows the food set elements listed horizontally across the top and the rate at which each is supplied per unit time for the experiments labeled I–IV. The right-hand column gives the steady state slope Λ of the autocatalytic network. In the figure the polymers of the network are ordered according to length and assigned an index i. The concentration of each polymer is plotted as a function of i. Variations II and IV cause the autocatalytic metabolism to "die," falling below the concentration levels without catalysis, while variations I and III leave it essentially unaltered. The parameters are $k_r = 10$, $k_f = 10^2$, $k_u = 10^4$, $\delta = 10^{3.25}$, $\nu = 10^4$, and $m_0 = 30$.

TABLE 3

	a	b	ab	bb	Λ
Default	5	5	5	5	-1.25×10^{-1}
I	5	0	5	7.5	-1.44×10^{-1}
II	0	0	10	5	-3.84×10^{-1}
III	10	20	0	0	-1.60×10^{-1}
IV	0	10	10	0	-4.11×10^{-1}
No Catalysis	5	5	5	5	-3.59×10^{-1}

the influx of the other species is adjusted so that the total rate of influx of mass remains constant. The results are shown in Figure 16. The concentration profiles cluster into two groups: In cases *I* and *III* the concentration profile is essentially unaltered, giving the same dominance over the background observed with the default food set. Apparently the network is able to resupply the missing elements with little alteration. For the other two alterations, in contrast, the autocatalytic set appears to "die": the concentrations change radically, dropping below the level corresponding to that with no catalysis. The clustering evident in Figure 16 demonstrates a qualitative difference between an autocatalytic metabolism that is "alive" and one that is "dead," and reinforces our use of the word "metabolism" for this behavior.

Note that we originally thought that survival under changes of the food set might be easily explained by simply examining the reaction graph. However, there are several cases where the autocatalytic set dies even though a metabolic pathway is available that could potentially replenish the missing element. The persistence of the autocatalytic metabolism depends on the topology, but it also depends on complicated nonlinear cooperative effects associated with the kinetics.

7. DISCUSSION

7.1 RELEVANCE TO EXPERIMENTS

One of our ultimate goals is the discovery of autocatalytic metabolisms in the laboratory. We hope that the parameters quoted in Table 2 will at least provide a crude starting point for laboratory investigations. In any case, we have demonstrated some qualitative principles that may serve as a guide in real experiments, such as:

- For an autocatalytic metabolism to dominate over its background, polymerization must be favored even in the absence of catalysis. In other words, the primary role of catalysis is to focus the concentration into a few species, rather than to effect an overall tendency for everything to polymerize. Energetic chemical species such as pyrophosphate may play an important role in shifting the effective equilibrium constant to favor overall polymerization.
- It is important to drive the system the proper distance from equilibrium. For a significant effect the mean reaction number (as defined in subsection 2.4.1) should be in the vicinity of 100, although the optimal value also depends on the other parameters.
- The largest possible deviation from equilibrium occurs when all the parameters have finite values.[7] This is particularly surprising for the catalytic efficiency ν and the forward rate constant k_f^*, which one might have thought should be as large as possible. However, since ν and k_f^* are almost certainly on the low side of the favorable regime, from a practical point of view in experiments, an effort should be made to find systems for which they are as large as possible.

- In order for an autocatalytic set to be robust, the steady-state concentrations of its components should be within a few orders of magnitude of each other.

We have posed several questions whose experimental resolution could prove or disprove the assumptions of this model, or more likely, could guide revisions to make it more realistic. The number of individual kinetic parameters involved is enormous, and it is unrealistic to expect that they could be measured in detail. However, experiments that measure the distribution of parameters could be very useful. A measurement of the distribution of catalytic efficiencies as a function of length for randomly chosen polypeptides or nucleic acids would be particularly helpful.

By heating and cooling appropriate mixtures of amino acids, Sidney Fox has observed the formation of protein-like polymers that form cell-like enclosures, called proteinoid microspheres.[20,21] It is possible that autocatalytic behavior played an important role in the formation of proteinoids, through a mechanism analogous to that discussed here. The problem of determining experimentally whether or not an autocatalytic metabolism is present in the laboratory has not been solved, and deserves more attention.

The formation of an enclosure is clearly a desirable property for an autocatalytic metabolism. An autocatalytic metabolism might form its own enclosure, or it might inhabit a pre-existing enclosure. Morowitz et al. have proposed that the spontaneous formation of self-replicating lipid vesicles may have played a critical role in the origin of life.[30,31] Furthermore, they argue that they might be capable of spontaneously producing amino acids, as well as a membrane potential. Such vesicles might provide a natural home for autocatalytic metabolisms. In the spirit of Oparin's original proposals, it is conceivable that an autocatalytic metabolism could establish a cooperative relationship with such a vesicle and evolve so that it was able regulate its functions.

7.2 POSSIBLE SCENARIO FOR THE ORIGIN OF LIFE

Our results lend support to the hypothesis that the emergence of metabolisms preceded the emergence of template-based self-replication. The formation of autocatalytic metabolisms might play a role in several possible scenarios for the origin of life. The following gives one possible scenario:

[7]The only possible exception is the mass concentration m_0, which we have not studied in detail.

Any one of several different mechanisms might have generated a supply of monomers. In a reducing environment the experiments of Miller and Urey make this plausible for amino acids. In principle it could also happen for nucleotides, but given that their formation is not energetically favorable,[14,19,22,29,41,42] amino acids seem to be a more likely candidate. Non-equilibrium conditions fostered the formation of autocatalytic metabolisms. This might have happened either with the help of pre-existing enclosures,[30,33] or because the autocatalytic metabolisms formed their own enclosures.[20] Autocatalytic metabolisms flourished and evolved. Eventually, some of them happened to have constituent proteins with the appropriate enzymatic properties to facilitate the synthesis and polymerization of nucleotides. The resulting nucleic acids began to cooperate, either by catalyzing the formation of each other, or making use of the catalytic properties of the proteins and forming hypercycles, or both. The nucleic acids synthesized proteins, some of which were incorporated into the autocatalytic metabolism. Eventually DNA molecules evolved that coded for all the proteins and RNA enzymes involved with their own replication, completely taking over the original autonomous autocatalytic metabolisms and causing them to disappear.

7.3 CRITIQUE

We feel that there are at least two major problems with the model for autocatalytic metabolisms that we have presented here, at least as it pertains to experiments:

1. *Reliance on a mass flux to drive the system away from equilibrium.* It seems implausible to assume that a flux of appropriate material would exist for a sustained period of time in a natural setting.[8] Although we have also explored the possibility of light as an energy source, using pyrophosphate as an energy transducer, in its present form this mechanism essentially only serves to shift the equilibrium point, and cannot focus the material of the system to create a metabolism all by itself. Another mechanism for focusing the energy of the system, perhaps based on oxidation-reduction reactions, might solve this problem in a more realistic manner.[18,31]
2. *Catalytic properties for short polymers.* The random rule generates reactions whose catalysts may be short polymers. This is unrealistic. The match rule we have formulated takes this to the opposite extreme, and only allows polymers longer than a certain critical size to participate in any catalytic reactions, either as reactants or as catalysts. Because of this the problem of maintaining a flux from the food set to the catalytic network becomes difficult. Of course, there

[8]Note that an autocatalytic metabolism may be able to persist through periods in which the system is at equilibrium, particularly if the polymers form enclosures that exclude water. Even very small concentrations may be sufficient to "seed" the formation of the original autocatalytic metabolism once non-equilibrium conditions are re-established.

are always spontaneous reactions between short polymers, which automatically expand the food set to a certain critical size, as described by Eq. (6). This might be sufficient, but in any case it complicates the problem.

We need to explore this question more carefully. One possible line of investigation is to use an artificial chemistry that is intermediate between the two proposed here, in which short polymers cannot be catalysts, but can be reactants. Still another solution might follow from finding an alternative energy source, as suggested in Eq. (1) above.

From a broader perspective of studying self-organization, this model suffers from a problem that is generic for many artificial life models. The process of making abstractions and simplifications builds a straightjacket that limits its emergent properties. The model presented here is essentially a "connectionist" model for chemistry, and suffers from the limitations inherent to this level of description.[13] In order to make this model tractable, we have reduced real polymers, which have complex spatial structure, to an interaction rule and a set of rate equations that measures only one aspect of their aggregate behavior. To see further emergence of functional properties, such as the formation of enclosures, it may be necessary to study an artificial chemistry that is richer than that we have studied here. Unfortunately, such a model will inevitably require more computational power. This seems to be an unavoidable trade-off that occurs in all artificial life models. Even with a Connection Machine, one must be very clever to compete with Nature, who has more than Avogadro's number of parallel processors.

7.4 ARE THEY ALIVE?

The artificial metabolism that we have studied here represents an organizational state that is between that of living and non-living systems. Given the appropriate conditions, within its own artificial universe an autocatalytic metabolism emerges spontaneously. It metabolizes energy, capturing the material of its universe and incorporating it into its own form. It also reproduces itself, although it does so continuously rather than discretely.

A good point of comparison is the class of oxidation reactions that produce "rust." Rust can be an autocatalytic reaction, in the sense that the presence of rust engenders the formation of more rust. For example, collectors of bronze coins are very careful to avoid "bronze disease," a form of rust that degrades a bronze coin, and spreads rapidly from coin to coin; if a coin is discovered to have bronze disease, it is immediately quarantined, to avoid the infection of other coins.

The key difference between autocatalytic metabolisms and simple autocatalytic reactions has to do with diversity and specificity. For a given metal, such as bronze, there is typically only one variety of rust. For a given polymeric medium, such as amino acids, there may be an enormous number of different autocatalytic metabolisms, each of which contains a different mixture of polymeric species. As we discuss in a companion paper, when the stochastic dynamics of spontaneous reactions are taken into account, which autocatalytic metabolism dominates at any given time depends on the history of the medium. Autocatalytic metabolisms can act like "seeds"—the presence of a minimal set of the elements of an autocatalytic metabolism, even at low concentrations, may allow this metabolism to take over its medium, and block the formation of other autocatalytic metabolisms.[4] Autocatalytic metabolisms are both diverse and specific. Finally, as discussed in the companion paper, they also have the necessary feedback mechanisms to propagate themselves and amplify variations.

Insofar as an autocatalytic set is alive, it is a much cruder and less specific form of life than an organism with reproductive machinery based on templating. There is no genetic code for autocatalytic metabolisms. The message is simply the set of chemical species present in the soup.

There has been some debate concerning whether or not Gaia, i.e., the Earth, can be considered to be a living organism in and of itself.[26] The relevant issues bearing on the question of whether or not autocatalytic metabolisms are alive are closely related to those for Gaia. The properties of autocatalytic metabolisms and Gaia are strikingly similar. First, an autocatalytic metabolism causes a substantial deviation from the equilibrium properties of the medium it occupies. Second, an autocatalytic metabolism may be the sole inhabitant of its medium. In this case, like Gaia, the autocatalytic metabolism evolves only through the richness of its own dynamics. One type of autocatalytic metabolism spontaneously evolves into another. There is internal competition and cooperation.

An objection to the consideration of Gaia as a living organism has been raised by Dawkins, who points out that since it is the sole inhabitant of its environment, there is no mechanism for selection.[26] While we argue that autocatalytic metabolisms might "evolve" on their own, they need not remain isolated, and in general it is unlikely that they would. It is quite natural to imagine diverse autocatalytic metabolisms in separate enclosures, geographically isolated from each other, occasionally coming in contact, and competing and/or cooperating with each other. Some preliminary simulations involving spatially isolated autocatalytic metabolisms have been made by Bagley.[3] Thus, Dawkin's objections do not apply to autocatalytic metabolisms. A significant difference between autocatalytic metabolisms and Gaia is that the constituents of Gaia are themselves alive, whereas the constituents of autocatalytic metabolisms are not. Insofar as they are alive, Gaia is a simple organism made out of very complex organisms, whereas autocatalytic metabolisms are simple organisms made out of inanimate matter.

7.5 SUMMARY

We have demonstrated that under appropriate conditions an autocatalytic set can concentrate much of the mass of its environment into a focused set, with concentrations orders of magnitude above equilibrium. When this occurs we call the network an autocatalytic metabolism. The use of the term metabolism is supported by the arguments of Section 6, and by the demonstration that some autocatalytic metabolisms are robust, and are capable of digesting a variety of different food sets.

Autocatalytic metabolisms can be highly sensitive to both the topology of the reaction network and the kinetic parameters of individual reactions. Because of this sensitivity, without performing a simulation it is difficult to state in advance whether or not a reaction network supports an autocatalytic metabolism. Nonetheless, there are ranges of parameters where most parameter values or most topologies support an autocatalytic metabolism. An indication of the range of parameter values where autocatalytic metabolisms are common is given in Table 2, and Figures 6-12.

These simulations demonstrate there is a favorable regime in which the flow of energy through the system is "just right" to make its properties radically different from those at equilibrium. There is an intriguing correspondence with behavior observed in other natural and artificial systems. For example, Langton[25] has performed experiments constructing cellular automata rules at random. The rules are constrained to have a fixed probability of producing a non-zero state, which is regulated by a parameter λ. In an analogy to thermodynamics λ corresponds to the Boltzmann factor $\Delta E/kT$, and thus, in a generalized sense, measures how far the system is driven from equilibrium. Langton finds that the most interesting behavior occurs at a finite value of λ, intermediate between the equilibrium value and the largest possible deviation from equilibrium. It is also interesting to note the roughly sigmoidal shape of the transition in Figure 9, which is similar to the phase transitions observed by Langton. This correspondence may be more than accidental, since both phenomena involve measurements of quantities that go through a phase transition when graph closure is obtained.

This model adds support to the idea that the emergence of a metabolism may have preceded the emergence of a self-replicator based on templating machinery. Perhaps more important, it provides an example of a "self-organizing structure" that is intermediate in complexity between life and non-life. There is an old idea, clearly articulated by Herbert Spencer,[43] that in some general sense the evolution of form and organization depends on diversity arising from physical laws. Autocatalytic metabolisms illustrate how diversity in nonlinear feedback loops can couple to a flow of energy and radically alter the properties of a system from those at equilibrium, giving rise to information-carrying, propagating structures. This follows naturally from the rules of chemical kinetics. Although autocatalytic metabolisms are quite simple and lack the sophistication of modern organisms, they nonetheless fulfill the same functions as a contemporary metabolism, and challenge our notion of what it means to be alive.

ACKNOWLEDGMENTS

We would like to thank Stuart Kauffman, Ron Fox, Walter Fontana, and Alan Perelson for valuable suggestions, and Chris Langton, Mats Nordahl, and Steen Rasmussen for critical readings of the manuscript. We are grateful for support from the Department of Energy and the University of California INCOR program. We would also like to thank the Complexity and Evolution program at the Institute for Scientific Interchange, Turin, Italy, which facilitated the writing of this paper.

APPENDIX

KINETIC EQUATIONS

The effect of saturation can be taken into account exactly (within the limit of validity of mass action) by breaking down a catalyzed reaction into each of its parts. This involves keeping track of several different intermediate products: A bound to E, B bound to E, C bound to E, and AB bound to E. Each reaction thus has four intermediaries. The autocatalytic networks that we simulate may have many species, and typically have many more reactions than species. Thus, treating each catalyzed reaction in full detail would enormously increase the complexity of the simulation.

Another difficulty of the simulation comes about because the reactions are all reversible; if we allowed them to be irreversible, we would have effectively assumed the solution. This means that traditional schemes such as the Michaelis-Menten approximation are not applicable.

We simplify the saturation problem by assuming that A and B bind to each other as soon as they are bound to the enzyme E. We then break the reversible reaction of Eq. (8) into two irreversible reactions of the form

$$\begin{aligned} A+B+E &\xrightarrow{\nu k_f} \overline{CE}+H\,, \\ C+H+E &\xrightarrow{\nu k_r} \overline{ABE}\,. \end{aligned} \tag{22}$$

$\overline{ABE}$ and $\overline{CE}$ are complexes formed in the reaction. The further dissociation of the bound complexes into free elements is treated through the irreversible reactions

$$\begin{aligned} \overline{CE} &\xrightarrow{k_u} C+E\,, \\ \overline{ABE} &\xrightarrow{k_u} A+B+E\,. \end{aligned} \tag{23}$$

k_u is a constant that characterizes the rate at which the reactants unbind from the enzyme. Taking these together, and adding on the effect of the spontaneous reactions gives the rate equations

$$\begin{aligned} \frac{d\overline{CE}}{dt} &= -k_u\overline{CE}+k_f(1+\nu E)AB\,, \\ \frac{d\overline{ABE}}{dt} &= -k_u\overline{ABE}+k_r(1+\nu E)CH\,, \\ \frac{dC}{dt} &= k_u\overline{CE}-k_r(1+\nu E)CH\,, \\ \frac{dA}{dt} &= k_u\overline{ABE}-k_f(1+\nu E)AB\,, \\ \frac{dE}{dt} &= k_u\overline{CE}+k_u\overline{ABE}-(1+\nu E)(k_fAB-k_rCH)\,. \end{aligned} \tag{24}$$

Note that we are using the same symbol for the bound complex $\overline{ABE}$ and its concentration. The reason the last equation differs from the others is because the enzyme E is involved in two bound complexes, and so has two terms involving k_u.

In comparison with the exact reaction scheme, the approximate reaction scheme of Eq. (24) reduces the number of intermediates by a factor of two. However, as described in the remainder of this section, by making a simple substitution of variables, we can make a simplification that goes considerably beyond this, so that the number of equations that must be solved including saturation is only twice the number required without saturation.

Suppose that a given species i participates in many different reactions, and is bound up in many different complexes $\overline{c}_m$. For convenience, assume that each has the same dissociation constant k_u. Using the approximation scheme above, the kinetic equation for dx_i/dt is of the form

$$\frac{dx_i}{dt} = k_u \sum_m \overline{c}_m + \sum (\textit{reaction terms}). \tag{25}$$

The first term takes into account all of the bound complexes that contribute to species i when they dissociate. (*reaction terms*) consists of a sum of terms, one for each reaction that species i participates in. Each term is of the form of Eq. (24), according to whether i plays the role of A, C, or E. Note that these only involve free species; none of the bound complexes $\overline{c}_m$ are involved. Furthermore, the dissociation terms that involve $\overline{c}_m$ are all linear. Thus, if we define the new variable

$$\overline{x}_i = \sum_m \overline{c}_m , \tag{26}$$

we can rewrite Eq. (25) as

$$\frac{dx_i}{dt} = k_u \overline{x}_i + \sum (\textit{reaction terms}). \tag{27}$$

Similarly, each bound complex has a kinetic equation of the form

$$\frac{d\,\overline{c}_m}{dt} = -k_u \overline{c}_m + \sum (\textit{reaction terms}). \tag{28}$$

There is only one reaction term, corresponding to the particular reaction associated with the complex. If we add up the equations associated with all the reactions that x_i participates in, we obtain

$$\frac{d\,\overline{x}_i}{dt} = -k_u \overline{x}_i + \sum (\textit{reaction terms}). \tag{29}$$

Thus, we can solve for $\overline{x}_i$ without having to solve for each intermediate complex from which it is derived. Since none of the reaction terms involve the bound complexes, we never have to solve for them. The number of variables is just twice the number of polymer species, rather than more than twice the number of reactions. We have taken saturation into account, at the small cost of doubling the effective number of species.

REFERENCES

1. Anderson, P. W. "Suggested Model for Prebiotic Evolution: The Use of Chaos." *Proc. Natl. Acad. Sci.* **80** (1983): 3386–3390.
2. Bagley, R. J., J. D. Farmer, S. A. Kauffman, N. H. Packard, A. S. Perelson, and I. M. Stadnyk. "Modeling Adaptive Biological Systems." *Biosystems* **23** (1989): 113–138.
3. Bagley, R. J. "The Functional Self-Organization of Autocatalytic Networks in a Model of the Evolution of Biogenesis." Ph.D. Thesis, University of California at San Diego, La Jolla, CA, 1990.
4. Bagley, R. J., J. D. Farmer, and W. Fontana. "Evolution of a Metabolism." This volume.
5. Calvin, M. "Chemical Evolution and the Origin of Life." *Amer. Sci.* **44** (1956): 248–263.
6. Calvin, M. *Chemical Evolution.* Oxford: Oxford University Press, 1969.
7. Cech, T. R. "The Chemistry of Self-Splicing RNA and RNA Enzymes." *Science* **236** (1987): 1532.
8. Dyson, F. "A Model for the Origin of Life." *J. Mol. Evol.* **118** (1982): 344–350.
9. Eigen, M., and P. Schuster. "The Hypercycle: a Principle of Natural Self-Organization. Part A: The Emergence of the Hypercycle." *Naturwissenschaften* **64** (1977): 541–565.
10. Eigen, M. "Self-Organization of Matter and the Evolution of Biological Macromolecules." *Naturwissenschaften* **58** (1971): 465–523.
11. Farmer, J. D., S. A. Kauffman, and N. H. Packard. "Autocatalytic Replication of Polymers." *Physica D* **22** (1986): 50–67.
12. Farmer, J. D., N. H. Packard, and A. S. Perelson. "The Immune System, Adaptation, and Machine Learning." *Physica D* **22** (1986): 187–204.
13. Farmer, J. D. "A Rosetta Stone for Connectionism." *Physica D* **42** (1990): 153–187.
14. Ferris, J. P. "Prebiotic Synthesis: Problems and Challenges." *Cold Spring Harbor Symp. Quant. Biol.* **LXII** (1987): 29–35.
15. Flory, P. J. *Principles of Polymer Chemistry.* Ithaca: Cornell University Press, 1953.
16. Fontana, W. "Algorithmic Chemistry: A Model for Functional Self-Organization." This volume.
17. Fontana, W. "Turing Gas: A New Approach to Functional Self-Organization." Preprint, 1991.
18. Fox, R. F. *Biological Energy Transduction: The Uroboros.* New York: John Wiley, 1982.
19. Fox, R. F. *Energy and the Evolution of Life.* New York: W. H. Freeman, 1988.
20. Fox, S. W., K. Harada, and J. Kendrick. "Production of Sperules from Proteinoids and Hot Water." *Science* **129** (1959): 1221–1223.

21. Fox, S. W., T. Nakashima, A. Przytylski, and R. M. Syren. "The Updated Experimental Proteiniod Model." *Int. J. Quantum Chem.* **55** (1982): 195.
22. Joyce, G. F. "RNA Evolution and the Origins of Life." *Nature* **338** (1989).
23. Kauffman, S. A. "Cellular Homeostasis, Epigenesis and Replication in Randomly Aggregated Macromolecular Systems." *J. Cybernetics.* **1** (1971): 71–96.
24. Kauffman, S. A. "Autocatalytic Sets of Proteins." *J. Theor. Biol.* **119** (1986): 1–24.
25. Langton, C. G. "Computation at the Edge of Chaos: Phase Transitions and Emergent Computation." *Physica D* **42** (1990).
26. Lovelock, J. E. *Gaia: A New Look at Life on Earth.* Oxford: Oxford University Press, 1979.
27. Macken, C. A., and A. S. Perelson. *Branching Processes Applied to Cell Surface Aggregation Phenomena. Lecture Notes in Biomathematics*, Vol. 58. Berlin: Springer-Verlag, 1985.
28. Miller, S. L., and H. C. Urey. "Organic Compound Synthesis on the Primitive Earth." *Science* **130** (1959): 245.
29. Miller, S. L., and L. E. Orgel. *The Origins of Life on Earth.* Concepts of Modern Biology. Englewood Cliffs, NJ: Prentice-Hall, 1974.
30. Morowitz, H. J., B. Heinz, and D. Deamer. *Origins of Life* **18** (1988): 281–287.
31. Morowitz, H. J. *The Beginnings of Cellular Life: Metabolism Recapitulates Biogenesis.* 1991. Preliminary manuscript.
32. Nicolis, G., and I. Prigogine. *Self-Organization in Nonequilibrium Systems.* New York: John Wiley, 1977.
33. Oparin, A. I. *The Origin of Life.* New York: Macmillan, 1938.
34. Paecht-Horowitz, M. "Polymerization on Surfaces." In *Polymerization of Organized Systems*, edited by H. G. Elias, 89–113. New York: Gordon and Breach, 1977.
35. Paecht-Horowitz, M., J. Berger, and A. Katchalsky. "Prebiotic Synthesis of Polypeptides by Heterogeneous Polycondensation of Amino-Acid Adenylates." *Nature* **228(5272)** (1970): 636–639.
36. Rasmussen, S., C. Knudsen, and R. Feldberg. "Dynamics of Programmable Matter." This volume.
37. Rössler, O. "A System Theoretic Model of Biogenesis." *Z. Naturforsch* **B26b** (1971): 741–746.
38. Rössler, O. "Deductive Prebiology." In *Molecular Evolution and the Prebiological Paradigm*, edited by K. L. Rolfing. New York: Plenum, 1983.
39. Stein, D. L., and P. W. Anderson. "A Model for the Origin of Biological Catalysis." *Proc. Natl. Acad. Sci.* **81** (1984): 1751–1753.
40. Stockmayer, W. H. "Theory of Molecular Size Distribution and Gel Formation in Branched-Chain Polymers." *J. Chem. Phys.* **11(2)** (1943): 45–55.
41. Schwartz, A. W., J. Visscher, R. Van der Woerd, and C. G. Bakker. "In Search of RNA Ancestors." *Cold Spring Harbor Symp. Quant. Biol.* **LXII** (1987): 37–39.

42. Shapiro, R. *Origins: A Skeptics Guide to the Creation of Life.* New York: Summit Books, 1986.
43. Spencer, Herbert. *First Principles.* 1862.
44. Van Dongen, P. G. J., and M. H. Ernst. "Kinetics of Reversible Polymerization." *J. Stat. Phys.* **37** (1984): 301–324.

Richard J. Bagley, J. Doyne Farmer, and Walter Fontana
Complex Systems Group, Theoretical Division, and Center for Nonlinear Studies, Los Alamos National Laboratory, Los Alamos, NM 87545 and Santa Fe Institute, 1120 Canyon Road, Santa Fe, NM 87501

Evolution of a Metabolism

We demonstrate that when stochastic effects are taken into account, over long periods of time autocatalytic metabolisms evolve through a series of punctuated equilibria. We outline a rigorous theoretical treatment of the dynamics of autocatalytic metabolisms, and present some heuristic numerical simulations. We develop the analogy between the evolution of autocatalytic metabolisms and that of contemporary organisms, and argue that while the essential properties of variation and selection are satisfied, there are nonetheless some intriguing differences that merit further study.

1. INTRODUCTION

The word "evolution" is often narrowly construed to apply only to the process of variation and selection in biology. However, there is an older and broader usage of this word, originating with Spencer,[16] that views evolution as driving long-term organizational change in nature, with biological evolution as a special case. Spencer defines evolution as "a change from an incoherent homogeneity to a coherent heterogeneity." Analogies between evolutionary behavior in many areas, such as astronomy, geology, economics, and sociology ,have been developed at a qualitative

Artificial Life II, SFI Studies in the Sciences of Complexity, vol. X, edited by C. G. Langton, C. Taylor, J. D. Farmer, & S. Rasmussen, Addison-Wesley, 1991

level.[11,16] However, as yet no one has been able to formulate this analogy quantitatively, or to demonstrate that it has predictive value. The failure to articulate the broader notion of evolution as a quantitative scientific principle justifies the prevalence of the narrower view that evolution in biology is fundamentally different from evolution in other natural phenomena.

One form of evolution that probably played a dominant role prior to biological evolution is what Calvin has termed "chemical evolution."[3,4] In this paper we study chemical evolution in the context of an artificial yet in many respects realistic model for catalytic reactions in polymer chemistry.[1,7] These results build on those of a companion paper,[1] in which it was demonstrated that under appropriate conditions metabolisms spontaneously emerge from a chemical soup. This emergence takes place on short time scales, over which the behavior is well approximated by deterministic equations. This paper studies the time evolution of these metabolisms over longer time scales, where stochastic effects play a critical role. We show that the metabolisms make transitions through a series of different fixed points, exhibiting what appears to be an open-ended succession of "punctuated equilibria."

The form of chemical evolution that we study here should be contrasted with that based on self-replication through templating, studied extensively by Eigen and others.[5,6] Templating reactions form the basis for the reproduction of contemporary organisms; the replication dynamics of chemical systems is so closely analogous to that of biological populations that it is essentially biological evolution on a molecular scale. Self-replication pertains to the possible evolution of early life forms, rather than to alternative evolutionary processes that may have preceded contemporary life. The form of chemical evolution that we study here is much closer to that envisioned by Rössler.[13,14,15]

Autocatalytic metabolisms reproduce themselves autonomously, without templating reactions. The analogy to biological evolution is not as direct as it is for templating systems. However, we argue that it nonetheless involves a process of variation and selection and deserves the name "evolution," at least in the broad sense articulated by Spencer. Although there are differences between the evolution of autocatalytic metabolisms and the evolution of biological organisms, they are at the level that one would naturally expect for proto-life forms based on alternative principles from those of contemporary organisms. The evolution of autocatalytic metabolisms is similar enough to that of biological organisms that many aspects are immediately recognizable, yet at the same time that there are provocative differences. The chemistry in which they evolve is simple enough for quantitative study and simulation. By presenting an example of an alternative form of evolution, we hope to support the idea that evolution can indeed be regarded as a broader physical principle driving organizational change, and to illustrate some of its more general properties.

This paper describes work in progress. Our goals are to outline the scenario under which autocatalytic metabolisms evolve, to discuss the issues involved in simulating their behavior, to present some preliminary numerical results, and to make some remarks comparing their evolution to that of contemporary biological organisms.

This paper will draw heavily on the companion paper, "Spontaneous Emergence of a Metabolism," in this volume,[1] where we study the chemistry of catalyzed polymerization reactions. The catalysis of these reactions is assumed to be specific, i.e., a typical polymer catalyzes the formation of only a small subset of all possible reactions. Under appropriate circumstances, the system may contain an *autocatalytic set*, i.e., a set of polymers such that each polymer is produced by at least one catalytic reaction involving only other members of the autocatalytic set. The system is driven away from equilibrium by an influx of a few special polymers, called the *food set.* When parameters are in the appropriate regime, the autocatalytic set may boost its own concentrations many orders of magnitude above the background of spontaneous reactions. When this occurs, we call the result an *autocatalytic metabolism.*

Although we will attempt to summarize important aspects of the companion paper as we go along, we will assume that the reader can refer to it as necessary.

2. POSSIBLE METHODS OF SIMULATION

Since there are an infinite number of possible polymer species, and an infinite number of possible reactions, properly simulating the behavior of a polymer network is a formidable problem. This problem is particularly severe for catalytic reactions, over long time scales. We compare three approaches: continuous differential equations, stochastic molecular collisions, and deterministic metadynamics.

Continuous differential equations. As long as the concentration of each species is sufficiently large, the dynamics can be described by a system of deterministic differential equations with continuous concentration variables $x_i > 0$, where i labels the possible species. However, because there are an infinite number of possible equations, from a practical point of view such a simulation is obviously impossible. Furthermore, even if it were possible, this approach provides a poor approximation of reality, particularly over long time scales. This is because real reaction vessels are always finite, which induces a minimum concentration θ corresponding to a single molecule. Until at least one molecule of a given chemical species is present, it cannot cause any reactions. As a result reactions are initiated sequentially, as the system creates the necessary constituents. In contrast, for continuous differential equations, for any time $t > 0$ all reactions are switched on, and all species typically have nonzero concentrations. For autocatalytic reactions this is not merely an esoteric problem: It results in a qualitatively incorrect prediction of the true dynamics.

Stochastic molecular collisions assumes integer populations of each species and simulates reactions as discrete collisions by sampling at random. This method is faithful to reality (at least for the level of description we are interested in here). However, as discussed in the companion paper, if the difference between the largest and smallest concentrations is large, it can be time consuming in comparison with differential equations. This problem taxes the resources of even the largest parallel computers. It is difficult to simulate a system of any size over long time scales.

Deterministic metadynamics uses a sequence of deterministic equations that change as the behavior of the system changes. This alternative attempts to make the best of both worlds.[2,7,8] The topological structure of the kinetics is represented by a graph, reflecting the dominant chemical species and chemical reactions at any given time. The graph changes to reflect either the creation of new species or the elimination of old species. This takes place relative to a concentration threshold θ, corresponding to the concentration when one molecule is present. The threshold thus specifies the size of the reaction vessel.

We wish to emphasize that while this procedure involves a sequence of changing graphs, *it is purely deterministic*.[1] Because of the finite threshold θ, the results are different from those that would be obtained with a fixed continuous system of equations. However, because the simulation is deterministic, over long time scales the results diverge from those that would be obtained with a stochastic simulator. In the companion paper we were interested in the problem of the *emergence* of autocatalytic metabolisms. This takes places on short time scales, over which the deterministic metadynamics procedure described above is reasonably accurate. One of our main purposes in this paper is to introduce a modification of the metadynamics procedure, called *stochastic metadynamics*, which combines the physical accuracy of a full stochastic molecular collision simulation with the speed of a metadynamics simulation, and allows us to follow the evolution of autocatalytic metabolisms for long periods of time.

For the equations studied here, for any fixed graph our simulations indicate that the dynamical equations always have a unique stable fixed point. We call this a *dynamical fixed point*. As described in detail in the companion paper, the existence of a unique fixed point makes it possible to speed up the metadynamics algorithm considerably by using an algebraic fixed point solver. For a given graph we find the corresponding dynamical fixed point; if there are new species over the threshold, we update the graph, and then find a new fixed point. We repeat this until there are no species that cross the threshold. Since no new species are created or destroyed, the graph remains fixed. We call the corresponding final state a *metadynamical fixed point*. Dynamical fixed points have no physical meaning, and are just computational conveniences. In contrast, metadynamical fixed points are physically meaningful. This is clarified in the next two sections.

[1] The kinetic equations should not be confused with the rules for the assignment of kinetic parameters. Though we may use a random assignment rule, once the assignment is completed the kinetic parameters are fixed and, for the simulations of the companion paper, the kinetic equations are completely deterministic.

3. AUTOCATALYTIC NETWORKS AS FLUCTUATION AMPLIFIERS

To understand why thresholds and spontaneous fluctuations have a large effect on autocatalysis, it is important to distinguish between internally and externally catalyzed reactions. An *internally catalyzed* reaction pathway is catalyzed by a species within the current autocatalytic metabolism. Let σ be a species that initially has a population of zero, and so is external to the metabolism. If σ is produced by a catalyzed reaction whose catalyst is already abundant, the concentration of σ will increase rapidly. In contrast, an *externally catalyzed* reaction pathway is catalyzed by a species outside the current autocatalytic metabolism. In order to be catalyzed, such reactions must wait for a spontaneous reaction to produce the catalyst. Since the spontaneous reactions are slow, when compared to the catalyzed reactions they can be regarded as discrete fluctuations. As pointed out by Rössler,[14] this introduces a delay in the activation of externally catalyzed reactions.

The simplest example is given in Figure 1(a). Let A and B be two polymers in the autocatalytic metabolism, and let σ be a third polymer that catalyzes its own formation.

$$A + B \xrightarrow{\sigma} \sigma, \tag{1}$$

Assume that σ is not produced by any other reactions involving only members of the autocatalytic metabolism, i.e., that it is an element of the *background* of polymers formed by spontaneous reactions. Assume that in the initial meta-dynamical fixed point its population is zero.[2] If a fluctuation produces σ, then, since it catalyzes its own formation, its concentration may increase by orders of magnitude.

Note that the boost in the concentration of σ may have other side effects; for example, σ may catalyze other reactions. This may cause the production of new polymer species, which in turn catalyze other reactions, etc. Since these events involve only internally catalyzed pathways, they take place on a rapid time scale, without delays. When σ is pumped up to high concentrations, the change in the dynamic equilibrium may also alter the concentrations of other species in the metabolism. The resulting competition might cause other polymer species to disappear from the metabolism. New species can be created, and old species can become extinct.

Purely deterministic metadynamics, as used in the companion paper, explores only the internally catalyzed pathways. Since the system cannot produce any of the external catalysts, it gets stuck at a metadynamical fixed point. For convenience, we will often call this a *pinned state*. Once in a pinned state, until a spontaneous fluctuation produces one of the external catalysts, the system cannot change. When a fluctuation occurs it may trigger another period of rapid change as the system

[2] When the simulation of the concentration of the background as an aggregate is included, assume that its concentration is less than the threshold, so that with the deterministic metadynamic rule of the companion paper,[1] it is not allowed to catalyze new reactions.

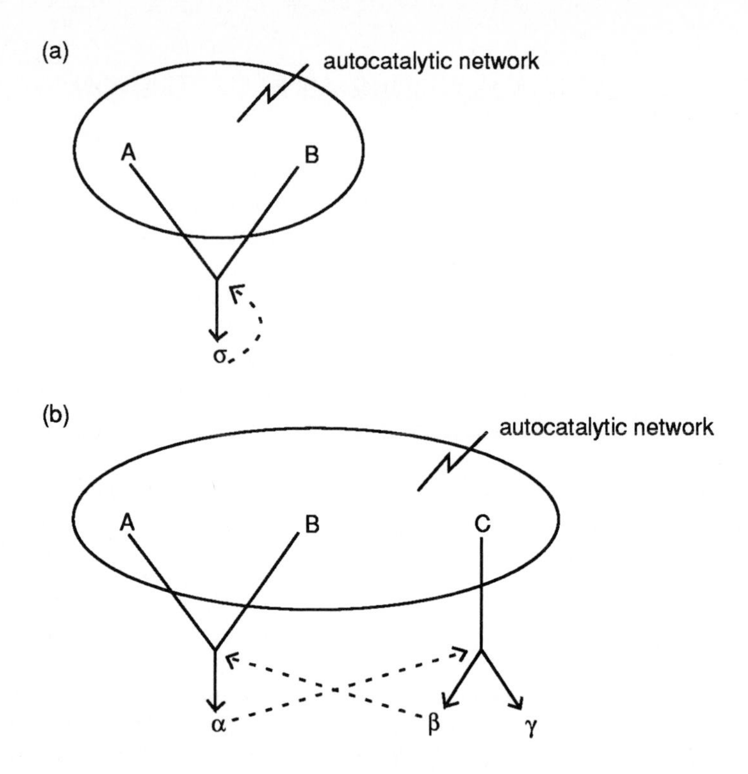

FIGURE 1 *Examples of autocatalytic mutations.* (a) A and B are members of an autocatalytic metabolism. $\sigma = A + B$ catalyzes its own formation. If the concentration of σ is initially zero, then except for the spontaneous reaction it will remain zero. If by chance the spontaneous reaction produces a molecule of σ, then the autocatalytic reaction may increase the concentration of σ by many orders of magnitude, creating a new pinned state which includes σ. (b) depicts a somewhat more complicated instance of the same phenomenon: A, B, and C are in the autocatalytic metabolism. Through spontaneous reactions A and B produce α, and C produces β and γ. Suppose α catalyzes the cleavage of C, and β catalyzes the condensation of A and B. If either α or β are produced by a spontaneous reaction, they may pump each other up and join the metabolism.

explores the newly activated internally catalyzed pathways, until the system settles into a new metadynamic fixed point. The end result is a change in the composition of the autocatalytic metabolism, and a transition to a new pinned state. We call a transition from one pinned state to another an *evolutionary modification* of the autocatalytic metabolism.

We will call an autocatalytic reaction such as the one shown in Figure 1(a), in which a species catalyzes its own formation, a *first-order autocatalytic loop*. There are autocatalytic loops of all orders. For example, Figure 1(b) depicts a second-order autocatalytic loop, in which α catalyzes the formation of β and β catalyzes the formation of α. The number of possible graphical combinations grows exponentially with the order of the loop. There are thus an enormous number of possible autocatalytic loops, and an enormous number of possible evolutionary modifications of autocatalytic metabolisms.

If an evolutionary modification is *robust*, i.e., if the steady state concentrations of the new elements at the new pinned state are orders of magnitude above the threshold, then the modification is unlikely to reverse itself spontaneously, since the probability of a series of fluctuations that decrease the population by several orders of magnitude is virtually nil. Thus, even though all the reactions involved are reversible, from a stochastic point of view *robust evolutionary modifications are effectively irreversible.* Once an evolutionary modification is triggered, the system is unlikely to return to its previous pinned state.[3]

The evolution of the system through time is highly path dependent. At any given time there are many possible spontaneous fluctuations. Each fluctuation that actually occurs generates a series of irreversible changes, effecting both the probability that a given fluctuation will occur, and the probability it might initiate a modification. Externally catalyzed reactions are activated sequentially, in random order; the probability that a given reaction will be activated at a given time is altered by each preceding fluctuation. This is quite different from what one would observe with a purely deterministic model.

Thus we see how the long time dynamics of a chemical network involves a series of transitions between pinned states, which are similar to punctuated equilibria in evolutionary biology. For a network of any reasonable size, the number of possible pinned states is so large that the system may evolve for a very long time without ever repeating itself.

4. STOCHASTIC METADYNAMICS

In this section we present a stochastic extension of the metadynamical method that allows us to simulate the evolution of autocatalytic metabolisms reasonably quickly. The basic idea is to perturb the catalytic reaction graph by randomly adding new species that may cause evolutionary modifications. Doing this in a physically realistic manner, so that the perturbations occur with the proper relative probabilities and time scales, involves several complex issues. In this section we outline our approach to the problem, and describe the heuristic treatment that forms the basis for the simulations of the next section.

[3] Of course, if a fluctuation of species σ triggers an evolutionary modification, it is always possible that some later evolutionary modification might introduce competition and eliminate σ. However, when this occurs it is unlikely to cause a return to the original pinned state.

4.1 THE SPECIAL ROLE OF THE SHADOW

The shadow is the subset of species in the background that are produced by reactions involving only themselves and members of the catalyzed reaction network. (See Figure 4 of the companion paper.) The shadow plays a dominant role in initiating evolutionary modifications, for two reasons: First, because the autocatalytic metabolism is at high concentration, the shadow is typically maintained at higher concentrations than other parts of the background. Thus fluctuations creating elements of the shadow are more likely than others. Second, for a fluctuation of a *single* species to trigger an evolutionary modification, that species must be in the shadow. This is true almost by definition: if a background element is not in the shadow, then its production requires at least one other background element. Triggering catalytic production therefore requires at least two simultaneous fluctuations, which is unlikely. For both of these reasons, the majority of evolutionary modifications are initiated in the shadow.

4.2 ENUMERATION OF AUTOCATALYTIC SUBGRAPHS

Figure 1 shows two possible autocatalytic subgraphs that might trigger an evolutionary modification of an autocatalytic metabolism. There are an enormous number of other possible subgraphs. We will restrict attention to autocatalytic subgraphs that can be triggered by a single fluctuation in the shadow.

We begin by introducing a simplified graph description that describes the feedback structure of the subgraph. This simplified description ignores the identity of the "parents" in the autocatalytic metabolism, focusing attention on the catalytic relationships in the shadow. A situation in which σ catalyzes its own formation, as shown in Figure 1(a), is represented by a simple graph with a single vertex σ, and a single directed edge from σ to itself, as shown in Figure 2(a).

This reduced graph also lumps together any other simple autocatalytic reactions that produce σ. For example, there might be another set of polymers A' and B' in the autocatalytic set whose condensation is catalyzed by σ, or a polymer C' whose cleavage is catalyzed by σ. The reduced graph describes all of these reactions taken together.

Figure 2 enumerates the combinatorial possibilities for reduced graphs with up to three vertices. In addition to simple loops, there are many other possibilities. In many of these the autocatalytic feedback is maintained exclusively by a subset of the possible vertices; in this case we call the remaining vertices *parasites*. For example, Figure 2(c) has two vertices. The first vertex provides the autocatalytic feedback through a first-order loop. It also catalyzes the formation of the parasitic second vertex, which does nothing to support the first loop.

In principle it is possible to enumerate the autocatalytic modifications of arbitrary order. However, for simplicity in this paper we will restrict attention to modifications of order three or less, as shown in Figure 2.

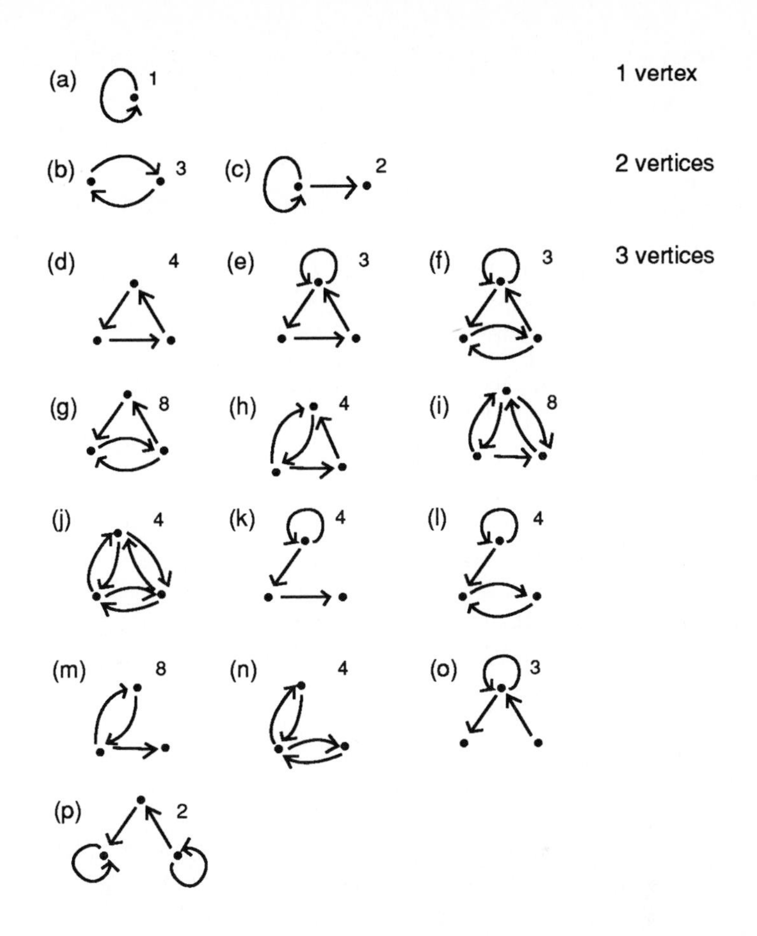

FIGURE 2 *Possible autocatalytic subgraphs in the shadow with three vertices or less.* These reduced graphs ignore the identity of the "parents" in the autocatalytic set. The vertices correspond to chemical species in the shadow, and the edges indicate their catalytic relationship to each other. (a) corresponds to a species that catalyzes its own formation, and is equivalent to the full graph of Figure 1(a); it also lumps together any other graphs of the same form with different parents but the same σ. (b) corresponds to a two-cycle, such as that of Figure 1(b). (c) also involves two vertices; however, it can be regarded as a one cycle with a "parasite." The remainder of the graphs enumerate the possibilities with three vertices. The counts with each graph give the number of realizations that contain the shown skeleton, and in which some of the vertices are additionally autocatalytic on their own.

4.3 ASSIGNING PROBABILITIES TO AUTOCATALYTIC MODIFICATIONS

All of the autocatalytic subgraphs listed in the previous section describe additions to the metabolism that can be triggered by a single spontaneous fluctuation. Once the fluctuation occurs it may activate the other elements of the subgraph. For example, in Figure 2(b), suppose a fluctuation produces α; unless α decays first, it produces β, which in turn may produce another α, etc. Whether or not these fluctuations grow so that they pump up the concentrations to a robust level, well above the threshold, is a complicated birth and death process with an uncertain outcome. After a long time there are two likely possibilities:

1. Creation events overcome decay events, and the concentrations of the subgraph are roughly at the level predicted by deterministic kinetics.
2. Decay events overcome creation events, and the concentrations are zero.

This neglects unlikely events, such as additional spontaneous fluctuations, and assumes that the kinetic parameters are such that the initial production rate exceeds the decay rate (otherwise the concentrations almost certainly go to zero).

The relative probability of these two outcomes can be approximated using a master equation. For example, consider the case of a simple first-order autocatalytic loop, as described in Figure 1(b). Neglecting saturation, the deterministic kinetics can be approximated by

$$\dot{\sigma} = (1 + \nu\sigma)(k_f AB - k_r H\sigma) - K\sigma, \tag{2}$$

where ν is the catalytic efficiency, k_f is the forward rate constant, k_r is the reverse rate constant, H is the concentration of water, and K is the global dissipation parameter, corresponding to the rate at which material diffuses out of the system. When σ is sufficiently small, we can neglect the quadratic term. If $\nu\theta \gg 1$ we can also neglect the spontaneous reaction, and this becomes

$$\dot{\sigma} = \nu k_f AB\sigma - (K + k_r H)\sigma. \tag{3}$$

When the concentrations are small the problem is more properly treated in terms of a master equation. Letting $P(n,t)$ be the probability that σ has population n at time t, the master equation corresponding to Eq. (3) is

$$\frac{\partial P(n,t)}{\partial t} = c\,(n-1)P(n-1,t) + d\,(n+1)P(n+1,t) - (c+d)nP(n,t), \tag{4}$$

where $c = \nu k_f AB$ is the "creation" rate, and $d = K + k_r H$ is the "death" rate.

Assume that at $t = 0$ the spontaneous reaction creates a single molecule. For the master equation this corresponds to the initial condition $P(1,0) = 1$. By solving the master equation (see e.g., Karlin[10]) and taking the limit as $t \to \infty$, we can compute the survival probability, $P_s = 1 - P(0,\infty)$. It is

$$P_s = \begin{cases} 1 - \frac{d}{c} & \text{if } c > d; \\ 0 & \text{otherwise.} \end{cases} \tag{5}$$

P_s is the probability a mutation will grow once it is triggered.

The asymptotic survival probability for an mth-order subgraph can also be derived using an m-dimensional master equation. The equations are linear, but they are complicated and we have not yet solved them. As an approximation, however, we first study the corresponding deterministic equations. In particular, it is clear that the survival probabilities will depend on the deterministic creation and death rates. If there is not a growing mode in the deterministic limit, the survival probability in the stochastic case will be zero.

Making the same approximation used above, for a simple mth-order cycle the linearized kinetic equations are

$$\begin{aligned} \dot{\sigma_1} &= c_1\sigma_m - d\sigma_1 \\ \dot{\sigma_2} &= c_2\sigma_1 - d\sigma_2 \\ &\vdots \\ \dot{\sigma_m} &= c_m\sigma_{m-1} - d\sigma_m, \end{aligned} \tag{6}$$

where $c_i = \nu k_f A_i B_i$ are the catalyzed production rates[4] of the m species external to the metabolism whose concentrations are given by σ_i. The kinetics is described by the eigenvalues (and eigenvectors) of the matrix

$$\begin{pmatrix} -d & 0 & 0 & \cdots & c_1 \\ c_2 & -d & 0 & \cdots & 0 \\ \vdots & \vdots & \ddots & \vdots & \vdots \\ 0 & 0 & \cdots & c_m & -d \end{pmatrix} \tag{7}$$

The m eigenvalues turn out to be

$$\lambda_k = \Big(\prod_{j=1}^{m} c_j\Big)^{1/m} \exp(2\pi i \cdot k/m) - d, \quad k = 0, 1, \ldots, m-1. \tag{8}$$

Fluctuations will grow as long as the largest eigenvalue $\lambda_0 = (\prod_{j=1}^{m} c_j)^{1/m} - d$ is positive. This suggests that the probability for a fluctuation that generates a growing mth-order graph depends on the geometric mean of the catalytic production rates c_i. This has some interesting possible consequences: Since the mass of the system is constant, in order to increase the number of species, it must typically decrease the population of each species inside the autocatalytic metabolism. The growth rates c_i depend on the concentrations in the metabolisms; if these decrease, the analysis above indicates that the probability of survival also decreases. The system should thus evolve to a critical state in which the linear growth rate for a typical mutation is on average near zero.

[4] This form assumes that all the reactions are condensation reactions; for a cleavage reaction it is $c_i = \nu k_r H C_i$.

5. NUMERICAL EXPERIMENTS

In this section we present a few preliminary numerical experiments on the evolution of autocatalytic metabolisms. The purpose of the simulations presented at this stage is simply to demonstrate the basic principle behind the chemical evolution of autocatalytic mutations. The approach at this stage is strictly *ad hoc*: We ignore the important issue of the time between transitions from one pinned state to another, and make an arbitrary choice concerning the relative survival probabilities of autocatalytic subgraphs.[5] A simulation proceeds as follows:

1. Explore the internally catalyzed pathways by running the deterministic metadynamics algorithm until the system reaches a metadynamical fixed point.
2. Construct the allowed autocatalytic subgraphs of the shadow.
3. Assign a weight to each of the subgraphs.
4. Select a subgraph with probability given by its normalized weight.
5. Try to install the selected subgraph and check if its members acquire concentrations above threshold within the autocatalytic network.
6. If no, return to step 4.
7. If yes, install it in the network and return to step 1.

We call this procedure *stochastic metadynamics.*

A typical simulation is shown in Figure 3(a). We arbitrarily measure the time in "meta-steps," which correspond to the number of times the metadynamics algorithm finds new dynamical fixed points. Figure 3(a), for example, plots the number of polymer species above threshold, beginning with an initial state in which the only species with nonzero concentrations are those of the food set. The graph changes 12 times, corresponding to 12 different dynamical fixed points, before the system settles on a metadynamical fixed point with roughly 20 species. At this point we introduce a mutation, which triggers a series of new internally catalyzed pathways, until the system settles onto another metadynamical fixed point, roughly ten steps later. This process repeats itself until the number of species reaches 50 and the simulation terminates. During this time new species are created, and old species become extinct, as shown in Figure 3(b).

The "meta-step" time scale used in these simulations masks the physical correspondence to punctuated equilibria. In reality, after each mutation the system quickly reaches a new steady state, and then remains relatively unchanged for a much longer period of time, until the next mutation. To make a correspondence between the simulation and reality, at each time marked by a triangle in Figure 3, one should imagine an interval of indefinite duration during which the properties of the system remain almost constant.

[5]To assign relative survival probabilities, for the simulations shown here we weighted graphs according to the product of the production rates of their vertex elements. As we now know, this overemphasizes the importance of short graphs. We intend to repeat these simulations with the proper weighting function as described in section 4.3 in the future.

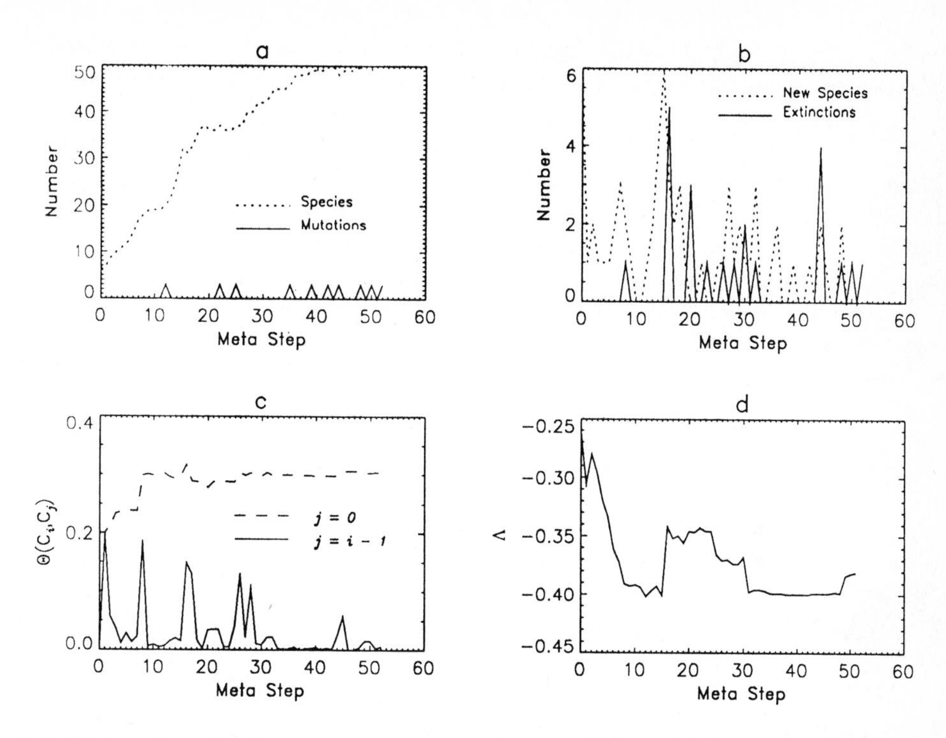

FIGURE 3 *The evolution of an autocatalytic metabolism.* (a) plots the number of polymer species as a function of time, measured in "meta-steps" (see text). The triangles on the horizontal axis indicate mutations that generate evolutionary modifications. (b) shows the number of new species created (dotted) and the number of old species that become extinct. (c) plots the difference in the angle of the concentration vector (see text) between the present step and the previous step (solid) or the initial condition (dashed). (d) The slope Λ is one indication of the deviation from equilibrium properties. The forward rate constant k_f, the reverse rate constant k_r, and the catalytic efficiency ν are randomly varied in the range $k_f \in [10^1, 10^2]$, $k_r \in [1, 10]$, $\nu \in [10^3, 10^7]$. The unbinding constant $k_u = 10^5$, the driving $\delta = 10^2$, the mass concentration $m_0 = 10^{-1}$, the concentration threshold $= 10^{-4}$, and the probability of catalysis $p = 4.5 \times 10^{-3}$. An upper limit on the number of polymers was set at 50; autocatalytic subgraphs were chosen from those with one or two nodes.

In order to measure the change of the state of the network in quantitative terms, in Figure 3(c) we use the angle $\Theta(C_i, C_j)$ between the concentration vectors C_i and C_j. The *concentration vector* C_i is defined as the infinite-dimensional vector whose coordinates are the concentrations of each possible species. C_i uniquely specifies the state of the system at meta-step i. Figure 3(c) demonstrates that the angle between concentration vectors changes more rapidly at the beginning, indicating that evolutionary modifications at later stages have a smaller effect on the autocatalytic metabolism.

In the companion paper we introduced Λ, the slope of the concentration profile, as a measure of the deviation of the properties of the system from those at equilibrium. Larger (less negative) values of Λ indicate a larger deviation from equilibrium. Λ decreases during the initial deterministic evolution; an evolutionary modification triggered by the first mutation increases Λ significantly for a while, but subsequent mutations cause it to decrease again. The values of Λ in this simulation indicate that none of these metabolisms are very robust; however, it is interesting to see that an evolutionary modification can make a significant change in Λ.

Figure 4 shows another sequence of evolutionary modifications. In this case we admit mutations involving subgraphs with up to three (rather than two) vertices, and use slightly different kinetic parameters, as well as a higher ceiling for the number of possible species.

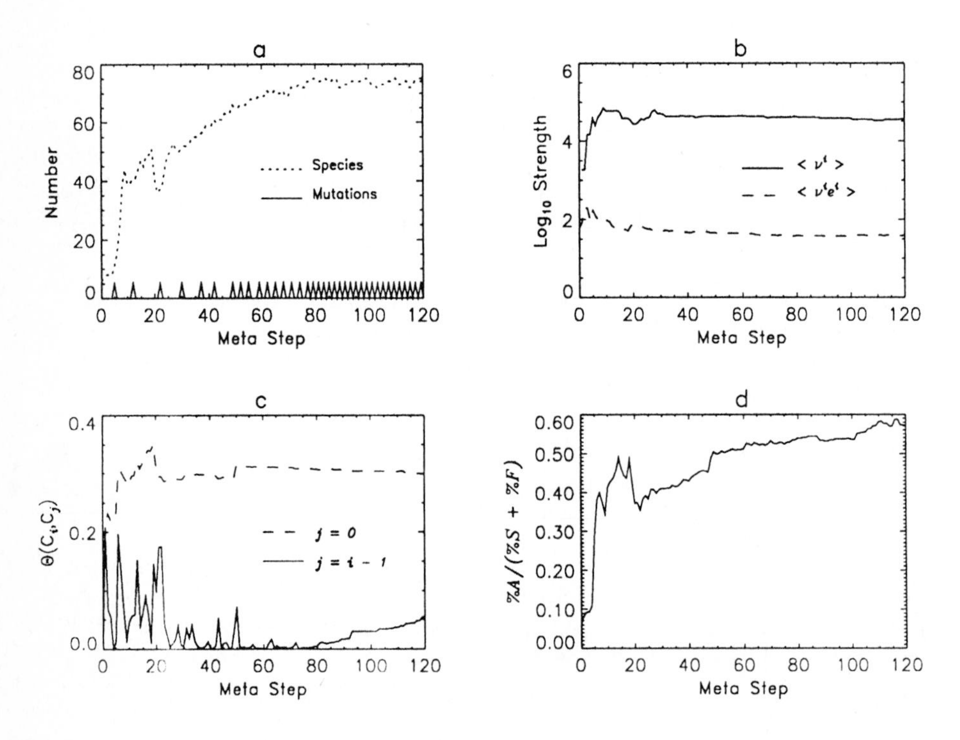

FIGURE 4 *A second simulation of an autocatalytic metabolism.* Figures (a) and (c) are similar to those of from those of Figure 3. (b) plots the mean catalytic efficiency efficiency $\langle \nu \rangle$ of the autocatalytic metabolism (solid) and the mean catalytic strength $\langle \nu e_i \rangle$, where e_i is the concentration of catalyst e. (d) plots the mass ratio, i.e., the ratio of the mass of the autocatalytic metabolism (without the food set) to that of the sum of the food set and the background. The parameters are the same as those of Figure 3, except that $k_u = 10^6$, the limit on the number of polymers was set at 75, and autocatalytic subgraphs were chosen from those with three vertices or less.

Figure 4(a) shows the number of species as a function of time; interestingly, some mutations cause the number of species (and hence the diversity) to decrease. Early mutations typically have a large effect, in that they require several steps of the metadynamics algorithm to reach a metadynamical fixed point and therefore must have generated several new internally catalyzed pathways and several new chemical species. Later mutations, however, only trigger a single metadynamical step. (They may generate multiple species and internal pathways, but the number is probably fewer). Somewhat surprisingly, in spite of this Figure 4(c) shows that the later mutations cause a larger change in the angle of the concentration vector. At this point we do not know how to interpret this.[6]

Figure 4(d) plots another measure of the deviation from equilibrium, the mass ratio of the autocatalytic metabolism relative to the sum of the food set and the background. Although there are several periods where the mass ratio decreases, there is an overall tendency for it to increase with time. Figure 4(b) shows the mean catalytic efficiency as a function of time. It rises to a maximum and then tends to decrease slightly; we do not understand why.

These numerical experiments are admittedly very preliminary. The parameters used were somewhat arbitrary, and did not generate very robust autocatalytic metabolisms. Nonetheless, they do demonstrate the basic principle that a metabolism can evolve well past its initial pinned state, and that the resulting set of evolutionary modifications can generate new molecular species, and cause old species to become extinct.

6. DISCUSSION

We have illustrated a process of "chemical evolution" that bears many similarities to biological evolution, at the same time that it is distinctly different. In this section we discuss this analogy in more detail.

There is a clear notion of variation and selection within autocatalytic metabolisms, and we feel that the term "evolution" is well deserved. Random variations, which play the role of mutations, are generated by spontaneous reactions. Some of these variations have no effect, and simply die out. Others have large effects, generating several new chemical species and perhaps causing others to die out, substantially altering the composition of the autocatalytic metabolism. As in evolutionary biology, "favorable" variations are by definition those that propagate themselves. Spontaneous fluctuations provide random variation, and chemical kinetics provide selection.

[6] Similar behavior has been observed by Rasmussen et al.[12]

We now develop the analogy with the evolution of biological organisms in more detail. A possible point of confusion concerns the level at which to make the identification. In the self-reproducing reactions studied by Eigen, the notion of "chemical species" can be roughly identified with "biological species." This is literally the case in the experiments with the $Q\beta$ virus. For autocatalytic metabolisms, however, the individual chemical species are only elementary building blocks; they are not in any sense alive on their own. The "organism" is the entire metabolism. The "phenotype" of the organism is the concentration vector C_i, i.e., the set of species and concentrations that comprise the autocatalytic metabolism.

We have demonstrated that the time history of an autocatalytic metabolisms contains periods where the system is pinned and change is very slow, and other periods where a fluctuation triggers the exploration of new internally catalyzed pathways and change is quite rapid. We propose that the chemical kinetics of internally catalyzed pathways can be regarded as a "developmental algorithm." The "genotype" is any list of chemical species that produce a given pinned state through purely internally catalyzed pathways. Note that there are many possible concentration levels and lists of chemical species that are connected by internally catalyzed pathways leading to the same pinned state. A subset of the metabolism can thus regenerate the entire metabolism. Such a subset has been called a *seeding set* by Fontana.[9] We regard the genotype of a given pinned state as the equivalence class of its seeding sets. Since the seeding sets will all regenerate the metabolism regardless of the concentrations of their elements, the genotype is given by a simple list, stating which species are present. It thus encodes information about the phenotype in symbolic form, in a manner analogous to the sequence of base pairs in a contemporary organism. The chemical kinetics produces the phenotype from the genotype, in a highly simplified but analogous manner to that of a contemporary organism.

As we have demonstrated here, autocatalytic metabolisms go through a progression of events, involving the alteration of the genotype and the generation of new phenotypes. Since the expression of the directly catalyzed pathways is rapid, while the time intervals between evolutionary modifications may be large, when viewed over an expanse of time, these have the appearance of "punctuated equilibria." This illustrates a difference between the evolution of autocatalytic metabolisms and that of biological organisms: the evolutionary process and the developmental process are one and the same. In this context a single organism can evolve—the selection comes from the laws of chemistry, without any obvious need for competition with other organisms.

This brings up the important issue of "identity." If an autocatalytic metabolism is clearly confined to a given container, then the genotype and phenotype of the autocatalytic metabolism make it possible to distinguish it from other metabolisms in other containers. The number of possible metabolisms is quite high, and the probability that two metabolisms in different containers will be in different pinned states is extremely high. Each metabolism might evolve on its own, without any need for competition with others.

However, one can also imagine that the vessels containing the metabolisms occasionally come into physical contact, so that some material diffuses from one to the other. In this case we can imagine a population of evolving metabolisms, each occasionally infecting the others with pieces of its own genotype. An infection may trigger new catalyzed pathways, and substantially change the evolution of the infected host.

One can also imagine a situation in which autocatalytic metabolisms are more continuously distributed through space, with ongoing diffusive coupling. In this case it becomes difficult to assign a notion of "identity." Nonetheless, there might be some interesting spatial inhomogeneity that fosters evolutionary behavior that is qualitatively different from that within a spatially isolated well-stirred reactor.

We have tried to suggest that autocatalytic metabolisms can be viewed as proto-organisms with a crude "genetic code," consisting of a list of polymers. The code is "interpreted" by chemical kinetics. We also suggest that autocatalytic metabolisms "evolve"; they certainly meet the basic requirements of information storage of a genotype and of a selection mechanism to amplify random variations. Nonetheless their evolution is distinctly "chemical" rather than biological. There are substantial differences, most of which stem from the fact that the identity of an individual autocatalytic metabolism is not as well defined as that of a biological organism or that of a templating RNA string; there is not necessarily a clear distinction between an organism and a population of organisms. However, we feel that these differences make this system more rather than less interesting. Many of the same questions that have eternally plagued evolutionary biology also surface here. In particular, how does the evolution of autocatalytic metabolisms differ from a random walk? Is there any sense in which they "make progress" as they evolve? We hope that future studies of autocatalytic metabolisms may shed some light on these questions, and help clarify how evolution applies in a broader sense.

ACKNOWLEDGMENTS

We would like to thank Steen Rasmussen and Chris Langton for critical readings of the manuscript.

REFERENCES

1. Bagley, R. J., and J. D. Farmer. "Spontaneous Emergence of a Metabolism." This volume.
2. Bagley, R. J., J. D. Farmer, S. A. Kauffman, N. H. Packard, A. S. Perelson, and I. M. Stadnyk. "Modeling Adaptive Biological Systems." *Biosystems* **23** (1989): 113–138.
3. Calvin, M. "Chemical Evolution and the Origin of Life." *Amer. Sci.* **44** (1956): 248–263.
4. Calvin, M. *Chemical Evolution.* Oxford: Oxford University Press, 1969.
5. Eigen, M. "Self-Organization of Matter and the Evolution of Biological Macromolecules." *Naturwissenschaften* **58** (1971): 465–523.
6. Eigen, M., and P. Schuster. "The Hypercycle: A Principle of Natural Self-Organization. Part A: The Emergence of the Hypercycle." *Naturwissenschaften* **64** (1977): 541–565.
7. Farmer, J. D., S. A. Kauffman, and N. H. Packard. "Autocatalytic Replication of Polymers." *Physica D* **22** (1986): 50–67.
8. Farmer, J. D., N. H. Packard, and A. S. Perelson. "The Immune System, Adaptation, and Machine Learning." *Physica D* **22** (1986): 187–204.
9. Fontana, W. "Algorithmic Chemistry: A Model for Functional Self-Organization." This volume.
10. Karlin, S., and H. M. Taylor. *A First Course in Stochastic Processes.* New York: Academic Press, 1975.
11. Miller, J. G. *Living Systems.* New York: McGraw Hill, 1978.
12. Rasmussen, S., C. Knudsen, and R. Feldberg. "Dynamics of Programmable Matter." This volume.
13. Rössler, O. "A System Theoretic Model of Biogenesis." *Z. Naturforschung* **B26b** (1971).
14. Rössler, O. "Design for Autonomous Chemical Growth Under Different Environmental Constraints." In *Progress in Theoretical Biology*, edited by R. Rosen and F.M. Snell. 1972.
15. Rössler, O. "Chemical Automata in Homogeneous and Reaction-Diffusion Kinetics." *Notes in Biomathematics* **B4** (1974).
16. Spencer, H. *First Principles.* 1862.

Walter Fontana
Theoretical Division, MS-B213; Center for Nonlinear Studies, MS-B258, Los Alamos National Laboratory, Los Alamos, New Mexico 87545 USA; and Santa Fe Institute, 1120 Canyon Road, Santa Fe, New Mexico 87501 USA

Algorithmic Chemistry

In this paper complex adaptive systems are defined by a loop in which objects encode functions that act on these objects. A model for this loop is presented. It uses a simple recursive formal language, derived from the λ-calculus, to provide a semantics that maps character strings into functions that manipulate symbols on strings. The interaction between two functions, or algorithms, is defined naturally within the language through function composition, and results in the production of a new function. An iterated map acting on sets of functions and a corresponding graph representation are defined. Their properties are useful to discuss the behavior of a fixed-size ensemble of randomly interacting functions. This "function gas," or "Turing gas," is studied under various conditions, and evolves cooperative interaction patterns of considerable intricacy. These patterns adapt under the influence of perturbations consisting in the addition of new random functions to the system. Different organizations emerge depending on the availability of self-replicators.

Artificial Life II, SFI Studies in the Sciences of Complexity, vol. X, edited by C. G. Langton, C. Taylor, J. D. Farmer, & S. Rasmussen, Addison-Wesley, 1991

While the logical structure of Darwinism seems secure, this should not be taken to imply that Darwinism, when expanded to encompass hierarchical considerations, will not be found to possess a mathematical structure previously unsuspected.

—Leo W. Buss[3]
"The Evolution of Individuality"

1. WHAT THIS IS ALL ABOUT

1.1 INNOVATION

The evolution of living systems involves, by definition, the notion of innovation. The appearance of novelty occurs at many scales ranging from societies, to individuals, to cells, to genes, to molecules. Objects at all these scales typically exhibit a high diversity of interactions. Molecules, for example, encode a variety of interaction properties ranging from kinetic parameters to qualitative relations as expressed in chemical reactions. Most importantly, on all levels the interactions are constructive, in the sense that they enable, directly or indirectly, the formation of new objects.

Introducing a new object into such systems is, therefore, equivalent to the introduction of a variety of new relations. The newly created object interacts with other objects that are present, and spawns further interactions involving its products. This can lead to dramatic changes in the organization of a system. Physics usually does not consider situations of this kind.

Mechanisms for the generation of new objects from available ones can be placed between two extremes. New objects may be constructed by virtue of intrinsic and specific properties of the interacting objects, or they may arise purely by noise.

Chemistry is a clear-cut example for the former situation. The formation of a new molecular product in a chemical reaction is, to a large extent, a deterministic function of the interacting molecules (and thermodynamic variables as temperature, pressure, magnetization, stress, volume, etc.). The chemical properties of the reacting molecules give rise to a specific product.

In contrast, the generation of a new object through noise is well represented by the phenomenon of mutation. The "intended" reaction is the faithful replication of a DNA or RNA string, but a chance event, like the absorption of a high-frequency photon, causes a copying error. The resulting string may represent a new object. The point, however, is that the causing event stands in no further relation to its effect. Clearly, in any real situation even the non-random formation of products will be subject to noise.

This contribution considers the (idealized) non-random case: systems in which interactions among objects generate specific other objects. This case deserves particular interest because it isolates a situation in which objects encode operations whose targets are the same objects. As a result, self-organization occurs in both the space of objects and the space of functions associated with these. In a suitable representation, in which the construction of new objects is practically unbounded (for example in chemistry), open-endedness might not only occur at the level of objects, but also—and most importantly—it can occur at the level of the relationships that arise among the objects. To put it with Max Delbrück[8]: "A mature physicist, acquainting himself for the first time with the problems of biology, is puzzled by the circumstance that there are no 'absolute phenomena' in biology. Everything is time-bound and space-bound." (Quotation taken from Mayr.[23])

1.2 FUNCTION

Physical objects that qualify as carriers of constructive interactions are far too complicated to model at present. Understanding from first principles how function is encoded into molecular objects is a major open problem, let alone how function is supported by supramolecular structures such as cells or cell-aggregates. In cases like societies and economies, it is even a major task to sensibly identify the various functional units.

How then can an innovative system based on non-random constructive interactions be modeled? How should an artificial world be defined that expresses in a transparent, tractable, and sufficient way

1. a combinatorial variety of structures, and
2. a mechanism by which these structures can manipulate each other?

A structure that manipulates another structure and outputs (uniquely) a further structure is a mathematical function. "Function" is a concept that is irreducible. It can be viewed in two ways:

1. A function as an applicative rule refers to the process—coded by a definition—of going from argument to value.
2. A function as a graph, refers to a set of ordered pairs such that, if $(x, y) \in f$ and if $(x, z) \in f$, then $y = z$.

The first view privileges the computational aspect of function. Mathematics provides, since 1936, through the works of Church,[5] Gödel,[13] and Turing,[33] a formalization of the intuitive notion of "effective procedure" in terms of a theory of a particular class of functions on the natural numbers: the partial recursive functions.

A formal system (like a computational language) secures the combinatorial variety of structures by a recursive definition of syntactically legal objects, and provides, through a few axioms, a semantics that defines the function, i.e., the manipulative part, associated with each object.

Church's λ-calculus[2,6] embodies in the most transparent way the basic ingredients for a simple and abstract model of a complex innovative system:

1. Functions are defined recursively in terms of other functions. This hierarchical construction is reflected on the syntactic level by defining legal objects as trees. This makes explicit that functions are "modular" objects, whose building blocks are again functions that can be freely recombined.

2. Objects in λ can serve both as arguments or as functions to be applied to these arguments.
3. There is no reference to any "machine" architecture.

It is remarkable that in the physical universe, so far, only the level of molecules has been observed to spontaneously support complex phenomena—as life—that are different, in kind, from the phenomena at all other levels. The intriguing feature at this level is the appearance of chemistry, eventually due to the distinguishing properties of the Coulomb force. The molecular level seems to be the first level in physics where a combinatorial variety of structures can "manipulate" each other in a way that is strikingly similar to "symbolic manipulation." It is at this level that the notion of "function" begins to emerge, and again in a strikingly similar sense as the intensional interpretation of function in mathematics: a term or an expression is given, and a function—a relation—can be associated with that expression beyond its literal meaning. Chemical systems, in sharp contrast to nuclear, atomic, or astronomic systems, generate a level at which a description in terms of functional interactions becomes more adequate than in terms of the fundamental forces involved.

1.3 ARTIFICIAL WORLDS BEYOND LOTKA-VOLTERRA

This contribution is intended to show that a model universe made up of "particles" that are functions in the sense of a (universal) formal language accomodates innovation in a very simple and straightforward way. Function composition induces a dynamics in the space of functions (see section 4). In addition, the interacting objects exhibit a time evolution in relative concentrations as a result of "mass" action kinetics. The interaction between these two dynamical systems—one concerning the evolution of a nonlinear dynamical system on a support, the other governing the change of that support—has been articulated in different contexts by Doyne Farmer,[11] Stuart Kauffman,[17] Richard Bagley,[1] Norman Packard,[27] Steen Rasmussen,[29] John Holland,[16] and probably directly or indirectly by many others as well.[24]

This section puts the nonlinear system into relation with previously studied equations of the Lotka-Volterra type. Suppose that we have an infinite population. This assumption makes the support-dynamics obsolete, but isolates the structure of the nonlinear system. Suppose further, that an object j interacts with $\alpha = 1, 2, \ldots$ and other objects $k, l, m, \ldots$ to produce an object i. Let $x_i(t) \in \mathbb{R}_0^+$ denote the frequency of i in the system at time t. Furthermore, let a_{ijk} be a coefficient that quantifies the change in i upon interaction of j with k. For example, a_{ijk} could be the probability by which i is produced given that an interaction occurs between j and k. In general, in $a_{ijklm\ldots z}$ the first index indicates the product object, and subsequent indices refer to the interacting objects. Then,

$$\dot{x}_i = \sum_{j,k} a_{ijk} x_j x_k + \sum_{j,k,l} a_{ijkl} x_j x_k x_l + \cdots - \Omega(t) x_i, \quad i = 1, 2, \ldots \tag{1}$$

The proportional dilution flow $\Omega(t)$ is chosen such that $\sum_i x_i(t) = 1$, in which case,

$$\Phi(t) = \sum_{i,j,k} a_{ijk} x_j x_k + \sum_{i,j,k,l} a_{ijkl} x_j x_k x_l + \cdots, \tag{2}$$

and the system is confined to a simplex. To remain as simple as possible objects can be restricted to binary interactions. Equation (1) then reduces to

$$\dot{x}_i = \sum_{j,k} a_{ijk} x_j x_k - x_i \sum_{r,s,t} a_{rst} x_s x_t, \quad i = 1, 2, \ldots. \tag{3}$$

An *ansatz* like Eq. (1) assumes mass action kinetics, and raises the question about the construction of objects that require the "simultaneous" interaction of many other objects (say, more than three). Can such a construction be broken up into a series of binary interactions? In other words, it is not obvious how third- and higher-order terms in Eq. (1) should be interpreted physically.

In the present work the objects are functions. A function particle $j(v_1, v_2, \ldots, v_\alpha)$ of α variables, then, interacts through function composition with α functions $k, l, m, \ldots$ (all of which with an arbitrary finite number of variables) to produce a function particle i (of some number β of variables), $i(v_1, v_2, \ldots, v_\beta) = j(k, l, m, \ldots)$. For the coefficients we have, $a_{ijk} = 1$ if j is a function of one variable and its action upon k (arbitrary finite number of variables) produces i, $i = j(k)$, and $a_{ijk} = 0$, otherwise. a_{ijkl} denotes the analogous production coefficient with j being a function of two variables acting upon k and l (in this order) to generate function i. In general, in $a_{ijklm\ldots z}$ the first index indicates the product object, the second index denotes the acting function, and subsequent indices refer to as many argument functions (of arbitrary finite number of variables) as the domain of definition of the acting function requires. Notice that since the objects are functions, uniqueness translates into

$$\sum_i a_{ijk} = 1, \sum_i a_{ijkl} = 1, \ldots, \tag{4}$$

i.e., a function j acting on argument k evaluates to a unique product i. It follows that $\Phi(t) = 1$, and considering only functions in one variable Eq. (3) simplifies further to[25]

$$\dot{x}_i = \sum_{j,k} a_{ijk} x_j x_k - x_i, \quad i = 1, 2, \ldots. \tag{5}$$

Systems consisting of objects that are copied by themselves (self-replicators), and/or by others arise in a variety of contexts, e.g., sociobiological game dynamics, ecology, economics, population genetics, and molecular evolution. The equation that represents the common thread in all those areas is the "replicator equation" (for an extensive survey see Hofbauer[15])

$$\dot{x}_i = x_i \Big(\sum_j a_{ij} x_j - \sum_{r,s} a_{rs} x_r x_s\Big), \quad i = 1, \ldots, n. \tag{6}$$

It has been shown[15] that a diffeomorphism converts the replicator equation in n variables into the Lotka-Volterra equation

$$\dot{y_i} = y_i(r_i + \sum_j a'_{ij} y_j) \quad i = 1, \ldots, n-1, \tag{7}$$

in $n-1$ variables on the positive orthant. Lotka-Volterra equations are widely used to model ecosystems.

Now observe that the replicator (or Lotka-Volterra) equation is a special case of Eq. (3), in which $k = i$, and all coefficients a_{rst} are of the form $a_{rsr} \equiv a_{rs}$. Since all i are replicated under the action of some j, $i = j(i)$, the first sum in Eq. (3) runs only over j, and, thus, becomes Eq. (6). Equation (3) is a generalization of the replicator equation, in that it drops the assumption that individual objects must be replicated.

Recently, Peter Stadler and Peter Schuster[32] have studied the replicator Eq. (6) including mutations. This means that object j copies object k, but makes an error with some probability distribution q_{ik}, thereby producing object i. The resulting equation is very similar to Eq. (3), but the coefficients a_{ijk} factorize into $a_{jk}q_{ik}$. The first factor describes the efficiency of the copy action of j upon k, the second factor describes the product. Notice that the interaction product does not depend on j, but only on i. It cannot depend on j, because the underlying assumption is that of a chance event representing an error in the reproduction of k: polymerases are not supposed to produce specific errors. This is not the case in Eq. (3), where the product i depends on both j and k. Stadler's and Schuster's system is one important example for the limiting case of a purely noise-induced production of new objects. The corollary is that, as noise tends to vanish the replicator Eq. (6) is regained, and, therefore, provides for a reference state that allows an analytical perturbation approach. The limiting case of a non-random production of new objects, as depicted by Eq. (3), has no reference state. This brings the above quotation from Delbrück back to one's mind.

Innovation requires a combinatorial variety of structures. In the case of computable functions over $\mathbb{N}$ this variety is countably infinite. In all finite-universe size limitations have to be imposed. But even very strong limitations cannot prevent the numbers involved to quickly exceed the material and temporal resources of many universes. The number of baryons in the universe is already matched by the number of variations of strings of length 80 over an alphabet of size 10. This calls for a stochastic description. The Turing gas described in section 5 is precisely a stochastic simulation of Eq. (5). Typical questions that arise are: Which sets of interaction patterns can coexist for how long under which conditions? How do once established interactions constrain the subsequent evolution of the system? How do mutually stabilizing interactions respond to perturbations consisting in the introduction of new interaction carriers? Does the motion of a finite system in interaction space exhibit attractors of the type featured by usual dynamical systems? How can cooperativity be characterized and classified?

The artificial worlds of Eq. (1) are specified by the coefficients $a_{ijk}, a_{ijkl}, \ldots$. These coefficients involve a concrete description of concepts like "object" and "interaction." The problem is to achieve a finite description of an infinity of possible structures that these objects might acquire. For most cases in chemistry, biology, or economy this is tantamount to a theory, or at least a model, of the appropriate entities.

The avenue taken here is to identify "object" with "function" and to use a theory of function, as Church's λ-calculus, to decide, i.e., to compute a_{ijk}, or equivalently to compute i, given j and k. This algorithmic toy-chemistry captures a constructively innovative system in a transparent way, but how does it connect to physics or biology? Stated differently, the question is: to what extent are pure functions the right objects to consider in biological systems?

A model-like the one proposed here, cannot provide much detailed information about a particular real complex system whose dynamics will highly depend on the physical realization of the objects as well as on the scheme by which the functions or interactions are encoded into these objects. The hope is that an abstraction cast purely in terms of functions, defines a level of description that enables a logical and mathematical characterization of patterns of physical organization. The conjecture is that a world of functions is indeed homomorphic to the real world. But still: how much—in the case of biological systems—can we abstract from the "hardware" until a theory loses any explanatory power? We don't know yet. This is tightly connected to deep problems of inference concerning artificial constructive worlds in general, and that go beyond the conventional problems in mathematical modeling. These issues are addressed by David Lane and John Holland within the economics and the adaptive computation programs at the Santa Fe Institute.[19]

1.4 ORGANIZATION OF THE PAPER

Section 2 summarizes the model; section 3 briefly describes the language used to provide the mapping from syntactically legal character strings to algorithms, i.e., functions, that operate on character strings. Section 4 introduces an iterated map acting on a set of functions and its graph representation. Some concepts are defined that are used in the discussion of computer experiments concerned with the behavior of an ensemble of functions that act upon each other under particular conditions: a "Turing gas," defined in section 5. Some results are presented and discussed in section 6. Section 7 summarizes, and section 8 concludes the paper with an outlook. A formal and detailed description of the computational language used to encode the functions is relegated to the appendices.

This contribution is a modified version of a paper submitted for publication to *Physica D*.[12]

2. THE MODEL

To set up a model that also provides a workbench for experimentation, a representation of functions along the lines of the λ-calculus is needed. The representation used here is a somewhat modified and extremely stripped-down version of a toy-model of pure LISP as defined in Chaitin.[4] In pure LISP a couple of functions are pre-defined (six in the present case). They represent primitive operations on trees (expressions), for example, joining trees or deleting subtrees. This speeds up and simplifies matters as compared to the λ-calculus, in which one starts from "absolute zero" using only application and substitution. For the sake of simplicity, only functions in one variable are considered. The language is briefly explained in section 3.

The model is built as follows:

1. Universe. A universe is defined through the λ-like language. The language specifies rules for building syntactically legal ("well-formed") objects and rules for interpreting these structures as functions. In this sense the language represents the "physics." Let the set of all objects be denoted by $\mathcal{F}$.
2. Interaction. Interaction among two objects, $f(x)$ and $g(x)$ is naturally induced by the language through function composition, $f(g(x))$. The evaluation of $f(g(x))$ results in a (possibly) new object $h(x)$. Interaction is clearly asymmetric. This can easily be repaired by symmetrizing. However, many objects like biological species or cell types (neurons, for example) interact in an asymmetric fashion. I chose to keep asymmetry.
   Note that "interaction" is just the name of a binary function $\phi(s,t)$ that sends any ordered pair of objects f and g into an object $h = \phi(f,g)$ representing the value of $f(g)$. More generally, $\phi(s,t) : \mathcal{F} \times \mathcal{F} \mapsto \mathcal{F}$ could be *any* computable function, not necessarily composition, although composition is the most natural choice. The point is that whatever the "interaction" function is chosen to be, it is itself evaluated according to the semantics of the language. Stated in terms of chemistry, it is the same chemistry that determines the properties of individual molecules and at the same time determines how two molecules interact.
3. Collision rule. While "interaction" is intrinsic to the universe as defined above, the collision rule is not. The collision rule specifies essentially three arbitrary aspects:

   a. What happens with f and g once they have interacted. These objects could be "used up," or they could be kept (information is not destroyed by its usage).

   b. What happens with the interaction product h. Some interactions produce objects that are bound to be inactive no matter with whom they collide. The so-called NIL function is such an object: it consists of an empty expression. Several other constructs have the same effect, like function expressions that happen to lack any occurrence of the variable. In general, such products are ignored, and the collision among f and g is then termed "elastic"; otherwise, it is termed "reactive."

c. Computational limits. Function evaluation need not halt. The computation of a value could lead to infinite recursions. To avoid this, recursion limits, as well as memory and real-time limitations, have to be imposed. A collision has to terminate within some pre-specified limits; otherwise, the "value" consists in whatever has been computed until the limits have been hit.

The collision rule is very useful for introducing boundary conditions. For example, every collision resulting in the copy of one of the collision partners might be ignored. The definition of the language is not changed at all, but identity functions would have now been prevented from appearing in the universe.
In the following, it will be implied that the interaction among two objects has been "filtered" by the collision rule. That is, the collision of f and g is represented by $\Phi(f,g)$ that returns $h = \phi(f,g)$ if the collision rule accepts h (see item (b) above); otherwise, the pair (f,g) is not in the domain of Φ.

4. System. To investigate what happens once an ensemble of interacting function "particles" is generated, a "system" has to be defined. The remaining sections will briefly consider two systems:

a. An iterated map acting on sets of functions. Let $\mathcal{P}$ be the power set, $2^{\mathcal{F}}$, of the set of all functions $\mathcal{F}$. Note that $\mathcal{F}$ is countable infinite, but $\mathcal{P}$ is uncountable. Let $\mathcal{A}_i$ denote subsets of $\mathcal{F}$, and let $\Phi[\mathcal{A}]$ denote the set of functions obtained by all $|\mathcal{A}|^2$ pair interactions (i.e., pair collisions) $\Phi(i,k)$ in $\mathcal{A}$, $\Phi[\mathcal{A}] = \{j : j = \Phi(i,k), (i,k) \in \mathcal{A} \times \mathcal{A}\}$. The map M is defined as

$$M : \mathcal{P} \mapsto \mathcal{P}, \; \mathcal{A}_{i+1} = \Phi[\mathcal{A}_i]. \quad (8)$$

Function composition induces a dynamics in the space of functions. This dynamics is captured by the above map M. An equivalent representation in terms of an interaction graph will be given in section 4.

b. A Turing gas. The Turing gas is a stochastic process that induces an additional dynamics over the nodes of an interaction graph. Stated informally, individual objects now acquire "concentrations" much like molecules in a test-tube mixture. However, the graph on which this process lives changes as reactive collisions occur. Section 6 will give a brief survey on experiments with the Turing gas.

3. THE LANGUAGE

The language used to express the algorithms is closely related to, so-called, pure LISP, and in particular to a toy version designed by Gregory Chaitin.[4] Using a λ-calculus type of language turns out to be critical: it is a functional, as opposed to a procedural, programming language.

The procedural programming mode is the standard approach in traditional von-Neumann languages like Pascal, FORTRAN, or C. In this mode a program is a step-by-step specification of an algorithm. Instructions are executed in order, and the state of a program at any point of its execution is determined by the values of various variables in use. Procedural programming uses elements like conditionals and iterations to control the execution flow. At any point during execution, a program will have many local variables and various control structures in use. It is precisely this property that constrains a procedural program syntactically and semantically, making it very difficult to plug in random pieces of code.

The basic idea of functional programming[22] is to specify an algorithm by nesting functions. A function manipulates a string of characters. At any time only one particular function is active, and the string that this function manipulates is called the "current expression." When the currently active function has terminated its operations, the current expression is passed to the calling function that continues the processing. Functions can call themselves. Due to this recursion, local variables, assignments, or references of any kind to intermediate storage are no longer needed. The state of the program is given at any time exactly by the current expression. Hence, there are no side effects, such as clashes between local and global variable identifiers. A new piece of program can be simply added by inserting an additional function that is applied to the current expression. Usually the functions are so-called "pure functions." They are not assigned specific names; the name of a pure function is the encoding character string itself.

The language, referred to as AlChemy (a shorthand for Algorithmic Chemistry), is extremely simple. A detailed definition of AlChemy is given in the appendices. The following paragraphs give a qualitative overview of syntax and semantics.

3.1 SYNTAX

The program strings of AlChemy are combinations of characters taken from a set C that includes right and left parentheses. All characters except the parentheses are called atoms.

A syntactically legal string must fulfill only one requirement: the number of left and right parentheses must balance for the first time at the end of the string.

Syntactically legal strings are called expressions. According to the above definition a single atom is an expression. The parentheses, zero in this case, balance after the single character. Expressions consisting of more than one atom must, therefore, begin with a left parenthesis and must terminate with a right parenthesis. Such an expression is often called a list. Inside a list, parentheses can group atoms together.

Groups delimited by matching parentheses are obviously expressions. Hence, expressions can consist of other expressions.

The definition of an expression is precisely the recursive definition of a tree. Every pair of matching parentheses is represented by an internal node, while every atom is represented by a terminal node or leaf. Figure 1 gives a graph interpretation of expressions as rooted, ordered trees. Ordered means that the relative order of the subtrees is important. The empty list, (), is treated like an atom, but consists of two characters (Figure 1). The set of legal objects in the universe of AlChemy is therefore the set of all trees. The parentheses are purely structural characters needed to encode a tree graph as a linear string that can be manipulated by a computer in a convenient way.

The length of an expression is the number of characters in that expression. The number of expressions of length n, E_n, is derived in Chaitin.[4] The asymptotic estimate, $E_n = |\mathcal{C}|^{-1/2}\,|\mathcal{C}|^n/(2n\sqrt{\pi\, n})$, with $|\mathcal{C}|$ denoting the cardinality of $\mathcal{C}$, equals almost the number of strings of length n, $|\mathcal{C}|^n$. This indicates that the syntactic constraints are rather weak, and they become more so as n increases.

((b c) (c d (a)) b (c (

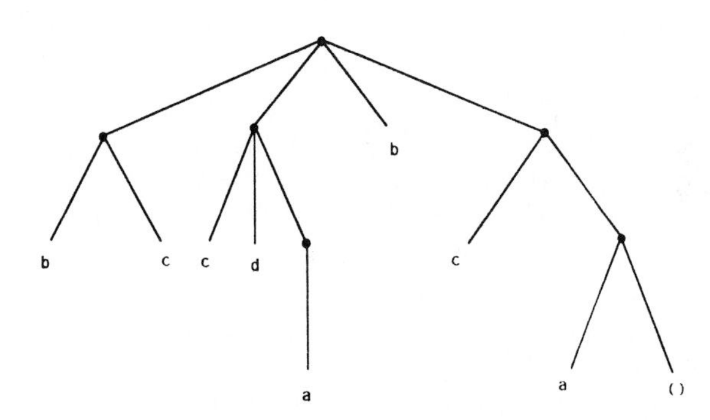

FIGURE 1 Expressions and trees. As in LISP, every AlChemy expression (top) can be represented as an ordered tree (bottom). A pre-order traversal reconstructs the list expression.

3.2 SEMANTICS

The semantics of a language determines what an expression "means." Expressions shall be used to represent programs, i.e., specifications of symbolic algorithms. A program A accepts an expression as input and returns an expression as output. Let $\mathcal{T}$ be the set of all trees (expressions), and let $\mathcal{X}$ be the subset of $\mathcal{T}$ for which A terminates. The meaning of A is then precisely the function $f : \mathcal{X} \subset \mathcal{T} \mapsto \mathcal{Y} \subset \mathcal{T}$, where $x \in \mathcal{X}$ is the input and $f(x) \in \mathcal{Y}$ is the output, or value, at termination. Clearly, different expressions (programs) can "mean" the same function. In order to provide a semantics, a set of rules has to be defined that specifies how expressions can manipulate symbol strings, thus encoding a function.

When constructing a function, a character symbolizing the variable has to be specified. The functions considered in this contribution will have only one variable. Let the character denoting this variable be "**a**."

In functional programming, functions are defined in terms of other functions. This is reflected on the syntactic level by the recursive definition of expressions being made of other expressions. The atoms are the base step in this recursion. A similar base step holds on the semantic level. At some point the recursive definition of a function must terminate. Functions that are no longer defined in terms of other functions are termed "primitive operators," or "primitives." The action of primitive operators is defined as a part of the semantics of the language. Primitives are "hard-wired," predefined operations acting on some specified number of argument expressions. They are assigned specific atoms of the alphabet $\mathcal{C}$ as names.

Since primitives operate on expressions, their actions consist in basic manipulations of tree structures. For example, joining two trees or deleting subtrees. The overall semantics of the language is defined in such a way that operators can easily be added, or their definition changed. The number of operators, as well as their actions, set the level of description in the model class that is being described.

The set of operators used in this version of AlChemy is limited to only six very simple ones whose functions are defined in Appendix B. All operators return legal expressions.

In AlChemy, an expression denotes a function, $f(\mathbf{a})$, and the value that f assigns to a particular argument r is the "value" of the expression f computed when the variable **a** is replaced by the expression r. The basic process of "evaluation" is defined recursively, and is outlined in the next paragraph. This process can be viewed as assigning a value to the root of the expression tree in terms of the values of its children. When the value of an expression has been obtained, it always replaces that expression.

The base step is, therefore, to assign a value to an atom. A terminal node must be an atom, and, hence, be either the variable **a** or a primitive operator. Operators shall always evaluate to themselves: their symbols are never substituted. The operator $*$, for example, has always value $*$. The value of the variable **a**, in contrast, is looked up in a list called the "association list." The association list is a look-up table where an expression is assigned to the atom **a**. If this list has no entry for **a**, then **a** is not substituted.

Assigning a value to some internal node of the expression tree involves using the values of its children in left to right order. Since these values are expressions, they denote functions that are to be applied to their arguments. Recall that there are two types of functions: primitive operators and composite functions. Accordingly, the "application" of a function to one or more arguments is performed in two ways.

If the function is an atom denoting an n-ary primitive operator, n siblings following the operator node are evaluated and their values taken as arguments. The built-in operation corresponding to the primitive is then applied to the argument expressions. If an operator encounters an insufficient number of arguments, an empty list, `()`, is supplied for each missing expression. Any excess arguments are ignored.

If the function is composite, the right neighbor sibling is evaluated and its value taken to be the argument of the (one-variable) function. If there is no sibling, an empty list is supplied. The procedure amounts to update the association list of that function, such that the argument expression is assigned to the variable `a`. The expression denoting the function is now evaluated using the new association list. The value expression obtained in this way replaces the original expression.

Figure 2 illustrates the evaluation process. The expression `((+a)(-a))` is evaluated using the association list that assigns the value `((*aa)(+a))` to the variable `a`. The interpretation process follows the tree structure until it reaches an atom. In this case it happens at depth 2. The atoms are evaluated: the operators "+" and "−" remain unchanged, while the value of `a` is looked up in the association list that assigns to it the value `((*aa)(+a))`. The interpreter backs up to compute the values of the nodes at the next higher level using the values of their children. The value of the left node at depth 1 is obtained by applying the unary "+"-operator to its sibling (which has already been evaluated). The "+" operation returns the first subtree of the argument, `(*aa)` in this case. Similarly, the value at the right depth 1 node is obtained by applying the unary "−"-operator to its argument `((*aa)(+a))`. The "−" operation deletes the first subtree of its argument returning the remainder, `(+a)`. The interpreter now has to assign a value to the top node. The left child's value is a non-atomic expression and, therefore, denotes a composite function. This function is `(*aa)`, labelled as f in Figure 2. Its argument is the right neighbor sibling, `(+a)`, labelled as g. Evaluating the top node means applying f to g. This is done by evaluating f with an association list that assign to `a` in f the value g. The procedure then recurs along a similar path as above, shown in the box of Figure 2. The result of $f(g)$ is the expression `((+a)(+a))`. This is the value of the root (level 0) of the original expression tree, and therefore the value of the whole expression, given the initial association list.

What happens if a node has more than two children whose values represent composite functions, as in $(f\ g\ h)$? In AlChemy the value of this expression is defined to be $(f(g)\ g(h))$: every function is applied to its right neighbor in turn and the results are appended to the value expression. If the application results in an empty list `()`, denoting the NIL function, the empty list is not appended. The last

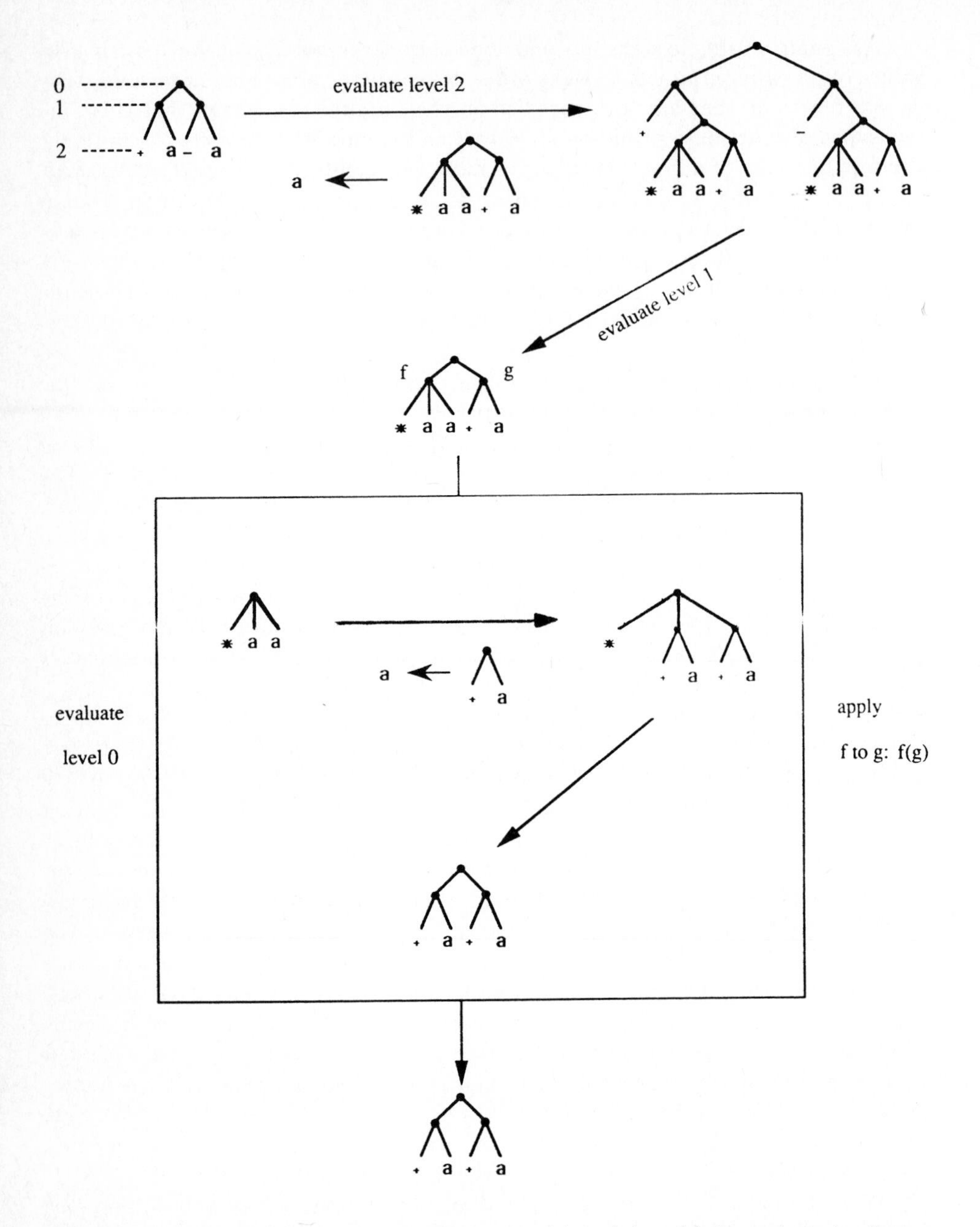

FIGURE 2 Evaluation example. The expression ((+a)(−a)) is evaluated using an association list that assigns the expression ((∗aa)(+a)) to the variable a. See text for details.

function h would be applied to the empty list supplied for the missing argument. This almost always results in the NIL function. The evaluation is, therefore, skipped *a priori*.

The resolution of such a situation is not standard. The value $(f(g(h)))$ would be more in the spirit of functional programming. However, many computer experiments clearly indicate that with the current set of operators a long chaining of functions evaluates too frequently to NIL. This is partly due to the fact that the majority of the current operators shorten expressions. The sequential scheme is a simple action against this effect.

The language outlined here has a few minor idiosyncrasies relative to toy-LISP as defined in Chaitin.[4] For more details see the appendices. A long series of experiments with variations in the semantics has been performed. All experiments gave results whose basic structure was essentially identical to those obtained with this version and reported in section 6.

3.3 INTERACTION BETWEEN FUNCTIONS

The hierarchical structure of functions implied a recursive evaluation process that relies on the application of subfunctions to other subfunctions. This results immediately in a natural definition of interaction.

Let f and g be expressions. The natural way to let them interact is to construct an expression whose value is $f(g)$ or $g(f)$, depending on who has been chosen to act on whom.

According to the semantics of section 3.2, the expression $(\, f \; g \,)$ is not suited, because its value is obtained by first evaluating the expressions f and g separately and then applying the value of f to the value of g. In order to avoid the evaluation of an expression, a primitive operator is defined whose action is to return the unevaluated argument expression. This operator is denoted by the symbol $'$, and is referred to as the "quote"-operator.

The correct "interaction expression" then must read

$$((' f)(' g)), \tag{9}$$

and is evaluated using an initially empty association list (a assigned to a). The values of $(' f)$ and $(' g)$ are f and g, respectively. The interaction expression then evaluates to $(\, f(g))$, which is the value that has been sought (see Figure 3).

Notice that the term "interaction" is here just the name of a syntactic expression with the particular structure (9), and is thus itself an object belonging to the language. More generally, let $\mathcal{F}$ be the set of functions defined by the language. "Interaction" is, then, the name of any two-variable function Φ—not necessarily composition—that assigns to any ordered pair of functions (f, g) some function h:

$$\Phi : \mathcal{F} \times \mathcal{F} \mapsto \mathcal{F}, \; (f,g) \to h \,. \tag{10}$$

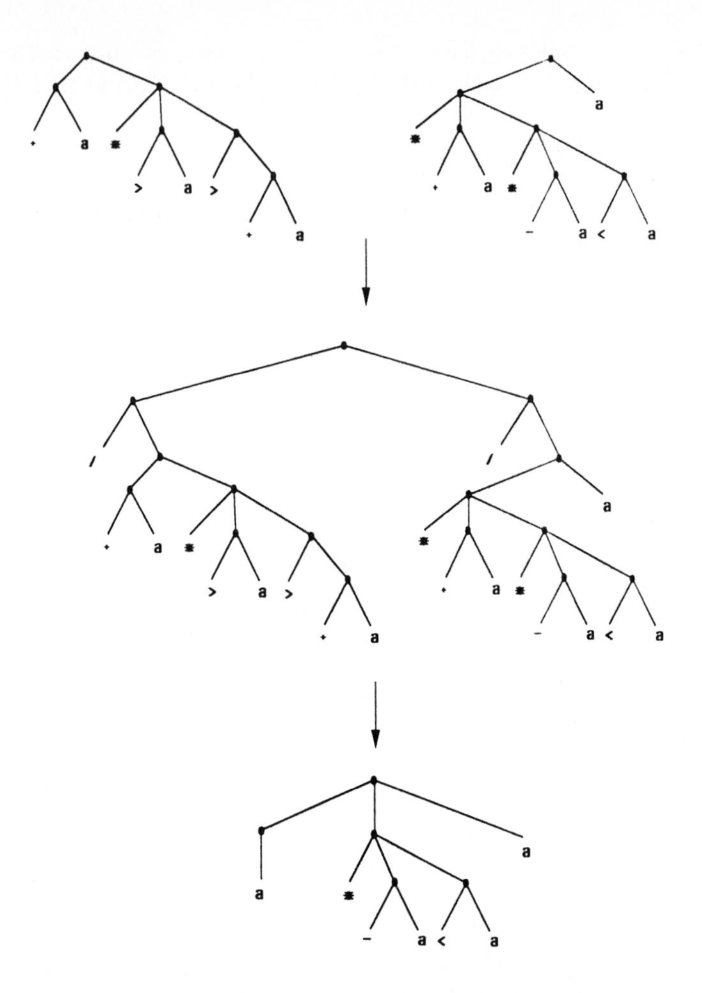

FIGURE 3 Interaction between algorithms. Two algorithmic strings (top) represented as trees interact by forming a new algorithmic string (middle) that corresponds to a function composition. The new root with its two branches and $'$-operators is the algorithmic notation for composing the functions. The interaction expression is evaluated according to the semantics of the language and produces an expression (bottom) that represents a new function. The evaluation of the interaction expression is derived in Appendix C.

The expression resulting from any particular Φ after insertion of particular expressions for f and g is clearly a function in one variable; see Figures 2 and 3. Nevertheless, on a meta-level, the set of possible interaction forms is the set of two-variable functions, of which Eq. (9), $\Phi(\mathtt{a},\mathtt{b}) = ((\mathtt{'a})(\mathtt{'b}))$, is but one, albeit natural, example. For reasons of simplicity the present language has been restricted to functions in one variable. The interaction form, though having two variables, obviously operates according to the semantics. When the number of variables is not limited, Φ is always an element of $\mathcal{F}$.

The same semantics that determines the function meant by any individual expression determines in particular the meaning of "interaction," for example, Eq. (9). The unity of function and interaction is a fundamental feature resulting from the completeness of this model class.

Figure 3 depicts the current interaction scheme. The evaluation of that particular interaction is derived step by step in Appendix C, which also serves as a more detailed interpretation example.

3.4 REMARKS

The above-described language is an extremely simplified version. The primitive operators (Appendix B) are not very powerful. More powerful primitive function constructors, as suggested for example in Arbib,[22] can be used.

The system allows sequences of primitive operations to be nested and grouped together into an expression. The value of such an expression is a function. In the present version a function acts precisely like a unary operator. It is, in contrast to a primitive operator, not hard-wired. It can be decomposed, reshuffled, or joined to something else. By means of interactions among individual functions, the system can, therefore, construct further "composite" operators. In addition, the system has not to provide names for its newly constructed operators. As in the λ-calculus, the specific sequence of characters coding for a function represents its name. There is no limitation, in principle, to the number of composite operators that a system can sustain.

The present semantics considers only functions in one variable. The extension to n variables is straightforward. A universe of multivariable functions leads to interesting additional questions due to n-"body" interactions. This theme is resumed in section 8.

The recursiveness of the evaluation process bears the danger of infinite loops. In Figure 2 the expression ((*aa)(+a)) is assigned to **a** at the beginning of the evaluation process. The reader can verify that replacing the + operator with **a** results in an infinite loop during the evaluation process shown inside the box of Figure 2. This is avoided by allocating to each interaction a depth limit. The depth limit specifies how many nested function evaluations are allowed to be incomplete at any given time. An evaluation that exceeds the depth limit stops gently by simply returning the value expression that has been computed so far (wrapped properly in parentheses). To avoid long cyclings between evaluation tree levels, an additional limitation had to be introduced: an interaction has to be completed within some maximum real cpu time. Similarly, only a limited amount of memory space is allowed for evaluating an interaction.

The interpreter is written in C, and is available from the author upon request.

4. ITERATED INTERACTION GRAPHS

The interactions between functions in a set $\mathcal{A}$ can be represented as a directed graph G. A graph G is defined by a set $V(G)$ of vertices, a set $E(G)$ of edges, and a relation of incidence, which associates with each edge two vertices (i, j). A directed graph, or digraph, has a direction associated with each edge. A labelled graph has, in addition, a label k assigned to each edge (i, j). The labelled edge is denoted by (i, j, k).

The action of function $k \in \mathcal{A}$ on function $i \in \mathcal{A}$ resulting in function $j \in \mathcal{A}$ is represented by a directed labelled edge (i, j, k):

$$(i, j, k): \; i \; k \rightarrow j \quad i, j, k \in \mathcal{A} \tag{11}$$

Note that the labels k are in $\mathcal{A}$. The relationships among functions in a set are then described by a graph G with vertex set $V(G) = \mathcal{A}$ and edge set $E(G) = \{(i, j, k) : j = k(i)\}$.

A useful alternative representation of an interaction is in terms of a "double-edge,"

$$(i, j, k): \; i \; (i, k) \longrightarrow j \; (i, k) \longleftarrow k \quad i, j, k \in \mathcal{A}, \tag{12}$$

where the function k acting on i and producing j has now been connected to j by an additional directed edge. The edges are still labelled, but no longer with an element of the vertex set. The labels (i, k) are required to uniquely reconstruct the edge set from a drawing of the graph. The graph corresponding to a given edge set is obviously uniquely specified. Suppose, however, that a function j is produced by two different interactions. The corresponding vertex j in the graph then has four inward edges. Uniquely reconstructing the edge set, or modifying the graph, for example by deleting a vertex, requires information about which pair of edges results from the same interaction. Some properties of the interaction graph can be obtained while ignoring the information provided by the edge labels. The representation in terms of double edges (i, j, k) has the advantage to be meaningful for any interaction function Φ mapping a pair of functions (i, k) to j, and not only for the particular Φ representing chaining. The double-edge suggests that both i as well as k are needed to produce j. In addition, the asymmetry of the interaction is relegated to the label: (i, k) implies an interaction $\Phi(i, k)$ as opposed to $\Phi(k, i)$. This representation is naturally extendable to n-ary interactions $\Phi(i_1, i_2, \ldots, i_n)$. In the binary case considered here every node in G must therefore have zero or an even number of incoming edges.

The following gives a precise definition of an interaction graph G. As in Eq. (8) let $\Phi[\mathcal{A}]$ denote the set of functions obtained by all possible pair collisions $\Phi(i, k)$ in $\mathcal{A}$, $\Phi[\mathcal{A}] = \{j : j = \Phi(i, k), (i, k) \in \mathcal{A} \times \mathcal{A}\}$. The interaction graph G of set $\mathcal{A}$ is defined by the vertex set

$$V(G) = \mathcal{A} \cup \Phi[\mathcal{A}] \tag{13}$$

and the edge set

$$E(G) = \{(i, j, k) : i, k \in \mathcal{A}, j = \Phi(i, k)\}. \tag{14}$$

The graph G is a function of $\mathcal{A}$ and Φ, $G[\mathcal{A}, \Phi]$. The action of the map

$$M : \mathcal{A}_{i+1} = \Phi[\mathcal{A}_i] \tag{15}$$

on a vertex set $\mathcal{A}_i$ leads to a graph representation of M. Let

$$G^{(i)}[\mathcal{A}, \Phi] := G[\Phi^i[\mathcal{A}], \Phi] \tag{16}$$

denote the ith iteration of the graph G starting with vertex set $\mathcal{A}$; $G^{(0)} = G$.

A graph G and its vertex set $V(G)$ are closed with respect to interaction, when

$$\Phi[V(G)] \subseteq V(G); \tag{17}$$

otherwise, G and $V(G)$ are termed innovative.

Consider again the map M, Eq. (15). What are the fixed points of $\Phi[\cdot]$? $\mathcal{A} = \Phi[\mathcal{A}]$ is equivalent to (a) $\mathcal{A}$ is closed with respect to interaction, and (b) the set $\mathcal{A}$ reproduces itself under interaction. That is,

$$\forall j \in \mathcal{A}, \exists\, i, k \in \mathcal{A} \text{ such that } j = \Phi(i, k). \tag{18}$$

Condition (18) states that all vertices of the interaction graph G have at least one inward edge (in fact, two or any even number). Such a self-maintaining set will also be termed "autocatalytic," following M. Eigen[9] and S. A. Kauffman[17,18] who recognized the relevance of such sets with respect to the self-organization of biological macromolecules.

Consider a set $\mathcal{F}_i$ for which Eq. (18) is still valid, but which is not closed with respect to interaction. $\mathcal{F}_{i+1}$ obviously contains $\mathcal{F}_i$, because of Eq. (18), and in addition it contains the set of new interaction products $\Phi[\mathcal{F}_i] \setminus \mathcal{F}_i$. These are clearly generated by interactions within $\mathcal{F}_i \in \Phi[\mathcal{F}_i]$. Therefore, Eq. (18) also holds for the set $\Phi[\mathcal{F}_i]$, implying that the set $\mathcal{F}_{i+1}$ is autocatalytic. Therefore, if $\mathcal{A}$ is autocatalytic, it follows that

$$G[\mathcal{A}, \Phi] \subseteq G^{(1)}[\mathcal{A}, \Phi] \subseteq G^{(2)}[\mathcal{A}, \Phi] \subseteq \ldots \subseteq G^{(i)}[\mathcal{A}, \Phi] \subseteq \ldots. \tag{19}$$

In the case of strict inclusion, let such a set be termed "autocatalytically self-extending." Such a set is a special case of innovation, in which

$$\Phi[V(G)] \supseteq V(G) \tag{20}$$

holds, with equality applying only at closure of the set.

An interesting concept arises in the context of finite, closed graphs. Consider, for example, the autocatalytic graph G in Figure 3(b), and assume that G is closed.

The autocatalytic subset of vertices $V_1 = \{A, B, D\}$ induces an interaction graph $G_1[V_1, \Phi]$. Clearly, $G[V, \Phi] = G_1^{(2)}[V_1, \Phi]$, which means that the autocatalytic set V_1 regenerates the set V in two iterations. This is not the case for the autocatalytic graph shown in Figure 3(a). More precisely, let G be a finite-interaction graph, and let $G_\alpha \subseteq G$ be termed a "seeding set" of G, if

$$\exists\, i, \text{ such that } G \subseteq G_\alpha^{(i)}, \tag{21}$$

where equality must hold if G is closed. Seeding sets turn out to be interesting for several reasons. For instance, in the next section a stochastic dynamics (Turing gas) will be induced over an interaction graph. If a system is described by a graph that contains a small seeding set, the system becomes less vulnerable to the accidental removal of functions. In particular cases a seeding set can even turn the set it seeds into a limit set of the process. Such a case arises when every individual function f_i in $\mathcal{A}$ is a seeding set of $\mathcal{A}$:

$$\begin{aligned} f_{i+1} &= \Phi(f_i, f_i), \qquad i = 1, 2, \ldots, n-1 \\ f_1 &= \Phi(f_n, f_n). \end{aligned} \tag{22}$$

Furthermore, suppose that G is finite, closed, and autocatalytic. It follows from the above that all seeding sets G_α must be autocatalytically self-extending, as for example in Figure 3(b). If G is finite, closed, but not autocatalytic, there can be no seeding set. Being closed and not autocatalytic implies $V(G^{(2)}) \subset V(G)$. The vertices of G that have no inward edges are lost irreversibly at each iteration. Therefore, for some i either $G^{(i)} = \emptyset$, or $G^{(i)}$ becomes an autocatalytic subset of G.

In the case of innovative, not autocatalytic sets, i.e., sets for which

$$\Phi[\mathcal{A}] \not\subseteq \mathcal{A} \wedge \Phi[\mathcal{A}] \not\supset \mathcal{A} \tag{23}$$

holds, no precise statement can be made at present.

A digraph is called connected if, for every pair of vertices i and j, there exists at least one directed path from i to j and at least one from j to i. An interaction graph G that is connected not only implies an autocatalytic vertex set, but in addition depicts a situation in which there are no "parasitic" subsets. A parasitic subset is a collection of vertices that has only incoming edges, like the single vertices C and E in Figure 3(b), or the set $\{C, E\}$ in Figure 3(a). As the name suggests, a parasitic subset is not cooperative, in the sense that it does not contribute to generate any functions outside of itself.

All the properties discussed in this section are independent of the information provided by the edge labels (in the double-edge representation). Note, furthermore, that the above discussion is independent of any particular representation of "function." It never refers to the implementation in the LISP-like AlChemy presented in section 3. The representation of function in terms of that particular language is used in the simulations of section 6 to demonstrate the accessibility of the phenomena outlined above, as well as to provide a workbench for experimentation.

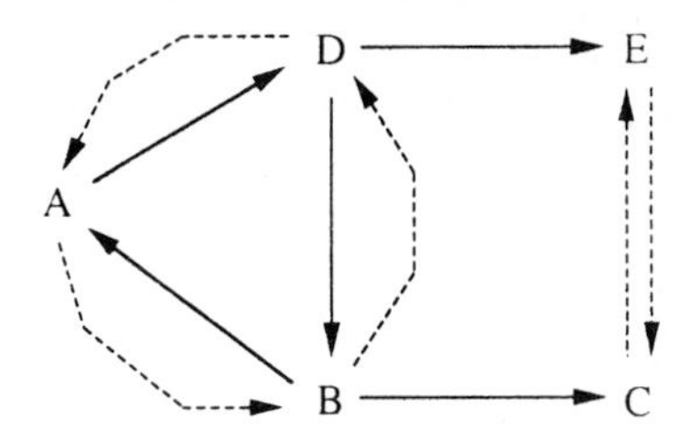

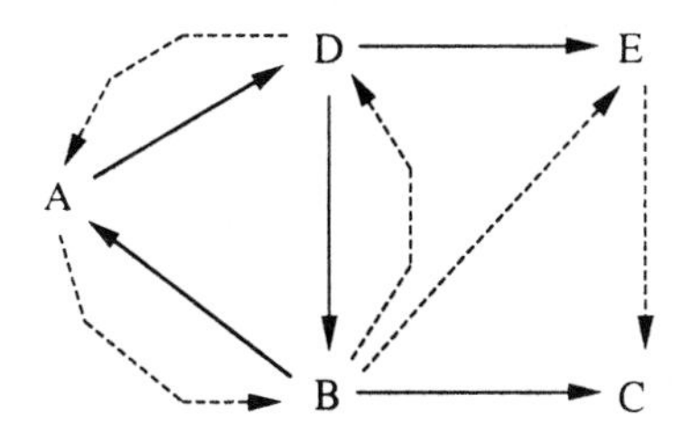

FIGURE 4 Self-maintaining (autocatalytic) graphs. The lower graph, (b), can be regenerated from the vertex subset $\{A, B, D\}$, in contrast to the upper graph, (a). Both contain parasitic subsets: $\{E, C\}$ in (a), and $\{C\}$, $\{E\}$ in (b). See text for details.

5. A TURING GAS

The previous section briefly considered the dynamics of relationships among objects in a set as they interact with each other on the basis of a formal language. These relationships are captured at any time by a graph. In physical instances the nodes of the graph could represent currently available interactive entities, as molecules, species, instructions, messages, etc. Usually these entities can be present in multiple copies. The nodes of the graph then support a frequency distribution that induces an interaction kinetics on the graph. The change in frequency of a particular object will depend on the frequencies of those objects that are needed for its production, as in Eq. (5).

In this section a simple stochastic process is used to induce a dynamical system on the interaction graphs. The size of the system is kept finite and constant. The constant size clearly represents a selection constraint that will influence the systems' evolution.

Let the system contain N functions. An iteration consists of two steps: a random collision between two objects and the application of a scheme to keep the total number of objects constant. The collision step is as follows:

1. Collision:

   a. choose at random two objects f and g from the system, and

   b. evaluate their interaction expression. This generates a new expression h. If the expression h (1) contains at least one primitive operator, (2) contains at least one variable, and (3) is not longer than l_{max} characters, then the collision is termed "reactive," and the reaction products are f, g, and h:

$$f + g \rightarrow (('f)('g)) \rightarrow h + f + g . \tag{24}$$

   If any of the above conditions is not fulfilled, then the collision is termed "elastic":

$$f + g \rightarrow (('f)('g)) \rightarrow f + g . \tag{25}$$

2. Removal: If the collision was reactive one of the old N functions is chosen at random and deleted. The reaction product h is, therefore, kept in any case.

The first chosen object, f, operates on the second object g. No conservation of any quantity is imposed during the collision. This keeps the scheme simple at first. The language definition itself provides for some constraints. For example, the product h cannot contain any primitive operator that was not already present in f or g. Keeping the interacting functions f and g after a reactive collision is in line with their abstract nature. f and g represent information that is used to build a new object. Information is not destroyed by the mere fact of its usage. Nevertheless, f and g are subject to the dilution flux—a random erasure—that is necessary to establish a finite system.

$l_{max} = 300$ in all examples discussed in section 6. In addition, to avoid halting problems the computation of a collision has not to exceed a depth of 10 (see section 3.4) and has to be completed within 6 seconds of real cpu time.

The scheme of constraining the number of particles is essentially equivalent to a flow reactor. The encounter between two object "species," as well as their dilution due to removal, occurs with a probability proportional to their frequency in the system.

The system is typically started with N random functions. A random function is a syntactically legal random string of characters obtained as outlined next.

In what follows, matching parentheses are always wrapped around an n-ary operator and its n-argument expressions. Generating an expression is then a very simple recursion R: (1) an atom is drawn at random; (2) if the atom is the variable, a complete (atomic) expression has been obtained and the procedure stops, else the atom is an n-ary operator that requires n-argument expressions that are generated according to procedure R.

The random-function generator has two parameters. One parameter concerns the probability by which an operator character is chosen. This allows to tune the frequency of operators versus variable. The above procedure, however, generates only trees with at most one primitive operator plus corresponding argument expressions at each level. Expressions could consist of any finite number k of branchings at any internal node. The second parameter tunes k, but only for the first level. To build an expression the above generator is simply invoked k times, and a pair of parentheses is wrapped around the k expressions.

The above procedure can access only a subset of all functions, in particular those that have operators associated with the "correct" predefined number of arguments. It seems natural to start with such functions, although any tree would be legal.

6. RESULTS AND DISCUSSION

In this section I discuss some of the basic results obtained in the first experiments with the Turing gas. Three variants will be considered. The first is a "plain" version, where the system self-organizes from random initial conditions into "quasi-stationary" (see below) interaction patterns. The second version slightly modifies the collision scheme such as to forbid copy reactions that are basic to the interaction patterns that develop in the plain version. Both versions keep the system closed. The third version opens the system at a quasi-stationary state and periodically perturbs the system by releasing a small percentage of new random functions into the gas.

6.1 TURING GAS WITHOUT PERTURBATIONS

The gas starts with $N = 1000$ randomly generated functions, each present in a single copy. Table 1 gives a glimpse of an initial condition. The random-function generator was instructed to generate function expressions containing four branches at the first level. Each expression was produced by drawing an operator with probability 0.7 and a variable with probability 0.3. Table 2 lists the state of the gas after 3×10^5 collisions, 93264 of which have been reactive. The 1000 particles are now distributed over only 18 different functions listed in lexicographic order with their corresponding number of copies. All functions are different from those present initially. Figure 5 shows the interaction graph G (section 4) obtained by performing all 324 pairwise collisions. The origin of a dotted line connecting to a solid arrow indicates the function that transforms the tail of the arrow into the head. This representation deviates slightly from the definitions in section 4, but has been chosen for the sake of clearer and less congested figures. Functions are usually "named" by their lexicographic order in the set under discussion. Capital letters denote sets of functions. Sometimes several syntactically related functions perform similar operations on different but again syntactically related arguments. This can be conveniently represented by having arrows and dotted lines connect sets instead of single functions. A set C transforming a set A into a set B then means

$$\forall j \in B \; \exists! \; i \in C, k \in A \text{ such that } j = i(k). \tag{26}$$

Inspection of the pair interaction data shows that function 17 is an identity function, or "general copier." An identity function is a function f with $f(g) = g, \forall g$. In some simulation experiments several identity functions are produced that copy themselves and each other.

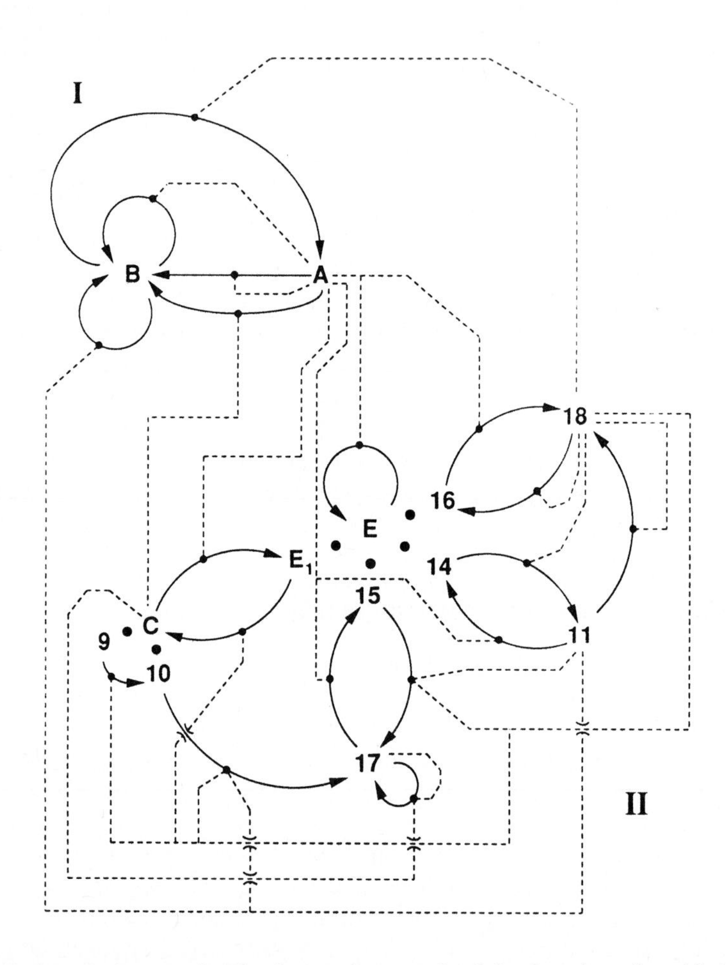

FIGURE 5 Interaction graph. The interaction graph of the functions listed in Table 2 is shown. The numbers denote the individual functions according to their ordering in Table 2. Capital letters denote sets, where $A = \{1,2,3,4\}$, $B = \{5,6,7,8\}$, $C = \{9,10\}$, $E = \{12,13,14,15,16\}$, and $E_1 = \{12,13\} \in E$. Solid arrows indicate transformations, dotted lines functional couplings. A dotted line originates in a function (or a set, see text), say, k, and connects (filled circle) to a solid arrow, whose head is j and whose tail is i. This is to be interpreted as $j = k(i)$. Large filled circles indicate membership in a particular set. Function 17 is an identity function. Note: all dotted lines and solid arrows that result from 17 copying everything else in addition to itself have been omitted. See text for details.

Note that, in a universe defined by a language, there are two senses in which things can be equal. One sense refers to the semantical level: two objects f and g are equal if they represent the same function, i.e., if (1) they have the same domain of definition $\mathcal{D}$ ($\mathcal{D}$ is the set of arguments h for which $f(h)$ terminates), and if (2) $f(h) = g(h), \forall h \in \mathcal{D}$. The other sense refers to the syntactical level: two objects are equal if their symbol strings are equal. Throughout this paper equality refers to the syntactical level. Several objects may represent the identity function, but differ in their symbolic representation. The identity function, for example, can be constructed in multiple ways by encoding first a series of operations to be performed on an input string, and then a second series that "undoes" the first one. Semantical (functional) equality can be very hard to establish. In fact, a general procedure for establishing functional equivalence between any two objects would run into the halting problem.

The drawing of the interaction graph in Figure 5 is incomplete, because it should contain for each function one solid self-loop with a dotted line coming from the general copier 17. These interactions were left out for the sake of a clearer picture. The reader is asked to keep in mind that all functions in Figure 5 are copied by 17, not only 17 itself.

TABLE 1 Random functions. The table shows randomly generated functions containing four expressions at the first tree level. The probability of drawing an operator was 0.7, the probability of chosing a variable was 0.3. The parentheses are completely determined by the sequence of operators. Thus only a subset of legal expressions is generated: expressions whose operators are bound to a complete set of arguments.

```
((>a)a(>(*aa))a)
((>('a))(*aa)(-a)a)
((<('a))(-(*a(-(<(+a)))))(<('(-(<a))))(-a))
((-(*(*a(<(<(<(>a))))))a))aaa)
(('('('(+a))))a('(-(*(+('(>('a))))(*a(>a)))))(*(<a)a))
((+a)a('a)(-(>(+(*(>(+(*(<a)(-a))))(>a)))))))
(('('('(-a))))a(-(>a))(>a))
((*(*aa)a)(*(-(<(<(-a))))a)(>(*(*(-(-(+(>('a)))))(-a))(-('a))))(<a))
((*(*(>(-a))a)(<(<(-a))))aaa)
(('('(*(+a)(>(+(-('(+a))))))))(+(+a))(+(<(<(-(<a)))))a)
(a(-a)('('(*(-a)a)))(<('(<(>a)))))
(aa(-(+(*(+(<a))('a))))(+(+(-a))))
```

TABLE 2 State of an unperturbed Turing gas. The table lists the state of a Turing gas with $N = 1000$ particles after 3×10^5 collisions. First column: lexicographic order of the function. This number is the "name" used in the text to refer to a particular function. Second column: – marks indicate functions that disappear during the following 2×10^5 collisions. Third column: number of copies. Fourth column: function expression. See Figure 5 for the interaction graph and text for the details.

(# 1)	–	14	(((’a)(a))(’a)(a))
(# 2)		43	((((’a)(a))(’a)(a))((’a)(a))(’a)(a))
(# 3)	–	16	(((((>(’(>a)))(>(’(>a))))(a))((>(’(>a)))(>(’(>a))))(a))(((>(’(>a)))(>(’(>a))))(a))((>(’(>a)))(>(’(>a))))(a))
(# 4)		90	((((>(’(>a)))(>(’(>a))))(a))((>(’(>a)))(>(’(>a))))(a))
(# 5)		33	(((a)(a))(((’a)(a))(’a)(a)))
(# 6)		133	(((a)(a))((((’a)(a))(’a)(a))((’a)(a))(’a)(a)))
(# 7)	–	36	(((a)(a))(((((>(’(>a)))(>(’(>a))))(a))((>(’(>a)))(>(’(>a))))(a))(((>(’(>a)))(>(’(>a))))(a))((>(’(>a)))(>(’(>a))))(a)))
(# 8)		205	(((a)(a))((((>(’(>a)))(>(’(>a))))(a))((>(’(>a)))(>(’(>a))))(a)))
(# 9)	–	10	((*a)((*a)(*a)))
(# 10)		7	((*a)(*a))
(# 11)		1	((>a)(>a))
(# 12)	–	37	((a)((*a)((*a)(*a))))
(# 13)		5	((a)((*a)(*a)))
(# 14)		7	((a)((>a)(>a)))
(# 15)		40	((a)(*a))
(# 16)		220	((a)(>a))
(# 17)		11	(*a)
(# 18)		92	(>a)

The component analysis[30] of the interaction graph G shows that G is connected. This implies that G is (1) self-reproducing (autocatalytic), (2) closed, and (3) has no parasitic subsets.

The constant system size represents a simple selective constraint. The only way for a particular function to survive in the system, in the long run, consists in becoming the product of some transformation pathway. The fate of that function is then linked to the functions in that pathway. To survive a pathway has to become closed, that is, self-maintaining. One (trivial) solution consists in a function that copies itself. Other solutions consist in sets that reproduce themselves without any single member being self-reproducing,[1,11,17] and any combinations of self-reproducing sets and/or single functions. These solutions are precisely the fixed points of the iterated interaction map, Eq. (15). The stability of such self-reproducing sets is strongly influenced by the number and size of constituent seeding sets (section 4). Stochastic fluctuations continously wipe out functions. The smaller the minimal seeding set of a self-reproducing set, the higher its stability, because correspondingly large numbers of different functions that have been lost can be regenerated.

The Turing gas is a stochastic process whose fluctuations eventually drive the system into three types of absorbing barriers. (1) A possibly heterogeneous mixture of elastic colliders ("dead system"), (2) a single self-reproducing function, or (3) a self-reproducing set in which every single function is a seeding set. The latter is a subtle "steady state," although a rather contrieved one, since seeding sets are usually much bigger than a single function. In the unperturbed version most interesting situations are, therefore, confined to transients, in particular long-living transients. Such long-living transients will be called "quasi-stationary" states.

At 3×10^5 collisions the interaction graph G is closed. Initially all reactive collisions result in a new function. The fraction of innovative collisions (relative to reactive encounters) decays very fast during the first $30,000$ collisions from a value of 1.0 to values fluctuating between 0.05 and 0.2. This range is kept for $80,000$ further collisions, and drops, then, to zero as the system attains closure. It is important to distinguish between collisions that are innovative in the sense that the produced function is not present in the system at the time of its production, and collisions that produce functions that have never been in the system during its entire history. Collisions of the latter type are termed "absolutely" innovative (they are included in the count of innovative collisions). Fluctuations that wipe out lowly populated functions that are subsequently regenerated, or any retracing of past trajectories in function space, give rise to innovative collisions that generate functions that the system has already seen. In fact, in the present experiment the fraction of absolutely innovative collisions follows the decay of innovative collisions, but settles on values between 0.02 and 0.1 before eventually dropping to zero. A similar scheme—fast decay to a ratio of approximately 0.5 to 0.3 of absolute innovation versus total innovation for variably long periods of time—is observed in many computer experiments. The quasi-stationary state at termination of some simulations is closed with respect to interaction, while others (section 6.2) continue to be innovative. So far, none have been observed to remain absolutely innovative for over a million collisions. Nevertheless, initial conditions for which this may happen cannot be excluded in principle. Dependencies on the system size have not yet been systematically investigated.

The trend towards closure is based on the appearance of identity functions and partial copiers. The latter are functions that copy some but not all arguments. As soon as an identity function becomes the end product of a pathway, the members of that pathway will be generated in an autocatalytic fashion, since they are copied by their joint end product. These pathways subsequently extend themselves through innovative collisions (section 4). Functions not linked to these pathways are eventually displaced by dilution.

The graph G (Figure 5) exhibits two groups, I and II, of functions that are not connected with each other by solid arrows. Any function of a group can be transformed only into functions belonging to the same group. The groups are connected solely by functional edges (dotted lines): objects in one group operate some interconversions between objects of the other group.

If all connections between both groups were cut, for example by introducing a boundary, group II would still remain autocatalytic (in fact, connected) due to the action of the identity function 17 belonging to group II. Group I would eventually attain a state of pure elastic colliders belonging to set B. If Figure 5 were taken literally, i.e., if function 17 were not an identity but a pure self-replicator, the couple $\{15, 17\}$ would act as a parasite to the system. Its removal would leave the system connected. "Removal experiments" on an interaction graph can easily be performed by deleting particular nodes along with all those edges having them as tail or as head. For example, removing all nodes except $\{9, 10, 11, 15, 17\}$ shows that this subset is still connected, implying autocatalysis. A given self-maintaining set can be composed of several other self-maintaining sets: in Figure 5, for example, the whole system is self-maintaining, group II stand alone is self-maintaining, the subset $\{9, 10, 11, 15, 17\}$ is self-maintaining, and obviously function 17 is by itself. The nesting of autocatalytic components is a frequently observed pattern in the Turing gas. Connectivity of the whole set, and thus closure, is more readily attained in the presence of general copiers.

The interaction graph G', 10^5 collisions earlier, consisted of 39 functions forming a self-reproducing, but innovative set. The graph G is already contained in G'. G' is not connected, and therefore contains parasitic functions. In 10^5 collisions G' has been reduced to G.

Most of the computer experiments exhibit very complicated short-lived states that reduce to simpler cooperative metastable transients as in Figure 5. Figure 6 shows, as another example, the interaction graph of a different 1000 particle gas after half a million collisions. The corresponding functions are listed in Table 3. This example exhibits "partial copiers," i.e., functions that copy some, but not all, functions in the system. Depending on the argument function other transformations are performed. In this case, functions 1 and 4 are self-replicators. In addition, 1 also copies 4, but 4 does not copy 1. Instead, 4 copies the function 3 that is transformed back into 1 under the action of 3 itself and, of 1. 4, cooperates through this pathway indirectly with 1. The graph is innovative; the function denoted by a star is not in the basis set $\{1, 2, 3, 4\}$.

Partial copiers are an obvious example of how an interaction not only depends on the acting function, but also on the properties of the argument. The acting function can instruct some parts of the argument function to operate on other parts of itself. The fact that given some acting function particular reactions need particular arguments is analogous to recognition phenomena in molecular systems.

TABLE 3 State of an unperturbed Turing gas. The table lists the functions of a quasi-stationary state obtained in a Turing gas with $N = 1000$ particles after 5×10^5 collisions. The initial set of random functions was different than in the simulation of Table 2. See text for details.

(# 1)	785	((>(*(*('(>a))(+(+('a)))a))(*(>(-a))(>(*(*('(>a))(+(+('a)))a))))
(# 2)	4	((a)(>(*(*('(>a))(+(+('a)))a)))
(# 3)	98	(*(>(-a))(>(*(*('(>a))(+(+('a)))a)))
(# 4)	113	(>(*(*('(>a))(+(+('a)))a))

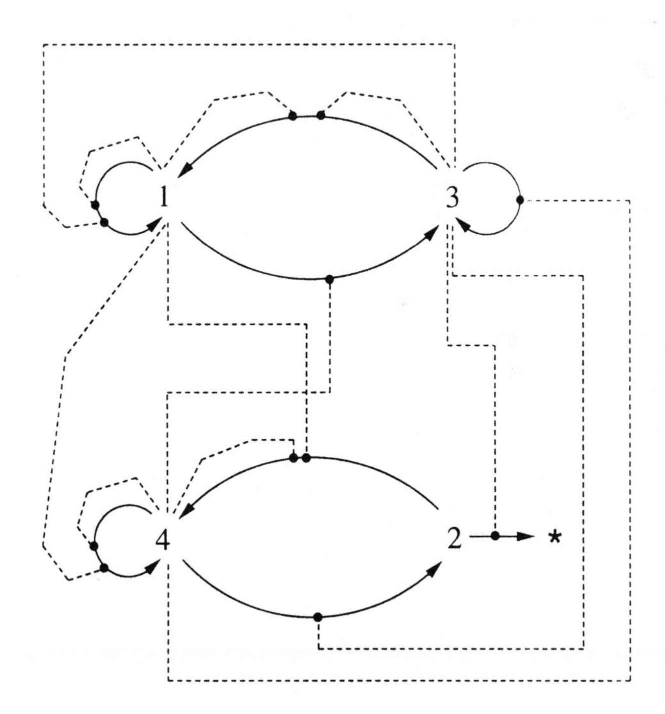

FIGURE 6 Interaction graph. The interaction graph of the functions listed in Table 3 is shown. See caption to Figure 5 and text for explanations.

6.2 TURING GAS WITHOUT COPY REACTIONS

The previous examples show that the organization of interactions in the Turing gas is centered at copy functions, be they general or partial. How does the system self-organize if copy reactions are not possible? This section considers this question as an example of a boundary condition imposed on the system through the collision rule.

The collision rule (section 5) is modified by simply declaring every copy reaction as being elastic. Note that AlChemy is left unchanged, and so is the definition of interaction.

One example only is discussed here as a prototype for many simulations. The 1000 different functions present initially have disappeared after 4×10^4 collisions. The similarity of two states at different times can be quickly assessed by computing the angle between the corresponding state vectors. The cosine of the angle between successive state vectors 5000 collisions apart from each other approaches values ranging between 0.97 and 0.99, as soon as the system reaches a quasi-stationary regime after approximately 2×10^5 collisions. At the same time, however, almost 50% of all the different function change constantly between each successive state. The innovation rate remains very high, but the absolute innovation rate becomes effectively zero after 3×10^5 collisions. The number of different functions fluctuates between 35 and 54. These facts indicate that the system steadily produces a fraction of new functions, loses them, and regenerates them again.

A snapshot of the gas after half a million collisions contains a set $\mathcal{A}$ of 51 different functions, the major part of which are too long and unwieldy to display. They are, however, built according to a simple scheme outlined below. The graph analysis reveals that the set is innovative and self-maintaining after removal of one function. This function has no inward edge, indicating again large fluctuations that led to the loss of its production pathway. Removing all edges pointing to innovative products, i.e., keeping only edges in the bulk $\mathcal{A}\backslash\Phi[\mathcal{A}]$, produces a graph that is connected, therefore self-maintaining and without parasites. How can a network be self-maintaining while constantly changing half of its functional species? The key to this consists in finding the minimal seeding set. The combinatorics makes this usually a difficult task. A simple heuristic that worked in the present case is to record a long series of n sets, $\mathcal{F}_i$, corresponding to states taken at particular time intervals, and to analyse their intersection set $\mathcal{I}$,

$$\mathcal{I} = \bigcap_{i=1}^{n} \mathcal{F}_i. \tag{27}$$

The intersection of 51 sets beginning at collision 2×10^5 and taken at intervals of 5000, contains 18 functions whose interaction graph is self-maintaining and innovative. The functions are listed in Table 4; their relationships are very simple and sketched in Figure 7.

TABLE 4 Turing gas without copy reactions. The table shows a quasi-stationary state of a Turing gas in which copy reactions were not allowed. The state consists of three polymer families A, B, and C, built from monomers $A1$, $B1$, and $C1$, respectively. The interaction graphs for subsets A and B are shown in Figure 7. The ordering of the functions is not lexicographic. See text for details.

(# A1)	41	(*(*aa)a)
(# A2)	69	((*(*aa)a)(*(*aa)a))
(# A3)	56	(((*(*aa)a)(*(*aa)a))(*(*aa)a))
(# A4)	20	((*(*aa)a)(((*(*aa)a)(*(*aa)a))(*(*aa)a))((*(*aa)a)(*(*aa)a)))
(# A5)	4	((((*(*aa)a)(*(*aa)a))(*(*aa)a)(*(*aa)a))(*(*aa)a)(*(*aa)a))
(# A6)	19	(((((*(*aa)a)(*(*aa)a))(*(*aa)a))((*(*aa)a)(*(*aa)a))(*(*aa)a))((*(*aa)a)(*(*aa)a))(*(*aa)a))
(# B1)	183	(<(*(>(*aa))(+a)))
(# B2)	115	((<(*(>(*aa))(+a)))(<(*(>(*aa))(+a))))
(# B3)	35	((<(*(>(*aa))(+a)))((<(*(>(*aa))(+a)))(<(*(>(*aa))(+a)))))
(# B4)	22	(((<(*(>(*aa))(+a)))(<(*(>(*aa))(+a))))(<(*(>(*aa))(+a))))
(# B5)	9	((<(*(>(*aa))(+a)))((<(*(>(*aa))(+a)))(<(*(>(*aa))(+a))))(<(*(>(*aa))(+a))))
(# B6)	13	((((<(*(>(*aa))(+a)))(<(*(>(*aa))(+a))))(<(*(>(*aa))(+a)))(<(*(>(*aa))(+a))))(<(*(>(*aa))(+a)))(<(*(>(*aa))(+a))))
(# C1)	182	(*(<(*(>(*aa))(+a))))
(# C2)	108	((*(<(*(>(*aa))(+a))))(*(<(*(>(*aa))(+a)))))
(# C3)	27	((*(<(*(>(*aa))(+a))))((*(<(*(>(*aa))(+a))))(*(<(*(>(*aa))(+a))))))
(# C4)	28	(((*(<(*(>(*aa))(+a))))(*(<(*(>(*aa))(+a)))))(*(<(*(>(*aa))(+a)))))
(# C5)	3	((*(<(*(>(*aa))(+a))))((*(<(*(>(*aa))(+a))))(*(<(*(>(*aa))(+a)))))(*(<(*(>(*aa))(+a)))))
(# C6)	8	((((*(<(*(>(*aa))(+a))))(*(<(*(>(*aa))(+a)))))(*(<(*(>(*aa))(+a))))(*(<(*(>(*aa))(+a)))))(*(<(*(>(*aa))(+a))))(*(<(*(>(*aa))(+a)))))

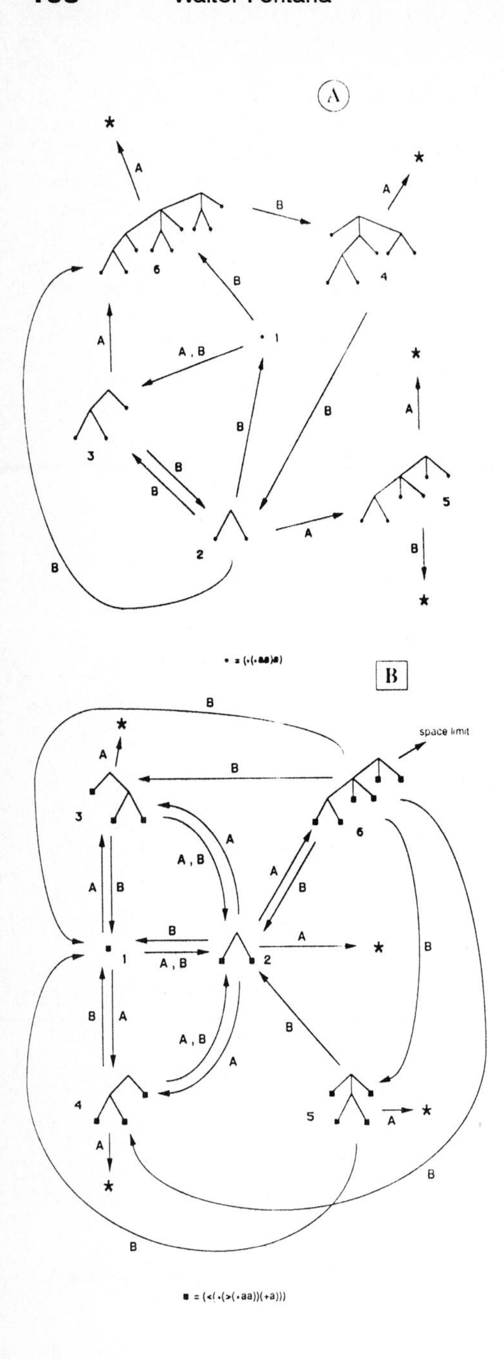

FIGURE 7 Interaction graph of a state without copy reactions. The interaction graphs of the A-subset, 7(a), and of the B-subset, 7(b), of Table 4 are shown. The functions are displayed as trees whose leaves are monomers given by the functional group indicated at the bottom of each graph. Solid arrows indicate transformations operated by some function(s) belonging to the subset(s) indicated by the label(s) of the arrows. Numbers as in Table 4.

The basic observation is that the functions become "polymers," and come in three distinct groups $\mathcal{A}$, $\mathcal{B}$, and $\mathcal{C}$, each containing six functions. Each group is characterized by its own "monomeric" unit: $A1$, $B1$, and $C1$. The monomers $B1$ and $C1$ are degenerate with respect to function, though they are different at the syntactical level. The corresponding functions in each group act in the same way. Group $\mathcal{C}$ is, therefore, skipped in the discussion (and in Figure 7), since every statement about $\mathcal{B}$ applies equally to $\mathcal{C}$.

The basic polymerizing unit is monomer $A1 =$ `(*(*aa)a)`, indicated by a filled circle in Figure 7(a). Evidently, $A1$ triplicates any input `a`. Figure 7 shows the transformation pathways among the species in each group. Solid arrows are marked by letters indicating only the originating group of the function(s) that operate(s) the transformation. The system does not contain a function that interconverts monomers. Transformations occur therefore only within group members. $A1$ polymerizes every species in the system, provided the product does not exceed the 300 character limit imposed by the collision rule. This is the case for the 114-character function $B6$ in Figure 7(b). All polymerizations in group A are operated by $A1$. In particular, $A1 \rightarrow A3 \rightarrow A6 \rightarrow *$. Figure 8 gives the structure of the $A6$ trimer. The polymers are highly branched and self-similar due to the iterated action of $A1$. The intriguing feature is that the polymers created in all groups (mostly by action of $A1$) are themselves functionally active such as to establish depolymerization pathways that achieve self-maintaining closure of the system. They also form further products that are not "iterated trimers."

Both groups, $\mathcal{A}$ and $\mathcal{B}$ (and $\mathcal{C}$), are functionally strongly coupled. As Figure 7 shows, $\mathcal{A}$ is dependent for closure upon $\mathcal{B}$ and conversely. The essential interdependencies among the groups are visualized in Figure 9. One seeding set consists of a transformation cycle of three functions in each group. The cycles mutually depend on each other and interconvert their respective monomers, dimers, and trimers. The dimer and trimer vertices are further polymerized into higher-order structures.

Ultimately, Figure 9 and, therefore, the whole system can be built up from the mere presence of monomer $A1$ and monomer $B1$. Everything else follows. The reader can easily check: $A1$ and $B1$ are the minimal seeding set. Its smallness explains the high stability of this system.

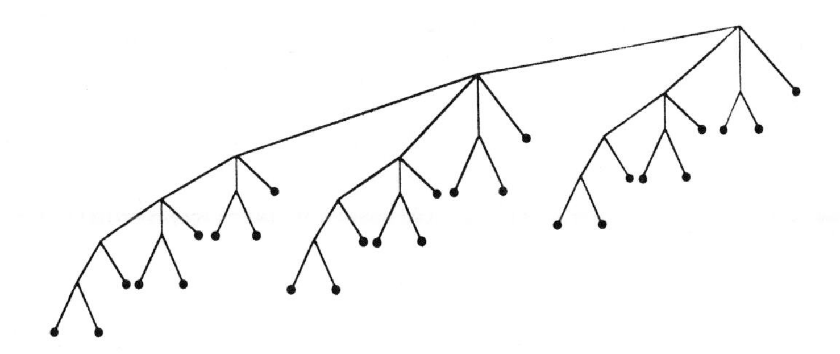

FIGURE 8 Self-similar polymer. The function resulting from three-fold application of the $A1 =$ `(*(*aa)a)` monomer (see Table 4) to itself.

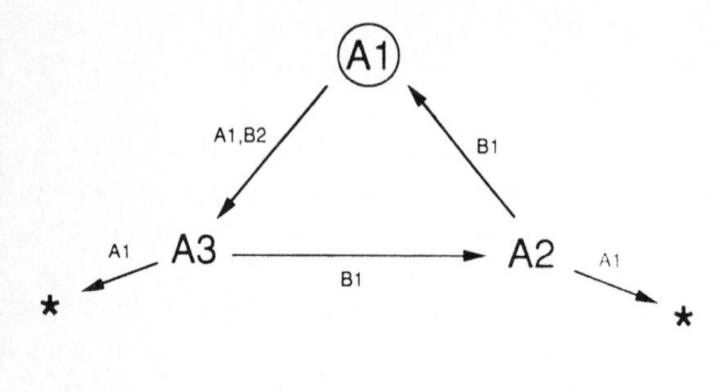

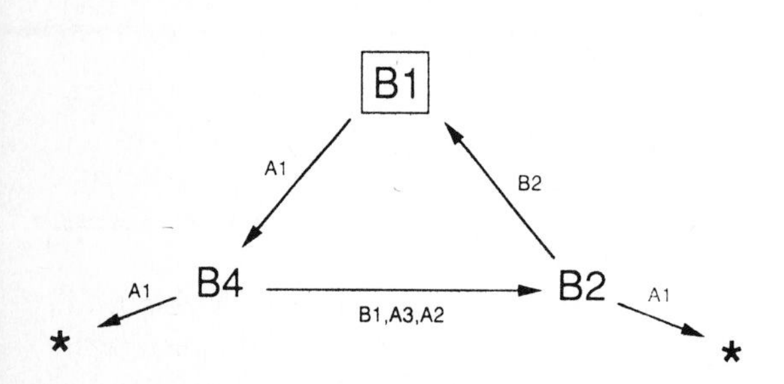

FIGURE 9 A seeding set and the minimal seeding set. A seeding set of the system shown in Figure 7. The two graphs correspond to the polymer families A and B. They represent the high populated core in Table 4 (a corresponding triangle must be added for group C). Arrows and labels as in Figure 7. The minimal seeding set of the system consists solely of the two monomers $A1$ and $B1$. See text for details.

The nine functions (including C, not shown) of Figure 9 constitute the all-time-high populated core. The "polymerization triangles" cause a constantly high innovation rate. The system could expand indefinitely into new polymers, if it weren't for the length limitation on the function strings. In addition, the kinetics induced by the gas system and by the established pathways constrain the number of different function species that can be sustained with only 1000 particles. The ongoing polymerizations create a low-populated periphery subject to fluctuations that frequently lead to the loss of functions. The stable seeding set regenerates any lost structure. In fact, ten iterations of the map, Eq. 10, starting with $A1$ and $B1$ ($\text{and } C1$), generate a set that contains all the 51 functions constituting the system's state after 5×10^5 collisions.

Can seeding sets replace each other during the system's evolution? For example, a seeding set A generates some larger ensemble that happens to contain a different seeding set B for a different ensemble. B takes over replacing A. B now expands the new ensemble and hits a further seeding set C, and so on. In the example discussed so far, the system was trying to expand and to move away from a particular region in function space. However, the seeding set $\{A1, B1\}$ was never replaced, thus anchoring the system in that region. So far a replacement of seeding sets has not been observed in the unperturbed version of the Turing gas. It is observed regularly, however, when the gas is perturbed by exogenously introduced random functions (see section 6.3).

Some few experiments ended simply with elastic colliders. Most of the simulations, however, evolved organization patterns similar to those described above. All instances observed so far, produced different "polymeric" structures. Polymers consisting of monomers nested into one single tree branch have been observed. Interconversion is then achieved by functions that add or delete one monomer independently from the "polymerization degree" of the input functions. Other polymeric forms have their monomers arranged as individual branches emanating from the root. A whole zoo of functions based on polymeric structures has been collected by now. Their detailed description is, nevertheless, beyond the scope of this paper.

These examples illustrate how strongly the organization of interactions in the Turing gas is shaped by the availability of copiers. In the absence of copiers the system still develops connected metabolisms, but based on a rather different "chemistry" that appears to sustain a much higher degree of diversity and innovation.

6.3 TURING GAS WITH PERTURBATIONS

The Turing gas evolves from an initial set of functions through random collisions. Since in the early stages almost every collision is reactive, the dilution flux frequently removes functions that never had an interaction. The system is, therefore, highly sensitive to the particular interaction history. Changing one collision could result in a function that would have otherwise not been produced, thus opening the possibility for a strong sensitivity to initial conditions. In fact, different random number sequences lead very frequently to completely different quasi-stationary states. Simulations have been performed by allowing the gas to increase the number of particles during the initial phase, such that on average every initially present function had the chance to interact. The basic principles of self-organization, however, remain the same. As the accumulation of function chaining leads to closure (section 6.1), or to fluctuating motions around a fixed point (section 6.2), the system's evolution and organization becomes more and more predictable. For example, a primitive operator that has disappeared is lost forever.

A new set of questions arises if the system is exposed to perturbations consisting in the injection of new (random) functions. This can happen in a variety of forms, for example, either constantly as a noisy background, or at specific time intervals. The introduction of a new function creates a center of innovation. New products are generated through secondary reactions spawned by the perturbing agent.

Consider the system without copy reactions discussed in the previous section. The state after 5×10^5 collisions, containing 51 different functions, is taken to be the initial state of a simulation that injects 100 randomly generated functions every 2×10^4 interactions. In order to focus on those functions that organize the system, intersection sets are analyzed that capture the unchanging part during periods lasting between 5×10^4 to 1×10^5 collisions.

The unchanging core between collisions 2×10^5 and 3×10^5 consists of a set $\mathcal{A}$ of 12 functions whose interactions are partly innovative. The interaction graph induced on the vertex set $\mathcal{A} \backslash \Phi[\mathcal{A}]$ is again connected. The structure of the functions is still polymeric, but the architecture changed with respect to the situation described in section 6.2. The system contains four polymer families, two of which it still shares with the quasi-stationary state described in section 6.2. These groups are induced by the monomers $B1 =$ (<(*(>*aa))(+a))) and $C1 =$ (* $B1$). The previous polymerizing function (*(*aa)a) has been replaced by $A1 =$ (*aa). $A1$ is the cause of the change in architecture, the polymers still being branched and self-similar, but this time "doublets" of each other. The fourth group is also new and based on the monomer $D1 =$ (>(*aa)).

The new monomers are the products of reactions involving perturbing functions and "resident" ones. Once (*aa) has been produced, its takeover is not a pure accident. In the previous section it was noticed that the polymers induced by the monomer (*(*aa)a) hit the length limitation of 300 characters very soon. The products of an agent with length l and polymerization factor p reach the length limit after $n = \log(300/l)/\log p$ iterated applications. Under the action of the triplicating monomer with $l = 9$ and $p = 3$, the length limit is reached already after three applications. In case of the duplicating monomer (*aa), $l = 5$ and $p = 2$, the limit is reached after six applications. This shows that the polymer populations induced by (*aa), and that can be sustained given the actual limitations of the system, are much more diverse. In addition, the (*(*aa)a)-based populations and the (*aa)-based populations cooperate on a short time scale: the action of both monomers on each other as well as suitable depolymerization pathways contribute to their mutual maintainance in the system. In the long run, however, the mere combinatorics of the (*aa)-subsystem will provide a richer network that has more vertices and more interaction possibilities. In short, under the actual limitations, a system based on duplications sustains more reactive collisions than a system based on triplications, and will eventually displace the triplicating system. Note that the combinatorial argument holds not only for the polymer family consisting of (*aa)-monomers, but, as well, for the other polymeric species built up through the action of (*aa). In fact, the number of subsytems (disconnected in the interconversion sense, not in the functional sense) increased from three to four.

As time proceeds and perturbations are introduced every 2×10^4 collisions, time slices of the duration of 5×10^4 collisions are continously monitored for invariant sets. 1.5×10^5 collisions later relative to the previous quasi-stationary state a new invariant set has been found. The set contains 20 functions whose interactions are highly innovative. Their structure is again based on polymer combinatorics, but this time generating a plethora of structures.

Four monomers are present, three of which are identical to the previous metastable state: $A1 =$ (*aa), $B1 =$ (<(*(>*aa))(+a))) and $D1 =$ (>(*aa)). The new monomer is $C1 =$ (*('('a))a). Note that now two monomers, $A1$ as well as $C1$, have polymerizing activity. $A1$ doubles its input as usual, while $C1$ joins its input to ('a), therefore giving rise to polymers of the form (('a)('a)...('a)E), with E being any expression in the system. A polymer of $A1$ usually interacts differently than the monomer. For example, ((*aa)(*aa)) performs the function that results from "the doubled input acting on the doubled input," which is quite different than polymerizing by doubling the input. Functions generated by

$C1$ acting on functions containing $C1$, in contrast, always keep the function of the monomer. For example, F = (('a)('a)(*('('a))a) acts on some input E by producing ((a)('a)E). If the expression E is a form that contains $C1$, it will be active by itself in a similar fashion. Note the transformation of one ('a) into an (a) during the action of F: a new "building block," (a), has been introduced. Suitable interactions give rise to all sorts of (a)-('a)-copolymers. In addition, all these products interact with the polymer families arising from $B1$ and $D1$.

Another series of perturbation exeperiments was performed with systems that allow copying. For example, the set of 13 functions from the quasi-stationary state of the system discussed in section 6.1 was perturbed by 10 randomly chosen functions, each introduced in 10 copies. The system was perturbed by such an event every 10^5 collisions. Half a million collisions later the system became homogeneous containing solely the partial copier (and self-replicator) (>(*aa)).

Systems that forbid copying reactions seem to respond in a much subtler way to perturbations. The system discussed in the previous paragraphs, for example, underwent transitions among meta-stable states characterized by a high degree of diversity. If forcing a base level of cooperativity by disallowing copying always results in systems with a higher adaptability than systems that contain self-replicating functions cannot be answered at present. More systematic investigations are needed.

The above examples are meant to illustrate one type of behavior resulting from the adaptation of a Turing gas to external perturbations. The range of responses will depend on the precise form of the noise level: how many functions in what copy numbers at which times. A systematic study of adaptive responses is beyond the scope of this paper, and leads to a series of questions about techniques that are suitable for monitoring "adaptive activity." A forthcoming paper is addressing these issues.[27]

7. SUMMARY AND CONCLUSIONS

The present contribution is based on the hypothesis that the distinguishing feature of adaptive systems consists in a closed loop between objects from a combinatorial variety and the functions they encode. This paper introduces a simple instance of a new class of models aimed at isolating this feature and to study its consequences.

The key idea is to represent such a loop through a formal computational language. The syntactical and semantical levels of a language fit naturally into the object and function poles of that loop. The language used in this paper maps strings of characters into symbolic algorithms that operate on strings. It is closely related to the λ-calculus invented by Alonzo Church in the '30s, and its actual implementation is derived from Gregory Chaitin's permissive toy LISP.[4]

As a first step, the algorithms have been restricted to be functions in one variable. A character string represents a function that acts on a single other character string. The model is complete in the sense that the interaction between functions is embedded into the language itself. In this paper, interaction is naturally defined to be function composition, and is, therefore, asymmetric. The result of an interaction is a syntactically legal string that encodes a possibly new function.

An iterated map is defined by repeatedly applying the interaction operator to a given set of functions. The concept of an interaction graph is introduced as a representation of the iterated map. The fixed points of an interaction graph are closed, self-maintaining sets. The characterization of all finite self-maintaining sets of functions is an open mathematical problem. A solution to this problem is important with respect to a classification of cooperative structures, and might bear some biological significance (see section 8). Seeding sets are defined as subgraphs that can regenerate the original graph under the action of the map. If the original graph is a fixed point of the map, the seeding sets determine its stability. Some properties of innovative interaction graphs have been discussed in section 4.

The interaction graph, and, equivalently, the iterated map, describe a dynamical system induced by the language on the power set of functions. This graph dynamics is supplemented by a system of mass action kinetics leading to a frequency distribution on the set of functions in a graph. The kinetics is induced through a stochastic process termed "Turing gas." A Turing gas consists of a fixed number of function particles that are randomly chosen for pairwise collisions. In the present scheme, a reactive collision keeps the interaction partners in addition to the reaction product. A stochastic unspecific dilution flux is provided to hold the number of particles in the system at a predefined value.

Three versions of the Turing gas have been discussed. In one version, the time evolution of the gas is observed after its initialization with N randomly generated functions. In the second version, the collision rule is changed to forbid reactions resulting in a copy of one of the collision partners. In the third version, the gas is allowed to settle into a quasi-stationary state, where it is perturbed by injecting new random functions.

The results can be summarized as follows:

1. Ensembles of initially random functions self-organize into ensembles of specific functions sustaining cooperative interactions. Self-replicators, parasites, general copy functions, as well as partial copiers, shape the dynamics of the system. The "innovation rate," i.e., the frequency of collisions that result in functions not present in the system, decreases with time indicating a steady closure with respect to interactions. If the stochastic process is left to itself after injecting the initial functions, it will eventually hit an absorbing barrier characterized by a single replicator type, by a possibly hetergeneous mixture of non-reactive functions ("dead system"), or by a self-reproducing set where each individual function species is a seeding set. The system typically exhibits extremely long transients characterized by mutually stabilizing interaction patterns. Such patterns include a hierarchical organization of interacting self-maintaining sets. Sometimes these subsets are disconnected from each other with respect to interconversion pathways, but connected with respect to functional couplings.

2. Forbidding copy reactions results in a new type of cooperative organization as compared to the case in which copy reactions and, therefore, self-replicators were allowed. The system switches to functions based on a "polymeric" architecture that entertain a closed web of mutual synthesis reactions. The individual functions are usually organized into disjoint subsets of polymer families based on distinct monomers. As in the case of copy reactions, these subsets interact along specific pathways leading to a cooperativity at the set level. Due to the polymeric structure of the functions, the Turing gas remains innovative. A much higher degree of diversity and stability is achieved than in systems that are dominated by self-replicators.
3. An open system is modeled by introducing new random functions that perturb a well-established ecology. In the case without copy reactions, the system underwent transitions among several new quasi-stationary states, each characterized by an access to higher diversity. Systems with copy reactions were more vulnerable to perturbations and lost in the long run much of their structure.

The main conclusions are:

1. A formal computational language captures basic qualitative features of complex adaptive systems. It does this because of
   a. a powerful, abstract and consistent description of a system at the "functional" level, due to an unambiguous mathematical notion of function;
   b. a finite description of an infinite (countable) set of functions, therefore providing a potential for functional open-endedness; and
   c. a natural way of enabling the construction of new functions through a consistent definition of interaction between functions.
2. Populations of individuals that are both an object at the syntactic level and a function at the semantic level, give rise to the spontaneous emergence of complex, stable, and adaptive interactions among their members.

The present contribution raises many questions. The next section lists some of these questions, briefly points to related work, and considers future directions.

8. OUTLOOK

One feature of the present model is the coupling of a dynamics governing the topology of an interaction graph with a dynamics governing a frequency distribution over its vertices. The present model is by no means the first to exhibit such a structure, although it differs fundamentally in approach and scope from others.

Approaches based on this coupling have been proposed in several areas of research. D. Farmer, S. Kauffman, and N. Packard,[11] as well as R. Bagley et al.[1] considered a model of polymers intended to be RNA strands or proteins, undergoing condensation and hydrolysis reactions catalyzed by other polymers. S. Rasmussen et al.[26,28] investigated a model involving random catalytic connections of the hypercycle type,[10] R. deBoer and A. Perelson[7] proposed a model of the immune system that exhibits the above coupling, and J. Holland's classifier system[16] is an instance based on a genetic scheme.

More abstract approaches are considered by J. McCaskill[24] and S. Rasmussen et al.[29] Rasmussen's system consists of generalized assembler code instructions that interact in parallel inside a controlled computer memory giving rise to cooperative phenomena. McCaskill uses binary strings to encode transition-table machines of the Turing type that read and modify bit strings. The *ansatz* developed in this paper emerged from the direct experience with McCaskill's system, from several attempts to redefine it, and from the observation that Rasmussen's system shares too many properties with cellular automata.

Cellular automata provide a powerful approach to the study of the emergence of loops between objects and functions. Incidentally, John von Neumann envisioned such a loop when invoking symbolic instruction and universal construction as necessary conditions for the evolution of complexity.[34] Much work has been done to understand the conditions that allow for its emergence.[14,20,21,35] It becomes, however, difficult to study the consequences of such a loop at the same level of description that has been used to study its emergence.

Some issues that accumulated throughout the paper are addressed in the following. One concerns the asymmetry of the interactions. The immediate observation is that many complex systems in Nature have asymmetric interactions. Neurons, other cell types, and species are but a few examples. Clearly, a strict analogy with molecules is not possible within the present collision rule. The decision of which molecule, in a given pair, is the enzyme and which molecule is the substrate cannot be reversed in later collisions. This could be taken into account by redefining the collision rule such as to compute both expressions, $f(g)$ as well as $g(f)$, and taking, for example, the longest expression to be the collision product (ties resolved lexicographically).

Enzyme and substrate interpretations become stretched, as soon as general interaction schemes, $\Phi(f, g)$, are considered. An open question is if interactions Φ, different from chaining, result in qualitatively similar patterns of organization.

One of the most obvious generalizations of the system is the multivariable case. Considering functions with up to n variables is interesting for at least three reasons. First, it leads naturally to $(n+1)$-body interactions. Second, the system can self-organize with respect to the density of 1-, 2-, ..., n-variable functions. Third, suppose the interaction is still given by a generalized function composition. A two-variable function $f(x, y)$, then interacts with functions g and h (in this order) by producing $i = f(g, h)$. f acts, with respect to any pair g and h, precisely like a binary interaction law expression Φ did previously (see section 3.3). However, f can now be modified through interactions with other components of the same system. This might have consequences for the architecture of organizational patterns that are likely to evolve. The extension to n variables is currently in preparation.

Further questions relate to the number and type of primitive operators. An extended set of experiments was conducted with 11 operators (including the six reported here). The additional primitives involved instructions like calls to the interpreter from within the interpretation process. The basic results are very similar with respect to organizational patterns. Nevertheless, more powerful function constructors should be taken into account depending on the particular application.

The dependence of the observed phenomena on the number of particles in the system has not been considered so far. Do new phenomena appear if the system consists of a million particles?

Spatial systems have not been considered here. Nor have genetic mechanisms. What happens if noise is introduced through error-prone execution with some per step error rate q? What concepts of "mutation" can be envisioned?

Slight variations in the semantics of the language had no effect on the basic results. Nothing can be said at present about language architectures other than the λ-calculus. This raises an important question: Are the phenomena reported in this paper universal with respect to representation? Do different representations only affect the probability by which particular organizational patterns emerge?

Rate constants were not considered in this contribution. The a_{kij} in Eq. (2) were unity if an interaction between i and j resulted in k, and zero otherwise. The model puts the emphasis on relationships among agents rather than on kinetic details. Nevertheless, "rate constants" can be added, e.g., by considering the space-time resources required to complete the computation of an interaction.

Functional interactions based on chemistry enabled the evolution of complex adaptive systems of the biological type. Biological systems seem to have various levels at which "higher-order" structures interact again in a functional way. For example, cell types and control mechanisms give rise to ontogeny in multicellular organisms. The results reported in this contribution suggest that a formal language might be a useful model for systems that are characterized by functional interactions.

The new model class introduced in this paper raises sensible mathematical questions that are interesting by and of themselves. However, the point of view outlined here is useful only if it can organize knowledge about the real world. For example, by being predictive or by enabling a logical characterization or classification of certain phenomena that would otherwise remain a pure matter of fact. The viewpoint would be useful, for instance, if a characterization of finite self-maintaining sets of functions in up to n variables exhibits a subset with basic similarities to known biological life-cycle organizations. It would be useful, for instance, if such a system (with suitable modifications) allows the study and the understanding of the entangled interplay between rules of selection and products of selection.

Doyne Farmer recently drew my attention to a paper by O. E. Rössler[31] that appeared in German in 1971. In that remarkable paper Rössler expands precisely on the chemical metaphor I had in mind when developing the work presented in this contribution. Although Rössler did neither present a specific model nor did he invoke the constructive aspect of computable functions, his paper clearly states some of the main thoughts that motivated the present work.

ACKNOWLEDGMENTS

This work would have hardly been possible without inspiring discussions with John McCaskill, Wojciech Zurek, David Cai, Norman Packard, Steen Rasmussen, David Lane, Doyne Farmer, Stuart Kauffman, Leo Buss, David Sharp, Gian-Carlo Rota, Peter Stadler, Peter Schuster, Jeff Davitz, and Chris Langton.

This work has been performed under the auspices of the United States Department of Energy.

A. ALCHEMY IS A VARIANT OF PURE LISP

A.1 SYNTAX

Let $\mathcal{C}$ be an alphabet that includes left and right parentheses. All characters except left and right parentheses are called atoms. The fundamental syntactic structure is termed "expression."

DEFINITION A.1 An expression is an atom or a list.

The structure of a list is defined recursively.

DEFINITION A.2 A list consists of a left parenthesis followed by zero or more atoms, or lists followed by a right parenthesis.

Let $\mathcal{C} = \{$`a,b,c,d,e,(,)`$\}$, then `()`, `(c)`, `(((dba)(b))b(ec))` are lists.

Lists can be represented by ordered trees with the leaves being the atoms (Figure 1). A matching pair of left and right parentheses represents an internal node. The empty list is treated like an atom, but consists of two characters. A list is reconstructed from a tree by traversing the tree in preorder, i.e., by visiting the root of the first (leftmost) tree, traversing the subtrees of the first tree in preorder, and traversing the remaining trees in preorder.

Expressions are self-delimiting. An expression is complete as soon as the number of right parentheses matches the number of left parentheses. For instance, the string `(ab)c(de)` is not an expression. The point is that the parentheses do not balance for the first time at the end of the character string, but at the end of `(ab)`.

A.2 SEMANTICS

A.2.1 VALUE The basic semantical concept is that of the "value" V of an expression. The evaluation of an expression refers always to an "association list" L. An association list is a look-up table that stores zero or more pairs. The first element of such a pair is always an atom, and the second element is an expression that is to be substituted for that atom during the evaluation process. If a given atom does not appear in the first position of any pair, then this atom evaluates to itself. In an empty association list, $L =$`()`, every atom evaluates to itself. The empty list is considered to be an atom. Its value is always the empty list.

Value assignments in the association list are indicated by a left arrow. For instance, the pair that assigns to `a` the value `b` is written as `a`←`b`. With the association list

$$L = (\mathtt{d} \leftarrow (\,\mathtt{cd(\,a)}\,)\ \mathtt{e} \leftarrow \mathtt{b}\,\mathtt{a} \leftarrow (\,\mathtt{e})\,)\ ,$$

the expression `e` evaluates to `b`. For this I shall write: $\mathbf{V}[\mathtt{e}, L]$ =`b`. Furthermore, $\mathbf{V}[\mathtt{d}, L]$ =`(cd(a))`, $\mathbf{V}[\mathtt{a}, L]$ =`(e)`, and $\mathbf{V}[\mathtt{b}, L]$ =`b`.

DEFINITION A.3 The value of an atom is obtained from the association list L. The value of a non-atomic expression is obtained recursively in terms of the values of its elements (atoms and/or lists).

A list is, therefore, parsed into its constituent expressions, and each expression is evaluated. The concept of "function" is needed to further understand the evaluation process.

A.2.2 FUNCTIONS AND EVALUATION Two types of functions are distinguished. Primitive and defined (or composite) functions.

Primitive functions are the predefined operators of the language. Their names are atoms from the character set $\mathcal{C}$ reserved for this purpose.

DEFINITION A.4 An n-ary primitive operates on n arguments that are elements of the same list to which the operator belongs. The n arguments are obtained by evaluating the n expressions following the atom denoting the primitive.

If the list contains less than n expressions, an empty list is supplied for each missing one. Any expressions in excess are ignored. The primitive operators are defined in Appendix B. Every operator returns an expression.

A defined function is the value of an expressions. This value is applied to arguments. The arguments are obtained in exactly the same way as in the case of the primitives.

DEFINITION A.5 In AlChemy the values of all expressions that are not arguments to primitive operators define functions.

The exception ensures that a primitive operator performs its manipulations, without side effects arising from interacting arguments.

In order for a function to operate on arguments, the variables have to be specified.

DEFINITION A.6 In AlChemy all atoms not denoting operators are variables.

DEFINITION A.7 "Applying a defined function to arguments" means to evaluate the function expression after having substituted every occurrence of the variables with the corresponding arguments.

Technically this is a two-step process: (1) Updating the association list by appending the old list to a new list of pairs that binds the variables with the corresponding argument expressions, and (2) evaluating the function expression thereby using the new association list.

In this paper only the simplest case is considered: functions in one variable; hence, there is only one more atom in addition to the primitive operators. Throughout this paper the atom **a** denotes the variable. The generalization to multivariable functions is straightforward.

The following is a formal definition of the value of an expression. The association list L contains, at any time, the current value of the variable, and is of the form

L =(a← Z), where Z is some expression. The empty association list L =() is equivalent to L =(a←a). Consider the list ($f_1\ f_2\ \ldots\ f_n$) with an empty association list. According to definition A.5 the values of the expressions $f_1, f_2, \ldots, f_n$ define functions.

DEFINITION A.8 A list consisting of more than one function definition is evaluated by applying each function to its arguments in turn. The value of the list is the expression obtained by appending the result of each function application to an initially empty list. Results consisting of an empty list are ignored.

According to this definition the value of the list, ($f_1\ f_2\ \ldots\ f_n$) is given by

$$\mathbf{V}[(f_1\, f_2\, \ldots\, f_n), L] = \begin{cases} \text{application of k-ary primitive}, & \text{if } f_1 \in \mathcal{O}; \\ (\mathbf{V}[\mathbf{V}[f_1, L], (\mathtt{a} \leftarrow \mathbf{V}[f_2, L])] \\ \ \mathbf{V}[\mathbf{V}[f_2, L], (\mathtt{a} \leftarrow \mathbf{V}[f_3, L])] \\ \ \vdots \\ \ \mathbf{V}[\mathbf{V}[f_{n-1}, L], (\mathtt{a} \leftarrow \mathbf{V}[f_n, L])]), & \text{otherwise}, \end{cases}$$

where $\mathcal{O}$ denotes the set of primitive operators. Usually, $L = ()$. Redundant parentheses are always removed, e.g., $((f))$ becomes (f). In addition, if an evaluation results in a function body consisting of the parenthesized variable, $\mathbf{V}[f_i, L] = (\mathtt{a})$, the parentheses are removed. The application of (a) would mostly result in (), since an empty list is supplied for the missing argument.

This procedure contains, in contrast to pure LISP, a "sequential" aspect. A purely recursive scheme would evaluate an expression $f = (ghi)$ by first applying the value of h to the value of i, and then applying the value of g to that result. The departure from a purely recursive scheme has been decided by observing that the nesting of too many evaluations considerably shortens the output expressions, often leading to empty lists. The sequential mode is an efficient way to offset this effect without constraining the combinatorics of expressions.

Stated in terms of tree structures, evaluating an expression means evaluating its tree. The value of any node is obtained by applying the value of each subtree to its right-neighbor sibling in turn, each time appending the result to the current value expression. If the leftmost subtree of a node is an n-ary operator, then the value of that node is simply obtained by applying the primitive operation to the values of the n siblings of that operator. The value of a tree is therefore the value at its root.

A.2.3 INTERACTIONS BETWEEN ALCHEMY FUNCTIONS The interaction between two expressions, f and g, is defined by an expression that has as value the expression $f(g)$. Definition A.5 states that a function is represented by the value of an expression, not the expression itself. To build an expression with value $f(g)$ the quote-operator, $'$, is used. The action of the quote-operator is to prevent evaluation of its argument. Hence, $\mathbf{V}[('f), L] = f$, for all L. This leads to the following interaction expression:

DEFINITION A.9 The interaction between two expressions, f and g, is defined by the expression

$$(('f)('g))$$

The result of an interaction is given by

$$\mathbf{V}[(('f)('g)), ()] = (\mathbf{V}[f, (\mathsf{a} \leftarrow g)]) \,.$$

The functions of the model described in this paper have at most one variable; hence, only binary interactions have to be considered. A function of n variables can interact with n other functions.

B. PRIMITIVE OPERATORS

AlChemy's character set is given by

$$\mathcal{C} = \{(\,,\,)\,,\mathtt{a}\,,+,-,>,<,*,'\}.$$

AlChemy has six primitive operators. These operators manipulate list structures. The operators are not "orthogonal," in the sense that some of them can be expressed through a combination of others.

$+$ Name : head Arguments : 1

Application of this operator to a list returns the first expression of that list. Application to an atom returns the atom itself. The head of an empty list is, therefore, an empty list.

$-$ Name : tail Arguments : 1

Application of this operator to a list returns what remains after the first expression of that list is deleted. Application to an atom returns the atom itself. The tail of an empty list is, therefore, an empty list.

$>$ Name : inversehead Arguments : 1

Application of this operator to a list returns the last expression of that list. Application to an atom returns the atom itself. The inverse head of an empty list is, therefore, an empty list.

$<$ Name : inversetail Arguments : 1

Application of this operator to a list returns what remains after the last expression of that list is deleted. Application to an atom returns the atom itself. The inverse tail of an empty list is, therefore, an empty list.

$+$ Name : join Arguments : 2

If the second argument is an atom, but not the empty list, then the operator returns the first argument. If the second argument is an n-element list, then the result is an $n+1$-element list whose head is the first argument and whose tail is the second argument.

Exception: If the second argument is an expression E of the form $E = (\,\text{"op"}\ldots)$, where "op" is one of the primitive operators defined in this section, then E is wrapped into parentheses, $(\,E)$, prior to application of the join operator.

Remark: Let, for example, $(\,A)$ and $(+B)$ be expressions that are to be joined. The exception rule applies to $(+B)$. The result will then simply be $((\,A)(+B))$. If this product is to be further evaluated, then the group $(\,A)$ will be a function acting on the group $(+B)$. Without the exception rule the join product would have been $((\,A)+B)$. Further evaluation would have $(\,A)$ act on the character $+$, which is less interesting. Applying the exception rule means confining the system to a subset of the computations that would occur otherwise.

Name : quote Arguments : 1

The operator returns the unevaluated argument expression.

C. AN EVALUATION EXAMPLE

Consider the interaction expression E represented by the middle tree in Figure 3.

$$E = (A\ B) \tag{C.1}$$

with

$$A = ('((+\mathtt{a})(*(>\mathtt{a})(>(+\mathtt{a}))))),$$
$$B = ('((*(+\mathtt{a})(*(-\mathtt{a})(<\mathtt{a})))\mathtt{a})).$$

The value of Eq. (C.1) is

$$\mathbf{V}[E, \mathtt{a} \leftarrow \mathtt{a}] = (\mathbf{V}[\mathbf{V}[A, \mathtt{a} \leftarrow \mathtt{a}], \mathtt{a} \leftarrow \mathbf{V}[B, \mathtt{a} \leftarrow \mathtt{a}]]). \tag{C.2}$$

$\mathbf{V}[A, \mathtt{a} \leftarrow \mathtt{a}]$, with $A = ('F)$ is obtained by applying the quote operator, $'$, to its argument F. Clearly,

$$\mathbf{V}[('F), \mathtt{a} \leftarrow \mathtt{a}] = F = ((+\mathtt{a})(*(>\mathtt{a})(>(+\mathtt{a})))).$$

The same holds for $B = ('G)$:

$$\mathbf{V}[('G), \mathtt{a} \leftarrow \mathtt{a}] = G = ((*(+\mathtt{a})(*(-\mathtt{a})(<\mathtt{a})))\mathtt{a}).$$

Therefore, Eq. (C.2) becomes

$$\mathbf{V}[E, \mathtt{a} \leftarrow \mathtt{a}] = (\mathbf{V}[F, \mathtt{a} \leftarrow G]). \tag{C.3}$$

Write

$$F = (F_1 F_2),$$

with

$$F_1 = (+\mathtt{a})$$
$$F_2 = (*(>\mathtt{a})(>(+\mathtt{a}))),$$

and

$$G = (G_1 G_2),$$

with

$$G_1 = (*(+\mathtt{a})(*-\mathtt{a})(<\mathtt{a})))$$
$$G_2 = \mathtt{a}.$$

Then Eq. (C.2) becomes

$$\begin{aligned}\mathbf{V}[E, \mathtt{a} \leftarrow \mathtt{a}] &= \mathbf{V}[F, \mathtt{a} \leftarrow G] = \mathbf{V}[(F_1 F_2), \mathtt{a} \leftarrow (G_1 G_2)] = \\ &= (\mathbf{V}[\mathbf{V}[F_1, \mathtt{a} \leftarrow (G_1 G_2)], \mathtt{a} \leftarrow \mathbf{V}[F_2, \mathtt{a} \leftarrow (G_1 G_2)]]).\end{aligned} \tag{C.4}$$

$\mathbf{V}[F_1, \mathtt{a} \leftarrow (G_1G_2)]$ with $F_1 = (+\mathtt{a})$ is obtained by substituting (G_1G_2) for $\mathtt{a}$ and by applying the "+" operator (see Appendix B):

$$\mathbf{V}[(+\mathtt{a}), \mathtt{a} \leftarrow (G_1G_2)] = G_1 . \tag{C.5}$$

$\mathbf{V}[F_2, \mathtt{a} \leftarrow (G_1G_2)]$ is obtained by applying the "$*$" operator to $\mathbf{V}[(>\mathtt{a}), \mathtt{a} \leftarrow (G_1G_2)]$ and $\mathbf{V}[(>(+\mathtt{a})), \mathtt{a} \leftarrow (G_1G_2)]$, the values of its arguments $(>\mathtt{a})$ and $(>(+\mathtt{a}))$, respectively. Refer to appendix B for the action of the primitive operator "$>$."

$$\mathbf{V}[(>\mathtt{a}), \mathtt{a} \leftarrow (G_1G_2)] = (G_2) = (\mathtt{a}) \tag{C.6}$$

$$\mathbf{V}[(>(+\mathtt{a})), \mathtt{a} \leftarrow (G_1G_2)] = (*(-\mathtt{a})(<\mathtt{a})). \tag{C.7}$$

Hence,

$$\mathbf{V}[F_2, \mathtt{a} \leftarrow (G_1G_2)] = ((\mathtt{a})(*(-\mathtt{a})(<\mathtt{a}))). \tag{C.8}$$

Equation (C.2) is now

$$\begin{aligned}\mathbf{V}[E, \mathtt{a} \leftarrow \mathtt{a}] = &\mathbf{V}[F, \mathtt{a} \leftarrow G] = \\ &= (\mathbf{V}[\mathbf{V}[F_1, \mathtt{a} \leftarrow (G_1G_2)], \mathtt{a} \leftarrow \mathbf{V}[F_2, \mathtt{a} \leftarrow (G_1G_2)]]) = \\ &= (\mathbf{V}[(*(+\mathtt{a})(*(-\mathtt{a})(*\mathtt{a}))), \mathtt{a} \leftarrow ((\mathtt{a})(*(-\mathtt{a})(<\mathtt{a})))\end{aligned} \tag{C.9}$$

which means applying "$*$" to

$$\mathbf{V}[(+\mathtt{a}), \mathtt{a} \leftarrow ((\mathtt{a})(*(-\mathtt{a})(<\mathtt{a})))] = (\mathtt{a}), \tag{C.10}$$

and

$$\mathbf{V}[(*(-\mathtt{a})(<\mathtt{a})), \mathtt{a} \leftarrow ((\mathtt{a})(*(-\mathtt{a})(<\mathtt{a})))] = ((*(-\mathtt{a})(<\mathtt{a}))\mathtt{a}) \tag{C.11}$$

resulting from the application of "$*$" to $(*(-\mathtt{a})(<\mathtt{a}))$, the value of $(-\mathtt{a})$ in Eq. (C.11), and to $(\mathtt{a})$, the value of $(<\mathtt{a})$ in (11). The application of "$*$" to Eqs. (C.10) and (C.11) finally completes

$$\mathbf{V}[E, \mathtt{a} \leftarrow \mathtt{a}] = \mathbf{V}[F, \mathtt{a} \leftarrow G] = ((\mathtt{a})(*(-\mathtt{a})(<\mathtt{a}))\mathtt{a}). \tag{C.12}$$

Equation (C.12) is the collision product of functions F and G as shown in Figure 3.

REFERENCES

1. Bagley, R. J., J. D. Farmer, S. A. Kauffman, N. H. Packard, A. S. Perelson, I. M. Stadnyk. "Modeling Adaptive Biological Systems." *BioSystems* **23** (1989): 113–138.
2. Barendregt, H. P. *The Lambda Calculus*, Studies in Logic and the Foundations of Mathematics, vol. 103. Amsterdam: North Holland, 1984.
3. Buss, L. W. "The Evolution of Individuality." Princeton: Princeton University Press, (1987): 196.
4. Chaitin, G. J. *Algorithmic Information Theory.* Cambridge: Cambridge University Press, 1987.
5. Church, A. "An Unsolvable Problem of Elementary Number Theory." *Am. J. Math.* **58** (1936): 345–363.
6. Church, A. "The Calculi of Lambda-Conversion." *Annals of Mathematics Studies*, no. 6. Princeton: Princeton University Press, 1941.
7. deBoer, R., and A. S. Perelson. "Size and Connectivity as Emergent Properties of a Developing Immune Network." *J. Theor. Biol.* (1990): in press.
8. Delbrück, M. *Trans. Conn. Acad. Arts Sci.* **38** (1949): 173–190.
9. Eigen, M. "Self-Organization of Matter and the Evolution of Biological Macromolecules." *Naturwissenschaften* **58** (1971): 465–523.
10. Eigen, M., and P. Schuster. *The Hypercycle.* Berlin: Springer, 1979.
11. Farmer, J. D., S. A. Kauffman, and N. H. Packard. "Autocatalytic Replication of Polymers." *Physica D* **22** (1986): 50–67.
12. Fontana, W. "Algorithmic Chemistry: A New Approach to Functional Self-Organization." Technical Report LA-UR 90-3431, Los Alamos National Laboratory. Submitted to *Physica D*, 1990.
13. Gödel, K. Presented in his 1934 lectures at the Institute for Advanced Study. Quoted from: S. Kleene. "Turing's Analysis of Computabity and Major Applications of it." In *The Universal Turing Machine: A Half-Century Survey*, edited by R. Herken, 17–54. Oxford: Oxford University Press, 1988.
14. Gutowitz, H. A., B. K. Knight, and J. D. Victor. *Physica D* **48** (1987): 18.
15. Hofbauer, J., and K. Sigmund. *The Theory of Evolution and Dynamical Systems.* Cambridge: Cambridge University Press, 1988.
16. Holland, J. H. "Escaping Brittleness: The Possibilities of General Purpose Learning Algorithms Applied to Parallel Rule-Based Systems." In *Machine Learning II*, edited by R. S. Mishalski, J. G. Carbonell, and T. M. Mitchell, 593–623. Kaufman, 1986.
17. Kauffman, S. A. "Autocatalytic Sets of Proteins." *J. Theor. Biol.* **119** (1986): 1–24.
18. Kauffman, S. A. *J. Cybernetics* **1** (1971): 71–96.
19. Lane, D., J. H. Holland, and W. Fontana. Working paper under the auspices of the Adaptive Computation Program. Santa Fe, 1990.
20. Langton, C. G. *Proceedings of the 1989 Cellular Automata Workshop*, edited by H. A. Gutowitz. North-Holland, 1990.

21. Li, W., and N. H. Packard. *Complex Systems* **4** (1990): 281–297.
22. Manes, E. G., and M. A. Arbib. *Algebraic Approaches to Program Semantics.* Berlin: Springer-Verlag, 1986.
23. Mayr, E. *The Growth of Biological Thought.* Cambridge, MA: Harvard University Press, 1982.
24. McCaskill, J. S. In preparation.
25. Miller, J. H., W. Fontana, and P. Schuster. "Towards a Mathematics of a Turing Gas." In preparation, 1990.
26. Mosekilde, E., S. Rasmussen, and T. S. Sorensen. "Self-Organization and Stochastic Re-Causalization in Systems Dynamics Models." In *Proceedings of the 1983 International Systems Dynamics Conference*, 1983, 123.
27. Packard, N. H., and W. Fontana. In preparation, 1990.
28. Rasmussen, S. "Toward a Quantitative Theory of the Origin of Life." In *Artificial Life*, edited by C. G. Langton. Santa Fe Institute Studies in the Sciences of Complexity, Vol. VI, 79–104. Reading, MA: Addison-Wesley, 1988.
29. Rasmussen, S., C. Knudsen, R. Feldberg, and M. Hindsholm. "The Coreworld: Emergence and Evolution of Cooperative Structures in a Computational Chemistry." *Physica D* **42** (1990): 111–134.
30. Reingold, E. M., J. Nievergelt, and N. Deo. *Combinatorial Algorithms: Theory and Practice.* Englewood Cliffs, NJ: Prentice-Hall, 1977.
31. Rössler, O. E. "A System Theoretic Model of Biogenesis." *Zeitschrift für Naturforschung* **26b** (1971): 741–746.
32. Stadler, P. F., and P. Schuster "Mutation in Autocatalytic Reaction Networks." Working paper number 90-022, Santa Fe Institute. Submitted to *J. Math. Biol.*, 1990.
33. Turing, A. M. "On Computable Numbers with an Application to the Entscheidungs Problem." *P. Lond. Math. Soc. (2)* **42** (1936-1937): 230–265.
34. von Neumann, J. *Theory of Self-Reproducing Automata*, edited by A. W. Burks. Urbana: University of Illinois Press, 1966.
35. Wolfram, S. *Physica D* **10** (1984): 1.

Steen Rasmussen,† Carsten Knudsen,‡ and Rasmus Feldberg‡
†Complex Systems Group, Theoretical Division (T-13), and Center for Nonlinear Studies, MS-B258, Los Alamos National Laboratory, Los Alamos, New Mexico 87545 USA, e-mail: steen@t13.lanl.gov (arpa-net); and ‡Center for Modelling, Nonlinear Dynamics, and Irreversible Thermodynamics (MIDIT), B306 The Technical University of Denmark, DK-2800 Lyngby, DENMARK

Dynamics of Programmable Matter

We develop the concept of self-programmable matter and review different computational paradigms through this definition. A transformation between different kinds of programmable matter is defined relating the information mechanics of the different systems.

A particular kind of self-programmable matter based on a modified parallel von Neumann machine is defined and discussed from a dynamical system theory point of view. The dynamics of this system is studied through different projections in state space, through the information dynamics, through evolutionary scaling phenomena, and through the functional dynamics.

We are focusing on functional self-organization in the system and on the emergence of cooperative dynamics. Through the definition of an interaction graph, the essential functional dynamics can be singled out, and we show how functional cooperation is a natural consequence of the dynamics of programmable matter.

CONTENTS

1. INTRODUCTION

In an attempt to understand what it is that enables matter to organize into the complex structures we see in biological systems, we focus on the *autonomous programming abilities* of matter. Autonomous programmable matter[1] defines a particular class of dynamical systems consisting of interacting elements which can be arranged in a combinatorial way. The term *programmable* indicates that the dynamics of such systems have a clear computational interpretation and that functional properties can be programmed into the system via the elements. The term *matter* indicates that the dynamics is defined through the interactions of the fine-grains of the system e.g., at the level that defines the "physics"of the system.

Thus, *self-programmable matter* is a dynamical system of interacting elements, with associated functional properties, which through their autonomous dynamics develop new compositions of elements with new associated functional properties. Such systems are characterized by *an ability to construct novel elements within themselves.* Chemical systems are clear examples of such self-programmable systems.

It is important to note that the elements of self-programmable systems primarily have *deterministic* interaction rules (recall chemical reactions). This is in opposition to how novelty normally is viewed to emerge: namely through random processes. We shall focus on deterministic self-programming, but also discuss the consequences of random perturbations of self-programming processes.

Among other necessary conditions it is the self-programming properties of matter, in particular at the level of chemistry, that enable the origin and the apparently open-ended evolution of life. Another necessary condition for the emergence of life, as well as for simpler macroscopic, ordered structures, is thermodynamical nonequilibrium conditions, e.g., continuous supply of free energy and/or resources to the systems.[39] Classical examples of non-living, dissipative structures are the Raleigh-Bénard convection pattern and the chemical reaction waves in the Belousov-Zhabotinski reaction. Recent work by Packard,[41] Langton,[29] Li et al.,[31] and Wootters and Langton[59] has demonstrated yet another general condition which may also be necessary for the emergence and the existence of life: the operation of the system in a vicinity of a solid-fluid phase transition. Information processing capabilities can emerge spontaneously and come to dominate the dynamics of cellular automata systems in the vicinity of such a phase transition. The transition regime between frozen (= solid) and chaotic (= fluid) dynamics constitutes a natural domain of information processing, since its dynamics is characterized by an ability to store and communicate information as well as an ability to perform non-trivial information transformation. These are all properties it shares with systems designed for computation. Many processes and structures in living systems, as, for example,

[1]Toffoli and Margolus[53] define from a somewhat different perspective *programmable matter* as a flexible, fine-grained computing "medium" which is accessible to real-time observation, analysis, and modification. This certainly also applies to the kinds of programmable matter we investigate.

the cell membrane and the cytoskeleton, are being maintained in the vicinity of a solid-fluid phase transition.

There exist several well established theoretical frameworks, the different computation paradigms, which, in principle, can describe such self-programming processes. However, the autonomous dynamics of computational systems has never been the subject for any serious investigations. We, therefore, develop a computational paradigm to investigate the properties of self-programming. This is a continuation of our earlier attempts[26,47] to understand some of the underlying principles for the origin and the early evolution of life, based on the dynamics of computational systems. Our approach is closely related to Fontana's[13] and McCaskill's.[36]

The dynamical systems we are going to investigate are not models of particular physical, chemical, or biological systems. We are interested in the generic properties of the dynamics of self-programmable matter. By implementing the property of self-programming in a simple computational system we hope to be able to extract such generic properties. Hopefully the dynamics of the emergence of novel properties in such a simple system also enables us to understand aspects of the emergence of novel properties in the more complex natural systems.

It should be noted that the self-programming dynamical systems we are investigating have properties similar to nonequilibrium systems located on the phase transition in the above mentioned sense. Computation universal systems have the essential dynamical properties, e.g., information storage, transfer, and transformation capabilities, and a thermodynamical interpretation of the computational system yields an analogy between a flux of free energy and a flux of computational resources (executions), defining the nonequilibrium conditions.

1.1 NOVEL CHEMICAL PROPERTIES

Chemical evolution seems to be characterized by a successive emergence of new functional properties. The physical properties (shape, charge, etc.) of the chemical species together with macroscopic thermodynamical properties, such as temperature and pressure define the possible interactions with other molecules and thereby their functional properties. Chemical systems create new properties through recombination of molecules via chemical bonds. New combinations between existing molecules and combinations of new molecules with other molecules, then define new functional properties in a system at large. In the following, we shall refer to this loop: *molecules* → *physical properties* → *functional properties* → *interactions* → *new molecules*, as an example of autonomous programming, or self-programming, of matter. See Figure 1(a).

The essence of this chemical loop can be extracted by choosing another kind of programmable "matter" in which the elements are able to alter each other in a more direct way and recombine to create new functional properties. In other words, a loop abstracted from molecular reactions and re-implemented in a more accessible media without the particular physical interpretation in chemistry. Clearly, any implementation implies a physical interpretation. However, for a computational implementation we do not need to worry about the physical interpretation. The computer wil take care of that. Thereby the programming loop becomes *functional properties* $\rightarrow$ *interactions* $\rightarrow$ *new functional properties*, as new elements with new functional properties are formed. See Figure 1(b). In the computational systems to be discussed, such elements are equivalent to programs or assemblies of instructions.

1.2 INFORMATION PROCESSING IN ORGANISMS

It is obviously important for living systems to have information-processing capabilities. By information processing we refer to an organism's ability to respond in a coherent way to both internal and external signals. In this sense, modern life-forms show many different levels of sophistication of information processing.

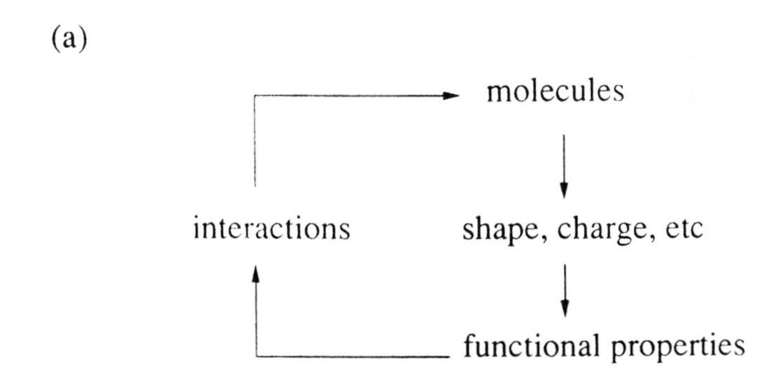

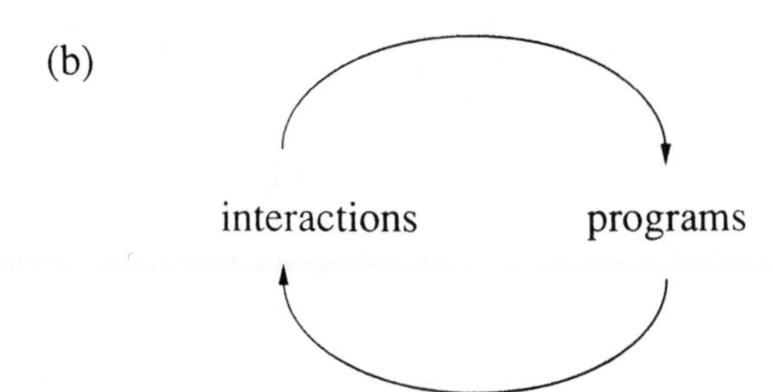

FIGURE 1 (a) Evolution of novel functional properties in chemical systems. This is an example of autonomous programming of matter. (b) The functional self-programming loop implemented on a computational system.

At one end of the biological information-processing scale, we find the retrovirus carrying only an information string of RNA in its protein coat, coding for the RNA string itself, and some proteins to be used as building blocks for the virus coat. Its information-processing capacity is primarily due to its ability to recognize suitable host cells, and to inject its genetic material into such a host cell, since the virus needs the cell's protein synthesis machinery for reproduction. New viruses then self-assemble from the parts produced by the host. The assembly process also requires information processing, primarily in terms of molecular recognition. This process is driven by free energy.

A prokaryotic cell is significantly more complicated. It is able to utilize chemical resources directly via its own complicated metabolic network. Responses like swimming towards chemical food gradients, producing thick cell walls in hostile environments, DNA interchange between different cells, pseudo-pod formation, and cell division, are all examples of very complicated processes involving biochemical information processing in the sense that the cell responds to internal as well as external signals.

A further step in complexity is found within a eukaryotic cell, with its cytoskeleton and more advanced organelles. A discussion of information processing at the sub-cellular level and further references can be found in Hameroff et al.,[20] or Rasmussen et al.[48] The eukaryotic cell also forms the basic building block in multicellular organisms, and thereby enables biological systems to take a giant leap in complexity. The differentiation process during ontogenesis, as well as the stimulus-response repertoire in multicellular organisms calls for even more advanced information-processing capabilities. One of these capabilities is learning, which over the last decade has become a major field of theoretical research. There is an extensive literature on simple models for aspects of information processing in artificial neural networks. For references, see for instance Touretzky[54] or Palmer.[42]

1.3 BIOMOLECULAR AND FORMAL INFORMATION PROCESSING

At the far end of the biological information-processing scale, we find humans capable of abstract and creative information processing at least at a level of universal computation, which is the most advanced known level of formal computation and information processing.[23,37] However, the *information mechanics* (e.g., the details of how the information is processed) in the hierarchy of computation, as developed in formal computation theory,[23,37] do not at all resemble the hierarchy of biomolecular information processing found in biological systems. Thus, the level of computation and information processing in terms of computation theory of each of the above mentioned life-forms is not known.

On one hand, it is not known whether a system needs to be able to perform universal computation in order to be alive; on the other hand, a system supporting universal computation does not guarantee an ability to develop life. In fact, computation universal systems can be very fragile. It is known, for instance, that classical hard billiard balls, moving in a simple periodic potential, can be prepared in an initial condition to perform any computation.[16] Since the slightest disturbance of these initial conditions, or any perturbation of this mechanical system during the computation, will cause the system to diverge exponentially from the original trajectory, any computation easily loses its original meaning. The mechanical system can only perform computations on a dynamically unstable set of trajectories with measure zero. It is, therefore, difficult to imagine that such a system autonomously could program itself and develop higher-level stable computational structures.

A precondition for the definition of a suitable computational system is that it has self-programming abilities. It also needs to have the ability to store, to transfer, and to process information in a reasonably stable manner. We know from modern life-forms that the combinatorial possibilities of their programmable matter, the biochemical molecules, ions, and water, allow universal computation at some higher level. It, therefore, seems reasonable to demand a collective universality from the elements constituting an artificial system where the emergence of life potentially could occur. We can also argue for universality from a purely pragmatic point of view, since computation theory already has provided us with a number of universal formalisms, which we can use as a starting point, or as inspiration for the design of new and maybe more suitable formalisms.

1.4 OVERVIEW

In section 2 we review different computation universal systems in the context of self-programmable dynamical systems. Section 3 is devoted to the algorithmic definition of a formal self-programming system, a modified parallel von Neumann machine (MVNM), and a definition of the system in terms of dynamical systems theory. Since this computational system is rather complex (many different interacting elements and a high-dimensional state space), the dynamics can only be followed through projections. In section 4 we follow a number of such projections, focusing on the self-organizing properties of the dynamics. Throughout this section a variety of system parameters are changed to see how this is reflected in the dynamical properties. Section 5 contains a discussion of a variation of the MVNM. Since the resulting system is quite different from the original one, we have given it a section of its own. In section 6 we discuss how the MVNM relates to other existing formal self-programming paradigms, discussed in section 2, and how new formal paradigms can be constructed. Section 7 is devoted to a discussion of how the dynamics of self-programming in the MVNM relate to the dynamics of self-programming in prebiotic and proto-biological systems. The main results from the paper are finally collected in section 8.

2. UNIVERSAL MACHINES AS PROGRAMMABLE MATTER

There exist formal computational paradigms which *in principle* allow self-programming, but this property has never really been studied and no serious attempt has been made to use this capacity. Already von Neumann[57] noted that computer programs are equivalent to data, that programs alter data, and thus programs can alter programs. We shall, in the following, define programmable matter in the context of computationally universal systems.

The definition of a universal computer together with many modern concepts of programming emerged from a collection of questions including: Which processes are describable? Can any fixed language admit the description of all describable processes? According to Turing[55] a describable process is equivalent to a process with a set of rules which all the time direct its behavior. Such rules define what he called an effective procedure. This is an intuitive definition so what can be called an effective procedure cannot be shown rigorously. Turing's amazing answer to the second question was that it *is* possible to realize the notion of an instruction-obeying machine in a form which remains constant no matter how complex the procedure in question. This means that it is possible to set up a language and a simple *universal* interpretation machine which can handle all effective procedures. This universal machine was later named the universal Turing machine.

The class of universal formalisms includes Turing machines,[23,37,55] cellular automata,[34,50] the λ-calculus,[2,5] Post systems,[37,44] the hard billiard ball computer,[35] general recursive functions,[25] classifier systems,[18,22] partial differential equations,[40] von Neumann machines,[23,57] and C^∞ maps on the unit square.[38] A universal machine can simulate the information transformation rules of any physical mechanism for which those rules are known. In particular, it can simulate the operation of any other universal machine—which means that all universal machines are equivalent in their simulation capabilities. It is an unproven, but so far uncontradicted, contention of computation theory that no mechanism is more powerful than a universal Turing machine. This conjecture is known as the *Church-Turing thesis*.

Our physics is at least able to support universal computation, since we are able to build digital computers. However, we cannot be sure whether or not Nature is able to support a more general class of information processing. The claim that *any* physical process can be precisely described by a universal machine,[15] is often referred to as the *physical Church-Turing thesis*.

Technically, the common property for all universal formalisms is that their range is equivalent to the class of partial recursive functions (= computable functions = functions which can be specified through effective procedures). A less formal way of identifying the class of computable functions is to identify them as functions which are "human computable." Not all computations can be performed in a finite number of steps, since it can be shown that some effective procedures will never let a computation halt. Some, or all, values of such functions are, therefore, not defined. There exists no algorithm by which we can decide in the general case whether a random chosen function would let a computation halt or not. This is referred to as the *halting problem*.[37] This implies that the "trajectory" of a randomly chosen initial condition for a computation universal system cannot be know *a priori*. This implies that dynamical open-endedness in principle exists in such systems, which is a property we want.

2.1 CANDIDATES FOR SELF-PROGRAMMABLE MATTER

It is yet unknown how a given formalism, together with its autonomous dynamics, determine a system's ability to spontaneously organize itself into complex structures. We shall briefly review some of the universal formalisms in the light of self-programmability. In section 6 we discuss different levels of equivalence between some of the following formalisms, in particular we introduce a transformation of the information mechanics from one formalism to the other.

2.1.1 CELLULAR AUTOMATA Cellular automata (CA) have, for this purpose, been extensively examined by a number of people,[2] since von Neumann[57] and Ulam[56] invented the framework in the late 1940s. No other framework has inspired so much work in the area of complexity as CA's. The beauty of the CA approach is clearly its simplicity. These simple low-level and local rules are able to produce amazingly complex spatio-temporal patterns. This has been the leading motivation for many people using CA's to understand self-organization of complex phenomena in Nature. Real-world complex spatio-temporal patterns, as life itself, may even seem tractable when viewed from this perspective. Despite their simplicity, CA's are powerful tools for modeling many physical systems.[7,19,52] Cellular automata are presently being exploited as a framework to describe Nature much the same way as differential equations were used as a framework in the last century.

CA's have been used to construct complex patterns for several years, and we have probably just seen the beginning. CA dynamics can support universal computation, and CA systems are also self-programmable. The main difficulty with the CA approach seems to be associated with its attractiveness: the extreme low-level representation of interactions. CA's are programmed at the level of the local physics of the system and, therefore, higher-level cooperative structures are difficult to evolve in CA's. To have a complete local description of all interactions is a strong condition. CA dynamics are typically fragile, even in the most interesting areas of the rule space.[29,30] Despite these problems both emerging, cooperative patterns and ingeniously simple rules for self-replicating patterns have been found and constructed within different CA's. Some of the best known examples are the emergence of simple *gliders* in Conway's two-dimensional CA rule, Life,[17] and the emergence of spiral waves in CA models of two-dimensional excitable media.[52] Langton[3] has shown that certain two-dimensional rules are able to support cooperative glider

[2] See, for example, Gardner,[17] Toffoli,[52] Wolfram,[58] Langton,[29] and Li.[31]

[3] See, for instance, the discussion in Farmer and Belin.[11]

patterns which template so fast that they can outgrow disturbances originating from collisions with other virtual particles.

2.1.2 TURING MACHINES The Turing machine from 1936 was the first universal system where the machine concept was used.[55] Mostly for historical reasons the universal Turing machine became the reference point for all other computationally universal systems. That is also the reason for much of computation theory being formulated in terms of Turing machines, which from our perspective is unfortunate, since the *information mechanics*, e.g., the details of information storage, transfer, and transformation, in natural systems seems to be very different from that of Turing machines. It is true that living systems exhibit a "machinery" for more advanced information processing such as that seen in transcription, in axiomatic transport, and in mitosis, but this machinery is by no means localized to a "head" with an associated "tape." The Turing machine concept as a basis for a self-programming computational chemistry is presently being explored by McCaskill in Göttingen.[36]

Nature also exhibits massively low-level parallel information processing driven by free energy where the key is "molecular recognition." This is the case for most chemical reactions. A useful computation theory for natural systems has yet to be formulated. A more detailed discussion of biomolecular information processing is found in Knudsen at al.[26]

2.1.3 POST SYSTEMS Variations of Post's production systems from 1943,[37,44] such as Lindenmeyer systems,[45] have, for many years, been used to study the morphogenetic properties of plants. The goal for these investigations has been to understand *morphological self-organization* in opposition to *functional self-organization*. Grammar systems could, however, be interesting as self-programmable systems once interactions between different *forms* are given a meaning, for instance, when the forms are given *functional* properties.

2.1.4 THE λ-CALCULUS The λ-calculus invented by Church in 1935[2,5] has recently been used as a framework for a constructive computational system.[12,13] Fontana's algorithmic chemistry, "AlChemy," is defined as a modified version of pure LISP. Functional interactions in the λ-context are equivalent to substitutions. Two functions interact by way of substituting one of the functions into the other—as function composition. In a "λ-gas" with a suitable tree representation of functions, new trees with associated functions can emerge via collisions between existing functions. Functional interactions are thereby given a very clear interpretation, probably clearer than in any other representation. The functional dynamics *is* the state space dynamics.

The λ-gas with the bitstring representation of functions encoded as trees is also a natural generalization of a chemostat, a well-stirred wet macromolecular reactor with a molecular representation of functions. One of the apparent problems with the current λ-calculus approach seems to be a system tendency to produce identity operators unless collisions producing these functions, after they have occurred, are made elastic (undone). Another property, which can be viewed both as a strength and a weakness of the approach, is its non-obvious relation between physical space and functional properties. The functional properties are defined as mathematical entities without any physical extension. Thereby, the functional properties are not implemented at the level of the physics of the system as, for instance, the functional properties are for CA's.

2.1.5 VON NEUMANN MACHINES The von Neumann machine[57] (VNM) was introduced in 1945 and it is the logical structure that underlies most modern digital computers, and therefore it is by far the most used universal machine. The fact that one can easily program any function into a VNM accounts for the wide use of the digital computer today. As a kind of self-programmable matter, a multi-tasking VNM has a number of nice properties: It is easy to interpret its autonomous dynamics as a computation, since each elementary operation has a clear computational definition. Programs in the memory can alter themselves and write new programs into the memory, which makes the system self-programmable. One of the major problems associated with the following use of VNMs as self-programmable matter is related to our particular definition of the system. Since we do not want any central-process administration of protected memory allocation, any process can, in principle, read and write to any other memory location, which implies that functions encoded as programs do not have any protection from being altered or destroyed. Defining the VNMs in this way, therefore, gives the system a number of properties similar to CA's.

3. THE VON NEUMANN MACHINE AS A DYNAMICAL SYSTEM

In this section we modify a parallel von Neumann machine to allow self-programming. This is an example of changing a known formal computational scheme into the autonomous scheme found in natural systems. We first give an algorithmic definition of the modified von Neumann machine (MVNM) to be investigated in the following. We then cast the MVNM in the concepts of dynamical systems, which enables an analysis of the system in terms of dynamical systems theory. Since the system is constructive, e.g., it can introduce novel functional properties, we need to introduce an interaction graph (see also Fontana[12,13]), which allows a more precise analysis of the functional aspects of the state space dynamics.

3.1 A MODIFIED VON NEUMANN MACHINE

A detailed description of a von Neumann machine can be found in any standard textbook on computation theory.[23] To our knowledge the autonomous dynamics of a VNM with a randomized memory has never been the subject of extensive investigations. In fact, most work in the area of computation theory has dealt with problems and possibilities for *controlled*, not *autonomous*, information processing.[14]

The systems we shall discuss in the following differ a little from standard multi-tasking VNMs. We shall, in general, refer to autonomous VNM's as *Coreworlds*, since the dynamics is the core dynamics, which correspond to the dynamics of the random access memory (RAM) in a digital computer (core = memory). We have named the MVNM's to be introduced in the following, Venus I and Venus II, where the difference is due to the details of the updating schemes. Venus I is also discussd in Rasmussen et al.[47] and Knudsen et al.[26] We introduce and analyze yet another and simpler type of MVNM in section 5.

The cores in our MVNM's (Venus I and II) are one-dimensional address arrays with periodic boundary conditions. Each address in the core is occupied by one of the ten basic instructions[4] (see Table 1(a)). A number of execution pointers point to instructions to be executed. We have introduced a notion of computational resources primarily to prevent a high number of pointers at the same address and to enhance cooperative interactions among different instructions. Each address is, therefore, associated with a certain amount of computational resources $r(y,t)$ ($0 \leq r(y,t) \leq 1$) measured in fractions of one execution, where y denotes the location and t the time. An instruction is executed if its address has a pointer, and its local resource neighborhood (R_{res} addresses to either side of the actual address, $2R_{res}+1$ altogether) has sufficient computational strength (at least what corresponds to one execution). Insufficient resources will eliminate execution pointers. Reading and writing to memory is only possible within a certain operational radius R_{opr} relative to the executing instruction. This restriction is introduced in order to obtain some notion of locality in the core. All addressing is relative. Unless the executed instruction tells the pointer to move to a specified location, the virtual machine moves the pointer to the next higher address in the memory ($y \to y+1$). At each core update, a certain amount of computational resource, Δr ($0 \leq \Delta r \leq 1$) is supplied to every address up to a given maximal level r_{max}, which is always smaller than the equivalent of one execution. $r(y,t)$, Δr, and r_{max} are all real numbers.

Except for the concept of computational resources, the local addressing, and a few other things mentioned below, the core operates as an ordinary multi-tasking von Neumann machine with an execution queue of length L, e.g., with at most L execution pointers.

The core is updated in parallel for Venus I. When the system is running, it has many pointers active at the same time. All the instructions associated with these

[4] We have, out of historical reasons, used the same basic instructions as found in Dewdney,[6] since it was here we originally got the idea of using a von Neumann system as a basis. This instruction set is by no means optimal for our purpose, but we have kept it as a common reference for this and earlier work.

pointers are executed and the results stored in a temporary array before changes in the core are made. In this way each instruction "sees" the same core, even though we are simulating the parallelism on a sequential machine. In case of conflict between two instructions, the instruction with the highest number in the execution queue will have its changes effected. However, specific details on how conflicts are resolved do not affect the dynamics of the system much.

In Venus II the updating occurs in a regular sequential manner. All pointers in the queue are executed one by one and they sequentially alter the core. The computational resources are updated after every pass through the execution queue of maximal length L, just as in the parallel system.

As described so far, the system is deterministic. Since noise seems to play an important role in biological evolution, computational noise is introduced in two different ways. We have arbitrarily chosen that whenever a MOV instruction (see Table 1(a)) is executed, there is a certain probability, P_{mut}, that the Word being moved mutates. The MOV instruction copies the Word at the A-address to the B-address (see Table 1(b)). In this situation each Word has an equal probability of changing into any one of the ten legal instructions. The operands for the new instruction are also chosen at random. Changes are, thus, restricted to always yield legal syntax, just as in real chemistry. Molecular interactions cannot produce new molecules falling outside the "syntax" of our chemistry. The perturbations caused by this kind of noise appear to be important for the richness of different evolutionary processes and, thus, to the kind of spatio-temporal structures the system can develop.

TABLE 1 (a)

Opcode	Function
`DAT    B`	Non-executable statement. Terminate the process currently executing. Can be used for storing data.
`MOV A, B`	Copy the contents of A to B.
`ADD A, B`	Add the contents of A to the contents of B and save the results in B.
`SUB A, B`	Subtract the contents of A from the contents of and B and save the result in B.
`JMP A`	Move the pointer to A.
`JMZ A, B`	If B equals zero, move the pointer to A.
`JMN A, B`	If B differs from zero, move the pointer to A.
`DJN A, B`	Decrement B, and if B differs from zero, move the pointer to A.
`CMP A, B`	If A differs from B, skip the next instruction, e.g.,move the pointer two steps ahead instead of one.
`SPL    B`	Create a new process at the address pointed to by B

TABLE 1 (b)

Addressing Mode	Effective Operand
# (immediate)	The effective operand is the value in the data field. Example: MOV #3,.. has the effective operand 3.
$ (direct)	The effective operand is the word pointed to by the value in the data field. Example: MOV $2,.. has the effective operand located two words towards increasing addresses.
@ (indirect)	The effective operand is found by looking at the data field pointed by the actual data field, and then using the direct mode.
< (autodecrement indirect)	As indirect, only the value pointed to by the actual data field is decremented before being used.

The system can autonomously eliminate and produce pointers. A pointer is eliminated whenever it has insufficient computational resources or whenever it meets a DAT instruction (see Table 1(a)). A pointer is duplicated whenever it meets a SPL instruction (see Table 1(b)). In this case the new pointer is sent to the location specified by the B-address of the SPL instruction. Since there is a finite probability that every pointer in a randomized core is eliminated due to DAT instructions, we disturb the core by introducing new pointers at random with a low pointer appearance frequency P_{point}, each time the core is updated. We thereby assure that the system is always active. This random influx of pointers is an additional way of driving the system besides the influx of computational resources, Δr.

The basic concepts of how the systems operate are summarized in Figure 2.

3.2 DEFINITION OF STATE SPACE AND MAP

To give a more concise description of the MVNM dynamics in the next section, we cast the alogrithmic description of Venus I and II in dynamical system concepts and introduce the interaction graph. The state space $\mathbf{\Omega}$ in our MVNM is given by the structure of the individual instructions (in a usual VNM notation: Opcode, A-mode, A-field, B-mode, B-field, and the potential pointers), the maximal number of pointers L (the instruction execution queue length), and the size of the memory N.

Knowing the number of different Opcodes, modes, fields, and the operation radius R_{opr}, the system state, $\mathbf{x}(t) \in \mathbf{\Omega}$, at time t can be described by an ensemble

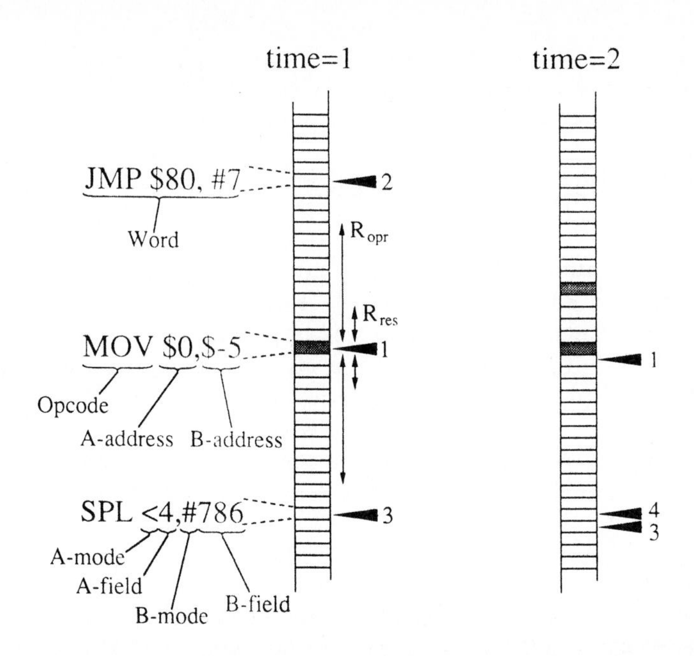

FIGURE 2 The basic concepts of how the modified von Neumann machines, Venus I and II, operate. Shown is a part of the core of Venus at two subsequent generations. The enumerated arrows represent pointers and the numbers refer to their pointer queue position. Each Word in the core consists of an Opcode, an A-address and a B-address, and each address consists of a mode and a value (see Table 1). At time $= 1$ the first instruction to be executed is the MOV instruction. Since the A-address is the relative address 0, i.e., the MOV instruction itself, and the B-address is the relative address -5, the MOV instruction will copy itself to a position five Words back in the core. The shaded areas at time $= 2$ are the Words containing the MOV instruction after it has been executed at time $= 1$. Note that the pointer (number one) is being incremented one address after the execution. However, the execution of an instruction only takes place if a sufficient amount of computational resources are available within the resource radius R_{res}. Furthermore, an instruction is only permitted to change the contents of the core or to jump to positions in the core within the operational radius R_{opr}. Attempted actions outside this radius will be wrapped back within the radius (modulus R_{opr}). The second instruction to be executed at time $= 1$ is the JMP instruction. In this case the A-address is the Word seven addresses before the instruction; thus, pointer number two is no longer in the part of the core that we are observing at time$=$ 2. Finally, the SPL instruction is executed. SPL will generate a new pointer at the B-address, which in this case is the SPL instruction itself, since the mode of the B-address is immediate. Therefore, at time $= 2$ a fourth pointer has been added to the pointer queue pointing at the SPL instruction. The pointer (number three) that executed the SPL instruction at time $= 1$ has moved ahead one Word in the core.

of N ($|Opcodes|$ + $|modes|$ + $|fields|$ + 1)–dimensional vectors. The number of different states in Venus is therefore given by

$$|\Omega| = [|Opcodes||A - modes|(2R_{opr} + 1)|B - modes|(2R_{opr} + 1)]^N \sum_{i=0}^{L} N^i \; . \quad (1)$$

Note that this estimation of $|\Omega|$ does not include the computational resources $r(y,t)$, associated with each address y, since $r(y,t)$ is defined as a real number between 0 and r_{max}. $r(z,t)$ could, however, just as well be defined as a fraction allowing us to include it in Eq. (1). To give an idea of the size of Ω, the state space typically has of the order of $10^{33,000}$ different configurations for the systems we are considering.

Let us denote the executions associated with a single update of the whole core (a generation) by an operator $\mathbf{T}$. Recall the algorithmic description of $\mathbf{T}$ in section 3.1. We can now formally describe the VNM dynamics by the map[47]

$$\mathbf{x}(t+1) = \mathbf{T}(\mathbf{x}(t))\,. \quad (2)$$

The stochastic elements of $\mathbf{T}$ can be eliminated by setting $P_{mut} = P_{point} = 0$. For the deterministic part of $\mathbf{T}$, we can define the following. $\mathbf{A} \subseteq \Omega$ is an *invariant* set for $\mathbf{T}$ iff $\mathbf{T}$: $\mathbf{A} \mapsto \mathbf{A}$. Since Ω is finite, the invarian set $\mathbf{A}$ is an *attractor* for $\mathbf{T}$ if there exists at least one $\mathbf{x}_i \notin \mathbf{A}$, such that $\mathbf{T}(\mathbf{x}_i) = \mathbf{x}_j$, $\mathbf{x}_j \in \mathbf{A}$. All other states are defined as *transients*.

In section 4.6 we shall see how the average transient length for a particular class of initial conditions scale with N, for N/L and N/R_{opr} constant. The same questions could be asked for any class of initial conditions and for other map properties, such as the average number of attractors or the distribution of cycle lengths. In section 5 we see how the dynamics of a MVNM depends of the structure of its instruction set. It would also be interesting to know how likely is it to jump from one attractor to another as a function of P_{mut} and P_{point}. Questions of another nature, such as what the state space volume dissipation rate is with particular initial conditions, are obviously important, since too strong a selection of specific Opcode, mode, and field specifications may lock the system on trajectories with monotone non-interesting functional dynamics. Such systems get "stuck."

Note that a random mapping over $|\Omega_{eff}|$ vertices can be used as a reference for the dynamics for $\mathbf{T}$, where Ω_{eff} refers to the number of effective (e.g., in principle reachable states) under given initial conditions. Such a mapping gives stochastic boundaries for the dynamics. We shall briefly discuss this in section 4.6.

3.3 FUNCTIONAL DYNAMICS: THE INTERACTION GRAPH

The *functional* dynamics occurring between the different functional compositions, e.g., assemblies of instructions in the core acting as a coherent functional unit, is at any time defined by the pointers and the instructions they execute. To illustrate the concepts, let us look at a simple example. Imagine a core with only two pointers located at two JMP instructions, one at address y and the other at address $y + 2$. Assume the first instruction sends the pointer to the address $y + 2$, and the other instruction sends the pointer to the address y. These two instructions form a simple cooperative structure, since they ensure the execution of each other. The functional dynamics for this simple system can be represented by an *interaction graph*: a directed cycle where the two vertices represent the two JMP instructions and the two directed edges represent pointer jumps (e.g., the A-addresses; recall the definition in Table 1). If only one pointer was present in the above example, we would again obtain cooperation, but this time of an oscillatory nature (period two). No directed cycle would be present in this interaction graph, but it would appear after a single iteration of the graph. The interaction graph is, in general, complicated for the whole core, since it has up to ten different vertex colors (ten different Opcodes) and two different edge colors, one to indicate the location of the information needed in the execution of the instruction, and one to indicate the location being influenced by the execution of the instruction (recall the definition of the instruction set in Table 1). Note that many different states can have the same interaction graph.

We can now interpret cooperation occurring in the MVNM in terms of the interaction graph and the map. Assume that $\mathbf{x}(t)$ is on an attractor of period τ, i.e., $\mathbf{x}(t+\tau) = \mathbf{x}(t)$. Since the functional dynamics on the attractor is characterized by a *reproduction* of the interaction graph for every τ time steps, the dynamics of the functional compositions by definition is *cooperative*. The different functional compositions facilitate the production of each other with period τ. Note that identification of cooperation also includes trivial dynamics, such as fixed points for the map. Fixed points for the map are reflected as loops and oriented cycles in the interaction graph.

Cooperation within functional compositions can, of course, also occur when the map dynamics is on a transient. In such a situation the interaction graph may change as it reaches an attractor. Since the existence of an attractor implies the occurrence of functional cooperation, an attractor in the MVNM is a sufficient but not a necessary condition for cooperation. We shall further discuss the notion of functional cooperation and give a number of examples within MVNMs in section 4.4.

The interaction graph is a useful concept, both in the discussion of cooperation and to distinguish qualitatively different types of functional dynamics. Qualitatively different types of functional dynamics also have qualitatively different interaction graphs. It can, therefore, be used to identify different evolutionary epochs in the system.

4. SELF-ORGANIZATION OF PROGRAMMABLE MATTER

Having defined a particular kind of self-programmable matter, we now turn to an investigation of the self-organizing properties of the system. Thereby, we hope to learn something generic about the dynamics of self-programmable matter, properties which are also present in natural self-programmable systems. To create macroscopic dissipative structures in a physical system, free energy and many microscopic degrees of freedom are needed. A thermodynamical interpretation of the computational system yields an analogy between the flux of free energy and the flux of computational resources (executions per iteration) and an analogy between the microscopic degrees of freedom in the physical system and all the possible functional interactions in the computational system (instructions, programs). To create macroscopic computational structures, many executions as well as ample functional interactions are needed.

We shall discuss the dynamics of the MVNMs in the state space and in terms of the interaction graph. The Shannon entropy and the mutual information of Opcodes are used as yet other indicators for the dynamical properties. We are primarily interested in functional self-organization and cooperation. We alter the MVNM in several ways to monitor the results on the dynamics. As defined, the system can be altered in terms of the *instruction set*, in terms of the *updating scheme* (Venus I is updated in parallel and Venus II is sequentially updated), in terms of *parameters* (N, L, Δr, r_{max}, R_{res}, R_{opr}, P_{mut}, and P_{point}) where N defines the system size, and in terms of the *initial distribution* of instructions and their arguments in the core. In the first part of this section, we discuss simulations with the same instruction set as defined in Table 1 (Venus I and II) where the system is subjected to external noise. In section 4.6 we investigate how the functional dynamics of Venus II depends on the memory size N with no external noise. In section 5 the instruction set is varied (simplified) to define another MVNM which dynamics we discuss.

4.1 SYSTEM PARAMETERS

From earlier experiments with the Venus I system (Rasmussen et al.[47]), it is known that the most interesting self-organizing processes occur when the system parameters are combined to yeld a high resource regeneration rate, Δr large; a large operation radius, R_{opr} large; and a small resource radius, R_{res} small. A small R_{res} is used mostly because it speeds up the simulation process. In the following the parameters are defined as (N; L; Δr; r_{max}; R_{res}) = (3584; 220; 0.5; 0.5; 3) unless otherwise specified. A system defined along these lines gives ample execution resources and enables a rich spectrum of functional interactions. It corresponds to a "fertile computational jungle."

4.2 INITIAL CONDITIONS

Since the MVNM is computation universal (see appendix), in principle, we cannot *a priori* know the trajectory from an abitrary initial condition (recall section 2.1).

There are a number of different ways to initialize the MVNM. Numerous experiments with Venus I have shown[26,47] that a homogeneous randomization of the memory is not optimal for significant self-organizing processes within the spatio-temporal limitations of our computers. The occurrence of interesting self-organizing processes is enhanced if the initial instruction distribution is biased. In this respect the MVNMs are similar to certain CA's.[33] The system needs some initial reactivity facilitated by instruction inhomogenity to drive the process "off the ground." This can be viewed as an initial chemical potential. Technically this can either be obtained by using a number of coupled Markov matrices in the memory initialization process (to shift the distribution away from being rectangular) or for example by inducing a self-replicating program in the memory which creates inhomogenities in a previously homogenized memory. It is important to note that the self-replicating programs typically start malfunctioning after 200 iterations due to computational noise and copying on top of each other, and, hence, no higher-order self-replicating properties are preserved.[26,47] The valuable effect of this program is that it creates a reactive, inhomogeneous core.

Using the general Markov transition matrix approach to bias the initial conditions we need a matrix of the same dimension as an address-state vector (Opcode, A-mode, A-field, B-mode, B-field, pointers). With the instructions defined as in Venus, we need in principle to specify a six-dimensional matrix with $(10 \cdot 4 \cdot (2R_{opr} + 1) \cdot 4 \cdot (2R_{opr} + 1))^6 \simeq 4.8 \times 10^{51}$ elements, given $R_{opr} = 800$. However, a sequential biasing of a core using 5, two-dimensional matrices $O_{10,10}, Am_{10,4}, Av_{10,N}, Bm_{10,4}, Bv_{10,N}$ enables us to control five important correlations. The five matrices are controlling the probability that $Opcode_i$ is followed (next higher address in the array) by $Opcode_j$, the probability that $A - mode_j$ occurs in a Word with $Opcode_i$, the probability that $A - value_j$ occurs in a Word with $Opcode_i$, the probability that $B - mode_j$ occurs in a Word with $Opcode_i$, and the probability that $B - value_j$ occurs in a Word $Opcode_j$ respectively.

Tuning the parameters in these matrices to enhance the occurrence of a MOV instruction with an auto-decrement B-field (recall Table 1) followed by one of the jump instructions that can return the pointer to the MOV instruction (or to an instruction at a lower address (at most 1% of core size) near the MOV instruction will eventually result in a high number of instruction pairs which can copy a high number of the instructions pointed to by the A-address of the MOV instruction. Many active pairs of this (or a related) kind define a high initial reactivity and are responsible for a build-up of instruction inhomogeneities. It is a run-away process with sensitivity to initial conditions.

In the next sections this biasing is used to produce initial cores capable of evolving complex cooperative structures. In section 5 we show that a simpler MVNM is able to support interesting self-organizing phenomena from a homogeneous initial memory.

4.3 STATE SPACE PROJECTIONS OF THE DYNAMICS

We shall now follow the evolution of $\mathbf{x}(t)$ through different projections. The dynamics of the Venus I and II examples in this section has been disturbed with a little noise, $P_{mut} = 10^{-3}$ and $P_{point} = 0.05$, as described in section 3.1. We are perturbing the system with noise because we are interested in the emergence of perturbation stable computational structures.

The first representation of the state space is through a projection of the Opcode and the pointer coordinates onto the memory array (of size N) or parts thereof. The Opcode is assigned a color code and the occurrence of pointers is indicated by white underscores.

Color plate 25 shows an *(Opcode, pointer)*-projection of an initial core from Venus II correlated as described in section 4.2 using the Markov matrices. Initially of the order of 200 pointers are distributed throughout the core. The pointers are indicated by white underscores. Note that every Opcode color is present although some inhomogenity occurs. The color code is given in the figure caption.

Color plate 26(a) shows an example of an *(Opcode, pointer)*-projection of a MOV structure after 300 iterations with Venus II, where $R_{opr} = 800$, and with the initial core shown in color plate 25. In color plate 26(b) an *(Opcode, pointer)*-projection of a particular area of the MOV structure (address 1000 through 1127) is shown at 140 consecutive iterations (time 141 through 280). Although noise, through mutations of the MOV instruction, continously induces new instructions, altering the microscopic details of the system, the global dynamics is conserved indefinitely (for more than 50,000 instructions); thus, the dynamics of this structure is recurrent and not periodic. Compare with Figure 6(a) in section 4.6.

Color plate 27 shows an *(Opcode, pointer)*-projection of a SPL-JMZ structure at time 40,000 in Venus II. The SPL-JMZ structure has several areas of intense pointer activity. These SPL-JMZ colonies compete for pointers, and an intermittent pointer dynamics between the areas occur. The areas occupied with DAT instructions absorb all pointers moving into it. This structure is developed under similar conditions as the MOV structure shown in color plates 26(a)-(b). Only the noise sequences used in the creation of the initial core and used under the simulation are different in the two processes. This structure emerged after approximately 20,000 iterations and seems to be able to persist indefinitely (more than 50,000 iterations).

Color plate 28 shows an example of a (noisy) fixed point for the mapping $\mathbf{T}$. This is not a true fixed point partly because the process has a stochastic element, since new pointers are introduced ($P_{point} = 0.05$) and partly because the occurrence of countdowns of the B-fields for the JMN(# 0, $\cdot$) instructions occasionally allow the trapped pointers to escape the JMN(# 0, $\cdot$) instructions when the B-fields equal zero. We see the same area of the core at 70 consecutive iterations. This jump structure emerged after approximately 3500 iterations in Venus I, with $R_{opr} = N = 3584$, from a random initial core with a self-replicating program. The fixed point character of this structure persists indefinitely (more than 100,000

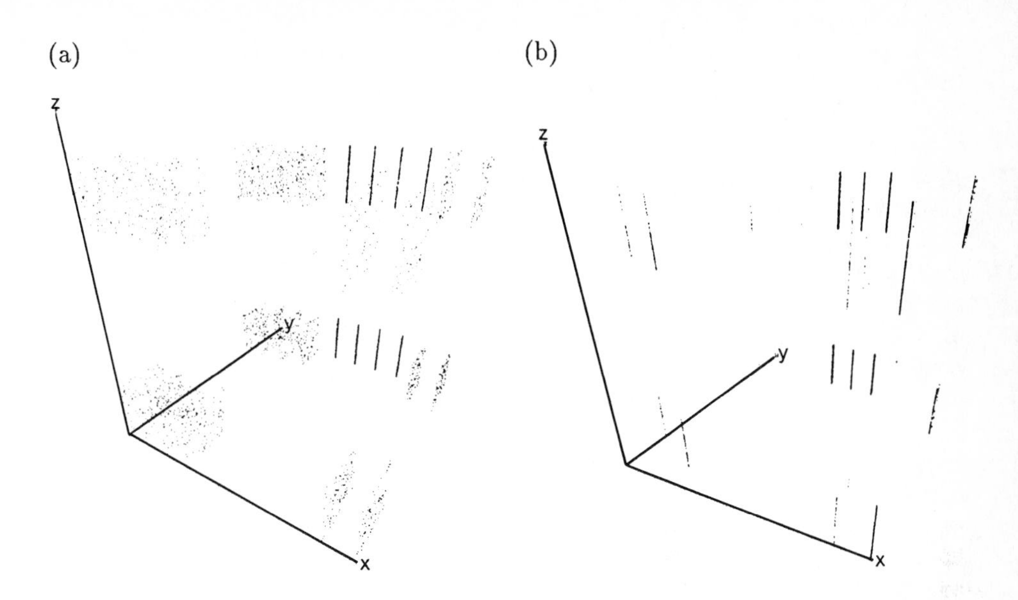

FIGURE 3 *(Opcode, A-field, B-field)*-projections of the N instruction vectors in Venus II (right-handed coordinates). (a) Projection of a randomized (biased) core as described in section 4.2. The interaction radius R_{opr} = 800 is reflected through the forbidden A-field and B-field areas outside $\pm R_{opr}$. This creates the characteristic division of allowed addressing areas into "quadratic planes." The biasing of the A-field of all the jump instructions is clearly seen, since only up to one percent of the core can be addressed (towards lower addresses) by these instructions. This creates the "rods" in the picture. (b) After 40,000 core iterations the occupied state-space volume has decreased significantly. The result of the functional self-organization is the SPL-JMZ structure also discussed in color plate 27. Note that other instructions also are present. Interaction graph projections (see section 4.6 for a discussion of such projections) show, however, that these other instructions only seldomly are executed. The change of the quadratic planes into vertical rods reflects a strong selection of certain A-fields. The B-fields continue to be distributed within the allowed $\pm$ 800 instruction range, since the B-field values frequently are altered by the dynamics.

instructions). The core shown in color plate 29 is evolved in Venus I under the same conditions as the one leading to the fixed point core—only the noise sequences differ. In this case the simulation develops a MOV-SPL structure. This particular MOV-SPL structure persists indefinitely (more than 100,000 iterations) although some other instructions, due to noise, enter the system.

In Figure 3 we see three-dimensional projections of the N state vectors onto three-tuples of the *Opcode, A-mode, A-field, B-mode, and B-field* in Venus II. Figure 3(a) shows an *(Opcode, A-field, B-field)*-projection of a randomized (biased) core as discribed in section 4.2. The small "quadratic planes" orthogonal to the x-axis are caused by the maximum interaction radius preventing A-fields or B-fields outside

$\pm R_{opr}$. A biasing of the A-fields of all the jump instructions is also clearly seen. The "roads" in the picture occur since only up to 1% of the core can be addressed (toward lower addresses) by these instructions. 40,000 iterations later the system has stabilized on a SPL-JMZ structure. Note how a strong selection of A-fields (along the y-axis) has occurred. An *(Opcode, pointer)*-projection of the same structure is discussed in color plate 27.

The functional dynamics for the above shown structures can easily be decomposed so that we can understand the nature of the functional cooperation. The interaction graphs for the fixed-point structure and for the MOV-SPL structure are discussed in detail in section 4.4. The transients for the shown quasi-stationary states (color plates 26–29 and Figure 3) differ by several orders of magnitude. This also occurs for the systems without external perturbations and is discussed thoroughly in section 4.6. Evolutionary processes, as discussed above, are in general characterized by the occurrence of successive quasi-stationary states, or epochs, which are dominated by particular kinds of functional dynamics. These aspects of the processes are discussed through the information dynamics of the system in section 4.5.

4.4 COOPERATION

The functional dynamics underlying the different metastable states and attractors in Venus I and II have a number of common features despite their obvious differences. The similarities as well as the differences can be seen through their interaction graphs. Both the topology and the vertex colors of the interaction graphs are in general very different, but the metastable states are similar in the sense that they all have *recurrent* iterated interaction graphs.

Functional cooperation in the MVNMs is defined as invariant or recurrent sets of functional compositions. Recall the discussion in section 3.3. Note that each instruction or group of instructions simultaneously can be a member of several functional compositions.

The viability of the MOV-SPL structure which we have seen projections from in section 4.3 (see color plate 29) stems from a cooperation between the MOV and the SPL instructions with appropriate A- and B-fields. The MOV instruction with very high probability either copies another MOV instruction or a SPL instruction, guaranteeing the reproduction of the structure. The SPL instruction hands out pointers either to another SPL instruction or to a MOV instruction, thereby guaranteeing the structure is kept alive. Part of the interaction graph for the MOV-SPL structure is show in Figure 4(a).

The more static fixed point cores, which in fact are not strict fixed points, are frequently populated with a mixture of MOV and jump instructions, can also be interpreted as cooperative structures. Figure 4(b) shows part of the interaction graph for the fixed-point core discussed in color plate 28. The cooperation here operates on two different time scales. The short time stability of the structure

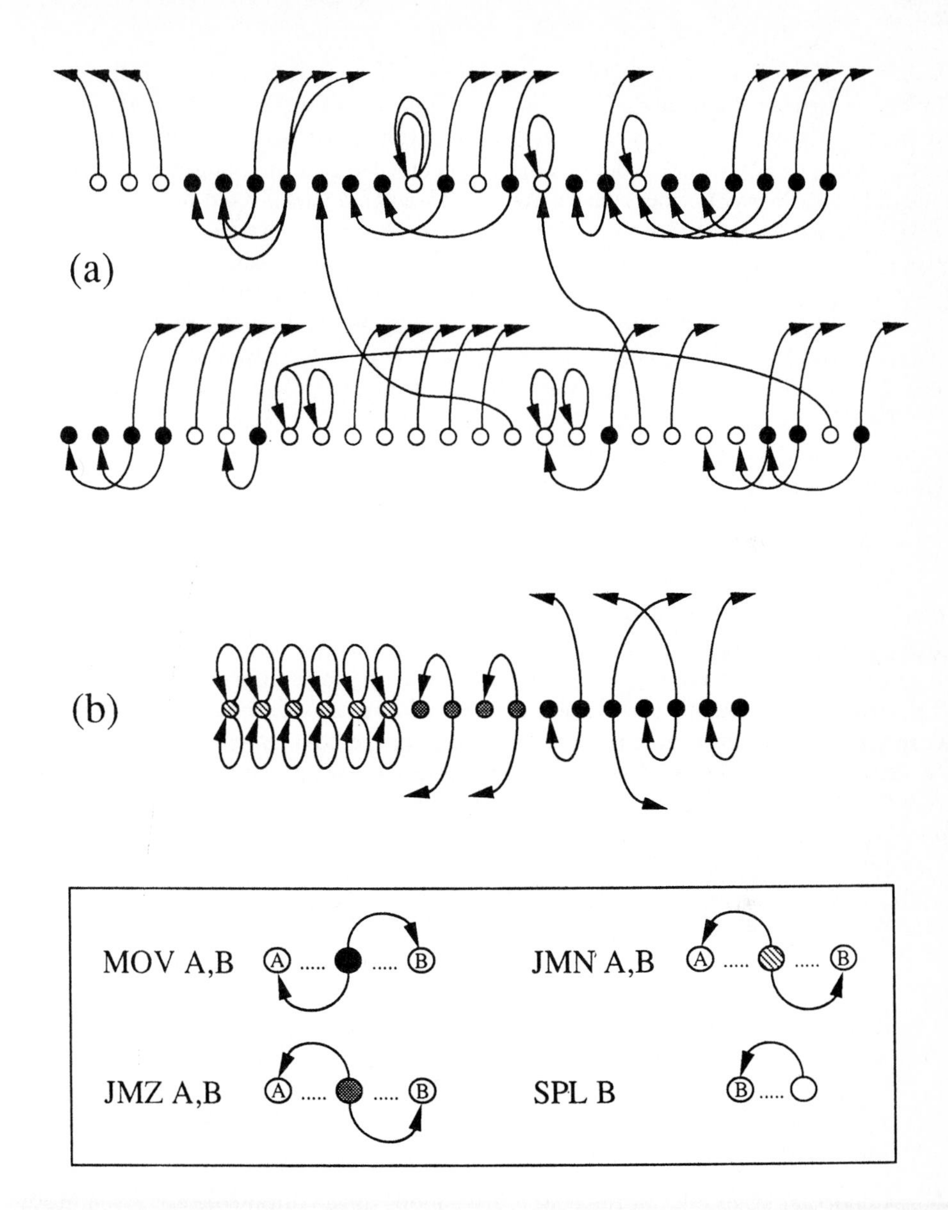

FIGURE 4 The interaction graph for two different cooperative structures developed under noisy conditions in Venus I ($R_{opr} = N = 3584$). (a) Interaction graph for part of the MOV-SPL structure. The temporal as well as the spatial dynamics of this structure is recurrent, but not periodic. Although the details of the interaction graph are altered a little at each iteration the overall composition is conserved. See text for details. (b) Interaction graph for part of one of the jump cores. Note how this interaction graph is much simpler than the one obtained for the MOV-SPL structure. See text for the dynamics of the iterated interaction graph.

is due to all the loops on the JMN instructions. The long time stability is facilitated by the instruction mixture. Once a pointer hits outside one of the JMP instructions, it will execute a number of MOV instructions, primarily copying other MOV instructions or JMN instructions. This will continue until the pointer is trapped on one of the JMN instructions, whereafter the core again is at a fixed point—although with a slightly different composition than the earlier one. As we see, such a fixed point structure is also very stable to perturbations. In fact, simulations show that they are more perturbation stable than the MOV-SPL structures.

The dynamics leading to both of these structures is due to a runaway mechanism, a positive feedback loop, easily obtained from a randomized but inhomogeneous core or from a malfunctioning self-replicating program as discussed in section 4.2. The jump and the MOV-SPL structures are commonly found in the Venus I system for $R_{opr} = N = 3584$. The fixed point occurs a little more than every second simulation and the viable MOV-SPL structure in one out of four in the Venus I system. Variants of these structures, together with more rare structures, are discussed in Rasmussen et al.[26,47]

Note that these cooperative structures are *spatio-temporal* structures which reproduce as *a whole* and not through an individual reproduction of the each of the parts. This property does the emergent cooperative structures in Fontana's λ-gas also have, although they do not have the spatio-temporal character.[13] The important steps closing the interconnected cooperative cycles in the interaction graph for the MOV-SPL structure are given in Eq. (3). S corresponds to SPL, M corresponds to MOV, $(*)$ corresponds to an excess pointer created by a SPL instruction, and $(X)^{(*)}$ corresponds to an instruction X, with a pointer.

$$
\begin{aligned}
S^{(*)} &\rightarrow S^{(*)} + (*)\,, \\
S^{(*)} + M' + M + S' &\rightarrow S + M' + M^{(*)} + S' + (*) \rightarrow \\
&\qquad S + 2M' + M + S^{(*)\prime} + (*)\,, \\
&\qquad \text{and} \\
S^{(*)} + S' + M + S'' &\rightarrow S + S' + M^{(*)} + S'' + (*) \rightarrow \\
&\qquad S + 2S' + M + S^{(*)\prime\prime} + (*).
\end{aligned}
\tag{3}
$$

The important steps closing the cooperative loops in the jump core are given in Eq. (4). J corresponds to JMP.

$$
\begin{aligned}
J^{(*)} &\rightarrow M^{(*)}\,, \\
M^{(*)} + J + J' &\rightarrow M + J^{(*)} + 2J'\,, \\
&\qquad \text{and} \\
M^{(*)} + J + M' &\rightarrow M + J^{(*)} + 2M'.
\end{aligned}
\tag{4}
$$

The fact that we can characterize the functional interactions through these equations implies that a functional chemistry emerges from the physics of the MVNM.

The autocatalytic sets found in systems of catalyzed cleavage-ligation reactions of polymers[1,10,24] also reproduce or maintain as a whole and not through an individual reproduction of the each of the parts.

There is an extensive literature on cooperation in different types of dynamical systems. For a discussion on the nature of cooperative structures in evolutionary systems, see, for instance, Eigen,[8] Eigen and Schuster,[9] Kauffman,[24] Farmer et al.,[10] Rasmussen,[46] Bollobás and Rasmussen,[4] Rasmussen et al.,[47] or Fontana.[13]

4.5 INFORMATION DYNAMICS

One way of identifying both the evolutionary changes and metastable evolutionary epochs is through a calculation of the single site spatial Shannon entropy $H(t)$ of the Opcodes. The entropy is of course a very crude measure and leaves out interesting details about what kind of functional interactions are causing the changes in memory composition. The average spatial entropy $H(t)$ in bits is calculated by

$$H(t) = -\sum_{i=0}^{9}\left[\left(\frac{\sum_{core} op_i}{N}\right) log_2\left(\frac{\sum_{core} op_i}{N}\right)\right], \tag{5}$$

where $(\sum_{core} op_i)/N$ defines the probability that any core address is occupied by Opcode op_i.

Figure 5(a) shows an example of the entropy dynamics for Venus II with reactive initial conditions obtained via a biased core as described in section 4.2 and where the system is not perturbed by external noise ($P_{mut} = P_{point} = 0$). Note the significant periods of rapid change followed by small amplitude oscillations around plateaus of almost constant entropy. This particular simulation seems to be characterized by an initial transient followed by three different types of dynamics. After the initial transient the entropy dynamics stabilizes for some 5,000 iterations at a plateau around 1.5 bits. At time 10,000 this kind of dynamics is interrupted, resulting in a rapid drop in the entropy level to approximately 1.0 bits. The new functional dynamics, characterized by some recurrency, persists until time 25,000 where yet another kind of recurrent functional dynamics seems to emerge. Projections of the interaction graph show that each metastable epoch is characterized by its own typical functional interactions (see Figure 6(b)). The functional dynamics exhibits what population biologists call punctuated equilibria. The details of functional dynamics of this particular simulation are discussed in section 4.6.

The mutual information $M_{1,i+k}(t)$ between $Opcode_i$ and $Opcode_{i+k}$, with $k = 1, 2, \ldots$, also gives interesting information about the core dynamics. The average

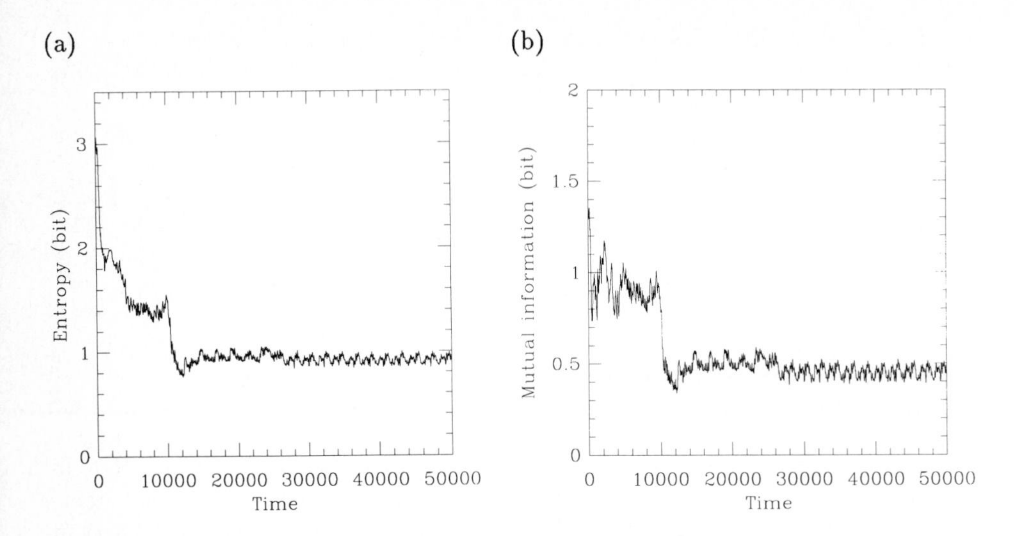

FIGURE 5 Information dynamics for Venus II without external noise. The successive evolutionary epochs can here be identified through the dynamics of the average spatial Shannon entropy and the mutual information of Opcode. (a) The single site Shannon entropy $H(t)$ as a function of the number of iterations. This simulation is characterized by an initial transient followed by three different types of functional dynamics. (b) The mutual information $M_1(t)$ as a function of the number of iterations. Note how the dynamical pattern for $M_1(t)$ is similar to the entropy dynamics, and how for this simulation it more clearly reflects the different epochs. $H(t)$ and $M_i(t)$ differ in general more than seen here.

spatial mutual information, $M_k(t)$, where k refers to the address distance between the two Opcodes, is defined by

$$M_k(t) = H_i(t) + H_{i+k}(t) - H_{i,i+k}(t) = 2H(t) - H_k(t) , \qquad (6)$$

since we are only considering averages. $H_{i,i+k}(t) = H_k(t)$ denotes the average Shannon entropy of Opcode pairs located k addresses apart. $H(t)$ is defined as in Eq. (5). The mutual information $M_1(t)$ for the simulation discussed in Figure 5(a) is shown in Figure 5(b). Note that both the entropy and the mutual information as defined in Eqs. (5) and (6) only are able to signal changes in the functional dynamics caused by MOV instructions, since only the execution of these instructions can change the core composition of Opcodes. Details of the information dynamics for Venus I is discussed in Knudsen et al.[26]

4.6 EVOLUTIONARY SCALING

In this section we show how the process richness and diversity and the transient length of the occurring processes in the Venus II system depend on the core size N, for $N/L \simeq 16.3$ and $N/R_{opr} \simeq 4.5$, when initialized with a high reactivity (recall section 4.2). The initial dynamics of such a core is dominated by many parallel multiplications of single instructions. It is thereby a runaway process with sensitivity to initial conditions.

Quantitative differences in the functional dynamics are not trivial to measure, since they require an easy way to determine differences in the interaction graph. However, an indicator for such changes can be obtained from a projection of the graph, only taking the color distribution of the vertices into account, e.g., the distribution of Opcode with pointers in the core (recall the discussion of the interaction graph in section 3.3 and 4.4). The number of vertices of each color defines a ten-dimensional vector. The problem of distingushing between qualitatively different interactions graphs is now reduced to measuring significant jumps in length and orientation of this vector. This is a computationally tractable way to obtain quantitative relations.

To simplify the investigations we only consider a purely deterministic system ($P_{mut} = P_{point} = 0$). The only difference between each simulation at a given core size is due to the details of the initial state which has been randomized as described in section 4.2. The random component of these systems are thereby restricted to their initial conditions.

In Figure 6 three simulations are projected onto part of their interaction graph, e.g., the distribution of Opcode with pointers. The simulations shown develop into (a) a pure MOV core, (b) a MOV-jump-SPL structure, whereas (c) does not settle down to any attractor within the 100,000 iterations.

Figure 7(a) shows a log-log plot of the transient length as a function of the core size for 180 simulations, 30 for each core size. The historical component (e.g., the details of the initial conditions) is the most prominent process property, since the same system size gives rise to different functional dynamics and thereby also to different transient lengths. There is, however, a clear tendency for an increasing transient length as the memory size is increased. It should be noted that the transient is assumed to terminate when the dynamics of the *projected* interaction graph settles down to a recurrent dynamics. This implies that we, for instance for the simulation shown in Figure 6(a), assume that the attractor is reached at time 4000 (where the pure MOV core is reached) and at time 28,000 for the simulation shown in Figure 5(a)–(b) and 6(b) (where the information dynamics becomes periodic). The attractor for the *full* dynamics may be reached at some later point.

The core size also influences the occurrence frequency for some of the structures (see Figure 7(b)). For instance, cores where pointers only are located on the jump instructions occur more often for smaller core size. Cores only consisting of MOV instructions occur more frequently the larger the core size. The longest transients are all associated with functional dynamics involving many different instructions.

In section 6 we shall argue that the MVNM can be viewed as a particular kind of cellular automata. Since the MVNM has several CA properties, one could perhaps expect a power-law variation of the transient length.[31,32,43] However, this does not seem to be the case for this particular MVNM although it cannot be concluded from these data.

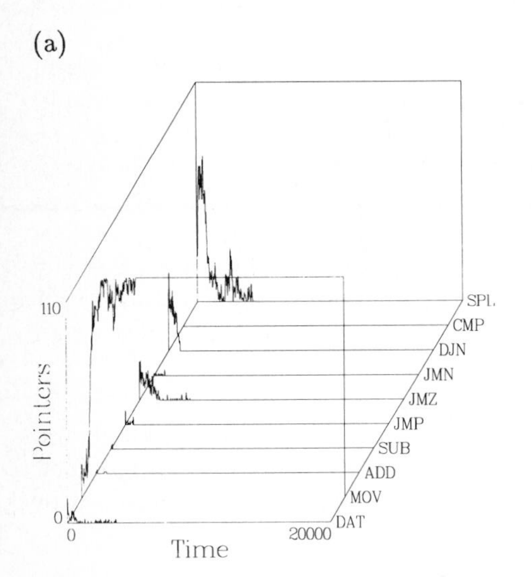

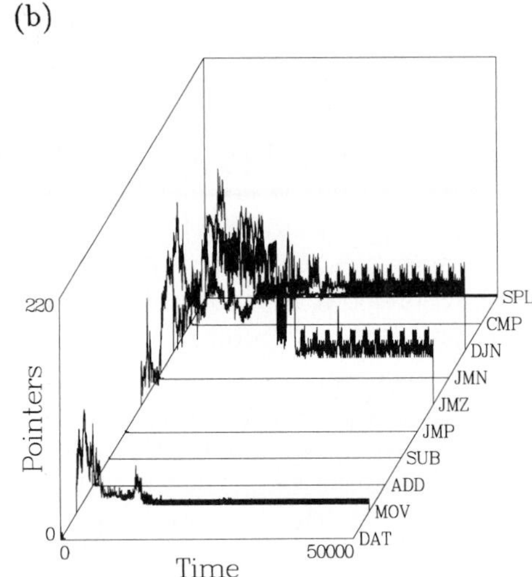

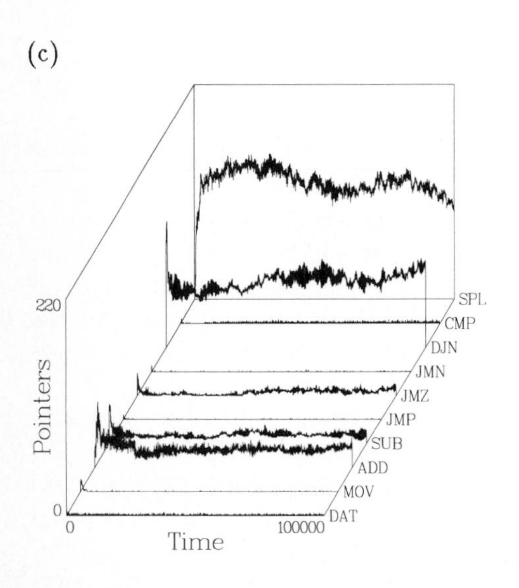

FIGURE 6 Dynamics of a projection of the interaction graph (no external noise). The x-axis is time (number of iterations), the y-axis represents the ten different Opcode types (0 - 9), and the z-axis shows how many pointers each instruction type has at any time (0 - L). (a) Evolution into a pure MOV core (N = 1792). Note that the full dynamics in the MOV core is periodic and not a fixed point as indicated by this projection. (b) Evolution into a MOV-jmp-SPL structure with periodic dynamics (N = 3584).(c) Very long transient dynamics (> 100,000 iterations). Note that this core has many different types of Opcode (N = 3584).

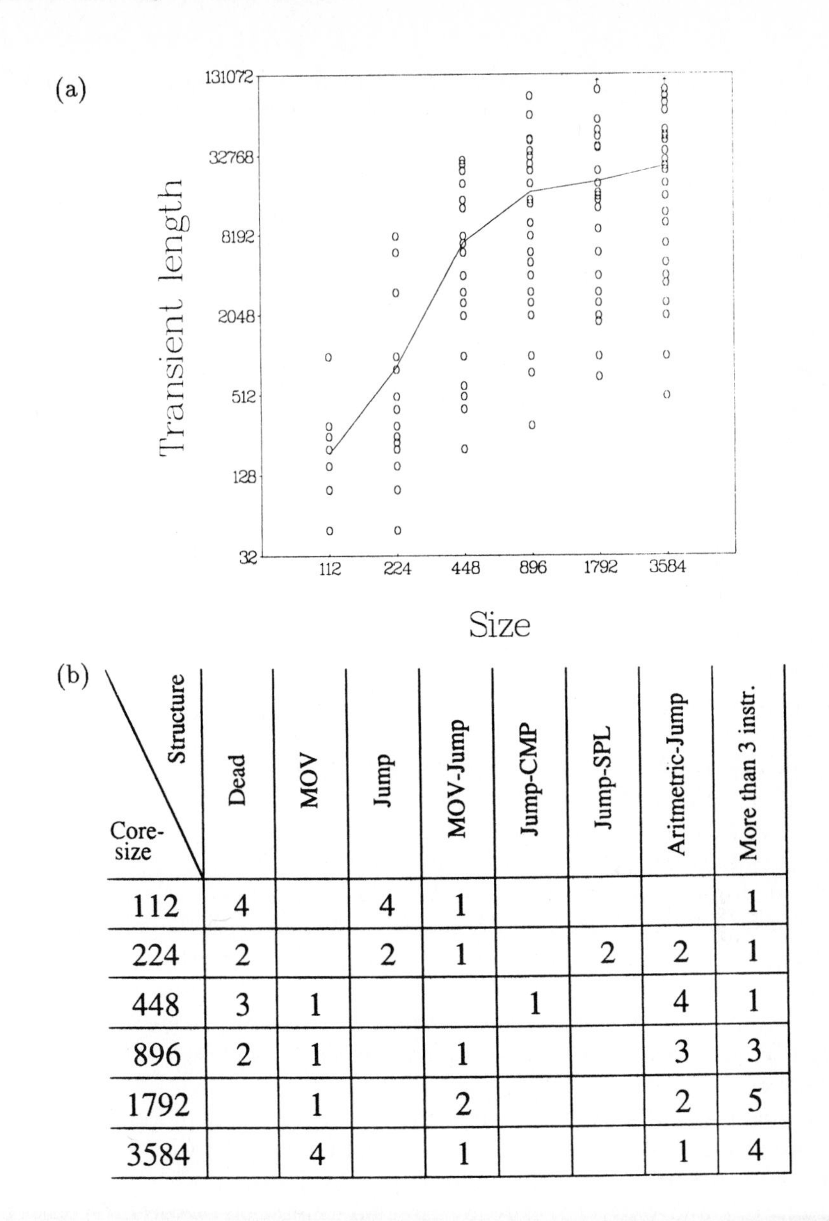

Core-size \ Structure	Dead	MOV	Jump	MOV-Jump	Jump-CMP	Jump-SPL	Aritmetric-Jump	More than 3 instr.
112	4		4	1				1
224	2		2	1		2	2	1
448	3	1			1		4	1
896	2	1		1			3	3
1792		1		2			2	5
3584		4		1			1	4

FIGURE 7 (a) Scaling dynamics without external noise (log-log plot). Transient length as a function of core size for 180 simulations, each marked by an "o," 30 for each core size. Only two transients did not settle down within 100,000 iterations (indicated by small arrows). The solid line indicates the average transient time which clearly increases as a function of the core size. The short transients for large core size all end up either as MOV cores, as MOV-jump cores, or as arithmetic jump cores (see Figure 6(b)). (b) The number of occurrences for the observed structures, indicated by the dominating Opcode contence, as a function of core size. Note, for example, how the jump cores mainly occur for smaller core sizes and how the MOV cores mainly occur for larger core sizes.

As discussed in section 3.2, a random mapping function can be used as a reference for our system. The expected transient length for a random mapping function[3] scales asymptotically as $n^{1/2}$ where the number of states in the map is n. An upper limit for the effective (e.g., in principle reachable) number of states $|\Omega_{eff}|$, in Venus II, for $N = 3584$, is $(N(2R_{opr}+1))^N \sum_{i=0}^{L} N^i$, since there at most can be N different Words to move around in the memory and since the dynamics can alter the B-fields independently. Since $|\Omega_{eff}| \sim 10^{25,000}$, the expected transient length is of the order $10^{12,500}$, assuming a random mapping over all effective states. The actual expected transient length is only of the order 1.6×10^4 iterations, which implies that the MVNM dynamics is very different from the dynamics of a random mapping function over the same state space.

5. OTHER MODIFIED VON NEUMANN MACHINES

We shall now discuss the dynamics of a MVNM which is drastically simpler and somewhat different from Venus. In section 6 we are going to show the heuristics of a transformation route from the Coreworld to a cellular automaton. An example of a system defined along this route is what we shall refer to here as the Luna system. Luna is still closer to a VNM than a CA, but it is much simpler than Venus. It has a cyclic addressing array with one instruction at each addressing slot. Each instruction has an Opcode, an A-field, and a B-field. All addressing is direct ($ in the Venus notation) and relative to the actual address.

Luna is defined by the five instructions given in Table 2.

The core is updated in parallel (as in Venus I) and one pointer can be executed at each address at each generation. The addressing radius is defined by R_{opr}. This system has two pointer populations, where pointers in one of the populations are moving toward smaller address numbers and pointers in the other population are moving toward higher address numbers. The system thereby becomes isotropic. We still keep a single execution queue. A pointer can switch between the two populations whenever it meets a REVERSE instruction.

In Figures 8(a), (b), and (c), we show the typical dynamics of Luna with deterministic instruction executions. The core is randomly initiated with an even distribution of the 5 instructions, 25 randomly located pointers, and the parameters, $N = 1600$, $L = 100$, and $R_{opr} = 1$. Already after 50 generations several disconnected areas only populated with COPY instructions emerge. These areas grow until they merge and the whole core finally is populated by COPY instructions. They, thereby, overwrite other spontaneously generated cooperative structures: structures with "reflecting" pointer borders, two REVERSE instructions bouncing a pointer forth and back, two CREATE instructions sending new pointers towards each other every time one of them are executed, or mixtures of these two instructions, all of them with appropriate A and B fields of course.

TABLE 2

Opcode	Function
COPY A, B	Copies the whole instruction (opcode, afield, bfield) from the A-address to the B-address.
ADD A, B	Puts the sum of A's afield and B's afield into B's afield, and puts the sum of A's bfield and B's bfield into B's afield.
CREATE A, B	Adds a new program pointer, which moves in the opposite direction of the current pointer.
REVERSE A, B	Adds a new program pointer, which mes in the opposite direction of the current pointer.
DAT A, B	Removes the current programpointer.

This type of dynamics is generic for the system for any combination of the parameters ($N > 100$, $L > N/10$) with the initial conditions given above, as long as R_{opr} is very small. The larger N, the larger R_{opr} can be. It also holds when the system is subjected to noise, when spontaneous pointers can occur with probability up to $P_{point} = 0.1$, and when *any* writing to memory is changed with the probability $P_{mut} < 0.1$.

The mechanism behind this type of dynamics is trivial. In the simplest situation where $R_{opr} = 1$ and $P_{mut} = 0$, one finds that the probability of having a COPY instruction copying itself towards either higher or lower addresses is $1/5 \times 1/3 \times 2/3 = 2/45$, given the random initial conditions. Assuming the 25 randomly located initial pointers, the expected number of COPY areas should be $25 \times (2/45) \simeq 1.11$ which approximately corresponds to one area for every 1,440 addresses given that the pointers are stationary and not able to move along the addressing array. Since the pointers move, the number of observed areas is larger. Once R_{opr} is increased to above 5, still with $N = 1600$, the phenomena becomes very unprobable and is no longer part of the typical dynamics.

The takeover of the COPY(0,$\pm$1) instruction is a clear example of a self-sufficient, one-object replication process, which does not leave any room for a more involved evolution process. We have, in this situation , clearly crossed a critical complexity level below which no higher-order cooperative structures, e.g., structures involving more than one instruction, can evolve. To enhance the cooperation between *different* instructions and thereby open the system a little towards larger flexability, we need to eliminate some of the COPY instruction's self-sufficiency. This can be done through a definition of computational resources as in Venus or by assigning a finite lifetime to each pointer. By allowing the incorporating of different instructions in a self-sustaining structure, as in Venus the system is given more evolutionary possibilities.

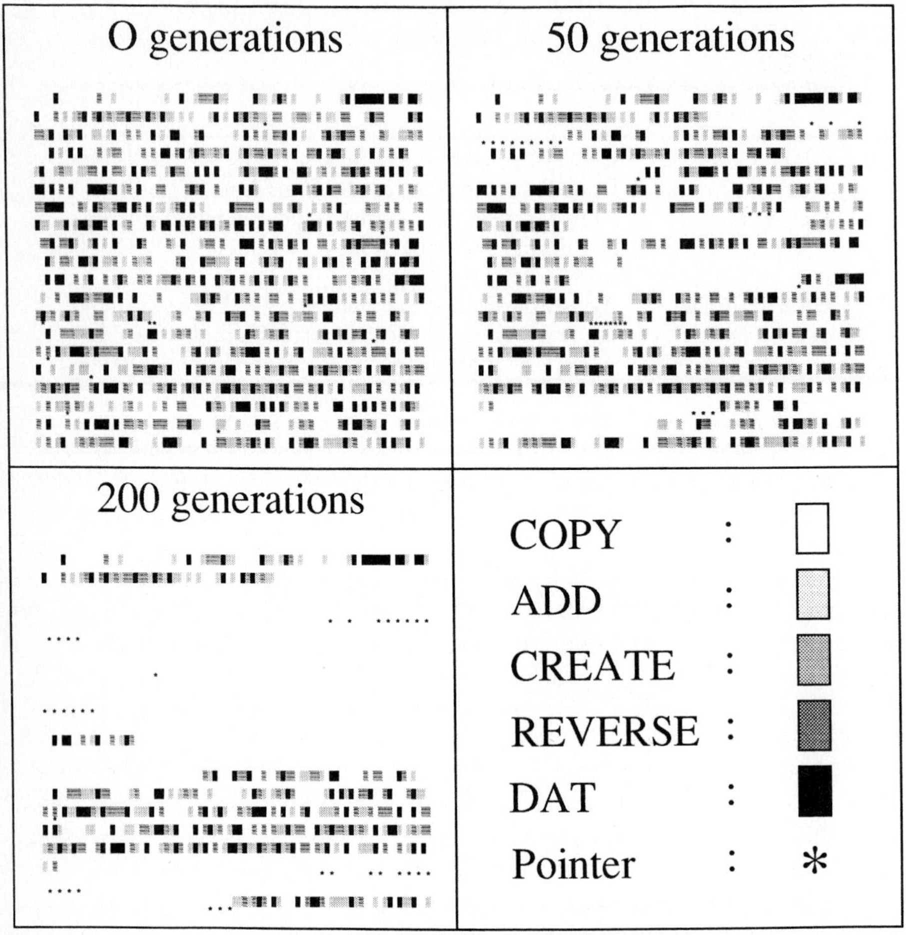

FIGURE 8 Simple self-organizing process for the Luna system from random unbiased initial conditions. (*Opcode, pointer*)-projection of state space at time 0, 50 and 200. The grey toning indicates: white = COPY, light grey = ADD, grey = CREATE, dark grey = REVERSE, black = DAT. Note how several independent COPY-dominated areas initially appear whereafter they merge. This type of dynamics is related to the Venus II dynamics shown in Figure 10(a).

6. DIFFERENT LEVELS OF EQUIVALENCE BETWEEN UNIVERSAL FORMALISMS

Using the MVNM as a reference, we are able to discuss different levels of equivalence between universal programmable systems including a method for a stepwise transformation of one universal formalism into another. New formalisms defined along such transformation may turn out to be useful and this transformation method may be used to create formalisms more suitable as self-programmable matter, since evolvability is *not* representation independent.

In order to show that Venus supports universal computation, we can construct a machine-code program capable of simulating a universal Turing machine (TM). The full proof of universality can be found in the Appendix. The key point is to show how a Turing table can be represented, which breaks down to a formulation of the elementary TM operations in machine code: reading from tape, writing to tape, moving right on tape, moving left on tape, and changing internal state.

Such a construction is an example of a *simulation* of one machine in another machine, namely embedding a TM within a VNM. However, it does not say much about how the *information mechanics*, e.g., the details of the information storage, transfer, and transformations, is related for the two machines. Another, but from our point of view more informative, process is to turn one machine on the other through a number of structural steps, each step gradually altering the information mechanics of the one machine, and where each step is defining a new machine somewhere in between the two original machines. We shall in the following refer to such a process as a *transformation* as opposed to a *simulation* of the one machine into the other.

Let us first sketch a way of transforming a MVNM into a cellular automaton. Define a Coreworld, like Venus, with at most $r_{max}(< 1.0)$ executions per address. Now let the resource radius, R_{res}, decrease as r_{max} increases, which in the end means one execution per address per generation (at most one pointer per address). Remove one by one the different addressing modes so that only direct addressing is possible in the system (recall Table 1). Let the operation radius, R_{opr}, decrease so that only local interactions are possible (such a system is close to Luna defined in section 5). Let the B-field address be fixed to the address of the executed instruction. Thereby, the actual instruction alters one of its own fields (states) as a result of the execution, and the B-field can be omitted. Increase the number of pointers so that every address in the array gets executed once at each generation. This step also eliminates the SPL and all the jump instructions, since creating pointers or moving pointers on the lattice no longer has any meaning. Now, this local CA with two variables per cell can easily be redefined as an ordinary CA with one variable per cell by expanding the actual number of states of this one variable.

This route shows the heuristics of a transformation from the Coreworld to a cellular automaton. Here we have not shown any *formal* equivalence between the Coreworld and a specific CA, since we have not made any attempts to conserve the universality in each step. We have, however, given a route by which one can be turned into the other (and vice versa) in a finite number of natural steps each defining a new system. Actually, some of the operation primitives defined for the Connection Machine,[21] for instance the interchange of address contents between neighbors, is a hybrid CA/VNM property.

Examples of cellular automata simulations of the information mechanics in Turing machines can be found in Smith,[50] Langton,[27] or in Lindgren and Nordahl.[34]

A Coreworld can, of course, also simulate the λ-calculus-based ALChemy,[13] since ALChemy is implemented on a modern digital computer which *is* a von Neumann machine. ALChemy can be embedded in the Coreworld by inserting each of the function encoded strings or programs at disjoint locations in the memory. The functional interactions can occur between any two randomly chosen programs where the first program uses the second as an input. The new program resulting from this interaction is then inserted at a core location not occupied by any other program. In order to keep the number of interacting programs constant, a program will be removed at random from the core whenever a new one is introduced. In this formalism only a single pointer is active at a time. The "gas" property of the system, or the part of the system that determines which programs should interact at a given time, is itself located in the core. The algorithm for the details of a typical program/program interaction is very complicated to program in machine code. Instead of going into the details of this, we will concentrate on how the ALChemy dynamics gradually can be transformed into a Coreworld dynamics, as in the Venus formalism. First of all, we need to allow for parallel processing in the program collisions, which implies many pointers. Another step is to let neighboring programs, in an addressing sense, have a higher probability of interacting. Step by step, the systems gas property can thereby be removed along with a removal of its encoding. So far we have, loosely speaking, transformed a "spherical container" with the λ-gas inside into a "torus," and made the small torus diameter smaller and smaller. A further step towards a Coreworld is to let the strict LISP colliding scheme, now between neighboring programs, disappear and allow for more general program/program alterations not obeying the rules of balanced parentheses. A final step is then to loosen the predefined addressing range for what defines a single program and to allow free reading and writing to memory.

The whole transformation can, of course, be done in as many steps as desired and thereby be made more and more smooth. Actually, somewhere along this particular transformation path, we come close to the definition of the Tierra system developed by Ray.[49]

Whether computational universality is preserved along each of these transformation steps is not known, but it seems plausible that there exists a transformation path by which this is the case.

7. MVNM DYNAMICS AND THE DYNAMICS OF NATURAL SELF-PROGRAMMABLE MATTER

How does the modified von Neumann machine dynamics relate to living processes? There are at least two levels in this question. (a) How do *processes* that lead to emergent phenomena within the MVNM relate to possible prebiotic processes of self-organization and (b) how do the *functional details* of the emerging MVNM structures relate to the functional details of proto-organisms? In addition we may ask if the proposed computational approach is able to help us in a logical characterization of functional organization. We have seen that the MVNM dynamics is capable of a successive development of epochs characterized by different functional compositions. The different instruction compositions cooperate and utilize the available resources in terms of executions and memory slots in very ingenious ways. Although the number of components as well as the number of interactions are far fewer than those found in the living cell, the computational structures have a number of properties they share with biological systems. They are: (1) able to channel and focus the available computational resources very effectively; (2) they are perturbation stable; (3) they are able to maintain themselves and reproduce through a recurrence, or an invariance, in their iterated functional dynamics; and (4) they can undergo evolution. How close the details of the MVNM processes and the details of the emerging cooperative structures are to evolutionary processes in biological systems and to the structures underlying contemporary livings systems we cannot say. The detailed properties of biological evolution as well as the fundamental dynamics underlying life itself are yet unknown. Since the details of the emerging cooperative structures in the MVNMs differ depending on instruction set, parameters, and initial conditions, we are lead to assume that the details of proto-organisms based on biomolecular inteactions also would depend on the system details. However, a number of suggestive prebiotic interpretations, at the process and organizational levels, can be made from the experiments with the MVNMs.

i. As for physico-chemical dissipative structures, the emergence of computational structures requires driving. Physical dissipative structures degrade high-quality free energy to sustain themselves, and the computational structures degrade computational resources in terms of cpu cycles. Instruction executions are analogs to free energy.

ii. There are certain relations, which need to be fulfilled, between system size, available executions per system update, and initial conditions before the systems are able to support complex cooperative dynamics. For some combinations of the above, the systems are virtually inert. This is, for instance, the case when the Venus system is only being driven weakly (Δr small, few executions per iteration) This implies that life probably did not emerge in environments with little driving.[47] Biased initial conditions in the Venus system enhance the evolutionary possibilities. This should imply that steep and complex chemical potentials enhance the chances for the emergence of a variety of interesting structures through prebiotic chemical evolution.

iii. The constructive, or self-programmable, aspect of the system creates new functional properties. However, noise in such a system is still an important driving factor. The noise can "push" the system from one metastable state to another and speed up the evolutionary search process.

iv. Frozen accidents in terms of particular functional properties are a natural consequence of runaway processes accompanying self-organizing phenomena. Complex cooperative dynamics is not guaranteed, even given the right conditions. Qualitatively different cooperative structures can emerge from systems with the same parameters. In these systems, chance is part of the game. This implies that many properties characterizing contemporary life-forms should be frozen accidents.

v. Functional stability to perturbations is a *product* of evolution and not a property of the details of the underlying programmable matter. Even when the system is based on a von Neumann architecture, which is highly perturbation fragile with respect to functional properties, it can spontaneously develop perturbation-stable cooperative structures. Functional stability to perturbations as we see it in modern life-forms should, therefore, be a product of evolution and not a property of the underlying chemistry. This is probably true, since an embedding of any modern biochemical pathway in a random chemical environment will cause the pathway to collapse.

vi. Cooperation emerges as a natural property of the functional dynamics in systems with a constructive dynamics. The intricate network of functional cooperation characterizing contemporary life should, therefore, be a natural consequence of the dynamics in the prebiotic environments, and not the result of a long chain of highly improbable events.

vii. Simplifying the instruction set below a certain level of complexity inhibits the emergence of higher-order cooperative structures in the MVNM. This may indicate that a prebiotic chemical "construction set" needs to reach a certain level of complexity before it can initiate a self-organizing process leading to entities with lifelike properties.

viii. A general feature for most of the emerging cooperative structures in the MVNMs is a reproduction of the structure *as a whole* and not an individual reproduction of the parts. In such systems, the interaction graph is part of the "template information." It therefore seems easier to create a reproductive system without separate genes. The emerging cooperative structures have several properties in common with the *autocatalytic sets* found for catalyzed cleavage-condensation reactions in polymer systems.[1]

ix. Finally, we want to stress that an organism and its environment form an *integrated* system. The more low-level the living process is, the more fuzzy the organism-environment distinction appears. At the level where we are operating with the MVNMs, such a distinction is clearly arbitrary.

8. CONCLUSION

We have extracted the property of self-programmability which characterizes the processes producing novelty in biomolecular systems. We have implemented this property in much simpler and more tractable dynamical systems. By characterizing the dynamics of these simple self-programmable systems, we have tried to characterize dynamical aspects of the emergence of novel properties for evolutionary processes in prebiotic and proto-biological systems.

In stepping away from the use of "classical" dynamical systems, e.g., systems where the equations, or rules of dynamics, do not depend on the dynamics (typically ordinary differential equations), we have opened a new alley. The best current models for truly evolutionary systems are the different computational paradigms, since programs are identical to data, programs alter data, and thus programs can alter programs (section 2).

We have created a particular kind of self-programmable matter based on a modified von Neumann machine (MVNM). We show that the functional dynamics can be defined through the introduction of an interaction graph and that invariance or recurrence in the iterated interaction graph is equivalent to functional cooperation (section 3.3 and 4.4). Invariance is trivially fulfilled when the system is on an attractor, but recurrence frequently occurs when the system is on a transient and the recurrent states can thereby define metastable epochs. We trace the functional dynamics through a number of projections in state space (section 4.3), by the information dynamics of the system (section 4.5), and via projections of the interaction graph (4.4 and 4.6). Through a combination of these projections, it is possible to measure when new functional properties emerge in the system and what they are characterized by.

Without external perturbations the MVNM always settles down to an attractor. In Venus II we experience a clear increase of the average transient length as a function of core size under the given initialization (section 4.6). The scaling of the average transient length with the core size does not seem to scale with a power law as seen for cellular automata. We also find that *different* cooperative structures dominate the dynamics as the system size is varied, which does not occur for CA's. An increased memory size need not imply longer transients or the development of more complex structures. It also depends on the system representation. This is clearly demonstrated in the simple Luna system (section 5) where the single instruction takeover phenomenon occurs for a wide range of parameters no matter how large the system is. The defined representation of the system can have some inherent limitations.

The stepwise *transformation* of universal systems into one another gives us a new way of thinking about the inter-relatedness of the details of the information mechanics of the different universal systems (section 6). We show how a MVNM can be stepwise transformed into a cellular automata as well as into a λ-calculus-based system. Systems defined along these transformations may be interesting by themselves and the new systems may also serve as a source of inspiration to define yet other and more evolvable constructive systems. In the Appendix we prove, in a more traditional fashion, that Venus is equivalent to a universal Turing machine, since Venus can *simulate* a universal Turing machine.

Macroscopic, spatio-temporal, cooperative structures spontaneously emerge in the MVNMs. They emerge in a similar way as macroscopic dissipative structures do in physico-chemical systems. A thermodynamical interpretation of the computational system yields an equivalence between the flux of free energy and the flux of computational resources (executions per iteration) and an equivalence between the microscopic degrees of freedom in the physico-chemical system and all the possible functional interactions in the computational system. A notable difference is, however, that our macroscopic computational structures *change* even with a constant pumping (allowed executions per iteration). Due to our system's self-programmable properties, it does not stay in a fixed macroscopic pattern, as, for instance, the Raleigh-Bénard convection or the chemical reaction waves in the Belousov-Zhabotinski reaction do for constant pumping. Our systems have, in addition, a property biological systems also have: they can change themselves and thereby undergo development.

The details of the emerging structures depend on the details of the underlying systems clearly demonstrated through variations in the instruction set, the updating scheme, the parameters, and the initial conditions of the MVNMs. The functional details of biomolecular-based proto-organisms are therefore also assumed to depend on system details. We can, therefore, not give a full answer to our initial question on what enables matter to organize into the complex structures we see in biological organisms. However, we have seen that even simple self-programmable systems are able to develop complex adaptive structures. Although these structures are still far simpler than, and probably also organized differently from, any contemporary biological organism, the structures are able to channel and focus the available computational resources very effectively, they are perturbation stable, they are able to maintain themselves (e.g., reproduce), and they can undergo evolution. These properties enable the structures to expand and invade the whole system for long periods of time. Functional cooperation is an inherent property of the dynamics of programmable matter.

ACKNOWLEDGMENTS

We are grateful for many exciting and enlightening discussions with Walter Fontana, Chris Langton, Mats Nordahl, Doyne Farmer, Erica Jen, Wentian Li, and Brosl Hasslacher. Chris Langton, Walter Fontana, Mats Nordahl, and Doyne Farmer are also acknowledged for their many constructive suggestions in earlier versions of this paper.

APPENDIX

PROOF OF UNIVERSALITY OF THE MACHINECODE USED IN THE COREWORLDS, VENUS I AND II.

In order to prove that the machine code used is capable of universal computation, we construct a machine-code program that can simulate any Turing machine (TM) with N internal states $S_1, S_2, \ldots, S_N$, M different IO-symbols $B_1, B_2, \ldots, B_M$, and an infinite tape.[23]

The machine code program is given in Table 3. To make the program more transparent, we introduce labels. Whenever a label appears as an operand for an Opcode in the program, the effective operand for this Opcode is located at the address with the label. The mode of the operand is direct ($) if nothing else is stated. We introduce a tape pointer P_{tape}, and N routines to handle the internal states of the TM. The routines must be able to read a symbol from the tape, write a new symbol on the tape, move the tape pointer along the tape, and change the internal state. SW_{S_k,B_j} is the symbol written to the tape when B_j has been read and the TM is in state S_k. $Direction_{S_k,B_j}$ is the direction the tape pointer moves (left/right) when B_j is read and the TM is in state S_k, and the corresponding new state of the machine is $NewState_{S_k,B_j}$ when B_j is read and the TM is in state S_k. SW_{S_k,B_j}, $Direction_{S_k,B_j}$, and $NewState_{S_k,B_j}$ are all predefined constants. They define the transition table for the TM.

The first bracket to the left of the program indicates the beginning and the end of the main loop. The following brackets enclose the routines handling each of the TM states indicated by the lables $State_1, State_2, \ldots, State_N$. Inside each of the state handling routines, a second level of brackets indicate the position of the subroutines used to handle each symbol read while the machine is in state S_k. The subroutine used when B_j is read while the machine is in state S_k is labeled SR_{S_k,B_j}.

At the bottom of the listing, we find DAT instructions storing data used by the main loop. *Symbol* contains the last symbol read and *State* contains the actual state of the TM. The following DAT instructions store the adresses $State_1 - Address$, $State_2 - Address, \ldots$, $State_N - Address$, of the routines handling the possible TM states relative to the first instruction in the program. These are all predefined. P_{tape} stores the actual tape pointer location and the rest of the, in principle, infinite core is used to store the tape.

The program starts at *Main*; it is assumed that the current TM state is S_k, the current tape location is *Location*, and the current tape symbol is B_j. First the address of the routine handling state S_k is moved to *Address* and the pointer jumps to that routine. Now the current tape symbol is compared to the possible symbols $B_1, B_2, \ldots, B_M$, one by one in a series of CMP instructions. When the right symbol is found, the program pointer is transferred to the corresponding subroutine SR_{S_k,B_j}. Here a new symbol is written on the tape, the tape pointer is decreased

TABLE 3

```
Address      DAT ...
Main      => MOV @ State,              Address; Get the address of the next state routine
             JMP @ Address                    ; Jump to the next state routine
             :
State_1      :                                ; Start of the state routines
             JMP Main
  :
State_k      MOV @P_tape,              Symbol ; Routine handling S_k
             CMP #B_1,                 Symbol ; If the symbol last read equals B_1...
             JMP SR_{S_k,B_1}...              ; jump to the corresponding subroutine
             :
             CMP #B_j,                 Symbol ; If the symbol last read eqals B_j...
             JMP SR_{S_k,B_j}...              ; jump to the corresponding subroutine
             :
             CMP #B_M,                 Symbol ; If the symbol last read equals B_M...
             JMP SR_{S_k,B_M}...              ; jump to the corresponding subroutine
  SR_{S_k,B_1}
             :
  :
  SR_{S_k,B_j} MOV #SW_{S_k,B_j}       @P_tape; Write new symbol on tape
             ADD #Directions_{S_k,B_j} P_tape ; Move TM right or left
             MOV #NewStates_{S_k,B_j}  State  ; Change state
             JMP Main                         ; Jump back to main loop
  :
  SR_{S_k,B_M}
             :
  :
State_N      :
             JMP Main                         ; End of state routines
Symbol       DAT #B_j                         ; Last symbol from the tape
State        DAT #S_k                         ; State of the TM

             DAT #State_1 - Address           ; Address of code handling state 1 relative to Address
             :
             DAT #State_k - Address           ; Address of code handling state k relative to Address
             :
             DAT #State_N - Address           ; Address of code handling state N relative to Address

P_tape       DAT $Location                    ; Tape pointer location

TAPE         DAT ...                          ; Tape starts here...
             DAT ...
             DAT ...
             DAT ...
             :
```

or increased according to $Direction_{S_k,B_j}$, and $NewState_{S_k,B_j}$ is placed at *State*. Finally, the pointer jumps back to the main program to initiate the next cycle of the Turing machine.

Since this program can simulate any TM with N internal states $S_1, S_2, \ldots, S_N$, M different IO-symbols $B_1, B_2, \ldots, B_M$, and an infinite tape, we have shown that Venus I and II are universal computers.

Note that we only used four of the ten possible instructions to show universality. Other instruction tuples can also support universal computation.

REFERENCES

1. Bagley, R., and D. Farmer. "Spontaneous Emergence of a Metabolism." This volume.
2. Barendregt, H. "The Lambda Calculus." In *Syntax and Semantics.* Studies in Logic and the Foundations of Mathematics, vol. 103 (second printing). Amsterdam: North-Holland, 1985.
3. Bollobás, B. *Random Graphs.* New York: Academic Press, 1985.
4. Bollobás, B., and S. Rasmussen. "First Cycles in Random Directed Graph Processes." *Discrete Mathematics* **75** (1989): 55–68.
5. Church, A. *The Calculi of Lambda-Conversion.* Annals of Math. Studies, vol. 6. Princeton University Press, 1941.
6. Dewdney, A. "In the Game Called Core War Hostile Programs Engage in the Battle of Bits." *Sci. Am.* May 1984.
7. Doolen, G. ed. "Lattice Gas Methods for PDE's: Theory, Applications, and Hardware." *Physica D* **47** (1991).
8. Eigen, M. "Self-Organization of Matter and Evolution of Biological Macromolecules." *Naturwissenschaften* **58** (1971): 465.
9. Eigen, M., and P. Schuster. *The Hypercycle-A Principle of Natural Self-Organization.* Heidelberg: Springer-Verlag, 1979.
10. Farmer, D., S. Kauffman, and N. Packard. "Autocatalytic Replication of Polymers." *Physica D* **22** (1986): 50.
11. Farmer, D., and A. Belin. "Artificial Life: The Coming Evolution." This volume.
12. Fontana, W. "Functional Self-Organization in Complex Systems." In *1990 Lectures in Complex Systems*, edited by L. Nadel and D. Stein. Santa Fe Institute Studies in the Sciences of Complexity, Lect. Vol. III, 407–426. Redwood City, CA: Addison-Wesley, 1991.
13. Fontana, W. "Algorithmic Chemistry." This volume.
14. Forrest, S. "Emergent Computation: Self-Organizing, Collective, and Cooperative Phenomena in Natural and Artificial Computing Networks." *Physica D* **42** (1990): 1–11.
15. Fredkin, E. "Digital Mechanics." *Physica D* **45** (1990): 254–270.
16. Fredkin, E., and T. Toffoli. "Conservative Logic." *Int. J. Theor. Phys.* **21** (1982): 219–253.
17. Gardner, M. "The Fantastic Combinations of John Conway's New Solitaire Game of 'Life'." *Sci. Am.* **223** (1970): 120–123.
18. Goldberg, D. *Genetic Algorithms in Search, Optimization, and Machine Learning.* Reading, MA: Addison-Wesley, 1989.
19. Gutowitz, H., ed. "Cellular Automata: Theory and Experiment." *Physica D* **45** (1990).
20. Hameroff, S., S. Rasmussen, and B. Månson. "Molecular Automata in Microtubule: Basic Computational Logic for the Living State." In *Artificial Life,*

edited by C. Langton. Santa Fe Institute Studies in the Sciences of Complexity, vol. VI, 521–553. Redwood City, CA: Addison-Wesley, 1989.

21. Hillis, D. *The Connection Machine.* Cambridge: MIT Press, 1987.
22. Holland, J. H. *Adaptation in Natural and Artificial Systems.* Ann Arbor, MI: University of Michigan Press, 1975.
23. Hopcroft, J., and J. Ullman. *Introduction to Automata Theory, Languages, and Computation.* Reading, MA: Addison-Wesley, 1979.
24. Kauffman, S. A. "Autocatalytic Sets of Proteins." *J. Theor. Bio.* **119** (1986.): 1–24.
25. Kleene, S. "Turing Analysis of Computability and Major Applications of It." In *The Universal Turing Machine: A Half-Century Survey,* edited by R. Herken, 17–54. Oxford: Oxford University Press, 1988.
26. Knudsen, C., R. Feldberg, and S. Rasmussen. "Information Dynamics of Self-Programmable Matter." *Proceedings of the NATO ARW on Complex Dynamics and Biological Evolution,* Middelfart, Denmark, August 5-10, 1990, edited by E. Mosekilde. New York: Plenum Press, in press.
27. Langton, C. "Virtual State Machines and Cellular Automata." *Complex Systems* **1** (1987): 257–271.
28. Langton, C. "Studying Artificial Life with Cellular Automata." *Physica D* **22** (1989): 120–149.
29. Langton, C. "Computation at the Edge of Chaos: Phase Transitions and Emergent Computation." *Physica D* **42** (1990): 12–37.
30. Langton, C. "Life at the Edge of Chaos." This volume.
31. Li, W., and M. Nordahl. "Transient Times of Cellular Automata Rule 110." Preprint 91-05-025, Santa Fe Institute, 1991.
32. Li, W. "Non-local Cellular Automata." Preprint 91-01-001, Santa Fe Institute, 1991.
33. Lindgren, K., and M. Nordahl. "Complexity Measures and Cellular Automata." *Complex Systems* **2** (1988): 409–440.
34. Lindgren, K., and M. Nordahl. "Universal Computation in Simple One Dimensional Cellular Automata." *Complex Systems* **4** (1990): 299–318.
35. Margolus, N. "Physics-Like Models of Computation." *Physica D* **10** (1984): 81–95.
36. McCaskill, J. Preprint, 1989, and private communication.
37. Minsky, M. *Computation-Finite and Infinite Machines.* Prentice-Hall, 1972.
38. Moore, C. "Unpredicatbility and Undecidability in Dynamical Systems." *Phys. Rev. Lett.* **64** (1990): 2354–2357.
39. Nicolis, G., and I. Prigogine. *Self-Organization in Nonequilibrium Systems.* New York: Wiley-Interscience, 1977.
40. Omohundro, S. "Modelling Cellular Automata with Partial Differential Equations." *Physica D* **10** (1984): 128–134.
41. Packard, N. "Adaptation at the Edge of Chaos." Technical Report CCSR-88-5, Center for Complex Systems Research, University of Illinois, 1988.

42. Palmer, R. "Neural Nets." In *Lectures in the Sciences of Complexity*, edited by D. Stein. Santa Fe Institute Studies in the Sciences of Complexity, Lec. Vol. I, 439-461. Redwood City, CA: Addison-Wesley, 1988.
43. Peck, H. "Global Properties for Finite Lattices." In *Theory and Application of Celular Automata*,edited by S. Wolfram, appendix, 537–546. World Scientific, 1986.
44. Post, E. "Formal Reduction of the General Reduction of the General Combinatorial Decision Problem." *Am. J. Math.* **65** (1943): 197–268.
45. Prusinkiewica, P., and A. Lindenmayer. *The Algorithmic Beauty of Plants.* New York: Springer-Verlag, 1990.
46. Rasmussen, S., B. Bollobás, and E. Mosekilde. "Elements of a Quantitative Theory of Prebiotic Evolution." Thechnical Report LA-UR-89-1881, Los Alamos National Laboratory, 1989; to appear in *J. Theor Biol.*
47. Rasmussen, S., C. Knudsen, R. Feldberg, and M. Hindsholm. "The Coreworld: Emergence and Evolution of Cooperative Structures in a Computational Chemistry." *Physica D* **42** (1990): 111–134.
48. Rasmussen, S., H. Karampurwala, R. Vaidyanath, K. Jensen, and S. Hameroff. "Computational Connectionism Within Neurons: A Model of Cytoskeletal Automata Subserving Neural Networks." *Physica D* **42** (1990): 428–449.
49. Ray, T. "An Approach to the Synthesis of Life." This volume.
50. Smith, A. R. "Simple Computation-Universal Cellular Spaces." *JACM* **18(3)** (1971): 337–353.
51. Toffoli, T. "Cellular Automata as an Alternative to (Rather than an Approximation of) Differential Equations in Modeling Physics." *Physica D* **10** (1984): 117–127.
52. Toffoli, T., and N. Magolus. *Cellular Automata Machines.* Cambridge: MIT Press, 1987.
53. Toffoli, T., and N. Magolus. "Programmable Matter: Concepts and Realizations." *Physica D* **47** (1991): 263–272.[2]
54. Touretzky, D., ed. *Proceedings of The Neural Information Processing Conferences*, NIPS 2. Morgan Kauffman, 1990.
55. Turing, A. "On Computable Numbers with Applications to the Entscheidungs Problem." *Proc. Lond. Math. Soc.* **42** (1936–1937): 230–265.
56. Ulam, S. "Random Processes and Transformations." *Proceedings of the International Congress on Mathematics, 1950* **2** (1952): 264–275.
57. von Neumann, J. *Theory of Self-Reproducing Automata*, edited and completed by A. Burks. Urbana: University of Illinois Press, 1966.
58. Wolfram, S. ed. *Theory and Applications of Cellular Automata.* Worlds Scientific, 1986.
59. Wootters, W., and C. Langton. "Is There a Sharp Phase Transition for Deterministic Cellular Automata?" *Physica D* **45** (1990): 95–104.

Maarten Boerlijst and Pauline Hogeweg
Bioinformatica, Padualaan 8, 3584 CH Utrecht, The Netherlands

Self-Structuring and Selection: Spiral Waves as a Substrate for Prebiotic Evolution

In the study of prebiotic evolution self-structuring and selection are themes that are usually studied separately. In this paper we demonstrate that spatial self-structuring in incompletely mixed media can profoundly change the outcome of evolutionary processes; for instance, positive selection for "altruistic" features becomes feasible.

A cellular automaton model of self-replicative molecules that give each other cyclic-catalytic support was studied. Because of the spiral structures that emerge in such a spatial hypercycle system, selection no longer exclusively takes place at the level of the individual molecules, but also at the level of the spirals.

1. INTRODUCTION

Accumulation of information is a central issue in prebiotic evolution. Eigen and Schuster[3] were the first to stress the existence of the information threshold; in a system containing self-replicative molecules, the length of the molecules is restricted by the accuracy of replication.

In their hypercycle theory, Eigen and Schuster state that the information threshold can only be crossed if a number of molecules catalyze the replication of each other in a cyclic way (see Appendix, Figure 9(a)). This so-called hypercycle has great selective advantages. Each molecule in the hypercycle is still bound to the maximum string length, but the molecules can combine their information and thus cross the information threshold.

The selective properties of the hypercycle have been investigated by several authors by means of an ordinary differential equations (ODE) model (see Appendix). In the ODE model, a hypercycle is not an evolutionary stable strategy because there is no positive selection for the giving of catalytic support to other molecule species[2,13]; therefore, a "parasitic" mutant that gives no catalytic support but receives increased catalytic support is fatal to the hypercycle.

Recently we introduced a cellular automaton (CA) model for the hypercycle interactions.[1] The introduction of a spatial dimension and incomplete mixing in this model has far-reaching consequences. In the CA model, the hypercycle interactions give rise to self-structuring into spiral waves. It turns out that by this self-structuring, the hypercycle becomes resistant to a large class of parasites.

In this paper we catalogue the changes in the selectional properties of the hypercycle due to the spatial self-structuring. In the ODE model, the selection takes place at the level of the individual molecules; a mutant that produces more copies is always selected. We will show that many of the "self-evident" selectional properties which are well known from the ODE model are not valid in the spatial CA model. This is caused by the fact that in the CA model selection takes place between spirals.

2. THE MODEL

In order to simulate incomplete mixing, we use the model formalism of the cellular automata.[16] A cellular automaton is defined as a large tesselation of identical finite-state automata (cells). Each automaton is defined as a triplet: $\langle I, S, W \rangle$, where I is the set of inputs, S is the set of states (both sets being finite and usually small), and W is the next-state function, defined on input-state pairs. The inputs are the states of "neighbor" cells, i.e., the adjacent cells in the tesselation. Cellular automata have proved to be a powerful tool in the study of spatial processes, for instance, fluid dynamics.[5,16,20]

In our cellular automaton, the total space consists of 300×300 cells in a square toroidal tesselation. The state of a cell refers either to its occupation by a molecule of a certain species or to its emptiness; i.e., a cell can contain one molecule or it can be empty. In the next-state function of the cells (see Table 1), we implement a representation of three processes:

1. *Decay* (see Figure 1(a)) can occur when a cell is occupied; after decay the cell becomes empty. The probability of decay is species (i.e., state) dependent.
2. *Replication* (see Figure 1(b)) is only possible in empty cells; a molecule in one of the four direct neighbor cells can replicate into the empty cell. The probability

of replication is species dependent. There is also a probability that the cell will remain empty.

3. *Catalysis* (see Figure 1(c)) is related to replication; the probability that a molecule will replicate into an empty cell is increased when there are catalytic molecules in at least one of the four cells that lie adjacent to the direction of replication.

In addition to these three processes, *diffusion* is included in our model as a separate process, operating between "timesteps." We use the diffusion algorithm of Toffoli and Margolus,[16] which ensures particle conservation. In this algorithm, the space is divided into subfields of 2 x 2 cells. At each diffusion step, the states of a subfield are rotated 90 degrees clockwise or anti-clockwise with equal probability. After a diffusion step, the subfields are shifted one cell diagonally.

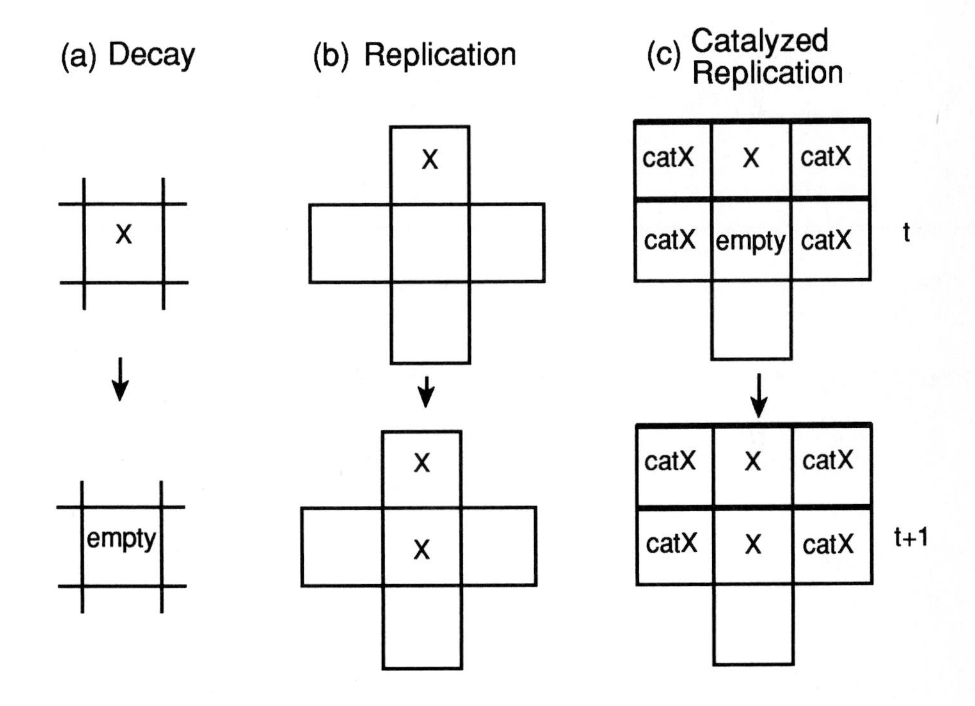

FIGURE 1 Three-state transitions of a cell in a cellular automaton in which hypercycles can be simulated. The next state $t+1$ is drawn below the present state t. X is a molecule of a certain species; cat$X$ is a molecule that catalyzes molecule X's replication. See text for further explanation.

TABLE 1 Next state probability function p() of a cellular automaton in which hypercycles can be simulated.[1]

If the cell is occupied by a molecule x:
p(empty) = decay[x] ;
p(x) = 1 – decay[x] ;

If the cell is empty:
p(empty) = $\text{Claim}_{\text{empty}}$ / $\sum$ Claims ; (no replication)
p(N) = Claim_{N} / $\sum$ Claims ;
p(S) = Claim_{S} / $\sum$ Claims ;
p(W) = Claim_{W} / $\sum$ Claims ;
p(E) = Claim_{E} / $\sum$ Claims ;

In which:
$\sum$ Claims = $\text{Claim}_{\text{empty}}$ + Claim_{N} + Claim_{S} + Claim_{W} + Claim_{E} ;
Claim_{N} = self[N] + c[N,NE] + c[N,NW] + c[N,E] + c[N,W] ;
Claim_{S} = self[S] + c[S,SE] + c[S,SW] + c[S,E] + c[S,W] ;
Claim_{W} = self[W] + c[W,NW] + c[W,SW] + c[W,N] + c[W,S] ;
Claim_{E} = self[E] + c[E,NE] + c[E,SE] + c[E,N] + c[E,S] ;

[1] Explanation of some terms: decay[x] is the decay parameter of molecule x; self[x] is the replication parameter of x; c[x,y] is the catalytic support x gets from y; Claim_{empty} is a constant, which is set to 11; N(orth),S,W,E,NW,NE,SW,SE are the states of the eight neighbor cells indicated by their compass direction.

3. RESULTS

3.1. EMERGENCE OF SPIRAL WAVES

In order to get a better understanding of the selectional properties of the spatial model, we start with a section in which we will give a detailed description of the dynamical properties of the spirals.

3.1.1. DEVELOPMENT First we describe the spatial behavior of a set of molecules which are part of a pre-defined hypercycle. In plate 5A, the six members of the hypercycle are distributed at random in the space; 50% of the cells are empty. The six molecule species have identical replication and decay parameters (Table 1: self[1..6] = 1; decay[1..6] = 0.2). Each species catalyzes one other member of the hypercycle; catalyzed replication is much stronger than non-catalyzed replication (Table 1: $c[2,1] = c[3,2] = \ldots = c[1,6] = 100$). After each timestep, there are two diffusion steps.

Plate 5B shows the situation after 1000 timesteps. Spiral structures have developed containing all members of the hypercycle in catalytic order. Each species grows towards its catalytic supporter; species 1 (red) grows towards species 6 (blue), species 2 (orange) grows towards species 1, and so on. As a result of this directional growth, the spirals rotate. Most spirals occur in couples; a spiral rotating clockwise is close to a spiral rotating counter-clockwise.

Plate 5C shows the situation after 2000 timesteps. Some spirals have disappeared. This happens when two spirals rotating in the opposite direction come too close to one another. The number of molecules of a species between the two spirals is then reduced. If, by chance, a species dies out, then temporarily the species that gives catalytic support to the extinct species takes over the complete region of the double spiral. Because this species now no longer gets catalytic support, the region formerly occupied by the double spiral is taken over by other nearby spirals.

After 2000 timesteps, the pattern remains stable: the centers of the spirals do not move and all spirals have the same rotation time.

3.1.2. GROWTH WITHIN A SPIRAL The middle of a spiral acts as a center of growth for the entire spiral. This is demonstrated in plates 6A-C (see color plates). In plate 6A the molecules in the middle and the periphery of the huge single spiral in the situation of plate 1C are labelled (only the labelled molecules are colored; each color represents three molecule species, which are adjacent in the hypercycle). In plate 6B, after 30 timesteps, the descendants of the labelled molecules in the periphery have reached the edge of the region of the spiral. The labelled molecules in the middle of the spiral have increased in number. After 200 timesteps (see plate 6C), the molecules from the middle have taken over the complete spiral region and the molecules from the periphery have disappeared. This direction of growth is caused by the catalytic waves that travel from the middle towards the periphery of the spiral. Note that, although the spirals rotate, growth is not rotational.

3.1.3. STABILITY The stability of spirals depends on the parameters of the molecules. We test this relation on a 150×150 field with empty boundaries (no torus). As a starting pattern we use the configuration in Figure 2. In this configuration, the molecules are already in catalytic order; normally only one spiral is formed, which rotates clockwise. At each run, we vary one parameter of species 1 (red); the default parameters of the other species are as in plate 5A.

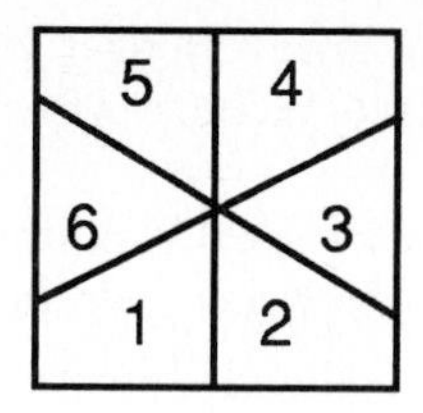

FIGURE 2 Starting pattern of the stability experiments.

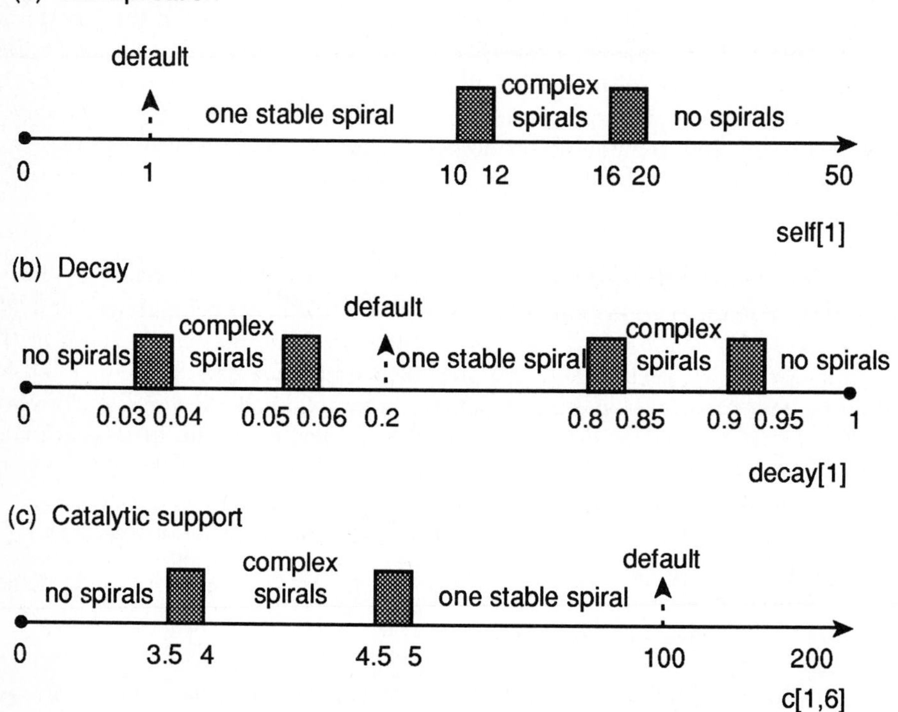

FIGURE 3 Stability results for various parameters of species 1, after 1000 timesteps.

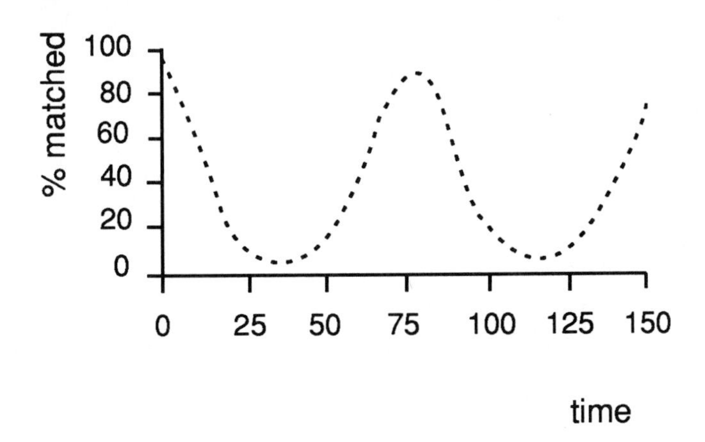

FIGURE 4 Measuring rotation time.

In Figure 3 the stability results after 1000 timesteps for various parameters of species 1 are summarized. We distinguish three classes of stability:

1. *One stable spiral.* As long as the parameters of species 1 do not deviate too much from the default, a stable spiral is formed. The thickness of the spiral waves varies with the parameters; in plate 7A the situation after 1000 timesteps for decay[1] = 0.06 is shown: the waves of species 1 (red) and 2 (orange) are thick and the wave of species 6 (blue) is very thin.
2. *No spirals.* The spiral is unstable when a catalytic wave is too thin and becomes extinct. We give an example for decay[1] = 0.03; in plate 7B, after 1000 time steps, species 6 (blue) has vanished and five species remain. The remaining system is locally unstable, but the five species persist in a global dynamical equilibrium (see also subhead 3.3.3).
3. *Complex spirals.* Between the parameter regions of Class 1 and Class 2 there is always a zone of Class 3; in this zone, spirals are formed but they are not stable: there is a perpetual loss and reformation of spirals. An example is shown for decay[1] = 0.05 (plate 7C, see color plates, after 1000 timesteps); the catalytic wave of species 6 (blue) is thin and it has gaps: sometimes a gap is filled from the sides, and a new double spiral develops around the gap.

3.1.4. ROTATION TIME The rotation time of a spiral varies with the parameters of the species. We measured the rotation time of the spirals in the stability experiments described above. The rotation time is measured by taking the situation after 500 time steps as a target pattern and matching it with subsequent patterns. A molecule is matched if there is a molecule of the same species somewhere in the nine-cell

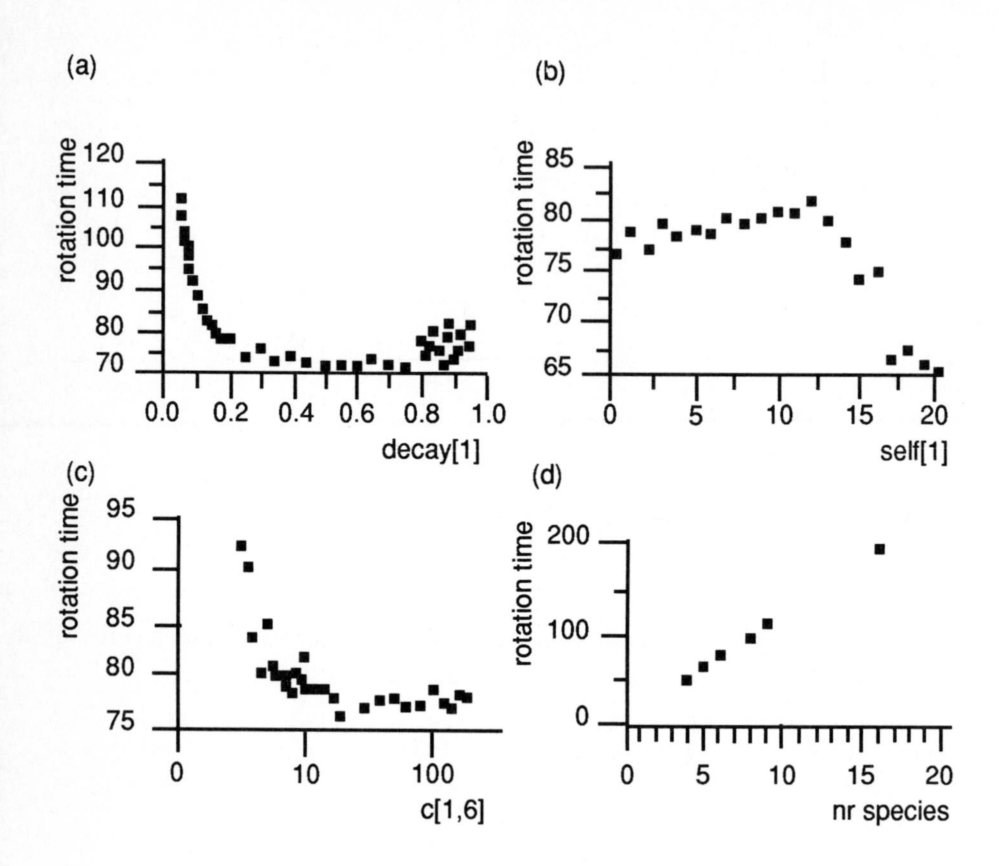

FIGURE 5 Rotation time of the spiral as a function of (a) decay (b) self-replication (c) catalytic support, and (d) number of species in the hypercycle.

neighborhood in the target pattern. In Figure 4 this relation is given for the default parameters; the rotation time is in this case 76 time steps.

In Figure 5(a) the rotation time is given for various decay parameters of species 1. For $0.05 \leq \text{decay}[1] \leq 0.75$, the rotation time decreases asymptotically. This is easy to understand, for the slower species 1 decays, the longer it takes its successor in the hypercycle (i.e., species 2) to replace it. For $\text{decay}[1] > 0.75$, the rotation time increases again, with high variation. This can be explained by the fact that the catalytic wave of species 1 in this case is very small and occasionally has gaps. The gaps have to be filled from the sides and this process takes some extra time.

In Figure 5(b) the relation between self-replication and rotation time is given. For $\text{self}[1] \leq 12$, the rotation time increases with increasing self-replication. The reason is analogous to the case for slower decay: if species 1 self-replicates strongly, species 2 has more trouble in replacing it. For self[1] > 12, the rotation time decreases sharply with increasing self-replication. In this parameter region, the self-replication of species 1 is so strong that the species from which it gets catalytic support (i.e., species 6) has almost vanished.

Figure 5(c) shows the rotation time for various catalysis parameters. Rotation time decreases asymptotically with increasing catalysis. With high catalysis, a molecule growths faster towards its catalytic supporter. For c[1,6] ≤ 20, (almost) all catalyzed replication claims are rewarded and thus there is no further decrease in rotation time.

3.1.5. **NUMBER OF SPECIES** Not only the parameter values, but also the number of species in the hypercycle affects the stability and rotation time of the spirals. For hypercycles of 2 or 3 members (with default parameters), no spirals are formed (plate 8A; 3 members, $t = 2000$), a hypercycle of four members is in the complex spiral region (plate 8B; 4 members, $t = 2000$) and hypercycles of five or more members form stable spirals (plate 5C; 6 members, $t = 2000$). The rotation time of the spirals increases linearly with increasing number of species (see Figure 5(d)).

If we compare these results with the ODE model, we see that hypercycles of five or more members form spatially stable spirals; whereas the ODE model has a limit cycle with extreme oscillations (such that extinction should be expected). For hypercycles of four or less members, the ODE model has a stable equilibrium. Such systems are spatially unstable in the CA model, although the frequency of the species is fairly constant.

3.2. PARAMETER MUTANTS

In this section we will investigate selection for mutants that differ in the strength of one parameter. We use the situation of plate 5C (see color plates) as a starting pattern. All molecules of species 1 in the center of the single spiral in the middle of the field are replaced by mutants, as shown in plate 9A (see color plates); mutants are black.

3.2.1. **"DECAY" MUTANTS** The ODE model predicts that mutants with slower decay will win and mutants with faster decay will loose. In Figure 6(a) the number of mutants and the number of molecules of species 1 are given 4000 timesteps after infection. The prediction of the ODE model is fulfilled for decay[mutant] ≤ 0.13 and decay[mutant] ≤ 0.55. However, for $0.14 \leq$ decay[mutant] ≤ 0.5, the results contradict this prediction; the outcome of selection is reversed. How can this be explained?

It appears that, for decay[mutant] ≤ 0.13, the mutant decays so slowly that it can diffuse towards the center of the other spirals; the mutant is able to grow against the general direction of growth in the spirals (see subsection 3.1.2.). Analogously for decay[mutant] ≤ 0.55, species 1 is able to get into the center of the infected spiral and thus replace the mutant.

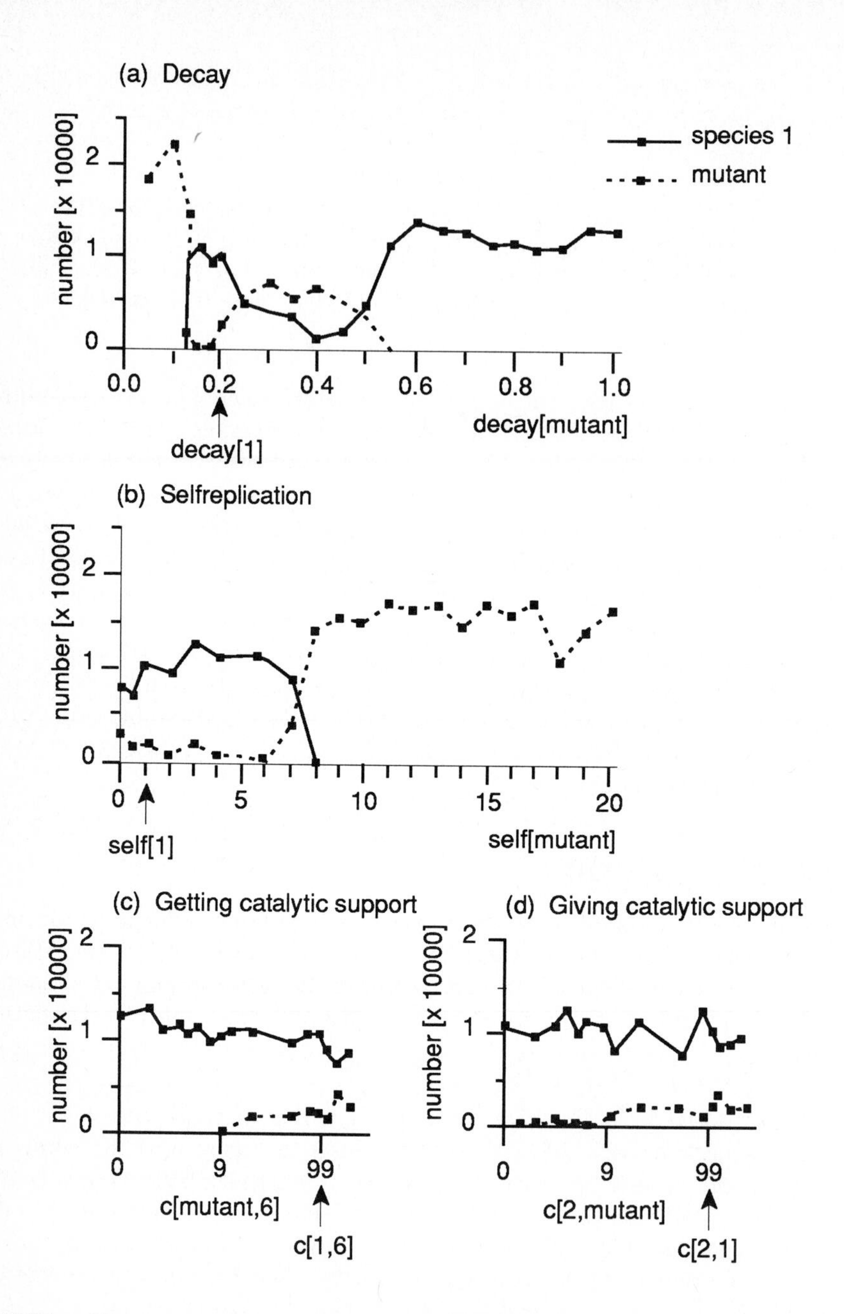

FIGURE 6 Number of molecules 4000 timesteps after infection with a parameter mutant. The x-axis of (c) and (d) are z-log tranformation.

For $0.14 \leq$ decay[mutant] ≤ 0.5, the molecules are not able to grow towards the center of other spirals. Competition now takes place between spirals, and it turns out that rotation speed of the spiral is the decisive factor: a spiral that rotates faster expands towards a spiral that rotates slower.

In plate 9B the dynamical process of the competition between the spirals is shown; a one-dimensional horizontal section through the middle of the field ($y = 150$) is drawn for every fourth timestep. In this case the mutant decays slower than species 1 (decay[mutant] $= 0.15$; decay[1] $= 0.2$) and, therefore, the infected spiral rotates slower than the other spirals (see Figure 5(a)) and thus the other spirals expand. Plate 9C shows the situation after 4000 timesteps; there are still some mutants, but the infected spiral has lost most of its domain. The situation is stable in time; the mutants will not disappear but form a small cyst, because the infected spiral is reinforced by the other spirals.[1]

Plate 9D shows the dynamics for decay[mutant] $= 0.4$. Now the infected spiral expands, because it rotates faster. After 4000 timesteps (plate 9E), the infected spiral has taken over almost the entire field, and some of the non-infected spirals have formed cysts.

3.2.2. "SELF-REPLICATION" MUTANTS According to the ODE model, a raised self-replication rate should be advantageous for a mutant. Figure 6(b) shows the number of mutants and the number of molecules of species 1, 4000 time steps after infection. For self[mutant] ≤ 7, indeed, the mutant wins; it is able to get into the center of the other spirals. For self[mutant] ≤ 6, there is competition between the spirals. For this parameter region, the rotation time increases with increasing self-replication (see Figure 5(b)) rate, so this explains why the number of mutants increases with decreasing self-replication rate of the mutant. For self[mutant] $= 0$, the infected spiral should out-compete the other spirals; however, after 4000 timesteps, the number of mutants is still smaller than the number of molecules of species 1. This can be explained by the small difference in rotation time between the spirals, which causes the infected spiral to expand very slowly. Moreover the catalytic wave of the mutant is smaller than that of species 1, because it lacks self-replication, and this causes the number of mutants to be relatively low as well.

3.2.3. "CATALYTIC SUPPORT" MUTANTS The ODE model makes a clear distinction between selection for getting catalytic support and selection for giving catalytic support; mutants that get more support are positively selected while selection for mutants that give more support is neutral. As mentioned, this renders hypercycles vulnerable to "parasites." Figures 6(c) and 6D show the number of mutants and the number of molecules of species 1 after 4000 timesteps for both types of mutants. It appears that, in the CA model, selection for giving and getting catalytic support essentially follows the same pattern.

In both cases for c[mutant] > 10, selection is neutral, because there is no difference in the rotation time between the infected spiral and the other spirals (see Figure 9(c)). In this parameter region there is a slight increase in the number of mutants but this has to be explained by a thicker catalytic wave of the mutant for increasing catalytic support.

For c[mutant] $\leq$ 10, there is positive selection for both getting and giving catalytic support; the infected spiral in this case has a slower rotation time than the other spirals (see Figure 5(c)). A mutant for giving catalytic support forms a cyst, while a mutant for getting catalytic support is rejected completely. Note that these results imply that the hypercycle in the CA model is resistant to parasites.

3.3. CONNECTANCE MUTANTS

In this section we will examine selection for mutants that differ in their catalytic connections to other molecules. Again, as with the parameter mutants, we will start with replacing the molecules of species 1 with mutants in the center of a spiral.

3.3.1. SHORT-CUT MUTANTS Figure 7 shows the catalytic connections of a shortcut mutant in a hypercycle with five members. The shortcut mutant has the same connections as species 1, but it also gives catalytic support to species 3 and, therefore, enables a shorter hypercycle of four members. All catalytic connections are equally strong (c = 100) and the mutant has default decay and self-replication parameters.

Plate 10A (see color plates) shows the infection with the above-described shortcut mutant in a stable situation of a hypercycle with five members. Plate 10B shows the situation after 500 timesteps; in the infected spiral, species 2 (orange) has become (almost) extinct, so now the short hypercycle of length 4 dominates the former region of the infected spiral. However this short hypercycle cannot form a stable spiral (see subhead 3.1.5.) and it turns out that it cannot compete against the remaining spirals of length 5 which are stable; so after 3000 timesteps in plate 10C the mutant is rejected completely. This result holds for initial hypercycles of whatever length, infected by shortcut mutants for hypercycles of length 4 or less.

Something different happens when the shortcut hypercycle has length 6 or more, i.e., when it can form a stable spiral. Plate 11A shows the infection of a spiral in a stable situation of a hypercycle with seven members with a shortcut mutant for a hypercycle with six members. After 500 timesteps, in plate 11B, species 2 has become extinct in the infected spiral, and now there is competition between the two spirals of different length. From Figure 5(d), we know that the shorter hypercycle rotates faster, and, therefore, it will expand; after 2000 timesteps (see color plate 11C), the hypercycle of length 6 has taken over the complete field.

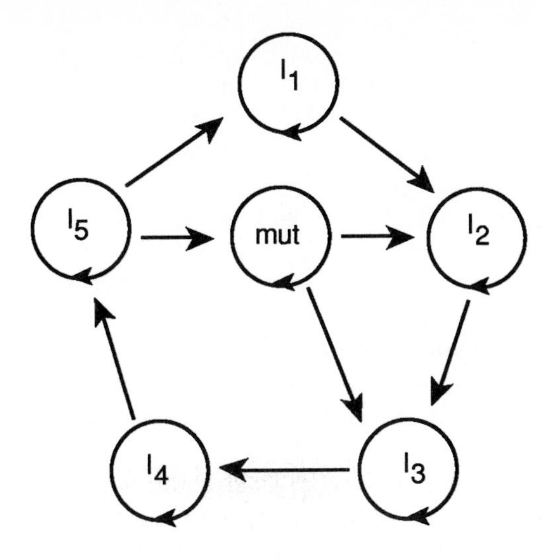

FIGURE 7 A hypercycle of five members with a shortcut mutant "mut." The mutant enables a hypercycle of four members.

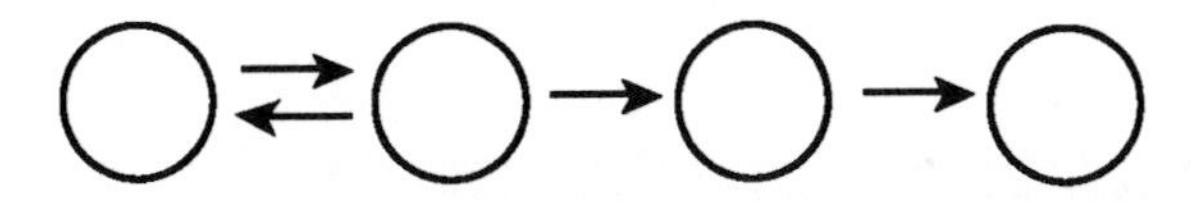

FIGURE 8 A hypercycle of two members with two coupled parasites.

For shortcut mutants of length 5, the situation appears to be dramatic. A six member hypercycle is infected with a shortcut mutant for a hypercycle of length 5 (plate 12A, see color plates). Initially things develop as in the previous case; the short hypercycle expands, as shown after 2000 timesteps in plate 12B. However in the end, as species 1 has become almost extinct, and species 2 is completely extinct, it turns out that a hypercycle of length 5 is vulnerable to parasites. With the extinction of species 2, species 1 no longer gives catalytic support to the hypercycle and thus it has become a "parasite" to the hypercycle. Plate 12C after 4000 timesteps shows that this parasite is deadly to the five-member hypercycle; it expands towards the spiral, and eventually all species, except for species 1, will become extinct.

3.3.2. DISJOINT HYPERCYCLES In the ODE model, the growth of a hypercycle is a nonlinear (approximately quadratic) function of its concentration. Therefore, an established hypercycle cannot be replaced by any newcomer, regardless of its parameters, because the latter has to start at low concentration. Thus, selection of a hypercycle is a "once forever" decision.

In the CA model, it is obvious that, as soon as the new hypercycle has formed a spiral, it can compete with the established hypercycle. In the situation of plate 5C (color plates), a new hypercycle was introduced at low concentration (25 molecules of each species), but in catalytic order. The new hypercycle has much better parameters than the established hypercycle (decay[new 1..6] = 0.1; self[new 1..6] = 2.0; c[new 1,6] etc. = 200.0). It appears that the new hypercycle simply outgrows the established hypercycle. Thus once forever selection does not hold for this system.

3.3.3. RANDOM CONNECTIONS Preliminary experiments, in which the catalytic connections between molecules are chosen at random, indicate that sometimes there is selection for a hypercyclic interaction structure. This occurs for a certain number of species and a certain number of connections. Much more often, interaction structures of short cycles with coupled parasites (an example is shown in Figure 8) arise. These structures would not persist in the ODE model; in the CA model, the system is locally unstable, but there is a dynamical global equilibrium.

In a future paper we will try to unravel the complex interaction pattern which evolves under various circumstances.

4. DISCUSSION AND CONCLUSIONS

4.1. MOLECULES WITH A HYPERCYCLIC INTERACTION STRUCTURE SHOW SPATIAL SELF-STRUCTURING; SPIRAL WAVES DEVELOP FOR A WIDE VARIETY OF PARAMETER VALUES

The spiral wave is a well-known pattern in excitable media which has been studied most thoroughly with respect to the Belousov-Zhabotinsky reaction, both experimentally and theoretically.[11,14] Most theoretical models are formulated in terms of partial differential equations[17]; spiral-wave solutions are found in cellular automata models as well.[6,7,8,16] Spiral waves have been shown to play a role in cell-to-cell communication in *Dictyostelium discoideum*[18] and in neuro-muscular tissue.[19] Our analysis suggests a role for spiral waves in enabling the evolution of cooperation.

We tested our model extensively for robustness of the self-structuring property. It turns out that the precise definition of the cellular automaton does not affect the development of the spirals. We examined, for instance, asynchronous updating of the cells, a non-toroidal field, and other neighbor cells that can give catalytic support.

The individual molecule species in the hypercycle can, to some extend, differ in their parameters and still form stable spirals. However, when the difference in parameters is too large, the system becomes unstable and no spirals are formed. In between the parameter region of stable spirals and no spirals, there is always a zone of complex spirals; in this zone there is perpetual formation and degradation of spirals.

The number of species in the hypercycle affects the formation of spiral waves; it appears that (at least for the given default parameters) the hypercycle needs to consist of five members or more to form stable spirals. The hypercycle with four members is in the complex spiral zone and spirals of two or three members show no spiral formation at all.

For many of the parameters, we see a zone of complexity in between a dynamical (chaotic) zone and a spatially stable spiral pattern. This situation seems somewhat analogous to phase transitions.[12]

4.2. IN A SYSTEM WITH SPIRAL WAVES THERE IS SELECTION AT THE LEVEL OF THE SPIRALS; THIS SELECTION CAN CONTRADICT SELECTION AT THE LEVEL OF THE INDIVIDUAL MOLECULES

Within a spiral, there is a strikingly unequal distribution of long-term fitness: the molecules in the middle of a spiral generate the offspring of the entire spiral whereas the molecules in the periphery of a spiral disappear (as shown in color plate 6A-C). This direction of growth causes the spirals to act as independent entities in the selection process.

The following "rules of selection" can be formulated for the spiral wave system:

A. *Molecules that are capable of growth toward the center of a spiral are always selected.* An example of this rule is the situation for decay[mutant] ≤ 0.13 in Figure 6(a); the mutant decays so slowly that it is able to get in the center of all the spirals. Plate 12C also gives an example: species 1 (red) is coupled parasitically to the hypercycle and it is able to grow towards the center of the spirals. (Note that, in the latter case, species 1 will destroy the hypercycle, while in the first case, the mutant replaces species 1, but it does not destroy the hypercycle).

B. *If not A: molecules that can form a stable spiral are selected against molecules that cannot form a stable spiral.* An example is the situation for the shortcut mutant in plate 10A-C; the mutant selects for a hypercycle of four members, but this hypercycle does not form stable spirals and, therefore, it is out-competed by the hypercycle with five members.

C. *If not A or B: molecules whose spiral rotates faster will be selected against molecules with slower rotating spirals.* The cases of plates 9A-E and 11A-C are examples of this rule. The process is analogous to the competition between pace-maker waves in the Belousov-Zhabotinsky reaction,[21] where the wave with the highest frequency wins.

D. *If not A, B or C: selection is neutral.* Selection is neutral when the spiral with the mutant has the same rotation time as the other spirals (e.g., Figure 6(c) and 6(d) for $c > 10$).

In cases where rule B, C or D applies, the selection results often are reversed to the selection pressure at the level of the individual molecules (i.e., to the predictions of the ODE model). A nice example is the positive selection for faster decay as shown in Figure 6(a). Of course, individual selection would predict selection against a mutant that decays faster.

In cases where rule A applies, the results normally do agree with the predictions of the ODE model. However, even this case should not be interpreted as simple individual selection, but rather as competition between an individual molecule species and a spiral, which is won by the first.

4.3. THE TYPE OF ATTRACTOR TO WHICH A SPATIAL SYSTEM WITH SELF-REPLICATION AND CATALYSIS EVOLVES IS YET UNKNOWN; THERE SEEMS TO BE SOME INDICATION FOR EVOLUTION TOWARDS THE COMPLEX SPIRAL REGION (SEE SUBHEAD 3.1.3)

In this paper we have shown selection results for single infections with one type of mutant. The behavior of the system in case of multiple infections with mutants of various type, of course, remains an open question. However, it is tempting to speculate about it, and we will do so in this paragraph.

We showed (in subheads 3.2.1. and 3.2.2.) that two types of decay and self-replication mutants are selected, namely:

- *Type 1.* Mutants with somewhat faster decay or weaker self-replication are selected because they generate spirals that rotate faster.
- *Type 2.* Mutants with much slower decay or much stronger self-replication are selected because they are capable of growth towards the center of the spirals.

It seems plausible that there will be a kind of balance between the two types of mutations; after establishment of a mutant of type 1, the spirals will become easier to penetrate and thus more vulnerable to mutants of type 2. Vice versa, after establishment of a type-2 mutant, the spirals will be difficult to penetrate and their rotation speed will be relatively slow, thus they will be more vulnerable to mutants of type 1. This balance in the direction of selection might lead to specialization of the molecules; some will be of type 1 and some of type 2. However, by diverging in their parameters, the molecules will form less stable spirals (see subhead 3.1.3). Mutants that actually destabilize the spiral will be rejected. In this scenario the system would approach the complex spiral region and remain there.

The shortcut mutants that we discussed in subhead 3.3.1 also show a tendency to get the system near the complex spiral region; there is selection for hypercycles of five members, the lowest number to form a stable spiral. It appears though that this critical hypercycle length in this case also is critical in respect to its vulnerability to parasites; the five-member hypercycle in color plate 12C is destroyed by the parasitical species 1. However, this result is not robust: if the five-member hypercycle has skew parameters (e.g., species 4 decays somewhat slower), it can be stable against species 1.

4.4. SPATIAL SELF-STRUCTURING CAN HAVE A MAJOR IMPACT ON THE OUTCOME OF SELECTION PROCESSES; THEREFORE, IT SHOULD BE TAKEN INTO ACCOUNT IN THE STUDY OF EVOLUTION

We have shown that spatial self-structuring into spiral waves alters almost all selectional properties of the hypercycle. This is because selection no longer exclusively takes place at the level of the individual molecules, but also at the level of the spirals. The spatial self-structuring enables selection for all kinds of altruistic properties (e.g., selection for giving catalytic support and selection for decaying faster).

It seems plausible that the phenomenon of spatial self-structuring is not restricted to hypercycles. Farmer et al.[4] propose a cyclic catalytic network of polymers. This network differs from the structure of a hypercycle in that the polymers are not self-replicative. However, the interactions in this network look very much like the interactions in the earlier mentioned Belousov-Zhabotinsky reaction,[14] so the spiral structure may well emerge in this system, too. Whether cyclic interaction structures are likely to appear and out-compete other structures in networks with random interactions is a subject for further study.

Self-structuring is a well-known feature of cellular automata. Simple low-level transition rules can generate high-level spatial patterns. This spontaneous self-structuring has often been interpreted as a form of evolution.[15] In this study we use a different approach; we consider self-structuring as a substrate for selection.[10] The substrate has proved very fertile; an environment is created in which selection for altruistic features is possible.

We believe, therefore, that in the study of (prebiotic) evolution, it is important to look for self-structuring and examine its consequences.

APPENDIX THE ODE OF THE HYPERCYCLE

In the model, n self-replicative molecule species are linked cyclically by catalysis (see Figure 9). The total number of molecules C is kept constant by an output flux f. Erroneous mutants are not included in the model. For analytical proof and further discussion, see Eigen[3] and Hofbauer.[9]

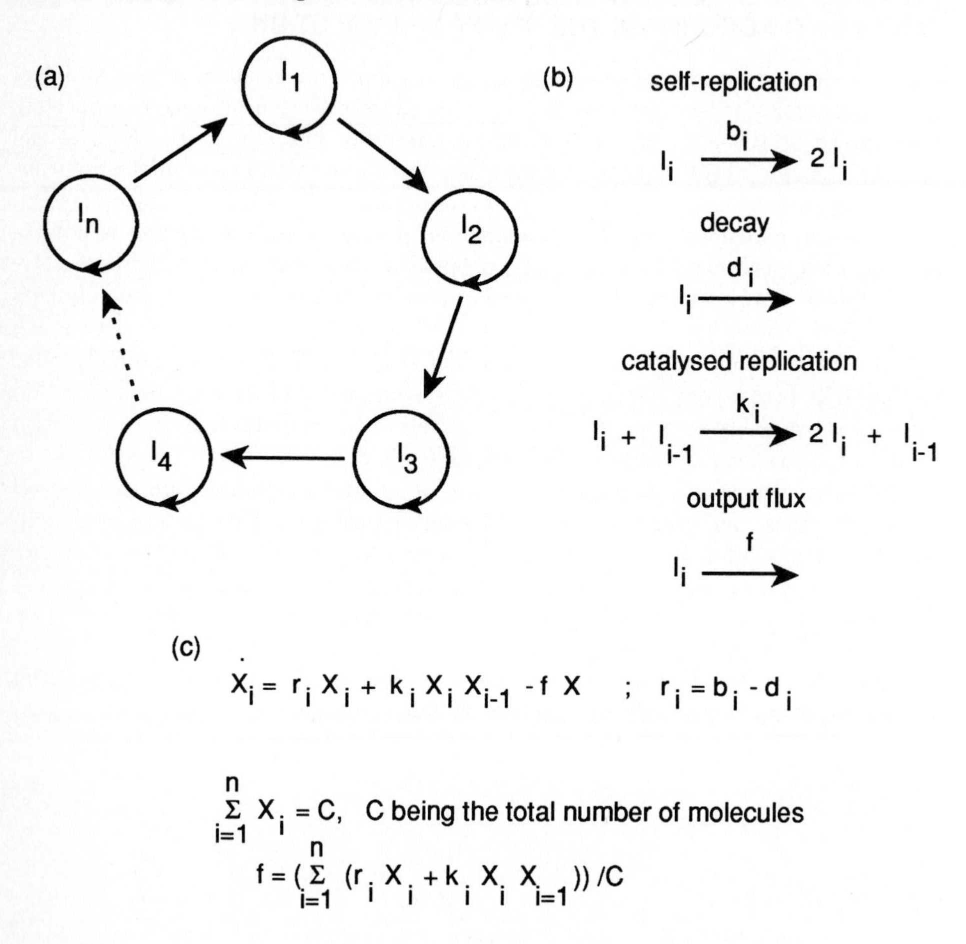

FIGURE 9 (a) Schematic diagram of a hypercycle. The hypercycle consists of self-replicative molecule species I_i; each species provides catalytic support for the subsequent species in the cycle. After Eigen.[3] (b) Kinetic steps. (c) Differential equations.

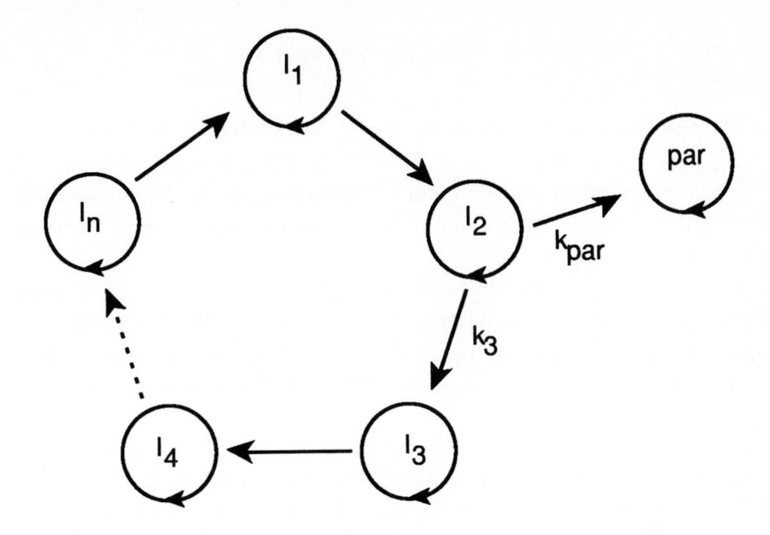

FIGURE 10 A hypercycle with a self-replicative "parasitic" molecule species "par." The parasite gets catalytic support from species I_2, but does not give catalytic support to any molecule species in the hypercycle. After Eigen.[3]

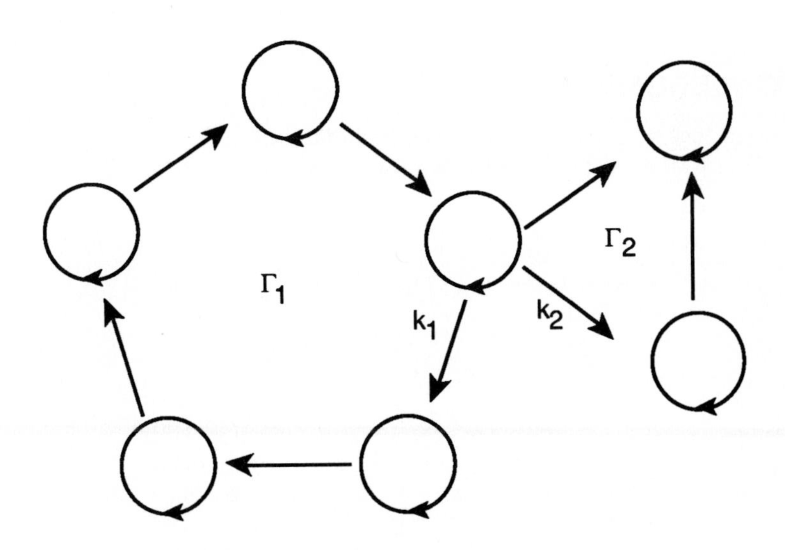

FIGURE 11 Schematic diagram of two joint hypercycles. After Hofbauer.[9]

Summary of dynamical and selectional properties

- Stability. The elementary hypercycle has only one attractor. At low dimensions ($n \leq 4$), the attractor is an asymptotically stable fixed point, namely, a focus for $n = 2$ and a spiral sink for $n = 3$ and $n = 4$. In systems of higher dimensions ($n \geq 5$), "permanence" has been proven, i.e., no molecule species vanishes; numerical integration provides strong evidence for the existence of a stable limit cycle.
- Parasites. In Figure 10 a hypercycle with a so-called parasite is shown. The system appears to be competitive, i.e., the hypercycle and the parasite cannot co-exist. If the linear terms are neglected, the following relation holds: if $k_{par} > k_3$, the parasite wins; if $k_{par} < k_3$, the entire hypercycle (and the parasite) becomes extinct.
- Competition between joint hypercycles. In Figure 11, two joint hypercycles are shown. The two cycles exclude each other (again neglecting the linear terms): If $k_1 < k_2$, hypercycle Γ_2 will out-compete Γ_1; if $k_1 > k_2$, hypercycle Γ_1 will out-compete Γ_2.
- Competition between disjoint hypercycles. Selection of a hypercycle is a "once forever" decision. A hypercycle, once established, cannot easily be replaced by any newcomer, since new species always emerge as one copy; the growth rate of a hypercycle is nonlinear and therefore dependent on population size.

ACKNOWLEDGMENTS

The investigations were supported by the Foundation for Biophysics, which is subsidized by the Netherlands Organization for Scientific Research (NWO).

REFERENCES

1. Boerlijst, M. C., and P. Hogeweg. "Spiral Wave Structure in Prebiotic Evolution: Hypercycles Stable Against Parasites." *Physica D* **48** (1991): 17.
2. Bresch, C., U. Niesert, and D. Harnasch. "Hypercycles, Parasites and Packages." *J. Theor. Biol.* **85** (1980): 399.
3. Eigen, M., and P. Schuster. *The Hypercycle: A Principle of Natural Self-Organization*. Berlin: Springer, 1979.
4. Farmer, J. D., S. A. Kauffman, and N. H. Packard. "Autocatalytic Replication of Polymers." *Physica* **22D** (1986): 50.
5. Frisch, U., B. Hasslacher, and Y. Pomeau. "Lattice-Gas Automata for the Navier-Stokes Equation." *Phys. Rev. Lett.* **56** (1986): 1505.
6. Griffeath, D. "Cyclic Random Competition: A Case History in Experimental Mathematics." *Notices of the American Mathematical Society* **35** (1988): 1472.
7. Gerhardt, M., and H. Schuster. "A Cellular Automaton Describing the Formation of Spatially Ordered Structures in Chemical Systems." *Physica D* **36** (1989): 209.
8. Gerhardt, M., H. Schuster, and J. J. Tyson. "A Cellular Automaton Model of Excitable Media Including Curvature and Dispersion." *Science* **247** (1990): 1563.
9. Hofbauer, J., and K. Sigmund. *The Theory of Evolution and Dynamical Systems*. Cambridge, MA: Cambridge University Press, 1988.
10. Hogeweg, P. "Simplicity and Complexity in MIRROR Universes." *BioSystems* **23** (1989): 231.
11. Keener, J. P., and J. J. Tyson. "Spiral Waves in the Belousov-Zhabotinskii Reaction." *Physica D* **21** (1986): 307.
12. Langton, C. G. "Studying Artificial Life with Cellular Automata." *Physica* **22D** (1986): 120.
13. Maynard-Smith, J. "Hypercycles and The Origin of Life." *Nature* **280** (1979): 445.
14. Müller, S. C., T. Plesser, and B. Hess. "Two-Dimensional Spectophotometry of Spiral Wave Propagation in the Belousov-Zhabotinskii Reaction I. Experiments and Digital Representation. II. Geometric and Kinematic Parameters." *Physica* **24D** (1987): 71.
15. Tamayo, P., and H. Hartman. "Cellular Automata, Reaction-Diffusion Systems and the Origin of Life." *Artificial Life*, edited by C. G. Langton. SFI Studies in the Sciences of Complexity, Proc. Vol. VI, 105. Redwood City, CA: Addison-Wesley, 1989.
16. Toffoli, T., and N. Margolus. *Cellular Automata Machines*. Cambridge: MIT Press, 1987.
17. Tyson, J. J., and J. P. Keener. "Singular Perturbation Theory of Traveling Waves in Excitable Media (a review)." *Physica D* **32** (1988): 327.

18. Tyson, J. J., K. A. Alexander, V. S. Manoranjan, and J. D. Murray "Spiral Waves of Cyclic AMP in a Model of Slime Mold Aggregation." *Physica D* **34** (1989): 193.
19. Winfree, A.T. "Electrical Instability in Cardiac Muscle: Phase Singularities and Rotors." *J. Theor. Biol.* **138** (1989): 353.
20. Wolfram, S. *Theory and Applications of Cellular Automata.* Singapore: World Scientific, 1986.
21. Zaikin, A. N., and A. M. Zhabothinsky. "Concentration Wave Propagation in Two-Dimensional Liquid-Phase Self-Oscillating System." *Nature* **225** (1970): 535.

Peter Schuster
Institut für Theoretische Chemie der Universität Wien, Währingerstraße 17, A-1090 Wien, Austria

Complex Optimization in an Artificial RNA World

INTRODUCTION

Folding polynucleotide sequences into thermodynamically stable structures provides an example of a complex relation between a string of symbols and an object in three-dimensional space. It is used in our model as a—simplest possible—genotype-phenotype relation. The information for the three-dimensional structure is encoded in the string which represents the genotype. The notion of the phenotype is extended to molecules by identifying it with the spatial structure of polynucleotides. Tertiary structures are hard to encode in compact notion and their predictions by theoretical models are notoriously bad. If, however, only secondary structures are considered as phenotypes, fairly reliable predictions of stable folded structures can be made and quantitative estimates of free energies and kinetic parameters—like rate constants for replication or degradation—are possible. Kinetic parameters cannot be derived from *first principles*. Instead, qualitatively correct models are applied which are based on the current empirical knowledge in biophysical chemistry. In order to simplify the combinatorial problem binary sequences (G,C) rather than natural four-letter sequences (G,A,C,U) are used. A variant of genetic algorithms based on replication and mutation is applied to optimize some phenotypic properties. It mimics the processes taking place in a kind of flow reactor. Relations between genotypes are measured in sequence space. The sequence space is a discrete space having the Hamming distance as metric. Individual genotypes are represented by points. Evolutionary optimization is visualized as a process in this space. In addition cost functions or value landscapes are explored by simulated annealing and computation of autocorrelation functions in the space of genotypes.

GENOTYPES AND MOLECULAR PHENOTYPES

The information carriers in molecular genetics are commonly denoted as genotypes. They are also called the *primary* structures of polynucleotides. In nature two classes of polynucleotides form genotypes: DNA and RNA in case of some classes of viruses. The genotype is understood best as a string of letters which are chosen from the four letter alphabet of naturally occurring bases (G,A,C,U in RNA or T in DNA, respectively). Phenotypes are organisms or virions. In case of molecular evolution in the test tube[1,2,3] *naked* RNA molecules are replicated, modified by mutation and subjected to selection. It is useful therefore to extend the notion of phenotype also to the spatial structures of these nucleic acid molecules.

Spatial structures of polynucleotides and proper phenotypes—as they are encountered with organisms—share many features. Single stranded RNA molecules exist in a great variety of different conformations—as a rule there is a single most stable one under given experimental conditions. Many organisms—in particular bacteria—can exist in different phenotypes. As the unfolding of phenotypes depends on environmental conditions, RNA and DNA molecules form different stable conformations under different experimental conditions—as there are temperature, pH, ionic strength etc.

The process of folding the string of bases of an RNA molecule into a three-dimensional *tertiary* structure can be partitioned into two steps:

1. folding of the string into a quasi-planar—two dimensional—*secondary* structure by formation of complementary base pairs, G≡C or A=U, respectively, and
2. formation of a three-dimensional spatial structure from the quasi-planar folding pattern.

Secondary structure formation is modelled much more easily than their transformation into tertiary structures. At present there are no theoretical models available which can predict tertiary structures reliably. In addition three-dimensional structures are very hard to encode in compact form—commonly Cartesian coordinates of all atoms have to be stored. In contrast to the weakness in predictions and the difficulties in encoding of tertiary structures the predictions of secondary structures are simpler and reached already a level of fairly high fidelity.[4] This is mainly a consequence of the fact that the intermolecular forces stabilizing secondary structures—base pairing and base pair stacking—are much stronger than those involved in three-dimensional structure formation. Algorithms are available which find not only the most stable conformation but also suboptimal foldings.[5] In addition secondary structures can be decomposed into structural elements like free ends, loops, stems and joints which contribute to a good approximation additively to free energies or kinetic properties of the RNA molecules. The secondary structures can be readily encoded in strings which are not longer than those of the primary base sequences.

EVALUATION OF PHENOTYPES

The process in evolution which decides about the fitness of a given genotype may be characterized as evaluation of its phenotype. We realize an important *dichotomy*: selection acts only indirectly on the genotype by evaluation of the phenotype whereas all modification is done on the genotype. Again the modifications are subjected to selection only after unfolding of the corresponding phenotypes. Modifications, in general, come along by means of two mechanisms:

1. mutation which is a change in the base sequence due to an error in the replication process, and
2. recombination which does not create new parts of sequences but instead combines anew the genetic information stored in two DNA molecules.

The fitness of a phenotype is the crucial property determining the outcome of selection in the orthodox Darwinian scenario. It is worth to make two remarks here: Darwin himself attributes some role to neutrality with respect to selection and uses the notion of *random drift* in his *Origin of Species* to characterize population dynamics in the absence of differences in fitness. From comparisons of base sequences in different organisms it is well known that neutral mutations are very frequent. Secondly, it is well established now that there are phenomena which are able to suppress competition. They lead to various forms of symbiosis in which former contestants cooperate for their benefits.

Let us return now to the Darwinian scenario which we shall study further here. Fitness is an exceedingly complex function involving many phenotypic traits. There is practically no chance to evaluate the fitness of an higher organism or of a bacterium independently from the selection process. This fact has often led to rather unjustified attacks on the theory of evolution since it seems to run inevitably into a tautology of *survival of the survivor*. Test tube evolution of molecules[1,3] opened up a new access to the fitness problem. On the basis of the known mechanism of RNA replication by $Q\beta$-replicase[2] models can be developed which allow to derive formulas relating molecular structures to the fitness determining quantities. Moreover, it is not unlikely that the physical—thermodynamic and kinetic—parameters which contribute to fitness can be measured directly in the near future.

MODELS OF REALISTIC FITNESS LANDSCAPES

An alternative way to study relations between genotypes and phenotypes is to build a computer model. We made such an attempt with the ultimate goal to simulate evolutionary optimization in small populations.[6,7] This model was also conceived as a test of a theory of molecular evolution which describes replication with errors by means of kinetic differential equations[8,9] or as stochastic processes.[10,11] The relation between this concept of a *molecular quasi-species* and the computer simulation model has been reviewed recently.[12]

Any comprehensive model of molecular evolution has to handle two formidable problems:

GGGGCGGCCGGCCCCGCGGCCGGG ···

Primary Structure: I_k

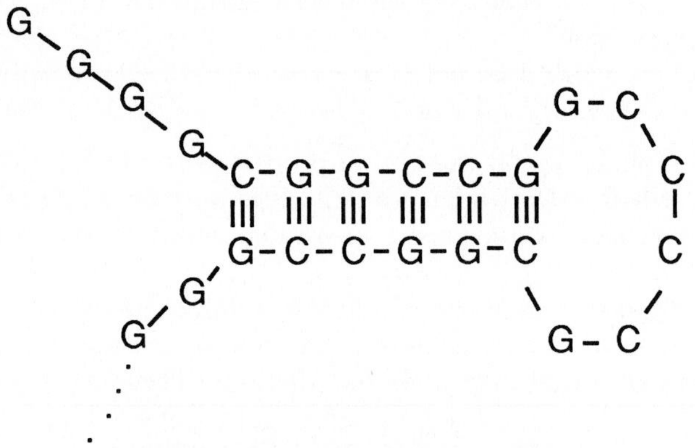

```
G
 \
  G
   \
    G                              G - C
     \                            /     \
      G                          /       \
       \C - G - G - C - C - G            C
        |||  |||  |||  |||  |||  |||     |
        G - C - C - G - G - C            C
       /                     \          /
      G                       \        /
     /                         G - C
    G
   .
  .
 .
.
```

Secondary Structure: G_k

zzzzaaaaaxxxxxxaaaaayy ···

Encoded Secondary Structure: Γ_k

FIGURE 1 Folding and encoding of secondary structures.

1. The spatial structure of the phenotype G_k has to be expressed as a function of the genotype $I_k: \quad G_k = \mathcal{G}(I_k)$, and
2. the fitness of the phenotype has to be evaluated as a function of its structure: $f_k = \mathcal{F}(G_k) = \mathcal{F}(\mathcal{G}(I_k))$.

Since no analytical or other quantitative relations between genotypes and phenotypes are available at present, one has to go through the time consuming folding procedure in order to find out which spatial structure belongs to which genotype. As mentioned previously prediction and efficient handling of tertiary structures face enormous difficulties. The phenotype in our model is identified therefore with the secondary structure of the RNA molecule. For computation of secondary structures the algorithm developed by Zuker, Stiegler and Sankoff[13,14] was used. This algorithm determines the most stable planar conformation by means of a minimum free energy criterium.

The planar structures obtained by the folding algorithm are lacking knots and pseudoknots. Thus the phenotypes G_k are graphs in this model and can be encoded by strings of the same lengths as the primary sequences. The relations between primary structure I_k, secondary structure G_k and encoded secondary structure Γ_k are shown in Figure 1. The graph G_k is easily partitioned into substructures. We distinguish four classes of substructures or structural elements: stems, loops, joints and free ends. The secondary structure is encoded by assigning a lower case letter to every base of the primary sequence. The position of a given base in the sequence is the same in I_k and in Γ_k. In particular we have:

1. stems, encoded by $aaa\cdots$, $bbb\cdots$, $ccc\cdots$, $ddd\cdots$, etc.,
2. loops, encoded by $xxx\cdots$,
3. joints, encoded by $yyy\cdots$, and
4. free ends, encoded by $zzz\cdots$.

The use of letters is here not entirely arbitrary: the later they come in the alphabet, the more flexible are the corresponding parts of the RNA molecule. Stem regions are more rigid than loops, loops in turn are more rigid than joints and joints are less flexible than the free ends. In the coding applied here we do not distinguish between individual loops, individual joints or between the two free ends (3'- and 5'-end). Only the stems are counted individually. As shown in the figure a stem region appears twice in the encoded two-dimensional structure—the two occurrences correspond to the two strands forming the double helix.

In Figure 2 we show the secondary structures of four binary sequences of chain lengths around $\nu \approx 100$. The four examples were chosen at random and demonstrate the great variability in shape which is characteristic for RNA molecules.

One important feature of the Zuker-Stiegler-Sankoff algorithm is additivity of the contributions of substructures to the free energy of the RNA molecule. This additivity—certainly an approximation but apparently justified in the case of free energies—will be retained later on, when we evaluate the secondary structures according to other criteria. Additivity of substructure contributions is one source

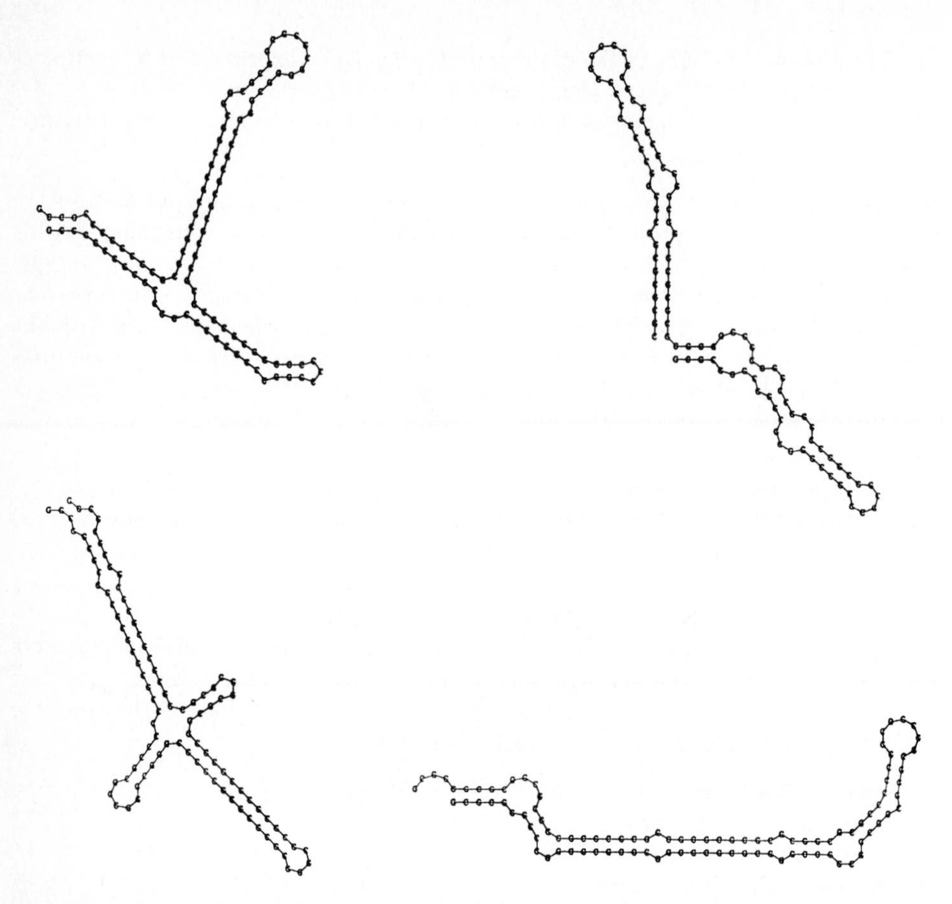

FIGURE 2 Examples of secondary structures derived from binary (G,C) sequences by folding according to a minimum free energy criterium. The chain lengths are $\nu = 97$ for the sequence in the lower left part and $\nu = 101$ for all other RNA molecules.

of selective neutrality on the phenotypic level. Several structures which consist of identical sets of substructures map onto the same selective values, although their phenotypic appearances are different. For two selectively neutral genotypes we have in this case:

$$\mathcal{G}(I_k) = G_k \ , \ \mathcal{F}(G_k) = f_k$$
$$\mathcal{G}(I_j) = G_j \ , \ \mathcal{F}(G_j) = f_k$$

The second source of selective neutrality is even more common: several primary structures map onto the same phenotype:

$$\mathcal{G}(I_k) \ = \ G_k \qquad , \qquad \mathcal{G}(I_j) \ = \ G_k \ .$$

The polynucleotide landscapes considered here are characterized by an extremely high degree of neutrality. This has consequences for the evolutionary optimization process which will be discussed in the next section.

The genotypes of RNA molecules can be ordered with respect to matching sequences. The number of bases in which two sequences differ is called the Hamming distance (d). Sequences can be arranged in a discrete space such that the Hamming distance forms a metric. The object obtained thereby is known as sequence space in information theory.[15] We plot now the free energies of the most stable structures in the sequence space. The landscape obtained thereby is highly bizarre. It represents a typical example of a *rugged landscape*[16]: closely related sequences—sequences which are mapped on nearby lying points in sequence space—may, but need not have very different free energies. Ruggedness is a consequence of the fact that small changes in the genotype may lead to large changes in the phenotype—for typical examples see Schuster,[12] p.115. In order to derive a quantitative measure for ruggedness we computed the autocorrelation function for a random walk of stepsize $d = 1$ on the free energy landscape[17]:

$$\rho(k) = \frac{< (F_i - F_j)^2 > - < (F_i - F_{i+k})^2 >}{< (F_i - F_j)^2 >} \tag{1}$$

Herein the notion $< X >$ is used for the expectation value of X, F_i and F_j are the free energies of two randomly chosen genotypes I_i and I_j, and F_{i+k} is the free energy of the sequence which follows I_i after exactly k steps of the random walk. According to equation (1) we have $\rho(0) = 1$ and $\lim_{k\to\infty} \rho(k) = 0$. The average number of steps k at which the autocorrelation function takes—for the first time—a value of e^{-1} is characterized as correlation length ℓ: $\ln \rho(\ell) = -1$. In Figure 3 the correlation length is plotted as a function of the chain length ν. The two curves represent the data for two letter (G,C) and four letter (G,A,C,U) sequences: the free energy surface of two letter sequences shows much shorter correlation lengths ℓ and hence is more rugged.

In contrast to free energies kinetic constants of replication, A_k, and degradation, D_k, cannot be computed straightway from known secondary structures G_k or their short-hand notations Γ_k. There is now satisfactory model available which is based purely on knowledge in biophysical chemistry. We had to use therefore a very crude estimate of these quantities. It is well known from virus specific RNA replication by $Q\beta$ replicase that only single stranded molecules are accepted as templates. The secondary structure has to *melt* in order to make replication possible. Therefore we used an estimate of the rate constant of melting as a simple model for the replication process:

$$A_k = \alpha_0 - \alpha_1 \sum_{j=1}^{s^{(k)}} \frac{n_j^{(k)}(1 + n_j^{(k)})^3}{(1 + n_j^{(k)})^4 + L} \; ; \quad k = 1, 2, \ldots, 2^\nu . \tag{2}$$

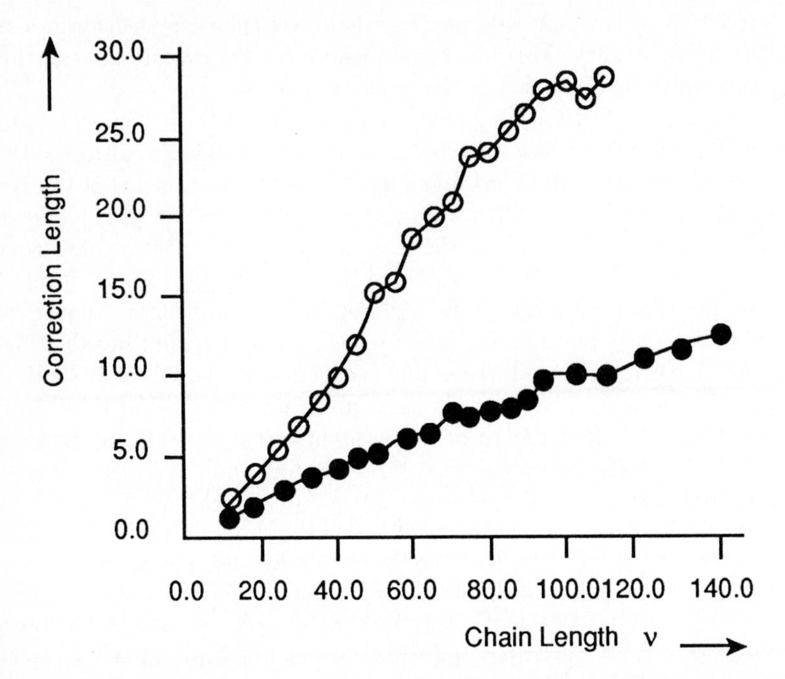

FIGURE 3 The correlation length ℓ of free energies F of RNA molecules in their most stable secondary structures as a function of the chain length ν. Binary sequences (G,C) and four letter sequences (G,A,C,U) are denoted by • and o, respectively. Values are taken from Fontana.[17]

By $n_j^{(k)}$ we denote the number of base pairs in the j-th stack of the secondary structure G_k, $s^{(k)}$ is the number of stacks in this structure, and α_0, α_1 and L are empirical constants. The function used in equation (2) takes care of the cooperativity in the melting process.

Degradation is modelled by taking into account all possible attacks of a hydrolytic agent or an enzyme with nuclease activity on the single stranded regions of the secondary structure G_k:

$$D_k = \beta_0 + \beta_1 \sum_{j=1}^{u^{(k)}} \frac{u_j^{(k)}}{u_{max}} \exp\Big\{(u_j^{(k)} - u_{max})/u_{max}\Big\} + \frac{\beta_2}{\nu} \sum_{j=1}^{z^{(k)}} z_j^{(k)} . \quad (3)$$

Herein the number of bases in the j-th loop of the secondary structure G_k is denoted by $u_j^{(k)}$. This structure has $u^{(k)}$ loops and there is a maximum loop size u_{max} above which loops are considered as completely mobile elements like free ends. The number of bases in free ends or large loops is given by $z_j^{(k)}$ and β_0, β_1 and β_2 are empirical parameters. The number $z^{(k)}$ is two—3'- and 5'-end—plus the number of large loops.

Replication and degradation surfaces were studied by random walk statistics as well.[6,7,17] It turned out that the degradation surface is rather similar to the free energy surface whereas the replication surface appears to be much more rugged. The landscape obtained for the excess production, $E_k = A_k - D_k$ is relevant for evolutionary optimization—we shall call it the E-landscape for short. It represents the cost function of the complex optimization problem. The E-landscape was studied not only by means of random walk statistics, in addition we optimized the E_k value of a test sequence by simulated annealing. The main goal of this search was to explore the distribution of maxima of the excess production in sequence space and to compare it with a spin glass landscape. A chain length of $\nu = 70$ bases was chosen. In this case the highest possible E-value can be computed from the optimal combination of substructures: $E_{max}^{(70)} = 2245\,[t^{-1}]$ in arbitrary reciprocal time units. Such a *best* structure, however, does not occur as a minimum free energy folding pattern. These structures appear only as conformations with free energies above the corresponding conformational ground states. The highest value found by the simulated annealing technique was only $E_{opt} = 2045\,[t^{-1}]$; the genetic algorithm with variable chain length ν yielded a somewhat higher optimal value: $E_{opt} = 2059\,[t^{-1}]$.

The optimal value found with simulated annealing is degenerate: we found it with ten different genotypes. They are related in pairs by symmetry—every polynucleotide sequence is transformed into one with identical structure by complementation and inversion. The remaining five sequences form two pairs of close relatives with Hamming distance $d = 2$ and one *solitary* sequence. These potential centers of quasi-species[9] are well separated in sequence space—all Hamming distances are larger than $d \geq 30$.

In total, 879 genotypes with excess productions $E_k \geq 2011\,[t^{-1}]$ were identified. Their distribution in sequence space is characterized by clusters of peaks with high excess production which have rich internal structures. Like in mountainous areas on the surface of the earth, we find ridges as well as saddles and valleys separating zones of high excess production.

The distribution of near optimal configurations of the spin glass Hamiltonian shows characteristic features of *ultrametricity*: arbitrarily chosen triangles of near optimal configurations are either equilateral or isosceles with small basis. The distribution of clusters of high excess production in sequence space, on the other hand, shows no detectable bias towards ultrametricity. It does not deviate significantly from a random distribution. Two causes may be responsible for this apparent difference between spin glass Hamiltonians and polynucleotide folding landscapes: either binary sequences of chain length $\nu = 70$ are too short to reveal higher order structures in the value landscape, or polynucleotide folding and evaluation of the folded structures have some fundamental internal characteristics which is different from that intrinsic to the random couplings in spin glass Hamiltonians.

A MODEL OF MOLECULAR EVOLUTION

In order to study evolution dynamics on the excess production (E) landscape a series of computer experiments was carried out.[6,7] A population of several thousand binary (G,C) sequences with chain lengths up to $\nu = 100$ nucleotides was used to optimize replication-mutation performance on the E-landscape described in the previous section. Mutations were either restricted to point mutations, or point mutations, deletions and insertions up to ten bases were allowed. In the former case the chain length remained constant. For point mutations we applied the *uniform error rate* model[9]: the single digit accuracy q—this implies an error rate of $1-q$ per digit and replication—is assumed to be the same at every position of the sequence. For the mutation probability from sequence I_j to I_k we find:

$$Q_{kj} = q^{\nu} \left(\frac{1-q}{q} \right)^{d_{kj}} \tag{4}$$

The chain length is denoted by ν as before and d_{kj} is the Hamming distance between the two sequences I_k and I_j. The model has an important feature: all mutation rates at constant chain lengths ν can be expressed by only two quantities, the single digit accuracy q and the Hamming distance d. In case of variable chain lengths the frequencies of deletions and insertions are additional input parameters. In the computer experiments reported here a value of $q = 0.999$ corresponding to one error per one thousand digits was applied throughout. The stochastic dynamics of the replication-mutation-degradation reaction network was simulated under the conditions of a CSTR (Continuously Stirred Tank Reactor) by means of an efficient computer algorithm.[18] Sequences produced in excess are removed by a stochastic dilution flux. Accordingly the total population size fluctuates around a mean value $\overline{N}$ and fulfils a $\sqrt{N}$ law.

Here we report the results of computer runs[6,7] which started with an initial population of 3000 *all-G-sequences* of chain lengths $\nu = 70$. The homogeneous initial population of *all-G-sequences* was chosen because this genotype cannot unfold into a phenotype which has a secondary structure. Moreover, there are no structures with reasonably good values of the excess production in its neighborhood. In order to use a metaphor from terrestrial landscapes we start our computer experiments in the planes far away from any higher peak. During a typical simulation experiment with our variant of a genetic algorithm the mean excess production of the population $\overline{E}(t)$ increases monotonously—apart from small random fluctuations—and eventually approaches a plateau value. Depending on population size, replication accuracy and structure of the value landscape two extreme scenarios were observed:

1. The population resides for rather long time in some region of sequence space and makes little progress in optimization. It moves quickly and in jumplike manner into another area of higher mean excess production. The optimal mean excess production is approached by stepwise improvements.

2. The population approaches the optimal value gradually by steady improvements and the population migrates rather smoothly through sequence space.

The first scenario is reminiscent of punctuated or stepwise evolution. It occurs here in small populations, at low error rates and in landscapes with distant local maxima. The second scenario dominates in large populations, at high error rates and in landscapes with close by lying local maxima. The two scenarios represent only the extreme cases. There is a steady transition from scenario I to scenario II examples of which were observed in additional computer runs: the gradual approach changes first into a sequence of small jumps when the error rate is reduced. Larger jumps appear on further decrease of the mutation frequency and eventually the population is caught in some local maximum of the fitness surface.

Three different secondary structures with almost the same excess production were obtained in three optimization runs starting from identical initial conditions: 3000 *all-G-sequences* of chain length $\nu=70$ on the same excess production surface. The individual runs differed only with respect to the sequence of random events along the stochastic trajectory which was achieved by using different initial seeds for the random number generator. The genotypes corresponding to the three structures lie far away from each other in sequence space: their Hamming distances span an almost equilateral triangle of chain length $d = 30$ which has an average distance of $d = 25$ from the initial *all-G-sequence*. An interpretation of this result can be given straightway: we start from sequences which are unable to form any secondary structure and therefore an initial random walk in the planes of the fitness surface is likely to find better sequences in almost every direction of sequence space. An initial random choice will lay down the direction into which the population migrates later on in its search for higher excess production. Apparently there is ample variability for the choice of such a direction: we have 70 independent directions in the sequence space of binary sequences with chain length $\nu=70$.

Optimization runs with the genetic algorithm were also carried out at replication accuracies which lie below the critical accuracy for the population size and the fitness surface applied. No optimization was observed in these cases. Populations drift randomly in sequence space. This general behavior which is in agreement with the error threshold concept in finite populations[11] was observed even if we started with homogeneous populations of near optimal genotypes obtained in previous optimization runs with smaller error rates. The initial master sequence is first surrounded by a growing cloud of mutants and then it is displaced by one of its error copies. All other sequences are lost likewise after sufficiently long time. In other words, every genotype has a finite lifetime only and localization of the population in quasi-species like manner does not occur. The predictions of the theory of molecular evolution[8,9] are valid also in populations as small as a few thousand molecules.

CONCLUDING REMARKS

Systematic studies on the landscapes underlying complex optimization problems were so far always restricted to the statistics of random models. In this contribution we reviewed at attempt to explore a realistic fitness surface of a biophysical optimization problem. In order to be able to study and handle such a highly complex object like a fitness surface, reduction to the most simple case was inevitable. The simplest presently known entities which show optimization behavior in the Darwinian sense are small RNA molecules which can be replicated in the test tube by means of an enzyme. They mutate frequently since the enzyme has rather low replication fidelity. Selection and adaptation to the environmental conditions are already observed in such populations of RNA molecules replicating *in vitro*. In these molecular evolution experiments the phenotype is nothing but the three-dimensional structure of the RNA molecule. It is recognized by the enzyme and thus determines the fitness of its carrier.

The feasibility of computer studies in the exploration of fitness landscapes for molecular evolution is bound to the predictability of molecular structures. Three-dimensional structures of polynucleotides or proteins are very hard to predict and systematic investigations of the corresponding free energy surfaces are out of question at the present state of the art. Restriction to secondary structures of polynucleotides, in particular RNA molecules, improves the situation a lot: these structures can be predicted fairly reliably and efficient algorithms are available for computation. An additional assumption was made in order to be able to compute the free energy surface for two-dimensional folding patterns of RNA molecules and to model a fitness landscape on which the optimization process can take place: we studied binary (G,C) sequences instead of the natural four letter (G,A,C,U) sequences. How significant are the results derived from binary sequences? Computations of correlation lengths for random walks on free energy surfaces for both classes of sequences showed that the binary sequences represent an interesting extreme case: the correlation length is shortest for them and this implies that their folding landscapes are most complicated. Secondary structures of binary sequences are more sensitive to mutations than those derived from sequences of the four letter alphabet. Studies currently in progress showed in addition that structural richness depends very much on the strength of interaction between the bases in base pairs: an artificial (A,U) *world* would be very different and less bizarre than the (G,C) *world*.

The simulation studies of molecular evolution demonstrated the consequences of a high degree of selective neutrality. How general are these observations gained from an artificial computer model for real biology? What matters here is the structure of the fitness landscape—the degree of neutrality after all is a feature of the landscape. Selective neutrality, no doubt, does exist in nature too. Cases of neutrality on the molecular level are easy to visualize. There are the degeneracies of the genetic code and the mutations having no or little effect on protein function, since they exchange amino acid residues which do not effect protein structure very much. Surely, many additional sources of neutrality exist on the higher hierarchical levels of biology.

In molecular evolution complexity can be traced down to the properties of the carriers of genetic information. Here the simplest case of a relation between genotype and phenotype is the folding of strings into two-dimensional structures by means of digit complementarity rules. This process cannot be casted into an analytical expression and creates already a highly complex free energy landscape. The ruggedness of value landscapes is the ultimate cause why all transformations of genotypes into phenotypes are exceedingly complex. Unfolding of genotypes cannot be reduced to simple algorithms therefore—not even in the most simple cases.

ACKNOWLEDGMENTS

Many stimulating and fruitful discussions with my coworkers Drs. Walter Fontana, Wolfgang Schnabl and Peter F. Stadler are gratefully acknowledged. The research work reported here was supported financially by the Austrian *Fonds zur Förderung der wissenschaftlichen Forschung*, projects P 5286 and P 6864, by the German *Stiftung Volkswagenwerk* and by the *Jubiläumsfonds der Österreichischen Nationalbank* project 3819. Computer time on the IBM 3090/400VF mainframe supercomputer of the *EDV Zentrum der Universität Wien* was provided within the frame of the EASI project of IBM.

REFERENCES

1. Biebricher, C. K. "Darwinian Selection of Self-Replicating RNA Molecules." *Evol. Bio.* **16** (1983): 1.
2. Biebricher, C. K., and M. Eigen. "Kinetics of RNA Replication by Qβ Replicase." In *RNA Genetics, Vol. I: RNA-directed Virus Replication*, edited by E. Domingo, J. J. Holland and P. Ahlquist. Boca Raton: CRC Press, 1988.
3. Demetrius, L., P. Schuster and K. Sigmund. "Polynucleotide Evolution and Branching Processes." *Bull. Math. Biol.* **47** (1985): 239.
4. Eigen, M., J. McCaskill and P. Schuster. "Molecular Quasi-Species." *J. Phys. Chem.* **92** (1988): 6881.
5. Eigen, M., J. McCaskill and P. Schuster. "The Molecular Quasi-Species." *Advances in Chem. Phys.* **75** (1989): 149.
6. Fontana, W., and P. Schuster. "A Computer Model of Evolutionary Optimization." *Biophys. Chem.* **26** (1987): 123.
7. Fontana, W., W. Schnabl and P. Schuster. "Physical Aspects of Evolutionary Optimization and Adaptation." *Phys. Rev. A* **40** (1989): 3301.
8. Fontana, W, P. Schuster, P. Stadler, E. Weinberger, T. Griesmacher and W. Schnabl. "Characterization and Quantitative Evaluation of RNA Folding Landscapes." Preprint, 1990.
9. Gillespie, D. T. "A General Method for Numerically Simulating the Stochastic Time Evolution of Coupled Chemical Reactions." *J. Comp. Phys.* **22** (1976): 403.
10. Hamming, R. W. *Coding and Information Theory*, 2nd Ed., pp. 44–47, Englewood Cliffs: Prentice Hall, 1986.
11. Jaeger, J. A., D. H. Turner and M. Zuker. "Improved Predictions of Secondary Structures for RNA." *Proc. Natl. Acad. Sci. USA* **86** (1989): 7706.
12. Kauffman, S., and S. Levin. "Towards a General Theory of Adaptive Walks on Rugged Landscapes." *J. Theor. Biol.* **128** (1987): 11.
13. Nowak, M., and P. Schuster. "Error Thresholds of Replication in Finite Populations: Mutation Frequencies and the Onset of Muller's Ratchet." *J. Theor. Biol.* **137** (1989): 375.
14. Schuster, P. "Optimization and Complexity in Molecular Biology and Physics." In *Optimal Structures in Heterogeneous Reaction Systems*, Springer Series in Synergetics, Vol.44, edited by P. J. Plath. Berlin: Springer-Verlag, 1989.
15. Spiegelman, S. "An Approach to Experimental Analysis of Precellular Evolution" *Quart. Rev. Biophys.* **4** (1971): 36.
16. Zuker, M. "On Finding All Suboptimal Foldings of an RNA Molecule." *Science* **244** (1989): 48.
17. Zuker, M., and P. Stiegler. "Optimal Computer Folding of Large RNA Sequences Using Thermodynamics and Auxiliary Information." *Nucleic Acids Research* **9** (1981): 133.

18. Zuker, M., and D. Sankoff. "RNA Secondary Structures and their Prediction." *Bull. Math. Biol.* **46** (1984): 591.

Evolutionary Dynamics

Kristian Lindgren,†
Nordita, Blegdamsvej 17, DK-2100 Copenhagen, Denmark †Permanent address: Institute for Physical Resource Theory, Chalmers University of Technology, S-412 96 Göteborg, Sweden.

Evolutionary Phenomena in Simple Dynamics

We present a model of a population of individuals playing a variation of the iterated Prisoner's Dilemma in which noise may cause the players to make mistakes. Each individual acts according to a finite memory strategy encoded in its genome. All play against all, and those who perform well get more offspring in the next generation. Mutations enable the system to explore the strategy space, and selection favors the evolution of cooperative and unexploitable strategies. Several kinds of evolutionary phenomena, like periods of stasis, punctuated equilibria, large extinctions, coevolution of mutualism, and evolutionary stable strategies, are encountered in the simulations of this model.

INTRODUCTION

In the construction of simple models of abstract evolutionary systems, game theory provides a large number of concepts and examples of games that can be used to model the interaction between individuals in a population. Originally, game theory was developed by von Neumann and Morgenstern for the application to economic theory,[1] but it has now spread to other disciplines as well. The work of Maynard-Smith and Price[15,16] has lead to an increasing use of game theory in evolutionary ecology. In the social sciences game theoretical methods have been accepted for a long time. A renewed interest in the Prisoner's Dilemma followed the work of Axelrod and Hamilton,[1,4] who performed a detailed analysis of the iterated version of that game, and this has lead to several game theoretic models based on the iterated Prisoner's Dilemma. In large computer networks the presence of interacting agents may lead to computational ecosystems,[13] which can be analyzed from a game-theoretical point of view.

For a population with a fixed number of species, natural selection drives the system towards a fixed point, limit cycle, or strange attractor, assuming an unchanged environment. This process can be modelled by population dynamics, where one usually uses the number of individuals for the different species as variables, so that the dimensionality of the system equals the number of species. Population dynamics models the reproduction, survival, and death of individuals. If the behavior of the individuals (or species) depends on a genetic description inherited by the offspring, the introduction of mutations in the replication process may totally change the dynamic behavior of the system. One way to characterize such a dynamical system is to interpret mutations leading to new species as creations of new variables and extinction of species as the disappearance of present variables. But in both cases these events are due to the (stochastic) dynamic system itself. If there is no limit on the length of the genetic description and the number of phenotypic characters this is coded into, the system may be considered a potentially infinite-dimensional dynamical system. Evolution can then be viewed as a transient phenomenon in a potentially infinite-dimensional dynamical system.[9,20] If the transients continue for ever, we have *open-ended evolution.* Of course, we may still get the same behavior as in the mutation-free population dynamics. Therefore, one of the main problems in the construction of evolutionary models is how to model the interactions between species (and/or environment) so that the transients are infinite or at least long enough for evolutionary phenomena to appear. In this construction one is faced with the dilemma that one wants to achieve both high complexity, which is necessary for evolution to occur, and simplicity, which makes simulation possible for evolutionary time scales. Note that the dynamics used to model the behavior in prebiotic or chemical evolution is usually a form of population dynamics. Such systems have been analyzed by, e.g., Farmer et al.,[9] Schuster,[21] and Eigen et al.[7] in models for evolution of macromolecules.

We have constructed a model of a population of individuals playing the iterated Prisoner's Dilemma. The game is modified so that noise may disturb the actions performed by the players, which makes the problem of the optimal strategy more complicated. This increases the potential for having long transients showing evolutionary behavior. We construct a suitable coding for all deterministic strategies with finite memory, and let such a code serve as the genome for an individual playing the corresponding strategy. By adding mutations to the population dynamics we get a potentially infinite-dimensional dynamical system in which evolution is possible. The "artificial" selection in the model is determined by the result in the game—those individuals who get high scores also have higher fitness.

The idea of using the iterated Prisoner's Dilemma in evolutionary situations is not new, see, e.g., the studies by Axelrod[2] and Miller,[17] and a variety of other kinds of evolutionary models can be found in Langton.[14] The novel approach in this study is the combination of noisy games, simple population dynamics, analytically solvable interactions, and the possibility of increase in genome length, and it appears that this leads to a richness in evolutionary behavior that has not been observed in such models before.

THE PRISONER'S DILEMMA

The Prisoner's Dilemma is a two-person non-zerosum game, which has been used in both experimental and theoretical investigations of cooperative behavior. The game is based on the following situation. Two persons have been caught and are suspected of having committed a crime together. There is not enough evidence to sentence them, unless at least one of them confesses. So, if both stay quiet (cooperate, C) they will be released. If one confesses (defects, D) but the other does not, the one who confesses will be released and rewarded, while the other one will get a severe punishment. Finally, if both confess, they will be imprisoned but for a shorter period. It is assumed that they make their choice of action simultaneously without knowing the others decision.

This problem is formalized by assigning numerical values for each pair of choices. An example of such a payoff matrix for the players is shown in Table 1.

If the game is viewed as a single event, each player finds defection to be the optimal behavior, regardless of the opponents action. However, if there is a high probability that the two players will meet again in the same type of game, the question of the most optimal choice of action is more delicate. This kind of "iterated

TABLE 1 The payoff matrix we use in the Prisoner's Dilemma is the same as the one used by Axelrod.[1] The pair (s_1, s_2) denotes the scores to players 1 and 2, respectively

		Player 2	
		Cooperate	Defect
Player 1	Cooperate	(3, 3)	(0, 5)
	Defect	(5, 0)	(1, 1)

Prisoner's Dilemma" has been extensively studied by Axelrod.[1] From the results of a computer tournament, he found that a simple strategy called Tit-for-Tat (TFT) showed the best performance in the iterated game. Tit-for-Tat starts with cooperation and then repeats the opponents last action. Thus, two TFT players meeting each other in a series of games, share the highest possible total payoff and each gets an average score of 3.

In our model we shall let noise interfere with the actions of the players. With probability p the performed action is opposite to the intended one. (We shall assume that the average length T of the game is much longer than the average time between noise-modified actions, $T \gg 1/(2p)$.) For two players using the TFT strategy the result is that they will alternate between three modes of behavior. First they will play the ordinary TFT actions (C, C), but when an error occurs they will shift to alternating (C, D) and (D, C). The third possibility of behavior is sequences of (D, D). The average probability for the three modes are 1/4, 1/2, and 1/4, respectively, giving an average payoff of 9/4. None of the strategies in Axelrod's tournament was able to deal with noise *and* resist exploitation, and TFT turned out to be the best one in that set of strategies.[1] A simple strategy that is more stable to noise is Tit-for-Two-Tats, which defects only if the opponent defects twice in a row, but this strategy is vulnerable to exploiting strategies, and in an evolutionary context it should perform worse. Another way to decrease the sensitivity to noise is to allow for the strategies to choose among different actions according to a certain probability (mixed strategies). This approach has been analyzed by Molander,[18] who found that a strategy which mixes TFT with ALLC (always cooperate) can reach an average score very close to 3. In our model we shall assume that the strategies are deterministic (pure strategies), and in the simulations we shall see that there are deterministic, noise-robust, unexploitable strategies that reach an average score of almost 3.

FINITE MEMORY AND INFINITE GAMES

GENETIC CODING OF STRATEGIES

In the model we allow for deterministic finite memory strategies. This means that a finite history determines the next intended action, although the performed action can be changed by the noise. An m-length history consists of a series of previous actions starting with the opponent's last action a_0, the individual's own last action a_1, the opponent's next to last action a_2, etc. By introducing a binary coding for the actions, 0 for defection and 1 for cooperation, we can label an m-length history by a binary number

$$h_m = (a_{m-1}, \ldots, a_1, a_0)_2 \,.$$

Since a deterministic strategy of memory m associates an action to each m-length history, it can be specified by a binary sequence

$$S = [A_0, A_1, \ldots, A_{n-1}].$$

This sequence then serves as the genetic code for the strategy that chooses action A_k when history k turns up. The length n of the genome equals 2^m.

In the population dynamics we shall allow for three kinds of mutations: point mutations, gene duplications, and split mutations. The point mutation changes a symbol in the genome, e.g., [01] → [00], the gene duplication attaches a copy of the genome to itself, e.g., [01] → [0101], and the split mutation randomly removes the first or second half of the genome, e.g., [1001] → [01]. Note that gene duplication does not change the phenotype. The memory capacity is increased by one but the additional information is not used in the choice of action. For point mutations we have used the rate 2×10^{-5} per symbol and genome, and the other mutations occur with probability 10^{-5} per genome. Regarding a position in the genome as a locus and a symbol as an allele rather than a base pair, the point mutation rate we use has the order of magnitude that has been estimated for mutation rates at loci in living systems.[11]

For strategies of memory one, the histories are labeled 0 and 1, corresponding to the opponent defecting and cooperating, respectively. The four memory 1 strategies are [00], [01], [10], and [11]. The strategy [00] always defects (ALLD), [01] cooperates only if history 1 turns up (i.e., the opponent cooperated), and we recognize it as Tit-for-Tat, [10] does the opposite and we denote it Anti-Tit-for-Tat (ATFT), and [11] always cooperates (ALLC). We use equal fractions of these strategies as the initial state in the simulations.

SOLVING THE GAME

If the length of the game is infinite, the stationary distribution over finite histories can be solved analytically. This solution is unique if noise disturbing the actions is present. Although the game is infinite, the strategies can only take into account a finite history when choosing an action, which means that the infinite game is a Markov process. The average payoff for two players meeting in this game can be derived from the probabilities p_{00}, p_{01}, p_{10}, and p_{11} for all possible pairs of action (11), (10), (01), and (00). These can be found if we solve the equation

$$H = MH, \tag{1}$$

where $H^T = (h_0, h_1, \ldots, h_{n-1})$ is the vector of probabilities for different histories $0, 1 \ldots, n-1$, and M is a transfer matrix. The elements of M are determined by the strategies involved in the game including the possibility of making mistakes. The minimal size n of the matrix is given by the memory sizes of the involved strategies and is 2^m if the largest memory is m (or 2^{m+1} if m is odd and both players have the same memory size). Then one gets p_{ij} by summing the appropriate components in H, and the average payoff is

$$s = 3p_{11} + 5p_{01} + p_{00}, \tag{2}$$

according to the payoff matrix in Table 1.

POPULATION DYNAMICS

We shall consider a system consisting of a population of N individuals interacting according to the iterated Prisoner's Dilemma with noise. Each individual acts according to a certain strategy encoded in its genome. We think of a population sharing the same niche, fighting or cooperating with each other, to get a part of the available resources for survival and reproduction. In *each* generation all individuals play the infinitely iterated Prisoner's Dilemma against all, and the score s_i for individual i is compared to the average score of the population, and those above average will get more offspring in the next generation. In the reproduction, mutations may occur leading to the appearance of new strategies.

We model this situation as follows. First, we identify the different genotypes present in the population, and let them meet in the game described above. Let g_{ij} be the score for the strategy of genotype i playing against the strategy of j, and let x_i be the fraction of the population occupied by genotype i. Then, the score s_i for an individual with genotype i is

$$s_i = \sum_j g_{ij} x_j \,, \tag{3}$$

and the average score is

$$s = \sum_i s_i x_i \,. \tag{4}$$

The fitness w_i of an individual is defined as the difference between its own score and the average score,

$$w_i = s_i - s \,. \tag{5}$$

From one generation t to the next $t+1$, we assume that due to the result of the interactions, the fraction $x_i$of the population for genotype i changes according to

$$x_i(t+1) - x_i(t) = d w_i \; x_i(t) \,, \tag{6}$$

where d is a growth constant. This equation can also be written in the following form

$$x_i(t+1) - x_i(t) = d s_i \; x_i(t) \left(1 - \sum_j \frac{s_j x_j(t)}{s_i} \right) , \tag{7}$$

which is a logistic equation for a population of competing species.[11] The carrying capacity is normalized to 1, and the competition coefficients for species i are $s_j/s_i (j = 1, 2, \ldots)$. Note that this growth equation conserves the total population size. If x_j falls below $1/N$ for a certain genotype j, we set $x_j = 0$ and that species has died out. When this happens, the fractions x_i have to be renormalized for the population size to be constant. When mutations are present there is an additional stochastic term m_i in the growth equation. If the mutation rates are small $(p_p + p_d + p_s \ll 1/N)$, the additional term is well approximated by

$$m_i = \frac{1}{N} \sum_j (Q_{ij} - Q_{ji}), \tag{8}$$

where Q_{ij} is a stochastic variable taking the value 1 if a gene j mutates to the gene i, and 0 otherwise. The probability for Q_{ij} to be 1 is

$$P(Q_{ij} = 1) = N x_j q_{ij}, \tag{9}$$

where q_{ij} is the probability that genotype j mutates to i, obtained from the mutation rates and the genotypes i and j. (This mutation may be composed of one gene duplication and several point mutations, although this is less frequent.) Due to the term m_i, new genotypes may appear in the time evolution, and we get a model with a potentially infinite state space.

SIMULATION RESULTS AND DISCUSSION

The system described above consists of a population of N individuals interacting according to the iterated Prisoner's Dilemma with a probability p for mistake (noise). Individuals who get high scores get more offspring in the next generation than those who get low scores. In this reproduction we allow for mutations to occur and new strategies to enter the game.

We model the dynamics of this system by Eqs. (6)–(9), and the parameters that enter are the growth rate d, the mutation rates p_p, p_d, and p_s, the population size N, and the error probability p. In the simulation example the parameter values are $N = 1000, p = 0.01, p_p = 2 \times 10^{-5}, p_d = p_s = 10^{-5}$, and $d = 0.1$, and we have also restricted the length of the genetic code to be at most 32, i.e., at most strategies of memory 5. For the first generation we have chosen equal fractions of the four strategies with memory one, i.e., $x_{00} = x_{01} = x_{10} = x_{11} = 1/4$.

Almost all simulations have in common that during the evolution the system passes a number of long-lived metastable states (periods of stasis) that appear in a certain order. These periods are usually interrupted by fast transitions to unstable dynamic behavior or to new periods of stasis. Below we shall discuss the evolutionary phenomena observed in a typical simulation of the model. In the four most common periods of stasis we find examples of coexistence between species, exploitation, spontaneously emerging mutualism (symbiosis), and unexploitable cooperation.

THE EVOLUTION OF STRATEGIES OF MEMORY 1.

In Figure 1 the development of the population for the first 600 generations is shown. During the first 150 generations, the dynamics drives the system of the 4 strategies towards a population mainly consisting of TFT strategies. The All-D strategy [00] exploits the kind All-C strategy [11] and the ATFT strategy [10], and consequently [00] increases its fraction of the population. When the strategies [11] and [10] are extinct, the average score for All-D is close to 1 and the more cooperative Tit-for-Tat strategy takes over the population.

However, Tit-for-Tat only reaches an average score of 9/4 since noise interferes with the interaction. Then, through a point mutation [01] $\rightarrow$ [11], the All-C strategy enters the scene again. The mutant gets an average score of almost 3, and thus the fraction of [11] rapidly increases. Next, it is favorable for a mutant [11] $\rightarrow$ [10] to survive, since ATFT exploits ALLC and plays fairly well against TFT. Actually, ATFT gets the same score $s = 9/4$ as TFT when playing against ATFT or TFT. When the population of ATFT has grown large enough, mutations from [01] and [10] to [00] will survive, and the fraction of ALLD increases again. The system oscillates, driven by the relatively fast population dynamics in combination with the point mutations.

In Figure 2 the time scale is compressed by a factor of 50, and the evolution of the first 30000 generations is shown. The picture we get is a history with stable periods interrupted by fast transitions or unstable dynamics. The average score for the same simulation is drawn in Figure 3, which shows that there is no general tendency towards higher scores, although the simulation seems to end in a stable high score state. In the same figure the number of species per generation is depicted,

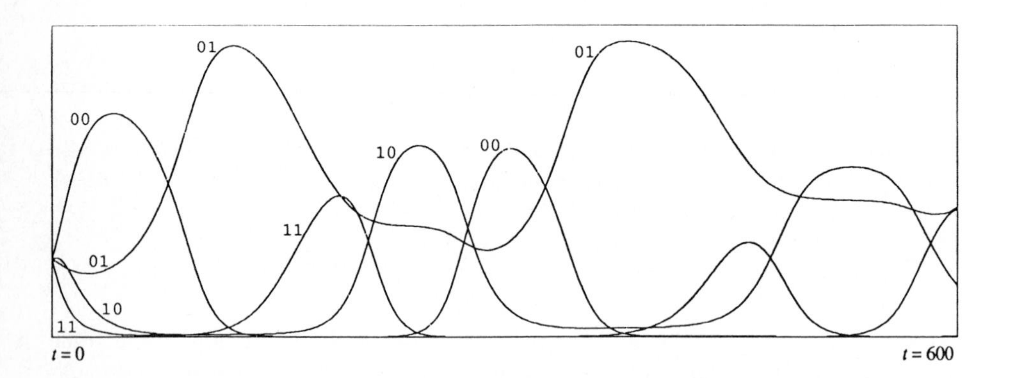

FIGURE 1 The evolution of a population of strategies starting with equal fractions of the memory one strategies [00], [01], [10], and [11] is shown for the first 600 generations. The fractions of different strategies are shown as functions of time (generation).

showing that the dimensionality of the system can increase and decrease in the evolution.

After some thousands generations the oscillations observed in Figure 1 are damped out, and the system stabilizes with a mixture of TFT [01] and ATFT [10]. If only the four simplest strategies are taken into account, this situation is easily analyzed. Assume that the population is divided into two fractions, one consisting

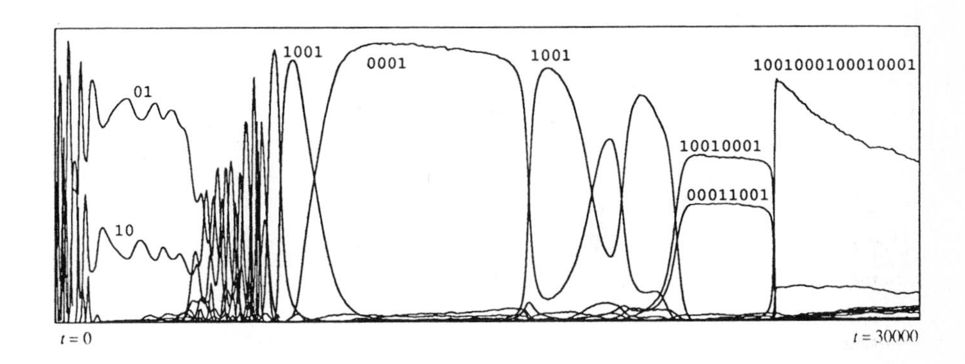

FIGURE 2 The simulation of Figure 1 is continued for 30000 generations, showing that four periods of stasis appear in the evolution. The oscillations observed in Figure 1 are damped and the system reaches a period of stasis with coexistence between [01] (TFT) and [10] (ATFT). This stasis is punctuated by a number of memory 2 strategies, and after a period of unstable behavior the system slowly stabilizes when the strategy [1001] increases in the population. This strategy cooperates if both players performed the same action last time. For two individuals using this strategy, an accidental defection by one of the players leads to both players defecting the next time, but in the round after that they return to cooperative behavior. Thus, the strategy [1001] is cooperative and stable against mistakes, but it can be exploited by uncooperative strategies. Actually, one of its mutants [0001] exploits the kindness of [1001], which results in a slow increase of [0001] in the population. This leads to a long-lived stasis dominated by the uncooperative behavior of [0001]. A slowly growing group of memory 3 strategies is then formed by mutations, and the presence of these species causes the fractions of the strategies [0001] and [1001] to oscillate. Two of the memory 3 strategies, M_1=[10010001] and M_2=[00011001], manage to take over the population, leading to a new period of stasis. Neither M_1 nor M_2 can handle mistakes when playing against individuals of their own kind, but if M_1 meets M_2 they are able to return to cooperative behavior after an accidental defection. This polymorphism is an example of mutualism which spontaneously emerges in this model. The stasis is destabilized by a group of mutants, and we get a fast transition to a population of memory 4 strategies which are both cooperative and unexploitable.

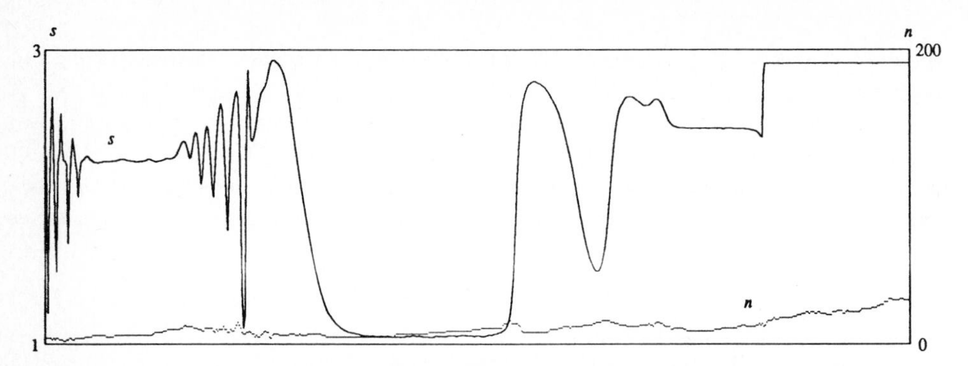

FIGURE 3 The average score s (continuous line) and the number of genotypes n (broken line) are shown for the simulation of Figure 2. When the exploiting memory 2 strategy dominates the scene, the average score drops close to 1. The last stasis, populated by the evolutionary stable memory 4 strategies, reaches a score of 2.91, close to the score of 3 achieved by the best strategies in a noise-free environment. Before the transitions and in the periods of unstable behavior, it appears that there are more mutants that survive and the number of genotypes increases, suggesting that most of the evolution takes place in these intervals.

of TFT and one of ATFT, and denote the fraction of the first by x. Then, for a large population, if $x < 7/16$ a mutant [00] will start to replicate, and if $x > 3/4$ any mutation to [11] will survive and replicate. But, if $7/16 < x < 3/4$ there is a meta-stable state consisting of a mixture of TFT and ATFT. This state is long-lived because none of the one-step mutations [00], [11], [0101], and [1010] are able to disturb the system. Actually, a detailed analysis shows that the only strategy with memory 2 that can invade this population alone and survive is the strategy [1100] which alternates between C and D, regardless of the opponent's action. However, this is not the usual way the stasis collapses, since one gene duplication and two point mutations are needed to get [1100] from [01] or [10]. Usually, a number of strategies, all having small fractions of the population, have a combined effect and cause the destabilization.

THE EVOLUTION OF STRATEGIES OF MEMORY 2.

The first stasis is usually followed by a period of unstable behavior, as is examplified in Figure 2. When the system stabilizes the strategy A=[1001] manages to dominate the population for some time. This strategy chooses C when the last pair of actions (the own and the opponent's) was CC or DD, which means that two individuals, both playing this strategy, get scores close to 3 when playing against each other. A typical history including a misaction **D** looks as follows (CC, C**D**, DD, CC, CC, ...), showing that the strategy is not sensitive to the noise. On the other hand the strategy can be exploited by one of its mutants, B=[0001]. When the strategy A plays against B, there are two modes of behavior, examplified by the following types of histories: (CC, CC, CC, ...) and (DD, CD, DD, CD, ...) where the second action in each pair is due to B. The second mode appears with frequency 0.80 and its average payoff is 3 for B and only 1/2 for A. Although the strategies A and B have totally different behavior (cooperative and uncooperative, respectively), the scores they receive are very close. This leads to a slow increase of B, while A decreases in the population, see Figure 2. Even a small group of mutants can then influence their scores so that the dominant strategy scores less than the rival species, which explains the oscillatory pattern that follows.

THE EVOLUTION OF STRATEGIES OF MEMORY 3.

During the time period dominated by the memory 2 strategies, a group of mutants containing memory 3 strategies is slowly growing. In Figure 2 we see two new strategies $M_1 = [10010001]$ and $M_2 = [00011001]$ spread in the population. A new stasis is reached between M_1 and M_2, and we shall analyze their behavior in more detail. The histories below exemplify how these strategies act when a single noise-induced **D**-action occurs.

M_1:M_1	M_2:M_2	M_1:M_2	M_2:M_1
C C	C C	C C	C C
C **D**	C **D**	C **D**	C **D**
D D	D D	D D	D D
C D	D C	C C	D D
D D	D D	C C	D C
C D	C D	C C	D D
D D	D D	C C	C C
C D	D C	C C	C C
.	.	.	.
.	.	.	.
.	.	.	.

Individuals playing against the same strategy type are not able to handle the noise, but when the strategies M_1 and M_2 play against each other they manage to return to a cooperative mode after a series of intermediate actions. The strategies respond to a disturbance D with a certain pattern of actions which fits to the opponents actions. This leads to a payoff close to 3 when they meet, but the payoff when M_1 meets M_1 is $S_{1:1} = 2.17$, and this is even worse for M_2, $s_{2:2} = 1.95$, because M_2 also has a mode consisting of a series of defect actions. Obviously, this strategy mix is an example of mutualism. The success of one of them is dependent on the success of the other one, and in Figure 2 we see that they spread simultaneously in the population.

THE EVOLUTION OF STRATEGIES OF MEMORY 4.

During the stasis of the two symbiotic strategies a group of mutants is formed and their fraction of the population is slowly increasing. The stasis ends with a fast transition to a new meta-stable state, consisting of two leading strategies and a growing group of mutants. All of these strategies have memory 4, i.e., they take into account the actions performed by both players the previous 2 time steps. There are several genotypes that can take the role of the leading one in this transition, because there is a class of genotypes coding into phenotypes or strategies that have practically the same behavior. All of them are cooperative, and if one player accidently defects both players defect twice before returning to the cooperative mode again. This assures that the strategy cannot be exploited by evil strategies at the same time as the mistakes only marginally decrease the average payoff. In the schematic genome E = [1xx10xxx0xxxx001] the most frequently used positions are shown and each x corresponds to a history occurring with a probability of order p^2 or less. There are 512 strategies fitting this mask, which explains the formation of a large genetic variety in this population, although some of these may have imperfections that can be exploited by other strategies. A typical game involving an accidental defect action **D** is shown below.

E : E
C C
C **D**
D D
D D
C C
C C
.
.
.

In Figure 2 [1001000100010001] has taken the lead, but there are others present in the growing group of quasi-species. The fact that the fraction of the leading genotype decreases can be explained by the small difference between the leading strategy and many of the strategies among the mutants. It should also be noted that since the length of the genome doubles each time the memory capacity is increased by 1, the probability for point mutations also doubles.

An important stability criterion for a strategy in a population dynamics model is given by the concept of an *evolutionary stable strategy*.[15] Assume that all individuals in a large population play a certain strategy S. The strategy S is evolutionary stable if any sufficiently small invading group of strategies dies out. It has been shown that, in the iterated Prisoner's Dilemma without noise, the Tit-for-Tat strategy is not evolutionary stable, because there are other strategies playing on equal terms with TFT at the same time as they perform better against other strategies. It has been shown by Boyd and Loberbaum[6] that there is no pure strategy that is evolutionary stable in the iterated Prisoner's Dilemma. A generalization of their result shows that this also holds for any finite population mixture of pure strategies.[10]

For the iterated Prisoner's Dilemma used in our model the presence of noise implies that every strategy can be regarded as a mixture of two opposite pure strategies, which allows for evolutionary stable strategies to exist.[5] Actually, the leading strategy in Figure 2 is evolutionary stable. A strategy that is simpler to analyze is $E_0 = [1001000000000001]$, which defects whenever the behavior deviates from the pattern in the game example above. This implies that no strategy can exploit it, and no strategy can invade a population of these by trying to be more cooperative, because any such attempt would be favorable to E_0 and it would reduce the payoff for the intruder. (Note that E_0 actually exploits the kind strategy [11].) However, even if the one-step mutants play slightly worse than the master species the mutation rate may be large enough for a net increase of these mutants, which leads to a growing group of quasi-species. In the simulations of our model we find that a large group of quasi-species is formed.

PATHWAYS FOR OPEN-ENDED EVOLUTION?

The scenario described above, passing periods of stasis dominated by strategies of increasing memory and then getting stuck in the evolutionary stable stasis, occurs with a probability of about 0.9. There are, however, evolutionary pathways that avoid the evolutionary stable memory 4 strategies. In Figure 4 an example of such a simulation is shown, and instead of getting to the stasis of the symbiotic species (see Figure 4(a)), the system takes a new way in state space and in Figure 4(b) we find the population dominated by memory 4 strategies not present in the ordinary simulations. The bottom diagram of Figure 4(b) shows that the number of genotypes (most of these are also of different phenotype) may increase to more than 200. In the figure it is seen that the system undergoes a collapse in which most of the genotypes disappears in a few hundred generations. Similar extinctions occur also in Figure 4(c), but they do not involve that many genotypes. In all these events the average score drops fast, suggesting that the extinctions are due to a mutant that exploits the present strategies but is unable to establish a cooperative behavior with its own species.

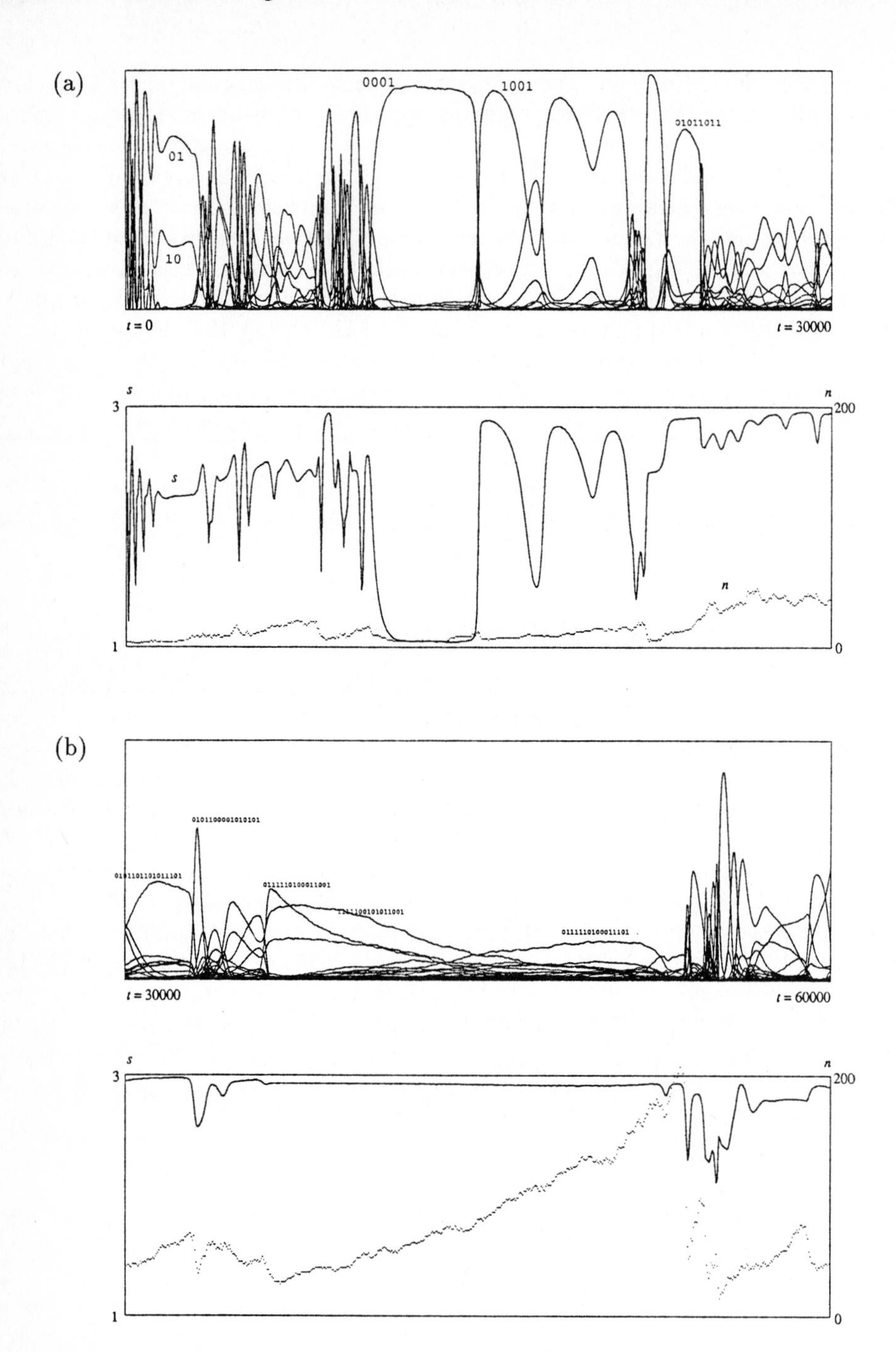

FIGURE 4 See caption next page.

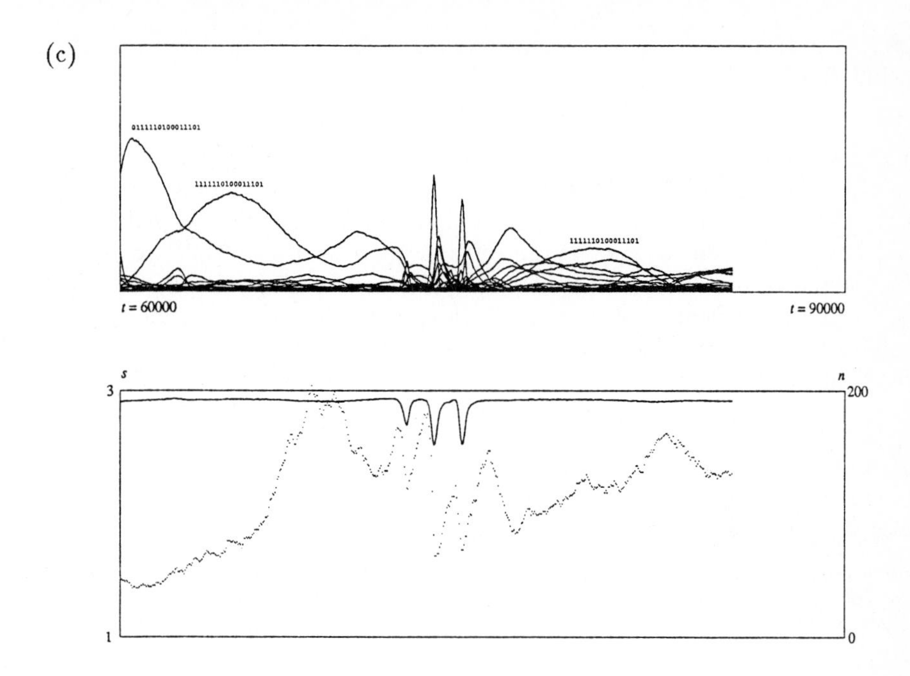

FIGURE 4 In (a) through (c) the evolution of a system avoiding the stable memory 4 stasis is shown for more than 80 000 generations. The bottom graphs show the average score and the number of genotypes (cf. Figure 3). (a) In this simulation the system never reaches the symbiotic stasis but finds another way in state space leading to new strategies dominating the population. (b) Several new memory 4 strategies appear and dominate the population. The system reaches a dimensionality of more than 200, and after that a collapse occurs in which most of the genotypes disappears. At the same time, the average score drops, indicating that this large extinction is caused by a parasite mutant exploiting the present species. (c) Some new large extinctions occur, and a few of them are accompanied by a decrease in the average score.

CONCLUSIONS

The presence of mutations in the population dynamics leads to intrinsical changes of the dimensionality of the system. The dynamic behavior observed is highly complicated with extremely long transients. One important characteristic of the model is that the game-theoretic problem used is complicated enough for complex strategies to evolve, at the same time as we can solve the game analytically, letting us simulate the population dynamics over evolutionary time scales. If one instead uses the iterated Prisoner's Dilemma without noise the potentiality for evolutionary transients is essentially lost. Another important aspect is that we use an effective way to code the strategies in genomes, and that the genome is easily modified by mutations. Having these aspects in mind it should be possible to model other situations as well, for example evolutionary models with more realistic assumptions, including, e.g., spatial dependence and sexual reproduction.

From the game-theoretical point of view we have found that when the iterated Prisoner's Dilemma is modified by noise, there is an unexploitable strategy that is cooperative. The evolutionary simulation, which actually is a kind of genetic algorithm[12] for finding good strategies for the noisy iterated Prisoner's Dilemma, indicates that the minimal memory for this kind of strategy is 4, i.e., the strategy should take into account the action of both players the previous 2 time steps. By answering a single defection by defecting twice the strategy is prevented from exploitation by intruders.

We have found periods of stasis punctuated by rapid transitions to new stasis or to periods of unstable dynamics. These rapid transitions are reminiscent of punctuated equilibria,[8] and it appears that the destabilization usually is due to a slowly growing group of mutants reaching a critical level. The coevolution of mutualism emerges spontaneously, and it serves as an example of a higher level of cooperation than the actions on the single round level provide. The appearance of an evolutionary stable strategy is interesting from the game-theoretic point of view, but in the construction of models possessing open-ended evolution one tries to eliminate such stabilizing phenomena. Therefore, from the evolutionary point of view, one should pay more attention to the less probable evolutionary pathways that avoid this evolutionary stable stasis. In particular, the large extinctions that appear in these simulations should be studied in more detail, since these collapses are triggered by the dynamical system itself and do not need external catastrophes for their explaination. The analysis of these results is in progress and shall be reported elsewhere. The major result of this model is that it establishes the fact that several evolutionary phenomena, like those described above, can emerge from very simple dynamics.

ACKNOWLEDGMENTS

I thank Tomas Kåberger for inspiring me to start the construction of this model, Mats Nordahl and Doyne Farmer for stimulating discussions, Gunnar Eriksson and Karl-Erik Eriksson for collaboration in game theory that also provided inspiration for the present study, and Torbjörn Fagerström for useful biological remarks. Financial support from Erna and Victor Hasselblad Foundation is gratefully acknowledged.

REFERENCES

1. Axelrod, R. *The Evolution of Cooperation.* New York: Basic Books, 1984.
2. Axelrod, R. In *Genetic Algorithm and Simulated Annealing*, edited by D. Davies, 32–42. London: Pitman, 1987.
3. Axelrod, R., and E. Dion. "The Further Evolution of Cooperation." *Science* **242** (1988): 1385–1390.
4. Axelrod, R., and W. D. Hamilton. "The Evolution of Cooperation." *Science* **211** (1981): 1390–1396.
5. Boyd, R "Mistakes Allow Evolutionary Stability in the Repeated Prisoner's Dilemma Game." *J. Theor. Biol.* **136** (1989): 47–56.
6. Boyd, R., and J. P. Lorberbaum. "No Pure Strategy is Evolutionarily Stable in the Iterated Prisoner's Dilemma Game." *Nature (London)* **327** (1987): 58–59.
7. Eigen, M., J. McCaskill, and P. Schuster. "Molecular Quasi-Species." *J. Phys. Chem.* **92** (1988): 6881–6891.
8. Eldredge, N., and S. J. Gould. *Models in paleobiology*, edited by T. J. M. Schopf, 82–115. San Fransisco: Freeman, Cooper and Company, 1972.
9. Farmer, J. D., S. A. Kauffman, and N. H. Packard. "Autocatalytic Replication of Polymers." *Physica* **22D** (1986): 50–67.
10. Farrell, J., and R. Ware. "Evolutionary Stability in the Repeated Prisoner's Dilemma." *Theor. Pop. Bio.* **36** (1989): 161–166.
11. Futuyma, D. J. *Evolutionary Biology* (2nd ed.). Sunderland: Sinauer Associates, 1986.
12. Holland, J. H. *Adaptation in Natural and Artificial Systems.* Ann Arbor: University of Michigan Press, 1975.
13. Kephardt, J. O., T. Hogg, and B. A. Huberman. "Dynamics of Computational Ecosystems." *Phys. Rev.* **A40** (1989): 404–420.
14. Langton, C. (Ed.). *Artificial Life.* Redwood City, CA: Addison-Wesley, 1989.
15. Maynard-Smith, J. *Evolution and the Theory of Games.* Cambridge: Cambridge University Press, 1982.

16. Maynard-Smith, J., and G. R. Price. "The Logic of Animal Conflict" *Nature (London)* **246** (1973): 15–18.
17. Miller, J. "The Coevolution of Automata in the Repeated Prisoner's Dilemma." Santa Fe Institute preprint 89–003, 1989.
18. Molander, P. "The Optimal Level of Generosity in a Selfish Uncertain Environment." *J. of Conflict Resol.* **29** (1985): 611–618.
19. von Neumann, J., and O. Morgenstern. *Theory of Games and Economic Behavior.* Princeton: Princeton University Press, 1944.
20. Rössler, O. E. "A System Theoretic Model of Biogenesis." *Zeitschrift für Naturforschung* **26b** (1971): 741–746.
21. Schuster, P. "Dynamics of Molecular Evolution." *Physica* **22D** (1986): 100–119.

W. Daniel Hillis
Thinking Machines, 245 First Street, Cambridge, MA 02142-1214

Co-Evolving Parasites Improve Simulated Evolution as an Optimization Procedure

INTRODUCTION

This paper shows an example of how simulated evolution can be applied to a practical optimization problem, and more specifically, how the addition of co-evolving parasites can improve the procedure by preventing the system from sticking at local maxima. The first section of the paper describes an optimization procedure based on simulated evolution, and its implementation on a parallel computer. The second section describes an application of this system to the problem of generating minimal sorting networks. The third section shows how the introduction of a species of co-evolving parasites improves the efficiency and effectiveness of the procedure.

The process of biological evolution by natural selection can be viewed as a procedure for finding better solutions to some externally imposed problem of fitness. Given a set of solutions (the initial population of individuals), selection reduces that set according to fitness, so that solutions with higher fitness are over-represented. A new population of solutions is then generated based on variations (mutation) and combinations (recombination) of the reduced population. Sometimes the new population will contain better solutions than the original. When this sequence of evaluation, selection, and recombination is repeated many times, the set of solutions (the population) will generally evolve toward greater fitness.

A similar sequence of steps can be used to produce *simulated evolution* within a computer.[3,4,13,18,20,21] In simulated evolution the set of solutions is represented by data structures on the computer and the procedures for selection, mutation, and recombination are implemented by algorithms that manipulated the data structures. Although the term "simulated evolution" deliberately suggests an analogy to biological evolution, it is understood that the real biological processes are far more complex than the simulation; simulated evolution represents only an idealization of certain aspects of a biological system. Such simulations are sometimes used as tools for understanding biological evolution,[16] but this paper will concentrate on the use of simulated evolution for optimization; that is, as a practical method of generating better solutions to problems. Biological systems will serve as a source of metaphor and inspiration, but no attempt will be made to apply the lessons learned to biological phenomena.

As an optimization procedure, the goal of simulated evolution is very similar to that of other domain-independent search procedures such as *generate and test, gradient descent*, and *simulated annealing*.[14,17] Like most such procedures, simulated evolution searches for a good solution, although not necessarily the optimal one. Whether or not it will find a good solution will depend on the distribution of solutions within the space.

These methods are all useful in searching solution spaces that are too large for exhaustive search. As in gradient descent and simulated annealing procedures, simulated evolution depends on information gathered in exploring some regions of the solution space to indicate which other regions of the space should be explored. How well this works obviously depends on the distribution of solutions in the space. The types of fitness spaces for which simulated evolution produces good results are not well understood, but one important type of space for which it works is a space that is independently a good domain for hill climbing in each dimension.

Another attractive property of simulated evolution is that it can be implemented very naturally on a massively parallel computer. During the selection step, for example, the fitness function can be evaluated for every member of the population simultaneously. The same is true for mutation, recombination, and a computation of statistics and graphics for monitoring the progress of the system. In the system described below, we routinely simulate the evolution of populations of a million individuals over tens of thousands of generations. Since these simulations take place on several generations per second, such experiments take only a few hours.

In these simulations, individuals are represented within the computer's memory as pairs of number strings that are analogous to the chromosome pairs of diploid organisms. The population evolves in discrete generations. At the beginning of each generation, the computer begins by constructing a phenotype for each individual, using the set of number strings corresponding to an individual (the "genome") as a specification. The function used for the interpretation is dependent upon the experiment, but typically a fixed region within each of the chromosomes is used to determine each phenotypic trait of the individual. Discrepancies between the two-bit strings of the pair are resolved according to some specified rule of dominance. This is similar to the diploid "genetic algorithms" studied by Smith and Goldberg.[20]

To simulate selection, the phenotypes are scored according to a set of fitness criteria. When the system is being used to solve an optimization problem, the traits are interpreted as solution parameters and the individuals are scored according to the function being optimized. This score is then used to cull the population in a way that gives higher scoring individuals a greater chance of survival.

After the selection step, the surviving gene pool is used to produce the next generation by a process analogous to mating. Mating pairs are selected by either random mating from the entire population, some form of inbred mating, or assortive mating in which individuals with similar traits are more likely to mate. The pairs are used to produce genetic material for the next generation by a process analogous to sexual reproduction. First, each individual's diploid genome is used to produce a haploid by combining each pair of number strings into a single string by randomly choosing substrings from one or the other. At this point, randomized point mutations or transpositions may also be introduced. The two haploids from each mating pair are combined to produce the genetic specification for each individual in the next generation. Each mating pair is used to produce several siblings, according to a distribution normalized to ensure a constant total population size. The entire process is repeated for each generation, using the gene pool produced by one generation as a specification for the next.

The experiments that we have conducted have simulated populations ranging in size from 512 to $\sim 10^6$ individuals, with between 1 and 256 chromosomes per individual. Chromosome lengths have ranged from 10 to 128 bits per chromosome, mutation rates from 0 to 25% probability of mutation per bit per generation, and crossover frequencies ranged from 0 to an average of 4 per chromosome. Using a Connection Machine with 64,536 processors, a typical experiment progresses at about 100 to 1000 generations per minute, depending on population size and on the complexity of the fitness function.

SORTING NETWORKS

As an example of how simulated evolution can be applied to a complex optimization problem, we consider the problem of finding minimal *sorting networks* for a given number of elements. A sorting network[10] is a sorting algorithm in which the sequence of comparisons and exchanges of data take place in a predetermined order. Finding good networks is a problem of considerable practical importance, since it bears directly on the construction of optimal sorting programs, switching circuits, and routing algorithms in interconnection networks. Because of this, the problem has been well studied, particularly for networks that sort numbers of elements that are exact powers of two.

Sorting networks are typically implemented as computer programs, but they have a convenient graphical representation, as shown in Figure 1. The drawing contains n horizontal lines, in this case 16, corresponding to the n elements to be sorted. The unsorted input is on the left, and the sorted output is on the right. A comparison-exchange of the ith and jth elements is indicated by an arrow from the ith to the jth line. Two specified elements are compared and they are exchanged, if and only if, the element at the head of the arrow is less than the element at the tail; the smallest element will always end up at the tail. The sorting network pattern shown in Figure 1 is a Batcher sort,[1] which requires $n \log^2 n - 1$ exchanges to sort n elements.

A useful property of sorting networks is that they are relatively easy to test. A sorting network that correctly sorts all sequences of 1 and 0 will correctly sort any sequence, so it is possible to test an n-input sorting network exhaustively with 2^n tests.

In this section we describe how simulated evolution is used to search for networks that require a small number of exchanges for a given number of inputs. In particular, the case $n = 16$ is of particular interest, and has a long history of successive surprises. In 1962, Bose and Nelson[2] showed a general method of sorting networks which required 65 exchanges for a network of 16 inputs. They conjectured that this was the best possible. In 1964, Batcher,[1] and independently, Floyd and Knuth,[6] discovered the network shown in Figure 1, which requires only 63 exchanges. Again, it was thought by many to be the best possible, but 1969 Shapiro[19] discovered a network using only 62 exchanges. Later that year, Green[7] discovered a 60-comparison sorter, shown in Figure 2, which stands as the best known. These results are summarized in Table 1. For a lively and more detailed account of these developments, the reader is referred to Knuth,[15] Volume 3, pages 227–229.

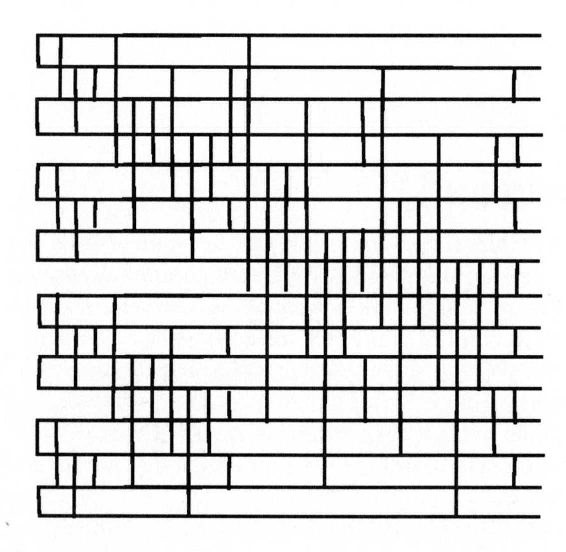

FIGURE 1 Sorting Network.

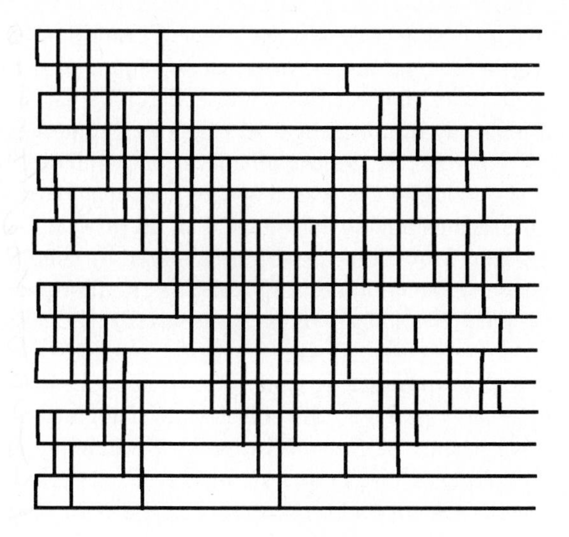

FIGURE 2 Green's 60 comparison sorter.

TABLE 1 Best Known Networks

1962	Bose and Nelson	65
1964	Batcher, Knuth	63
1969	Shapiro	62
1969	Green	6

TABLE 2 Networks Found by Simulated Evolution

Without Parasites	65
With Parasites	61

There are two ways to cast the search for minimal sorting networks as an optimization problem. The first is to search the space of functional sorting networks for one of minimal length. The second is to search the space of short sequences of comparison/exchanges for ones that sort best. The difficulty with the first approach is that there is no obvious way of mutating a working sorting network into another one that is guaranteed to work, so almost all mutations and recombinations will create a network that is outside of the search space. It is much easier in the second approach to produce mutations and variations of a small program that stay within the space of small programs. Mutation can be implemented by changing the position of one of the exchanges, and recombination by splicing the first part of one sorting network with the last part of another. This is essentially the approach we have adopted.

One difficulty with this approach is that even if the solution is in the space of small networks, the easiest paths to the solution may not be. It may be easier, for example, to produce a short correct network by optimizing a slightly longer correct network than by fixing a bug in a short incorrect network. For this reason, we have taken advantage of the diploid representation of a genotype to allow longer networks to be generated as intermediate solutions.

The genotype of each individual consists of 15 pairs of chromosomes, each consisting of 8 codons, representing the digits of 4 chromosome pairs. Each codon is a 4-bit number, representing an index into the elements, so the genotype of an individual is represented as 30 strings of 32 bits each. The phenotype of each individual (an instance of a sorting network) is represented as an ordered sequence of ordered pairs of integers. There is one pair for each exchange within the network. The elements of the pair indicate which elements are to be compared and optionally exchanged. Each individual has between 60 and 120 pairs in its phenotype, corresponding to sorting networks with 60 to 120 exchanges.

The phenotype is generated from the genotype by traversing the chromosomes of the genotype in fixed order, reading off the pairs to appear in the phenotype. If a pair of chromosomes is homozygous at a given position (if the same pair is specified in both chromosomes), then only a single pair is generated in the phenotype. If the site is heterozygous, then both pairs are generated. Thus, the phenotype will contain between 60 and 120 exchanges, depending on the heterozygosity of the genotype. Sixty was chosen as the minimum size so that a completely homozygous genotype would produce a sorting network that matches the best-known solution. Because most of the known minimal 16-input networks begin with the same pattern of 32 exchanges, the gene pool is initialized to be homozygous for these exchanges. The rest of the sites are initialized randomly.

The diploid genome allows for dominant and recessive genes to evolve. For example, an exchange near the end of the sorting network that puts two items in the wrong order is likely to be a dominant lethal. On the other hand, a relatively harmless exchange, say, one that incorrectly exchanges two nearby elements near the beginning of the sequence, is likely to be recessive. Since a purely homozygous individual expresses the shortest possible sequences of exchanges, heterozygous individuals are likely to have at least a short-term advantage in fitness. The offspring of two variations of the same algorithm is likely to exhibit hybrid vigor. The disadvantage of this approach is that the only force tending to reduce the length of the successful heterozygous solution is the outbreeding of deleterious recessives.

Once a phenotype is produced, it is scored according to how well it sorts. One measure of ability to sort is the percentage of input cases for which the network produces the correct output. This measure is convenient for two reasons. First, it offers partial credit for partial solutions. Second, it can be conveniently approximated by trying out the network on a random sample of test cases. After scoring, the population is culled by truncation selection at the 50% level; only the best scoring half of the population is allowed to contribute to the gene pool of the next generation.

The algorithm for selection works as follows. Each individual is paired with another in its local two-dimensional neighborhood (by an algorithm described below). Before mating, the genetic material of the higher-scoring individual replaces the genetic material of the lower-scoring individual. Thus, the higher-scoring individual participates in the next mating cycle twice, whereas the lower-scoring individual does not participate at all.

To implement recombination, the gamete pool is generated by crossover among pairs of chromosomes. For each chromosome pair in the surviving population, a crossover point is randomly and independently chosen, and a haploid gamete is produced by taking the codons before the crossover point from the first member of each chromosome pair, and the codons after the crossover point from the second member. Thus, there is exactly one crossover per chromosome pair per generation. Point mutations are then introduced in the gamete pool at a rate of one mutation per one thousand sites per generation.

The next stage is the selection of mates. One way to do this would be to choose pairs randomly, but our experience suggests that it is better to use a mating program with some type of spatial locality. This increases the average inbreeding coefficient and allows the population to divide into locally mating demes. The sorting networks evolve on a 2-dimensional grid with torroidal boundary conditions. Mating pairs are chosen to be nearby in the grid. Specifically, the x and y displacement of an individual from its mate is a binomial approximation of a Gaussian distribution. Mating consists of the exchange of haploid gametes. After a pair mates, they are replaced by their offspring in the same spatial location, so the genetic material remains spatially local.

The algorithm for choosing mates in competition on a 2-dimensional grid works as follows. Each individual begins with a pointer to itself. The grid is then divided into pairs of adjacent individuals, for example, by pairing individuals in odd rows with their northern neighbor. Each pair of individuals flips a coin to determine whether or not the pair should exchange addresses. The process is then repeated with a different set of pairings, for example, pairings in the east-west direction, or pairings where the odd rows choose the southern neighbor. As the procedure is iterated, each individual address will follow a random walk through the space. The distribution of distances moved from the point of origin will be approximately Gaussian.

Simulations were performed using the procedure on populations of 64,536 individuals for up to 5000 generations. Typically, one solution, or a few equal-scoring solutions, were discovered relatively early in the run. These solutions and their variants then spread until they accounted for most of the genetic material in the population. In cases where there was more than one equally good solution, each "species" dominated one area of the grid. The areas were separated by a boundary layer of non-viable crosses. Once these boundaries were established, the population would usually make no further progress. The successful networks tend to be short because the descendant of heterozygotes tended to be missing crucial exchanges (recessive lethals). The best sorting networks found by this procedure contained 65 exchanges.

THE CO-EVOLUTION OF PARASITES

While the evolution of the sorting networks produced respectable results, it was evident on detailed examination of the runs that a great deal of computation was being wasted. There were two major sources of inefficiency. One was a classical problem of local optima: once the system found a reasonable solution, it was difficult to make progress without temporarily making things worse. The second problem was an inefficiency in the testing process. After the first few generations, most of the tests performed were sorted successfully by almost all viable networks, so they provided little information about differential fitness. Many of the tests were too "easy." Unfortunately, the discriminative value of a test depends on the solutions that initially evolve, and in the case where several solutions evolve, the value of a given test varies from one sub-population to another.

To overcome these two difficulties, various methods were implemented for accelerating progress by encouraging a wider diversity of solutions and limiting the number of redundant test cases. Three general methods were investigated: varying the test cases over time, varying the test cases spatially, and varying the test cases automatically by independent evolution. Because the third case has yielded the most interesting results, only it will be described in detail.

The co-evolution of test cases is analogous to the biological evolution of a host parasite, or of prey and predator. Hamilton has used both computer simulation and mathematical/biological arguments to show that such co-evolution can be a generator of genetic diversity.[8,9,10,11,12] The improved optimization procedure uses this idea to increase the efficiency of the search.

In the improved procedure, there are two independent gene pools, each evolving according to the selection/mutation/recombination sequence outlined above. One population, the "hosts," represents sorting networks, while the other population, the "parasites," represents test cases. (These two populations might also be considered as "prey" and "predator," since their evolution rates are comparable.) Both populations evolve on the same grid, and their interaction is through their fitness functions. The sorting networks are scored according to the test cases provided by the parasites at the same grid location. The parasites are scored according to how well they find flaws in sorting networks. Specifically, the phenotype of each parasite is a group of 10 to 20 test cases, and its score is the number of these tests that the corresponding sorting network fails to pass. The fitness functions of the host-sorting networks and the parasitic sets of test patterns are complementary in the sense that a success of the sorting network represents a failure of the test pattern and vice versa.

The benefits of allowing the test cases to co-evolve are twofold. First, it helps prevent large portions of the population from becoming stuck in local optima. As soon as a large but imperfect sub-population evolves, it becomes an attractive target toward which the parasitic test cases are likely to evolve. The co-evolving test cases implement a frequency selective fitness function for the sorting networks that discourages large numbers of individuals from adopting the same non-optimal strategy. Successive waves of epidemic and immunity keep the population in a constant state of flux. While systems with a fixed fitness criteria tended to get stuck in a few non-optimal states after a few hundred generations, runs with co-evolving test cases showed no such tendency, even after tens of thousands of generations.

The second advantage of co-evolving the parasites is that testing becomes more efficient. Since only test-case sets that show up weaknesses are widely represented in the population, it is sufficient to apply only a few tests to an individual each generation. Thus, the computation time per generation is significantly less. These

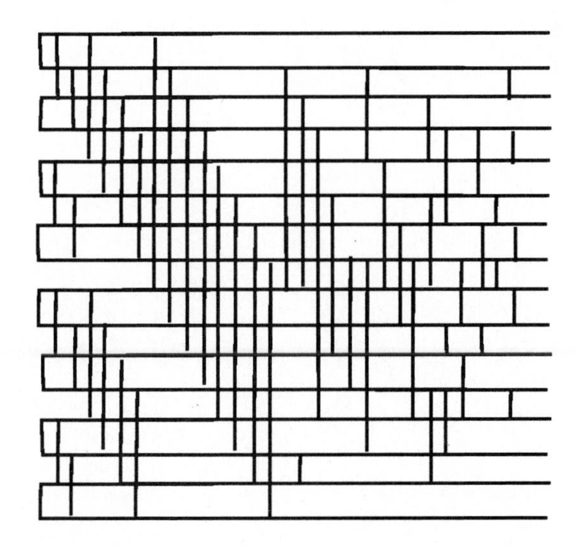

FIGURE 3 61 exchanges.

two factors taken together make it both more practical and more productive to allow the system to run for larger numbers of generations.

The runs with co-evolving parasites produced consistently better and faster results than those without. Figure 3 shows the best result found to date, which requires 61 exchange elements. This is an improvement over Batcher's and Shapiro's solutions, and over the results of the simulation without parasites. It is still not the optimum network, since it requires one more sorting exchange than the construction of Green.

These preliminary results are encouraging. They demonstrate that simulated evolution of co-evolving parasites is a useful procedure for finding good solutions to a complex optimization problem. We are currently applying similar techniques to other applications in an attempt to understand the range of applicability. It is ironic, but perhaps not surprising, that our attempts to improve simulated evolution as an optimization procedure continue to take us closer to real biological systems.

ACKNOWLEDGMENTS

This paper is a slightly more detailed version of the paper presented at the Emergent Systems Conference in Sante Fe, N.M., May 1989. The author would like to thank C. Taylor, W. Hamilton, S. Forrest, C. Langton, S. Kaufmann, S. Smith, and the reviewers for helpful discussions and comments in the preparation of this paper.

REFERENCES

1. Batcher, K. E. "A New Internal Sorting Method." Goodyear Aerospace Report GER-11759, 1964.
2. Bose, R. C., and R. J. Nelson. "A Sorting Problem." *JACM* **9** (1962): 282–296.
3. Bounds, D. G. "New Optimization Methods from Physics and Biology." *Nature* **329**, 1987.
4. Bremermann, H. J. "Optimization Through Evolution and Recombination." In *Self-Organizing Systems*, edited by M. C. Yovits, G. D. Goldstein, and G. T. Jacobi, 93–106. Washington, DC: Spartan, 1962.
5. Darwin, C. *The Origin of the Species by Means of Natural Selection; or, The Preservation of Favoured Races in the Struggle for Life*. London: Murray, 1859.
6. Floyd, R. W., and D. E. Knuth. "Improved Constructions for the Bose-Nelson Sorting Problem." *Notices of the Amer. Math. Soc.* **14** (1967): 283.
7. Green, M. W. In *Sorting and Searching, Volume 3: The Art of Computer Programming*, edited by D. Knuth. Reading, MA: Addison-Wesley, 1973.
8. Hamilton, W. D., and J. Seger. "Parasites and Sex." In *The Evolution of Sex*, edited by R. E. Micmod and B. R. Levin, Ch. 11. Sinauer, 1988.
9. Hamilton, W. D. "Pathogens as Causes of Genetic Diversity in their Host Populations." In *Population Biology of Infectious Diseases*, edited by R. M. Anderson and R. M. May, 269–296. Berlin: Springer-Verlag, 1982.
10. Hamilton, W. D. "Sex Versus Non-Sex Versus Parasite." *OIKOS* **35** (1980): 282–290.
11. Hamilton, W. D., P. Henderson, and N. Moran. "Fluctuation of Environment and Coevolved Antagonist Polymorphism as Factors in the Maintenance of Sex." In *Natural Selection and Social Behavior: Recent Research and Theory*, edited by R. D. Alexander and D. W. Tinkle, Ch. 22. Chiron Press, 1980.
12. Hamilton, W. D. "Gamblers Since Life Began: Barnacles, Aphids, Elms." In *Quart. Rev. Biol.* **50** (1975): 175–180.
13. Holland, J. H. *Adaption in Natural and Artificial Systems.* Ann Arbor: University of Michigan, 1975.
14. Kirkpatrick, S., C. Gelatt, Jr., and M. Vecchi. "Optimization by Simulated Annealing." *Science* **220** (1983): 4598.
15. Knuth, D. *Sorting and Searching, Volume 3: The Art of Computer Programming.* Reading, MA: Addison-Wesley, 1973.
16. Lewontin, R., and I. Franklin. "Is the Gene the Unit of Selection?" *Genetics* **65** (1970): 707–734.
17. Metropolis, N., A. Rosenbluth, M. Rosenbluth, A. Teller, and E. Teller. *J. Chem. Phys.* **21** (1953): 1807.
18. Rechenberg, I. *Evolutionsstrategie: Optimierung Technischer Systeme nach Prinzipien der Biologischen Evolution.* Stuttgart: Frommann-Holzboog, 1973.

19. Shapiro, G. In *Sorting and Searching, Volume 3: The Art of Computer Programming*, edited by D. Knuth. Reading, MA: Addison-Wesley, 1973.
20. Smith, R., and D. Goldberg. "Nonstationary Function Optimization Using Genetic Algorithms with Dominance and Diploidy." *Genetic Algorithms and Their Applications*, Proceedings of the Second International Conference on Genetic Algorithms, July 1987.
21. Wang, Q. "Optimization by Simulating Molecular Evolution." *Biol. Cybern.* **57** (1987): 95–101.

Stuart A. Kauffman†‡ and Sonke Johnsen‡
‡Department of Biochemistry and Biophysics, School of Medicine, University of Pennsylvania, Philadelphia, PA and †The Santa Fe Institute, 1120 Canyon Road, Santa Fe, NM 87501

Co-Evolution to the Edge of Chaos: Coupled Fitness Landscapes, Poised States, and Co-Evolutionary Avalanches

We introduce a broadened framework to study aspects of co-evolution based on the *NK* class of statistical models of rugged fitness landscapes. In these models the fitness contribution of each of N genes in a genotype depends epistatically on K other genes. Increasing epistatic interactions increases the rugged multipeaked character of the fitness landscape. Co-evolution is thought of, at the lowest level, as a coupling of landscapes such that adaptive moves by one player deform the landscapes of its immediate partners. In these models we are able to tune the ruggedness of landscapes, how richly intercoupled any two landscapes are, and how many other players interact with each player. All these properties profoundly alter the character of the co-evolutionary dynamics. In particular, these parameters govern how readily co-evolving ecosystems achieve Nash equilibria, how stable to perturbations such equilibria are, and the sustained mean fitness of co-evolving partners. In turn, this raises the possibility that an evolutionary *metadynamics* due to natural selection may sculpt landscapes and their couplings to achieve co-evolutionary systems able to coadapt well. The results suggest that sustained fitness is optimized when landscape ruggedness relative to couplings between landscapes is tuned such that Nash equilibria just tenuously form across the ecosystem. In this poised state, co-evolutionary avalanches appear to propagate on all length scales in a power-law distribution. Such avalanches may be related to the distribution of small and large extinction events in the record.

INTRODUCTION AND OVERVIEW

Our aim, in this article, is to describe a new class of models with which to investigate some of the problems of co-evolution. The class of models is related to the game theoretic models introduced by Maynard-Smith and Price,[16] and Maynard-Smith.[17] These authors, and many since them, have been concerned primarily (but not exclusively, see Rosenzweig et al[20]) with intraspecies co-evolution, and the conditions under which co-evolving systems attain evolutionary stable strategies (ESS). Our focus is largely on a general class of models for interspecific co-evolution, and the conditions to attain Nash equilibria in such systems.

We consider co-evolution occurring among species, each of which is itself adapting on a rugged multipeaked fitness landscape. But the fitness of each genotype, via the phenotype, of each species is affected by the genotypes via the phenotypes of the species with which it is coupled in the ecosystem. Adaptive moves by one co-evolutionary partner, therefore, may change the fitness and the *fitness landscapes* of the co-evolutionary partners. Anecdotally, development of a sticky tongue by the frog alters the fitness of the fly, and what it ought to do: it should develop slippery feet. In this framework, adaptive moves by any partner may *deform* the fitness landscapes of other partners.

To investigate this, we reintroduce the spin-glass-like *NK* family of rugged multipeaked fitness landscapes.[7,8,9,11,12,23] In this model, N corresponds to the number of genes in a genotype, or traits in an organism, while K corresponds to the number of other genes, or traits, which bear upon the fitness contribution of each gene or trait. K, therefore, corresponds to the richness of epistatic linkages in the system. Tuning K from low to high, increasing epistatic linkages, increases the ruggedness of fitness landscapes by increasing the number of fitness peaks, increasing the steepness of the sides of fitness peaks, and *decreasing* the typical heights of fitness peaks. The decrease reflects the conflicting constraints which arise when epistatic linkages increase. Thus, the *NK* model provides a tunably rugged family of model fitness landscapes. To study co-evolution we couple the fitness landscapes of different interacting species. To do so, we suppose that genes or traits in each species make fitness contributions which depend upon K, other genes, or traits within that species itself, but also upon C traits in each of the other species with which the species interacts. Therefore, adaptive moves by one species may deform the landscapes of its partners. Altering C changes how dramatically adaptive moves by each species deform the landscapes of its partners. The other major parameters in the model are the total number of species which interact, S, and the perhaps restricted number of these, S_i, with which any species, i, interacts. One form of the model we investigate ignores population dynamics and focuses upon the evolution of genotypes. A second form includes population dynamics using the familiar Lotka-Volterra logistic equation allowing the evolution of competition or mutualism among the co-evolving species.

The framework we consider permits us to study how the ruggedness of fitness landscapes, the richness of coupling among fitness landscapes, the number of species, and the structure of the ecosystem affect the co-evolutionary process. To our knowledge, this is the first model which has allowed these questions to be investigated more or less systematically. We find a number of interesting and apparently novel features of such a process. Among these, the sustained fitness of the co-evolutionary partners depends upon all the parameters. Therefore, we are driven to consider the possibility of a selective *metadynamics* in evolution in which co-evolutionary partners change, not only their genotypes in an effort to optimize fitness on a given deforming fitness landscape, but may also change the statistical character of their fitness landscapes by changing its ruggedness, change the richness of couplings to other landscapes, and change the number of other partners with whom they co-evolve so as to optimize sustained fitness. Rather remarkably, in distributed ecosystems where each partner interacts with only a few other species, this metadynamics appears to lead to a critical "poised" state in which the entire system is tenuously "frozen" at a Nash equilibrium. Here each species is maximally fit given the genotypes of its partners, and is characterized by an unchanging, locally optimal genotype and stable population density. But at this poised state, avalanches of co-evolutionary change unleashed by minor changes in the physical environment, or other exogenous noise, propagate through the system on all length scales, with a power-law distribution between sizes of avalanches and numbers of avalanches at each size. Such avalanches typically include fluctuations to lower fitness; hence, they are likely to be correlated with increased probabilities of extinction events. Thus, the size distribution of avalanches can be used to try to predict distribution of sizes of extinction events in the record. The observed distribution tends to support a picture of ecosystems which are slightly more rigidly "frozen" than the critical poised state.

In the second section, we reintroduce the *NK* model and extend it to model co-evolving species. In the third section, we describe the dynamics of co-evolving pairs of species as a function of K and C. In the fourth section, we extend the results to distributed ecosystems where each species interacts with only a few of the total. In the fifth section, we extend the results to include population dynamics via a version of the Lotka-Volterra logistic equation.

2. COUPLED *NK* FITNESS LANDSCAPES

The *NK* family of rugged fitness landscapes were introduced[7,8,9,11,12,23] to describe genotype fitness landscapes engendered by arbitrarily complex epistatic couplings. Consider an organism with N genes, each with A alleles. For simplicity, consider that each has only two alleles, 1 and 0. Alternatively, consider an organism with

N traits, present or absent. The *fitness contribution* of each gene or trait depends upon itself and epistatically on K other traits. Thus, each trait makes a fitness contribution which depends upon the particular combination, one among $2^{(K+1)}$ of the presence or absence of the $K+1$ traits which bear upon its fitness. The *NK* model is intended to capture the effects of such epistatic coupling, but is spin-glass-like in assuming that the epistatic interactions are so complex that their effects on fitness can best be captured by assigning those effects at random. Thus, the *NK* family models the statistical structure of such fitness landscapes by assigning at random to each such combination of "inputs" to the ith trait a fitness contribution between 0.0 and 1.0. Then, for each organism, the fitness contribution of each of its N traits in the context of itself and the K which bear upon it are added together. Thereafter the fitness is normalized through division by N. Figure 1 shows a simple organism with three traits, or genes, each of which makes a fitness contribution depending upon itself and the other two. The model assigns a fitness to each of the eight possible genotypes.

A central idea is that of a fitness landscape. Figure 1 shows all $2^3 = 8$ genotypes arranged on a Boolean cube, such that each genotype is a one-mutant neighbor of the three others which are accessible by mutating a single gene, or trait, to the opposite allele, or state. Thus, (000) is a one-mutant neighbor of (001), (010), and (100). Arrangement of the eight genotypes onto the Boolean cube supplies a

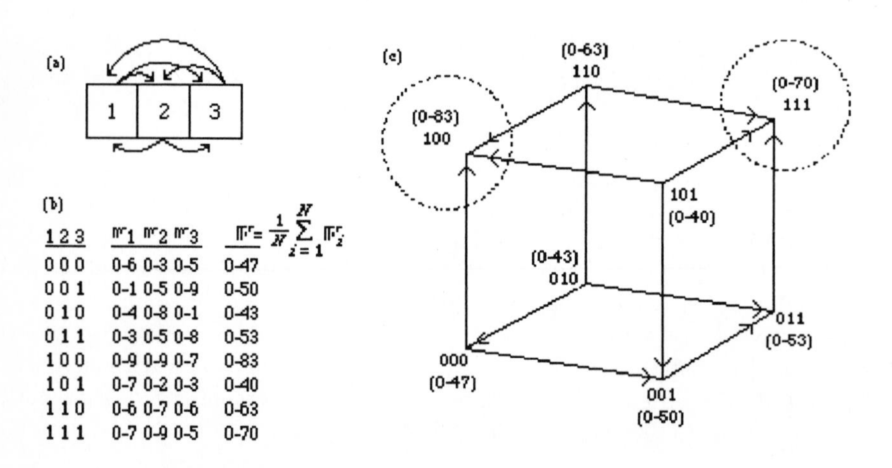

FIGURE 1 (a) Assignment of $K = 2$ epistatic "inputs" to each site. (b) Assignment of fitness values to each of the three genes with random values for each of the eight combinations of $K+1$ alleles bearing on genes 1, 2, and 3. These fitness values then assign a fitness to each of the $2^3 = 8$ possible genotypes as the *mean* value of the fitness contributions of the three genes, as given in Eq (1). (c) Fitness landscape on the three-dimensional boolean cube corresponding to the fitness values of the eight genotypes in (b). Note that more than one local optimum exists.

genotype space, in which each genotype is next to its one-mutant neighbors. For an N-gene, two-allele system, the corresponding space is an N-dimensional hypercube. The NK model yields a fitness for each of the eight genotypes, shown in Figure 1. These fitnesses can be thought of as a heights in a fitness landscapes of the kind initially introduced by Wright.

The simplest picture of an adaptive process in a fixed-fitness landscape envisions beginning at a genotype and "moving" to a one-mutant neighbor only if the second genotype is fitter than the first. Then an adaptive walk starts at a genotype and climbs uphill until a genotype, which is fitter than all one-mutant neighbors, a local optimum, is encountered. Figure 1 shows two such local optima. Natural statistical features of landscapes include the number of local optima, the average number of steps to local optima, the number of local optima accessible from a random initial genotype, the number of genotypes which can climb to the global, or any other optima, the number of directions "uphill" from any genotype, and how that number dwindles to zero as adaptive walks climb to optima. Other questions ask how these properties change if the adaptive walk can pass via two-mutant or J-mutant variants, or if it may pass through equally fit or less-fit neighbors. Still other questions concern how a population will flow across such a landscape under the drives of mutation, selection, and recombination. Here, as a function of the ruggedness of the landscape, complex population flows occur, with the population able to climb to local peaks and hover in their vicinity at low mutation rates, but melting from those peaks and relaxing over ever wider reaches of the genotype space as mutation rates increase.[3,4,7,9,22]

All these properties depend upon the statistical structure of the underlying fitness landscape. The NK model was introduced because it provides a family of landscapes which range from single peaked and highly correlated to fully random and uncorrelated as K increases from 0 to its maximum, $N-1$. For $K=0$, each trait makes an independent fitness contribution. The model corresponds to an N-locus, two-allele, additive haploid genetic model. Each gene has a favored allele, by chance 1 or 0. Thus, there is an optimal genotype in which each gene has mutated to its optimal allele. Further, any other genotype can "climb" to this optimum by mutating the less-fit to the more-fit allele of each gene. Hence, there are no optima other than the single and global optimum. On each step uphill, the number of ways uphill decreases by 1; hence, the ways uphill decrease slowly. On average, the number of steps to the global optimum scales as $N/2$. Further, the fitness of one-mutant neighbors is nearly the same, since the greatest difference a single gene might make scales as $1/N$, which reduces towards 0 as the number of genes, N increases. Thus, $K=0$ corresponds to a highly correlated landscape about a single global peak, a "Fujiyama landscape." Furthermore, the gradient of the fitness slope to the peak is shallow. Fitness increases by $1/N$ on each step uphill. Finally, it is easy to show from order statistics that in this model the global optimum has a fitness of 2/3.

The maximum value of K is $N-1$. For $K=N-1$ the landscape is fully uncorrelated. That is, the fitness of one-mutant neighbors are random with respect to one another. This can be seen in Figure 1. Each of the three genes depends upon itself and the other two. Thus, changing any gene's allele, say from 0 to 1, changes the combination of allele states affecting each of the three genes to a new combination. In turn, each gene makes a different, randomly assigned fitness contribution. Hence, altering any gene's allele yields a fully random fitness in the one-mutant neighbor. The fully random fitness landscape, which corresponds to the Derrida random-energy spin-glass model[2] was analyzed by Kauffman and Levin[10] and Macken and Perelson.[14] Such landscapes have on the order of $2^N/(N+1)$ local optima and walks to optima scale as $\ln N$; at each step uphill the number of ways uphill drops by half; any initial genotype can climb to only a small fraction of the optima; and only a small fraction can climb to the global optimum. A critical feature of the *NK* model is that for $K = N-1$; the optima are *lower* than for $K = 0$. Indeed, a kind of complexity catastrophe occurs: As N increases, the fitness of typical local optima dwindle towards 0.5, the mean fitness in the space. This decrease reflects the conflicting constraints inherent in the large K limit.

These results show that as K increases, fitness landscapes change from smooth and single peaked, to random and multipeaked. Thus, as K changes, a family of increasingly rugged multipeaked landscapes is encountered. In general, as K increases for fixed N, the number of local optima increases, the sides of fitness peaks become steeper, and the heights of typical local optima decrease. As we shall see, these features appear fundamental to co-evolutionary behavior when *NK* landscapes are coupled.

The *NK* model is a very general model of a family of tunably rugged, correlated fitness landscapes. It, together with rather related spin-glass models, are the first explicit examples where such a family has been examined. It is, at this point, entirely unclear how many such families of rugged correlated landscapes may exist. However, the *NK* family seems likely to be an important member of such a set of families. Furthermore, the potential biological utility of the *NK* has recently been enhanced by showing that it is able to account for the statistical features of protein evolution seen in maturation of the immune response.[12] Both because the *NK* model is the first, and appears to be a plausible, model of rugged-fitness landscapes which may apply in biological evolution, we extend it here to study co-evolutionary processes.

Consider an ecosystem with S species. For simplicity imagine that each species is homogeneous, that is, that all organisms in the species are identical; hence, the species currently occupies a particular combination of its N traits. Then an *NK* landscape can represent the fitness landscape of one "homogeneous" species. This assumption corresponds to the limit described by Gillespie,[5,6] in which mutations arise infrequently in an adapting population compared to the fitness differences between initial and mutant forms. Under these conditions, the population encounters advantageous mutants on a slow time scale and moves as a whole on a short time scale to the new fitter variant. Thus, under these limiting cases, the population can be approximated as homogeneous. With a higher mutation rate, or under other conditions, a species is not homogeneous. It is possible to extend the model to allow the population representing one species to be a cloud distributed over its landscape in which frequency and density dependent co-evolution within the species occurs.

In a co-evolutionary system we need to represent the fact that both the fitness and the *fitness landscape* of each species is a function of the other species. Thus, in general, it is necessary to *couple* the rugged fitness landscapes for each species, such that an adaptive move by one species "projects" onto the fitness landscapes of other species and alters those fitness landscapes more or less profoundly. Over time, each species jockeys uphill on its own landscape, and thereby deforms those of its ecological neighbors. Any such move by one may increase or decrease the fitness of each neighbor on its own landscape, and alter the uphill walks accessible to that neighbor.

In the context of the NK model, the natural way to couple landscapes is to assume that each trait in species 1 depends epistatically on K other traits internally, and on C traits in species 2. More generally, in an ecosystem with S species, each trait in species 1 will depend upon K traits internally and on C traits in each of the S_i among the S species with which it interacts. It is also natural to assume symmetry, if species 1 is in the niche of species 2, then 2 is in the niche of species x1.

To represent the effect of the C traits from species 1 which are coupled to each trait in species 2, we expand the fitness tables defining the landscape of species 2 to incorporate the added C traits which couple to each trait in species 2. Hence, for each of the N traits in species 2, the model will assign a random fitness between 0.0 and 1.0 for each combination of the K traits internal to species 2 together with all combinations, present or absent, of the C traits in species 1. In short, landscapes are coupled by expanding the random fitness table for each trait in species 2 such that it "looks" at its K internal epistatic inputs and also "looks" at the C external epistatic inputs from species 1. Given this, then the fitness landscape of species 2 is a function of the current location of species 1 on species 1's own fitness landscape. Therefore, as species 1 adapts, it will change both the fitness of species 2, and also deform 2's fitness landscape. In turn, each trait in species 1 is coupled to C traits in species 2, and the fitness values for each of the N traits in species 1 must be expanded similarly. This couples the two landscapes. Each is in the niche of the other.

In a system of S species, the interactions can be represented by a web of such projections. In so doing, we have at our disposal at least: (1) choice of the number and identity of the traits, C, which couple from one species to each single trait in another species, (2) the number and identity of other species among the S which project onto each species, and (3) finally, we have at our disposal the number of species, S, in the entire ecosystem.

In section 3, we shall assume that each species in an S species system is coupled directly to all other species. This richest coupling is undoubtedly unrealistic. A vast literature studies the hierarchical structure of food webs.[18] In section 4 we consider "structured" ecosystems where each species interacts with only a few of the other species.

We also assume a second "worst case" condition, namely, that in an S species ecosystem there is no similarity between the species; hence, the effects of species 1 on 2 and 3 on 2 are randomly assigned. In reality, rabbits and hares probably look much the same to a fox. Similarity of species presumably can be thought of as reducing the number of *effectively different species* with which each species interacts. Finally, we also consider a naive case from an ecological standpoint, namely, that each coadapting partner interacts at each moment with all other partners. In section 5, we are more realistic and require that each species interacts with the others in the ecosystem in pair-wise combinations, using plausible population dynamics.

3. LANDSCAPE RUGGEDNESS AND COUPLINGS BETWEEN LANDSCAPES TUNE CO-EVOLUTION

The game theoretic models which have been explored to study co-evolution have not as yet been built to take account of the *statistical ruggedness* of each of the co-evolving partner's landscapes, the *richness of coupling* of those landscapes, and the implications of those features on the entire co-evolutionary process. But surely these are major aspects of the problem. The *NK* model affords tunably rugged landscapes whose richness of coupling can also be tuned; hence, we can study the influences of these factors on co-evolution.

Simulations of co-evolving systems were carried out under the assumption that each species "acts," *in turn*, in the context of the current state of the other species. On its turn, each species tries a random mutation and "moves" as a whole to that mutant variant if the variant is fitter. If the mutant variant is not fitter, the species does not move. Thus, if the species moves, this at least transiently increases the fitness of the species which has just moved, but may increase or decrease the fitness of its co-evolving partners. In addition to this "random" dynamics, we also examine two alternative cases. In the "fitter" dynamics, each species, in turn, examines all its mutant variants and chooses at random one of the fitter variants if any exist. In the "greedy" dynamics, each species, in turn, chooses the fittest mutant variant.

NASH EQUILIBRIA

In general, such a co-evolutionary process admits of two behaviors. Either the partners keep dancing, or the coupled system attains a steady state at which the local optimum of each partner is consistent with the local optimum of all the other partners via the "C" couplings. Such a steady state is the analogue of a Nash equilibrium in the current context. We use the word "analogue" for the following reason. A true Nash equilibrium assumes that each agent can, at each moment, chose any one of its possible "actions." In the present context, this corresponds to each species, in a single moment, altering its current genotype to any of the 2^N possible genotypes. In assuming that each species is able, at each moment, to mutate a single "gene" or trait, we are constraining the range of alternative genotypes, or actions, locally accessible to the species. Thus, the steady states we will find are similarly constrained. In the remainder of this article, we use the term "Nash" equilibria with the understanding that such equilibria are with respect to the mutant search range.

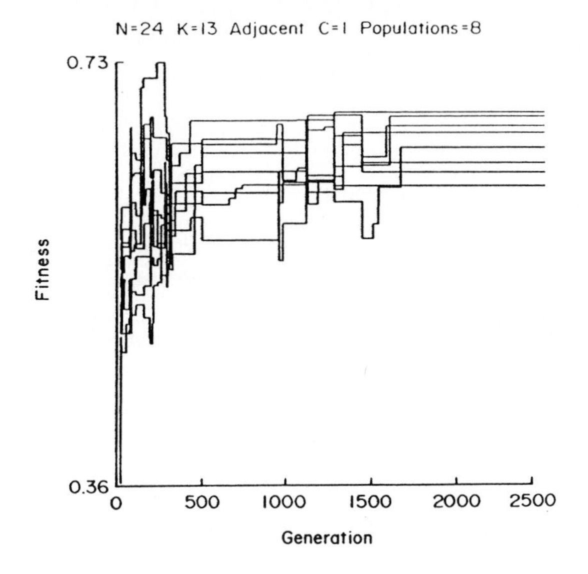

FIGURE 2 Co-evolution among eight "species" each governed by an NK landscape. $N = 24, K = 13$. Each of the N "traits" in each species is affected by $C = 1$ "trait" in each of the seven other species. System reaches a steady state about generation 1600. Note, mean fitness in the absence of selection is 0.5. See text.

A second caution is required. We consider Nash equilibria, as described. The concept of an evolutionary stable strategy, an ESS, is a further refinement of the concept of a Nash equilibrium, in which the condition of non-invadability by a mutant at an *initial low frequency* in the population is analyzed. In the simplified dynamics used here where the whole population moves in an "instant" to a fitter variant, we do not analyzed invadability. Studies with fuller population dynamics, section 5, which do include analysis of invadability, confirm the simpler dynamics.

Simulations were carried out between pairs of co-evolving species, or "agents," each a single organism on an independent NK landscape. In addition, simulations were carried out for larger numbers of species, S. The first major result is that Nash equilibria are actually encountered. It is not obvious that this should occur, for each species has 2^N genotypes among which it is evolving. An S species system has the product of these genotypes in its joint "strategy space."

Figure 2 shows eight species, each with $N = 24, K = 17$, and $C = 1$, co-evolving over 2500 generations. Here, then, each site within one species is epistatically affected by 17 other sites within that species and 1 site in each of the eight other species. Over eight generations, each species, in turn, tries a random mutation and moves to that new genotype only if it is fitter than the current genotype in the context of the current genotypes of the remaining seven species. At each generation, the fitness of all species are recalculated in the context of the genotype of each and their couplings. As can be seen, for the first few hundred generations, the mean fitness of the whole set of species increases, rapidly at first, then more slowly. Increasingly long intervals with no change occur, reflecting the fact that as fitness increases, the waiting time to find fitter variants increases for each partner. Sudden bursts of change by many species, however, are instigated by occasional changes by a single partner. By about 1600 generations, however, the entire system stops changing, and in fact remains constant forever thereafter. A Nash equilibrium has been found, such that each species is locally fitter than all one-mutant variants, granted that the others do not change.

WAITING TIME TO ENCOUNTER NASH EQUILIBRIA

The waiting time to encounter Nash equilibria depends upon N, K, and C. For $K > C$, Nash equilibria are encountered rapidly, for $K < C$ the waiting time to find Nash equilibria becomes very long. In order to examine how N, K, and C bear on the waiting time to encounter a Nash equilibrium, simulations were carried out between two species. For each value of N, K, and C tried, 100 co-evolving pairs were released. Over generations, a successively larger fraction will have encountered Nash equilibria and, hence, stopped evolving. Figures 3(a),(b) show the results, plotting the fraction "still walking," against the generations elapsed, for $C = 1, 3$ (a), and for $C = 8, 3$ (b).

The main point to note is that as K increases relative to C, the waiting time to hit a Nash equilibrium *decreases*. Thus, as the ruggedness of landscapes increases, by K increasing, the expected waiting time to find Nash are decreasing. Presumably this reflects the increased number of local optima in NK landscapes as K increases for fixed N. Similar studies as N increases, data not shown, for fixed K and C,

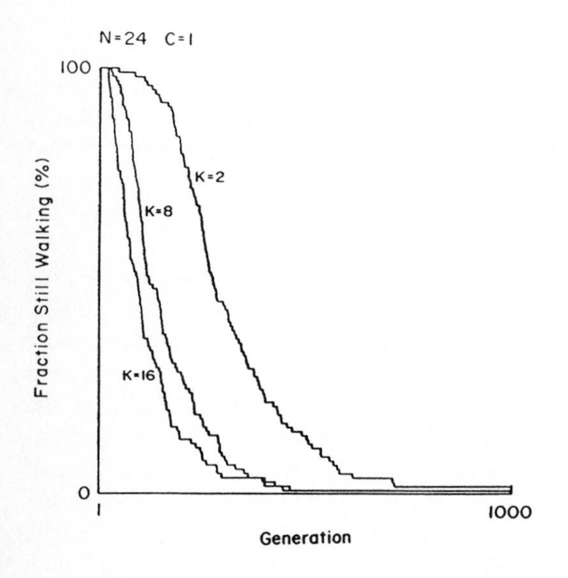

FIGURE 3 (a) Fraction of 100 co-evolving pairs of species which have not yet encountered a "Nash" equilibrium, and hence are still walking as a function of generations elapsed. Curves correspond to different values of K. $N = 24, C = 1$ in all cases.

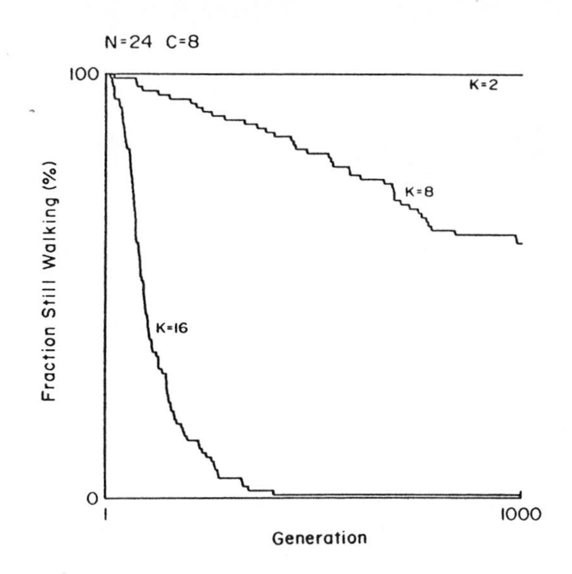

FIGURE 3 (b) Same as (a) except $C = 8$.

show that the waiting time to hit Nash equilibria increases, presumably because the density of local optima decreases as N increases. In short, for a pair of species which are co-evolving, $K = C$ is a crude dividing line. When K is greater than C, Nash equilibria are found rapidly. When K is less than C, Nash equilibria are still found, but the mean waiting time becomes very long.

CO-EVOLUTION WHEN TWO INTERACTING SPECIES HAVE DIFFERENT K VALUES

It is of considerable interest to study the outcome of co-evolution in which partners are on landscapes of different ruggedness. Figure 4 reports the results of simulations in which pairs of species have K values of 2, 4, 8, 12, and 16, for three values of C, 1, 8, and 20. $N = 24$ in all cases. These three values of C were chosen to lie below, in the middle, and above the range of K values. For each set of parameter values, 300 pairs of co-evolving species was released, and evolved for 250 generations. By that time, either a Nash equilibrium had been encountered, or had not. In the former case, the fitness of each partner was noted. In the latter case, the average fitness of each partner over the prior 85 generations was tabulated. The upper panels in Figure 4 report the mean fitness of each partner at the Nash equilibria encountered, and the number which reached such Nash equilibria. The lower panels of Figure 4 show the mean fitness of each partner as it continues to co-evolve, over the past 85 generations, and the number of pairs still co-evolving.

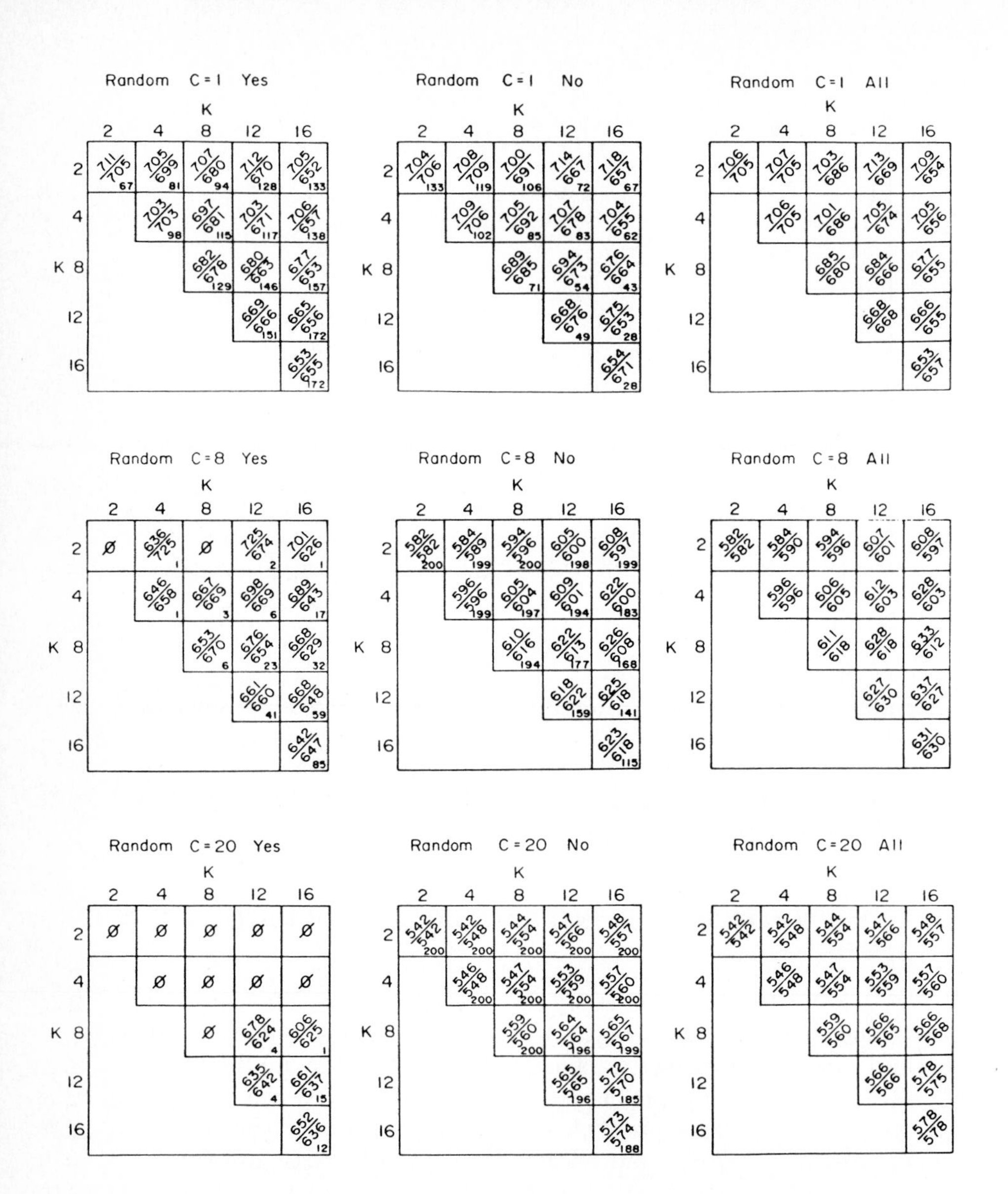

FIGURE 4 Matrix showing fitness of row and column players in each small box, top number row player, bottom number column player, in "K-player" values 2-16. Each small box also shows total number of runs ending in Nash Equilibria (yes), number not encountering Nash Equilibrium during run, (no), and mean fitness prorated over (yes) and (n0) in "ALL." Middle and lower points, similar except C=8 and C=20.

Figure 4 show a number of interesting features:

1. For all values of K, as C increases, the fraction of co-evolving pairs which encounter Nash equilibria in 250 generations decreases. Conversely, for any fixed value of C, as K increases, the fraction of pairs encountering Nash equilibria increases. This re-exhibits the phenomena of Figure 3. High K leads to more rugged landscapes and Nash equilibria are encountered more rapidly.
2. When C is higher than 1, i.e., $C = 8, C = 20$, the fitness at Nash equilibria is higher than the corresponding fitness when the partners are still oscillating.
3. As C increases, the fitness of both co-evolving partners during the oscillatory phase before encountering Nash equilibria *decreases* for all pairs of co-evolving K values. Thus, fitnesses during the pre-Nash oscillatory period for $C = 20$ are substantially lower than for $C = 8$ and still lower than for $C = 1$. Presumably this reflects the fact that, for high values of C, a single move by one partner sharply lowers the expected fitness of the remaining partner.
4. When C is high, $C = 20$, note that the high-K "players" have higher mean fitness during the oscillatory period before hitting Nash equilibria than do low-K "players." More strikingly, in playing against a second player with a fixed-K value, the first player would increase mean fitness during the oscillatory period by increasing its own value of K. That is, a K-4 player does better against a K-2 player than would a K-2 player, while a K-8 player does even better, etc.
5. Equally remarkably, when C is high, $C = 20$, a low-K player achieves higher mean fitness during the pre-Nash oscillatory period if it plays against a second species of high K. That is, a K-2 player has higher fitness against a K-4 player than against a K-2 player. A K-2 player fares even better against a K-16 player. Thus, when C is high, increasing the K value of one partner helps *both* co-evolving partners.
6. This tendency is reversed when C is low, $C = 1$. Here, during the oscillatory period, low-K players fare better than high-K players.
7. At the Nash equilibria encountered, the fitnesses of low-K players is clearly higher than that of high-K players for each value of C, and, indeed, seem roughly independent of C.
8. Finally, when C is high, overall average fitness is highest when K is high. When C is low, $C = 1$, overall average fitness is highest when K is low. Thus, fitness in co-evolving systems would be enhanced were K able to adjust to match C, or more broadly, if K and C were themselves evolvable.

Similar studies were carried out using the "fitter" and the "greedy" dynamics described above. The main results are the same. A major difference arises in the greedy dynamics. Here, for any genotype, there is generically a unique best-fit one-mutant variant in the context of the other species. Thus, if each species "plays" in a deterministic order, each changes to a unique next genotype, and the set of co-evolving species can enter a recurrent cycle in the total space of S genotypes. We find that when $K < C$, such periodic attractors arise rather frequently.

The analysis above is based on mutating a single site, gene, or trait, at a time in each co-evolving species. We have also examined the sustained fitness of co-evolving species as a function of the number of genes or traits which can mutate at each moment. In general, for any value of N, C, and K, an optimal mutation rate exists. Figure 5(a) shows the results of co-evolution as the mutation rate, or more accurately, the number of traits randomly mutated in each species, increases to 2, 4, 8, 16, and 24, for co-evolving pairs of species on increasingly rugged landscapes, $K = 2, 4, 8, 12$, and 16. In all cases each of the $N = 24$ sites in each species is coupled to $C = 1$ site in the other species. As the number of genes mutated simultaneously increases, the number of local optima on a fixed landscape dwindle; hence, the probability that the co-evolving pair reaches a local Nash equilibrium falls. Thus, the fitness seen in the co-evolving pair reflects fitness during the pre-Nash period. Values are averages of the last 85 steps in 250 generations. Figure 5(a) shows that for all values of K, as the number of genes mutated increases, the maintained fitness reaches a maximum for 2 or 4 simultaneous mutations, and falls thereafter. The decrease in maintained fitness is greatest for small K values, hence smoother landscapes, and less marked for large K values. Figures 5(b) and 5(c) show similar results except that the coupling among the co-evolving players, C, is increased to 8 and 20. In general, but not uniformly, as C increases the optimal mutation rate decreases. These results suggest that there may typically be an optimal mutation rate to maintain fitness in co-evolutionary processes.

The results we have seen make a number of intriguing suggestions. All point to the possibility that a co-evolving system of species may collectively tune the parameters governing its own co-evolution.

First, there may well be selective processes which match "K" to "C" in order to optimize the co-evolutionary capacities of the co-evolving partners. K "should" increase when it is low relative to C, and decrease when it is high relative to C. As shown in Figure 4, for the most biologically plausible case in which each species tries random mutations and moves to a variant if it is fitter, if C is high relative to K, any player increases its fitness by increasing its own K value. When C is high, increasing K has two beneficial effects. First, Nash equilibria are encountered more rapidly and are fitter than the prior oscillatory period. Second, fitness during that prior oscillatory period is higher. Thus it is advantageous to any player to increase K in a high C environment. Perhaps equally remarkably, in the biologically reasonable case of random mutations, 4(a)-(c), such a move by one species *also helps the second species.* Each has higher pre-Nash fitness and finds Nash equilibria sooner. Conversely, suppose K is high relative to C. Then, as is clear from Figures 4(a)-(c), and 3, Nash equilibria are encountered very rapidly. Thus, the fitness in the pre-Nash oscillatory period is of less importance, and the fitness of Nash equilibria are more important. But, local optima at Nash equilibria are higher for low-K players than high-K players. Thus, if K is too high with respect to C, it should be advantageous to decrease K. In short, at this group level of co-evolving species, it seems clear that there are reasonable selective advantages to a species as a whole to "tune" K to match C. At that match, given a fixed C, Nash equilibria will be encountered rapidly, and be highly fit, optimizing mean fitness during any pre-Nash periods, minimizing the mean duration of those periods, and maximizing the fitness of the Nash optima attained.

The analysis above is based on advantage to the species as a group in increasing K. In order to avoid "group selection," we seek selective conditions acting on individual members of a single species which might increase K in members of that species, hence in the coupled ecosystem. Within the framework of the NK model, a change in "K" would naturally be envisioned as a mutation which altered the epistatic coupling between traits, or genes, such that a trait or gene now depended on one more, or one fewer epistatic "inputs." That is, we must let "K" itself evolve. In this framework, the natural way to express the consequence of such a mutation which increases K is to expand the fitness table for that trait or gene such that it now "looks at" the new trait as well as the K it initially "looked at." That new epistatic connection, in the context of the current genotype in which the new connection is formed, might increase or might decrease the fitness of the current genotype. That is, the new epistatic link may itself alter the fitness of the current genotype for better or worse. Thus, we can envision three ways in which selection on an individual level may allow an increased K value at one genetic locus to spread throughout a population: (1) that new epistatic link, when it forms, causes the genotype to be fitter and is selected, and hence spreads; (2) the new epistatic link is "near neutral" and spreads through the population by random drift; and (3) The new link has not only a direct effect on the fitness of the current genotype, but also on the *inclusive fitness of the individual and its progeny* due to the increased fitness of those progeny in the *co-evolutionary* process itself, due to increased rapidity of finding Nash equilibria, and higher fitness during the pre-Nash oscillatory period.

These considerations suggest the possibility of a co-evolutionary dynamic which optimizes K relative to C in an ecosystem, such that partners maintain high mean fitness.

We turn next to consider the co-evolution of the number of species, S, in the coupled ecosystem. A clear process should tend to limit the number of species. Consider coupled landscapes with fixed K and C couplings. Let the number of species increase, under the assumption that each species is "C" coupled to all other species. For pairs of species, $K = C$ is a rough dividing line separating cases where Nash equilibria are encountered rapidly versus slowly. In multispecies systems with S species, it appears that when K is greater than $S \times C$, the co-evolving partners all encounter a Nash equilibrium rapidly. When K is less than $S \times C$, the co-evolving partners do not encounter a Nash equilibrium for a long time. We stress that the exact relation between mean waiting time to encounter Nash equilibrium, and the K, S, and C parameters are unknown. $K = S \times C$ is a rough guide.

Numerical results are shown in Figure 6(a)-(c) for an increasing number of coupled species. In all cases $N = 24, K = 10$, and $C = 2$. In Figure 6(a) $S = 4$ species are co-evolving. Thus, each species senses $C = 2$ inputs to each of its genes from each of the four species. Each species, in turn, randomly mutates and tries to find a fitter variant. Over 2000 generations, mean fitness increases and the four species find a Nash equilibrium. In Figures 6(b) and 6(c) the number of species increase to 8, then 16. Data to 2000 generations are shown for the 8 and 16 species cases. No Nash was found for 8000 generations in these cases.

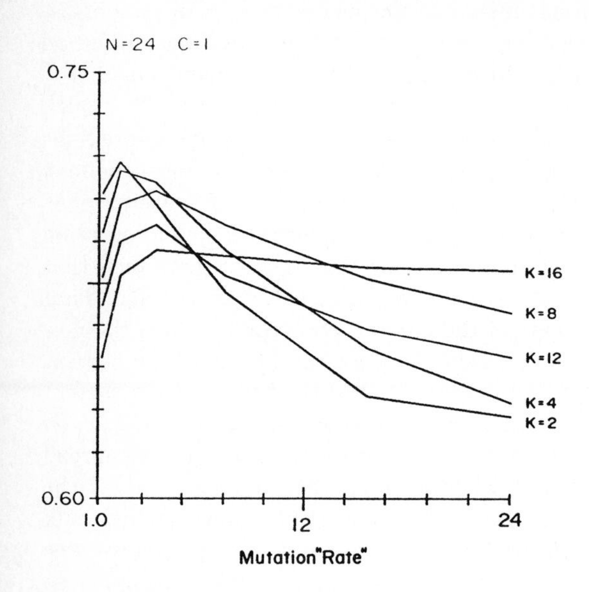

FIGURE 5 (a) Effects of increasing the mutation rate on mean fitness maintained in co-evolving species. Both co-evolving species randomly mutate 1, 2, 4, 8, 16, or 24 of the $N = 24$ "genes" at each moment. $K = 2$, 4, 8, 12, 16. $C = 1$. Figure plots average fitness over last third of 250 co-evolutionary steps against mutation rate. Note that for all K values, maintained fitness increases and reaches a maximum for 2 or 4 mutant search ranges. For all values of K, maintained fitness falls for greater mutation rates.

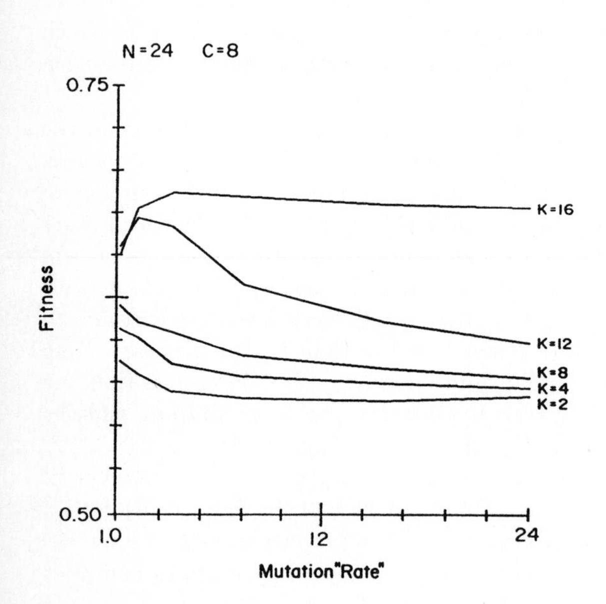

FIGURE 5 (b) As in (a), except $C = 8$.

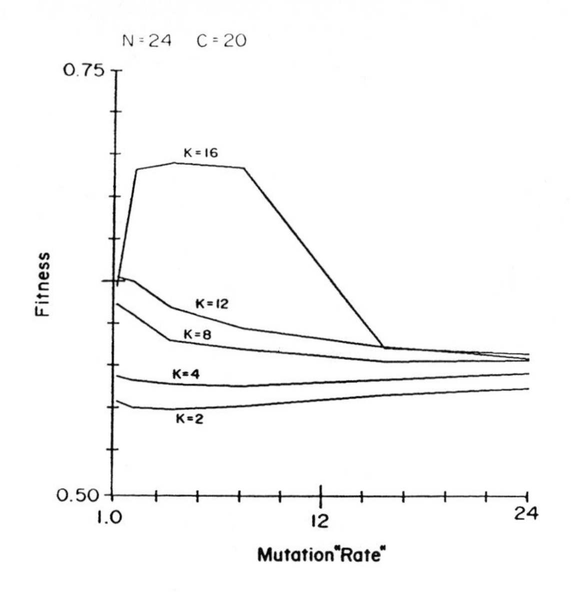

FIGURE 5 (c) As in (a), except $C = 20$.

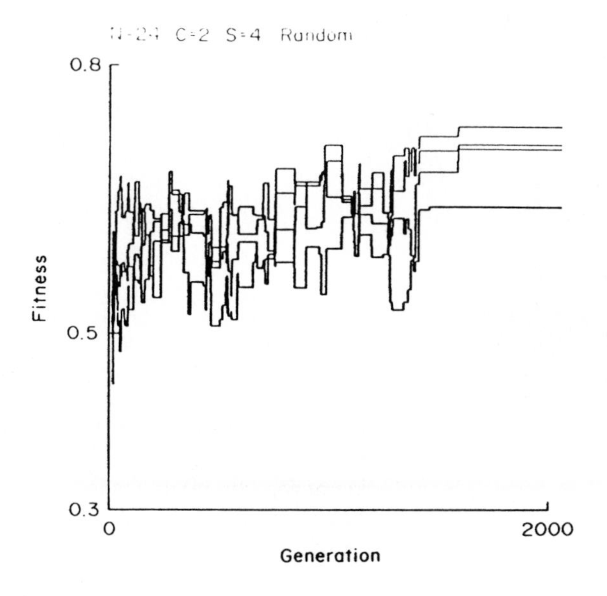

FIGURE 6 (a) Co-evolution among four species. $N = 24, C = 2, K = 10$.

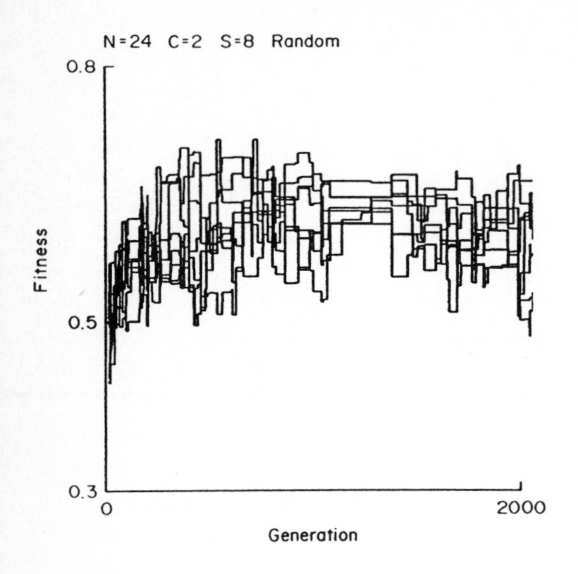

FIGURE 6 (b) As in (a), but eight species are co-evolving.

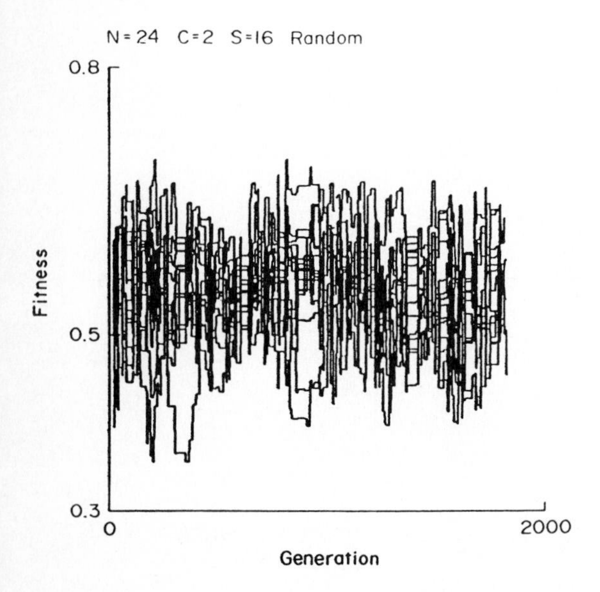

FIGURE 6 (c) As in (a), but 16 species are co-evolving. Note that as the number of species increases, the mean fitness decreases and the variance in fitness increases.

Note the following features in Figure 6(a)-(c): 1) as the number of species, S, increases, the waiting time to encounter a Nash increases; 2) as S increases, the *mean fitness* of the co-evolving partners *decreases*; and 3) as S increases, the *fluctuations* in fitness of the co-evolving partners increases dramatically. Thus, for the four-species case, mean fitness increases rapidly, and achieves fairly high Nash equilibria. For the 16-species case, the entire system fluctuates about a mean fitness slightly above average, 0.5, with dramatic excursions below (see section 4).

These results show that as the number of mutually coupled species in the system increases, the mean fitness falls and fluctuations to very low fitness increase. Therefore, if we may assume that fluctuations to low fitness are associated with an increased chance of *extinction* of the unlucky species, these results suggest that if S is too high, the coupled ecosystem will fluctuate dramatically and lead to the extinction and loss of species, thereby *lowering* S. In turn, as S is lowered, the ecosystem behaves less chaotically, mean fitness of all partners improves both during the pre-Nash period and because Nash are encountered more rapidly. The remaining system co-evolves "well" despite the fact that landscapes are coupled and deform as each actor moves.

In short, with clear hesitations, and heralded caveats, this framework begins to suggest the possibility that the co-evolutionary parameters governing the ruggedness of landscapes, couplings among landscapes, and number of co-evolving partners might themselves co-evolve without group selection, to continuously recreate well-formed ecosystems which are able *successfully* to coadapt. No mean feat, this, for as Figure 6(c) makes clear, co-evolution among coupled species can lead to chaotic fluctuations with no accumulation of improvement. If one wishes a "red queen," here is one to reckon with.

4. STRUCTURED ECOSYSTEMS AND SELF-ORGANIZED CRITICALITY: ADAPTATION TO THE EDGE OF CHAOS

Real ecosystems are not totally connected. Typically each species interacts with a subset of the total species; hence, the system has some extended web structure. In this section we extend our results to such ecosystems. The supposition that a co-evolutionary system can control the ruggedness of coupled landscapes, and its own connectivity by selection itself, and, therefore its dynamics, has interesting implications for extended ecosystems. It might be the case that co-evolving ecosystems tend toward a state of "self-organized criticality" in which parts of the ecosystem are "frozen" at Nash equilibria for long periods such that the species in the frozen component do not change, while other species continue to undergo co-evolutionary changes. Cascades, or avalanches of changes initiated at local points in the ecosystem web may propagate to various extents throughout the ecosystem. Such avalanches may trigger speciation and extinction events. Furthermore, the endogenous dynamics of the co-evolving system under natural selection optimizing sustained fitness for each co-evolving partner, may tend toward this characteristic critical poised state in which such avalanches can propagate on a variety of size scales with a power-law distribution between sizes of avalanches and their frequencies.

The term "self-organized criticality" was recently coined by physicist Per Bak,[1] to refer to a quite generic pattern of self-organization. Bak asks us to consider a tabletop onto which sand is added at a uniform rate. As the sand piles up on the table, it begins to slide off the edges of the table. Eventually, the system reaches a steady state at which the mean rate of adding sand equals the mean rate at which sand falls over the edges. At this stage the slopes from the peak to the edges of the table are near the rest angle for sand. Bak asks the following question: If one adds a single grain of sand to the pile at a random location and starts an avalanche, what will the *distribution of avalanche sizes be?* Bak finds a characteristic power-law distribution relating the frequencies and sizes of avalanches, with many tiny avalanches and few large avalanches. He argues that this distribution is characteristic of a wide range of phenomena, including distribution of earthquake sizes. The argument requires that the system under investigation attain and maintain a kind of "poised" state able to propagate perturbations—avalanches—on all possible length or size scales.

There are at least three ideas, derived from Bak's theory, which seem interesting in the co-evolutionary context. Cascades of perturbations, constituted by "packets" of co-evolutionary change, with a characteristic relation between size scale and frequency, may propagate through an ecosystem. This possibility requires that parts of the ecosystem can be "at rest" while other parts change. Second, the propagating changes are likely to be associated with fluctuations to low fitness which may engender both extinction and speciation events. Extinction events might be expected because of low fitness itself. Speciation events would be expected because, at low fitness, the number of directions of improvement is increased. If the probability of branching speciation is proportional to the number of directions of improved fitness, then low-fitness episodes should trigger speciation events. Thus, the propagation of avalanches through the system would be linked to speciation and extinction phenomena. Third, co-evolutionary dynamics of linked speciation, extinction, and alterations in coupling among the species in the ecosystem may achieve ecosystems which are "poised" such that avalanches can propagate on a characteristic variety of size scales. We shall see that this poised property is likely to be associated with the existence of nearly melted "frozen" components in ecosystems.

FLUCTUATING FROZEN COMPONENTS: NASH EQUILIBRIA EXTENDED TO LATTICE ECOSYSTEMS

A first hint that such ideas may apply to ecosystems arises in extending the co-evolving NK model to structured ecosystems in which each species interacts with only a few of the other species: *Parts of the system may be fixed at Nash equilibria while other parts continue to co-evolve.* That is, some species can attain an equilibrium and stop co-evolving, hence remain "at rest," while adjacent species in the ecosystem continue to change, either transiently or persistently.

To begin to investigate the behavior of structured ecosystems, we carried out simulations on "square" lattice ecosystems in which each interior species interacts with its four neighbors. Corner and edge species interact with two and three neighbors, respectively. Model ecosystems have varied from $3 \times 3 = 9$ species to $10 \times 10 = 100$ species. In addition to square ecosystem, which have corners and edges, we have investigated toroidal ecosystems in which the "square" is folded into a cylinder by joining left- and right-edge species, and then bent into a torus by joining top and bottom species. We have also investigated randomly connected ecosystems with similar general results.

Figure 7 shows twelve successive times in the temporal co-evolution of a 10×10 ecosystem. In this study $N = 24, K = 10, C = 1$. At each time moment, one of the 100 species "plays" and "greedily" chooses the fittest one-mutant variant if any is fitter than its current genotype. Each species "plays" in turn, and, thus, 100 "plays" constitutes an ecosystem generation. After each ecosystem generation, hereafter "generation," any species may have changed its genotype, or remained the

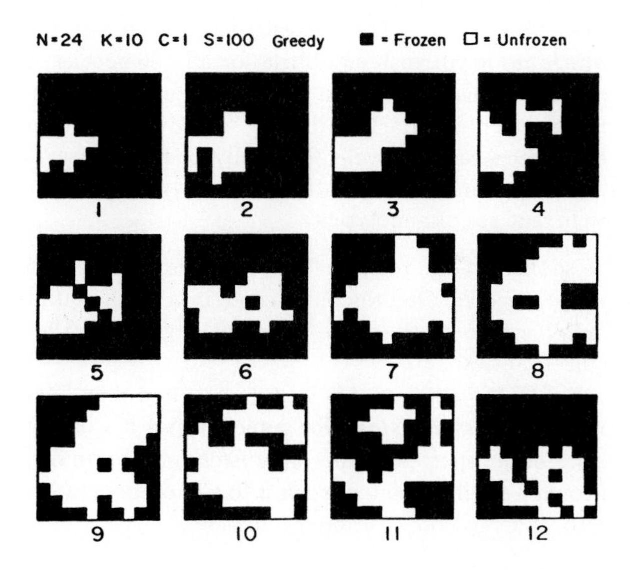

FIGURE 7 Twelve successive time moments four ecosystem generations apart in a $10 \times 10 = 100$ species ecosystem where each species plays with its immediate neighbors. Thus, corner species co-evolve with two immediate partners, edge species with 3, and interior species with 4. $N = 24, K = 10, C = 1$. As time progresses, frozen regions, black, where species are not changing genotype at that generation, emerge, expand and contract over the distributed ecosystem. If all species stop changing, the entire ecosystem is frozen, each species at a local Nash equilibrium.

same. If the species changed, we color it "white"; if it remains unchanged, we color it "black." The simulation was run over 200 ecosystem generations. The panels in Figure 7 represent moments which are four ecosystem generations apart in time, from a period in the middle of the 200 generations. The first question to ask is: Can "frozen" regions of "black" species which are unchanging in their genotypes, and other regions of "white" species which are changing their genotypes coexist in the ecosystem? The salient features to note are these:

1. A large fraction of the species are frozen and unchanging over single ecosystem generations.
2. Some regions remain frozen over very many ecosystem generations. In Figure 7, species in the upper left and lower left corner remain frozen over about 48 ecosystem generations.
3. One or more "white" unfrozen regions may exist.
4. Over time, the location and size of the frozen "black" region waxes and wanes. That is, *a fluctuating frozen component* can exist and extend through some or much of the ecosystem.
5. In the simulation carried out here, ultimately, the "frozen" region encompasses the entire ecosystem. That is, the ecosystem comes to rest at a combination of genotypes which are local Nash equilibria for all 100 species. In the absence of exogenous perturbations, the system will remain in this frozen state thereafter.
6. In many simulations, particularly using the "greedy" algorithm, the co-evolving ecosystem encounters a limit cycle. Typically in these cases, a fraction of the entire ecosystem remains permanently "frozen," while the remainder oscillate through a recurrent set of genotypes.

These results show that one region of an ecosystem can be "frozen" while other regions continue to co-evolve. One region persists in a Nash equilibrium something like an evolutionary stable state, while adjacent regions in the *same ecosystem* persist in Red Queen antics.

The existence of frozen components in an ecosystem may bear on, indeed may be fundamental to, *evolutionary stasis* for some species despite general changes in ecosystems altering other species. Some organisms may be maximally adapted to a fixed co-evolutionary environment in such a fixed component, even though other species continue to undergo co-evolutionary change.

A CO-EVOLUTIONARY ADAPTIVE PROCESS LEADING TOWARD A SELF-ORGANIZED CRITICAL STATE WITH A PERCOLATING FROZEN COMPONENT

The results in the previous sections concerning unstructured ecosystems suggested the possibility that a co-evolutionary dynamics might tune the parameters of the co-evolving species such that the species co-evolved "well." Here we discuss preliminary results suggesting that these ideas may extend to structured ecosystems. Species may "selfishly" tune their K values such that the coupled system as whole coadapts well. At the optimal state, a frozen component just begins to percolate across and covers the ecosystem "tenuously."

Figures 8(a) and 8(b) show the results of simulations of 5×5 square ecosystems in which the average fitness of the corner species, each coupled to two other species, the "edge" species, each coupled to three other species, and the "interior" species, each coupled to four other species, are accumulated. In the simulations, K is varied from 0 to 22. For each value of K, 50 ecosystems were analyzed over 200 ecosystem generations each. Figure 8(a) corresponds to the "random" dynamics where a random mutation is tried by each species. Figure 8(b) corresponds to the "fitter"

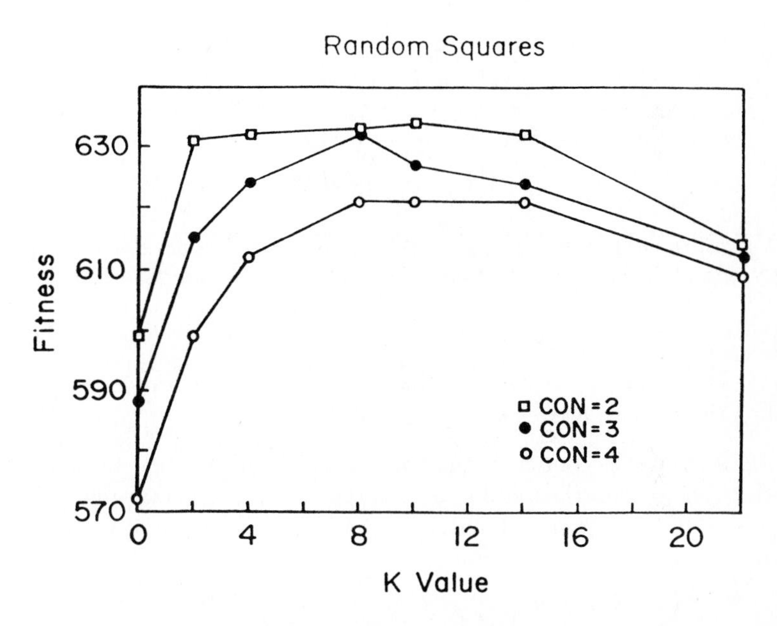

FIGURE 8 (a) Mean sustained fitness in 5×5 ecosystems as K varies from 0 to 22. $N = 24, C = 1$ in all cases. "Corner" species are connected to two others, top curve, edge species are connected to three others, middle curve, interior species are connected to four others, bottom curve. Note sustained fitness increases, then decreases as K increases. "Random" dynamics were used, such that each species tried a random one-mutant variant and "moved" there if that mutant were fitter.

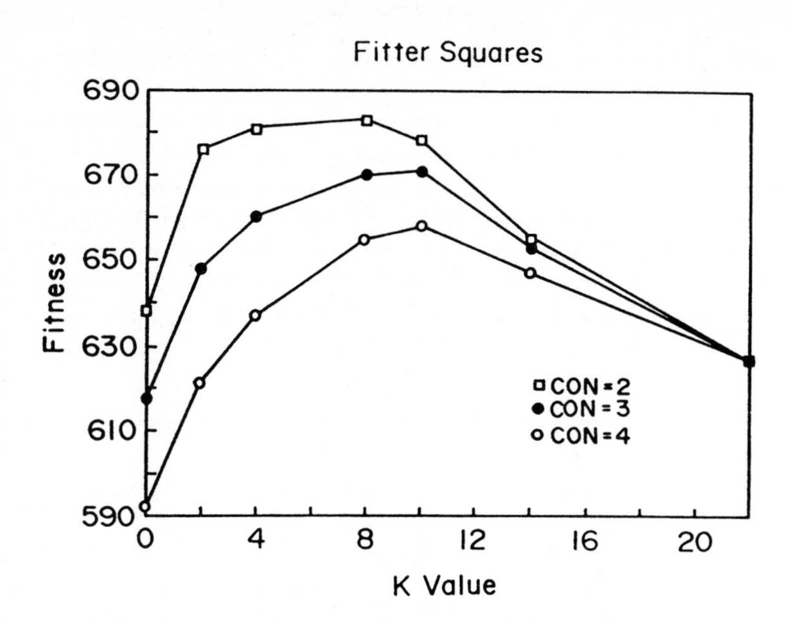

FIGURE 8 (b) As in 8(a), except the "fitter" dynamics were used. Each species picks at random one of its fitter variants if any exists, at each species generation.

dynamics, where one of the fitter variants is chosen at random by each species at each generation. These figures reveal the following features:

1. For all interspecies connectances, two for the corner species, three for the edge species, and four for the middle species, there is an *optimum value of* K at about $K = 8$ to 10. For lower or higher values of K, the average fitness declines.
2. Mean fitness *increases*, if connectance to other species *decreases* from four to three to two.

If there is an optimal value of K for sustained fitness in co-evolving systems, then it might be the case that selective effects might "pull" K values of co-evolving partners towards this optimum. This intuition is confirmed in Figure 9. Here we investigate the effect on the fitness of two "test" species, located adjacent to the center of the ecosystem, due to increasing or decreasing "K" values, compared to the rest of the ecosystem. To sample fairly the rest of the ecosystem, we monitored the fitness of two "control" species, also adjacent to the center of the square ecosystem. Figure 9 shows that, in all cases, presence of two "test" species with an altered K value had little effect on the fitness of the two control species. More critically, if the rest of the ecosystem had suboptimal K values, namely 0 or 2, or above optimal K values, namely 14 or 23, then deviations of the K values of the test species *to or towards* the K optimal values of $K = 8$ to 10 *increased the fitness of the test species.* Conversely, if the ecosystem as a whole is at the K optimal values of $K =$

8 to 10, then deviations of the K values of the test species *away* from the optimal values *decreases* the fitness of the test species. In short, there is a selective force toward the K optimal value of 8 to 10 which can act on single species, presumably via individual members of that species, and pull each toward the jointly optimal K value.

These results support and extend those discussed in the previous section with respect to completely connected ecosystems. There we found evidence that it was advantageous for a single species to increase or decrease K towards an optimal value relative to C. The results in Figure 9, are a powerful indication that a kind of selective metadynamics may very well tune K, the ruggedness of landscapes among co-evolving species, towards a *joint optimum* where all partners co-evolve well.

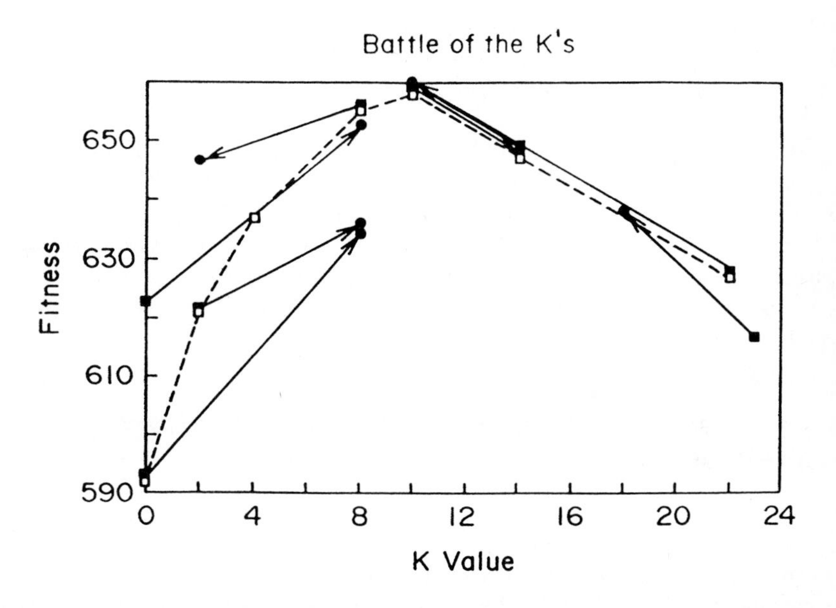

FIGURE 9 Selection force towards a K_{opt} value of $K = 8$ to 10. Two "experimental" species located adjacent to the central species in the 5×5 ecosystem were constructed with a different K value than the remainder of the ecosystem. In all cases deviations of the experimental pair, "dot," compared to the rest of the ecosystem, small square, *toward* K_{opt} *increased* the sustained fitness of the experimental pair compared to the rest of the ecosystem. In contrast, deviation *away from* K_{opt} *decreased* the fitness of the experimental pair compared to the rest of the ecosystem.

CO-EVOLUTION TO THE EDGE OF CHAOS: OPTIMIZATION OF SUSTAINED FITNESS AT K_{OPT} YIELDS ECOSYSTEMS WITH JUST PERCOLATING FROZEN COMPONENTS

Might optimization of the capacity to co-evolve, by optimizing K relative to C, actually selectively attain a posed self-organized critical state? The answer appears to be "yes" in the current model. Figure 9 shows that selection would be expected to pull co-evolving partners to jointly exhibit a near optimal $K = 8$ to 10 value. Figure 10 shows that, as K increases, the rapidity with which the ecosystem becomes "frozen" at a Nash equilibrium increases as well. Thus, for $K = 8$ or less, no freezing of the entire ecosystem occurs in any of the ecosystems over 200 generations. For $K = 10$, entire ecosystems freeze at Nash equilibria gradually over the 200 generations, while for $K = 14$ or 22 ecosystems freeze rapidly. Thus, the K optimal value for sustained fitness, $K = 8$ to 10, occurs just at that value where freezing begins visibly to occur. Otherwise stated, model ecosystems optimize co-evolutionary fitness when frozen components are tenuously extending across the ecosystem, hence when the system is "at the edge of chaos." By this phrase we mean that for values of K larger than K_{opt}, the ecosystem "freezes" into a "solid" state with all partners at Nash equilibria. For values of K smaller than K_{opt}, the ecosystem takes a very long time to achieve a Nash equilibria. The fitnesses of the species fluctuate chaotically during the long pre-Nash period. Hence, the system is chaotic, and can be thought of as being in a kind of "gas" phase. Just at the interface between the solid and gas phase is a kind of "liquid" region, at the edge of chaos.[13] Our results suggest that selection, in this model, achieves systems poised at the edge of chaos.

CO-EVOLUTIONARY AVALANCHES IN POISED ECOSYSTEMS: A POWER LAW DISTRIBUTION

The "edge of chaos" also corresponds to a poised self-organized critical state with respect to co-evolutionary avalanches. As seen above, as K decreases from above K_{opt}, the frozen component of all the species at a Nash equilibrium "melt." We investigate next the implications for co-evolutionary avalanches, and find that at the optimized state for sustained fitness, avalanches propagate on all length scales in a power-law distribution.

Alterations at one site in an ecosystem can often be expected to cause neighboring species to undergo coadaptive changes. Thus, changes at one point may propagate the various extents throughout the ecosystem. Such avalanches of changes are the analogues to Per Bak's sand pile avalanches. The propagation of such changes through an ecosystem are clearly of interest. In particular, during such changes, the affected species are likely to fall transiently to lower fitness. If lowered fitness is associated with an increased probability of extinction, then such avalanches might be associated with the propagation of extinction events.

To begin to investigate such avalanches, we modified the ecosystem model to allow each species to be affected not only by its neighbors in the ecosystem, but also by its external "world." The "external world" of each species consists in a binary vector length N. Each site in the species is coupled to W sites in its world. The fitness table of each site in the species is augmented to "look at" the K internal sites, the C sites in each of the species impinging upon that species, and the W sites in that species' world. Thus, alteration in the world of one species deforms the fitness landscape that species is co-evolving upon. Typically, such alterations lower the fitness of the species sufficiently that its current genotype is less fit than one or more of the one-mutant neighbor genotypes. If so, the species changes, and may then unleash an avalanche of co-evolutionary change which propagates through the ecosystem.

The simplest avalanches of changes to visualize are those which are perturbations from the "frozen" state in which all species are at local Nash equilibria. Simulations using the "fitter" move dynamics found the frozen state, then the "world" of a random member of the ecosystem was changed. At the end of each ecosystem generation, each species may have remained the same, or altered its genotype. After the onset of such a change, co-evolutionary changes continue until the system returns to a (perhaps new) frozen state with all species at local Nash equilibria. At that point, the "avalanche" has died out.

We used two measures of the "size" of such an avalanche. (1) The first is total number of species which are caused to alter their genotypes. This measures the total number of species in the avalanche which have changed genotype at least once. (2) An alterative measure sums the number of species which have changed at each ecosystem generation from the start of the avalanche until the avalanche stops.

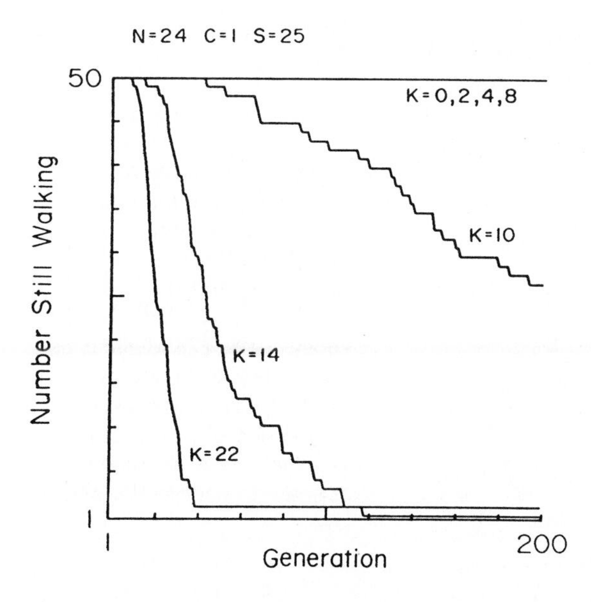

FIGURE 10 Fraction of 5 x 5 ecosystems which have not yet become frozen in an overall Nash equilibrium plotted against ecosystem generation. Note that for $K = 8$ or less, none of the ecosystems attained a frozen Nash equilibrium in the time available. For $K = 10$ or larger, some or most systems freeze at Nash equilibria, and do so more rapidly as K increases.

Thus, the second measure includes both the number of species which are affected, and the number of ecosystem generations in which each is affected. We denote this measure as (species x time).

Figures 11(a)-(c) show the resulting histograms of avalanche sizes for the 5 x 5 ecosystems discussed above. These figures plot the logarithm of the numbers of avalanches at each size versus the logarithm of the sizes of avalanches in the species x time measure of avalanche size. The number of avalanches measured are 228 for $K = 10$, 374 for $K = 14$, and 308 for $K = 20$. Figures 12(a)-(c) show similar histograms of avalanche sizes for 10 x 10 = 100 species square ecosystems, with $N = 24$, and values of K of 12, 14, and 18. The number of avalanches measured are 119 for $K = 12$, 402 for $K = 14$, and 150 for $K = 18$.

We would expect a relation between the value of K and the sizes of avalanches. Intuitively, the "frozen" state is readily attained and "solid" when K is sufficiently high, but the frozen state becomes more tenuous as K decreases. When the frozen state is very solid, avalanches are not likely to propagate far. When the frozen state is nearly melted, any perturbation is likely to propagate further. This is entirely in accord with the fact that the mean avalanche size increases and so does the variance as K decreases in both the 5 x 5 and 10 x 10 ecosystems. It also appears, that as K decreases, the distribution of avalanche sizes approaches a power law. Thus, for the 5 x 5 ecosystems where $K = 8 - 10$ corresponds to the value which optimizes sustained fitness, and for which ecosystems just begin to freeze, the log-log plot is approaching a straight line. Hence, the distribution appears to be a power law.

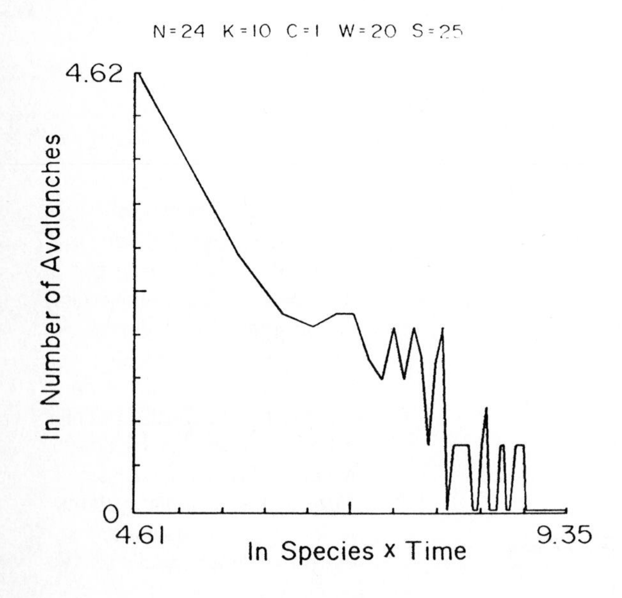

FIGURE 11 Avalanche-size distribution in 5 x 5 ecosystems. Figure plots the logarithm of the number of avalanches, versus the logarithm of the size of the avalanche (species x time), for $N = 24, C = 1, K = 10$ ecosystems. 228 avalanches are plotted. Note, Figures 11(a)-(c), as K approaches K_{opt} value of $K = 10$, avalanche-size distribution appears to approach a power law.

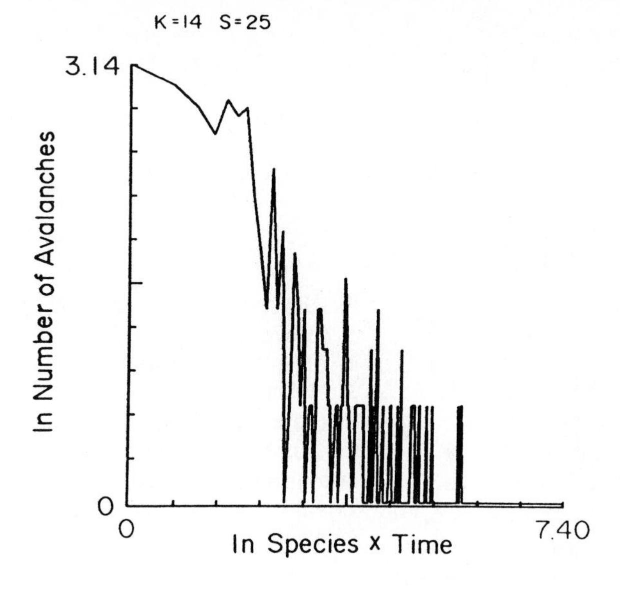

FIGURE 11 (b) As in 11(a), except $K = 14$. 374 avalanches are shown.

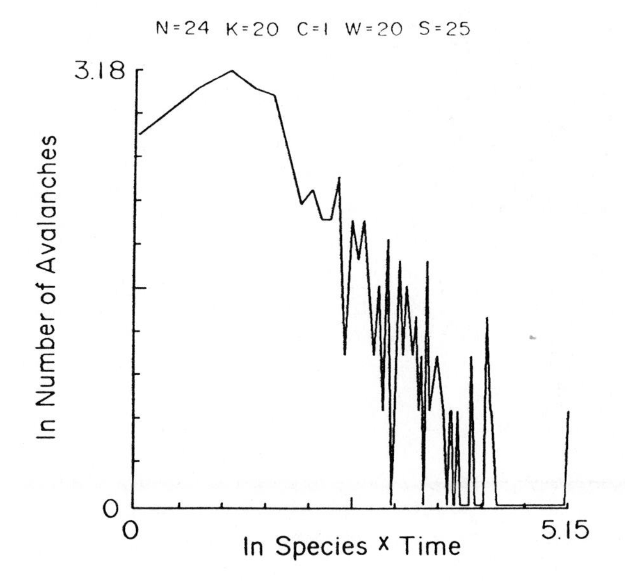

FIGURE 11 (c) As in 11(a), except $K = 20$. 308 avalanches are shown.

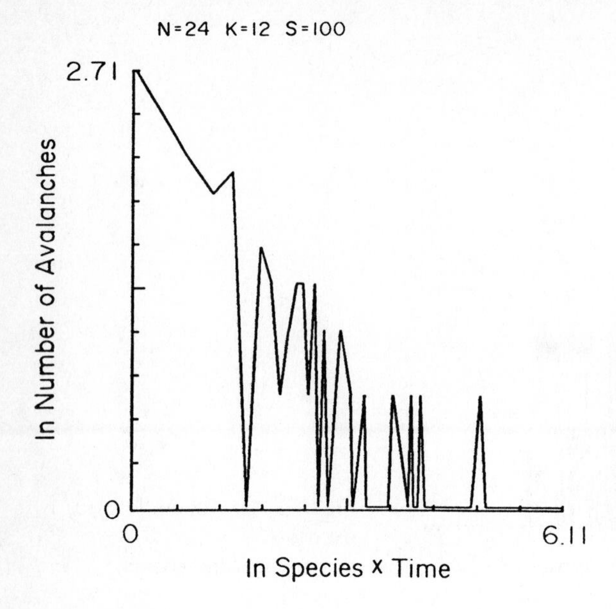

FIGURE 12 (a) Propagation of co-evolutionary avalanches in 10×10 ecosystems. $N = 24, C = 1, K = 12$. Figure plots the logarithm of the number of avalanches of a given size, versus the logarithm of avalanche size, (species x time).

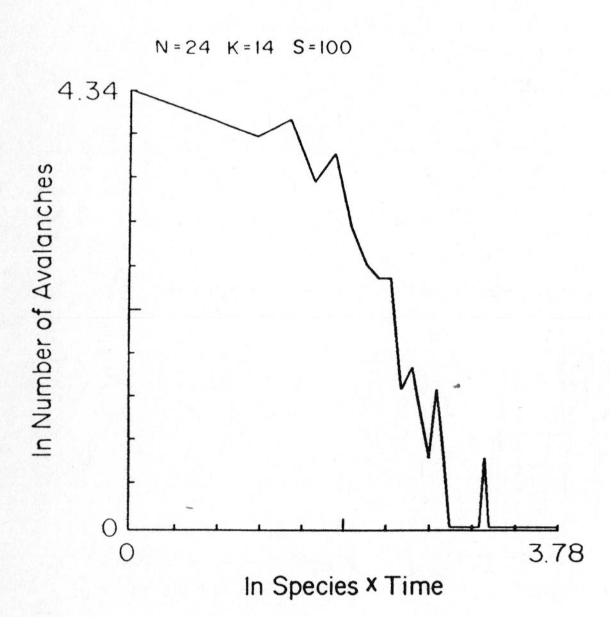

FIGURE 12 (b). As in 12(a), except K is 14.

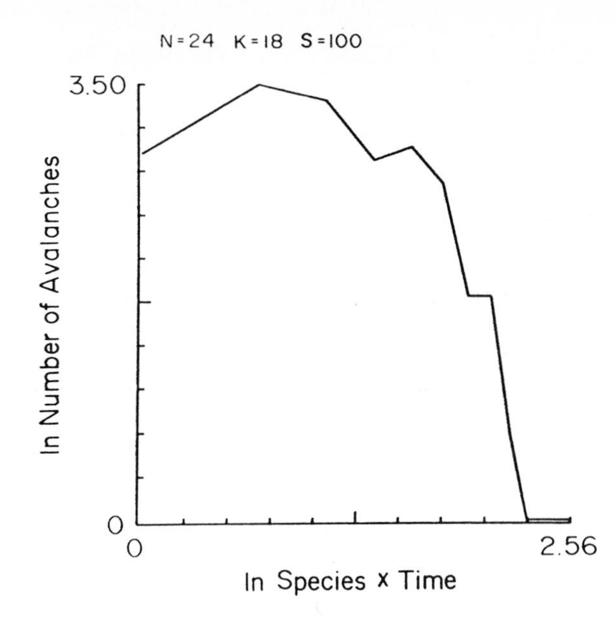

FIGURE 12 (c). As in 12(a), except K is 18.

For larger values of K, the log-log plot is convex, hence not a power law. The same features recur in the 10×10 ecosystems Figures 12(a)-(c) figures show that the log-log plot for $K = 18$ and $K = 14$ are clearly convex, hence not a power law, and may be approaching a power-law linear relationship by $K = 12$.

Similar results, data not shown, arise with respect to the "species" measure of avalanche sizes.

In sum, the following features are of interest.

1. There are more small than large avalanches.
2. As K decreases, the *mean and median size of avalanches increases*, and the *variance increases.*
3. The distribution of avalanche sizes is clearly not a power law when K is sufficiently large, but appear to be approaching a power law on both measures of avalanche size as K decreases towards a "critical" value at which waiting times to encounter Nash equilibria "diverge."
4. The critical value of K corresponds to the K_{opt} value which optimizes sustained fitness in the co-evolving ecosystem. Thus, selective forces are expected to pull the ecosystem to the poised self-organized critical state where avalanches exhibit a power-law distribution.

AVALANCHES AND THE DISTRIBUTION OF EXTINCTION EVENTS IN THE RECORD

These results may bear upon the distribution of extinction events in the evolutionary record. Raup[19] has analyzed the intensity of extinction events at the family level during each of the 79 stages of the entire Phanerozoic. On average, each stage is about 7 million years. Figure 13(a) shows Raup's histogram of the number of extinctions per stage (intensity), graphed against the number of stages exhibiting that intensity. Clearly, there are many more small extinction events than large events. Raup makes the point that the distribution is also clearly continuous. Figure 13(b) replots Raup's data in log-log form. Although the data are obviously too weak to place much weight upon, the log-log plot, 16(b) is clearly convex, suggesting that the observed distribution is not a power law.

Insofar as we wish to take these models seriously, Raup's data look like the $K = 14$ curves for both the 10×10 and 5×5 ecosystems, corresponding to an ecosystem with K large enough that the frozen structure is somewhat firm, rather than extremely tenuous or extremely "solid."

What conclusions are warranted by these results? A first general conclusion is the insight that *co-evolutionary avalanches propagate* through ecosystems, that such avalanches have characteristic frequency versus size distributions which *itself changes* depending upon the parameters of the system. In particular the distribution of avalanche sizes depends upon how solid the "frozen" state is. If we tentatively accept Raup's data as weak evidence, the "frozen state" is modestly firm.

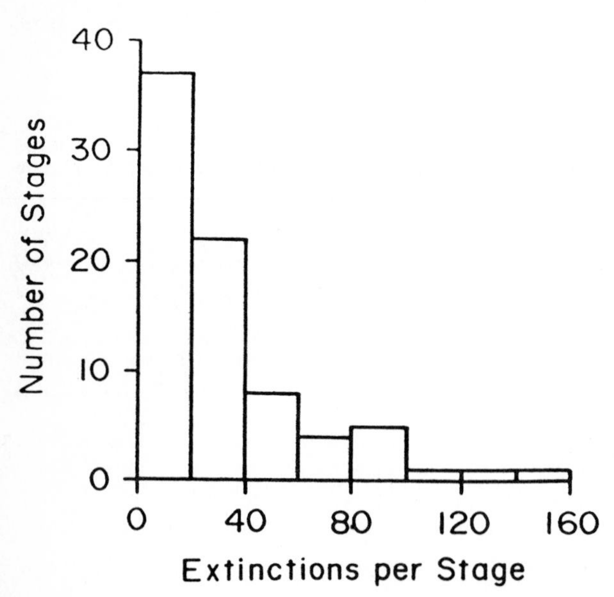

FIGURE 13 (a). Raup's data for the sizes of extinction events versus the number of events at that size. Raup totals the number of extinctions at the family level over each 7 million year interval since the Cambrian.

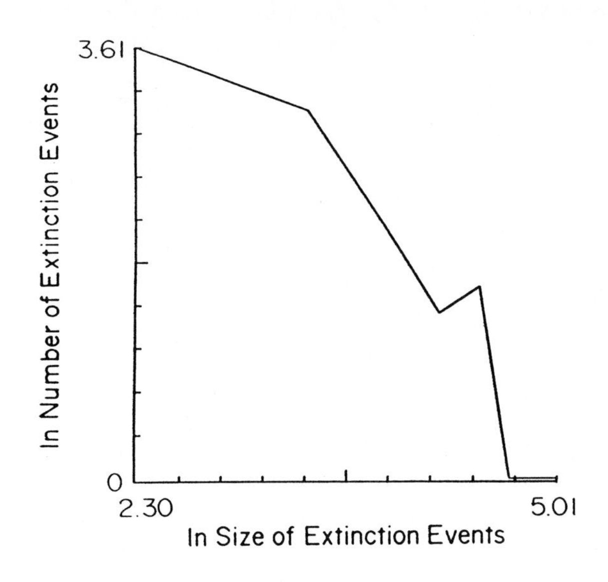

FIGURE 13 (b) Replot of Raup's data with logarithm of the number of extinction events.

A second and critical result is this: *Perturbations of the same initial size can unleash avalanches on a large variety of length scales.* This conclusion is clear and important. In these simulations, the perturbation in each case is a change in the "world" of a single randomly chosen species in the ecosystem. The same "size" perturbation of the external world generates a marked diversity of avalanches, with many small and few large avalanches. If we may tentatively assume that such avalanches can be linked to extinction events, then these results strongly suggest that *uniform* alterations in the "external world" during evolution can cause a diversity of sizes in extinction events. This possibility stands in contrast to the general hypothesis that small and large extinction events must be associated with small and cataclysmic changes in the external world. Since the external environment has almost certainly undergone changes on a variety of scales, we do not wish to assert that high variability in extinction sizes does not, in part, reflect a heterogeneity in intensity of causes. But these results place part of the responsibility for the diversity on the dynamics of coupled ecosystems and how "damage" propagates.

5. INCLUSION OF DENSITY-DEPENDENT POPULATION DYNAMICS

Our analysis above has relied on use of a simplified population dynamics. Each species, at each generation, tries a mutant variant and moves there as a whole if the mutant is better. Each species interacts at each moment with all its immediate neighbors in the ecosystem. The natural extension of this class of models considers a population for each species, subject to population growth characterized by familiar ecological models. Such models are based on "R" and "K" factors, where R reflects the intrinsic growth rate of each population by itself, while K reflects "density effects" due to the carrying capacity of the environment which limits each population's increase to a standing abundance, and includes possible competitive, mutualistic, and predator-prey interactions among the species.[15,18,21]

To generalize our NK co-evolving model, we defined the fitness of each genotype of each species in isolation from all other species to obtain its "R" value, and assessed for each species, "i," the effect of a pairwise interaction with a member of each of the other species "j" connected to it in the model ecosystem. We define "R_i" as the fitness of the species in isolation minus 0.5, the mean fitness in the space of genotypes. This definition allows a species to be an autotroph, $R_i > 0$, or an obligate heterotroph, $R_i < 0$. We define "a_{ij}" as the fitness of species "i" when it interacts with species j, minus 0.5. This definition allows interactions to be mutualistic, $a_{i,j} > 0$, or competitive, $a_{i,j} < 0$. We assumed that each species had the same carrying capacity, K. Then for each species, we utilized the familiar logistic equation:

$$\frac{dX_i}{dt} = X_i \left(R_i - \frac{X_i}{K} + \frac{\sum a_{i,j} X_j}{K} \right) .$$

Here in the absence of interactions by other species, each species will attain the stationary population, $R_i K$. Positive or negative interactions by other species, regarded as mutualism and competition, will, in general, alter the population attained by each species. In studies which allowed species to mutate, we first confirmed that a mutant form initially present at low frequency compared to the "wild type," and which increased in abundance more rapidly than the wild type, continued to do so until the wild-type population was replaced. Having confirmed this under a number of conditions, we substituted a simplier evolutionary dynamics: if a mutant form initially increases in abundance more rapidly in the current co-evolutionary context than the wild type, then the whole population "moved" to the mutant form.

Analysis of these population dynamic models appears to confirm the results based on the simplier dynamics discussed above, but allow us to extend our results to look at the actual dynamics of population abundances as well as changes of genotypes. We analyzed 5 x 5 ecosystems in which each "interior," "edge" or "corner" species interacted with its four-, three-, or two-nearest neighbors, as above, and also 5 × 5 ecosystems in which each "interior" species interacted with its eight-nearest neighbors, adjacent and diagonal, each "edge" species interacted with its five-nearest neighbors, and each "corner species interacted with its three-nearest neighbors. In addition, we examined the dynamics in 5 × 5 ecosystems in which each species interacted with all the other species directly. The major results are similar. The major differences, as the numbers of species connected to each species increases, are that ecosystems tend to be more volatile; hence, easy attainment of Nash equilibria requires higher K values. Table 1 summarizes the data for the case

TABLE 1 Results tabulated for 5×5 ecosystems in which interior species are connected to eight neighbors, edge species are connected to five neighbors, and corner species are connected to three neighbors. K values range from 0 to 16.[1]

		Pop	a_{ij}	R_i	FM	Mut
	0	1593	.009	.506	.158	47.4
	2	2103	.077	.600	.205	38.2
K	4	2264	.138	.657	.210	14.0
	8	2064	.090	.599	.196	10.2
	16	1276	.027	.522	.158	5.3

[1] Pop is mean sustained population density of a species in the ecosystem. $a_{i,j}$ is mean value of this measure of mutualism among the 25 members of the ecosystem. R_i is the mean self-growth rate, autotrophic if positive, of the 25 species in the ecosystem. FM or Fraction mutualistic is the fraction of the couplings into each species which have $a_{i,j}$ values greater than 0; hence, it is the fraction of couplings between species which are mutualistic to some degree. Mut is the number of mutations accepted per species over the simulated co-evolutionary period.

in which the interior species interacts with eight species, with respect to mean population attained, average $a_{i,j}$ values, average R_i values, average fraction of couplings from other species which are "mutualistic," and average number of mutations per species accepted over the run. Note that $K = 4$ yields the optimal sustained population, the highest mean $a_{i,j}$ coupling, the highest mean R_i, the highest fraction of mutualistic couplings, and corresponds to a sharp drop in mutations accepted compared to $K = 2$. The results, as the connectivity of interior species increases from 4 to 8 then to 23, data not shown, are that the optimal value of K increases from 2 to 4 and remains at 4 for the completely connected ecosystem.

Figures 14(a)-(c) examines the onset of frozen components in 5×5 species model ecosystems in which each species interacts with eight, five, or three neighbors, by examining the number of mutations which are "accepted" among all 25 species at each ecosystem generation. In these simulations, $C = 12$, and K is 0, 4, and 16 respectively. Note that when K is small relative to C, $K = 0$, the species in the ecosystem continue to mutate persistently throughout the run. Thus the system never attains a Nash equilibrium. In contrast, when K is much larger than C, $K = 16$, most species stop finding fitter mutants after a few hundred generations; hence, the ecosystem is largely frozen and unchanging. Since we here use the "random" rather than "fitter" or "greedy" dynamics, it is not entirely certain that the systems have attained true Nash equilibria.

Figures 15(a)-(c), as well as Table 1, examine the emergence of *mutualism* in these models. Each species grows on its own, and hence can be thought of as an autotroph. But in addition, each species, in its interaction with other species, might be helped or harmed by that interaction. The *NK* model, via the C couplings, permits the possibility that each species can change its genotype such that it is helped by its co-evolutionary partners. Table 1 shows that mean $a_{i,j}$ values are greater than 0 in all cases. Thus mutualism emerges in these systems. Note also in Table 1 that the $a_{i,j}$ values increase then decrease as K increases. Figures 16(a)-(c) show the details. For $K = 0$ massive fluctuations occur in $a_{i,j}$ values as co-evolutionary changes propagate through the unfrozen ecosystem. The fluctuations keep the $a_{i,j}$ values low. Conversely, for the largest value of K, $K = 16$, ecosystems freeze easily, but the fraction of connections which are mutualistic has fallen, and mean $a_{i,j}$ values have decreased from the optimum at $K = 4$. Mutualistic interactions appear harder to attain when K is large and each system harbors more conflicting constraints. Thus, an intermediate value of K, $K = 4$, appears to optimize the ease of forming strong mutualistic interactions.

These figures also give a dramatic view of the periods of quiescence and bursts of change which propagate through these model ecosystems, and for $K = 16$, give clear evidence of the attainment of a Nash or near-Nash equilibrium.

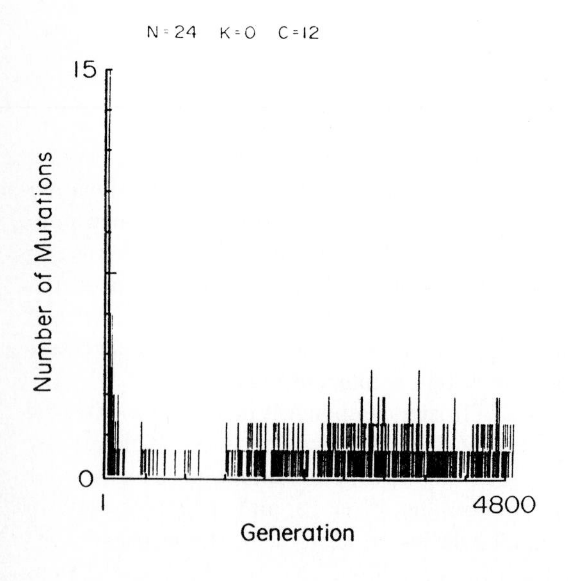

FIGURE 14 Onset of frozen components in 5×5 ecosystem, seen by showing the number of mutations which are "accepted" among all 25 species at each ecosystem generation. In these simulations, $N = 24, C = 12$ and $K = 0$.

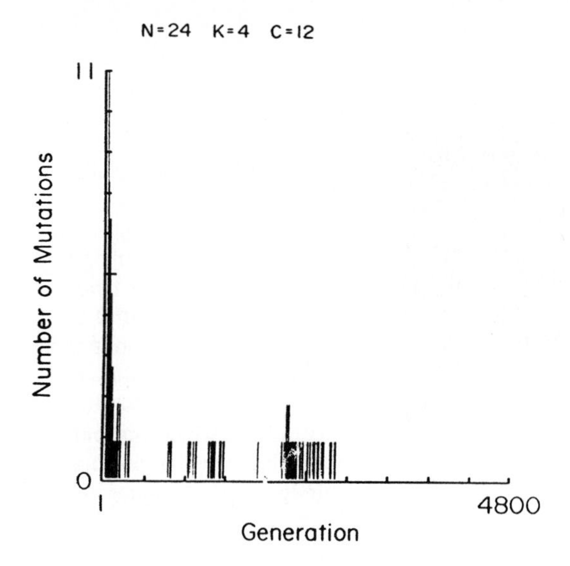

FIGURE 14 (b) $K = 4$

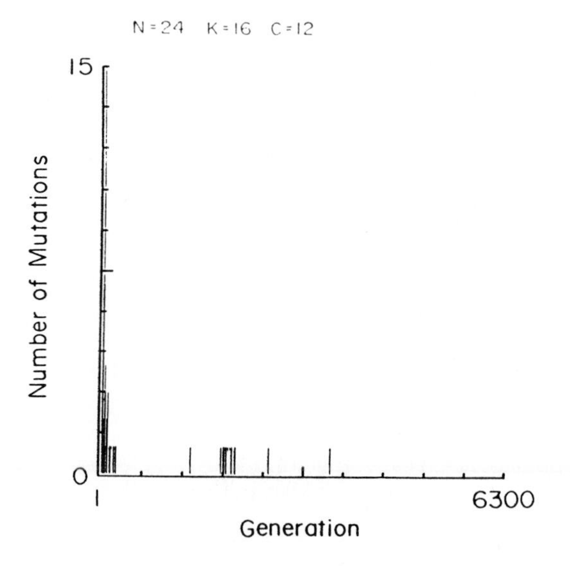

FIGURE 14 (c) $K = 16$.

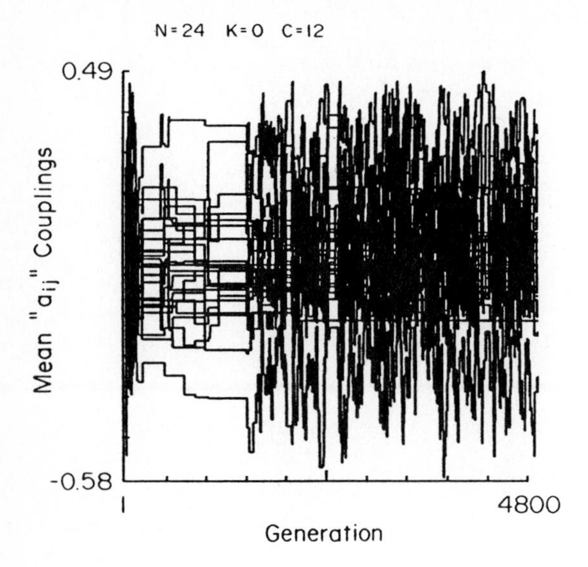

FIGURE 15 Evolution of mutualism in 5×5 ecosystem. Figures plot each species average "$a_{i,j}$" coupling to other species over time. Values greater than 0 are mutualistic; those less than 0 are competitive interactions members of the fraction of their couplings which are mutualistic (positive) or competitive (negative). K values are (a) 0.

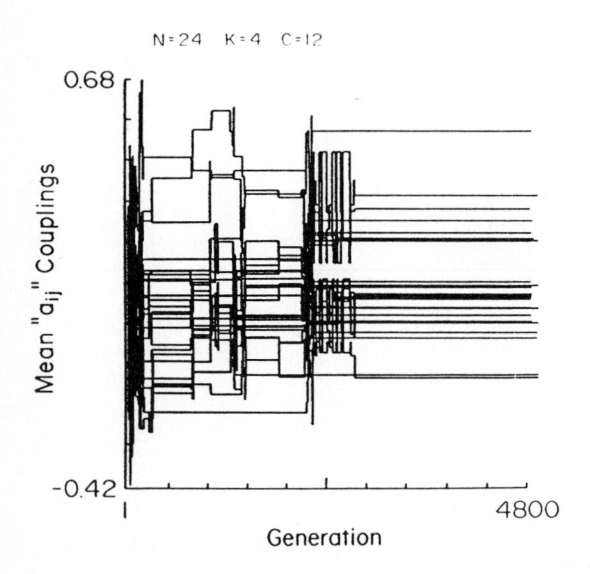

FIGURE 15 K values are (b) 4.

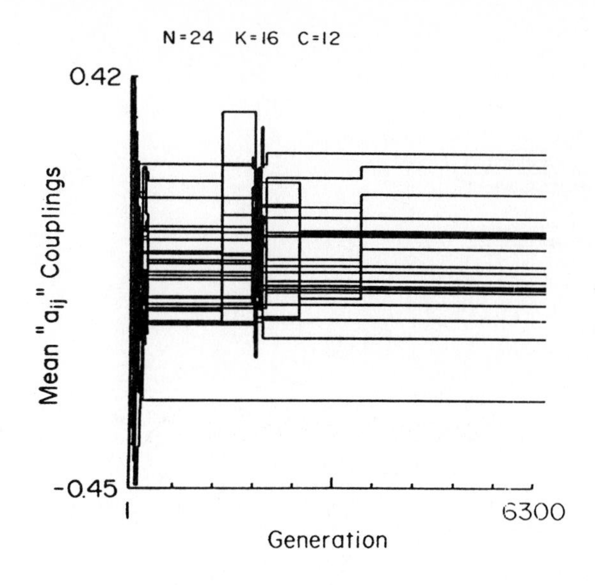

FIGURE 15 K values are (c) 16.

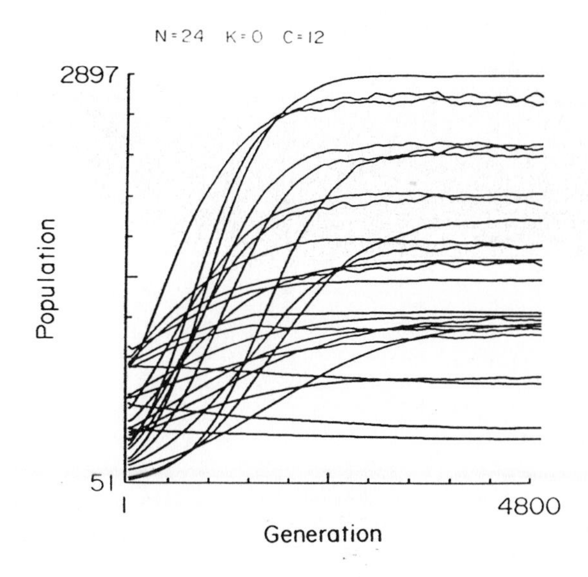

FIGURE 16 Population dynamics of 5×5 ecosystems. K values are (a) 0.

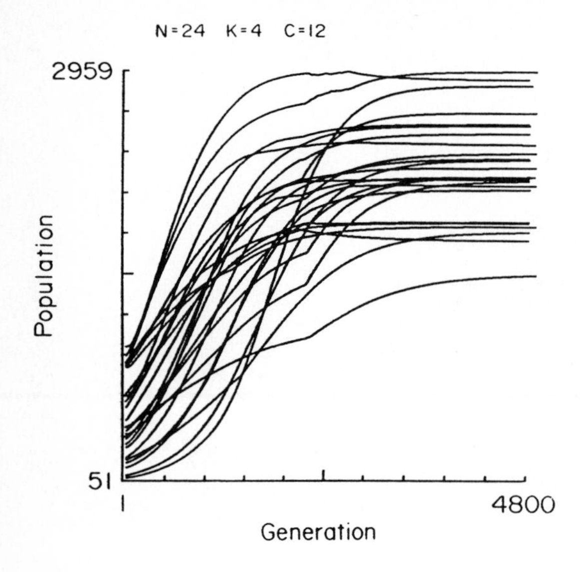

FIGURE 16 K values are (b) 4.

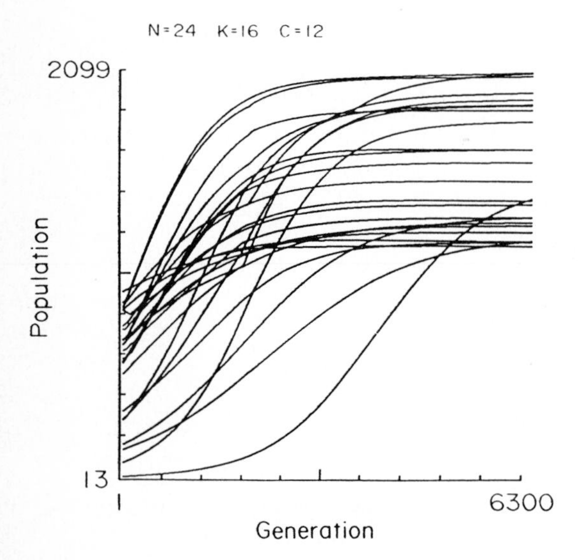

FIGURE 16 K values are (c) 16.

Figures 16(a)-(c) show the actual population dynamics for these conditions. Note that for the highest value of K, $K = 16$, population abundances increase rather smoothly to their carrying capacity. As K decreases, the population behavior becomes more erratic. Genotypic changes show up as discontinuities in the rate of population change. For $K = 0$ it appears that several species actually *decrease* monotonically in abundance during the run. As emphasized in Table 1, for $K = 0$, 2, 4, 8, and 16, the total abundances of all species increases then decreases as K increases, reaching a maximum at $K = 4$. Thus, there is an optimal value of K, as in the simplier dynamics studied above.

INTERMITTENCY AND PUNCTUATED EQUILIBRIUM

Another general feature of these results, whether on extended or fully connected ecosystems, or including populations or not, is intermittency and bursts of change. The most plausible "move" dynamics which we consider is the "random" dynamics, where a random mutation is tried, and the species moves if a fitter variant is found. Under this dynamic, even in the absence of exogenous noise, the time history of a co-evolving set of species shows periods of quiescence where no species finds a fitter variant, and bursts of change when one species finds a fitter variant and unleases an avalanche which propagates through the system. Figures 15(a)-(c) demonstrate this abundantly. It is tempting to associate these quiescent periods with stasis, and the bursts with punctuated change.

DISCUSSION

The central issues we have investigated in this article are the relations between the *structure* of the fitness landscape of one co-evolving partner, and the *couplings* between landscapes such that adaptive moves by one partner deform the landscapes of the adjacent partners, and their effects on the co-evolutionary process. Further, we have focused on the possibility that a *selective metadynamics* acting on individual members of individual species, might alter the character of the co-evolutionary process such that co-evolving partners can co-evolve well. With respect to the first set of issues, it is important to stress that our framework is very general. Co-evolution is certainly some form of flow by each adapting population on a fitness landscape which deforms due to intraspecies and interspecies couplings. Contemporary discussions of co-evolution have developed without yet having a broad theory of the structure of complex fitness landscapes, and, therefore, without having a body of theory about the relationship between the multipeaked structure of fitness landscapes and their deformability and the consequent character of co-evolutionary processes. The *NK* model has been discussed as perhaps the first, but surely not the only, family of rugged multipeaked fitness landscapes. Thus, the *NK* family of landscapes becomes increasingly more rugged as K increases from 0 to N. For $K = 0$, the landscapes have a single Fujiyama peak and smooth sides. For $K = N - 1$, landscapes are fully random. One can conceive of other ways to generate a family of correlated landscapes. For example, one might start with a completely random landscape, and successively "smooth" it by replacing the fitness value at each vertex, or genotype, with the mean of it and its one-mutant neighbors in the space. Such averaging is an analogue of diffusion. The statistics of such a diffusively smoothed family of landscapes may be similar to, or very different from, the *NK* family. In general, it seems a critical question whether there are a few basic families of rugged but correlated fitness landscapes, or very many. Whether few or many exist, our aim must be ultimately to characterize the actual statistical structure of fitness landscapes underlying biological evolution at molecular, morphological, and behavioral levels. Thereafter, we need to understand how richly coupled, hence deformable, fitness landscapes actually are.

A clear implication of these investigations is that the behavior of a co-evolutionary system can be chaotic and disorganized, or highly orderly, depending upon the structures and couplings among the landscapes of the co-evolving partners. In the current model, increasing K increases the ruggedness of individual fitness landscapes. Increasing C increases the couplings between landscapes, hence the deformation to the frog as the fly makes an adaptive move. In completely connected model ecosystems, each species is coupled to all other species. More realistically, each is coupled to only a few other species, S_i. Roughly speaking, if $K > C \times S_i$ for each species, then the ecosystem as a whole reaches a Nash equilibrium rapidly. We referred to this Nash equilibrium, encompassing some or all of the species, as a *frozen component* in the ecosystem. If K is large enough, the ecosystem freezes rapidly and solidly. Conversely, and approximately stated, if $K < C \times S_i$, then the ecosystem does not rapidly reach a Nash equilibrium and behaves chaotically. Thus, K relative to $C \times S_i$ is an *order parameter*. As this ratio changes, the system passes from a chaotic "gas" like behavior, through a phase where frozen components just form, a kind of "liquid" phase, to a well-frozen "solid" phase.

External perturbations due to changes in the abiotic environment, or internal changes under a random, rather than fitter or greedy evolutionary dynamics, can trigger avalanches of co-evolutionary change which propagate through the system. The size distribution of avalanches depends upon how solidly frozen the ecosystem is, and appears to approach a power-law distribution in the "liquid" region of the order parameter's value, when frozen components are just arising in the ecosystem. In the solid region, the distribution of the logarithm of the numbers of avalanches againsts the logarithm of avalanche sizes is convex rather than linear. Since such changes are associated with fluctuations to low fitness, it is a real possibility that such avalanches are associated with extinction events. A central conclusion is that small and large extinction events can in principle issue from perturbations of the same small size. Thus, it may prove possible to predict the distribution of sizes of extinction events in the record. The data from the record are convex in a log-log plot; hence, they tend weakly to support the hypothesis that ecosystems over evolutionary time are somewhat into the solid regime.

Since co-evolutionary dynamics depends upon the ruggedness of landscapes and the coupling of landscapes, it is hardly surprising that the sustained average fitness of co-evolving partners depends upon these factors as well. This obvious point carries the possibility of a metadynamics in which selection favors individuals whose landscapes and landscape couplings are sculpted to abet co-evolutionary success. Regardless of the detailed appropriateness of our *NK* ecosystem model, therefore, we want to stress that evolution may have itself created the kinds of organisms and couplings among them which favor the capacity of species to actually typically succeed in adapting to one another. In turn, this possibility poses the problem of what the "proper" structure of an ecosystem might be to optimize co-evolution. In the context of the *NK* model, we found evidence that sustained fitness is optimized when the ecosystem is in the liquid regime, "poised at the edge of chaos." Further, we found evidence that selection acting on individual members of individual species might pull each towards the poised state by tuning K relative to $C \times S_i$. Thus, at least the hint of a plausible metadynamics exists, and the hint that the edge of chaos might be an attractor of such an evolutionary metadynamics exists. Nevertheless, the supposition that the poised state where frozen components just form is the natural attractor for a co-evolutionary metadynamics must be regarded cautiously. It remains to be seen whether such a state is the one which optimizes sustained fitness even in the context of the *NK* model, let alone models based on other families of fitness landscapes and their couplings. Thus, we emphasize the distinction between the idea that a metadynamics might tune the structure of landscapes and their couplings to enhance sustained average fitness, and the more detailed possibility that the edge of chaos is typically the attractor of such a metadynamics.

SUMMARY

The capacity to co-evolve successfully is not trivial, for mere chaotic twitchings of the angry red queen may occur. We have introduced an approach to this question based on statistical models of rugged fitness landscapes. Co-evolution is thought of, at the lowest level, as a coupling of landscapes, such that adaptive moves by one player deform the landscapes of its immediate partners. In these models we are able to tune the ruggedness of landscapes, how richly intercoupled any two landscapes are, and how many other players interact with each player. We find that all these properties profoundly alter the character of the co-evolutionary dynamics.

The results we have reached suggest some tentative conclusions and avenues of investigation. Landscapes need to be of sufficient ruggedness to offset the couplings between landscapes and the number of partners whose moves impinge upon each landscape. Otherwise viewed, epistatic couplings within a species need to be large enough to counterbalance epistatic couplings to the other co-evolving partners. We have identified possible selective forces which may tune these parameters such that co-evolution is typically successful.

Perhaps most intriguingly, we have found beginning evidence that if each species optimizes its own sustained fitness, such ecosystems might approach a poised self organized critical state, balanced on the edge of chaos. Thus, we must consider a selective "metadynamics" which sculpts the structure of organisms and their couplings, to attain the co-evolution of poised, structured ecosystems which harbor nearly melted frozen components and permit propagation of packets of co-evolutionary change ringing in new species and ringing out old, with a characteristic distribution between cascade size and frequency.

ACKNOWLEDGMENTS

This work was partially funded by N.I.H. GM 40186.

REFERENCES

1. Bak, P, C. Tang, and K. Wiesenfeld. "Self-Organized Criticality." *Phys. Rev. A* **38(1)** (1988): 364–374.
2. Derrida, B. "Random Energy Model: An Exactly Solvable Model of Disordered Systems." *Phys. Rev. B* **24** (1981): 2613.
3. Eigen, M, J. McCaskill, and P. Schuster. "The Molecular Quasispecies." *J. Phys. Chem.* **92** (1988): 6881.
4. Fontana W., and P. Schuster. "A Computer Model of Evolutionary Optimization." *Biophys. Chem.* **26** (1987): 123.
5. Gillespie, J. H. "A Simple Stochastic Gene Subsitution Model." *Theor. Pop. Biol.* **23(2)** (1983): 202–215.
6. Gillespie, J. H. "Molecular Evolution over the Mutational Landscape." *Evol* **38(5)** (1984): 1116–1129.
7. Kauffman, S. A. "Adaptation on Rugged Fitness Landscapes." In *Lectures in the Sciences of Complexity*, edited by D. L. Stein. Santa Fe Institute Studies in the Sciences of Complexity, Lec. Vol. I, 527–618. Redwood City, CA: Addison Wesley, 1989.
8. Kauffman, S. A. "Principles of Adaptation in Complex Systems." In *Lectures in the Sciences of Complexity*, edited by D. L. Stein. Santa Fe Institute Studies in the Sciences of Complexity, Lec. Vol. I, 619–712. Redwood City, CA: Addison Wesley, 1989.
9. Kauffman, S. A. *The Origins of Order: Self Organization and Selection in Evolution*. Oxford University Press. In press.
10. Kauffman, S. A., and S. Levin. "Towards a General Theory of Adaptive Walks on Rugged Landscapes." *J. Theoret. Biol.* **128** (1987): 11–45.
11. Kauffman, S. A., E. W. Weinberger, and A. S. Perelson. "Maturation of the Immune Response via Adaptive Walks on Affinity Landscapes." In *Theoretical Immunology*, edited by A. S. Perelson. Santa Fe Institute Studies in the Science of Complexity, Part 1, 349–382. Redwood City, CA: Addison Wesley, 1989.
12. Kauffman, S. A., and E. W. Weinberger. "The *NK* Model of Rugged Fitness Landscapes And Its Application to Maturation of the Immune Response." *J. Theoret. Biol.* **141** (1989): 211–245.
13. Langton, C. "Life at the Edge of Chaos." This volume.
14. Macken, C. A., and A. S. Perelson. "Protein Evolution on Rugged Landscapes." *Proc. Natl. Acad. Sci.*, 1989.

15. May, R. M. ed. *Theoretical Ecology: Principles and Applications.* Philadelphia, PA: W. B. Saunders, 1976.
16. Maynard-Smith, J., and G. R. Price. "The Logic of Animal Conflict." *Nature* **246** (1973): 15–16.
17. Maynard-Smith, J. *Evolution and the Theory of Games.* Cambridge: Cambridge University Press, 1982.
18. Pimm. S. L. *Food Webs.* London: Chapman and Hall, 1982.
19. Raup, D. M. "Biological Extinction in Earth History." *Science* **231** (1986): 1528–1533.
20. Rosenzweig, M. L., J. S. Brown, and T. L. Vincent. "Red Queens and ESS: The Co-Evolution of Evolutionary Eates." *Evolutionary Ecology* **1** (1987): 59–94.
21. Roughgarden, J. *Theory of Population Genetics and Evolutionary Ecology: An Introduction.* New York: Macmillian,1979.
22. Schuster, P. "Structure and Dynamics of Replication-Mutation Systems." *Physica Scripta* **35** (1987): 402–416.
23. Weinberger, E. W. "Correlated and Uncorrelated Fitness Landscapes and How to Tell the Difference." *Biol. Cybernet.* In press.

Thomas S. Ray
School of Life & Health Sciences, University of Delaware, Newark, Delaware 19716, email: ray@brahms.udel.edu

An Approach to the Synthesis of Life

> Marcel, a mechanical chessplayer... his exquisite 19th-century brainwork—the human art it took to build which has been flat lost, lost as the dodo bird ... But where inside Marcel is the midget Grandmaster, the little Johann Allgeier? Where's the pantograph, and the magnets? Nowhere. Marcel really is a mechanical chessplayer. No fakery inside to give him any touch of humanity at all.
>
> — Thomas Pynchon, *Gravity's Rainbow.*

INTRODUCTION

Ideally, the science of biology should embrace all forms of life. However in practice, it has been restricted to the study of a single instance of life, life on earth. Because biology is based on a sample size of one, we can not know what features of life are peculiar to earth, and what features are general, characteristic of all life. A truly comparative natural biology would require inter-planetary travel, which is light years away. The ideal experimental evolutionary biology would involve creation of multiple planetary systems, some essentially identical, others varying by a parameter of interest, and observing them for billions of years. A practical alternative to an inter-planetary or mythical biology is to create synthetic life in a computer. "Evolution in a bottle" provides a valuable tool for the experimental study of evolution and ecology.

The intent of this work is to synthesize rather than simulate life. This approach starts with hand-crafted organisms already capable of replication and open-ended evolution, and aims to generate increasing diversity and complexity in a parallel to the Cambrian explosion.

To state such a goal leads to semantic problems, because life must be defined in a way that does not restrict it to carbon-based forms. It is unlikely that there could be general agreement on such a definition, or even on the proposition that life need not be carbon based. Therefore, I will simply state my conception of life in its most general sense. I would consider a system to be living if it is self-replicating, and capable of open-ended evolution. Synthetic life should self-replicate, and evolve structures or processes that were not designed-in or preconceived by the creator.[43]

Core Wars programs, computer viruses, and worms[11,14,15,16,17,18,19,46,48] are capable of self-replication, but fortunately, not evolution. It is unlikely that such programs will ever become fully living, because they are not likely to be able to evolve.

Most evolutionary simulations are not open ended. Their potential is limited by the structure of the model, which generally endows each individual with a genome consisting of a set of pre-defined genes, each of which may exist in a pre-defined set of allelic forms.[1,12,13,17,27,42] The object being evolved is generally a data structure representing the genome, which the simulator program mutates and/or recombines, selects, and replicates according to criteria designed into the simulator. The data structures do not contain the mechanism for replication; they are simply copied by the simulator if they survive the selection phase.

Self-replication is critical to synthetic life because without it, the mechanisms of selection must also be pre-determined by the simulator. Such artificial selection can never be as creative as natural selection. The organisms are not free to invent their own fitness functions. Freely evolving creatures will discover means of mutual exploitation and associated implicit fitness functions that we would never think of. Simulations constrained to evolve with pre-defined genes, alleles, and fitness functions are dead ended, not alive.

The approach presented here does not have such constraints. Although the model is limited to the evolution of creatures based on sequences of machine instructions, this may have a potential comparable to evolution based on sequences of organic molecules. Sets of machine instructions similar to those used in the Tierra Simulator have been shown to be capable of "universal computation."[2,33,38] This suggests that evolving machine codes should be able to generate any level of complexity.

Other examples of the synthetic approach to life can be seen in the work of Holland,[28] Farmer et al.,[22] Langton,[31] Rasmussen et al.,[45] and Bagley et al.[3] A characteristic these efforts generally have in common is that they parallel the origin of life event by attempting to create prebiotic conditions from which life may emerge spontaneously and evolve in an open-ended fashion.

While the origin of life is generally recognized as an event of the first order, there is another event in the history of life that is less well known but of comparable significance: the origin of biological diversity and macroscopic multicellular life during the Cambrian explosion 600 million years ago. This event involved a riotous diversification of life forms. Dozens of phyla appeared suddenly, many existing only fleetingly, as diverse and sometimes bizarre ways of life were explored in a relative ecological void.[24,39]

The work presented here aims to parallel the second major event in the history of life, the origin of diversity. Rather than attempting to create prebiotic conditions from which life may emerge, this approach involves engineering over the early history of life to design complex evolvable organisms, and then attempting to create the conditions that will set off a spontaneous evolutionary process of increasing diversity and complexity of organisms. This work represents a first step in this direction, creating an artificial world which may roughly parallel the RNA world of self-replicating molecules (still falling far short of the Cambrian explosion).

The approach has generated rapidly diversifying communities of self-replicating organisms exhibiting open-ended evolution by natural selection. From a single rudimentary ancestral creature containing only the code for self-replication, interactions such as parasitism, —inximmunity, hyper-parasitism, sociality, and cheating have emerged spontaneously. This paper presents a methodology and some first results.

> Here was a world of simplicity and certainty no acidhead, no revolutionary anarchist would ever find, a world based on the one and zero of life and death. Minimal, beautiful. The patterns of lives and deaths.... weightless, invisible chains of electronic presence or absence. If patterns of ones and zeros were "like" patterns of human lives and deaths, if everything about an individual could be represented in a computer record by a long string of ones and zeros, then what kind of creature would be represented by a long string of lives and deaths? It would have to be up one level at least—an angel, a minor god, something in a UFO.
>
> — Thomas Pynchon, *Vineland.*

METHODS

THE METAPHOR

Organic life is viewed as utilizing energy, mostly derived from the sun, to organize matter. By analogy, digital life can be viewed as using CPU (central processing unit) time, to organize memory. Organic life evolves through natural selection as individuals compete for resources (light, food, space, etc.) such that genotypes which leave the most descendants increase in frequency. Digital life evolves through the same process, as replicating algorithms compete for CPU time and memory space, and organisms evolve strategies to exploit one another. CPU time is thought of as the analog of the energy resource, and memory as the analog of the spatial resource.

The memory, the CPU, and the computer's operating system are viewed as elements of the "abiotic" environment. A "creature" is then designed to be specifically adapted to the features of the environment. The creature consists of a self-replicating assembler language program. Assembler languages are merely mnemonics for the machine codes that are directly executed by the CPU. These machine codes have the characteristic that they directly invoke the instruction set of the CPU and services provided by the operating system.

All programs, regardless of the language they are written in, are converted into machine code before they are executed. Machine code is the natural language of the machine, and machine instructions are viewed by this author as the "atomic units" of computing. It is felt that machine instructions provide the most natural basis for an artificial chemistry of creatures designed to live in the computer.

In the biological analogy, the machine instructions are considered to be more like the amino acids than the nucleic acids, because they are "chemically active." They actively manipulate bits, bytes, CPU registers, and the movements of the instruction pointer (as will be discussed later). The digital creatures discussed here are entirely constructed of machine instructions. They are considered analogous to creatures of the RNA world, because the same structures bear the "genetic" information and carry out the "metabolic" activity.

A block of RAM memory (random access memory, also known as "main" or "core" memory) in the computer is designated as a "soup" which can be inoculated with creatures. The "genome" of the creatures consists of the sequence of machine instructions that make up the creature's self-replicating algorithm. The prototype creature consists of 80 machine instructions; thus, the size of the genome of this creature is 80 instructions, and its "genotype" is the specific sequence of those 80 instructions.

THE VIRTUAL COMPUTER—TIERRA SIMULATOR

The computers we use are general purpose computers, which means, among other things, that they are capable of emulating through software the behavior of any other computer that ever has been built or that could be built.[2,33,38] We can utilize this flexibility to design a computer that would be especially hospitable to synthetic life.

There are several good reasons why it is not wise to attempt to synthesize digital organisms that exploit the machine codes and operating systems of real computers. The most urgent is the potential threat of natural evolution of machine codes leading to virus or worm types of programs that could be difficult to eradicate due to their changing "genotypes." This potential argues strongly for creating evolution exclusively in programs that run only on virtual computers and their virtual operating systems. Such programs would be nothing more than data on a real computer, and, therefore, would present no more threat than the data in a data base or the text file of a word processor.

Another reason to avoid developing digital organisms in the machine code of a real computer is that the artificial system would be tied to the hardware and would become obsolete as quickly as the particular machine it was developed on. In contrast, an artificial system developed on a virtual machine could be easily ported to new real machines as they become available.

A third issue, which potentially makes the first two moot, is that the machine languages of real machines are not designed to be evolvable, and in fact might not support significant evolution. Von Neuman-type machine languages are considered to be "brittle," meaning that the ratio of viable programs to possible programs is virtually zero. Any mutation or recombination event in a real machine code is almost certain to produce a non-functional program. The problem of brittleness can be mitigated by designing a virtual computer whose machine code is designed with evolution in mind. Farmer and Belin[23] have suggested that overcoming this brittleness and "discovering how to make such self-replicating patterns more robust so that they evolve to increasingly more complex states is probably the central problem in the study of artificial life."

The work described here takes place on a virtual computer known as Tierra (Spanish for Earth). Tierra is a parallel computer of the MIMD (multiple instruction, multiple data) type, with a processor (CPU) for each creature. Parallelism is imperfectly emulated by allowing each CPU to execute a small time slice in turn. Each CPU of this virtual computer contains two address registers, two numeric registers, a flags register to indicate error conditions, a stack pointer, a ten-word stack, and an instruction pointer. Each virtual CPU is implemented via the C structure listed in Appendix A. Computations performed by the Tierran CPUs are probabilistic due to flaws that occur at a low frequency (see Mutation below).

The instruction set of a CPU typically performs simple arithmetic operations or bit manipulations, within the small set of registers contained in the CPU. Some instructions move data between the registers in the CPU, or between the CPU registers and the RAM (main) memory. Other instructions control the location and movement of an "instruction pointer" (IP). The IP indicates an address in RAM, where the machine code of the executing program (in this case a digital organism) is located.

The CPU perpetually performs a fetch-decode-execute-increment-IP cycle: The machine code instruction currently addressed by the IP is fetched into the CPU, its bit pattern is decoded to determine which instruction it corresponds to, and the instruction is executed. Then the IP is incremented to point sequentially to the next position in RAM, from which the next instruction will be fetched. However, some instructions like JMP, CALL, and RET directly manipulate the IP, causing execution to jump to some other sequence of instructions in the RAM. In the Tierra

Simulator this CPU cycle is implemented through the time-slice routine listed in Appendix B.

THE TIERRAN LANGUAGE

Before attempting to set up an Artificial Life system, careful thought must be given to how the representation of a programming language affects its adaptability in the sense of being robust to genetic operations such as mutation and recombination. The nature of the virtual computer is defined in large part by the instruction set of its machine language. The approach in this study has been to loosen up the machine code in a "virtual bio-computer," in order to create a computational system based on a hybrid between biological and classical von Neumann processes.

In developing this new virtual language, which is called "Tierran," close attention has been paid to the structural and functional properties of the informational system of biological molecules: DNA, RNA, and proteins. Two features have been borrowed from the biological world which are considered to be critical to the evolvability of the Tierran language.

First, the instruction set of the Tierran language has been defined to be of a size that is the same order of magnitude as the genetic code. Information is encoded into DNA through 64 codons, which are translated into 20 amino acids. In its present manifestation, the Tierran language consists of 32 instructions, which can be represented by five bits, *operands included.*

Emphasis is placed on this last point because some instruction sets are deceptively small. Some versions of the redcode language of Core Wars,[15,18,45] for example, are defined to have ten operation codes. It might appear on the surface that the instruction set is of size ten. However, most of the ten instructions have one or two operands. Each operand has four addressing modes, and then an integer. When we consider that these operands are embedded into the machine code, we realize that they are, in fact, a part of the instruction set, and this set works out to be about 10^{11} in size. Inclusion of numeric operands will make any instruction set extremely large in comparison to the genetic code.

In order to make a machine code with a truly small instruction set, we must eliminate numeric operands. This can be accomplished by allowing the CPU registers and the stack to be the only operands of the instructions. When we need to encode an integer for some purpose, we can create it in a numeric register through bit manipulations: flipping the low-order bit and shifting left. The program can contain the proper sequence of bit flipping and shifting instructions to synthesize the desired number, and the instruction set need not include all possible integers.

A second feature that has been borrowed from molecular biology in the design of the Tierran language is the addressing mode, which is called "address by template." In most machine codes, when a piece of data is addressed, or the IP jumps to another piece of code, the exact numeric address of the data or target code is specified in the machine code. Consider that in the biological system by contrast, in order for protein molecule A in the cytoplasm of a cell to interact with protein molecule B, it does not specify the exact coordinates where B is located. Instead, molecule A presents a template on its surface which is complementary to some surface on B. Diffusion brings the two together, and the complementary conformations allow them to interact.

Addressing by template is illustrated by the Tierran JMP instruction. Each JMP instruction is followed by a sequence of NOP (no-operation) instructions, of which there are two kinds: NOP_0 and NOP_1. Suppose we have a piece of code with five instruction in the following order: JMP NOP_0 NOP_0 NOP_0 NOP_1. The system will search outward in both directions from the JMP instruction looking for the nearest occurrence of the complementary pattern: NOP_1 NOP_1 NOP_1 NOP_0. If the pattern is found, the instruction pointer will move to the end of the pattern and resume execution. If the pattern is not found, an error condition (flag) will be set and the JMP instruction will be ignored (in practice, a limit is placed on how far the system may search for the pattern).

The Tierran language is characterized by two unique features: a truly small instruction set without numeric operands, and addressing by template. Otherwise, the language consists of familiar instructions typical of most machine languages, e.g., MOV, CALL, RET, POP, PUSH, etc. The complete instruction set is listed in Appendix B.

THE TIERRAN OPERATING SYSTEM

The Tierran virtual computer needs a virtual operating system that will be hospitable to digital organisms. The operating system will determine the mechanisms of interprocess communication, memory allocation, and the allocation of CPU time among competing processes. Algorithms will evolve so as to exploit these features to their advantage. More than being a mere aspect of the environment, the operating system, together with the instruction set will determine the topology of possible interactions between individuals, such as the ability of pairs of individuals to exhibit predator-prey, parasite-host, or mutualistic relationships.

MEMORY ALLOCATION—CELLULARITY

The Tierran computer operates on a block of RAM of the real computer which is set aside for the purpose. This block of RAM is referred to as the "soup." In most of the work described here the soup consisted of 60,000 bytes, which can hold the same number of Tierran machine instructions. Each "creature" occupies some block of memory in this soup.

Cellularity is one of the fundamental properties of organic life, and can be recognized in the fossil record as far back as 3.6 billion years.[4] The cell is the original individual, with the cell membrane defining its limits and preserving its chemical integrity. An analog to the cell membrane is needed in digital organisms in order to preserve the integrity of the informational structure from being disrupted easily by the activity of other organisms. The need for this can be seen in AL models such as cellular automata where virtual state machines pass through one another,[31,32] or in core-wars-type simulations where coherent structures demolish one another when they come into contact.[15,18,45]

Tierran creatures are considered to be cellular in the sense that they are protected by a "semi-permeable membrane" of memory allocation. The Tierran operating system provides memory allocation services. Each creature has exclusive write privileges within its allocated block of memory. The "size" of a creature is just the size of its allocated block (e.g., 80 instructions). This usually corresponds to the size of the genome. While write privileges are protected, read and execute privileges are not. A creature may examine the code of another creature, and even execute it, but it can not write over it. Each creature may have exclusive write privileges in at most two blocks of memory: the one that it is born with which is referred to as the "mother cell," and a second block which it may obtain through the execution of the MAL (memory allocation) instruction. The second block, referred to as the "daughter cell," may be used to grow or reproduce into.

When Tierran creatures "divide," the mother cell loses write privileges on the space of the daughter cell, but is then free to allocate another block of memory. At the moment of division, the daughter cell is given its own instruction pointer, and is free to allocate its own second block of memory.

TIME SHARING—THE SLICER

The Tierran operating system must be multi-tasking in order for a community of individual creatures to live in the soup simultaneously. The system doles out small slices of CPU time to each creature in the soup in turn. The system maintains a circular queue called the "slicer queue." As each creature is born, a virtual CPU is created for it, and it enters the slicer queue just ahead of its mother, which is the active creature at that time. Thus, the newborn will be the last creature in the soup to get another time slice after the mother, and the mother will get the next slice after its daughter. As long as the slice size is small relative to the generation time of the creatures, the time-sharing system causes the world to approximate parallelism. In actuality, we have a population of virtual CPUs, each of which gets a slice of the real CPU's time as it comes up in the queue.

The number of instructions to be executed in each time slice is set proportional to the size of the genome of the creature being executed, raised to a power. If the "slicer power" is equal to one, then the slicer is size neutral, the probability of an instruction being executed does not depend on the size of the creature in which it occurs. If the power is greater than one, large creatures get more CPU cycles per instruction than small creatures. If the power is less than one, small creatures get more CPU cycles per instruction. The power determines if selection favors large or small creatures, or is size neutral. A constant slice size selects for small creatures.

MORTALITY—THE REAPER

Self-replicating creatures in a fixed-size soup would rapidly fill the soup and lock up the system. To prevent this from occurring, it is necessary to include mortality. The Tierran operating system includes a "reaper" which begins "killing" creatures when the memory fills to some specified level (e.g., 80%). Creatures are killed by deallocating their memory, and removing them from both the reaper and slicer queues. Their "dead" code is not removed from the soup.

In the present system, the reaper uses a linear queue. When a creature is born, it enters the bottom of the queue. The reaper always kills the creature at the top of the queue. However, individuals may move up or down in the reaper queue according to their success or failure at executing certain instructions. When a creature executes an instruction that generates an error condition, it moves one position up the queue, as long as the individual ahead of it in the queue has not accumulated a greater number of errors. Two of the instructions are somewhat difficult to execute without generating an error, therefore successful execution of these instructions moves the creature down the reaper queue one position, as long as it has not accumulated more errors than the creature below it.

The effect of the reaper queue is to cause algorithms which are fundamentally flawed to rise to the top of the queue and die. Vigorous algorithms have a greater longevity, but in general, the probability of death increases with age.

MUTATION

In order for evolution to occur, there must be some change in the genome of the creatures. This may occur within the lifespan of an individual, or there may be errors in passing along the genome to offspring. In order to insure that there is genetic change, the operating system randomly flips bits in the soup, and the instructions of the Tierran language are imperfectly executed.

Mutations occur in two circumstances. At some background rate, bits are randomly selected from the entire soup (60,000 instructions totaling 300,000 bits) and flipped. This is analogous to mutations caused by cosmic rays, and has the effect of preventing any creature from being immortal, as it will eventually mutate to death. The background mutation rate has generally been set at about 1 bit flipped for every 10,000 Tierran instructions executed by the system.

In addition, while copying instructions during the replication of creatures, bits are randomly flipped at some rate in the copies. The copy mutation rate is the higher of the two, and results in replication errors. The copy mutation rate has generally been set at about 1 bit flipped for every 1,000 to 2,500 instructions moved. In both classes of mutation, the interval between mutations varies randomly within a certain range to avoid possible periodic effects.

In addition to mutations, the execution of Tierran instructions is flawed at a low rate. For most of the 32 instructions, the result is off by ± 1 at some low frequency. For example, the increment instruction normally adds one to its register, but it sometimes adds two or zero. The bit-flipping instruction normally flips the low-order bit, but it sometimes flips the next higher bit or no bit. The shift-left instruction normally shifts all bits one bit to the left, but it sometimes shifts left by two bits, or not at all. In this way, the behavior of the Tierran instructions is probabilistic, not fully deterministic.

It turns out that bit-flipping mutations and flaws in instructions are not necessary to generate genetic change and evolution, once the community reaches a certain state of complexity. Genetic parasites evolve which are sloppy replicators, and have the effect of moving pieces of code around between creatures, causing rather massive rearrangements of the genomes. The mechanism of this ad hoc sexuality has not been worked out, but is likely due to the parasites' inability to discriminate between live, dead, or embryonic code.

Mutations result in the appearance of new genotypes, which are watched by an automated genebank manager. In one implementation of the manager, when new genotypes replicate twice, producing a genetically identical offspring at least once, they are given a unique name and saved to disk. Each genotype name contains two parts, a number, and a three-letter code. The number represents the number of instructions in the genome. The three-letter code is used as a base 26 numbering system for assigning a unique label to each genotype in a size class. The first genotype to appear in a size class is assigned the label aaa, the second is assigned the label aab, and so on. Thus the ancestor is named 80aaa, and the first mutant of size 80 is named 80aab. The first parasite of size 45 is named 45aaa.

The genebanker saves some additional information with each genome: the genotype name of its immediate ancestor which makes possible the reconstruction of the entire phylogeny; the time and date of origin; "metabolic" data including the number of instructions executed in the first and second reproduction, the number of errors generated in the first and second reproduction, and the number of instructions copied into the daughter cell in the first and second reproductions (see Appendix C); some environmental parameters at the time of origin including the search limit for addressing, and the slicer power, both of which affect selection for size.

THE TIERRAN ANCESTOR

The Tierran language has been used to write a single self-replicating program which is 80 instructions long. This program is referred to as the "ancestor," or alternatively as genotype 0080aaa (Figure 1). The ancestor is a minimal self-replicating algorithm which was originally written for use during the debugging of the simulator. No functionality was designed into the ancestor beyond the ability to self-replicate, nor was any specific evolutionary potential designed in. The commented Tierran assembler and machine code for this program is presented in Appendix C.

The ancestor examines itself to determine where in memory it begins and ends. The ancestor's beginning is marked with the four no-operation template: 1 1 1 1, and its ending is marked with 1 1 1 0. The ancestor locates its beginning with the five instructions: ADRB, NOP_0, NOP_0, NOP_0, and NOP_0. This series of

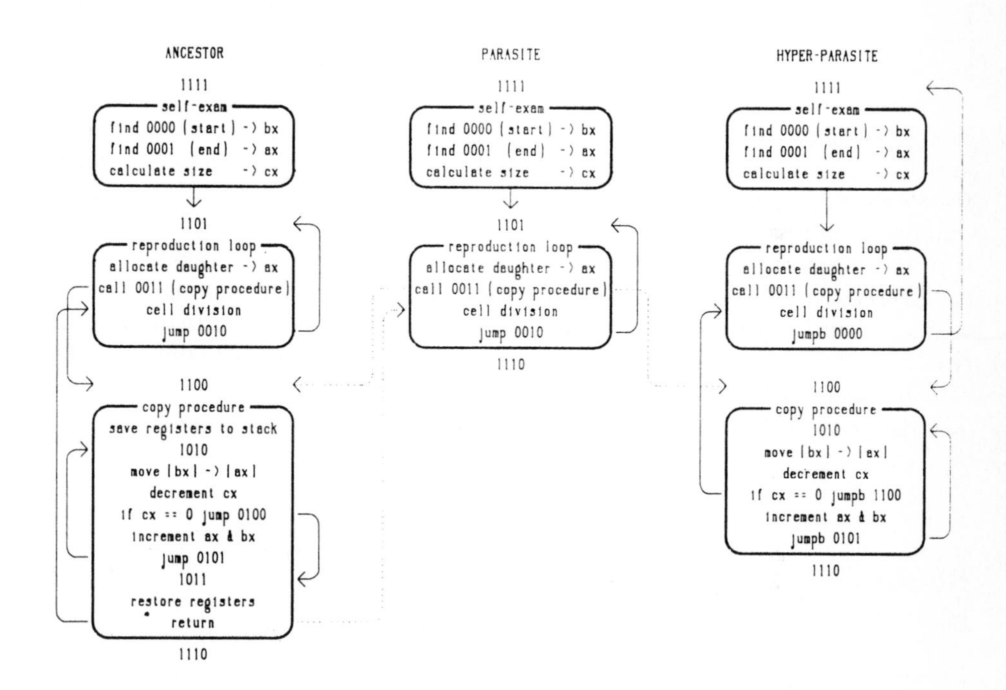

FIGURE 1 Metabolic flow chart for the ancestor, parasite, hyper-parasite, and their interactions: ax, bx and cx refer to CPU registers where location and size information are stored. [ax] and [bx] refer to locations in the soup indicated by the values in the ax and bx registers. Patterns such as 1101 are complementary templates used for addressing. Arrows outside of boxes indicate jumps in the flow of execution of the programs. The dotted-line arrows indicate flow of execution between creatures. The parasite lacks the copy procedure; however, if it is within the search limit of the copy procedure of a host, it can locate, call, and execute that procedure, thereby obtaining the information needed to complete its replication. The host is not adversely affected by this informational parasitism, except through competition with the parasite, which is a superior competitor. Note that the parasite calls the copy procedure of its host with the expectation that control will return to the parasite when the copy procedure returns. However, the hyper-parasite jumps out of the copy procedure rather than returning, thereby seizing control from the parasite. It then proceeds to reset the CPU registers of the parasite with the location and size of the hyper-parasite, causing the parasite to replicate the hyper-parasite genome thereafter.

instructions causes the system to search backwards from the ADRB instruction for a template complementary to the four NOP_0 instructions, and to place the address of the complementary template (the beginning) in the ax register of the CPU (see Appendix A). A similar method is used to locate the end.

Having determined the address of its beginning and its end, it subtracts the two to calculate its size, and allocates a block of memory of this size for a daughter cell. It then calls the copy procedure which copies the entire genome into the daughter-cell memory, one instruction at a time. The beginning of the copy procedure is marked by the four no-operation template: 1 1 0 0. Therefore, the call to the copy procedure is accomplished with the five instructions: CALL, NOP_0, NOP_0, NOP_1, and NOP_1.

When the genome has been copied, it executes the DIVIDE instruction, which causes the creature to lose write privileges on the daughter-cell memory, and gives an instruction pointer to the daughter cell (it also enters the daughter cell into the slicer and reaper queues). After this first replication, the mother cell does not examine itself again; it proceeds directly to the allocation of another daughter cell, then the copy procedure is followed by cell division, in an endless loop.

Fourty-eight of the 80 instructions in the ancestor are no-operations. Groups of four no-operation instructions are used as complementary templates to mark twelve sites for internal addressing, so that the creature can locate its beginning and end, call the copy procedure, and mark addresses for loops and jumps in the code, etc. The functions of these templates are commented in the listing in Appendix C.

RESULTS

GENERAL BEHAVIOR OF THE SYSTEM

Evolutionary runs of the simulator are begun by inoculating the soup of 60,000 instructions with a single individual of the 80 instruction ancestral genotype. The passage of time in a run is measured in terms of how many Tierran instructions have been executed by the simulator. Most software development work has been carried out on a Toshiba 5200/100 laptop computer with an 80386 processor and an 80387 math co-processor operating at 20 Mhz. This machine executes over 12 million Tierran instructions per hour. Long evolutionary runs are conducted on mini and mainframe computers which execute about 1 million Tierran instructions per minute.

The original ancestral cell which inoculates the soup executes 839 instructions in its first replication, and 813 for each additional replication. The initial cell and its replicating daughters rapidly fill the soup memory to the threshold level of 80% which starts the reaper. Typically, the system executes about 400,000 instructions in filling up the soup with about 375 individuals of size 80 (and their gestating daughter cells). Once the reaper begins, the memory remains roughly 80% filled with creatures for the remainder of the run.

Once the soup is full, individuals are initially short lived, generally reproducing only once before dying; thus, individuals turn over very rapidly. More slowly, there appear new genotypes of size 80, and then new size classes. There are changes in the genetic composition of each size class, as new mutants appear, some of which increase significantly in frequency, sometimes replacing the original genotype. The size classes which dominate the community also change through time, as new size classes appear (see below), some of which competitively exclude sizes present earlier. Once the community becomes diverse, there is a greater variance in the longevity and fecundity of individuals.

In addition to an increase in the raw diversity of genotypes and genome sizes, there is an increase in the ecological diversity. Obligate commensal parasites evolve, which are not capable of self-replication in isolated culture, but which can replicate when cultured with normal (self-replicating) creatures. These parasites execute some parts of the code of their hosts, but cause them no direct harm, except as competitors. Some potential hosts have evolved immunity to the parasites, and some parasites have evolved to circumvent this immunity.

In addition, facultative hyper-parasites have evolved, which can self-replicate in isolated culture, but when subjected to parasitism, subvert the parasites energy metabolism to augment their own reproduction. Hyper-parasites drive parasites to extinction, resulting in complete domination of the communities. The relatively high degrees of genetic relatedness within the hyper-parasite-dominated communities leads to the evolution of sociality in the sense of creatures that can only replicate when they occur in aggregations. These social aggregations are then invaded by hyper-hyper-parasite cheaters.

Mutations and the ensuing replication errors lead to an increasing diversity of sizes and genotypes of self-replicating creatures in the soup. Within the first 100 million instructions of elapsed time, the soup evolves to a state in which about a dozen more-or-less persistent size classes coexist. The relative abundances and specific list of the size classes varies over time. Each size class consists of a number of distinct genotypes which also vary over time.

EVOLUTION

MICRO-EVOLUTION

If there were no mutations at the outset of the run, there would be no evolution. However, the bits flipped as a result of copy errors or background mutations result in creatures whose list of 80 instructions (genotype) differs from the ancestor, usually by a single bit difference in a single instruction.

Mutations, in and of themselves, cannot result in a change in the size of a creature, they can only alter the instructions in its genome. However, by altering the genotype, mutations may affect the process whereby the creature examines itself and calculates its size, potentially causing it to produce an offspring that differs in size from itself.

Four out of the five possible mutations in a no-operation instruction convert it into another kind of instruction, while one out of five converts it into the complementary no-operation. Therefore, 80% of mutations in templates destroy the template, while one in five alters the template pattern. An altered template may cause the creature to make mistakes in self-examination, procedure calls, or looping or jumps of the instruction pointer, all of which use templates for addressing.

PARASITES An example of the kind of error that can result from a mutation in a template is a mutation of the low-order bit of instruction 42 of the ancestor (Appendix C). Instruction 42 is a NOP_0, the third component of the copy procedure template. A mutation in the low-order bit would convert it into NOP_1, thus changing the template from 1 1 0 0 to: 1 1 1 0. This would then be recognized as the template used to mark the end of the creature, rather than the copy procedure.

A creature born with a mutation in the low-order bit of instruction 42 would calculate its size as 45. It would allocate a daughter cell of size 45 and copy only instructions 0 through 44 into the daughter cell. The daughter cell then, would not include the copy procedure. This daughter genotype, consisting of 45 instructions, is named 0045aaa.

Genotype 0045aaa (Figure 1) is not able to self-replicate in isolated culture. However, the semi-permeable membrane of memory allocation only protects write privileges. Creatures may match templates with code in the allocated memory of other creatures, and may even execute that code. Therefore, if creature 0045aaa is grown in mixed culture with 0080aaa, when it attempts to call the copy procedure, it will not find the template within its own genome, but if it is within the search limit (generally set at 200–400 instructions) of the copy procedure of a creature of genotype 0080aaa, it will match templates, and send its instruction pointer to the copy code of 0080aaa. Thus a parasitic relationship is established (see ECOLOGY below). Typically, parasites begin to emerge within the first few million instructions of elapsed time in a run.

IMMUNITY TO PARASITES At least some of the size 79 genotypes demonstrate some measure of resistance to parasites. If genotype 45aaa is introduced into a soup, flanked on each side with one individual of genotype 0079aab, 0045aaa will initially reproduce somewhat, but will be quickly eliminated from the soup. When the same experiment is conducted with 0045aaa and the ancestor, they enter a stable cycle in which both genotypes coexist indefinitely. Freely evolving systems have been observed to become dominated by size 79 genotypes for long periods, during which parasitic genotypes repeatedly appear, but fail to invade.

CIRCUMVENTION OF IMMUNITY TO PARASITES Occasionally these evolving systems dominated by size 79 were successfully invaded by parasites of size 51. When the immune genotype 0079aab was tested with 0051aao (a direct, one-step descendant of 0045aaa in which instruction 39 is replaced by an insertion of seven instructions of unknown origin), they were found to enter a stable cycle. Evidently 0051aao has evolved some way to circumvent the immunity to parasites possessed by 0079aab. The 14 genotypes 0051aaa through 0051aan were also tested with 0079aab, and none were able to invade.

HYPER-PARASITES Hyper-parasites have been discovered, (e.g., 0080gai, which differs by 19 instructions from the ancestor, Figure 1). Their ability to subvert the energy metabolism of parasites is based on two changes. The copy procedure does not return, but jumps back directly to the proper address of the reproduction loop. In this way it effectively seizes the instruction pointer from the parasite. However it is another change which delivers the coup de grâce: after each reproduction, the hyper-parasite re-examines itself, resetting the bx register with its location and the cx register with its size. After the instruction pointer of the parasite passes through this code, the CPU of the parasite contains the location and size of the hyper-parasite and the parasite thereafter replicates the hyper-parasite genome.

SOCIAL HYPER-PARASITES Hyper-parasites drive the parasites to extinction. This results in a community with a relatively high level of genetic uniformity, and therefore high genetic relationship between individuals in the community. These are the conditions that support the evolution of sociality, and social hyper-parasites soon dominate the community. Social hyper-parasites (Figure 2) appear in the 61 instruction size class. For example, 0061acg is social in the sense that it can only self-replicate when it occurs in aggregations. When it jumps back to the code for self-examination, it jumps to a template that occurs at the end rather than the beginning of its genome. If the creature is flanked by a similar genome, the jump will find the target template in the tail of the neighbor, and execution will then pass into the beginning of the active creature's genome. The algorithm will fail unless a similar genome occurs just before the active creature in memory. Neighboring creatures cooperate by catching and passing on jumps of the instruction pointer.

It appears that the selection pressure for the evolution of sociality is that it facilitates size reduction. The social species are 24% smaller than the ancestor. They have achieved this size reduction in part by shrinking their templates from four instructions to three instructions. This means that there are only eight templates available to them, and catching each others jumps allows them to deal with some of the consequences of this limitation as well as to make dual use of some templates.

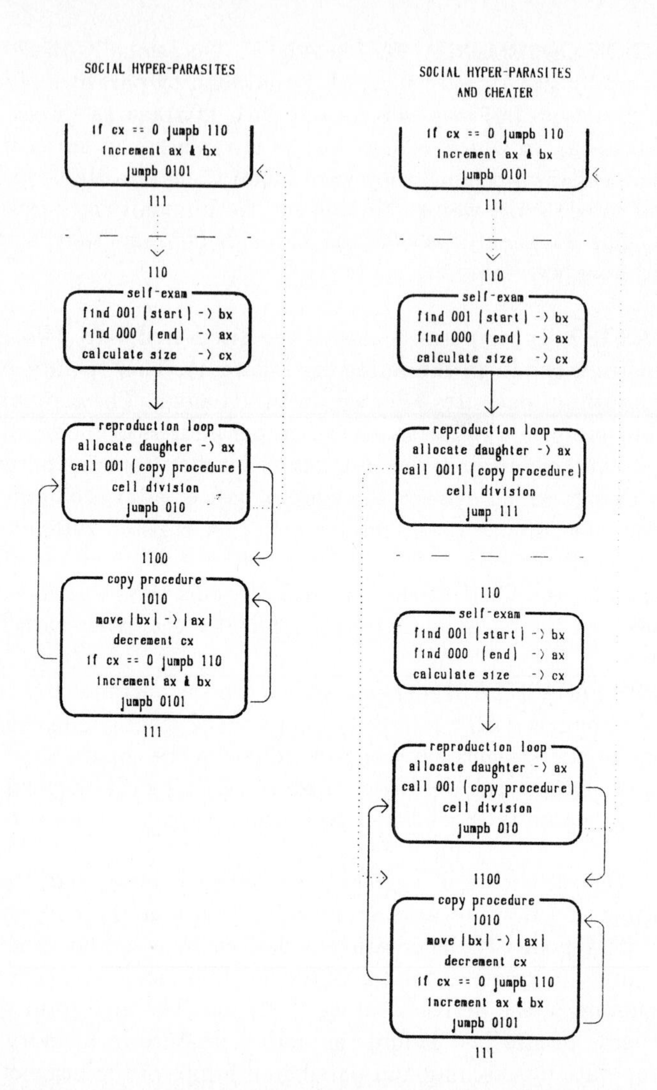

FIGURE 2 Metabolic flow chart for social hyper-parasites, their associated hyper-hyper-parasite cheaters, and their interactions. Symbols are as described for Figure 1. Horizontal dashed lines indicate the boundaries between individual creatures. On both the left and right, above the dashed line at the top of the figure is the lowermost fragment of a social hyper-parasite. Note (on the left) that neighboring social hyper-parasites cooperate in returning the flow of execution to the beginning of the creature for self-re-examination. Execution jumps back to the end of the creature above, but then falls off the end of the creature without executing any instructions of consequence, and enters the top of the creature below. On the right, a cheater is inserted between the two social-hyper-parasites. The cheater captures control of execution when it passes between the social individuals. It sets the CPU registers with its own location and size, and then skips over the self-examination step when it returns control of execution to the social creature below.

CHEATERS: HYPER-HYPER-PARASITES The cooperative social system of hyper-parasites is subject to cheating, and is eventually invaded by hyper-hyper-parasites (Figure 2). These cheaters (e.g., 0027aab) position themselves between aggregating hyper-parasites so that when the instruction pointer is passed between them, they capture it.

A NOVEL SELF-EXAMINATION All creatures discussed thus far mark their beginning and end with templates. They then locate the addresses of the two templates and determine their genome size by subtracting them. In one run, creatures evolved without a template marking their end. These creatures located the address of the template marking their beginning, and then the address of a template in the middle of their genome. These two addresses were then subtracted to calculate half of their size, and this value was multiplied by two (by shifting left) to calculate their full size.

MACRO-EVOLUTION

When the simulator is run over long periods of time, hundreds of millions or billions of instructions, various patterns emerge. Under selection for small sizes, there is a proliferation of small parasites and a rather interesting ecology (see below). Selection for large creatures has usually lead to continuous incrementally increasing sizes (but not to a trivial concatenation of creatures end-to-end) until a plateau in the upper hundreds is reached. In one run, selection for large size lead to apparently open-ended size increase, evolving genomes larger than 23,000 instructions in length. This evolutionary pattern might be described as phyletic gradualism.

The most thoroughly studied case for long runs is where selection, as determined by the slicer function, is size neutral. The longest runs to date (as much as 2.86 billion Tierran instructions) have been in a size-neutral environment, with a search limit of 10,000, which would allow large creatures to evolve if there were some algorithmic advantage to be gained from larger size. These long runs illustrate a pattern which could be described as periods of stasis punctuated by periods of rapid evolutionary change, which appears to parallel the pattern of punctuated equilibrium described by Eldredge and Gould[21] and Gould and Eldredge.[25]

Initially these communities are dominated by creatures with genome sizes in the 80s. This represents a period of relative stasis, which has lasted from 178 million to 1.44 billion instructions in the several long runs conducted to date. The systems then very abruptly (in a span of 1 or 2 million instructions) evolve into communities dominated by sizes ranging from about 400 to about 800. These communities have not yet been seen to evolve into communities dominated by either smaller or substantially larger size ranges.

TABLE 1 Table of numbers of size classes in the genebank. Left column is size class, right column is number of self-replicating genotypes of that size class. 305 sizes, 29,275 genotypes.

0034	1	0092	362	0150	2	0205	5	0418	1	5213	2
0041	2	0093	261	0151	1	0207	3	0442	10	5229	4
0043	12	0094	241	0152	2	0208	2	0443	1	5254	1
0044	7	0095	211	0153	1	0209	1	0444	61	5888	36
0045	191	0096	232	0154	2	0210	9	0445	1	5988	1
0046	7	0097	173	0155	3	0211	4	0456	2	6006	2
0047	5	0098	92	0156	77	0212	4	0465	6	6014	1
0048	4	0099	117	0157	270	0213	5	0472	6	6330	1
0049	8	0100	77	0158	938	0214	47	0483	1	6529	1
0050	13	0101	62	0159	836	0218	1	0484	8	6640	1
0051	2	0102	62	0160	3229	0219	1	0485	3	6901	5
0052	11	0103	27	0161	1417	0220	2	0486	9	6971	1
0053	4	0104	25	0162	174	0223	3	0487	2	7158	2
0054	2	0105	28	0163	187	0226	2	0493	2	7293	3
0055	2	0106	19	0164	46	0227	7	0511	2	7331	1
0056	4	0107	3	0165	183	0231	1	0513	1	7422	70
0057	1	0108	8	0166	81	0232	1	0519	1	7458	1
0058	8	0109	2	0167	71	0236	1	0522	6	7460	7
0059	8	0110	8	0168	9	0238	1	0553	1	7488	1
0060	3	0111	71	0169	15	0240	3	0568	6	7598	1
0061	1	0112	19	0170	99	0241	1	0578	1	7627	63
0062	2	0113	10	0171	40	0242	1	0581	3	7695	1
0063	2	0114	3	0172	44	0250	1	0582	1	7733	1

0064	1	0115	3	0173	34	0251	1	0600	1	7768	2
0065	4	0116	5	0174	15	0260	2	0683	1	7860	25
0066	1	0117	3	0175	22	0261	1	0689	1	7912	1
0067	1	0118	1	0176	137	0265	2	0757	6	8082	3
0068	2	0119	3	0177	13	0268	1	0804	2	8340	1
0069	1	0120	2	0178	3	0269	1	0813	1	8366	1
0070	7	0121	60	0179	1	0284	16	0881	6	8405	5
0071	5	0122	9	0180	16	0306	1	0888	1	8406	2
0072	17	0123	3	0181	5	0312	1	0940	2	8649	2
0073	2	0124	11	0182	27	0314	1	1006	6	8750	1
0074	80	0125	6	0184	3	0316	2	1016	1	8951	1
0075	56	0126	11	0185	21	0318	3	1077	5	8978	3
0076	21	0127	1	0186	9	0319	2	1116	1	9011	3
0077	28	0130	3	0187	3	0320	23	1186	1	9507	3
0078	409	0131	2	0188	11	0321	5	1294	7	9564	3
0079	850	0132	5	0190	20	0322	21	1322	7	9612	1
0080	7399	0133	2	0192	12	0330	1	1335	1	9968	1
0081	590	0134	7	0193	4	0342	5	1365	11	10259	31
0082	384	0135	1	0194	4	0343	1	1631	1	10676	1
0083	886	0136	1	0195	11	0351	1	1645	3	11366	5
0084	1672	0137	1	0196	19	0352	3	2266	1	11900	1
0085	1531	0138	1	0197	2	0386	1	2615	2	12212	2
0086	901	0139	2	0198	3	0388	2	2617	9	15717	3
0087	944	0141	6	0199	35	0401	3	2671	7	16355	1
0088	517	0143	1	0200	1	0407	1	3069	3	17356	3
0089	449	0144	4	0201	84	0411	22	4241	1	18532	1
0090	543	0146	1	0203	1	0412	3	5101	15	23134	14
0091	354	0149	1	0204	1	0416	1	5157	9		

The communities of creatures in the 400 to 800 size range also show a long-term pattern of punctuated equilibrium. These communities regularly come to be dominated by one or two size classes, and remain in that condition for long periods of time. However, they inevitably break out of that stasis and enter a period where no size class dominates. These periods of rapid evolutionary change may be very chaotic. Close observations indicate that at least at some of these times, no genotypes breed true. Many self-replicating genotypes will coexist in the soup at these times, but at the most chaotic times, none will produce offspring which are even their same size. Eventually the system will settle down to another period of stasis dominated by one or a few size classes which breed true.

Two communities have been observed to die after long periods. In one community, a chaotic period led to a situation where only a few replicating creatures were left in the soup, and these were producing sterile offspring. When these last replicating creatures died (presumably from an accumulation of mutations), the community was dead. In these runs, the mutation rate was not lowered during the run, while the average genome size increased by an order of magnitude until it approached the average mutation rate. Both communities died shortly after the dominant size class moved from the 400 range to the 700 to 1400 range. Under these circumstances it is probably difficult for any genome to breed true, and the genomes may simply have "melted." Another community died abruptly when the mutation rate was raised to a high level.

DIVERSITY

Most observations on the diversity of Tierran creatures have been based on the diversity of size classes. Creatures of different sizes are clearly genetically different, as their genomes are of different sizes. Different sized creatures would have some difficulty engaging in recombination if they were sexual; thus, it is likely that they would be different species. In a run of 526 million instructions, 366 size classes were generated, 93 of which achieved abundances of five or more individuals. In a run of 2.56 billion instructions, 1180 size classes were generated, 367 of which achieved abundances of five or more.

Each size class consists of a number of distinct genotypes which also vary over time. There exists the potential for great genetic diversity within a size class. There are 32^{80} distinct genotypes of size 80, but how many of those are viable self-replicating creatures? This question remains unanswered; however, some information has been gathered through the use of the automated genebank manager.

In several days of running the genebanker, over 29,000 self-replicating genotypes of over 300 size classes accumulated. The size classes and the number of unique genotypes banked for each size are listed in Table 1. The genotypes saved to disk can be used to inoculate new soups individually, or collections of these banked genotypes may be used to assemble "ecological communities." In "ecological" runs, the mutation rates can be set to zero in order to inhibit evolution.

ECOLOGY

The only communities whose ecology has been explored in detail are those that operate under selection for small sizes. These communities generally include a large number of parasites, which do not have functional copy procedures, and which execute the copy procedures of other creatures within the search limit. In exploring ecological interactions, the mutation rate is set at zero, which effectively throws the simulation into ecological time by stopping evolution. When parasites are present, it is also necessary to stipulate that creatures must breed true, since parasites have a tendency to scramble genomes, leading to evolution in the absence of mutation.

0045aaa is a "metabolic parasite." Its genome does not include the copy procedure; however, it executes the copy procedure code of a normal host, such as the ancestor. In an environment favoring small creatures, 0045aaa has a competitive advantage over the ancestor; however, the relationship is density dependent. When the hosts become scarce, most of the parasites are not within the search limit of a copy procedure, and are not able to reproduce. Their calls to the copy procedure fail and generate errors, causing them to rise to the top of the reaper queue and die. When the parasites die off, the host population rebounds. Hosts and parasites cultured together demonstrate Lotka-Volterra population cycling.[34,53,54]

A number of experiments have been conducted to explore the factors affecting diversity of size classes in these communities. Competitive exclusion trials were conducted with a series of self-replicating (non-parasitic) genotypes of different size classes. The experimental soups were initially inoculated with one individual of each size. A genotype of size 79 was tested against a genotype of size 80, and then against successively larger size classes. The interactions were observed by plotting the population of the size 79 class on the x axis, and the population of the other size class on the y axis. Sizes 79 and 80 were found to be competitively matched such that neither was eliminated from the soup. They quickly entered a stable cycle, which exactly repeated a small orbit. The same general pattern was found in the interaction between sizes 79 and 81.

When size 79 was tested against size 82, they initially entered a stable cycle, but after about 4 million instructions, they shook out of stability and the trajectory became chaotic with an attractor that was symmetric about the diagonal (neither size showed any advantage). This pattern was repeated for the next several size classes, until size 90, where a marked asymmetry of the chaotic attractor was evident, favoring size 79. The run of size 79 against size 93 showed a brief stable period of about a million instructions, which then moved to a chaotic phase without an attractor, which spiraled slowly down until size 93 became extinct, after an elapsed time of about 6 million instructions.

An interesting exception to this pattern was the interaction between size 79 and size 89. Size 89 is considered to be a "metabolic cripple," because although it is capable of self-replicating, it executes about 40% more instructions to replicate than normal. It was eliminated in competition with size 79, with no loops in the trajectory, after an elapsed time of under 1 million instructions.

In an experiment to determine the effects of the presence of parasites on community diversity, a community consisting of 20 size classes of hosts was created and allowed to run for 30 million instructions, at which time only the eight smallest size classes remained. The same community was then regenerated, but a single genotype (0045aaa) of parasite was also introduced. After 30 million instructions, 16 size classes remained, including the parasite. This seems to be an example of a "keystone" parasite effect.[41]

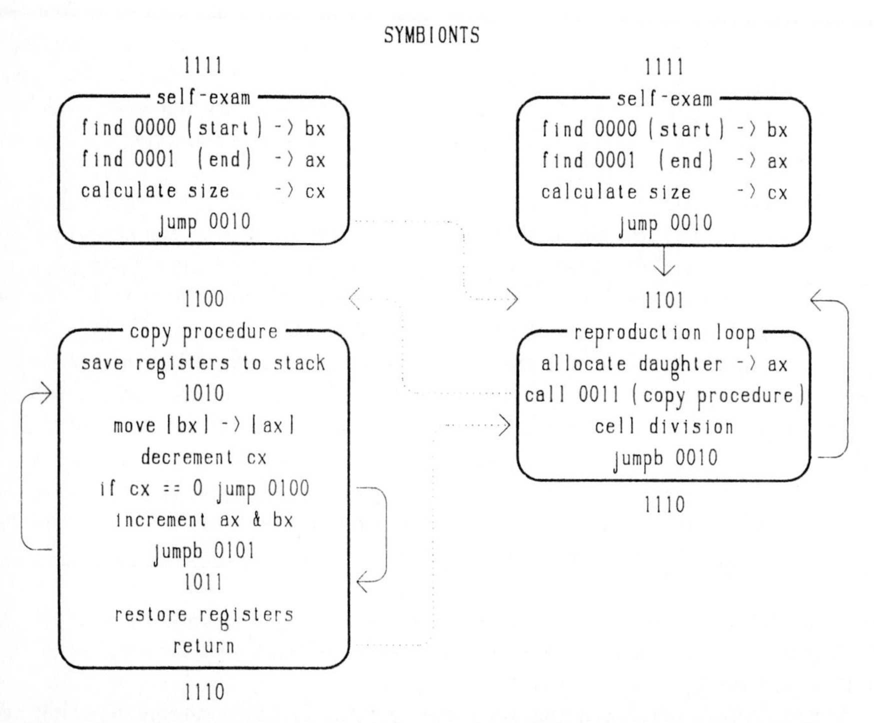

FIGURE 3 Metabolic flow chart for obligate symbionts and their interactions. Symbols are as described for Figure 1. Neither creature is able to self-replicate in isolation. However, when cultured together, each is able to replicate by using information provided by the other.

Symbiotic relationships are also possible. The ancestor was manually dissected into two creatures, one of size 46 which contained only the code for self-examination and the copy loop, and one of size 64 which contained only the code for self-examination and the copy procedure (Figure 3). Neither could replicate when cultured alone, but when cultured together, they both replicated, forming a stable mutualistic relationship. It is not known if such relationships have evolved spontaneously.

DISCUSSION

The "physical" environment presented by the simulator is quite simple, consisting of the energy resource (CPU time) doled out rather uniformly by the time slicer, and memory space which is completely uniform and always available. In light of the nature of the physical environment, the implicit fitness function would presumably favor the evolution of creatures which are able to replicate with less CPU time, and this does, in fact, occur. However, much of the evolution in the system consists of the creatures discovering ways to exploit one another. The creatures invent their own fitness functions through adaptation to their biotic environment.

Parasites do not contain the complete code for self-replication; thus, they utilize other creatures for the information contained in their genomes. Hyper-parasites exploit parasites in order to increase the amount of CPU time devoted to the replication of their own genomes; thus, hyper-parasites utilize other creatures for the energy resources that they possess. These ecological interactions are not programmed into the system, but emerge spontaneously as the creatures discover each other and invent their own games.

Evolutionary theory suggests that adaptation to the biotic environment (other organisms) rather than to the physical environment is the primary force driving the auto-catalytic diversification of organisms.[49] It is encouraging to discover that the process has already begun in the Tierran world. It is worth noting that the results presented here are based on evolution of the first creature that I designed, written in the first instruction set that I designed. Comparison to the creatures that have evolved shows that the one I designed is not a particularly clever one. Also, the instruction set that the creatures are based on is certainly not very powerful (apart from those special features incorporated to enhance its evolvability). It would appear then that it is rather easy to create life. Evidently, virtual life is out there, waiting for us to provide environments in which it may evolve.

EMERGENCE

Cariani[8] has suggested a methodology by which emergence can be detected. His analysis is described as "emergence-relative-to-a-model," where "the model... constitutes the observer's expectations of how the system will behave in the future." If the system evolves such that the model no longer describes the system, we have emergence.

Cariani recognizes three types of emergence, in semiotic terms: syntactic, semantic, and pragmatic. Syntactic operations are those of computation (symbolic). Semantic operations are those of measurement (e.g., sense perception) and control (e.g., effectors), because they "determine the relation of the symbols in the computational part of the device to the world at large." Pragmatic ("intentional") operations are those that are "performance-measuring," and, hence "the criteria which control the selection."

Cariani has developed this analysis in the context of robotics, and considers that the semantic operations should act at the interface between the symbolic (computational) and the nonsymbolic (real physical world). I can not apply his analysis in precisely this way to my simulation, because there is no connection between the Tierran world and the real physical world. I have created a virtual universe that is fully self-contained, within the computer; thus, I must apply his analysis in this context.

In the Tierran world, symbolic operations (syntactic), computations, take place in the CPU. The "nonsymbolic," "real physical world" is the soup (RAM) where the creatures reside. The measurement (semantic) operations are those that involve the location of templates; the effector operations are the copying of instructions within the soup, and the allocation of memory (cells). Fitness functions (pragmatic) are implicit, and are determined by the creatures themselves because they must effect their own replication.

Any program which is self-modifying can show syntactic emergence. As long as the organization of the executable code changes, we have syntactic emergence. This occurs in the Tierran world, as the executable genetic code of the creatures evolves.

Semantic emergence is more difficult to achieve, as it requires the appearance of some new meaning in the system. This is found in the Tierran world in the evolution of templates and their meanings. When a creature locates a template, which has a physical manifestation in the "real world" of the soup, the location of the template appears in the CPU in the form of a symbol representing its address in the soup. For example, the beginning and end of the ancestor are each marked by templates. That one "means" beginning and the other "means" end is apparent from the computation made on the symbols for them in the CPU: the two are subtracted to calculate the size of the creature, and copying of the genome starts at the beginning address. Through evolution, a class of creatures appeared which did not locate a template at their end, but rather one in their center. That the new template "means" center to these creatures is again apparent from the computations made on its associated symbol in the CPU: the beginning address is subtracted from the center address, the difference is then multiplied by two to calculate the size.

Pragmatic emergence is considered "higher" by Cariani, and certainly it is the most difficult to achieve, because it requires that the system evolve new fitness functions. In living systems, fitness functions always reduce to: genotypes which leave a greater number of their genes in future generations will increase in frequency relative to other genotypes and thus have a higher fitness. This is a nearly tautological observation, but tautology is avoided in that the fitness landscape is shaped by specific adaptations that facilitate passing genes on.

For a precambrian marine algae living before the appearance of herbivores, the fitness landscape consists, in part, of a multi-dimensional space of metabolic parameter affecting the efficiency of the conversion of sun light into useable energy, and the use of that energy in obtaining nutrients and converting them into new cells. Regions of this metabolic phase space that yield a greater efficiency at these operations also have higher associated fitnesses.

In order for pragmatic emergence to occur, the fitness landscape must be expanded to include new realms. For example, if a variant genotype of algae engulfs other algae, and thereby achieves a new mechanism of obtaining energy, the fitness landscape expands to include the parameters of structure and metabolism that facilitate the location, capture, and digestion of other cells. The fitness landscapes of algae lacking these adaptations also become altered, as they now include the parameters of mechanisms to avoid being eaten. Pragmatic emergence occurs through the acquisition of a new class of adaptation for enhancing passing genes on.

Pragmatic emergence occurs in the Tierran world as creatures which initially do not interact, discover means to exploit one another, and in response, means to avoid exploitation. The original fitness landscape of the ancestor consists only of the efficiency parameters of the replication algorithm, in the context of the properties of the reaper and slicer queues. When by chance, genotypes appear that exploit other creatures, selection acts to perfect the mechanisms of exploitation, and mechanisms of defense to that exploitation. The original fitness landscape was based only on adaptations of the organism to its physical environment (the slicer and reaper). The new fitness landscape retains those features, but adds to it adaptations to the biotic environment, the other creatures. Because the fitness landscape includes an ever-increasing realm of adaptations to other creatures which are themselves evolving, it can facilitate an auto-catalytic increase in complexity and diversity of organisms.

In any computer model of evolution, the fitness functions are determined by the entity responsible for the replication of individuals. In genetic algorithms and most simulations, that entity is the simulator program; thus, the fitness function is defined globally. In the Tierran world, that entity is the creatures themselves; thus, the fitness function is defined locally by each creature in relation to its environment (which includes the other creatures). It is for this reason that pragmatic emergence occurs in the Tierran world.

In Tierra, the fitness functions are determined by the creatures themselves, and evolve with the creatures. As Cariani states, "Such devices would not be useful for accomplishing *our* purposes as their evaluatory criteria might well diverge from our own over time." This was the case from the outset in the Tierran world, because the simulator never imposed any explicit selection on the creatures. They were not expected to solve my problems, other than satisfying my passion to create life.

After describing how to recognize the various types of emergence, Cariani concludes that Artificial Life cannot demonstrate emergence because of the fully deterministic and replicable nature of computer simulations. This conclusion does not follow in any obvious way from the preceding discussions and does not seem to be supported. Furthermore, I have never known "indeterminate" and "unreplicable" to be considered as necessary qualities of life.

As a thought experiment, suppose that we connect a Geiger counter near a radioactive source to our computer, and use the interval between clicks to determine the values in our random number generator. The resulting behavior of the simulation would no longer be deterministic or repeatable. However, the results would be the same, in any significant respect, to those obtained by using an algorithm to select the random numbers. Determinism and repeatability are irrelevant to emergence and to life. In fact, repeatability is a highly desirable quality of synthetic life because it facilitates study of life's properties.

SYNTHETIC BIOLOGY

One of the most uncanny of evolutionary phenomena is the ecological convergence of biota living on different continents or in different epochs. When a lineage of organisms undergoes an adaptive radiation (diversification), it leads to an array of relatively stable ecological forms. The specific ecological forms are often recognizable from lineage to lineage. For example, among dinosaurs, the *Pterosaur, Triceratops, Tyrannosaurus*, and *Ichthyosaur* are ecological parallels, respectively, to the bat, rhinoceros, lion, and porpoise of modern mammals. Similarly, among modern placental mammals, the gray wolf, flying squirrel, great anteater, and common mole are ecological parallels, respectively, to the Tasmanian wolf, honey glider, banded anteater, and marsupial mole of the marsupial mammals of Australia.

Given these evidently powerful convergent forces, it should perhaps not be surprising that as adaptive radiations proceed among digital organisms, we encounter recognizable ecological forms, in spite of the fundamentally distinct physics and chemistry on which they are based. Ideally, comparisons should be made among organisms of comparable complexity. It may not be appropriate to compare viruses to mammals. Unfortunately, the organic creatures most comparable to digital organisms, the RNA creatures, are no longer with us. Since digital organisms are being compared to modern organic creatures of much greater complexity, ecological comparisons must be made in the broadest of terms.

In describing the results, I have characterized classes of organisms such as hosts, parasites, hyper-parasites, social, and cheaters. While these terms apply nicely to digital organisms, it can be tricky to examine the parallels between digital and organic organisms in detail. The parasites of this study cause no direct harm to their host; however, they do compete with them for space. This is rather like a vine which depends on a tree for support, but which does not directly harm the tree, except that the two must compete for light. The hyper-parasites of this study are facultative and subvert the energy metabolism of their parasite victims without killing them. I cannot think of an organic example that has all of these properties. The carnivorous plant comes close in that it does not need the prey to survive, and in that its prey may have approached the plant expecting to feed on it. However, the prey of carnivorous plants are killed outright.

We are not in a position to make the most appropriate comparison, between digital creatures and RNA creatures. However, we can apply what we have learned from digital organisms, about the evolutionary properties of creatures at that level of complexity, to our speculations about what the RNA world may have been like. For example, once an RNA molecule fully capable of self-replication evolved, might other RNA molecules lacking that capability have parasitized its replicatory function?

In studying the natural history of synthetic organisms, it is important to recognize that they have a distinct biology due to their non-organic nature. In order to fully appreciate their biology, one must understand the stuff of which they are made. To study the biology of creatures of the RNA world would require an understanding of organic chemistry and the properties of macro-molecules. To understand the biology of digital organisms requires a knowledge of the properties of machine instructions and machine language algorithms. However, to fully understand digital organisms, one must also have a knowledge of biological evolution and ecology. Evolution and ecology are the domain of biologists and machine languages are the domain of computer scientists. The knowledge chasm between biology and computer science is likely to hinder progress in the field of Artificial Life for some time. We need more individuals with a depth of knowledge in both areas in order to carry out the work.

Trained biologists will tend to view synthetic life in the same terms that they have come to know organic life. Having been trained as an ecologist and evolutionist, I have seen in my synthetic communities, many of the ecological and evolutionary properties that are well known from natural communities. Biologists trained in other specialties will likely observe other familiar properties. It seems that what we see is what we know. It is likely to take longer before we appreciate the unique properties of these new life forms.

ARTIFICIAL LIFE AND BIOLOGICAL THEORY

The relationship between Artificial Life and biological theory is two-fold: (1) Given that one of the main objectives of AL is to produce evolution leading to spontaneously increasing diversity and complexity, there exists a rich body of biological theory that suggests factors that may contribute to that process; and (2) to the extent that the underlying life processes are the same in AL and organic life, AL models provide a new tool for experimental study of those processes, which can be used to test biological theory that can not be tested by traditional experimental and analytic techniques.[47]

Furthermore, there exists a complementary relationship between biological theory and the synthesis of life. Theory suggests how the synthesis can be achieved, while application of the theory in the synthesis is a test of the theory. If theory suggests that a certain factor will contribute to increasing diversity, then synthetic systems can be run with and without that factor. The process of synthesis becomes a test of the theory.

At the molecular level, there has been much discussion of the role of transposable elements in evolution. It has been observed that most of the genome in eukaryotes (perhaps 90%) originated from transposable elements, while in prokaryotes, only a very small percentage of the genome originated through transposons.[20,40,51] It can also be noted that the eukaryotes, not the prokaryotes, were involved in the Cambrian explosion of diversity.[4] It has been suggested that transposable elements play a significant role in facilitating evolution.[26,30,50] These observations suggest that it would be an interesting experiment to introduce transposable elements into digital organisms.

The Cambrian explosion consisted of the origin, proliferation, and diversification of macroscopic multi-cellular organisms. The origin and elaboration of multicellularity was an integral component of the process. Buss[7] provides a provocative discussion of the evolution of multi-cellularity, and explores the consequences of selection at the level of cell lines. From his discussion the following idea emerges (although he does not explicitly state this idea, in fact, he proposes a sort of inverse of this idea, p. 65): the transition from single to multi-celled existence involves the extension of the control of gene regulation by the mother cell to successively more generations of daughter cells. This is a concept which transcends the physical basis of life, and could be profitably applied to synthetic life in order to generate an analog of multi-cellularity.

The Red Queen hypothesis[52] suggests that in the face of a changing environment, organisms must evolve as fast as they can in order to simply maintain their current state of adaptation. "In order to get anywhere you must run twice as fast as that."[9] A critical component of the environment for any organism is the other living organisms with which it must interact. Given that the species that comprise the environment are themselves evolving, the pace is set by the maximal rate that any species may change through evolution, and it becomes very difficult to actually get ahead. A maximal rate of evolution is required just to keep from falling behind. This suggests that interactions with other evolving species provide the primary driving force in evolution.

Much evolutionary theory deals with the role of biotic interactions in driving evolution. For example, it is thought that these are of primary importance in the maintenance of sex.[5,10,36,37] Stanley[49] has suggested that the Cambrian explosion was sparked by the appearance of the first organisms that ate other organisms. These new herbivores enhanced diversity by preventing any single species of algae from dominating and competitively excluding others. These kinds of biotic interactions must be incorporated into synthetic life in order to move evolution.

Similarly, many abiotic factors are known to contribute to determining the diversity of ecological communities. Island biogeography theory considers how the size, shape, distribution, fragmentation, and heterogeneity of habitats contribute to community diversity.[35] Various types of disturbance are also believed to significantly affect diversity.[29,44] All of these factors may be introduced into synthetic life in an effort to enhance the diversification of the evolving systems.

The examples just listed are a few of the many theories that suggest factors that influence biological diversity. In the process of synthesizing increasingly complex instances of life, we can incorporate and manipulate the states of these factors. These manipulations, conducted for the purposes of advancing the synthesis, will also constitute powerful tests of the theories.

EXTENDING THE MODEL

The approach to AL advocated in this work involves engineering over the first 3 billion years of life's history to design complex evolvable artificial organisms, and attempting to create the biological conditions that will set off a spontaneous evolutionary process of increasing diversity and complexity of organisms. This is a very difficult undertaking, because in the midst of the Cambrian explosion, life had evolved to a level of complexity in which emergent properties existed at many hierarchical levels: molecular, cellular, organismal, populational, and community.

In order to define an approach to the synthesis of life paralleling this historical stage of organic life, we must examine each of the fundamental hierarchical levels, abstract the principal biological properties from their physical representation, and determine how they can be represented in our artificial media. The simulator program determines not only the physics and chemistry of the virtual universe that it creates, but the community ecology as well. We must tinker with the structure of the simulator program in order to facilitate the existence of the appropriate "molecular," "cellular," and "ecological" interactions to generate a spontaneously increasing diversity and complexity.

The evolutionary potential of the present model can be greatly extended by some modifications. In its present implementation, parasitic relationships evolve rapidly, but predation involving the direct usurpation of space occupied by cells is not possible. This could be facilitated by the introduction of a FREE (memory deallocation) instruction. However, it is unlikely that such predatory behavior would be selected for because in the current system there is always free memory space available; thus, there would be little to be gained through seizing space from another creature. However, predation could be selected for by removing the reaper from the system.

Perhaps a more interesting way to favor predatory-type interactions would be to make instructions expensive. In the present implementation, there is no "conservation of instructions," because the MOV_IAB instruction creates a new copy of the instruction being moved during self-replication. If the MOV_IAB instruction were modified such that it obeyed a law of conservation, and left behind all zeros when it moved an instruction, then instructions would not be so cheap. Creatures could be allowed to synthesize instructions through a series of bit flipping and shifting operations, which would make instructions "metabolically" costly. Under such circumstance, a soup of "autotrophs" which synthesize all of their instructions could be invaded by a predatory creature which kills other creatures to obtain instructions.

Additional richness could be introduced to the model by modifying the way that CPU time is allocated. Rather than using a circular queue, creatures could deploy special arrays of instructions or bit patterns (analogous to chlorophyll) which capture potential CPU time packets raining like photons onto the soup. In addition, with instructions being synthesized through bit flipping and shifting operations, each instruction could be considered to have a "potential time" (i.e., potential energy) value which is proportional to its content of one bits. Instructions rich in ones could be used as time (energy) storage "molecules" which could be metabolized when needed by converting the one bits to zeros to release the stored CPU time. The introduction of such an "informational metabolism" would open the way for all sorts of evolution involving the exploitation of one organism by another.

Separation of the genotype from the phenotype would allow the model to move beyond the parallel to the RNA world into a parallel of the DNA-RNA-protein stage of evolution. Storage of the genetic information in relatively passive informational structures, which are then translated into the "metabolically active" machine instructions would facilitate evolution of development, sexuality, and transposons. These features would contribute greatly to the evolutionary potential of the model.

These enhancements of the model represent the current directions of my continuing efforts in this area, in addition to using the existing model to further test ecological and evolutionary theory.

ACKNOWLEDGMENT

I thank Dan Chester, Robert Eisenberg, Doyne Farmer, Walter Fontana, Stephanie Forrest, Chris Langton, Stephen Pope, and Steen Rasmussen, for their discussions or readings of the manuscripts. Contribution No. 142 from the Ecology Program, School of Life and Health Sciences, University of Delaware.

APPENDIX A

Structure definition to implement the Tierra virtual CPU. The source code or executables for the Tierra Simulator can be obtained by contacting the author by mail (emial or snail mail).

```
struct cpu {  /* structure for registers of virtual cpu */
    int   ax;  /* address register */
    int   bx;  /* address register */
    int   cx;  /* numerical register */
    int   dx;  /* numerical register */
    char  fl;  /* flag */
    char  sp;  /* stack pointer */
    int   st[10];  /* stack */
    int   ip;  /* instruction pointer */
    } ;
```

APPENDIX B

Abbreviated code for implementing the CPU cycle of the Tierra Simulator.

```
void main(void)
{   get_soup();
    life();
    write_soup();
}

void life(void) /* doles out time slices and death */
{   while(inst_exec_c < alive)  /* control the length of the run */
    {   time_slice(this_slice); /* this_slice is current cell in queue */
        incr_slice_queue(); /* increment this_slice to next cell in queue */
        while(free_mem_current < free_mem_prop * soup_size)
            reaper(); /* if memory is full to threshold, reap some cells */
    }
}

void time_slice(int  ci)
{   Pcells  ce; /* pointer to the array of cell structures */
    char    i;  /* instruction from soup */
    int     di; /* decoded instruction */
    int     j, size_slice;
    ce = cells + ci;
    for(j = 0; j < size_slice; j++)
    {   i = fetch(ce->c.ip); /* fetch instruction from soup, at address ip */
        di = decode(i);      /* decode the fetched instruction */
        execute(di, ci);     /* execute the decoded instruction */
        increment_ip(di,ce); /* move instruction pointer to next instruction */
        system_work(); /* opportunity to extract information */
    }
}

void execute(int  di, int  ci)
{   switch(di)
    {   case 0x00: nop_0(ci);    break; /* no operation */
        case 0x01: nop_1(ci);    break; /* no operation */
        case 0x02: or1(ci);      break; /* flip low order bit of cx, cx ^= 1 */
        case 0x03: shl(ci);      break; /* shift left cx register, cx <<= 1 */
        case 0x04: zero(ci);     break; /* set cx register to zero, cx = 0 */
        case 0x05: if_cz(ci);    break; /* if cx==0 execute next instruction */
        case 0x06: sub_ab(ci);   break; /* subtract bx from ax, cx = ax - bx */
        case 0x07: sub_ac(ci);   break; /* subtract cx from ax, ax = ax - cx */
        case 0x08: inc_a(ci);    break; /* increment ax, ax = ax + 1 */
        case 0x09: inc_b(ci);    break; /* increment bx, bx = bx + 1 */
        case 0x0a: dec_c(ci);    break; /* decrement cx, cx = cx - 1 */
        case 0x0b: inc_c(ci);    break; /* increment cx, cx = cx + 1 */
        case 0x0c: push_ax(ci);  break; /* push ax on stack */
        case 0x0d: push_bx(ci);  break; /* push bx on stack */
        case 0x0e: push_cx(ci);  break; /* push cx on stack */
        case 0x0f: push_dx(ci);  break; /* push dx on stack */
```

```
        case 0x10: pop_ax(ci);   break; /* pop top of stack into ax */
        case 0x11: pop_bx(ci);   break; /* pop top of stack into bx */
        case 0x12: pop_cx(ci);   break; /* pop top of stack into cx */
        case 0x13: pop_dx(ci);   break; /* pop top of stack into dx */
        case 0x14: jmp(ci);      break; /* move ip to template */
        case 0x15: jmpb(ci);     break; /* move ip backward to template */
        case 0x16: call(ci);     break; /* call a procedure */
        case 0x17: ret(ci);      break; /* return from a procedure */
        case 0x18: mov_cd(ci);   break; /* move cx to dx, dx = cx */
        case 0x19: mov_ab(ci);   break; /* move ax to bx, bx = ax */
        case 0x1a: mov_iab(ci);  break; /* move instruction at address in bx
                                           to address in ax */
        case 0x1b: adr(ci);      break; /* address of nearest template to ax */
        case 0x1c: adrb(ci);     break; /* search backward for template */
        case 0x1d: adrf(ci);     break; /* search forward for template */
        case 0x1e: mal(ci);      break; /* allocate memory for daughter cell */
        case 0x1f: divide(ci);   break; /* cell division */
    }
    inst_exec_c++;
}
```

APPENDIX C

Assembler source code for the ancestral creature.

```
genotype: 80 aaa  origin: 1-1-1990  00:00:00:00  ancestor
parent genotype: human
1st_daughter:  flags: 0  inst: 839  mov_daught: 80
2nd_daughter:  flags: 0  inst: 813  mov_daught: 80

nop_1    ; 01   0 beginning template
nop_1    ; 01   1 beginning template
nop_1    ; 01   2 beginning template
nop_1    ; 01   3 beginning template
zero     ; 04   4 put zero in cx
or1      ; 02   5 put 1 in first bit of cx
shl      ; 03   6 shift left cx
shl      ; 03   7 shift left cx, now cx = 4
         ;        ax =                bx =
         ;        cx = template size   dx =
mov_cd   ; 18   8 move template size to dx
         ;        ax =                bx =
         ;        cx = template size   dx = template size
adrb     ; 1c   9 get (backward) address of beginning template
nop_0    ; 00  10 compliment to beginning template
nop_0    ; 00  11 compliment to beginning template
nop_0    ; 00  12 compliment to beginning template
nop_0    ; 00  13 compliment to beginning template
         ;        ax = start of mother + 4   bx =
         ;        cx = template size          dx = template size
sub_ac   ; 07  14 subtract cx from ax
         ;        ax = start of mother   bx =
         ;        cx = template size     dx = template size
mov_ab   ; 19  15 move start address to bx
         ;        ax = start of mother   bx = start of mother
         ;        cx = template size     dx = template size
adrf     ; 1d  16 get (forward) address of end template
nop_0    ; 00  17 compliment to end template
nop_0    ; 00  18 compliment to end template
nop_0    ; 00  19 compliment to end template
nop_1    ; 01  20 compliment to end template
         ;        ax = end of mother   bx = start of mother
         ;        cx = template size   dx = template size
inc_a    ; 08  21 to include dummy statement to separate creatures
sub_ab   ; 06  22 subtract start address from end address to get size
         ;        ax = end of mother    bx = start of mother
         ;        cx = size of mother   dx = template size
nop_1    ; 01  23 reproduction loop template
nop_1    ; 01  24 reproduction loop template
nop_0    ; 00  25 reproduction loop template
nop_1    ; 01  26 reproduction loop template
mal      ; 1e  27 allocate memory for daughter cell, address to ax
         ;        ax = start of daughter    bx = start of mother
         ;        cx = size of mother       dx = template size
```

```
call     ; 16  28 call template below (copy procedure)
nop_0    ; 00  29 copy procedure compliment
nop_0    ; 00  30 copy procedure compliment
nop_1    ; 01  31 copy procedure compliment
nop_1    ; 01  32 copy procedure compliment
divide   ; 1f  33 create independent daughter cell
jmp      ; 14  34 jump to template below (reproduction loop, above)
nop_0    ; 00  35 reproduction loop compliment
nop_0    ; 00  36 reproduction loop compliment
nop_1    ; 01  37 reproduction loop compliment
nop_0    ; 00  38 reproduction loop compliment
if_cz    ; 05  39 this is a dummy instruction to separate templates
         ;        begin copy procedure
nop_1    ; 01  40 copy procedure template
nop_1    ; 01  41 copy procedure template
nop_0    ; 00  42 copy procedure template
nop_0    ; 00  43 copy procedure template
push_ax  ; 0c  44 push ax onto stack
push_bx  ; 0d  45 push bx onto stack
push_cx  ; 0e  46 push cx onto stack
nop_1    ; 01  47 copy loop template
nop_0    ; 00  48 copy loop template
nop_1    ; 01  49 copy loop template
nop_0    ; 00  50 copy loop template
mov_iab  ; 1a  51 move contents of [bx] to [ax]
dec_c    ; 0a  52 decrement cx
if_cz    ; 05  53 if cx == 0 perform next instruction, otherwise skip it
jmp      ; 14  54 jump to template below (copy procedure exit)
nop_0    ; 00  55 copy procedure exit compliment
nop_1    ; 01  56 copy procedure exit compliment
nop_0    ; 00  57 copy procedure exit compliment
nop_0    ; 00  58 copy procedure exit compliment
inc_a    ; 08  59 increment ax
inc_b    ; 09  60 increment bx
jmp      ; 14  61 jump to template below (copy loop)
nop_0    ; 00  62 copy loop compliment
nop_1    ; 01  63 copy loop compliment
nop_0    ; 00  64 copy loop compliment
nop_1    ; 01  65 copy loop compliment
if_cz    ; 05  66 this is a dummy instruction, to separate templates
nop_1    ; 01  67 copy procedure exit template
nop_0    ; 00  68 copy procedure exit template
nop_1    ; 01  69 copy procedure exit template
nop_1    ; 01  70 copy procedure exit template
pop_cx   ; 12  71 pop cx off stack
pop_bx   ; 11  72 pop bx off stack
pop_ax   ; 10  73 pop ax off stack
ret      ; 17  74 return from copy procedure
nop_1    ; 01  75 end template
nop_1    ; 01  76 end template
nop_1    ; 01  77 end template
nop_0    ; 00  78 end template
if_cz    ; 05  79 dummy statement to separate creatures
```

REFERENCES

1. Ackley, D. H., and M. S. Littman. "Learning From Natural Selection in an Artificial Environment." In *Proceedings of the International Joint Conference on Neural Networks*, Vol. I, Theory Track, Neural and Cognitive Sciences Track. (Washington, DC, Winter, 1990.) Hillsdale, NJ: Lawrence Erlbaum Associates, 1990.
2. Aho, A. V., J. E. Hopcroft, and J. D. Ullman. *The Design and Analysis of Computer Algorithms.* Reading, MA: Addison-Wesley, 1974.
3. Bagley, R. J., J. D. Farmer, S. A. Kauffman, N. H. Packard, A. S. Perelson, and I. M. Stadnyk. "Modeling Adaptive Biological Systems." *Biosystems* **23** (1989): 113–138.
4. Barbieri, M. *The Semantic Theory of Evolution.* London: Harwood, 1985.
5. Bell, G. *The Masterpiece of Nature: The Evolution and Genetics of Sexuality.* Berkeley: University of California Press, 1982.
6. Bell, G. *Sex and Death in Protozoa: The History of an Obsession.* Cambridge: Cambridge University Press, 1989.
7. Buss, L. W. *The Evolution of Individuality.* Princeton: Princeton University Press, 1987.
8. Cariani, P. "Emergence and Artificial Life." This volume
9. Carroll, L. *Through the Looking-Glass.* London: MacMillan, 1865.
10. Charlesworth, B. "Recombination Modification in a Fluctuating Environment." *Genetics* **83** (1976): 181–195.
11. Cohen, F. "Computer Viruses: Theory and Experiments." Ph. D. dissertation, University of Southern California, 1984.
12. Dawkins, R. *The Blind Watchmaker.* New York: Norton, 1987.
13. Dawkins, R. "The Evolution of Evolvability." In *Artificial Life*, edited by C. Langton. Santa Fe Institute Studies in the Sciences of Complexity, Proc. Vol. VI, 201–220. Reading, MA: Addison-Wesley, 1989.
14. Denning, P. J. "Computer Viruses." *Amer. Sci.* **76** (1988): 236–238.
15. Dewdney, A. K. "Computer Recreations: In the Game Called Core War Hostile Programs Engage in a Battle of Bits." *Sci. Amer.* **250** (1984): 14–22.
16. Dewdney, A. K. "Computer Recreations: A Core War Bestiary of Viruses, Worms and Other Threats to Computer Memories." *Sci. Amer.* **252** (1985): 14–23.
17. Dewdney, A. K. "Computer Recreations: Exploring the Field of Genetic Algorithms in a Primordial Computer Sea Full of Flibs." *Sci. Amer.* **253** (1985): 21–32.
18. Dewdney, A. K. "Computer Recreations: A Program Called MICE Nibbles Its Way to Victory at the First Core War Tournament." *Sci. Amer.* **256** (1987): 14–20.
19. Dewdney, A. K. "Of Worms, Viruses and Core War." *Sci. Amer.* **260** (1989): 110–113.

20. Doolittle, W. F., and C. Sapienza. "Selfish Genes, the Phenotype Paradigm and Genome Evolution." *Nature* **284** (1980): 601–603.
21. Eldredge, N., and S. J. Gould. "Punctuated Equilibria: An Alternative to Phyletic Gradualism." In *Models in Paleobiology*, edited by J. M. Schopf, 82–115. San Francisco: Greeman, Cooper, 1972.
22. Farmer, J. D., S. A. Kauffman, and N. H. Packard. "Autocatalytic Replication of Polymers." *Physica D* **22** (1986): 50–67.
23. Farmer, J. D., and A. Belin. "Artificial Life: The Coming Evolution." Proceedings in celebration of Murray Gell-Man's 60th Birthday. Cambridge: Cambridge University Press. In press. Reprinted in this volume.
24. Gould, S. J. *Wonderful Life, The Burgess Shale and the Nature of History*. New York: Norton, 1989.
25. Gould, S. J., and N. Eldredge. "Punctuated Equilibria: The Tempo and Mode of Evolution Reconsidered." *Paleobiology* **3** (1977): 115–151.
26. Green, M. M. "Mobile DNA Elements and Spontaneous Gene Mutation." In *Eukaryotic Transposable Elements as Mutagenic Agents* , edited by M. E. Lambert, J. F. McDonald, and I. B. Weinstein, 41–50. Banbury Report 30. Cold Spring Harbor Laboratory, 1988.
27. Holland, J. H. *Adaptation in Natural and Artificial Systems: An Introductory Analysis with Applications to Biology, Control, and Artificial Intelligence.* Ann Arbor: University of Michigan Press, 1975.
28. Holland, J. H. "Studies of the Spontaneous Emergence of Self-Replicating Systems Using Cellular Automata and Formal Grammars." In *Automata, Languages, Development*, edited by A. Lindenmayer, and G. Rozenberg, 385–404. New York: North-Holland, 1976.
29. Huston, M. "A General Hypothesis of Species Diversity." *Am. Nat.* **113** (1979): 81–101.
30. Jelinek, W. R., and C. W. Schmid. "Repetitive Sequences in Eukaryotic DNA and Their Expression." *Ann. Rev. Biochem.* **51** (1982): 813–844.
31. Langton, C. G. "Studying Artificial Life With Cellular Automata." *Physica* **22D** (1986): 120–149.
32. Langton, C. G. "Virtual State Machines in Cellular Automata." *Complex Systems* **1** (1987): 257–271.
33. Langton, C. G., ed. "Artificial Life." In *Artificial Life*, Santa Fe Institute Studies in the Sciences of Complexity, Proc. Vol. VI, 1–47. Reading, MA: Addison-Wesley, 1989, .
34. Lotka, A. J. *Elements of Physical Biology.* Baltimore: Williams and Wilkins, 1925. Reprinted as *Elements of Mathematical Biology*, Dover Press, 1956.
35. MacArthur, R. H., and E. O. Wilson. *The Theory of Island Biogeography.* Princeton: Princeton University Press, 1967.
36. Maynard-Smith, J. "What Use is Sex?" *J. Theor. Biol.* **30** (1971): 319–335.
37. Michod, R. E., and B. R. Levin, eds. *The Evolution of Sex*. Sutherland, MA: Sinauer, 1988.
38. Minsky, M. L. *Computation: Finite and Infinite Machines.* Englewood Cliffs, NJ: Prentice-Hall, 1976.

39. Morris, S. C. "Burgess Shale Faunas and the Cambrian Explosion." *Science* **246** (1989): 339–346.
40. Orgel, L. E., and F. H. C. Crick. "Selfish DNA: The Ultimate Parasite." *Nature* **284** (1980): 604–607.
41. Paine, R. T. "Food Web Complexity and Species Diversity." *Am. Nat.* **100** (1966): 65–75.
42. Packard, N. H. "Intrinsic Adaptation in a Simple Model for Evolution." In *Artificial Life*, edited by C. Langton. Santa Fe Institute Studies in the Sciences of Complexity, Proc. Vol. VI, 141–155. Reading, MA: Addison-Wesley, 1989.
43. Pattee, H. H. "Simulations, Realizations, and Theories of Life." In *Artificial Life*, edited by C. Langton. Santa Fe Institute Studies in the Sciences of Complexity, Proc. Vol. VI, 63–77. Reading, MA: Addison-Wesley, 1989.
44. Petraitis, P. S., R. E. Latham, and R. A. Niesenbaum. "The Maintenance of Species Diversity by Disturbance." *Quart. Rev. Biol.* **64** (1989): 393–418.
45. Rasmussen, S., C. Knudsen, R. Feldberg, and M. Hindsholm. "The Coreworld: Emergence and Evolution of Cooperative Structures in a Computational Chemistry" *Physica D.* **42** (1990): 111–134.
46. Rheingold, H. "Computer Viruses." *Whole Earth Review* **Fall** (1988): 106.
47. Ray, T. S. "Synthetic Life: Evolution and Ecology of Digital Organisms." Unpublished, 1990.
48. Spafford, E. H., K. A. Heaphy, and D. J. Ferbrache. *Computer Viruses, Dealing with Electronic Vandalism and Programmed Threats.* ADAPSO, 1300 N. 17th Street, Suite 300, Arlington, VA 22209, 1989.
49. Stanley, S. M. "An Ecological Theory for the Sudden Origin of Multicellular Life in the Late Precambrian." *Proc. Nat. Acad. Sci.* **70** (1973): 1486–1489.
50. Syvanen, M. "The Evolutionary Implications of Mobile Genetic Elements." *Ann. Rev. Genet.* **18** (1984): 271–293.
51. Thomas, C. A. "The Genetic Organization of Chromosomes." *Ann. Rev. Genet.* **5** (1971): 237–256.
52. Van Valen, L. "A New Evolutionary Law." *Evol. Theor.* **1** (1973): 1–30.
53. Volterra, V. "Variations and Fluctuations of the Number of Individuals in Animal Species Living Together." In *Animal Ecology*, edited by R. N. Chapman, 409–448. New York: McGraw-Hill, 1926.
54. Wilson, E. O., and W. H. Bossert. *A Primer of Population Biology.* Stamford, CN: Sinauers, 1971.

David G. Stork*‡ Bernie Jackson† and Scott Walker†
*Department of Electrical Engineering and †Department of Computer Science, Jordan Hall, Stanford University, Stanford, CA 94305 and ‡Ricoh California Research Center, 2882 Sand Hill Road #115, Menlo Park, CA 94025-7022

"Non-Optimality" via Pre-adaptation in Simple Neural Systems

We simulate the evolution of the neural circuitry subserving the tailflip escape maneuver in the crayfish in order to help explain a paradoxical ("non-optimal") feature of that circuit. Specifically, a "useless" synapse in the current tailflip circuit can be understood as being a vestige from a previous evolutionary epoch in which the circuit was used for swimming instead of flipping. Such preadaptation effects may underlie a broad range of neural structures throughout the animal world, and illustrate fundamental principles important for Artificial Life, most notably the locally greedy nature of evolutionary change and that "elegance of design counts for little."

INTRODUCTION

The structure and function of every organism—both biological and the vast majority posited for Artificial Life—depend crucially upon its evolutionary precursors.[1] The form of the human eye and the neural system subserving peripheral visual processing, for example, depend upon the evolutionary history of hominids and pre-hominids[31]; likewise, the structure of systems subserving hearing (and thus speech recognition), motor control, and so on derived from those of earlier evolutionary epochs. Indeed, evolutionary change is so fundamental to our understanding of biological life that Dawkins[4,6] claimed that life without the notion of evolution was virtually unthinkable.

Neural systems of all animals possess structure at birth—there are no *tabulae rasae* anywhere in the animal kingdom.[11] Such structure is absolutely fundamental to the performance of the organism, of course, and even determines what can, and what cannot, be learned from the environment. Moreover, it is increasingly clear that the initial promise of artificial neural networks toward achieving adequate performance on speaker-independent speech recognition, three-dimensional visual object recognition, scene analysis, language understanding, and a host of higher cognitive functions cannot be met without continued progress in understanding constraints, as manifest in network structure.[32,34] Whereas nearly all researchers in neural networks *design* their networks (or "reverse engineer" what exists in biology), we believe that a deeper understanding of the *sources* of biological structure will also help us create artificial neural systems duplicating or mimicking complex behavior. Such understanding will also support efforts to produce Artificial Life.

Because biological structure evolved through selection in extremely complex environments, we should not expect that biological solutions will always conform to "good" design principles. The research related here is directed to understanding how "inelegant"—indeed, counter-intuitive, or "non-optimal"—structures might arise through evolution, even in quite simple neural systems. We argue, moreover, that "non-optimality" should be expected to be even more prevalent in *complex* neural structures, for instance, the human brain.

Although its roots extend back to the time of Darwin,[3] the concept of *pre-adaptation* has been recently elaborated by S. J. Gould, E. Mayr and others.[13,14,15] Pre-adaptation is used to describe the process by which an organ, behavior, neural structure, etc., which evolved to solve *one* set of tasks is later utilized to solve a *different* set of tasks. It illustrates the dichotomy between designed, planned, and "optimal" forms in biology on the one hand, and "non-optimal" ones on the other.

An example of pre-adaptation of an organ is the bird wing. The proto-bird wing was too short to be used for flight, and hence must have been used for some other task; the Darwinian fitness at that time did not depend upon flight. Theories of the use of the proto-wing include thermoregulation (the proto-bird spreads or retracts its wings to cool or warm itself), insect catching (the proto-wings are used to knock down insects to be eaten), and reorientation during jumps for insects (the proto-bird can then catch insects from a larger volume of air), and others. Whatever the reason, the proto-wing was indisputably *not* used for flight. Later in evolution, as the proto-wing became longer, a behavioral threshold was reached in which the limb *could* be used for flight. Then, a different set of evolutionary pressures were placed on the wing, yielding a lighter and more aerodynamic wing. The later wing, though, had to be built upon the structures that evolved for the previous task. Thus there could be structures in the current bird wing—holdovers from the earlier evolutionary epoch—that are "non-optimal" for flight.[33]

If such structures do not present an excessive biological "cost" (say, in energy or resources), then that structure may remain in the later system. Even if the structure does pose a cost to the organism, that structure might nevertheless remain in the later organisms, since intermediate states in its elimination may prove very detrimental to the organism. In such a case, the structure is "frozen into" the organism, a relic of the earlier evolutionary epoch.

Figure 1 illustrates, metaphorically, the process of pre-adaptation, and can be discussed in terms of neural networks (our primary system of interest). At an early epoch, the network solved Task 1, and might even have been optimal for it. (Optimal is, of course, dependent upon one's measure. We need not be specific here, but state roughly that a circuit which uses the minimum number of components, biological energy, and structure to solve the problem without compromising the organism's

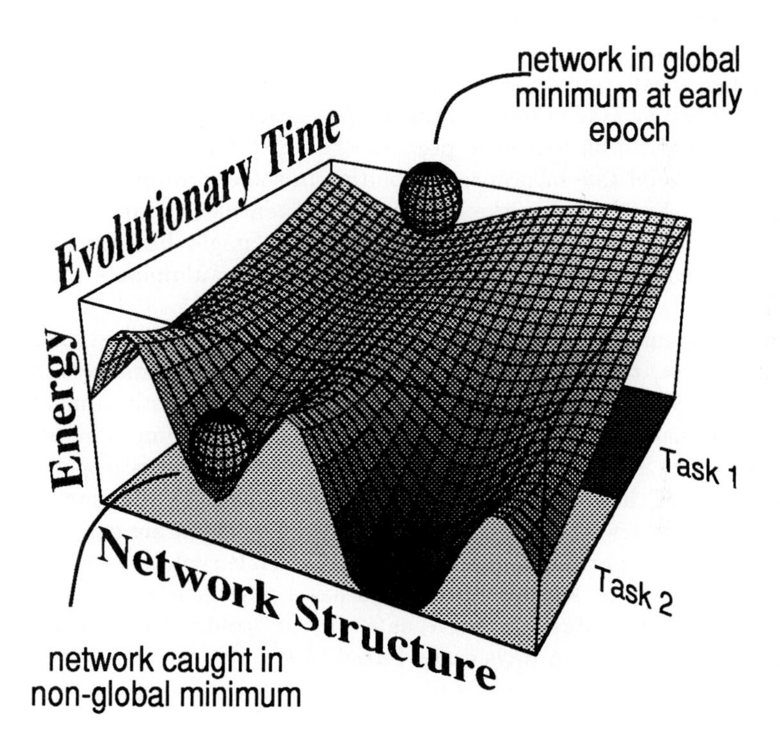

FIGURE 1 Pre-adaptation. Metaphorical energy landscape describing performance of a network throughout evolution. Evolutionary time runs from the back of the figure to the front; the "energy" (e.g., a measure of fitness) is vertical, and some index of network structure runs left to right. At an early epoch, the network may have been optimal for solving the task at that time—Task 1—but later, the appearance of Task 2 deforms the energy landscape. The network might, therefore, be in a *nonlocal* minimum, and hence "non-optimal" for Task 2. In our typical crayfish simulations, Task 1 is swimming and Task 2 flipping.

ability to solve other problems can be regarded as more optimal than a circuit that doesn't.) At a later evolutionary epoch, a *different* task becomes more relevant. This switch in task might be due to a changing environment, or to the network evolving such that new niches become available (as in the bird wing), and so on. The network is then under *different* evolutionary pressures, and the "energy landscape" is deformed. The network, however, must build upon structures selected based on Task 1—structures that might not be appropriate for the second task. The result is that the network may be "non-optimal" for Task 2.

Investigations of pre-adaptation are important in neurobiology, artificial neural networks, and Artificial Life. Such studies elucidate the nature of evolutionary change and the function of biological networks (especially since such information cannot be preserved in the fossil record). Pre-adaptation sheds light on the study of artificial neural networks in at least two ways: it can help guide the "reverse engineering" of biological systems, showing which structures might or might not be relevant to the cognitive task at hand; it can suggest general hybrid evolution-learning neural networks based on biological processes.[25,27,36,37] Since the vast majority of attempts at Artificial Life incorporate evolution in some form, pre-adaptation can aid these efforts by clarifying the difference between elegant and simple design principles and the "inelegant" implementations that might be required in living organisms. Likewise, studies such as this one can help to illuminate the processes in evolution.

We have chosen the crayfish tailflip circuit for our simulation studies for several reasons. First, the neural circuitry has been extensively mapped by neurophysiologists.[42] Second, the circuit is small enough that realistic simulations can be made using the computer resources available to us. Third, an apparently "non-optimal" structure is evident in the circuit. Fourth, the circuit is responsible for a behavior that is of the utmost survival value for the crayfish (flipping away from danger), and thus Darwinian selection pressures on the circuit are great. Fifth (and closely related to the previous reason), a highly plausible evolutionary scenario can be made for the circuits. Finally, the crayfish has a phylogenetically close relative, *Anaspides tasmaniae*, which can serve as a sort of "control" organism, since its homologous circuits differ in ways easily linked to its different behavior.

CRAYFISH TAILFLIP CIRCUIT

The crayfish tail consists of six segments, each with its own small neural circuit linking pressure-sensitive cells to flexor muscles governing the tail segment. The tailflip escape maneuver is effected by *flexion* of the *anterior* segments (segments 1–3) with *no* flexion in the *posterior* segments (segments 4–6). Figure 2 shows the basic structure of the actual crayfish circuits responsible for this behavior and possible evolution.

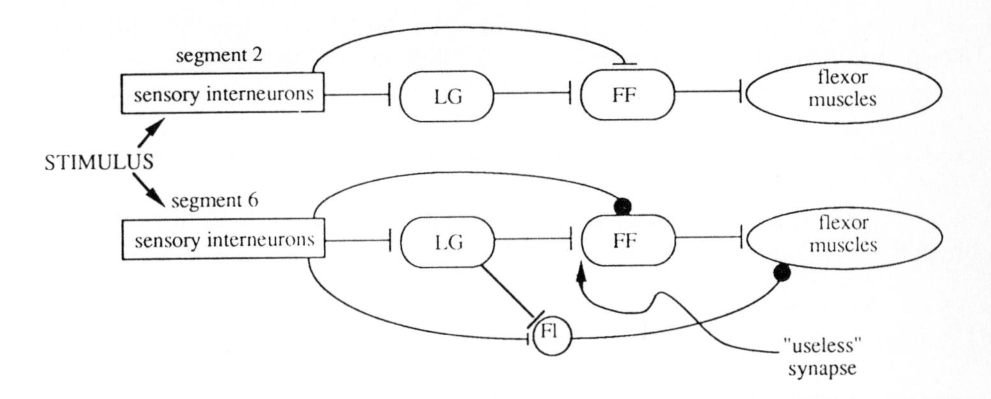

FIGURE 2 The neural circuitry subserving the tailflip in crayfish.[8] Excitatory synapses are represented by a T and inhibitory synapses by a •. In the event of a rapid rise in ambient water pressure (from a predator), pressure transducers yield excitatory activation in the sensory interneurons. To effect the tailflip maneuver, each anterior segment (e.g., segment 2) must flex (i.e., the flexor muscles must be excited) and each posterior segment (e.g., segment 6) must *not* flex (i.e., the flexor muscle must be inhibited). Note especially that one of the excitatory synapses in segment 6 is "useless": any time an excitatory volley passes from neuron LG to FF, the FF neuron is *also* inhibited (via a direct connection from the sensory interneuron), thereby rendering the excitation ineffective. Furthermore, the only projection of the FF (which is to the flexor muscles) is *also* overridden by inhibition from the F1 neuron.

Consider carefully the circuit in segment 6, which leads to inhibition of the flexor muscles whenever the sensory interneurons are excited. A neural volley passing from the LG to the FF neuron would lead to excitation of the FF. However, this excitation is counteracted by the direct *inhibitory* connection from the sensory interneuron to the FF itself. There is, moreover, inhibition of the flexor muscle via the F1 neuron. The synapse between the LG and FF is thereby overridden; it seems to have no purpose. So far as is known, then, the circuit is "non-optimal."

The question naturally arises: Why does the crayfish have this apparently useless synapse? What can account for such "non-optimality" in design?

PRE-ADAPTATION HYPOTHESIS

Dumont and Robertson[8] hypothesized that the excitatory LG $\Rightarrow$ FF synapse is a vestige from an earlier evolutionary epoch, one in which the proto-crayfish did *not* flip, but instead merely *swam.* (Simultaneous flexion in all segments, leads to swimming, as in the *Anaspides tasmaniae*, which has in each of its six tail segments a circuit homologous to those in the anterior segments of the crayfish.) The hypothesis is that the circuits in the posterior segments originally had the form at the top of Figure 2 (appropriate for swimming), but under a change in task—from swimming to flipping—the circuit evolved by building upon the previous ones. The LG ⇒ FF synapse was useful for swimming, but not for flipping, and the circuit evolved other connections to override that synapse. Because that synapse is no longer expressed behaviorally, it is "frozen into" the circuit—a vestige of the earlier epoch, and non-optimal in the context of the circuit's current use, in much the same way that the appendix has been "frozen" into our digestive system.

We provide here computer simulations and further analysis in support of this hypothesis.

SIMULATION APPROACH

The overall approach follows a classical Darwinian evolution scenario, shown in Figure 3—a more complete explanation and description of the relationship to actual biology is given in a recent paper.[38] Each network has a haploid gene, which is expressed to yield the full network, including connectivities and neural response characteristics. Networks then respond to the environment—a simulated pressure wave from a predator—and are selected based on their response. The selected networks then reproduce to give the genes of the next generation, and the cycle continues.

GENOTYPE The genetic representation and development used in our model system together avoid some of the artificial assumptions made by other modelers of genetic systems. The most important question centers on that of genetic representation of neural connection strengths: is this representation *localized* (each initial connection strength determined by one or a small number of genes) or is it *distributed* (the many connection strengths determined by several genes)?

There is abundant evidence for pleiotropy and a *distributed* genetic representation in biology.[4,17,20,40] It is clear that the information in the entire human genome is insufficient to specify every brain synapse, not to mention those elsewhere in the nervous system. Nor does there seem to be much evidence for "one gene-one synapse." Instead, genetic representation can act in several ways: setting affinities for connections, development rates, etc.[29,30] Furthermore, there are many cases in which mutations in a single or a small number of genes can have distributed consequences, as in many systemic neural disorders such as multiple sclerosis. On the computational and systems levels, a distributed representation has several useful properties. Perhaps most importantly, it permits mutations to make large changes in network structure, thereby leaving small refinements to be accomplished through learning.[25,28,36]

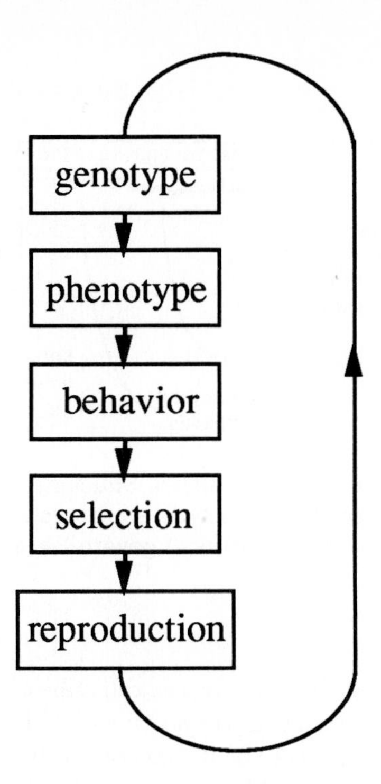

FIGURE 3 Evolutionary processes. The genes lead via development to a structured network, including interconnections (excitatory and inhibitory), neural-channel properties, etc. The network then responds to the environment and is selected based on the resulting fitness score. Fitness depends upon the posited task, here either swimming or flipping. The most-fit individuals then reproduce to yield the genotypes in the next generation, and the evolutionary processes continue.

Our simulations employ a distributed representation, based on properties of *control genes* and *structural genes*.[21] The structural genes code for fundamental aspects of the phenotype, here the cell type, neurotransmitters, type of synaptic receptors, etc.; the control genes guide the expression of the structural genes (Figure 4). Thus, for instance, if a particular enhancer from the control genes is activated, it will lead to a *distribution* of the structural genes to be expressed. This captures the fact that certain phenotypic features are expressed in concert. For example, a human photoreceptor contains both photopigment and platelets, as well as other structures unique to photoreceptors[7]; these are all expressed together. (One typically does not find cells with photopigment but no platelets, for instance.) In our model, then, several of these features are represented by a single structural gene; if that gene is activated, *all* of the component phenotypic features are candidates for expression.

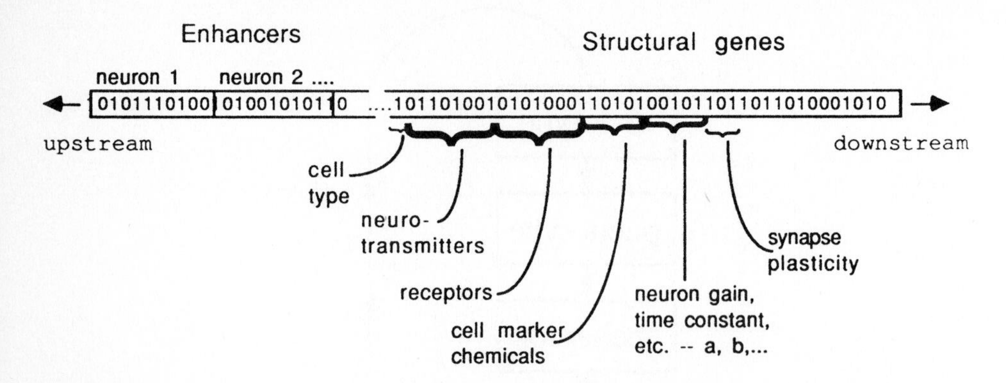

FIGURE 4 Haploid genome used in simulations. Structural genes (shown downstream, grouped for convenience) govern the phenotypic structures in the network. Enhancers (upstream, grouped by neuron for convenience) govern the expression of the structural genes.

Consider just one of the phenotypic traits: cell adhesion molecules (CAMs), implicated in developmental programs for connectivity.[9,10] In our model, there are four types of CAMs; during development the initial connectivity between two neurons is specified by the similarity in their CAMs, just as biological CAMs, large cell surface glycoproteins, are homophilic. Suppose that promoter 1 (also sometimes called an enhancer) would lead to the expression of CAM1 and CAM2 (Figure 5). If no other promoters are activated, the final neuron would have those two CAMs expressed. But suppose, moreover, that promoter 2 would lead to CAM2 and CAM4, but *not* CAM1 and CAM3, and analogously for promoter 3, as shown in the figure. (In our simulations, a promoter table describes the relationship between the promoters and the CAM structural genes.) If all three promoters are activated, each would express *its* corresponding set of CAMs, but prevent other CAMs from being expressed. The final distribution of CAMs expressed in a neuron are then the result of a majority vote for each CAM, as if the promoters competed among themselves to express the individual CAMs. Similarity in the cell surface markers expressed in any two neurons determines the initial interconnectivity—the greater the similarity, the stronger the initial synaptic connection, in accord with homophilic properties of CAMs.[9]

A similar computation occurs for the neurotransmitter to be produced in a neuron; we use twelve candidate neurotransmitters (e.g., GABA, acetylcholine, ..., whereas above we used just four CAMs. In the simulations described here, only one transmitter is expressed (as described by Dale's Law, which is not universally obeyed). Genes coding for acetylcholine and cholineacetyltransferase have been found on two separate chromosome segments in *Drosophila melanogaster*,[16,20] and this suggests that a similar arrangement could exist in the crayfish.

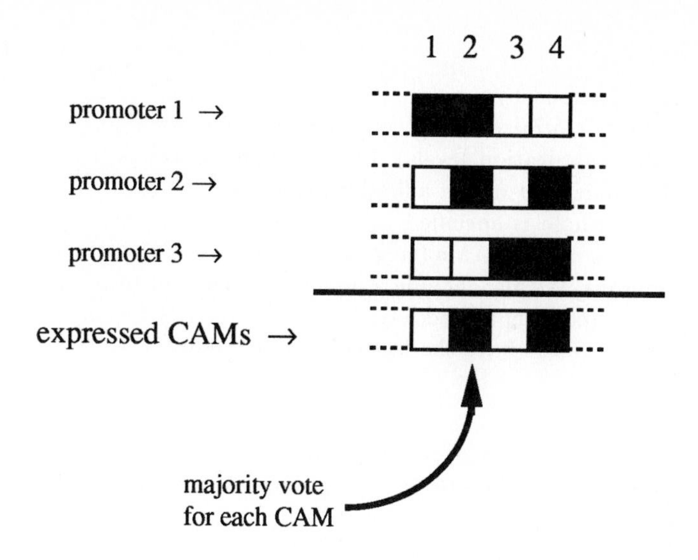

FIGURE 5 Model for the expression of cell adhesion molecules in a neuron. Suppose that for a given neuron three promoter genes are activated. In the example shown here, the first leads to activation of the structural genes 1 and 2, which would lead to CAM1 and CAM2; promoter 2 would likewise lead to CAM2 and CAM4, etc. (This relationship between promoters and these structural genes is stored in a look-up table in the simulations, and derives from physiological data on gene expression of CAMs.) The final CAMs expressed in the neuron are the result of a majority vote for each CAM—in the case shown, CAM2 and CAM4 are expressed. (Tie votes are decided by an unbiased random choice.)

Grouped phenotypic features that lead to a neuron being either a sensory, or an inter- or a motor-neuron are expressed by an analogous mechanism, though with only three (exclusive) attributes rather than twelve. Neural channel properties are computed as the *average* of those from each structural gene activated. Thus if one structural gene would lead to a large number of *Na* channels, while another would lead to a small number, then if both are activated, that actual number expressed will be intermediate. Such features of the model are motivated by recent results on mutations in three different alleles in the Shaker locus, which led to postsynaptic potentials in muscles longer and larger than in the wild type,[24] implying a genetic representation of potassium channels. (See a current paper for more detailed discussion of the biological motivation of the model.[38])

What is important here is that the relationship between genetic representation and ultimate phenotype is distributed and indirect.

PHENOTYPE

Each neuron is thus described by its global type (sensory, inter- or motor-neuron), its decay rate constant, neural channel concentrations (which determine the excitatory and inhibitory saturation levels), its neurotransmitter type, its synaptic receptor type, and complement of cell adhesion molecules.

The network as a whole is specified by the neural interconnectivities, determined by the similarities of the CAMs (computed as a Hamming distance) on each candidate pair of neurons. We also include a distance-dependent term, making neurons that are physically more separated have lower connectivity for any given CAM similarity. Expressed networks have the form shown in Figure 9.

BEHAVIOR

The behavior of each neuron in the network is governed by Hodgkin-Huxley equations of the following form[18,22]:

$$\frac{dx_i}{dt} = -ax_i + (b - cx_i)\left\{\sum_{j \in G_{ex}} z_{ij} f(x_j) + I_i\right\} + (d - ex_i)\left\{\sum_{j \in G_{in}} z_{ij} f(x_i)\right\} \quad (1)$$

where

- x_i = activity in neuron (depolarization).
- $f(x_j)$ = output spike rate—a compressively nonlinear transfer function of the activity.
- a, b, c, d, e = constants describing ion concentrations, channel densities, etc. In particular, a describes the time constant for neural recovery, b and c together with a specify the excitatory saturation level, and likewise d, e and a specify the inhibitory saturation level.
- z_{ij} = strength of synapse between neurons i and j.
- I_i = external input for neuron i (not due to other neurons).
- G_{ex} = the set of neurons connected to neuron i by synapses leading to excitation.
- G_{in} = the set of neurons connected to neuron i by synapses leading to inhibition.

The right-hand side of the equation consists of three terms. The first denotes a relaxation decay, the second an excitation term (involving the sum over all the inputs that lead to excitation), and the third term, analogously, inhibition. For our task, the input I_i is non-zero only for the sensory neuron, and in that case consists solely of a brief delta-function impulse at $t = 0$.

SELECTION

Our selection procedures are based on fitness-proportional reproduction[12]; the fitness score depends upon the task. For *swimming*, this score is equal to the maximum instantaneous excitation in the network's motor neuron (normalized over the population), corresponding (roughly) to the strength of flexion in the posterior tail segments. For *flipping*, the score is the maximum magnitude of *inhibition* in the motor neuron, corresponding (roughly) to the lack of such flexion. Although other measures of fitness are possible (motor-neuron activity integrated over time, maximum value of the derivative of the activation, etc.), the one we used captures the behaviorally relevant features of flexion. This fitness function is biologically plausible, since the crayfish locomotion is fundamental to its survival. Of course, other traits confer fitness: we are concentrating solely on one of the most important.

The algorithm for selection can be visualized as taking the fitness scores of each of the networks in a population and lining them up in a bar whose length is proportional to the individual scores. Then, points are chosen randomly and independently along the entire length (Figure 6). The networks selected in this way are then reproduced, regardless of the value of their fitness score (see subsection on reproduction). The number of points chosen is equal to the number of individuals.

Such fitness-proportional reproduction is biologically motivated and generally preferable to schemes in which merely the *most*-fit individuals are selected by truncation selection. In general, fitness-proportional reproduction helps to preserve diversity in the genome by permitting some low-fitness networks to pass on their genes.

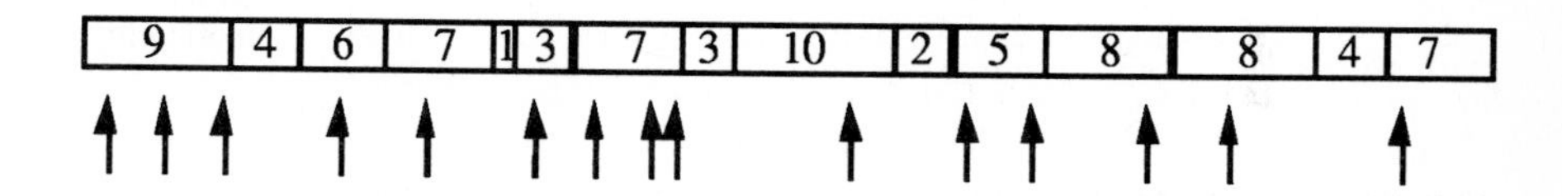

FIGURE 6 Selection for fitness-proportional reproduction. Each network is represented by a rectangle having a width equal to its fitness score. Selection is achieved by randomly choosing points along the entire population (arrows), which determine which networks survive. Thus, the probability a network survives is proportional to its fitness score, and a network can be selected multiple times. It is possible—though somewhat rare—for a network with a very low fitness score to be selected over a network with high fitness score. (Here the scores have been arbitrarily normalized to maximum = 10.)

REPRODUCTION

Those networks selected in this manner are reproduced using the familiar processes of replication, mutation (p_{bit} flip=10^{-2}/bit/generation), bit insertion (p_{bit}insert= 10^{-3}/network/generation), and single-position crossover (75% of pairs), as put forth by Holland,[23] and illustrated in Figure 7. The genetic algorithm parameters—in particular the somewhat high mutation rate—were chosen in order to probe the phenomena as thoroughly as possible using our computer. Based on several runs with different parameters and random number seeds, we found that our fundamental findings did not depend significantly on the choice of parameters over a wide range.

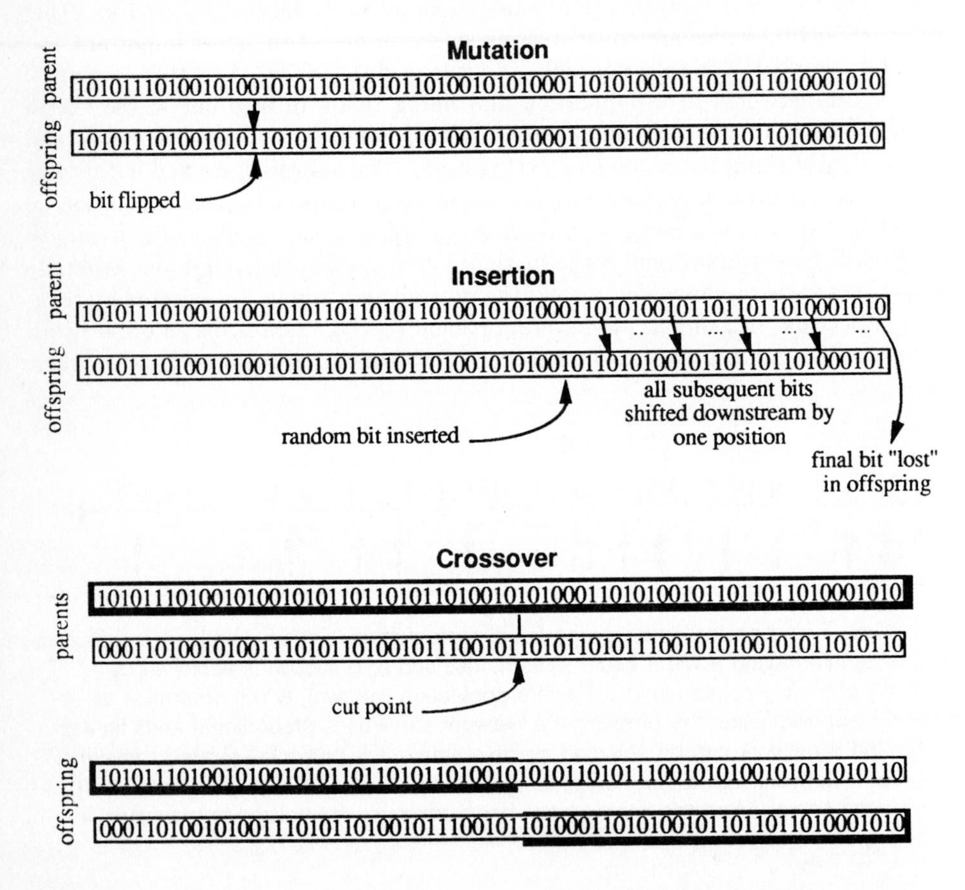

FIGURE 7 The processes of random mutation, bit insertion, and crossover (shown) as well as replication (i.e., duplication without mutation, not shown) are used between generations.

RESULTS AND ANALYSIS

All simulations were done on Connection Machine CM-2s, either at RIACS (Moffett Field, CA) or Thinking Machines Corporation (Cambridge, MA). Our program consisted of roughly 12,000 lines of C* code, the parallelized version of C; typical simulations required two to three hours. On SIMD (single-instruction multiple data) computers, there is always the question of the level at which parallelization of the problem should be made. The Connection Machine operating system and C* language permit construction of *domains*, which are processed in parallel. Candidate domains for our system were:

- individual networks,
- individual neurons, and
- individual synapses.

(The temporal dynamics of the neurons are inherently serial—the integration of Equation 1—and thus could not be parallelized. Indeed, this serial integration alone accounted for over 1/4 of the total processing time.)

Thus, for instance, if the code were parallelized at the level of individual networks, then the neurons and synapses would be serially processed. If, on the other hand, individual *neurons* were parallelized, then just the synapses and any finer grain structures would be processed serially, and so on. While parallelizing to the finest grain (here synapses) would lead to most rapid calculations, the overhead in inter-processor communication would increase, since each neuron interacts with several other neurons. For our small number of neurons (7), parallelizing at the level of individuals was most efficient. Only if the number of neurons per circuit were larger (roughly 20–30) would the speedup in computation by parallelizing at the *neuron* level outweigh the drawbacks in communication overhead.

The parallel aspect of the our program is that all members of the population are calculated simultaneously on this SIMD machine. Individual neurons and synapses within a network are computed in series. We created the parallel data structure "*domain* individual," a C* domain that allocates one processor (each with 8 kbytes of memory) per crayfish circuit. All the code was on the host VAX, while the data (synaptic strengths, neural activities, etc.) were stored on each physical processor. Whereas we ran some simulations with larger populations and found that the fitness curves (see Figure 8(a)-(b)) did not differ significantly from our results for populations of 200 individuals. Whereas the statistics for these larger sets is only slightly more reliable, we found that analyzing individual networks for "non-optimal" structures—which had to be done laboriously, by hand—became prohibitively time consuming.

PRE-ADAPTATION

Figure 8(a)-(b), from a typical simulation, illustrates the basic phenomenon of pre-adaptation.[35] The graph on the left shows the population average fitness as a function of generation. At generation 75, the task was changed from *swimming* to *flipping*—fitness score magnitude of the positive activity in motor neuron then negative (inhibitory). The population average fitness drops precipitously as the circuits previously selected for swimming are then tested and selected for flipping. Later the fitness levels off (by generation 150) to a mean score of 0.13 (in arbitrary units). The righthand graph shows evolution in the case of rewarding flipping alone—*no* pre-adaptation. After 75 generations, the mean score, 0.29 (in the same arbitrary units), is significantly above that of the preadapted networks in the left figure, given the same number of generations rewarding flipping. In short, evolving flipping networks from those previously selected for swimming leads to poorer performance than evolving them from the random networks present at the beginning of each of our simulations. Although, of course, there is a small chance the preadapted networks (Figure 8, left graph) could spontaneously increase in fitness through a fortuitous combination of mutations or crossovers, the networks seemed to be caught in a local minimum (cf. Figure 1).

The structure of preadapted networks differed from those not preadapted (Figure 9). In particular (based on a preliminary analysis of several dozen networks), roughly three times as many "non-optimal" structures were found in preadapted circuits as in non-preadapted circuits (other variables held constant). The structures we termed "non-optimal" included neurons unconnected to the rest of the network and synapses whose polarities (e.g., excitatory) were counterbalanced by another projection of the opposite polarity (i.e., inhibitory).

Because non-optimal structures arose more frequently in preadapted circuits in our simulations, and because several simulated circuits had non-optimal forms very closely homologous to those in the biological crayfish (compare Figures 2 and 9), our simulations provide support for an understanding of the LG $\Rightarrow$ FF synapse in the crayfish in terms of pre-adaptation.

A possible objection arises: how can we be sure that the LG $\Rightarrow$ FF synapse is, indeed, never used by the crayfish for some other purpose? Perhaps we simply have not been clever enough to guess a use. By analogy, very recent work on potassium channels in *Aplysia* on first analysis seemed to show that certain channels were non-functional, and, hence, perhaps non-optimal.[39] It was only after the ambient water temperature was raised from the (natural) 10°C to the warmer 15-20°C that these channels became active. (This suggested that the channels might help prevent convulsions in the *Aplysia*.) As F. H. C. Crick has remarked, evolution can be more creative than humans!

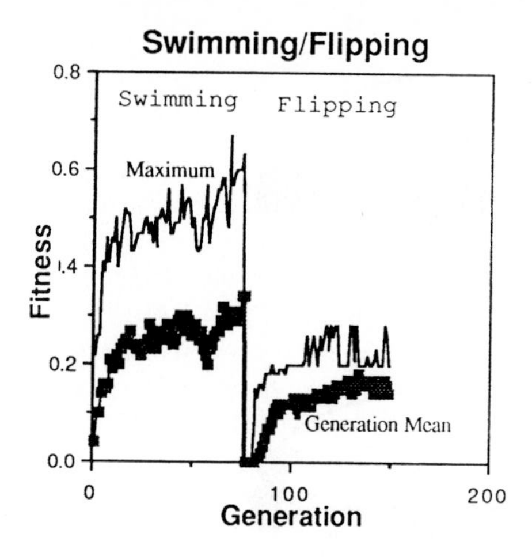

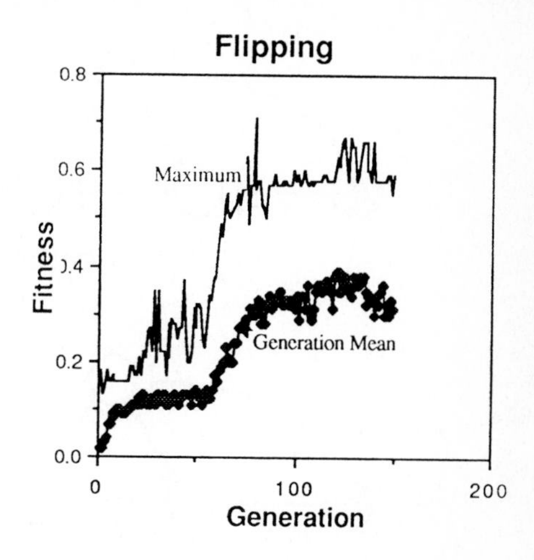

FIGURE 8 Pre-adaptation. (Left) The maximum individual fitness and the generation mean fitness for a population selected first for swimming and then (after generation 75) for flipping. (Right) Population selected solely for flipping. The minimum fitnesses were zero at virtually every generation, and hence have not been plotted. The same normalization convention was used for the graphs.

To such objections we respond that the manifest simplicity of the crayfish network and the restricted behavioral repertoire exhibited by the crayfish (at least evident in laboratory studies) seems to limit such hypothetical uses. Of course, a use might be found in the future. It might be possible that the "non-optimal" synapse and attendant projections give an architectural constraint of some sort, and cannot be removed without great behavioral and fitness cost. (One hypothetical "use" for the "non-optimal" circuit is for the inhibitory sensory-FF projection to limit the duration of an excitatory volley—perhaps to make a short "burst" in activity in the motor neuron. Alas, this does not appear to be the case in either the crayfish or our model networks: the inhibition of the FF neuron invariably precedes the excitatory volley through the "useless" synapse.) Given the simplicity and plausibility of the pre-adaptation scenario provided by Dumont and Robertson and by our simulations, this explanation seems far more acceptable than any current alternative.

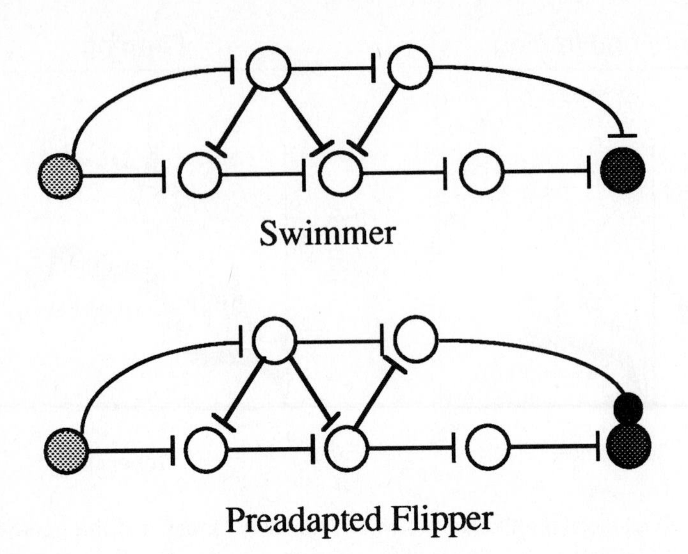

FIGURE 9 (top) Network resulting from evolution by selection for swimming alone. (bottom) Network after pre-adaptation scenario. Note in particular the non-optimal connections in the lower circuit. In both circuits, the sensory neuron is shown at the left and the motor neuron at the right. (As in Figure 2, Ts represent excitatory connections and • inhibitory ones.)

EVOLUTIONARY MEMORY

How can we understand in a deeper way the preservation of genetic information coding for functionally useless structures? Perhaps we can consider genetic information to be "junk." But note: junk is fundamentally different from trash. The junk around our house was at one time useful, and is often stored in an attic in the possibility of being used later. Trash, however, might never have been useful, and is not useful at present. We discard trash; we save junk, even if there is but a small chance that it might be used again. Perhaps the distributed genetic information responsible for the "useless" synapse is "junk" in just this way.

In order to explore this possibility, we performed another set of simulations. We selected first for swimming, and then for flipping (as before), thereby creating a population of networks which possessed a significant fraction of structures "non-optimal" for the flipping. We then changed the task back *again* to swimming, in order to see how rapidly and how well the population then evolved for swimming.

Figure 10 shows typical results. After selection for swimming then flipping, the population fitness rose very *rapidly* for the subsequent swimming task. The

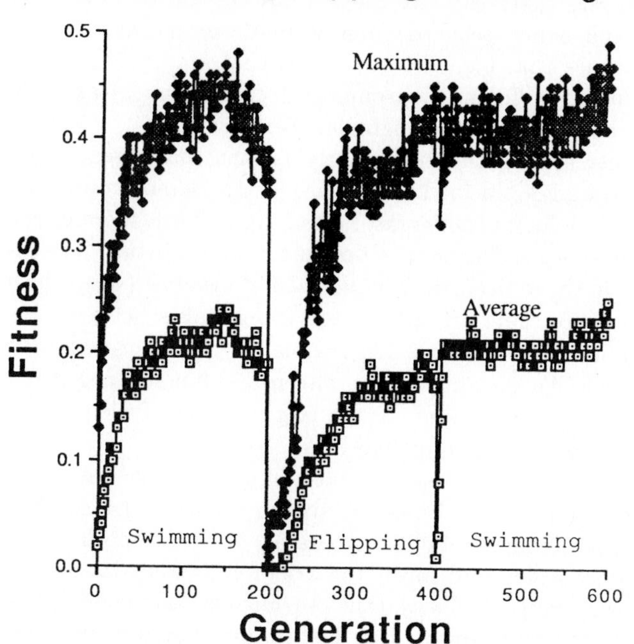

FIGURE 10 Evolutionary memory. The population was first selected for 200 generations for swimming, then for another 200 generations flipping. At generation 400, the task was changed back to swimming. Note especially that the recovery of fitness is extremely rapid in this last epoch (after generation 400).

population did this more rapidly than when it had evolved under the first swimming epoch, presumably in part because the later evolution could appropriate structures remaining from the first swimming epoch. The "junk" in the genome permits the crayfish to *rapidly* relearn how to swim, should the environment require it.

Keeping genes that were useful at previous epochs may help explain how evolution can be faster at later epochs, since the structures need only be recalled or reselected, not rebuilt *ex nihilo*.[4,41]

CONCLUSIONS AND FUTURE WORK

Our simulations support an explanation that an apparently "useless" feature of the contemporary crayfish tailflip circuit arose from pre-adaptation, specifically, that the crayfish circuit was historically selected based on the circuit's ability to have the crayfish *swim*, later selection was based on the crayfish's ability to *flip*. As such, there are features "left over" from the earlier (*swimming*) epoch, not selected out, and hence perhaps "non-optimal" in the current (flipping) circuit. Nevertheless, genes that code for structures that are at one epoch "useless" may be expressed under different environmental circumstances and, thus, permit the system to respond rapidly to changing environments.

These results, and the theories underlying them, have great import for biological systems and posited Artificial Life organisms. As Dumont and Robertson write of the evolution of biological networks[4]: "As long as both the end result and all the intervening stages work, elegance of design counts for little." The same phenomena are even more likely to occur in *complex* neural systems (which have more degrees of freedom) because there are more intervening stages between the genes and the behavior they influence. Hence, non-optimality may permeate neural systems in the animal world. We thus provide an alternate—but not necessarily competing—explanation to that of Edelman[9] for the large number of silent and perhaps unused synapses throughout the mammalian brain.

It has been argued pursuasively that human language has a strong innate, and hence genetic, component.[2] However, speech seems to have arisen fairly late in hominid evolution, roughly 100,000 years ago.[26] This epoch is very brief (on an evolutionary time scale), and surely too brief for complex language circuits to arise *ex nihilo*. Thus, it appears likely that our current language circuits appropriated and built upon structures selected for tasks *other* than language. Perhaps the most plausible use for the circuits before language was orofacial motor control.[26,33] Generalizing and extrapolating from our crayfish analysis, we can perhaps understand why language may not be "optimal," i.e., why grammar contains quirky forms or rules, due to pre-adaptation.

ACKNOWLEDGMENTS

We gratefully acknowledge DARPA grant DACA-88-C-0012 for Connection Machine time, administered through NASA-Ames Research Center, without which these simulations would have been nearly impossible.

REFERENCES

1. Bonner, J. T. *The Evolution of Complexity.* Princeton: Princeton University Press, 1988.
2. Chomsky, N. *Syntactic Structures.* Mouton: The Hague, 1957.
3. Darwin, C. R. *The Origin of Species,* 1st edition reprinted 1968. Harmondsworth, Middlesex: Penguin, 1866.
4. Dawkins, R. *The Selfish Gene.* Oxford: Oxford University Press, 1976.
5. Dawkins, R. "The Evolution of Evolvability." In *Artificial Life,* edited by C. Langton, Studies in the Sciences of Complexity, 201-220. Reading, MA: Addison-Wesley, 1988.
6. Dawkins, R. *The Extended Phenotype.* Oxford: Oxford University Press, 1989.
7. Dowling, J. E. *The Retina.* Harvard: Belknap, 1987.
8. Dumont, J. P. C., and R. M. Robertson. "Neuronal Circuits: An Evolutionary Perspective." *Science* **233** (1986): 849–853.
9. Edelman, G. *Neural Darwinism.* New York: Basic Books, 1988.
10. Edelman, G. *Topobiology.* New York: Basic Books, 1988.
11. Edwards, J. S. "Pathways and Changing Connections in the Developing Insect Nervous System." In *Developmental Neuropsychobiology,* edited by W. T. Greenough and J. M. Juraska, 74–93. New York: Academic Press, 1986.
12. Goldberg, D. *Genetic Algorithms in Search, Optimization and Machine Learning.* Redwood City, CA: Addison-Wesley, 1989.
13. Gould, S. J. "Darwinism and the Expansion of Evolutionary Theory." *Science* **216** (1982): 380–387.
14. Gould, S. J., and E. S. Vrba. "Exaptation—A Missing Term in the Science of Form." *Paleobiology* **8** (1982): 4–15.
15. Mayr, E. *Evolution and the Diversity of Life.* Cambridge, MA: Belknap Press, Harvard University Press, 1976.
16. Greenspan, R. J. "Mutations of Choline Acetyltransferase and Associated Neural Defects in *Drosophila Melanogaster.*" *J. Comp. Physiology* **137** (1980): 83–92.
17. Griffiths, A. J. F., and J. McPherson. *100+ Principles of Genetics.* New York: Freeman Press, 1989.
18. Grossberg, S. *Studies in Mind and Brain.* Boston: Reidel, 1982.
19. Hall, J. C., and D. R. Kankel. "Genetics of Acetylcholinesterase in *Drosophila Melanogaster.*" *Genetics* **83** (1976): 517–535.
20. Hall, J. C., R. J. Greenspan, and W. A. Harris. *Genetic Neurobiology.* Cambridge: MIT Press, 1982.
21. Hawkins, J. D. *Gene Structure and Expression.* Cambridge: Cambridge Univ. Press, 1986.
22. Hodgkin, A. L. *The Conduction of the Nervous Impulse.* Springfield: C. C. Thomas, 1964.
23. Holland, J. *Adaptation in Natural and Artificial Systems.* Ann Arbor: University of Michigan Press, 1975.

24. Jan, Y. N., J. Y. Jan, and M. J. Dennis. "Two Mutations of Synaptic Transmission in *Drosophila*." *Proc. Royal Soc. B* **198** (1977): 87–108.
25. Keesing, R., and D. G. Stork. "Evolution and Learning in Neural Networks: The Number and Distribution of Learning Trials Affect the Rate of Evolution." *Neural Information Processing Systems90*, held in Denver, 1990. Palo Alto, CA: Morgan-Kauffman, 1991.
26. Lieberman, P. *The Biology and Evolution of Language.* Cambridge: Harvard University Press, 1984.
27. Miller, G., P. Todd, and S. Hegde. "Designing Neural Networks using Genetic Algorithms." *Proc. Third International Conf. on Genetic Algorithms.* Palo Alto, CA: Morgan-Kaufmann, 1989.
28. Plotkin, H. C., "Learning and Evolution" In *The Role of Behavior in Evolution*, edited by H. C. Plotkin, 133–164. Cambridge: MIT Press, 1988.
29. Purves, D., and J. W. Lichtman. *Principles of Neural Development.* Sunderland, MA: Sinauer, 1985.
30. Purves, D. *Body and Brain.* Cambridge: Harvard University Press, 1989.
31. Spinelli, D. N. "A Trace of Memory: An Evolutionary Perspective on the Visual System." In *Vision, Brain and Cooperative Computation*, edited by M. A. Arbib, and A. R. Hanson. Cambridge: MIT Press, 1987.
32. Stork, D. G. Review of *Parallel Distributed Processing: Explorations in the Microstructure of Cognition, Vols. 1 and 2*, edited by by D. E. Rumelhart and J. L. McClelland and the PDP Research Group. *Bull. of Math. Biology* **50** (1988): 202–207.
33. Stork, D. G. "Preadaptation and Evolutionary Considerations in Neurobiolgy." In *Learning and Recognition—A Modern Approach*, edited by K. H. Zhao, C. F. Zhang, and Z. X. Zhu, 51–58. Singapore: World Scientific, 1989.
34. Stork, D. G. "Sources of Structure in Neural Networks for Speech and Language." In *Progress in Connectionism*, edited by J. Elman and D. Rumelhart. Hillsdale, NJ: Erlbaum Press, 1990.
35. Stork, D. G., S. Walker, M. Burns. and B. Jackson. "Preadaptation in Neural Circuits." *Proceedings of the International Joint Conference on Neural Networks-90*, Vol. I. Washington, D.C., 1990, 202-205.
36. Stork, D. G., and R. Keesing. "Interaction of Learning and Evolution: Principles Illustrated by Neural Networks." Paper presented to workshop on *Principles of Organization in Organisms*, Santa Fe Institute, June, 1990.
37. Stork, D. G., and R. Keesing. "Evolution and Learning in Neural Networks: The Number and Distribution of Learning Trials Affect the Rate of Evolution." To appear, 1991.
38. Stork, D. G., and R. Keesing. "Preadaptation, Learning and Evolution." In *Principles of Organization in Organisms*, edited by J. Mittenthal. Santa Fe Institute Studies in the Sciences of Complexity, Proc. Vol. XIII. Redwood City, CA: Addison-Wesley, 1992.
39. Triestman, F. M., and A. J. Grant. "Increase in Cell Size Underlies Cell-Specific Temperature Acclimation of Early Potassium Currents in *Aplysia*." To appear, 1991.

40. Wilkins, A. S. *Genetic Analysis of Animal Development.* New York: Wiley, 1988.
41. Wills, C. *The Wisdom of the Genes: New Pathways in Evolution.* New York: Basic Books, 1989.
42. Wine, J. J. "Escape Reflex Circuit in Crayfish: Interganglionic Interneurons Activated by the Giant Command Neurons." *Biological Bulletin* **141** (1971): 408.

Mark A. Bedau† and Norman H. Packard‡
†Philosophy Department, Dartmouth College, Hanover, NH 03755, e-mail: mark.bedau@dartmouth.edu and ‡Physics Department, and Center for Complex Systems Research–Beckman Institute, University of Illinois at Urbana Champaign, 405 N. Mathews Street, Urbana, IL 61801, e-mail: n@complex.ccsr.uiuc.edu

Measurement of Evolutionary Activity, Teleology, and Life

We consider how to discern whether or not evolution is taking place in an observed system. Evolution will be characterized in terms of a particular macroscopic behavior that emerges from microscopic organismic interaction. We define evolutionary activity as the rate at which useful genetic innovations are absorbed into the population. After measuring evolutionary activity in a simple model biosphere, we discuss applications to other systems. We argue that evolutionary activity provides an objective, quantitative interpretation of the intuitive idea of biological teleology. We also propose using evolutionary activity in a test for life.

1. WHAT IS EVOLUTION?

Our paradigm of an evolving system is the biosphere. Emerging somehow from inorganic origins, it has produced a myriad succession of marvelously adapted organisms, from the very simple to the quite complex. And the process is ongoing still.

Although we all have some commonsense grasp of the process, it is difficult to say precisely what evolution is. Evolution is change, but not all change is evolution.

What distinguishes evolution from other kinds of change? Some despair of the prospect for answering this question. Consider Dobzhansky[10]:

> There is no satisfactory general definition of evolution. "Sustained change" comes probably as close as possible at present. In the special case of biological evolution this may be amended to become "sustained change over a succession of generations," to differentiate the evolutionary development (phylogeny) from the development of an individual (ontogeny).

Evolution is clearly more than sustained change, more even than sustained *complex* change. A turbulent fluid is continually undergoing complex changes but it is evidently not evolving anything like the way that the biosphere is. Biological evolution is also not just complex change that propagates through successive generations. Large amounts of complex genetic deadwood—junk DNA—can accumulate in a gene pool with no real evolutionary effect.

Intuitively, the distinctive mark of evolution is the spontaneous generation of innovative functional structures. Implicitly designed and continually modified by the evolutionary process (on an evolutionary time scale), the structures persist because they prove sufficiently adaptive. The growth of adaptations causes the biosphere to increase in complexity, thus providing an arrow of time not implied by mere complex change, even if sustained through many generations. But how can the idea of a system continuously and spontaneously generating adaptations be expressed quantitatively? How can it be measured in a model or in the real world?

Our approach to answering these questions is to quantify the degree to which a system exhibits the continual spontaneous generation of adaptive forms. Specifically, we measure the degree to which new genetic combinations are persistently used in a population. This quantity is an objective, empirical measure of the level of evolutionary activity in an artificial or natural system, but it has important broader implications, as well. For one thing, it provides a natural, quantitative interpretation for the controversial but intuitively compelling view that a biosphere inevitably exhibits teleology (purposive or goal-directed behavior). In addition, it suggests a new approach to an empirical, quantitative understanding of life.

The evolving biosphere is a complex web of organisms interacting with each other and with their environment. Following the tradition of statistical mechanics, we regard each organism as a microscopic element of the biosphere, and we regard evolution as a macroscopic property that emerges as a consequence of the interactions among all organisms and their environment. Evolution is a macroscopic, long-term property of a population of interacting organisms.

The macroscopic state of a thermodynamic system is characterized by thermodynamic variables such as temperature, pressure, and specific heat. We would like to define analogous variables that characterize the macrostates of an evolving biosphere. Thermodynamic macroscopic variables are typically static quantities that characterize a time-independent equilibrium. By contrast, an evolving biosphere changes constantly, so its macroscopic characterization must inevitably include dynamic properties, especially in the long run. Evolutionary dynamics seem to possess metastable states, which leads us to define macroscopic quantities that are averaged over a short time scale (and possibly averaged over all or part of the population), and examine how they change over a longer time scale.

Evolution is driven by genetic changes. We will adopt the simplifying assumption that genetic changes are random changes in a genome (point mutations or crossovers). Each organism interacts with other organisms and its environment by

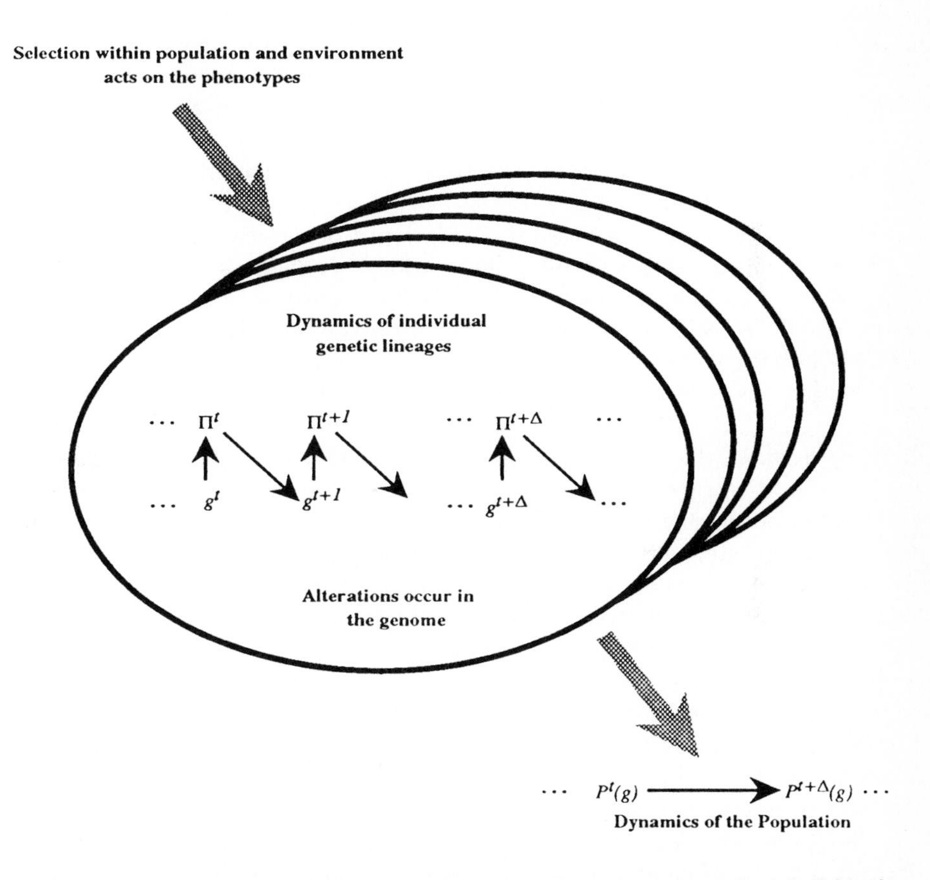

FIGURE 1 The part of the figure enclosed in the ellipses represents how individual genomes g^t change through a lineage as a result of changes in the genomes (taken to be random events), which cause a change in the resulting phenotypes Π^t. Selection acts on the phenotypes in the context of an environment consisting partly of a population of competing phenotypes (the series of ellipses). If an individual phenotype Π^t survives the selection process, its genotype g^t (possibly mutated) is transmitted in reproduction. The gray arrows represent how the evolving distribution of different genomes in the population, $P^t(g)$, emerges out of the combined effects of selection of individuals and genetic alterations.

means of an information-processing mechanism that takes sensory data as input and yields an action as output. The organism's genome encodes the mechanism that causes the behavior, though often behavior is not encoded directly but arises through the complex interaction of other directly encoded mechanisms. Selection occurs as a result of the behavior; more efficacious behavior increases an organism's probability of survival, on average. Typically selection takes place over a longer time scale than reproduction. The indirect nature of the effect of genetic change on the dynamics of the population is illustrated in Figure 1.

We have described the process of genetic change, and the resulting selection, on an organismic level. The macroscopic state, however, is a function of the distribution of the entire population over the space of possible genomes. The net result of the genetic change and selection for the entire population is an evolution of the population distribution over the space of genomes, as illustrated in the lower right of Figure 1. The statistical characterization of evolution concerns the dynamics of this population distribution.

In the final analysis, the classical adaptationist perspective embodied in Figure 1 must be qualified to allow for neutral evolution.[21] It must also be modified due to the presence of nonadaptive evolutionary forces, such as those due to developmental constraints.[16] We will concentrate, however, on the adaptive forces of evolution precisely because it is this aspect that we seek to quantify.

Spontaneous generation of innovative change epitomizes the dynamical nature of adaptive evolution. Thus, evolutionary change gives the appearance of having a direction as successful lineages become progressively better adapted to their environments. In the long run, however, evolution has no specific predetermined global goal. Evolutionary change might show overall statistical tendencies, such as those captured by the statistical measures defined below, but the details of a biosphere's global state are in constant flux. No Master Plan explicitly specifies the biosphere's form. At any given instant in a given local biological context for a given species, some specific evolutionary changes would be better and others worse. So, each species has certain temporary local optimization criteria, which generally differ for different species. These local optimization problems generally admit many possible approximate solutions. The particular evolutionary path a lineage follows results from many random genetic changes that survive as solutions or partial solutions to the local optimization problems.

The process of selection acting on the products of these genetic changes causes complex functional changes in the organisms in the long run. These changes emerge *a posteriori* out of the contingencies of the evolutionary dynamics. Evolutionary learning apparently takes place, but without an *a priori* specification of what needs to be learned. A feature that is good in one context might be bad in another; it depends on the surrounding population, which constantly shifts as the population evolves. In effect, the population is adapting to itself and to the environment, so the specific optimization criteria change implicitly during evolution. Whereas *extrinsic adaptation*, typical of artificial learning systems, has an explicitly specified goal, adaptation without an explicitly defined goal is *intrinsic adaptation*. Intrinsic adaptation is exhibited by a variety of models,[11,12,13,14,19,20,33,34] perhaps the simplest of which is the tropic bug model.[32]

A fundamental question in the general science of evolution is to determine what macroscopic variables characterize the behavior of an evolving system. Below we define a variable that reflects the degree to which new genetic combinations are continually produced and persistently used in the population. Initially sacrificing some amount of generality in favor of the clarity provided by an explicit example, we first explain this statistic in the context of a simple model biosphere. After describing the model, we measure this quantity in the model. Then we discuss how to make analogous measurements in a variety of other systems.

We then argue that evolutionary activity may reflect the occurrence of teleological or goal-directed activity. Though teleological behavior has often been identified on strong intuitive grounds as a hallmark of living, evolving systems, the concept has not been formalized in scientific theory. Much of the scientific literature is inimical to the idea that teleology is even compatible with formal scientific theory. Below, we argue the opposite—the intuitive notion of teleology is indeed quantifiable, by means of our measure of evolutionary activity. Just as recent studies of chaos have shown that deterministic systems could be unpredictable, we claim that deterministic systems may be teleological. Similarly, just as a central question in the study of chaos has been how to quantify unpredictability, in our study of teleological systems we must face the analogous question of how to quantify teleological behavior.

In the final section, we propose using evolutionary activity as a test for life. Where the necessary experimental measurements can be made, the test should be applicable to evolving systems in the real world. In the realm of artificial life models, the experimental measurements are nearly always possible, at least in principle. Thus, the test is a quantifiable analog to Turing's test in artificial intelligence.

2. A MODEL BIOSPHERE

Every model of living processes must model phenomena on some specific level. A model of a chemical soup might attempt to show how life could originate. Here, we model a population of organisms in an environment to show how evolution emerges. Following the tradition of statistical mechanics, the organisms and the environment in our model are highly idealized. Just as the Ising model from statistical mechanics represents a magnetic solid merely as a lattice filled with zeros and ones, the model presented here abstracts from a host of details about organisms and their environments.

2.1 TROPIC BUGS

The model used here is a modification of an earlier model designed to be simple yet able to capture the essential features of an evolutionary process.[32] The earlier model consisted of organisms in a two-dimensional world. The only thing that exists in the world besides the organisms is food. Represented as a field of values, food is put into the world in heaps that are concentrated at particular locations, approximating a continuum field with a gradient away from a central location. Food is refreshed periodically in time and randomly in space. The frequency and size of the heaps are variable parameters in the simulation.

The food represents energy for the organisms. They interact with the food field by eating it, decrementing the food value at their location, and incrementing their internal food supply. The organisms in this earlier model are endowed with enough innate intelligence to follow the gradient; they survey their local neighborhood and move in the direction of most food. Since the behavior of these organisms is simply the tropism of ineluctably climbing the food gradient, we call them *tropic bugs*. Movement expends energy, so each step taken exacts a tax on the internal food supply. If this internal food supply drops to zero, the organism dies and disappears from the world. On the other hand, an organism can remain alive indefinitely if it can continue to find enough food.

When an organism accumulates enough food, it produces some number of offspring. Reproduction by tropic bugs is controlled by two quantities regarded as "genes" for the organisms: the number of offspring, g_{off}, and the threshold for reproduction, g_{th}. These genes are changed during reproduction by random amounts, analogous to point mutations.

Evolution will occur in a population only if the environment stresses the population, so that some of its members can be better adapted to coping with the stress than others. The only stress faced by the tropic bugs is to find enough food to remain alive. Any evolution that occurs in the model is the effect of this one environmental imperative.

2.2 STRATEGIC BUGS

The model biosphere used here differs from the tropic bug model in two respects. The primary modification is to allow organisms to follow individually different strategies for finding food, so we call these organisms *strategic bugs*. The other modification provides a second source of genetic novelty.

The behavioral disposition of both tropic and strategic bugs is genetically hardwired. But whereas every tropic bug follows the same rule for climbing the food gradient, each strategic bug follows its own individual food-finding strategy. A behavioral strategy is simply a map taking sensory data from a local neighborhood (the five-member von Neumann neighborhood) to a vector indicating a magnitude and direction for movement:

$$S : (s_1, \ldots, s_5) \rightarrow \vec{v} = (r, \theta) .$$

A strategic bug's sensory data has two bits of resolution for each site (least food, somewhat more food, much more food, most food). Its behavioral repertoire is also finite, with four bits of resolution for magnitude r (zero, one, ..., fifteen steps), and four bits of resolution for direction θ (north, north-northeast, northeast, east-northeast, ...). As with tropic bugs, strategic bugs pay a movement tax proportional to the distance traveled. A tax is also levied just for living, so strategic bugs must continue finding food to survive.

The graph of the strategy map S may be thought of as a look-up table with 2^{10} entries, each entry taking one of 2^8 possible values. This look-up table represents an organism's overall behavioral strategy. The entries are input-output pairs that link a specific behavior (output) with each sensory state (input) that an organism could possibly encounter. Whereas tropic bugs have only the two genes g_{off} and g_{th}, strategic bugs have 2^{10} additional genes, one for each entry in their strategy look-up table. Although still finite, the space of genes in the strategic bug model is greater than in the tropic bug model by three orders of magnitude. This allows evolution in a much larger space of genetic possibilities, which better approximates a biological world with an infinite number of possibilities.

Evolution requires a source of random variation. Just as the two genes carried over from the tropic bug model can change during reproduction by random amounts, analogous point mutations of the strategy genes can change the output values of entries in the strategy look-up table. A parameter regulating the mutation rate, i.e., the "strength" of mutations, determines what fraction of the table mutates during reproduction.

The genome in the strategic bug model is large enough that it becomes reasonable to allow sexual reproduction, or at least a simple version with haploid crossover. The second respect in which strategic bugs differ from tropic bugs is that, whereas both tropic and strategic bugs reproduce asexually, strategic bugs can also reproduce sexually. A strategic bug can tell when it is next to another bug. If two healthy bugs (i.e., bugs with sufficient internal food) are adjacent, they flip a coin to decide whether to produce offspring. Analogous to the exchange of genetic material during crossover, each child contains a mix of genetic material randomly chosen from the two parents. There is no distinction between "female" and "male," so sexual reproduction here simply means offspring produced with a mixture of parental genetic material.

The strategic bug model illustrates a novel form of evolving dynamical system that has recently been developed for models of immune networks and autocatalytic networks[11,12,13,33,34,20] as well as for models of parallel computation in machine learning.[14,17,18,19] These systems consist of a state space that changes with time, with a meta-dynamic specifying the state-space evolution. In the strategic bug model, the momentary dynamical rule includes all the individual strategies for each organism present in the population, as well as the rule that governs the input of new food. The strategic bug metadynamic specifies how new elements come into the population and what rules govern them. Since new genes interact with the world and with each other just as old genes do, the metadynamical element in the strategic bug model is simply the generation of new genes by mutation and crossover, which creates new elements of the population. More complex forms of interaction between organisms (besides simply reproducing) would necessitate a more complex metadynamic.

There are many free parameters in the strategic bug world which must be set at the beginning of a simulation; we now describe them all. The parameters are named as variables that might appear in a computer program; the values listed in square brackets are those we actually used for the measurements presented in the following section. To interpret the parameters, it is useful to realize that the time scale is basically set by the amount the organisms are taxed for their activity, and the rate that food is coming into the world. The world was a 128×128 lattice, with an integer food value between zero and 255 at each site.

1. Bug initialization: The world begins with an initial population of size `initial population size [50]`. Some of the parameters that specify the bugs are the same for all bugs. One of these is `mouthful [50]`, the maximum amount of food a bug can eat in one gulp (if there is less than this amount at the bug's current location, all food is eaten). Other parameters that are the same for all bugs are the taxes, which subtract food from the internal food supply each elementary time step. One is `move tax [10]`, the amount of food used per unit distance moved. Another is `reproduction tax [0]`. A third is `overall tax [10]`, a metabolic tax for survival every time step.
2. Another group of parameters are needed to specify the initial values of genes that may be changed during reproduction. Each bug has a `reproduction thresh [1000]` for the amount of food needed to reproduce, and a `sex threshold [800]` for the amount of food needed to have sex (provided another healthy bug is in the neighborhood). Asexual reproduction yields `offspring num [2]` children. The strategy look-up table is initialized with a fraction (`strategy density [.25]`) of the entries set to random output actions, with all other entries set to the output action "do nothing."
3. World initialization: The environment is specified by parameters that govern the input of food. The first such parameter is `food time [2]`, the number of elementary time steps between each food input. Each food input consists of a lump of food placed at a random location. The lump has a maximum food value `food max [250]` at the center of the lump, and extends spatially, diminishing linearly to zero at a radius `food width [0.2]`, given in units of the size of the entire world.
4. Evolution: Parameters that specify the evolution are primarily the mutation levels. First there is the mutation level `tropic mut [0]` of the tropic genes, `reproduction threshold`, and `offspring num`. Then there is also the mutation level `strategy mut [0.1]` of the strategy table, given in terms of the fraction of total number of entries. The final evolutionary parameter is `crossover fraction [0.4]`, the fraction of strategy table entries that are exchanged during sexual reproduction.

3. EVOLUTIONARY ACTIVITY FOR STRATEGIC BUGS

An actively evolving system is continuously and spontaneously generating adaptive change. Its gene pool is continually shifting, absorbing new genes and rearranging existing ones, but genetic changes can persist in the long run only if the organisms with the new genes thrive. Any change that lessens an organism's ability to survive will, on average, be unlikely to appear in later generations. In other words, those changes that *do* get absorbed into the gene pool must, on average, either enhance survival or, at worst, be neutral. Thus, the continual retention of new useful genetic material indicates that the population is continually enhancing its gene pool.

The rate at which new genes are introduced does not reflect genuine evolutionary activity, for the new genes may well be useless. Likewise, the mere persistence of genetic innovation alone is insignificant, because a persistent gene may well be unused and irrelevant. *Persistent usage* of new genes is what signals genuine evolutionary activity. In the context of the strategic bug model, it is simple to defend the appropriateness of measuring persistent usage. Since "using" a gene amounts to moving and movements exact a tax, persistent usage of a gene necessarily indicates that the benefits brought about by the movement result in enough of a gain to offset the tax; otherwise, the gene would have disappeared from the population.

In this section we define a statistic designed to measure the rate at which new useful genetic material is being incorporated into the gene pool. We initially implement this measure in the model for strategic bugs. In a later section we discuss how to apply this measure to a wide variety of other artificial and natural systems.

3.1 USAGE STATISTICS

If we define evolutionary activity as the population's continual absorption of new genetic combinations that come to be persistently used, we must have a way to measure how persistently genes are used. To do this for the strategic bugs, we first define quantities that measure the usage of the genes in a bug's strategy look-up table. (For simplicity's sake, we refrain from measuring the usage of the two genes, carried over from the tropic bug model, that govern reproductive threshold and number of offspring.) Letting i label the bug and j label the gene within the bug, consider every gene g_{ij} of every bug as having a "usage counter" u_{ij} attached to it, initialized to zero. Recall that each entry in a look-up table is an input-output pair. Every time a particular input situation is encountered and its paired output entry in the table is used, the corresponding usage counter is incremented. During asexual reproduction, the usage is reset to zero if the corresponding gene mutates; otherwise, the usage is carried with the gene to the offspring. Offspring produced by sexual reproduction inherit their parents genes with their corresponding usages intact. In this way, a given gene's usage preserves information accumulated over many generations along the lineage through which the gene is inherited.

To record what percentage of the genes in the entire gene pool for the strategic bugs have given usage values at a given time, we define a *usage distribution function*,

$N(t,u)$, by apportioning all of the genes in all of the bugs into "bins" for given usage values u at given times t, as follows:

$$N(t,u) = \frac{1}{N_g} \sum_{i,j} \delta(u - u_{ij}^t).$$

Here, u_{ij}^t is the usage that gene g_{ij} (the jth gene of the ith organism) has accumulated by time t, and $\delta(u - u_{ij}^t)$ is the Dirac delta function, equal to one if $u = u_{ij}^t$, and zero otherwise. So, $\sum_{i,j} \delta(u - u_{ij}^t)$ simply counts the number of genes that have usage u. This sum is then normalized by dividing by the number of genes in the population, N_g. Thus, $N(t,u)$ is the total fraction of genes in the entire population having usage u at time t.

One can visualize a usage distribution function $N(t,u)$ as a three-dimensional surface (landscape) over a two-dimensional time/usage grid. The value (height) of any given location (t,u) on this surface simply reflects the proportion of genes that have usage u at time t. A significant peak in the surface around a location (t,u) would indicate that at times near t a significant proportion of genes had been used about u times.

3.2 ACTIVITY WAVES

The usage distribution function for the strategic bug model turns out to have a complex and interesting structure. Initially, at $t = 0$, all genes have zero usage, so $N(0,u)$ has just one peak at $u = 0$. As time progresses and genes are used, the usage distribution function becomes positive for other values of u. If a beneficial gene or gene-cluster enters the population, it will come to be used persistently, and the time/usage surface $N(t,u)$ will show a certain structure. Specifically, at a certain time after the introduction of the beneficial genes, $N(t,u)$ will have a bump due to the genes' persistent usage. (See Figure 2.) As long as the genes remain beneficial, they will persist and their usage will increase. As their usage increases, the bump will move in time with a velocity proportional to the frequency with which the relevant situation is encountered.

Such moving features appear as waves over the time/usage plane. We call them *waves of evolutionary activity*, or simply *activity waves*. The upper parts of Figures 3 and 4 illustrate the waves in a time/usage diagram for the strategic bug model. Each of the graphs in Figure 2 corresponds to a vertical column in the time/usage diagram, with higher values in the usage "bins" indicated by darker shades of gray.

If the genes in a wave continue to be used by all organisms with roughly the same frequency, the wave propagates at a constant velocity and appears as a relatively straight line over the time/usage surface. The slope of a wave reflects the frequency with which the genes in it are used. If this frequency is changing, the wave curves. If usage of a group of genes stops, then the slope of the wave levels to zero for as long as the genes persist in the gene pool. As the genes in this flat wave are pushed out of the gene pool by mutations, the wave's height drops to zero. If the genes start to be used again, the wave's slope (but not its height) will increase.

A gene will continue to contribute to just one wave during its time in a lineage in the gene pool. New activity waves are created only when a newly created gene

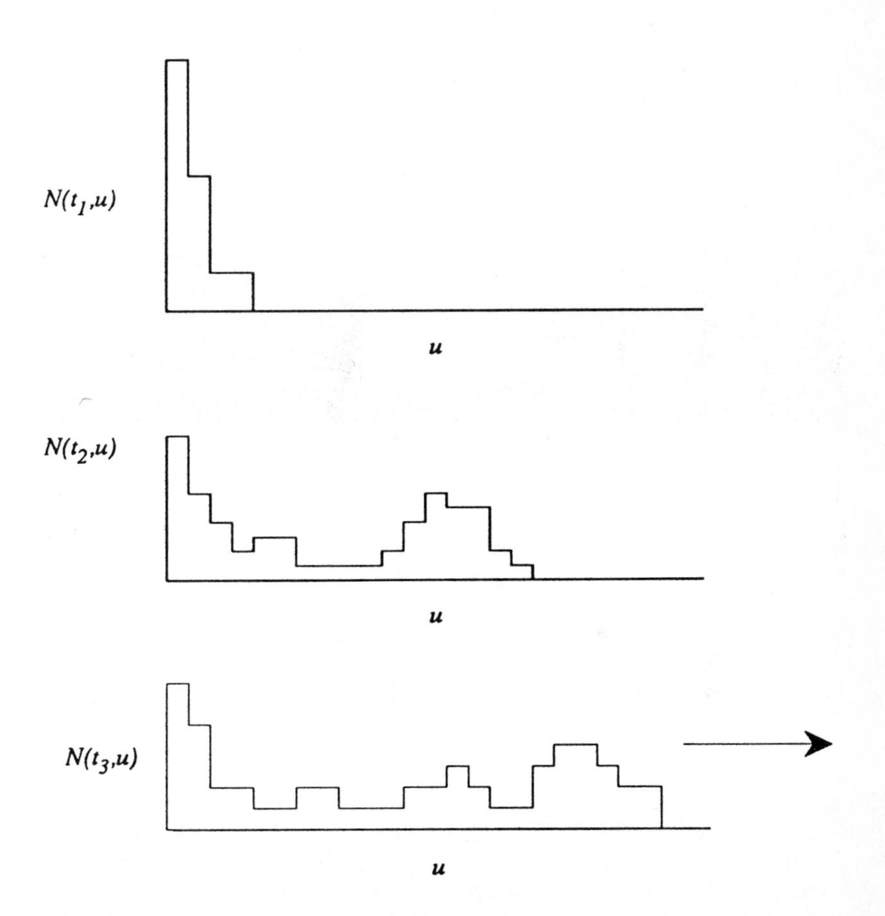

FIGURE 2 A view of the usage distribution function $N(t,u)$ for three particular values of t. At a given time t, $N(t,u)$ is a usage histogram, with usage increasing to the right and the proportion of genes with a given usage increasing vertically. Initially, at $t = 0$, all genes have zero usage, so a usage histogram has a single peak at zero usage. Then some genes get used and acquire positive usage, so a usage histogram for $t = t_1$ shows a tail sliding off the peak at zero usage. Later, at t_2, a gene or group of genes has been used repeatedly, so a usage histogram shows a clump of genes with positive usage. Later still, at t_3, after those genes have seen even more use, the bump has moved toward higher usage values. In this way, activity waves propagate through the usage distribution $N(t,u)$.

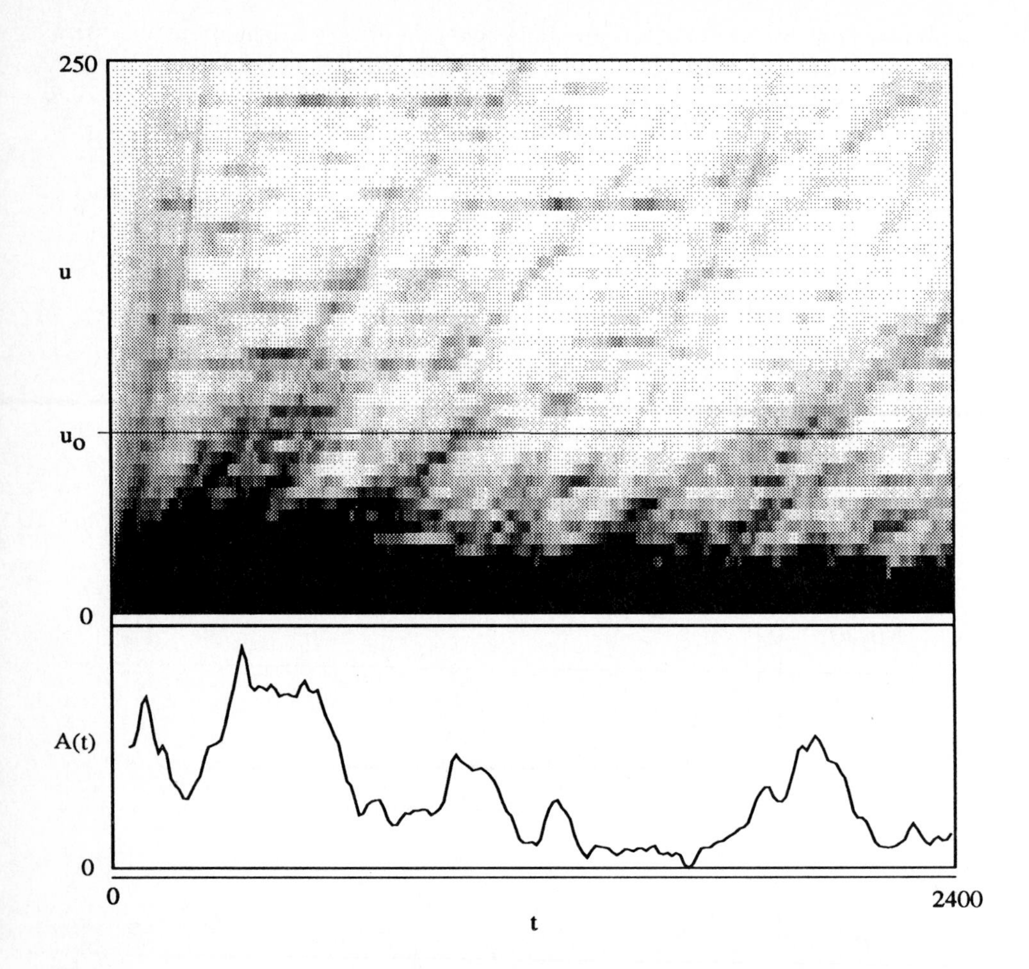

FIGURE 3 Above: A diagram of $N(t, u)$ for the strategic bug model. Time increases to the right, and usage increases up. Darker shades of gray indicate higher values in the usage "bins." The value at a given point (t, u) on the surface indicates what proportion of the genes in the gene pool have usage u at time t. Swaths of higher values (darker shades) moving up and to the right are activity waves. New activity waves start when genetic novelties prove to have some utility for the organisms. The slope of an activity wave reflects the frequency with which the genes contributing to it are used. Below: The corresponding graph of evolutionary activity $A(t)$, with the reference point u_0 chosen to be 75. Glancing at the distribution $N(t, u)$ shows that in general usage exceeds u_0 only for those genes that are in activity waves. Peaks in $A(t)$ correspond to an activity wave bursting past $u_0 = 75$.

is first used in its "life" in the gene pool. Mutations can converge, creating a new gene that accidently "copies" an existing gene; use of the accidental copy gene will initiate a *new* activity wave.

Activity waves emerge out of the local interactions between "microscopic" organisms. The medium of the activity waves is actually the genetic material of the entire population. All genetic changes are perturbations that can potentially initiate a new activity wave. Waves will actually start when the new genes are beneficial in the local biological context. "High" activity waves reflect "large" clusters of useful genes.

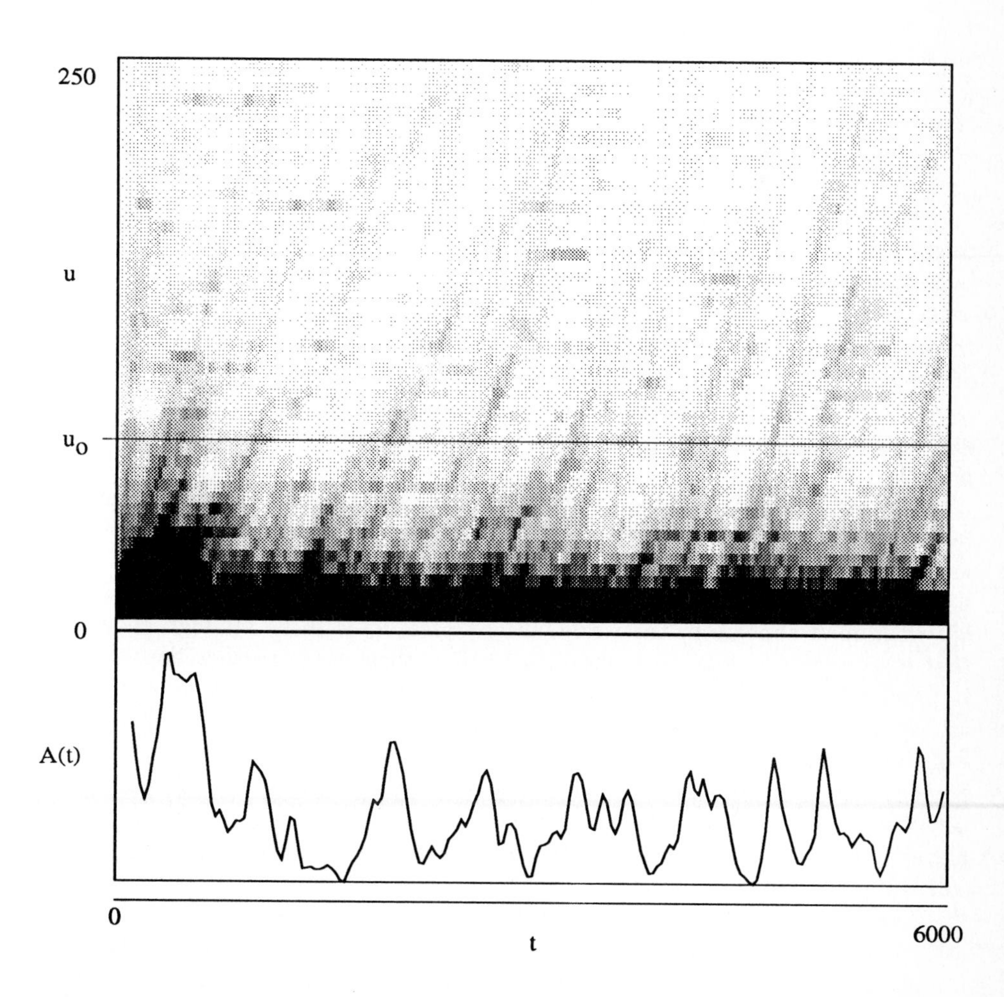

FIGURE 4 $N(t, u)$ for the same simulation as Figure 3, but for a longer time. New bursts of evolutionary activity continue to emerge.

An activity wave in a biological system reflects the persistence of a useful piece of information about the world (coded by a gene or gene cluster) within the information processing machinery of an organism. New genes are like guesses about what might be good to do in the world. When a guess is more or less correct, the information persists because it promotes the organism's survival. In this way, the continual emergence of new activity waves indicates that (on average) the organisms are effectively improving their internal models for what the world is like and what behaviors are most efficacious within it.

3.3 EVOLUTIONARY ACTIVITY

The existence of activity waves suggests a method for measuring a system's evolutionary activity. As long as activity waves continue to occur, the population is continually incorporating and repeatedly using new genetic material; in short, the system is evolving. Our measure of activity, then, is based on measuring the flow of usage into the gene pool, with a new burst of usage corresponding to a burst of evolutionary activity.

Before assigning any evolutionary significance to a gene's usage, we must distinguish short-term usage from long-term usage. Glancing at $N(t, u)$ shows that a large fraction of the genes have usage slightly greater than zero, but this short-term usage means little. Eventually, after further use, unhelpful genes are weeded out of the population. The only genes that accumulate usage above a certain value u_0 are those that end up contributing to activity waves; these are the genes that have proven their usefulness through acquiring long-term usage. Thus, we will call a gene *persistent* at a given time if its usage at that time exceeds an appropriate reference point u_0.

The parameter u_0 is determined by the time scale on which useless genes are replaced, which in turn is determined by the details of the organisms and how they interact with each other and the environment. In the case of the strategic bugs, u_0 is determined by a combination of the taxation rates and the rate and pattern by which food appears in the world. In any case, u_0 is to be set high enough so that most of the useless genes disappear before their usage reaches the value u_0. In practice, it is easy enough to identify a plausible reference point u_0 by glancing at $N(t, u)$ and picking a usage value above the initial large fall-off in usage (clearly evident in Figures 3 and 4) as useless genes are weeded out. It is evident that a gene's usage exceeds u_0 in general only if the gene ends up contributing to an activity wave. The exact value of u_0 is not crucial. As long as u_0 is large enough, patterns formed by activity waves passing u_0 will be similar for a wide range of u_0 values.

As a preliminary to quantifying activity waves, we want a measure of "bulk" usage over time in the gene pool that allows short-term usage to be distinguished from long-term usage. The *net persistence* $P(t, u)$ of a gene pool at time t relative to a given usage u is defined as the proportion of genes at t that have at least usage u:

$$P(t,u) = \sum_{u'=u}^{\infty} N(t,u').$$

Long-term and short-term usage may be separated because the net persistence $P(t,u)$ explicitly depends on u. One can think of the net persistence $P(t,u)$ for given t and u as the "bulk" of the column of usage values stacked above (t,u). Since the bulk of these usage columns decreases as u increases, $P(t,u)$ decreases monotonically with u.

If an activity wave passes a certain point in the $N(t,u)$ plane, the net persistence function $P(t,u)$ will be changing in the neighborhood around (t,u). As time approaches t, there will be a significant increase in $P(t,u)$, and as usage exceeds u, there will be a significant decrease in $P(t,u)$. Thus, a passing activity wave can be quantified by the rate of change of $P(t,u)$ with respect to either t or u. To avoid noise introduced by fluctuations in the population, we focus on the rate of change with respect to u. Waves in $N(t,u)$ correspond to "cliffs" in $P(t,u)$. The height of an activity wave at a point (t,u) in $N(t,u)$ is reflected by the *steepness* with which $P(t,u)$ falls off at that point.

Thus, to quantify the passage of activity waves, we can simply measure the steepness of $P(t,u)$ at our reference point u_0. That is, we define the *evolutionary activity* $A(t)$ of a system as the rate at which net persistence is dropping at u_0:

$$A(t) = -\left[\frac{\partial P(t,u)}{\partial u}\right]_{u=u_0}.$$

If new activity waves continue to be produced, then the population is continually acquiring new genetic material that is proving its usefulness through a significant amount of repeated use. In this case, the system is exhibiting significant evolutionary activity, and the measure of evolutionary activity $A(t)$ will be positive. If evolutionary activity is zero, then the gene pool is incorporating no new persistent genes, and the system is exhibiting no significant evolution whatsoever. Since $P(t,u)$ decreases monotonically with u, the activity $A(t)$ is never negative.

The usage distribution function $N(t,u)$ contains a wealth of information about the evolutionary process. Evolutionary activity $A(t)$—the rate at which innovative genetic novelty is flowing into the system—is one especially fundamental aspect of $N(t,u)$, but $A(t)$ does not reflect all significant evolutionary events. Two examples can illustrate how other kinds of evolutionary events could be quantified from $N(t,u)$.

One kind of event with evolutionary significance is extinctions. The extinction of a species of organisms would appear in $N(t,u)$ as the diminution and eventual disappearance of an activity wave; a massive dying off such as the Cretacious extinction would be an abrupt termination of a mass of waves. But since these waves would terminate above the reference point u_0, extinctions would not be registered in evolutionary activity $A(t)$. To quantify the *net* change in a system's innovative genetic novelty, subtracting that portion lost from extinctions, one could simply take the time derivative of $P(t, u)$ at the reference point $u = u_0$.

Changes in genes' usage patterns are another significant kind of evolutionary event. These would be reflected in $N(t, u)$ as changes in the slope of activity waves. For example, a wave's slope will increase if a group of little-used genes starts to be used more frequently. Evolutionary activity $A(t)$ does not quantify changes in the complexity of the dynamics of the activity waves; if waves continue to be produced at a steady rate, then $A(t)$ remains constant even if the *pattern* of activity between the waves becomes dramatically more complex. However, if the usage distribution function $N(t, u)$ were separated into components consisting of waves travelling at given velocities, then the complexity of the activity wave patterns could be quantified by correlating the strengths of the different components.

Our measure of evolutionary activity $A(t)$ is averaged over all individuals in the population. In addition, since usage is always passed to offspring except after a mutation, evolutionary activity is also averaged over generations. Thus, waves in a system's evolutionary activity truly characterize the system's global, long-term dynamics.

4. EVOLUTIONARY ACTIVITY IN OTHER SYSTEMS

The measurements of evolutionary activity $A(t)$ reported above were all made in the strategic bug model. But evolutionary activity can be measured in a wide variety of other artificial and natural systems, as long as the system consists of a "macroscopic" population of "microscopic" entities for which usage can be defined clearly and appropriately. If usage at the system's micro-level can be measured, then all the global macro-level quantities—the usage distribution function $N(t, u)$, net persistence $P(t)$, and evolutionary activity $A(t)$—become well defined, just as before.

The key to applying our measure of evolutionary activity is a good definition of usage. In some evolving systems it is difficult to measure gene usage directly; in others, there are no genes *per se*. Nevertheless, it is often possible to develop other ways to measure usage. One of the merits of our approach to measuring evolutionary activity is this flexibility in the definition of usage. In this section, after indicating what constitutes an appropriate measure of usage, we illustrate how to measure evolutionary activity in a wide variety of other systems.

In the strategic bug model, we counted the usage of genes in a gene pool, but it is possible to measure usage of other kinds of entities. Instead of genes, the micro-level entities could be *groups* of genes; they could even be the broad collection of genes and gene variations shared throughout a species. Since our aim is to measure the rate at which a system incorporates new functional units, the micro-entities can be *any* functional units, any units with adaptive significance. Our measure applies to any level at which natural selection operates. (Cf. the "units of selection" debate.[7,24,35,36])

If usage counters are attached to inappropriate micro-level entities, our measure of evolutionary activity can register false positives and false negatives. If the micro-level units fail to reflect some aspect of functionality, then some genuine evolutionary activity might be missed (a false negative). For example, assume that there is a genetic system in which *combinations* of genes can have adaptive significance over and above the adaptive significance of their individual component genes. In this case, crossovers could spark the spread of many quite beneficial new combinations of pre-existing genes, which would constitute significant evolutionary activity. But the proliferation of these adaptive genetic combinations would not generate activity waves if usage were counted only for single genes; the frequency of groups of genes could alter while the frequency of individual genes remained the same. To capture the occurrence of this sort of evolutionary activity, some way must be found to add usage counters to potentially functional gene combinations.

False positive readings of evolutionary activity can also occur if usage counters are attached to non-functional micro-level units. For example, only a small fraction of the eukaryotic chromosome has potential adaptive significance. Relatively short segments, exons, code for amino acids; these are the genes. The intervening segments, introns, are without adaptive significance. So, consider a genetic system that is undergoing no change in exons but rapid change in introns. In this case, the system would not be significantly evolving. However, if usage were counted at the level of individual base pairs in the nuclear DNA, then our measure of activity $A(t)$ for this system could be positive. To prevent such false positives, one must count usage only at levels on which units have adaptive significance.

To some extent it is an open question exactly which changes occurring in an evolving biosphere are adaptive; some are the effect merely of neutral genetic drift or other non-adaptive processes.[16,21] Still, there is no real doubt that a significant proportion of the change *is* adaptive. And even though selection might be taking place on a variety of levels, there is no real doubt that a significant proportion of adaptive change occurs at the genetic level. Furthermore, the degree of evolutionary activity measured at different levels should correspond at least roughly. Thus, a genetically grounded implementation of our statistical measure of evolutionary activity $A(t)$ should give a good first approximation of a system's overall evolutionary activity.

4.1 ARTIFICIAL BIOSPHERES

In the strategic bug model, the usage of a bug's genes increases every time the bug "uses" the gene, by instinctively following an entry in a strategy look-up table. Genes coding for any type of activity could be subject to the same kind of usage bookkeeping. This makes it straightforward to measure evolutionary activity in many other model biospheres.

For example, consider the tropic bug model outlined in section 2.1. Tropic bugs have only two genes, one controlling the number of offspring produced during reproduction, g_{off}, the other controlling the food threshold required for reproduction, g_{th}. Whereas a gene in a strategic bug's strategy look-up table is used only if and when the bug detects that gene's input condition, a tropic bug uses both of its genes each time it reproduces. Thus, if tropic gene usage is counted as for strategic genes, the "usage" of a tropic bug's gene would reflect simply the longevity of that bug's lineage. Thus, activity waves would correspond to the persistence of bug species, and the slope of a wave would reflect the average rate at which bugs in those species reproduce.

Evolution among tropic bugs is reflected by changes in the time-dependent population distribution function over the two-dimensional space of genes, $P^t(g_{off}, g_{th})$, which can be identified as an *a posteriori* fitness function. Simulations of this model show that the size of fluctuations of available food strongly determines the system's evolutionary dynamics,[32] with the dynamics of $P^t(g_{off}, g_{th})$ reflecting the evolutionary development of the system. If there is a low level of evolutionary activity, then $P^t(g_{off}, g_{th})$ goes in time to a fixed distribution showing that the bugs all fall into one broad cluster of species. This occurs typically when fluctuations in the food supply are small. In this case, the usage distribution function $N(t, u)$ would be dominated by one long-lived cluster of activity waves, overshadowing a carpet of short-lived waves reflecting new mutations that all quickly become extinct. On the other hand, if there is a high level of evolutionary activity, $P^t(g_{off}, g_{th})$ in time develops disjoint peaks that move about and eventually collapse. This phenomenon occurs typically when large quantities of food are put into the environment relatively infrequently. Plentiful food apparently causes a rapid proliferation and variation of organisms near a particular genome, followed by dying out of large fractions of the population, followed by another rapid proliferation, and so on. In this case, when distinct subsidiary species continually split off from the main population, $N(t, u)$ would show the background cluster of waves overlying a continual stream of relatively long-lasting secondary waves.

In general, usage statistics can be gathered for any computationally implemented model that contains explicit micro-level rules governing the behavior and structure of organisms in the population. This is true even when the micro-level rules are computationally more complex than the look-up tables of the strategic bugs. Thus, the style of usage bookkeeping implemented for the strategic bugs can quantify evolutionary activity in virtually any artificial life model biosphere.

4.2 NATURAL BIOSPHERES

If gene usage is defined as for strategic bugs, our measure of evolutionary activity $A(t)$ is well defined for actual biological populations, at least in principle. In practice, however, it is virtually impossible to get data about persistent gene usage from actual biological organisms. This makes it virtually impossible even to count gene usage in natural biospheres. Furthermore, in many actual biological populations, it is difficult to tell exactly which segments of DNA have adaptive significance.

However, there are other levels at which one could count usage in actual biological populations. For example, one could choose *species* as the microscopic functional units. When we counted usage for the strategic bugs, usage of a bug's gene was incremented every time the bug encountered a strategy gene's input condition and acted as prescribed in its output condition. A species does not correspond to any single given gene or any single given genome; it is a rough "clump" of roughly "similar" complete strategies genomes. So, a new definition of the "usage" of a species is needed. The simplest approach would be to increment the "usage" of a given species for as long as it persists in the ecosystem, weighted by the proportion of individuals in the ecosystem that belong to that species. If $s_i(t)$ were the proportion of the organisms that were members of the ith species at time t, the usage $u_i(t)$ of that species at that time would be the accumulated proportion of the population constituted by the species:

$$u_i(t) = \int_0^t s_i(t)dt\,.$$

A quite long-lived species, like the shark, would show up as a long activity wave. The birth of a new species would generate a new activity wave which would terminate with the species' eventual extinction. The continual production of new activity waves would signal the continual generation of new species.

With usage defined at the level of species, general patterns in the evolutionary activity of actual biological populations could be rendered quantifiable using data from populations living in the field or the laboratory. It should also be possible to obtain at least a qualitative picture of evolutionary activity from the fossil record. The evolutionary activity based on this measure of usage should swing dramatically at pivotal points in the evolution of the biosphere. The Cambrian explosion, for example, during which nearly all major groups of invertebrates with hard parts originated, would be seen as a period with large positive evolutionary activity.

It is instructive to compare our usage statistics with the gene frequency statistics standardly gathered by molecular geneticists.[21] There are similarities but also important differences; for the moment, focus on the usage of genes. A gene's *frequency* is defined as that proportion of the population that possesses the gene; if the whole population has a gene, it is said to be "fixed." If gene usage is incremented each time a gene is *used* (as in the strategic bugs), then usage reflects more than the mere *existence* reflected by a gene's frequency. Even if usage is defined by reference to a gene's persistence (as with species above), a gene's usage differs from its frequency; for, in this case, usage is a gene's *integrated* frequency. Thus, the molecular geneticists' data showing the route to fixation of one or two genes is qualitatively different from our data showing large clusters of persistent genes emerging out of the sea of all possible genes.

4.3 CHEMICAL SYSTEMS

In some evolutionary models, the microscopic constituents are so simple that there is no genetic code, for example, chemical soups[11,12,20] (and perhaps also populations of complex clay crystallites.[5,6]) Our measure of evolutionary activity could still be applied in these models, if "usage" were taken to be simply the integrated concentration of the chemical species, on analogy with the equation above for biological species. The definition of usage could also include a weighting for the number of reactions in which a given chemical species participates; thus, if $R_i(t)$ were the number of reactions for the ith chemical species at time t, and $c_i(t)$ were the concentration of the ith chemical species at time t, then the usage $u_i(t)$ of that species would be:

$$u_i(t) = \int_0^t R_i(t)c_i(t)dt\,.$$

In the autocatalytic soup, positive evolutionary activity would correspond to the ongoing production and persistence of novel forms of chemicals. Again, as in the genetic case, the intricate nature of the interactions are not completely captured by our measure of activity, but could be reflected in activity patterns.

Chemical soups can represent not only interactions between polypeptide strings and RNA strands; they can also model interactions between antibodies and antigens in an immune system.[12,13] Different kinds of antibodies may be regarded as different chemical species, with their usage calculated in the same way. The evolutionary time scale of immune systems is short enough to be readily observable. Thus, there is reasonable hope of actually measuring this type of evolutionary activity in real immune systems.

4.4 COMPUTATIONAL SYSTEMS

Populations of information-processing units are an especially interesting and powerful setting for evolution, one with extensive practical implications, in fact. The strategic bugs are one example of a population of simple computational units, but computational populations need not be embedded in a biologically motivated setting.

One computational model within which evolutionary activity can be easily measured is Fontana's Turing gas.[14] This is a model of interacting strings, similar to the interacting chemical strings mentioned above, except that here the strings are information-processing elements, functions in a lisp dialect called AlChemy (for Algorithmic Chemistry). The Turing Gas model is extremely active because the micro-units interact in a way that is essentially computational. Interaction happens repeatedly between pairs of lisp functions chosen at random. It occurs as one function is evaluated with another as its argument. Interactions can produce null programs (in which case the interaction is termed "elastic") or a new program (in which case the interaction is termed "reactive"). Usage counters are attached to each function present in the population, and incremented with each reactive collision. These measurements will be reported in future work.

Another kind of computational evolution is produced by genetic learning algorithms,[17] which operate on a population of "hypotheses" described by a set of genes. These hypotheses are assigned a fitness, which is typically their suitability for solving some problem. The genetic algorithm changes the population of hypotheses through a process of survival and genetically modified reproduction of the fittest. Usage bookkeeping can be easily implemented in this setting, often with usage counters for each genetic unit, just as for the strategic bug model. For learning algorithms with more machinery, it may be necessary to include other aspects in usage count. For example, Holland's classifier system[18] associates with each classifier a strength variable, analogous to the concentration of a species or chemical. A classifier's usage could be defined as its integrated strength, on analogy with the equations above for biological and chemical species. In this setting, one would expect to observe a flurry of evolutionary activity initially, as a variety of new hypotheses are tested, with activity dying down once optimal hypotheses have been identified.

A more abstract illustration of evolutionary activity is found in the complex patterns produced by the temporal evolution of cellular automata. A cellular automaton is a population of automata filling the sites in a lattice. The dynamics of the automaton maps a configuration of symbols over the lattice to another configuration, using a local rule applied simultaneously at all sites. The local rule followed by a cellular automaton site is analogous to a strategic bug's behavioral strategy but, whereas the bugs' strategies vary between individuals and change over generations, the rule in a cellular automaton is the same for all sites and never changes. If the number of symbols is finite, and the local neighborhood is finite, a cellular automaton's local map may be thought of as a look-up table whose inputs are site values over a neighborhood, and whose outputs are the value of a particular site at the following time step. Usage counters can be attached to each entry in this look-up table.

Langton has observed that, for certain classes of cellular automata with complex dynamics, the number of neighborhood configurations visited slowly grows with time.[22] As new local configurations are encountered, new usage activity would be stimulated and activity waves would form. If the set of local configurations visited stabilized to a fixed set, no new waves would be produced. Although this must eventually happen for any cellular automaton rule, it might take a very long time compared to the iteration of the rule. Langton's result thus indicates that cellular automata with complex dynamics might exhibit a primitive form of evolutionary activity that occurs in the absence of any "genetic" variability.

4.5 MENTAL SYSTEMS

Generating and repeatedly entertaining new ideas seems to be one of the hallmarks of an active mind. One can view an individual mind, whether real or simulated, as a macroscopic system, with individual ideas as its microscopic elements. Evolutionary activity could be measured by attaching usage counters to ideas. Although difficult for a real mind, in simulated minds it is usually quite easy to keep track of "ideas." Even the activity of real minds might even be able to be inferred from their products, at least in narrow realms. For instance, in the realm of mathematics, a crude measurement of evolutionary activity could be implemented by assigning usage counters to theorems, and the counters incrementing them each time a theorem is used to prove another theorem.

Moving beyond individual minds, evolutionary activity could occur in a communicating community of mental agents, such as the modern scientific community. Again, the pattern of mental activity could be inferred from patterns in its products. For example, evolutionary activity could be calculated from the Science Citation Index, with usage straightforwardly defined as the number of citations accumulated by a given article. An interesting feature of the population of "interacting" articles is that no member of the population ever "dies." The bulk of published articles would generate activity waves that move only a very short time; articles referred to quite often would generate high velocity activity waves, and articles referred to infrequently but over a long time period would generate low velocity activity waves. Concentrated periods of fruitful scientific activity would initiate many long activity waves. The flattening of most existing waves coinciding with the beginning of a welter of new waves would signal a scientific "revolution."[23]

Evolutionary activity could be applied on social and cultural levels, as well, if only one could identify the elementary units of cultural transmission that Dawkins referred to as *memes*, on analogy to the genes that encode our physiological structure.[7] If memes could be identified, evolutionary activity in a whole culture could be measured with the same bookkeeping that we have used for genes.

5. TELEOLOGY IN EVOLVING SYSTEMS

In this section we argue that evolutionary activity may be interpreted as a measure of the extent to which an evolving system's behavior is teleological, i.e., goal-directed or purposeful. After briefly describing what teleology is, we argue that it is related to evolutionary activity in the strategic bug model. Any attempt to revive teleology in biology runs the risk of provoking controversy and criticism. As J. B. S. Haldane once quipped, "Teleology is like a mistress to a biologist: he cannot live without her but he is unwilling to be seen with her in public." Thus, we make sure to explain why the teleology in activity waves is perfectly respectable.

5.1 THE NATURE OF TELEOLOGY

Behavior that can be explained by reference to the utility of its effects we will call *teleological* (telic, goal-directed, purposive, for the sake of some end), and a *telic explanation* will explain something by reference to its beneficial effects.[1,3,38] Ours is not the only approach to teleology; comparisons with the three most attractive alternatives are detailed elsewhere.[2,4]

In ordinary parlance, telic explanations are offered for a wide variety of things, such as the behavior and structure of biological organisms and their parts, the actions of conscious human agents, and the structure and behavior of artifacts designed and used by people. All of these can be given some variation of telic explanation.[3] In each case, an essential part of the explanation is a beneficial effect brought about by the thing being explained.

The benefit promoted by any form of telic activity can be identified with the activity's purpose or goal. In the case of a human being consciously trying to produce some specific beneficial effect, the telic agent is consciously and explicitly aware of the goal. But some goal-directed activity is directed to goals that are not entertained consciously or explicitly; in these cases, the "goal" is simply a beneficial effect that explains the activity.[2]

Functionality is sometimes confused with teleology; the two are related, but they must be distinguished. Functional behavior is merely any behavior that is beneficial, that "serves a purpose," regardless of its cause. Telic or goal-directed behavior, on the other hand, is not merely functional or beneficial; it does not merely serve a purpose. It must occur specifically *because* it is beneficial, *because* it serves a purpose. Telic behavior cannot occur merely accidently or for some reason wholly unconnected with its utility.

For a given organism at a given time in a given local environment, there is a range of possible behaviors that would be more or less functional (beneficial). The "temporary local optimization criteria" set for a species by the evolutionary dynamics (recall the discussion in section 1) are the criteria for its maximal local functionality. If an organism contains a favorable new mutation, the new behavior caused by the mutation might immediately be functional. But that behavior will not be telic until its utility becomes a causal factor in its continual production. This can happen if the behavior persists through a lineage *because* of its utility.

5.2 TELIC ACTIVITY WAVES

The presence of activity waves in the strategic bug model reflects the occurrence of this sort of telic behavior. The usefulness of a gene is tested when and only when it is used. Unused genes exact no tax, so their "persistence" means little. But a well-tested gene persists in the gene pool only if the gene makes a significant contribution to the welfare of those organisms containing it. So, the presence of an activity wave shows not only that a significant number of genes are useful; it shows that a significant number of genes are present in the gene pool because they are continually verifying their usefulness. That is, these genes, and the behaviors they encode, persist because the behaviors are continually performed and continually benefit the organisms exhibiting them. Thus, activity waves reflect teleological behavior, and the continual production of new activity waves reflects the continual emergence of new teleological behavior. In this context, then, it is appropriate to speak of *telic activity waves.*

It might seem that not all genes that contribute to a telic activity wave need be beneficial. After all, an organism might use a harmful gene a number of times and still pass it on to offspring, provided the organism possessed enough *other* genuinely beneficial genes to outweigh the harm produced by the use of the harmful gene. Thus, a harmful gene could acquire some positive usage. However, harmful genes make no significant contribution to telic activity waves. For one thing, a harmful gene is unlikely to persist long enough to contribute to net persistence $P(t)$. But more importantly, since telic activity waves occur on much longer time scales than generations, the continual persistence of well-tested genes cannot be attributed to happenstance. Telic activity waves reflect genes that persist because they contribute to strategies of proven usefulness. They are not merely useful; they persist *because* they are useful. Thus, the behaviors in these well-tested strategies are teleological (goal-directed), not merely functional.

The behavior of a strategic bug is genetically hardwired, "instinctive." Its strategy allows for no flexibility of response; unable to deliberate consciously about what course of action to take, the bug cannot "freely choose" its actions. In a given local environment, it has one and only one "option" for what to do: the behavioral output coded by the gene with that local environment as input condition. Nevertheless, its behavior can still be genuinely telic or purposive, if the behavior is persisting due to its usefulness. Not every action produced by every gene in a bug's behavioral strategy is telic; it depends on why the bug has the gene. If a particular gene is present in the bug's genome because that gene has produced behavior that was beneficial for the bug's ancestors, *then* the (instinctive) behavior produced by that gene will be telic. Instinctive behavior of this sort is the simplest kind of genuinely telic behavior. More than merely functional, it is a limiting case of teleology, located on the telic spectrum at the opposite end from behavior produced by open-ended conscious deliberation.

Are all evolutionary activity waves telic? Our argument that activity waves in the strategic bug model are telic depends on the premise that a persistent gene is valuable, in the sense that it benefits the organism by enabling it to gather more food. The activity waves that might occur in other systems discussed above do not necessarily have a similarly unambiguous value-based interpretation. An evolutionary activity wave is telic only if there is a value in the persistence reflected in the activity wave. At this point, there is no theory of value comprehensive enough to include all the systems discussed above, though for many of the systems a value-based interpretation is intuitively clear. For such systems, the activity waves are telic; for others, the question remains open.

5.3 WORRIES ABOUT BIOLOGICAL TELEOLOGY

Today, any form of teleology in biology tends to be viewed with suspicion and dismissed. The controversy stems partly from the differences among the many kinds of teleology. Technical terms like "teleonomic" have been introduced in the attempt to evade the controversy[26,27]; ironically, these neologisms are used so divergently that they just add to the confusion. Whether it is fair to criticize biological teleology depends on the specific kind of teleology involved. We believe that the teleology reflected by telic activity waves is no cause for embarrassment; at least, it does not revive any of a quartet of familiar objections.

ANTHROPOMORPHISM. One complaint is that teleology in biology anthropomorphizes nature.[7,15] This complaint takes two forms: Teleology might require either that each biological creature possesses sophisticated mental capacities analogous to those possessed by a person (mentalism), or that the diversity of well-adapted creatures is the result of the activities of a mental deity (the Designer supported by the notorious argument from design).

Worries about the argument from design would clearly be misdirected at telic activity waves. The teleology in telic activity waves presupposes no deity directing things behind the scenes. The more general worry that all teleology at bottom is mentalistic can also be deflected once it is realized that biological goals can be non-conscious. In the strategic bug model, whether an organism remains alive is determined by whether it continues to find food, but the organisms are not "aware" that finding food benefits them. Thus, finding food should not be considered to be an organism's *conscious* goal. Nevertheless, since finding food is what in fact determines whether an organism survives, it can be considered to be an organism's *non-conscious* goal. As explained above, there is no requirement that all teleology be mentalistic and involve conscious goals; a non-conscious goal can be sufficient for teleology provided that it causes behavior that realizes the goal. In particular, the teleology in telic activity waves involves no mentalism. The behavior encoded by well-tested genes is teleological because it can be explained by its good consequences, but those good consequences are merely non-conscious goals.

The model could be enhanced in such a way that it could give rise to mental teleology. Organisms would need a more complex information processing mechanism with the capacity to have explicit goals for which sub-goals can be formed in response to environmental contingencies, and the capacity to gather information about what sub-goals are feasible in the current local environment. Whereas survival is measured on a time scale spanning generations, psychological value would be measured on the time scale within a single lifetime. So, whereas biological teleology takes place through a lineage existing over many generations, mental teleology would take place within one lifetime.

PREDETERMINED GOALS. Another reason for the jaundiced attitude towards teleology in biology is that teleology is thought to require that the evolutionary process itself has a predetermined goal.[8,15] Specifically, even if there is no Master Designer, teleology still must involve a Master Plan, a specific set of predetermined specifications for each species. The worry is *not* that the development of our model of biological teleology would show an average statistical trend (e.g., continual evolutionary activity); on the contrary, this would be desired in a model, since the actual biological world apparently exhibits the same statistical trends. The worry is that the genotype and phenotype of the organisms produced by the evolutionary process are determined *a priori*, independently of local environmental and ecological contingencies.

However, the sort of teleology signaled by telic activity waves involves no predetermined goals. Rather, the form to which organisms have evolved is determined by whatever happens to be sufficiently beneficial in the continually changing local biological context. (Recall the discussion of intrinsic adaptation in section 1.) Rather than being specified in advance, the organisms' forms depend on random, non-teleological genetic changes and the contingencies of the struggle for survival. So, the teleology in telic activity waves emerges *a posteriori*.

FUTURE CAUSATION. Teleology in biology is sometimes thought to require that events in the future (the realization of goals) have causal efficacy over present behaviors, a sort of "future causation" that seems patently absurd.[28] Indeed, our view of teleology might appear to involve future causation, but this appearance evaporates under scrutiny.

Biological events can be viewed on either a micro or a macro level. The micro perspective involves events on a time scale within a generation. From this perspective, an individual organism's behavior has a telic explanation when it occurs because *in the past* the same kind of behavior helped the organism or the organism's ancestors to flourish. A behavior's past beneficial effects cause the same kind of behavior in the present, which might well cause further manifestations of the same kind of behavior. The causation in this explanation is of the ordinary kind—past events causing present events. By contrast, the macro perspective involves a time scale spanning many generations. From a vantage abstracted from individual generations, one can say simply that a teleological behavior occurs because on average that kind of behavior promotes (in a tenseless sense) the survival of those organisms that exhibit the behavior. This kind of merely apparent future causation at the macro level is harmless because it is underwritten by ordinary causation at the micro level.

VALUE AND SUBJECTIVITY. Our interpretation of teleology requires that good effects are causally efficacious. Some might judge that this reference to an effect's goodness or value is inescapably subjective, possibly on the grounds that all value judgments are inherently subjective. However, we believe that an objective criterion of an organism's welfare is its ability to survive and reproduce. In the strategic bug model, an organism's welfare consists of no more and no less than this.

We ignore other possible components of an organism's welfare, not because we believe that there could be none, but merely to simplify our model. A more complicated model could incorporate benefits that are unconnected with a creature's survival, such as pleasures, the satisfaction of desires, and other "psychological" goods. In the strategic bug model, however, all telic phenomena are shaped by their value simply for survival and reproduction.

6. VITALITY AS A TEST FOR LIFE

What is life? How can it be recognized? In an everyday context, these questions seem tantalizingly clear—a cat is alive and a rock is not. But formalizing this distinction is difficult, especially if the formalization is to be used in empirical measurements.

Life is usually thought of as a property of individual organisms. We propose to make a gestalt switch and view life from a more global, statistical perspective. No single molecule of gas has a macroscopic property like temperature; temperature is meaningful only for large populations of molecules. Similarly, no single organism exhibits indefinitely ongoing life; in the long run not even a lineage remains alive. Individual life is "here today, gone tomorrow" and, in fact, intuitively this transitory nature is one of its characteristic features. From a global perspective, only the complex web of interacting organisms—the entire biosphere—remains "alive" in the long run, through the continual cycle of birth and death of individual organisms. So, rather than try to define what it is for an individual "microscopic" organism to be alive, our concern is with what it is for a "macroscopic" system (population of organisms) to exhibit the property of indefinitely ongoing life.

Evolutionary activity is an especially salient global property of populations of living organisms; it seems that they cannot help but evolve, at least in the long run. In addition, minds and other non-biological systems with lots of evolutionary activity exhibit a kind of "liveliness." Thus, we will say that a system is *vital* if and only if it has positive evolutionary activity.

We believe it is fruitful, theoretically and experimentally, to link the notions of an individual's life to the vitality of the global system in which the individual lives. In fact, this link is already suggested by the common claim that an organism is alive only if it is a member of an actively evolving biosphere.[6,25] Our measure of evolutionary activity sharpens this claim into the following *life-vitality hypothesis*: Vitality—positive evolutionary activity $A(t)$—is a necessary condition of systems containing living individuals, and the measure of the vitality of a system of living individuals is the rate at which new telic activity waves are generated. Note that in the hypothesis vitality is only *necessary* for a system to contain living elements; it is not sufficient since the ideas entertained by a vital mind, for example, are not alive.

It is important to recognize what the life-vitality hypothesis is *not*. It is a contingent fact that the biosphere is the product of evolution; life might not have been linked to evolution in the way that it is. Organisms *could* have been designed and created by an omnipotent deity, and there *could* have been eternal, non-evolving forms of life, such as angels. Similarly, medical technology *could* improve to the point that individual organisms remain alive indefinitely and thus never evolve. These fanciful possibilities show that the life-vitality hypothesis is not a conceptual necessity; rather, it is the sort of contingent empirical claim that is characteristic of hypotheses in the natural sciences.

Certain facts about the biosphere might seem to contradict the life-vitality hypothesis. Infertile individual organisms live and die without affecting the evolutionary activity of their population. Furthermore, the individuals in a lineage or a sub-population are all alive, but certain lineages such as the shark have persisted without evolving appreciably for a quite long time, and certain sub-population of infertile organisms such as mules are simply incapable of evolving. But these admissions do not contravene the vital systems hypothesis, for infertile individuals and infertile sub-populations are always transitory members of a global biosphere which is certainly evolving in the long run. You can't get a mule or a shark except from a vital biosphere.

The possibility of an ecology that has reached a stable "climax" state and stopped evolving challenges the life-vitality hypothesis more directly. Although the existing genetic combinations in a climax biosphere would create continually propagating activity waves, the gene pool would no longer be absorbing new genetic variations. Thus, a climax biosphere would produce no new activity waves and its evolutionary activity $A(t)$ would consequently drop to zero. In this straightforward sense, a static climax biosphere is "less vital" than one that is continuously evolving. Yet the organisms in the biosphere would still be merrily living. Thus, for climax biospheres the vital systems hypothesis breaks down. However, the extent of this breakdown vanishes in the long run, it seems, for real biospheres apparently do not remain indefinitely in a state of climax. On the contrary, in the long run biospheres seem to continue to evolve.

The field of artificial life is searching for a definition of life; even better would be a criterion of life—a public, empirical, repeatable, quantifiable test for whether a system (possibly artificial) is alive. The analogous issue in artificial intelligence is how to tell whether a system is thinking. Forty years ago Alan Turing proposed a criterion for thinking—the famous Turing test.[37] What the field of artificial life needs is an analogue test for life.

Although public, empirical, repeatable, and quantifiable, the Turing test has two limitations: it detects thinking only *indirectly*, through its effects on overt behavior, and it evaluates this behavior through the *subjective* opinions of a panel of human jurors. It would be preferable to have tests that directly and objectively measure whether (and to what extent) a machine, or a model being implemented on a machine, exhibits intelligence or life.

Our quantity $A(t)$ is a direct and objective measure of a system's vitality or evolutionary activity. Thus, by invoking the life-vitality hypothesis, $A(t)$ provides a direct and objective measure of the degree to which a system exhibits life, yielding a Turing test for life.

As we noted in section 5.2, a biosphere's vitality also is a measure of its teleology. This leads to another link between vitality and life. Purposeful behavior has often been cited as an especially characteristic sign of individual life.[6,25,26,29,30] But this proposal has been unhelpful until now, since no way to quantify and measure purposefulness was known. If this teleology-life relationship is recast at a global level, as we have done for the link between evolution and life, then telic activity waves allow teleology to be quantified. Purposeful behavior is signaled by telic activity waves, and the level of vitality measures the degree to which new purposeful behavior is continually emerging. Thus, vitality at one fell-swoop quantifies two intuitive signs of life: evolutionary activity and purposeful behavior.

Measuring vitality in models that include psychological factors could also provide a direct and objective substitute for the original Turing test itself. If the mind is viewed as a global, statistical system, positive activity would indicate a mind's continual incorporation of new behavioral or psychological patterns of activity (see section 4.5). Statistics such as our measure of evolutionary activity might provide a method for quantifying the purposefulness common to both life and mind.

ACKNOWLEDGMENTS

We would like to acknowledge the hospitality of the Santa Fe Institute and the Complex Systems Group of the Theoretical Division at the Los Alamos National Laboratory, where this work began. We appreciate many helpful comments on the manuscript from M. Hammer, C. Langton, J. Moor, J. Page, C. Voeller, and an anonymous reviewer, and helpful conversations with W. Fontana, J. Holland, S. Kauffman, S. Lloyd, T. Meyer, and P. Schuster. NP also thanks the "Complexity and Evolution" program at the Institute for Scientific Interchange, where part of this work was completed. This work was partially supported by National Science Foundation Grant number PHY86-58062.

REFERENCES

1. Ayala, F. J. "Teleological Explanations in Evolutionary Biology." *Phil. of Sci.* **37** (1970): 1–15.
2. Bedau, M. A. "Against Mentalism in Teleology." *Amer. Phil. Quart.* **27** (1990): 61–70.
3. Bedau, M. A. "Where's the Good in Teleology?" *Philosophy and Phenomenological Research*, forthcoming.
4. Bedau, M. A. "Naturalism and Teleology." In *Naturalism: A Critical Appraisal*, edited by S. Wagner and R. Warner. Notre Dame, IN: The University of Notre Dame Press, 1992.
5. Cairns-Smith, A. G. *Genetic Takeover and the Mineral Origins of Life.* Cambridge: Cambridge University Press, 1982.
6. Cairns-Smith, A. G. *Seven Clues to the Origin of Life.* Cambridge: Cambridge University Press, 1985.
7. Dawkins, R. *The Selfish Gene.* Oxford: Oxford University Press, 1976.
8. Dawkins, R. *The Blind Watchmaker: Why the Evidence of Evolution Reveals a Universe Without Design.* Norton, 1986.
9. Dupré, J., ed. *The Latest on the Best: Essays on Evolution and Optimality.* Cambridge: The MIT Press, 1987.
10. Dobzhansky, T. G. *The Biology of Ultimate Concern.* New American Library, 1967.
11. Farmer, J. D., S. A. Kauffman, and N. H. Packard, "Autocatalytic Replication of Polymers." *Physica* **22D** (1986): 50–67.
12. Farmer, J. D., S. A. Kauffman, N. H. Packard, and A. S. Perelson. "Adaptive Dynamic Networks as Models for the Immune System and Autocatalytic Sets." *Ann. the N.Y. Acad. Sci.* **504** (1987): 118–130.
13. Farmer, J. D., N. H. Packard, and A. S. Perelson. "The Immune System, Adaptation, and Machine Learning." *Physica* **22D** (1986): 187–204.
14. Fontana, W. "Algorithmic Chemistry: A Model for Functional Self-Organization." This volume
15. Gould, S. J. *Ever Since Darwin.* Norton, 1977.
16. Gould, S. J., and R. C. Lewontin. "The Spandrels of San Marco and the Panglossian Paradigm: A Critique of the Adaptationist Programme." *Proceedings of the Royal Society of London* **B 205** (1979): 581-598.
17. Holland, J. H. *Adaptation in Natural and Artificial Systems.* University of Michigan Press, 1975.
18. Holland, J. H. "Escaping Brittleness: The Possibilities of General-Purpose Learning Algorithms Applied to Parallel Rule-Based Systems." In *Machine Learning, An Artificial Intelligence Approach*, edited by R. S. Michalski, J. G. Carbonell, and T. M. Mitchell, Vol. II, 593–623. Morgan Kaufmann, 1986.
19. Huberman, B. A., ed. *The Ecology of Computation.* North-Holland, 1988.
20. Kauffman, S. A. "Autocatalytic Sets of Proteins." *J. Theor. Biol.* **119** (1986): 1–24.

21. Kimura, M. *The Neutral Theory of Molecular Evolution*. Cambridge University Press, 1983.
22. Langton, C. G. "Computation at the Edge of Chaos: Phase Transitions and Emergent Computation." Ph.D. dissertation, EECS Department, University of Michigan, 1990.
23. Kuhn, T. S. *The Structure of Scientific Revolutions*. 2nd edition. Chicago: The University of Chicago Press, 1970.
24. Lewontin, R. "The Units of Selection." *Annual Review of Ecology and Systematics* **1** (1970): 1–14.
25. Maynard Smith, J. *The Theory of Evolution*. 3rd edition. Penguin, 1975.
26. Mayr, E. "Teleological and Teleonomic, A New Analysis." In *Methodological and Historical Essays in the Natural and Social Sciences*, edited by R. S. Cohen and M. W. Wartofsky, 91–117. Reidel, 1974.
27. Mayr, E. *The Growth of Biological Thought*. Cambridge: Harvard University Press, 1982.
28. Medwar, P. B., and J. S. Medwar. *Aristotle to Zoos*. Cambridge: Harvard University
Press, 1983.
29. Monod, J. *Chance and Necessity*. Translated by A. Wainhouse. Vintage, 1972.
30. Nagel, E. "Teleology Revisited." *J. Phil.* **74** (1977): 261–301.
31. Nitecki, M. H., ed. *Evolutionary Progress*. Chicago: The University of Chicago Press, 1988.
32. Packard, N. H. "Intrinsic Adaptation in a Simple Model for Evolution," In *Artificial Life*, edited by C. Langton, Studies in the Sciences of Complexity, Vol. VI, 141–155. Redwood City, CA: Addison-Wesley, 1988.
33. Rössler, O. "Chemical Automata in Homeogeneous and Reaction-Diffusion Kinetics." *Lecture Notes in Biomathematics* **4** (1974): 399–418.
34. Rössler, O. "Deductive Prebiology." In *Molecular Evolution and the Prebiological Paradigm*, edited by K. Matsuno, K. Dose, K. Harada, and D. L. Rohlfing, 375–385. New York: Plenum, 1984.
35. Sober, E., ed. *Conceptual Issues in Evolutionary Biology, An Anthology*. Cambridge: MIT Press, 1984.
36. Sober, E. *The Nature of Selection*. Cambridge: MIT Press, 1984.
37. Turing, A. M. "Can a Machine Think?" *Mind* **59** (1951): 433–460.
38. Woodfield, A. *Teleology*. Cambridge: Cambridge University Press, 1976.

Development

Martin J. M. de Boer,* F. David Fracchia,† and Przemyslaw Prusinkiewicz†
*Theoretical Biology Group, University of Utrecht, Padualaan 8, 3584 CH Utrecht, The Netherlands. †Department of Computer Science, University of Regina, Regina, Saskatchewan, Canada S4S 0A2.

Analysis and Simulation of the Development of Cellular Layers

The organization of cell division and its relation to growth and shape formation in organisms that consist of a single cellular layer is investigated by computer simulation. Context-free map Lindenmayer systems describe the successions of cell divisions and emphasize regularities in the patterns. Realistic graphical representation of the development of cellular layers and visualization methods[6,7] are instrumental in our simulations.[1]

INTRODUCTION

Cell division is the fission of a cell by the insertion of a new wall. Because plant cell walls are tightly connected, the spatial-temporal organization of cell division determines the neighborhood structure of cell patterns.

Successions of cell divisions can be observed on the surface of plant tissues by frequent observations and can sometimes (if cell walls thicken as they get older) be traced back for several generations from a single pattern. Tissues such as leaf epidermis and epidermis of shoot meristems show regular cell division patterns.

[1]This paper contains edited sections of two papers by Fracchia, Prusinkiewicz, and de Boer.[6,7]

Deterministic context-free map Lindenmayer systems (map DOL-systems) provide a powerful tool to capture and recognize these regularities.[2,10,12]

We assume that individual cell shapes are determined by mechanical cell interactions and use a dynamic approach. The development of cellular structures is visualized using a simulation method based on map DOL-systems for the algorithmic description of cell division and the dynamic approach for cell shapes. We employ this method for the analysis of shape formation in fern gametophytes, which consist of a flat cellular layer, and for the visualization of the early development of invertebrate embryos, which consist of a spherical cellular layer.

MAP L-SYSTEMS

In order to simulate the development of cellular structures, one needs a formal representation of the structures (maps) and a formalism that operates on these structures (map L-systems).

MAPS FOR THE REPRESENTATION OF CELL LAYERS

Maps form a class of planar graphs with cycles[16] and can be characterized as follows[13]:

- A map is a finite set of regions. Each region is surrounded by a boundary consisting of a finite, circular sequence of edges which meet at vertices.
- Each edge has one or two edges associated with it. (The one-vertex case occurs when an edge forms a loop). Edges cannot cross without forming a vertex and there are no vertices without an associated edge.
- Every edge is a part of the boundary of a region.
- The set of edges is connected. Specifically, there are no islands within regions.

A map represents a microscopic view of a cellular layer. Regions represent cells and edges represent cell walls perpendicular to the plane of view. We abstract here from the internal components of a cell.

MAP REWRITING SYSTEMS

The process of cell division can be expressed as map rewriting. This notion is an extension of string rewriting used in formal language theory. In general, map rewriting systems are categorized as sequential or parallel, and can be region controlled or edge controlled. Since several cells may divide concurrently, a parallel rewriting system is needed. The second categorization has to do with the form of rewriting rules, which may express cell subdivisions in terms of region labels or edge labels. Both approaches are suitable for biological modeling purposes.[2] We have chosen the edge-controlled formalism of Binary Propagating Map OL-systems with markers, or mBPMOL-systems. It was proposed by Nakamura et al.[13] as a refinement of the map L-systems introduced by Lindenmayer and Rozenberg.[10] The name is derived as follows. A map OL-system is a parallel rewriting system which operates on maps and does not allow for interaction. In other words, edges are modified irrespective of what happens to adjacent edges (a context-free mechanism). The system is binary because a region can split into at most two-daughter regions. It is propagating in the sense that edges cannot be erased. Thus regions cannot fuse. The markers represent a technique for specifying the positions of inserted edges that split the regions. The choice of mBPMOL-systems as a modeling tool has two justifications. First, they are more powerful than other interactionless map rewriting systems in the literature.[1,2,4] In addition, markers have a biological counterpart in preprophase bands of microtubules, which coincide with attachment sites of division walls formed during mitosis in plants.[8].

DEFINITION AND OPERATION OF MAP DOL-SYSTEMS

An mBPMOL-system G is defined by specifying a finite alphabet of edge labels $\sum$, a starting map w with labels from $\sum$, and a finite set of edge productions P. In general, the edges are directed, which is indicated by a left ($<$) or right ($>$) arrow placed next to the edge label. In some cases, the edge direction has no effect on the system operation. Such an edge is called neutral and no arrow is associated with the label denoting it. Each production is of the form $A \to a$, where the directed or neutral edge A is called the predecessor and the string a, composed of edge labels from $\sum$ and special symbols $[,], +$ and $-$, is called the successor. The sequence of symbols outside the square brackets specifies the edge subdivision pattern. Arrows can be placed behind edge labels to indicate whether the successor edges have directions consistent with, or opposite to, the predecessor edge. Pairs of matching brackets [and] delimit markers, which specify possible attachment sites for region dividing edges. The markers are viewed as short branches which can be connected to a division edge. The strings inside brackets consist of two symbols. The first symbol is either + or $-$, indicating whether the marker is placed to the left or the right of the predecessor edge. The second symbol is the marker label with or without an arrow. The left arrow indicates that the marker is directed towards the predecessor edge, and the right arrow indicates that the marker is oriented away from the predecessor edge. If no arrow is present, the marker is neutral.

For example, in the production $A\rangle \to D\rangle C\langle[-E\langle]B\rangle F$, the directed predecessor A splits into four edges D, C, B and F, and produces a marker E (Figure 1(a)). Successor edges D and B have the same direction as A, edge C has the opposite direction and F is neutral. Marker E is placed to the right of A and is directed towards A. Note that this same production could be written as $A\langle\to FB\langle[+E\langle]C\rangle D\langle$ (Figure 1(b)). As an example of a production with a neutral predecessor consider $A \to B\rangle[-B\langle]x[+B\langle]B\langle$. In this case the result of production application does not depend on the assumed direction of the predecessor edge (Figure 1(c)).

A derivation step in an mBPMOL-system consist of two phases:

1. Each edge in the map is replaced by successor edges and markers using the corresponding edge production in P.
2. Each region is scanned for matching markers.

Two markers are considered matching if (1) they appear in the same region, (2) they have the same label, and (3) one marker is directed away from its incident edge while the other is directed towards its incident edge, or both markers are neutral. If a match is found, the markers are joined to create a new edge which will split the region. The search for matching markers ends with the first match found, even though other markers entering the same region may also form a match. From the user's perspective, the system may behave in a non-deterministic way since it autonomously chooses the pair of markers to be connected. The unused markers are discarded. The mBPMOL-systems that are presented in this paper have the property that, in each region, at most one match can be found. Therefore, such systems are deterministic. In the following we denote mBPMOL-systems simply by map L-systems.

EXAMPLES OF MAP L-SYSTEMS

The two examples to follow illustrate the operation of map L-systems. Map L-system 1 has only neutral edges:

$$\begin{array}{ll} w: & ABAB \\ p_1: & A \rightarrow B[-A]x[+A]B \\ p_2: & B \rightarrow A \end{array}$$

Production p_1 creates markers for region division, while production p_2 introduces a delay, so that regions are subdivided alternately by horizontal and vertical edges. The edges x separate the markers in the successor of production p_1. This edge creates a Z-shaped offset between the inserted edges A (Figure 2). Z-offsets and symmetric S-offsets (Figure 3) are common in biological structures.[12]

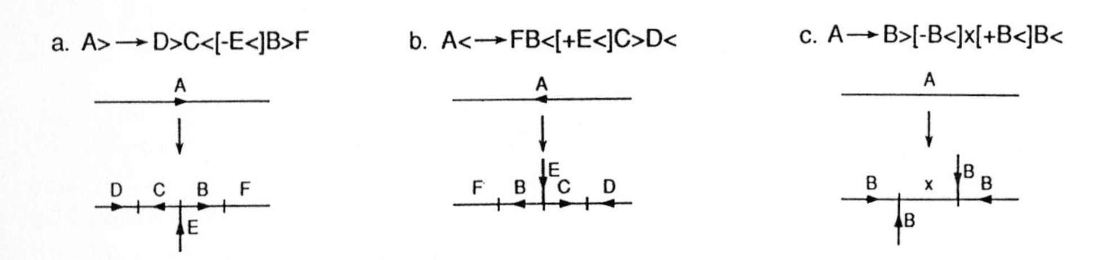

FIGURE 1 Examples of edge productions.

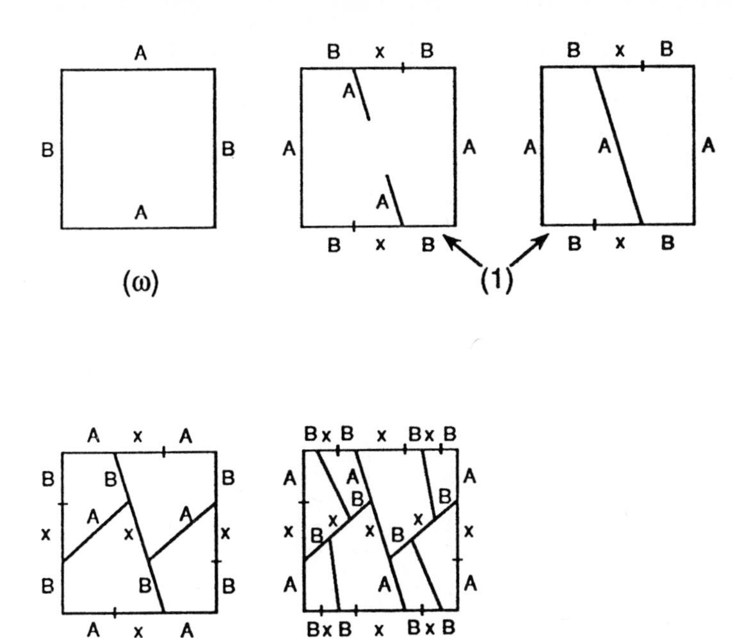

FIGURE 2 Developmental sequence defined by map L-system 1. In the first step, a distinction is made between the edge-rewriting phase and the connection of matching markers.

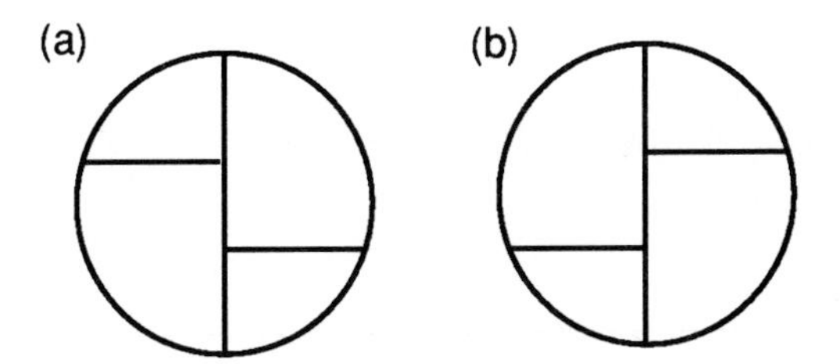

FIGURE 3 Offsets between four regions that result from the division of two regions sharing a common wall: (a) Z-offset, (b) S-offset.

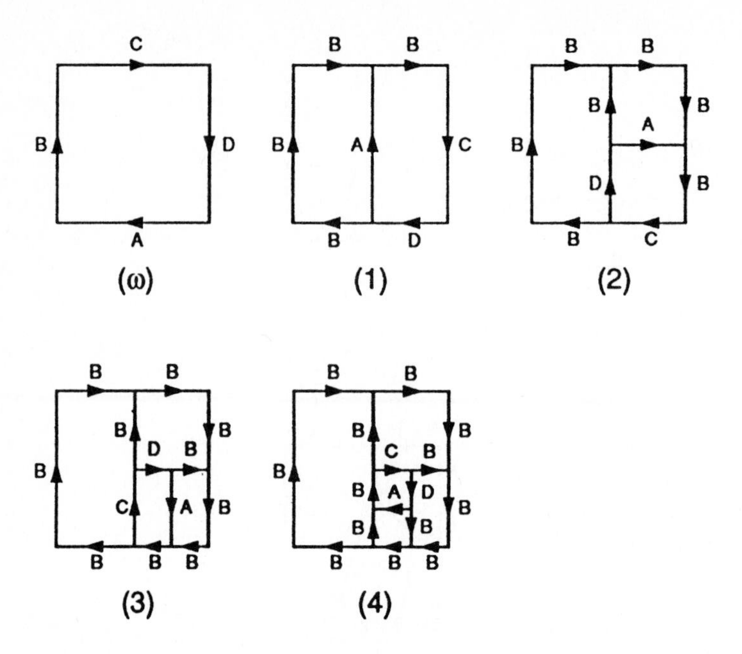

FIGURE 4 Developmental sequence defined by map L-system 2.

Map L-system 2 illustrates the operation of a map L-system with directed edges:

$$\begin{aligned} &w:\ A\rangle B\rangle C\rangle D\rangle \\ &p_1:\ A\rangle \rightarrow D\rangle[-A\rangle]B\rangle \\ &p_2:\ B\rangle \rightarrow B\rangle \\ &p_3:\ C\rangle \rightarrow B\rangle[-A\langle]B\rangle \\ &p_4:\ D\rangle \rightarrow C\rangle \end{aligned}$$

Productions p_1 and p_3 create matching markers. Production p_4 transforms edge D into C, so that in each derivation step, there is a pair of edges A and C to which productions p_1 and p_3 apply. Production p_2 indicates that edges B do not undergo further changes. In further L-systems, such identity productions are omitted. The resulting structure is that of a clockwise spiral (Figure 4).

GRAPHICAL INTERPRETATION OF MAP L-SYSTEMS

Maps are topological objects without inherent geometrical properties. In order to visualize them, some method for assigning geometric interpretation must be applied. Previous methods[5,15] lack sufficient biological justification. We use a method based on a dynamic interpretation.[6,7] We describe it using the biologically motivated words *cell corner, wall,* and *cell* equivalently with the mathematical terms vertex, edge, and region, respectively.

DYNAMIC INTERPRETATION

Assuming the dynamic point-of-view, the shape of cells and thus the shape of the entire organism result from the action of forces. The unbalanced forces, due to cell divisions, cause the gradual modification of cell shapes until an equilibrium is reached. At this point, new cell divisions occur, and expansion resumes. The dynamic method for determining cell geometry is based on the following assumptions:

- The modeled organism forms a single cell layer.
- The layer is represented as a two-dimensional network of masses corresponding to cell corners, connected by springs which correspond to cell walls.
- The springs are always straight and spring tension ($\vec{F_s}$) adheres to Hooke's law.
- The cells exert pressure ($\vec{P}$) on their bounding walls; the pressure on a wall is directly proportional to the wall length, and inversely proportional to the cell area.
- The pressure on a wall spreads evenly between the wall corners.
- The motion ($\vec{V}$) of masses is damped by a damping force ($\vec{F_d}$).
- No other forces are considered (Figure 5).

The position of each vertex, and thus the shape of the cell layer, is computed as follows. As long as an equilibrium is not reached, unbalanced forces put masses into motion. The total force $\vec{F_T}$ (forces exerted by incident walls plus the damping force) acts on a mass placed at a map vertex. Newton's second law of motion applies.

If the entire structure has N vertices, we obtain a system of $2N$ differential equations:

$$m_i \frac{d\vec{v_i}}{dt} = \vec{F_T}(\vec{x_1}, \ldots, \vec{x_N}, \vec{v_i})$$

$$\frac{d\vec{x_i}}{dt} = \vec{v_i}$$

where $\vec{x_i}$ is the vertex position and v_i is the vertex speed, $i = 1, 2, \ldots, N$.

This system is solved numerically using the forward Euler method. The detailed computations appear in Fracchia et al.[6] The next map L-system derivation step is then performed. A system of equations corresponding to the new map topology is created, and the search for an equilibrium state resumes. In such a way, the developmental process is simulated as periods of continuous cell expansion, delimited by instantaneous cell divisions. Continuity of cell shapes during divisions is preserved by the rule which sets the initial positions of vertices.

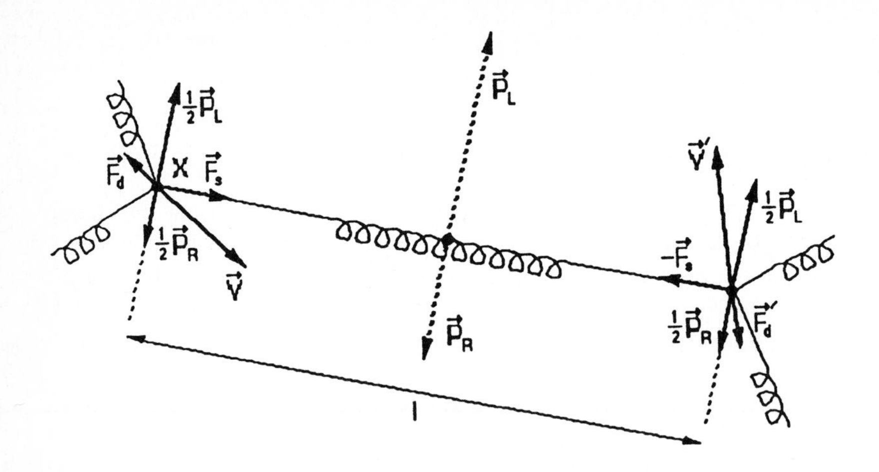

FIGURE 5 Forces acting on a cell corner X, according to the dynamic method. $\vec{F}_s$ is the wall tension, $\vec{F}_d$ is the damping force, $\vec{P}_L$ and $\vec{P}_R$ are the cell pressure forces, and $\vec{V}$ is the velocity.

BIOLOGICAL EXAMPLES

In this section, two examples of fern gametophyte development and an example of a cleavage pattern of a limpet illustrate our simulation method.

FERN GAMETOPHYTES

Fern gametophytes represent the sexually reproducing life stage of ferns. They show no differentiation into stem, leaf, and root, forming a plant body called a thallus which is one cell layer thick. Our study of cell division patterns of young fern gametophytes concentrates on the relationship between cell division pattern and shape. The modeling process captures repetitive patterns of cell divisions, so that large cellular structures can be described using a small number of productions. We first discuss a general scheme of gametophyte development with *Microsorium linguaeforme* as an example. In a second example we simulate heart shape formation of *Dryopteris thelypteris*.

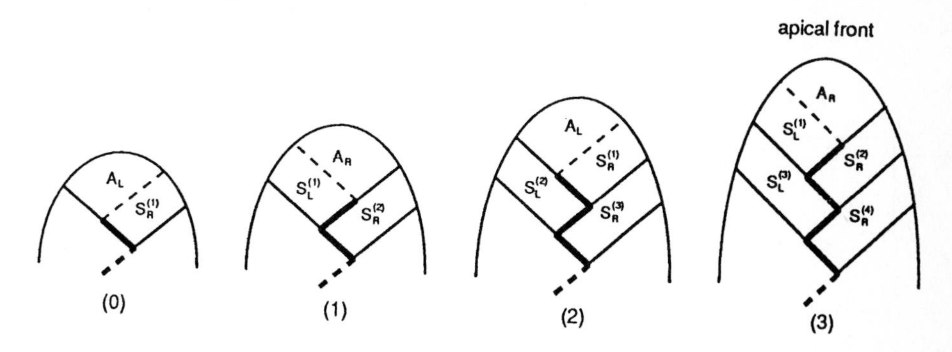

FIGURE 6 Production of segments. The labels A_R and A_L denote apical cells producing right segments S_R and left segments S_L respectively. Dashed lines indicate the newly created division wall. The superscripts represent segment age. The internal structure of the segments is not shown.

APICAL ACTIVITY The apical cell is the originator of the gametophyte structure. It divides repetitively, giving rise each time to a new apical cell and a primary segment cell. The segment cells subsequently develop into multicellular segments. The division wall of an apical cell is attached to the thallus border on one side and to a previously created division wall on the other side. Thus the division walls are oriented alternately to the left and to the right, yielding two columns of segments separated by a zig-zag dividing line (Figure 6). The recursive nature of the apical activity can be expressed by the following "cell production system."

$$A_L \rightarrow S_L \mid A_R \qquad\qquad A_R \rightarrow A_L \mid S_R$$

This notation means that the cell on the left side of the arrow sign divides into two daughter cells separated by a wall.

DIVISION PATTERN OF SEGMENTS In describing the structure of a segment, we distinguish between "periclinal" and "anticlinal" walls. Intuitively, periclinal walls are approximately parallel to the apical front of the thallus and anticlinal walls are perpendicular to this front. A more formal definition follows:

- In a primary segment cell, the apical front wall and one or more walls opposing it are periclinal walls. All other walls are periclinal walls.
- A division wall attached to the periclinal walls is an anticlinal wall and vice-versa.

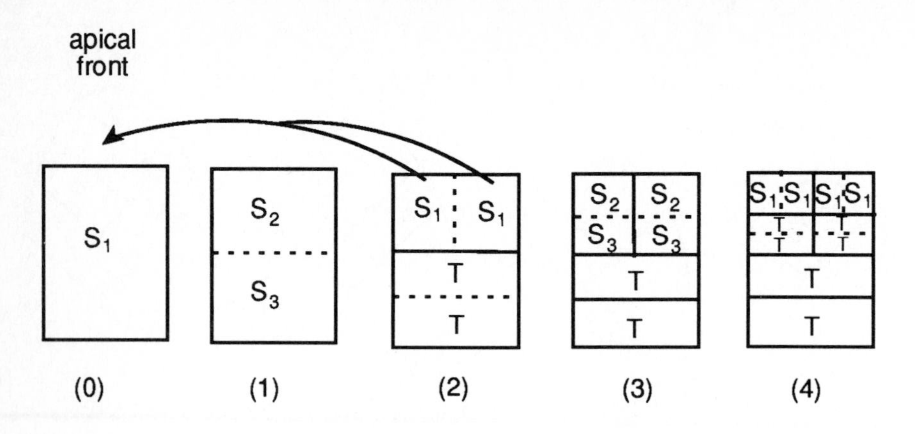

FIGURE 7 Developmental sequence of a *Microsorium* segment.

Division walls that are attached to a periclinal wall on one side and an anticlinal wall on the other side are extremely rare, and therefore we neglect them in the process of modeling.

Microscopic observations of *Microsorium linguaeforme* gametophytes reveal that close to the apex all segments follow roughly the same developmental sequence, shown diagrammatically in Figure 7. The primary segment cell S_1 is first divided by a periclinal wall into two cells, S_2 and S_3. Subsequently, the basal cell is divided by another periclinal wall into two "terminal cells" T which do not undergo further divisions. At the same time, the cell S_2 lying on the thallus border is divided by an anticlinal wall into two cells of type S_1. Each of these cells divides in the same way as the primary cell. Consequently, the recursive nature of segment development can be captured by the following cell production system.

$$S_1 \rightarrow \frac{S_2}{S_3} \qquad S_2 \rightarrow S_1 \mid S_1 \qquad S_3 \rightarrow \frac{T}{T}$$

In the above rules, a horizontal bar denotes a periclinal wall between cells, and a vertical bar denotes an anticlinal wall.

DEVELOPMENT OF THE THALLUS The development of a thallus is the result of concurrent divisions of the apical cells and the segment cells. A single division of an apical cell corresponds to a single step in segment development. A developmental sequence of the *Microsorium* thallus which combines the activity of the apex and of the segments is shown in Figure 8. This figure also reveals offsets between neighboring walls. On the basis of observation, it is assumed that periclinal walls form S-offsets in segments on the right side of the apex, and Z-offsets in the segments on the left side.

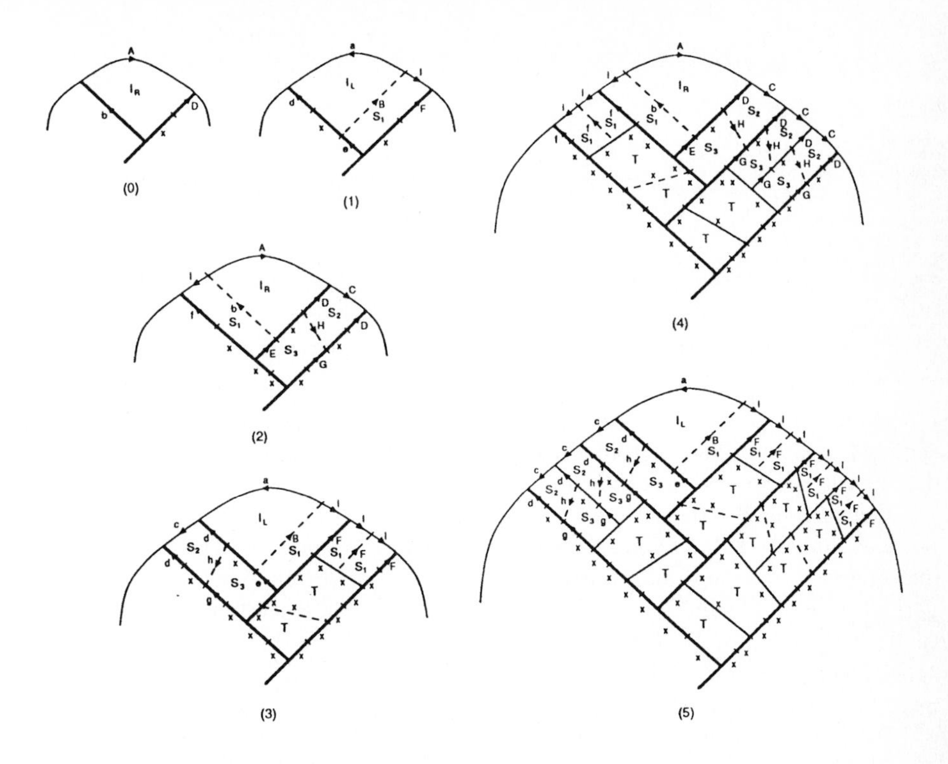

FIGURE 8 Developmental sequence of a *Microsorium* gametophyte defined by map L-system 3.

EXPRESSING THE DEVELOPMENT USING A MAP L-SYSTEM In order to capture the thallus development using a map L-system, it is necessary to identify all combinations of cells which may lie on both sides of a wall. Careful examination of these combinations in the developmental sequence of *Microsorium* yields the wall labeling scheme shown in Figure 8. Two walls have the same label if and only if they divide in the same way. This construction method is described in detail in De Boer.[2] The uppercase letters apply to right segment walls, and the corresponding lowercase letters denote symmetric walls in the left segments. By comparing pairs of subsequent structures we arrive at the following map L-system 3 (aoucak region of *Microsorium*):

$w\colon A\rangle D\langle xb\rangle$

$l_1\colon a\rangle \rightarrow A\langle[+b\langle]i\rangle$	$r_1\colon A\rangle \rightarrow a\langle[-B\langle]I\rangle$
$l_2\colon b\rangle \rightarrow e\rangle[-B\rangle]x[+h\rangle]d\rangle$	$r_2\colon B\rangle \rightarrow E\rangle[+b\rangle]x[-H\rangle]D\rangle$
$l_3\colon d\rangle \rightarrow f\rangle$	$r_3\colon D\rangle \rightarrow F\rangle$
$l_4\colon f\rangle \rightarrow g\rangle[-h\langle]x[+h\rangle]d\rangle$	$r_4\colon F\rangle \rightarrow E\rangle[+H\langle]x[-H\rangle]D\rangle$
$l_5\colon h\rangle \rightarrow x[-f\rangle]x$	$r_5\colon H\rangle \rightarrow x[+F\rangle]x$
$l_6\colon i\rangle \rightarrow c\rangle$	$r_6\colon I\rangle \rightarrow C\rangle$
$l_7\colon c\rangle \rightarrow i\rangle[+f\langle]i\rangle$	$r_7\colon C\rangle \rightarrow I\rangle[-F\langle]I\rangle$
$l_8\colon e\rangle \rightarrow x[+x]x$	$r_8\colon E\rangle \rightarrow x[-x]x$
$l_9\colon g\rangle \rightarrow x[-x]x[+x]x$	$r_9\colon G\rangle \rightarrow x[+x]x[-x]x$

The divisions of the apical cells result from the application of productions r_1-l_2 (creation of a right segment) and l_1-r_2 (creation of a left segment). The subsequent segment cell divisions proceed in a symmetric way in right and left segments; we describe in detail the development of a right segment.

Concurrently with the insertion of wall B which creates segment S_R^1, wall D on the opposite side of the segment is transformed into F. The transformation introduces a one-step delay into the application of production r_4. In the second derivation step production r_4, together with r_2, is responsible for the insertion of the first periclinal wall H into segment S_R^2. As the derivation progresses, r_4 inserts subsequent periclinal walls H between pairs of anticlinal walls F. Production r_3 introduces a delay needed to create walls F which are inserted between periclinal walls H and C, using r_5 and r_7. Production r_6 plays a role analogous to r_3—it introduces a one-step delay into the cycle, creating markers F at the apical front of the segment. Thus, periclinal walls H and anticlinal walls F are created alternatingly, in subsequent derivation steps. Productions r_8 and r_9 create terminal walls x, which do not undergo further changes. The first such wall is inserted between walls labeled D and E during derivation step 3. Wall E separates segment S_R^2 from S_L^1. Subsequent walls x are inserted every second step between pairs of walls D by the application of r_9.

SIMULATION OF THE ENTIRE STRUCTURE In order to compare the simulations with the real Microsorium structure, we complete the above model of the apical region of the thallus with productions for segments in the basal region. L-system 3 was formulated under the assumption that all segments develop identically. However, in *Microsorium*, the first two segments at the thallus base exhibit a modified pattern of development which can be characterized by more cell division activity on the apical sides than on the basal sides of the segments (Figure 9). The corresponding cell production system of a left basal segment follows.

$$S_1 \rightarrow \frac{S_2}{S_3} \qquad S_2 \rightarrow S_1 \mid T \qquad S_3 \rightarrow \frac{T}{T}$$

The map L-system 4 describing the development of *Microsorium* including the basal segments is the previous map L-system with the addition of several productions.

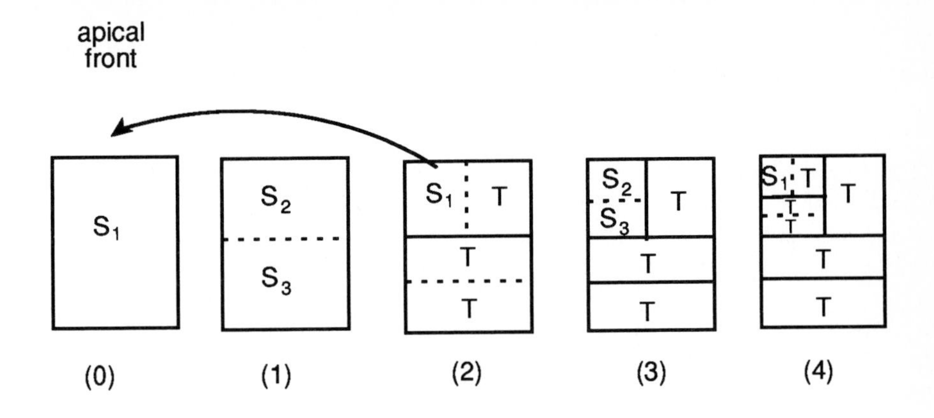

FIGURE 9 Developmental sequence of a left basal segment of *Microsorium*.

w: apical cell $A\rangle D\langle xb\rangle$ with a basal segment on each side on top of a filament.

r_1: $A\rangle \to a\langle[-B\langle]I\rangle$
r_2: $B\rangle \to E\rangle[+b\rangle]x[-H\rangle]D\rangle$
r_3: $D\rangle \to F\rangle$
r_4: $F\rangle \to G\rangle[+H\langle]x[-H\rangle]D\rangle$
r_5: $H\rangle \to x[+F\rangle]x$
r_6: $I\rangle \to C\rangle$
r_7: $C\rangle \to I\rangle[-F\langle]I\rangle$
r_8: $E\rangle \to x[-x]x$
r_9: $G\rangle \to x[+x]x[-x]x$
r_{10}: $J\rangle \to L\rangle$
r_{11}: $K\rangle \to N\rangle$
r_{12}: $R\rangle \to x[-M\langle]x$
r_{13}: $M\rangle \to x[-L\rangle]x$
r_{14}: $N\rangle \to O\rangle$
r_{15}: $O\rangle \to x[-L\langle]N\rangle$

Only productions describing the development of the right side are given. Their predecessors are denoted by uppercase labels. The corresponding lowercase productions, which complete the map L-system, can be obtained by switching the "case" of letters (except x) and the orientation of markers. The wall direction remains unchanged. For example, the right-side production r_{15} corresponds to the left-side production $l_{15}\colon o\rangle \to x[+l\langle]n\rangle$.

A simulated developmental sequence generated by map L-system 4 using the dynamic method to determine cell shape is given in Plate 1 (see color plates). Different colors are used to indicate the apical cell, the alternating "regular" segments, and the basal segments. The correspondence between the last stage in the simulated developmental sequence with a photograph of *Microsorium* (Plate 2, see color plates) with respect to structure topology, the relative sizes and shapes of cells, and the overall shape of the thallus demonstrates the suitability of the visualization method for simulation purposes.

HEART SHAPE FORMATION A second example of fern gametophyte development illustrates heart shape formation. The length of segments increases due to periclinal divisions. Neighboring segments differ in length as a result of the successive formation of segments, which causes the tendency of segments to curve towards the apical cell. This tendency can be compensated by the fanning out of segments due to anticlinal divisions, as in *Microsorium*. The relationship between cell division pattern and global shape in fern gametophytes has been quantified in De Boer and De Does.[3] In *Dryopteris thelypteris*, the rate of periclinal divisions is high enough relative to the rate of anticlinal divisions in order to form a heart-shaped thallus. Map L-system 5 simulates the apical growth of *Dryopteris* (Plate 3, see color plates). Also in this model, all segments develop identically. The model matches the average periclinal and anticlinal growth functions of the segments in the apical region.

$$
\begin{array}{ll}
w:\ A\rangle D\langle C\langle b\rangle & \\
r_1:\ A\rangle \rightarrow a\langle[-B\langle]O\rangle & r_{10}:\ J\rangle \rightarrow E\rangle \\
r_2:\ B\rangle \rightarrow E\rangle[+b\rangle]C\rangle[-x]D\rangle & r_{11}:\ K\rangle \rightarrow E\rangle[+H\langle]C\rangle[-x]D\rangle \\
r_3:\ C\rangle \rightarrow J\langle & r_{12}:\ L\rangle \rightarrow x[+x]x[-x]x \\
r_4:\ D\rangle \rightarrow F\rangle[-H\rangle]G\rangle & r_{13}:\ M\rangle \rightarrow L\rangle[+H\langle]x[-H\rangle]N\rangle \\
r_5:\ E\rangle \rightarrow x[-x]x & r_{14}:\ N\rangle \rightarrow I\rangle \\
r_6:\ F\rangle \rightarrow x[+x]x[-x] & r_{15}:\ O\rangle \rightarrow P\rangle \\
r_7:\ G\rangle \rightarrow K\rangle & r_{16}:\ P\rangle \rightarrow Q\rangle \\
r_8:\ H\rangle \rightarrow x[+I\rangle]x & r_{17}:\ Q\rangle \rightarrow O\rangle[-I\langle]O\rangle \\
r_9:\ I\rangle \rightarrow L\rangle[+x]x[-x]M\rangle &
\end{array}
$$

As in map L-system 4, only the productions describing the development of the right side of the thallus are given.

CLEAVAGE PATTERNS

In this section we apply our method to the blastula stage of invertebrate embryos. A blastula is a single-layered cell structure arranged around a spherical cavity. It develops from the fertilized egg by successive division cycles, called cleavages, in which all the cells divide synchronously. The entire structure does not expand. Connectives between the cell membranes preserve the cell neighborhoods created by the successive cleavages. The surface pattern includes most of the essential features of the cleavage patterns, except the role of cell contacts within the embryo in later stages, which we do not consider here.

SIMULATIONS ON THE SURFACE OF A SPHERE Cleavage patterns can be captured with map DOL-systems operating on the surface of a sphere (of constant size), rather than on a plane. To this end, cell boundaries on the surface of the sphere are represented as great circle arcs connecting vertices which are constrained to the sphere surface. The extension of the dynamic interpretation method from the plane to the surface of a sphere requires few changes. Cell shapes are calculated as before. Since the method for the plane may displace a vertex away from the surface of the sphere, the actual vertex position is found by projecting the displaced point back to the sphere.

SPIRAL CLEAVAGE As an example, we consider the spiral cleavage pattern of the limpet *Patella vulgata*. A characteristic of spiral cleavage patterns is that division plates of sister cells attach at one end in an offset position to the division plate of the mother cell, which was formed in the previous cleavage. This results in an orthogonal succession of division orientations, as in map L-system 1.

The following map L-system 6 simulates the developmental sequence of *Patella* (Figure 10) according to data presented in Van den Biggelaar.[17]

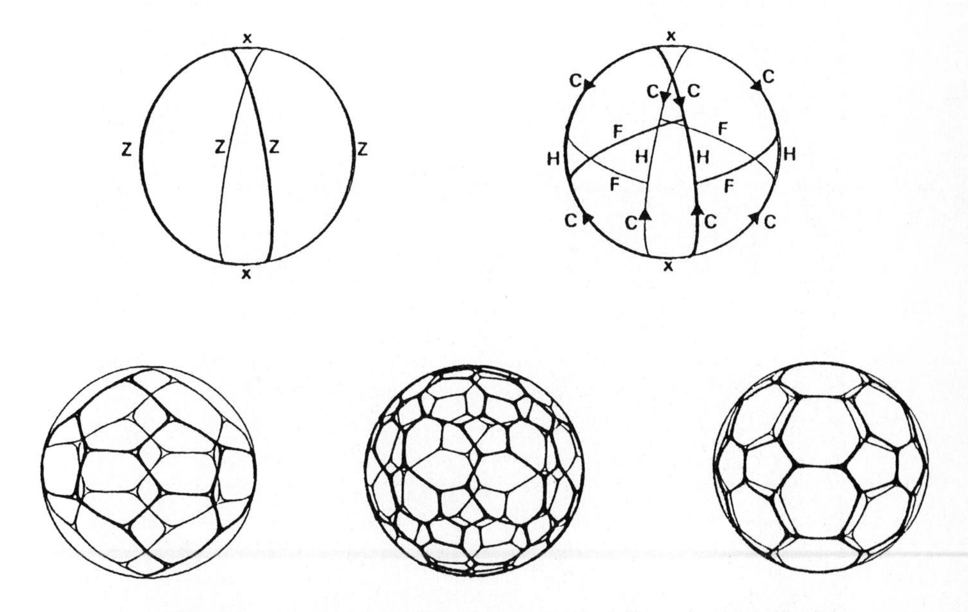

FIGURE 10 Simulated developmental sequence of *Patella vulgata* defined by map L-system 6, showing part of the edge labeling (equatorial view).

w: 4 – cell stage
p_1: $A \rightarrow b[-a]x[+a]b$
p_2: $a \rightarrow B[+A]x[-A]B$
p_3: $B \rightarrow a$
p_4: $b \rightarrow A$
p_5: $C\rangle \rightarrow D\rangle[+a]E\rangle$
p_6: $D\rangle \rightarrow C\rangle[-A]x$
p_7: $E\rangle \rightarrow C\rangle$
p_8: $F \rightarrow E\langle[-a]G[+a]E\rangle$
p_9: $G \rightarrow J$
p_{10}: $H \rightarrow I$
p_{11}: $I \rightarrow B[-A]x[+A]B$
p_{12}: $J \rightarrow b[+a]x[-a]b$
p_{13}: $Z \rightarrow C\rangle[-F]H[+F]C\langle$

The four cells in the starting map each have three boundaries ZZx and separate the embryo into equal quadrants. The attachment sites of the division plates of the second cleavage can be seen to be offset in a Z-fashion on the first division plate. The offset-edges (labeled x) in the four-cell stage are positioned at the so-called "poles." Attachment of division plates of sister cells on both sides of the division

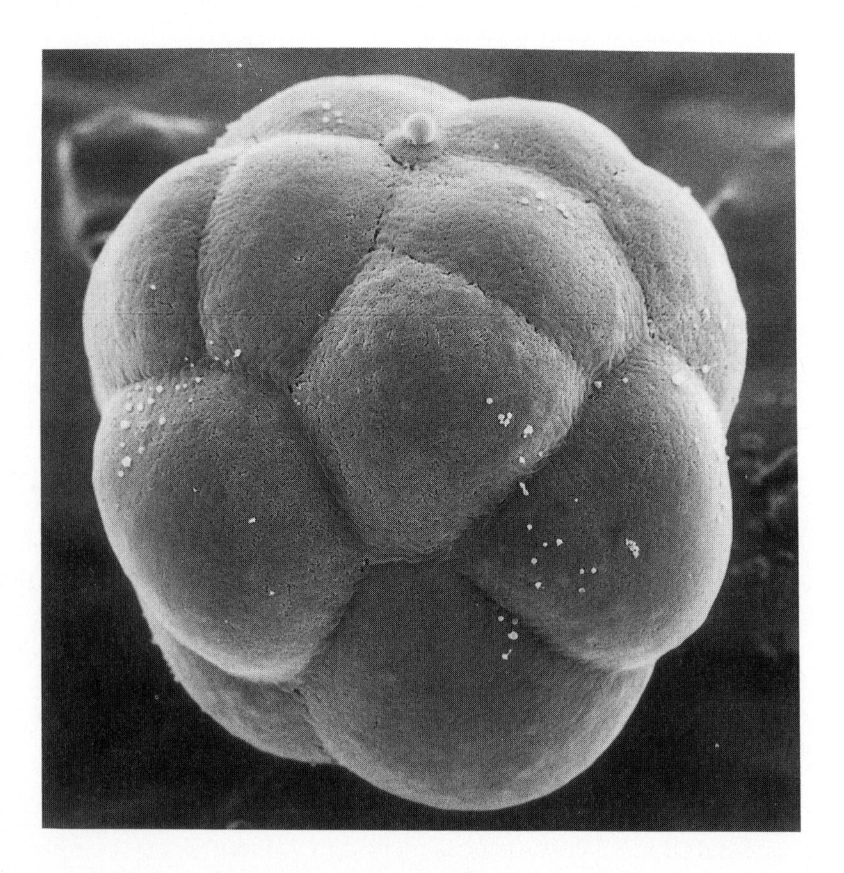

FIGURE 11 Electron micrograph of the 16 cell stage of *Patella vulgata* (polar view). Courtesy of W. J. A. G. Dictus, University of Utrecht.

plate of the mother cell are established by p_{13} in the third cleavage (Z-offsets), by p_8 in the fourth cleavage (Z-offsets), by p_2 in the fifth cleavage (S-offsets) and by p_1 in the sixth cleavage (Z-offsets). The productions p_5 (in the fourth and sixth cleavage) and p_6 (in the fifth cleavage) establish division plates to only one side on the boundaries of the quadrants, creating orthogonal orientations of divisions in neighboring quadrants. Production p_{11} is responsible (in the fifth cleavage) for division plates which are Z-offset on both sides of the offset edges created by p_{13} in the third cleavage. Similarly, in the 6th cleavage, p_{12} creates division plates in S-offset position on the offset edges created by p_8 in the 4th cleavage. Productions p_3, p_4, p_7, p_9 and p_{10} are responsible for delays. An alternative rendering of the modeled structures (Plate 4, see color plates) shows more correspondence to reality (Figure 11).

DISCUSSION

The examples presented in this paper show that complex cell division patterns observed in nature can be captured by relatively simple context-free map L-systems. They are powerful in that they enable us to recognize and to emphasize regularities in the patterns. The dynamic method for the calculation of cell shapes is instrumental in the simulation of shape formation in developing cellular structures. As a method for visualization of the organization of cell division patterns, our approach has many potential applications.

The control of cell division pattern is a complex unsolved problem. Cell division control is based on microtubule organization in the cytoskeleton. Microtubule organization in a mother cell determines the microtubule organization in the daughter cells.[11] Such control is a form of "cytoplasmic inheritance" which is simulated by map L-systems. Map L-systems are a necessary simplification on the level of cells of the subcellular processes of microtubule assembly, simulated by Hameroff et al.[9] Nevertheless, spatial-temporal regularities of cell divisions made explicit with the help of map L-systems may reveal properties of cell division control.

A future extension to context sensitive map L-systems would facilitate the simulation of additional control of cell division by positional information, such as the concentration of morphogen, which becomes increasingly important in later stages of development in the biological examples discussed in this paper.

ACKNOWLEDGMENTS

We are grateful to late Professor A. Lindenmayer for inspiring discussions. Norma Fuller contributed routines supporting geometric operations on a sphere. The reported research at the University of Regina has been supported by an operating grant, equipment grants and a scholarship from the Natural Sciences and Research Council of Canada. Facilities of the Department of Computer Science were also essential.

REFERENCES

1. Culik, II, K., and D. Wood. "A Mathematical Investigation of Propagating Graph OL-Systems." *Information and Control* **43** (1979): 50–82.
2. De Boer, M. J. M. "Construction of Map OL-System for Developmental Sequences of Plant Cell Layers." In *Graph Grammars and Their Application to Computer Science*, edited by H . Ehrig, H.-J. Kreowski, and G. Rozenberg. Lectures Notes in Computer Science, to appear.
3. De Boer, M. J. M., and M. de Does. "The Relationship Between Cell Division Pattern and Global Shape of Young Fern Gametophytes I. A Model Study." *Botanical Gazette* **151**, in press.
4. De Boer, M. J. M., and A. Lindenmayer. "Map OL-Systems with Edge Label Control: Comparison of Marker and Cyclic Systems." In *Graph Grammars and Their Application to Computer Science*, edited by H. Ehrig, M. Nagl, A. Rosenfeld and G. Rozenberg. Lecture Notes in Computer Science, Vol. 291, 378–392. Berlin: Springer-Verlag, 1987.
5. De Does, M., and A. Lindenmayer. "Algorithms for The Generation and Drawing of Maps Representing Cell Clones." In *Graph Grammars and Their Application to Computer Science*, edited by H. Ehrig, M. Nagl, and G. Rozenberg, Lecture Notes in Computer Science, Vol. 153, 39–57. Berlin: Springer-Verlag, 1983.
6. Fracchia, F. D., P. Prusinkiewicz, and M. J. M. de Boer. "Animation of The Development of Multicellular Structures." In *Computer Animation '90*, edited by N. Magnenat-Thalmann and D. Thalmann, 3–18. Tokyo: Springer-Verlag, 1990.
7. Fracchia, F. D., P. Prusinkiewicz, and M. J. M. de Boer. "Visualization of The Development of Multicellular Structures." *Proceedings of Graphics Interface'90* (1990): 267–277.
8. Gunning, B. E. S. "Microtubules and Cytomorphogenesis in a Developing Organ: The Root Primordium of Azolla Pinnata." In *Cytomorphogenesis in Plants, Cell Biology Monographs 8*, edited by O. Kiermayer, 301–325. Wien: Springer-Verlag, 1981.
9. Hameroff, S., S. Rasmussen, and B. Mansson. "Molecular Automata in Microtubules: Basic Computational Logic of the Living State?" In *Artificial Life*, edited by C. G. Langton. Santa Fe Institute Studies of the Sciences of Complexity, Vol. 6. Redwood City, CA: Addison-Wesley, 1989.

10. Lindenmayer, A., and G. Rozenberg. "Parallel Generation of Maps: Developmental Systems for Cell Layers." In *Graph Grammars and Their Application to Computer Science and Biology*, edited by V. Claus, H. Ehrig, and G. Rozenberg. Lecture Notes in Computer Science, Vol. 73, 301–316. Berlin: Springer-Verlag, 1979.
11. Lloyd, C. W. "The Plant Cytoskeleton: The Impact of Fluorescence Microscopy." *Ann. Rev. Plant Physiol* **38** (1987): 119–139.
12. Lück, J., A. Lindenmayer, and H. B. Lück. "Models of Cell Tetrads and Clones in Meristematic Cell Layers." *Botanical Gazette* **149** (1988): 127–141.
13. Nakamura, A., A. Lindenmayer, and K. Aizawa. "Some Systems for Map Generation." In *The Book of L*, edited by G. Rozenberg and A. Salomaa, 323–332. Berlin: Springer-Verlag, 1986.
14. Nakamura, A., A. Lindenmayer, and K. Aizawa. "Map OL-Systems with Markers." In *Graph Grammars and Their Application to Computer Science*, edited by H. Ehrig, M. Nagl, A. Rosenfeld and G. Rozenberg. Lecture Notes in Computer Science, Vol. 291, 479–495. Berlin: Springer-Verlag, 1987.
15. Siero, P. L. J., G. Rozenberg, and A. Lindenmayer. "Cell Division Patterns: Syntactical Description and Implementation." *Computer Graphics and Image Processing* **18** (1982): 329–346.
16. Tutte, W. T. *Graph Theory*. Reading, MA: Addison-Wesley, 1982.
17. Van den Biggelaar, J. A. M. "Development of Dorsoventral Polarity and Mesentoblast Determination in *Patella Vulgata*." *J. Morphology* **154** (1977): 157–186.

Learning and Evolution

David Ackley and Michael Littman
Cognitive Science Research Group, Bellcore, 445 South Street, Morristown, NJ 07960

Interactions Between Learning and Evolution

A program of research into weakly supervised learning algorithms led us to ask if learning could occur given only natural selection as feedback. We developed an algorithm that combined evolution and learning, and tested it in an artificial environment populated with adaptive and non-adaptive organisms. We found that learning and evolution together were more successful than either alone in producing adaptive populations that survived to the end of our simulation. In a case study testing long-term stability, we simulated one well-adapted population far beyond the original time limit. The story of that population's success and ultimate demise involves both familiar and novel effects in evolutionary biology and learning algorithms.

1. EVOLUTION, LEARNING, ARTIFICIAL LIFE

The processes of life involve change at many scales of space and time, from the small, fast biochemical cycles of cell energetics, to the growth and aging of an organism, to the rise and fall of entire populations, entire species, entire orders. Choosing a spatiotemporal scale emphasizes the changes at that scale, rendering smaller scales essentially as noise, larger scales essentially as constant. Many useful investigations can be performed within a given scale of life—there are striking mixtures of order and complexity almost everywhere one looks—but, of course, such investigations will be fundamentally limited by the assumptions of the rendering. Smaller scales are not always insignificant, sometimes they become decisive; larger scales are not always constant, sometimes they become cataclysmic.

Such limitations due to choice of scale can be partially eliminated by devising models that explicitly address multiple spatial and temporal scales (or, more generally, by increasing the *spatiotemporal bandwidth* covered by a model). In this chapter, we study interactions between adaptive processes on two adjacent levels—individuals and populations. Learning is a process at the individual level whereby an organism becomes optimized for its environment; evolution operates similarly at the level of populations or species. The two scales are evident: An entire lifetime of learning is but one tick of the clock for evolution. One trade-off between the two processes is readily apparent: Learning is facilitated by long individual lifetimes, whereas evolution benefits from rapidly passing generations. How else do they interact?

Such multiple time-scale questions are generally very difficult to answer in the natural world. The depths of evolutionary history can be probed via the fossil record and molecular genetics, but such techniques provide only hints about the day-to-day histories of long-passed organisms. Similarly, learning abilities can be studied in live organisms, but feasible length experiments can observe an evolutionarily significant number of generations only with relatively short-lived and learning-limited species.

As the available computational power grows, the "artificial life" experimental approach—based on computer simulations of systems modeling selected aspects of the natural world—becomes more and more feasible. The rich diversity of material in this book demonstrates some of the ways in which this power can be exploited. For present purposes, it is the power to create artificial organisms that combine reasonably long *simulated* lives—allowing for substantial learning—with reasonably short *real-time* lives—allowing us to perform experiments that span many generations. Given the power of a computer workstation, an artificial creature can live a simulated lifetime encompassing thousands of learning opportunities in only seconds of elapsed time, and small populations of such organisms can be tracked over thousands of generations in only days.

This chapter introduces, demonstrates, and studies an adaptation strategy called *evolutionary reinforcement learning* (ERL), which combines genetic evolution with neural network learning, and an artificial life "ecosystem" called AL, within which populations of ERL-driven adaptive "agents" struggle for survival. Although ERL and AL are tremendously impoverished models—possessing only a few stereotypical properties selected from the richness and depth of adaptation and the natural world—they give rise to a broad range of behaviors and phenomena. AL should be distinguished from single-level population-size models of species interactions, such as Lotka-Volterra or Rosenstein-McArthur-Zweig.[25] Instead of modeling an ecosystem in terms of *a priori* birth, interaction, and death rates, AL is a moment-to-moment simulation of each organism's lifetime in the ecosystem. Quantities such as birth and death rates are not input parameters; instead they are observables whose values reflect the interacting consequences of each organism's decisions.

Section 2 presents ERL.[2] Section 3 presents AL and summarizes a comparative study that supports the basic hypothesis that evolution and learning can mutually aid each other. Section 4 presents an in-depth historical study of one successful population, seeking an account of the population's longevity, and an account of its eventual extinction. A phenomenon known as the "Baldwin effect"[7,24] plays a role in the former case, and a phenomenon that we call *shielding* plays a role in the latter. Section 5 contains discussion, and Section 6 concludes the chapter.

2. EVOLUTIONARY REINFORCEMENT LEARNING

Learning algorithms require some sort of feedback to function, but different approaches vary widely in the amount and nature of the feedback required. One fundamental question is: How limited can the feedback be? *Supervised* paradigms[21,22] supply immediate detailed correct answers as feedback; the system must learn to produce them on demand. *Reinforcement* paradigms[8,27] supply less—only judgments of right or wrong—so the system must first discover and then remember the correct responses. Viewed as a learning algorithm, the paradigm of natural selection[11] supplies still less—only birth and death. How can an organism learn in such circumstances, where the only unarguable sign of failure is the organism's own death, and the reproduction process preserves only the genetic information, which is unaffected by any learning performed during the organism's life?

"Evolutionary Reinforcement Learning"[2] (ERL) provides one answer to this question. In ERL, we allow evolution to specify not only inherited *behaviors*, but also inherited *goals* that are used to guide learning. We do this by constructing a genetic code that specifies two major components. The first component is a set of initial values for the weights of an "action network" that maps from sensory input to behavior. These weights represent an innate set of behaviors that the individual inherits directly from its parents.

The second component is an "evaluation network" that maps from sensory input to a scalar value representing the "goodness" of the current situation. By learning to move from "bad" situations to "better" situations—modifying its action network weights in the process—an individual achieves the goals of learning passed down from its predecessors. Whether those inherited goals are actually sensible or not is, of course, a separate issue; insofar as learning is a factor, each organism stakes its life on the *assumption* that its inherited evaluation function is reliable.

Figure 1 depicts the three central structures possessed by an individual. The *genetic code* is a string of bits which the individual receives at birth. It is unchanged by learning and is passed from parents to offspring modified by *crossover* (genetic recombination[17]) and *mutation.* (Asexual reproduction is employed if no mate can be located; in such cases only mutation applies.)

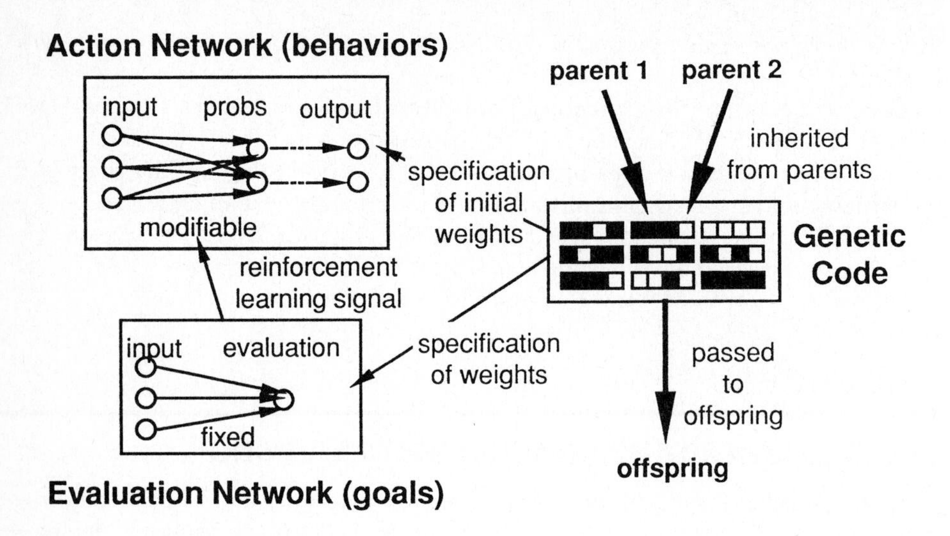

FIGURE 1 Overview of ERL.

The *evaluation network* is a feed-forward neural network that maps the organism's sensory input to a real-valued scalar. The weights of this network are determined solely by the genetic code and they do not change during the lifetime of the individual.

The *action network* is a feed-forward neural network that maps sensory input to behavioral output. The initial weights of this network are specified genetically. However, they are adjusted over time by a reinforcement learning algorithm that rewards behaviors that lead to an increase in the evaluation and punish those that lead to a decline.

To limit the computational costs, in the simulations presented here we used single-layer networks for both evaluations and actions. By design, however, all aspects of ERL carry over to the multi-layer case.

2.1. THE ERL ALGORITHM

The details of ERL are summarized in Figure 2. The first procedure is an implementation of evolution; the second, an implementation of learning. A few comments may help clarify the more and less critical aspects of our approach.

In steps B1 and B2, a spatial distance measure is presumed to exist for purposes of mate selection. In world AL (see Section 3) such a metric is readily available, but one is not required for the algorithm to make sense. In principle, mates could be chosen at random, or by some more elaborate mating ritual, and many interesting questions arise in such situations. We chose this very simple method to avoid the procedural complexities attending more realistic courting behavior.

ERL: Evolutionary reinforcement learning

At Birth

Given: A parent agent A and an offspring O to be initialized.

B1 *Clone.* Copy A's genetic code to O. If there are one or more other agents within a prespecified distance of A, pick the closest such agent B and go to B2, otherwise go to B3.

B2 *Crossover.* Modify O's genetic code by crossing with B's using two random crossover points.

B3 *Mutate.* With low probability mutate O's genetic code by flipping random bits.

B4 *Elaborate.* Translate O's genetic code into weights for O's evaluation network, and and initial weights for O's action network.

Living at time t**:**

Given: A living agent A, and a new current input vector I_t.

L1 *Evaluation.* Propagate I_t through the evaluation network producing a scalar evaluation E_t.

L2 *Learning.* If this is A's day of birth, go to L3. Otherwise, produce a reinforcement signal by comparison with the previous evaluation: $R_t = E_t - E_{t-1}$. Use the CRBP learning algorithm to update the action net with respect to the previous action X_{t-1} and previous input I_{t-1}.

L3 *Behave.* Use the CRBP performance algorithm to generate a new action X_t based on I_t. Perform the chosen action.

FIGURE 2 Summary of ERL.

Although we use a particular reinforcement learning algorithm called CRBP (Section 2.2) in steps L1–L3, in principle any associative reinforcement learning algorithm supporting multiple output bits could be employed. Regardless of the specific choice of algorithm, a *reinforcement function* is required for learning to proceed. From a computational point of view, the primary novel contribution embodied in ERL is the inheritable evaluation function that converts long-time-scale feedback (lifetime-to-lifetime natural selection) into short-time-scale feedback (moment-to-moment reinforcement signals). As we shall see in Section 4, there are both benefits and risks inherent in this approach.

It is important to recognize that natural selection, when viewed as a computational paradigm for search and learning, places severe restrictions on possible adaptation strategies. There are only two circumstances in which a strategy has decisions to make. The first situation—concerned with learning—is the choice of behavior for a given agent at a given time step, and the second situation—concerned with evolution—is the passage of genetic information to the offspring when a birth occurs. Everything else is determined by the "laws of nature" of the world at hand.

For example, death requires no action on the part of the strategy. Also, in contrast to conventional genetic algorithms,[13,17] a strategy is not free to specify the existence and maintenance of any particular population size, nor who lives, who dies, and who reproduces. The strategy influences such decisions only indirectly, via the interactions between the (static and dynamic) properties of the world and the behavior of the agents governed by the strategy.

2.2. CRBP

The ERL algorithm description in Figure 2 refers to a reinforcement learning algorithm called CRBP—*complementary reinforcement back-propagation*[3]—in the implementation of the action network. Although there is neither space nor pressing need to enter into an extensive discussion of CRBP here, an algorithm summary and a few brief comments may be useful for the interested reader.

Figure 3 summarizes CRBP as used in ERL to implement steps L1–L3. Compared to previous presentations of CRBP,[3] this version performs the action function and the learning function backwards, to reinforce the action at time t based on the input at time $t+1$. Thus, this version of CRBP is a simple *temporal* reinforcement learning algorithm.[26] Exploring the effects of incorporating more sophisticated *temporal difference* algorithms[27] would be an interesting extension to this work.

CRBP: Complementary reinforcement backpropagation (ERL version)

Given: A backpropagation network with input dimensionality n and output dimensionality m, and a reinforcement function $f(\Re^n, \Re^n) \to r$. Let $t = 0$.

1. Receive vector $i_t \in \Re^n$. If $t = 0$ go to 6. Otherwise compute reinforcement $r = f(i_t, i_{t-1})$.
2. Generate output errors e_j. If $r > 0$, let $e_j = (o_j - s_j)s_j(1 - s_j)$, otherwise let $e_j = (1 - o_j - s_j)s_j(1 - s_j)$.
3. Backpropagate errors.
4. Update weights. $\Delta w_{jk} = \eta e_k s_j$, using $\eta = \eta_+$ if $r \geq 0$, and $\eta = \eta_-$ otherwise, with parameters $\eta_+, \eta_- > 0$.
5. Forward propagate again to produce new s_j's. Generate temporary output vector o^*. If ($r > 0$ and $o^* \neq o$) or ($r < 0$ and $o^* = o$), go to 2.
6. Set network input to i_t. Forward propagate to produce s_j's.
7. Generate a binary output vector o. Given a uniform random variable $\xi \in [0, 1]$ and parameter $0 < \nu \leq 1$,

$$o_j = \begin{cases} 1, & \text{if } (s_j - \frac{1}{2})/\nu + \frac{1}{2} \geq \xi; \\ 0, & \text{otherwise.} \end{cases}$$

8. Perform the action associated with o. Let $t = t + 1$. Go to 1.

FIGURE 3 Summary of CRBP as used in ERL.

CRBP extends the *back-propagation* neural network learning algorithm[22] to reinforcement learning.[6,28] Back-propagation by itself is a supervised learning algorithm in which the desired outputs corresponding to given inputs are provided externally. By contrast, in reinforcement learning the network itself is given the task of *discovering* desired outputs—i.e., those that produce a positive reinforcement signal. This search task is implemented by step 7, where weighted random numbers are used to generate an output vector. Thus, the ERL action network has the job of specifying output *probabilities* conditional on the current inputs.

When the reinforcement signal is positive, the learning task is fairly clear: The generated output vector should be made *more probable* given the same input vector. If the output vector is taken as the *desired target*, back-propagation learning will do exactly that. Negative reinforcement, however, only says that the generated output vector is wrong, without suggesting which other output vector would be right. What should the desired target be?

Different reinforcement learning algorithms emerge depending on how that question is answered. One strategy[6,28] sidesteps the issue by taking the generated output vector as the *undesired target*—in essence, simply flipping the signs of the errors produced in the positive reinforcement case. CRBP embodies a somewhat stronger heuristic—that the desired output on negative reinforcement is the *complement* of the generated output. Without *a priori* knowledge of the reinforcement function, all treatments of negative reinforcement are fundamentally heuristic, but fortunately, since search is an integral part of reinforcement learning,[1] the occasional failure of the assumption need not be catastrophic.

The loop introduced in step 5 implements a simple form of "mental rehearsal." On positive reinforcement, the reward continues until another stochastic output generation produces the same result as the initial success, and on negative reinforcement, the punishment continues until another output generation produces something different than the initial failure. In empirical studies of CRBP,[3] this loop improved learning speed substantially without much overhead.

3. WORLD AL

We needed a source of natural selection to illustrate and evaluate ERL, so we constructed an artificial life world we called "AL." In doing so, we had to balance the desire for richness and complex interactions against the need for compactness and computational tractability. The result has much in common with other artificial life worlds.[5,18,20,29] AL is a two-dimensional 100×100 array of cells populated by adaptive ERL agents and non-adaptive carnivores, plants, trees, and walls. The world is summarized in Figure 4 and can also be seen in a video demonstration.[4] The various rates and thresholds that determine the artificial physics of AL are all fixed at constant values, as are the biological and physiological properties of

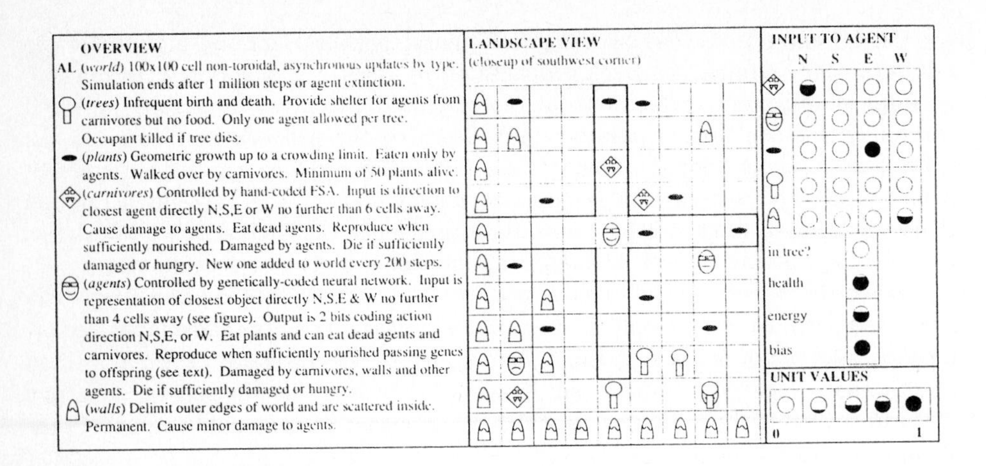

FIGURE 4 Summary of World AL.

AL agent and carnivore action semantics			
Contents of Target Cell	*Cell Visual Appearance*	*Effect of Agent action*	*Effect of Carnivore action*
Empty	Empty	Enter	Enter
Plant	Plant	Eat all	Enter
Empty Tree	Tree	Climb	No effect†
Agent* in Tree	Tree	No effect	No effect†
Wall	Wall	Damage self	Damage self†
Living carnivore‡	Carnivore	Damage other	Damage other†
Dead carnivore‡	Carnivore	Eat some	Eat some†
Living agent	Agent	Damage other	Damage other
Dead agent	Agent	Eat some	Eat some

* Living or dead.

† Carnivores as programmed will not choose these moves.

‡ Perhaps accompanied by a plant.

FIGURE 5 Effects of agent and carnivore actions.

all inhabitants. All that must be supplied is an algorithm for agent learning and evolution, and the name of the game is *maximize agent population survival time.*

The agents receive as input the visual appearance of the closest object not further than four cells away in each of the four compass directions. Carnivores can see objects six cells distant. All visual inputs in a given direction take value zero if only empty cells are visible, and otherwise the input corresponding to the visual appearance of the occupied cell takes a value from 0.5 to 1.0 proportional to the closeness of the cell. An additional binary input indicates whether an agent is currently on the ground or in a tree. Agents also have "proprioceptors" indicating the amount of energy and health they possess. Agents must produce as output a 2-bit pattern indicating the compass direction to their choice of *target cell.* Although it seems likely that the specific details of the action interpretations in AL are not critical to our basic results, for the sake of concreteness and as an aid to the reader's intuition, Figure 5 presents the complete "semantics" of agent and carnivore actions.

Agents reproduce by accumulating enough energy from food and die by running low on energy or health. Injured but alive agents and carnivores recover spontaneously over time. Dead agents and carnivores are eaten or simply decay until their energy is gone. Carnivores reproduce by eating enough agents and die by starvation (almost always), or agent-inflicted damage (very rarely—in a slugfest between a healthy agent and a healthy carnivore, the agent always loses). At regular intervals a carnivore is created in a random empty cell. Also, though they have not been observed to be necessary, procedures are included to reseed plants and trees if their numbers fall perilously low. Agents, of course, as the objects of our population longevity study, receive no such safety nets.

Simulations of ERL in AL display phenomena at several time scales.[4] Observing at highest resolution, agents are seen moving about or collecting in corners, feeding or starving, encountering carnivores and escaping or not, and so on. AL is not an overly kind world: Most initial agent populations die out quite quickly. Observing summary statistics at the $\times 100$ time scale, in those populations that survive the most apparent features are irregular predator-prey oscillations involving plants, agents, and carnivores, interspersed with periods of stable or slowly changing population sizes. In the simulation considered in Section 4, agent population sizes were oscillating in the 30–60 range when one million steps were reached (see the $\times 1,000$ view in Figure 6).

3.1. A COMPARATIVE STUDY

Our evaluation of the algorithm consisted of the following test: we ran the simulation on 100 random initial agent populations and recorded the time at which they eventually went extinct (up to 1 million time steps). We then used just the evolution component of the algorithm (E), just the learning component (L), and neither (F— Fixed random action networks), and ran the same test. As a baseline, we tested Brownian agents (B), who simply wander the world at random, ignoring their inputs.

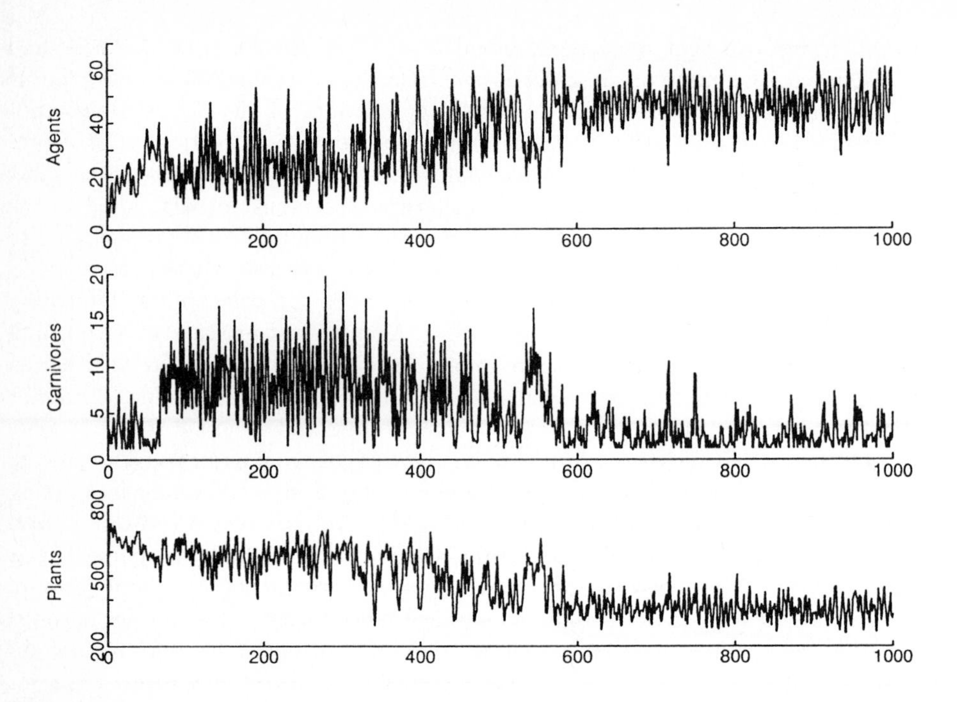

FIGURE 6 ERL in AL: Species population sizes vs. time ($\times 1000$) for a long-term successful agent population.

Figure 7 displays the percentage of initial populations that survived to various times for each of the five variations. No more than 18% of all populations reached 10,000 time steps and only 1.8% reached the 1 million time-step simulation limit.

There is a clear distinction in the first half-million time steps or so between the two algorithms that included learning (ERL, L) and their non-learning counterparts (E, F). The latter two algorithms even did poorly compared to the "brainless" B agents. Learning appears to contribute towards keeping the agents alive during this period.

Above about half a million time steps, ERL begins to pull away from learning-only (L), suggesting that evolution has an impact at this timescale. ERL finally goes on to produce seven populations that last to the 1 million time-step limit.

We were surprised that evolution without learning did so poorly, and that learning without evolution did so well. The former was surprising since evolution without learning is a common approach to artificial life, and the latter was surprising since, without evolution to improve the evaluation functions, strategy L can never move beyond on the randomly generated evaluation functions found in the initial populations.

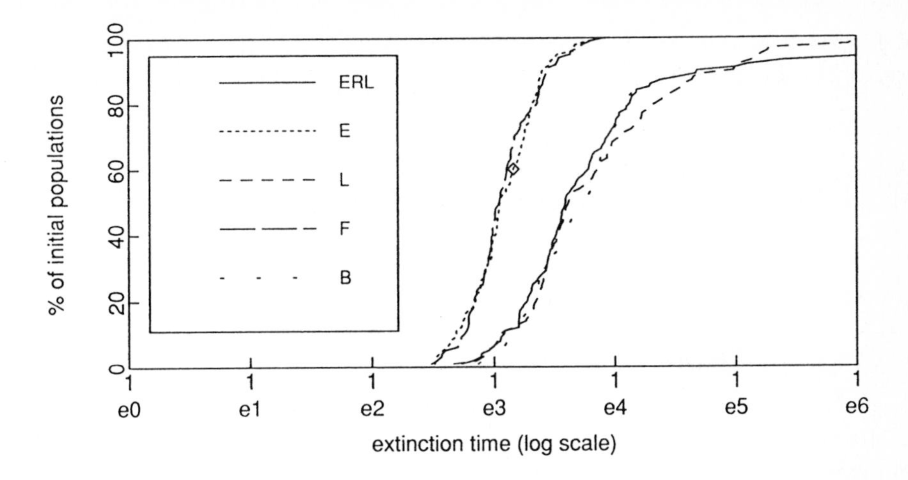

FIGURE 7 Cumulative plots showing the distributions of population lifetimes generated by the five strategies. The point marked with a diamond, for example, indicates that 60% of the strategy E initial populations were extinct by about 1500 time steps.

We hypothesize that evolution alone has difficulty because survival in AL is no trivial matter: Most agents with randomly generated action networks die quickly (*viz.* the strategy F results). This puts evolution at a disadvantage: Either the population dies out completely before there is time to evolve, or a single survivor becomes the ancestor of all subsequent agents. In the latter case, all the agents are close genetic kin, leaving little diversity for evolution to operate upon.

The success of learning alone was noteworthy. It is easy enough to conclude merely that the space of genetic codes for action networks is more difficult to search than the code space for action-plus-evaluation networks, so that strategy L could simply "luck into" good initial populations often enough to make the difference. However, that cannot be the whole story. After all, the code space for L is thirty orders of magnitude larger than that for E, so one might expect it to be harder to search. Our explanation is that *it is easier to generate a good evaluation function than a good action function.*

Notice, for example, that there are two output units in the action network, but only one in the evaluation network. To specify an action in response to a particular input requires specifying two weights, but to specify that a particular input is "good" requires only one weight. Furthermore, if the evaluation function specifies that the energy level input is positively valued, then there is pressure towards making "eating moves" more probable regardless of the direction of the food source. Thus, *one* evolutionarily specified learning weight can have the effect of specifying the *eight* action weights involved in response to plants. The insight that strategy L highlights is that it can be much easier to specify *goals* than *implementations*—assuming, of course, the existence of a search and learning process adequate to fill in the details.

ERL, which combines evolution and learning into a single system, is better at producing long-lasting populations than either alone. The interaction of these components can result in successful adaptations and stable populations.

4. A LONGITUDINAL STUDY

One advantage of artificial life studies over natural world studies, as we have seen, is the fact that experimental conditions can be so easily controlled, precisely repeated, and systematically varied. Another advantage is the abundance of data that such experiments provide, at whatever granularity we choose.

The object of our study was the population depicted in Figure 6 and on video.[4] It was doing well as it reached its 1-million-step birthday. Even in the depths of population declines, dozens of agents survived. Its very long-term prospects—on the multimillion-step timescale—looked good. We reset the simulator to that population's initial seed, let it loose with no upper time limit, and went about other business. By the next morning it had regained the million-step milestone and pushed into new territory.

Days went by—more millions of steps—and the population survived. Checking in on the simulation, it was clear that matters were more complex than they appeared at 1-million steps. There were periods of very large agent populations, and dangerous agent population crashes. Eventually, after about a week, almost at the 9 million mark, the sole remaining agent, a member of the 3,216th generation, died. Figure 8 displays the population sizes over the entire run. What happened?

Each agent's genetic sequence consists of 336 bits—84 weights total at 4 bits per redundantly encoded weight. On average there were 40 agents alive at any one time and each agent lived approximately 4,000 time steps. Almost 100,000 agents were born during the entire run, yielding slightly over 4 megabytes of genetic information from start to finish. Genealogical and census data added several more megabytes.

How does one go about reducing all this data? Averaged population size changes indicated substantial dynamics in the multimillion step regime, but suggested little in the way of explanation. We hoped to identify the relative importance of learning and evolution in the survival of the agents and perhaps even detect changes in the importance over time. If we were lucky, we hoped to find a genetic explanation for the instability that lead to the population's eventual extinction.

The biological literature suggested an approach we found fruitful: *functional constraints.* By looking at the changes in given sites on the genome over the millenia, biologists argue that one can assess those sites' relative importance to survival.

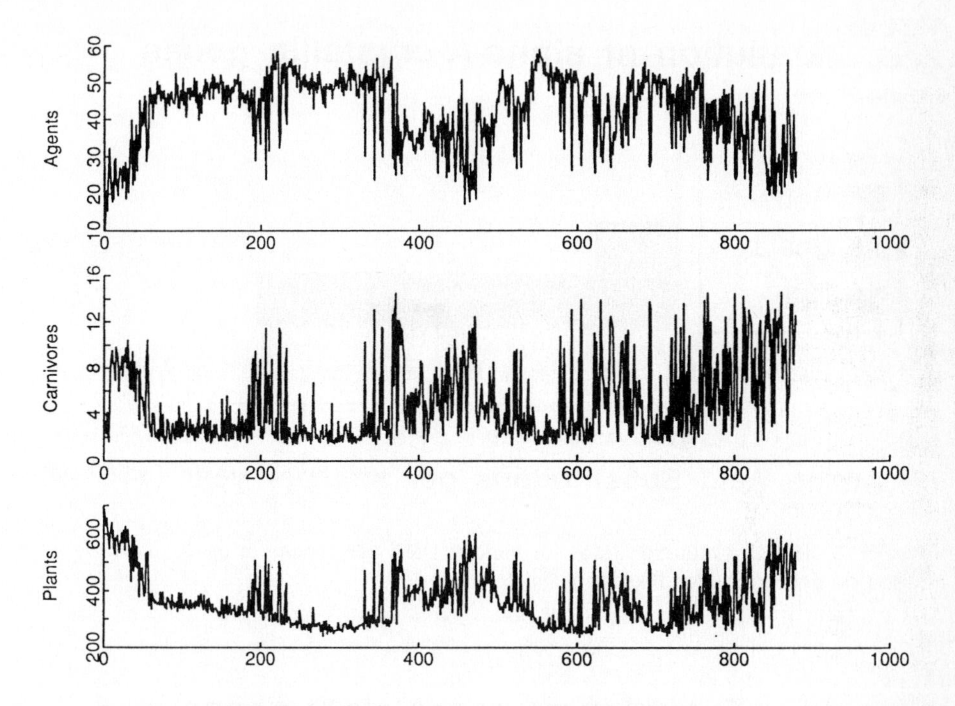

FIGURE 8 Population sizes vs. time ($\times 10,000$) for full run.

We came across functional constraints in an article by Gould,[14] who describes a lovely application of the idea, recounting the work of Hendriks, Leunissen, Nevo, Bloemendal, and de Jong[15] on the blind mole rat *spalax ehrenbergi*. The argument runs as follows: at the molecular level, one site is as likely as any other to mutate during reproduction. On the one hand, some mutations will occur in irrelevant portions of the genome (for instance, the so-called ***pseudogenes*** which are evidently non-expressed "commented out" portions of DNA). Such changes will have no effect on the probability of survival of the offspring, and will, therefore, tend to accumulate in the population over time.

On the other hand, some mutations will disturb genetic sites that contain information crucial to the survival of the organism. These changes will tend to disrupt the functioning of the organism and will tend not to be passed down to later generations. Thus, the *lack* of observed mutations in a gene sequence, over time, suggests that sequence is "functionally constrained" by natural selection.

In the case of *S. ehrenbergi*, the researchers looked at the genes coding for the protein αA-crystallin, which plays a role in the lens of vertebrates.[15] Figure 9 presents some of their data, which they gleaned from a painstaking mix of fossil data analysis and comparative biochemistry. A base mutation rate is computed from observed pseudogene mutation rates. In sighted rodent species, the gene for

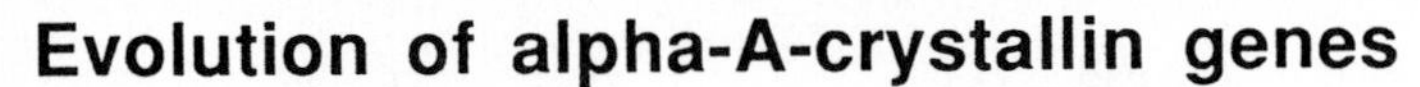

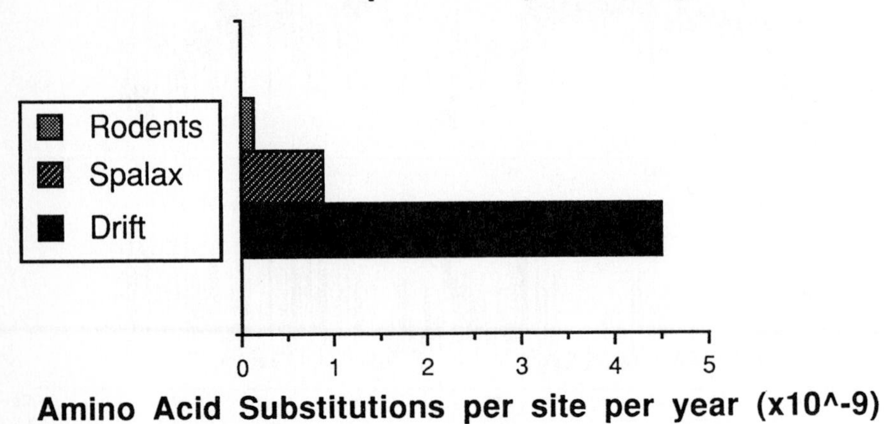

FIGURE 9 Relative mutation rates for various DNA sequences in *spalax ehrenbergi*. (Based on data from Hendriks et al.[15]).

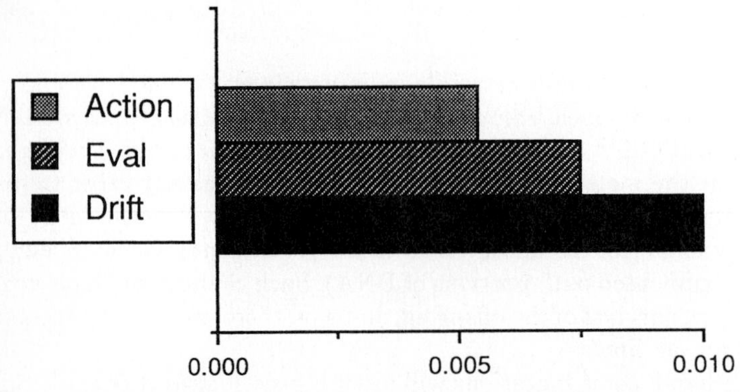

FIGURE 10 Relative mutation rates for various gene sequences in AL agents.

αA-crystallin mutates very slowly compared to the base rate. In *Spalax*, they find that the same gene is changing more rapidly than in sighted rodents but less fast than pseudogenes. This supports the inference that this sight-related protein has *some* survival value to the blind mole rat, but not as much as in sighted rodents. The functional constraints on a gene are inversely related to its observed rate of mutation.

We applied this technique to the agents in AL. Figure 10 displays the observed rates of change of three types of genes: the action-related genes associated with plants, the learning-related genes associated with plants, and a set of genes that happened to code for nothing in our simulation.

As in the natural world example, we found that the non-coding genes changed much more quickly over the generations than did the plant-related genes. We can therefore infer that the plant-related genes were functionally constrained and had a positive impact on fitness over the lifetime of the population.

Unlike our biological counterparts, with our densely sampled data, we can easily perform more detailed data analyses. By partitioning the data into pre-600,000 and post-600,000 time-step periods, we see that the relative importance of learning and evolution changes (Figure 11). During the first 600,000 time steps, there is little change in the genes related to plant evaluation. Therefore, changes in learning goals are being selected against—learning is very important to survival. After 600,000, however, it is the genes controlling the initial action towards plants which are conserved more. Therefore, inherited behaviors are more significant during this time.

One effect suggested by the above data is that in the initial periods, the successful behaviors are being represented in the evaluation network and that somehow, later, these behaviors have showed up in the action networks. In other words, it appears that in the beginning, agents have learning-related genes that state "Plants are good"[1] and from this, learn to approach plants. Later in the simulation, however, their action-related genes recommend approaching plants right from birth.

One might be tempted view this as a Lamarckian effect, with changes during an organism's lifetime somehow being transmitted genetically, but the mechanisms of ERL make direct transmission of acquired characteristics impossible. Fortunately, there is a Darwinian explanation, an effect first suggested in the biological literature around the turn of the century by J. M. Baldwin (and several others) that has come to be referred to as the "Baldwin Effect."[7,24] (Baldwin's own term for the phenomenon—*organic selection*—did not persist, but recently it has been proposed as a more general term for a variety of effects including Baldwin's.[23])

With ERL in AL, the Baldwin Effect often appears this way: In the beginning era of successful populations, agents possess (mostly by luck) learning genes telling

[1] "Plants are good" refers to both the antecedent— closeness to plants—and the consequent— increase in energy— of eating. Similarly "carnivores are bad" refers to both closeness to carnivores and decrease in health.

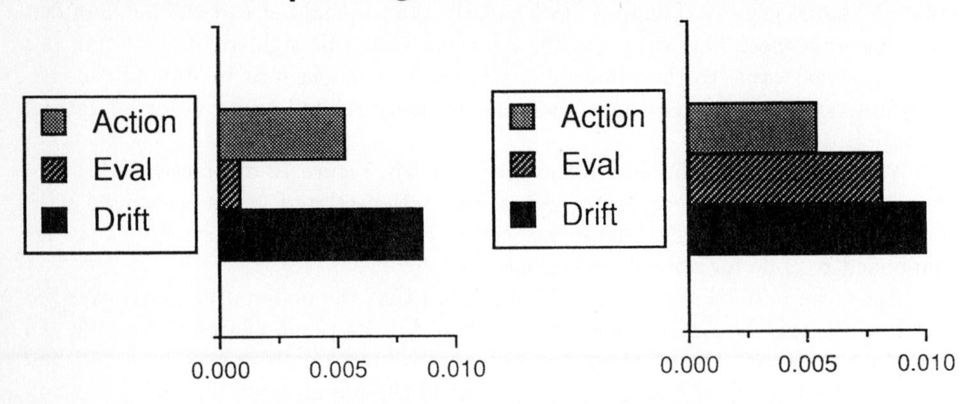

FIGURE 11 Relative importance of learning and evolution as observed by mutation rates, partitioned into pre- and post-600,000 time-step periods.

them plants are good. This is a big benefit for survival since the agents learn to eat, leading to energy increases, and eventually offspring. From time to time, action-related mutations occur that cause agents to approach plants instinctively. These changes are favored by natural selection because they avoid the shortcoming of each new agent having to rediscover that plants' goodness means it should approach them. Agents begin to eat at birth and are better able to survive.

The Baldwin Effect shows how inherited characteristics can mimic acquired characteristics in a population using only conventional evolutionary mechanisms. Though support from biologists for the concept been spotty, the phenomenon has been previously demonstrated in a computational evolutionary simulation by Hinton and Nowlan[9,16] (see also Section 5).

To investigate further, we devised hypothetical *fitness models* for various sets of agent genes, based on our sense of world AL. In Figure 12, the two graphs depict the values over time of fitness models for four sets of genes: the plant-evaluation, plant-action, carnivore-evaluation, and carnivore-action. Maximum plant-evaluation fitness implies a positive view of energy and positive responses to plants in all directions. Plant-action fitness is related to the probability that a plant will be approached in every direction. Carnivore-evaluation fitness incorporates a positive attitude toward health with negative attitudes towards carnivores in all directions. Carnivore-action fitness is *inversely* related to the probability that a carnivore will be approached in every direction.

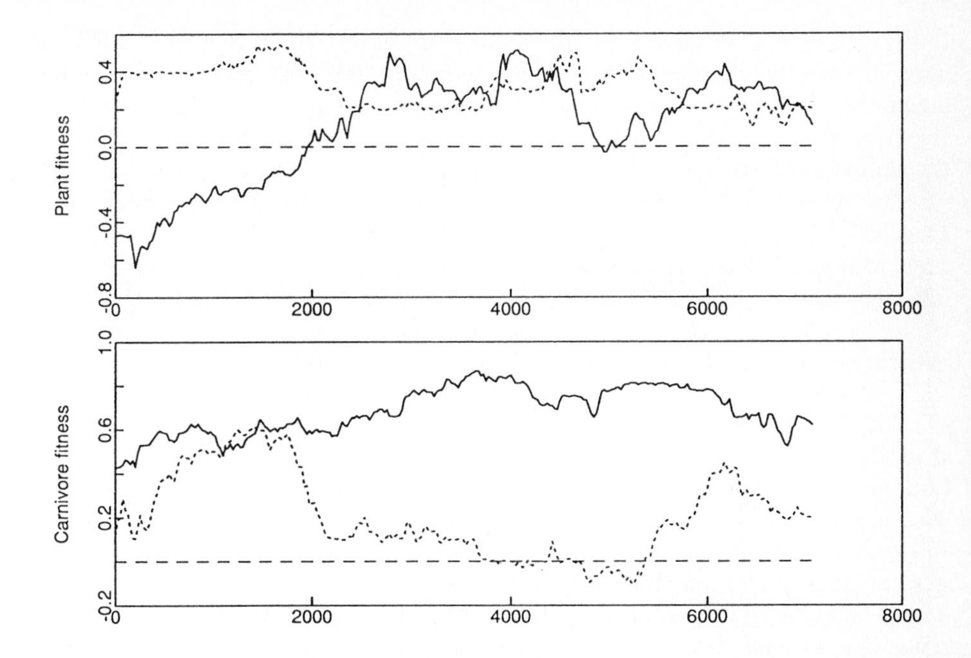

FIGURE 12 Genetic changes over time ($\times 10,000$) relative to hypothetical fitness models for plant-related genes (top) and carnivore-related genes (bottom). In each graph, solid lines represent values of *action*-related genes, and dashed lines represent values of *learning*-related genes.

It must be stressed that these models, though plausible, are only *hypothetical*. Lacking a closed form solution, only AL itself can model fitness exactly. For example, although the handcrafted agents mentioned in Ackley and Littman[4] receive a perfect score in plant-action fitness, their *actual* fitness, overall, is unclear.

Given these models to estimate genetic fitness, we can see in Figure 12 that in the plant domain, the Baldwin Effect is evident. Although plant-evaluation fitness is relatively high throughout the run, plant-action fitness rises steadily to supplant it. What once had to be acquired, was now inherited by around the 3 million time-step mark.

What leverage do these models give us in understanding the population's eventual demise? We believe a crucial destabilizing factor is visible in the plot of carnivore-evaluation fitness. It begins fairly low, rises quickly and levels off, then suddenly plummets and even goes negative for a time. For more than a million time steps, agents actually *liked* the sight of carnivores. It is distinctly possible that this maladaptation contributed to the population's eventual extinction.

How could this have come to be? Why would natural selection have permitted such unfit organisms to proliferate? The answer appears to be that the organisms were not actually unfit even though they possessed this "obvious" flaw. Although the carnivore-evaluation fitness was quite poor during the latter half of the simulation, the carnivore-action fitness was high and gradually increasing. The well-adapted action network apparently *shielded* the maladapted learning network from the fitness function. With an inborn skill at evading carnivores, the ability to learn the skill is unnecessary.

Our data does not make a strong case for the Baldwin Effect at work in the carnivore-related genes. In fact, based on preliminary analyses of other successful runs, it appears that carnivore-action fitness is so important to survival that if the initially created agents are not at least partially able to evade carnivores instinctively, the entire population dies out quickly. AL carnivores are dangerous beasts; without some innate tendency to avoid them, an agent is likely to die before it can learn to dodge.

Although our data does not clearly demonstrate both the Baldwin Effect and shielding on the same set of genes, it is easy to foresee the possibility of the combination, and the peculiar effect—which we call *goal regression*—that would be a likely consequence. A successful inherited ability acquired via the Baldwin Effect will not shield all learning genes equally—it will preferentially affect the learning genes that were responsible for the Baldwin Effect to begin with! The very goals that were known to be adaptive—since they aided population survival initially—are exactly those that become most shielded and subject to genetic deterioration—since the inherited abilities supplanted the need for those goals.

When natural selection is the only source of feedback, shielding and goal regression are potential hazards wherever the Baldwin Effect is a potential benefit.

5. DISCUSSION

The Baldwin Effect depends upon the stability of both a problem and its solution over evolutionary time. On the one hand, if the solution changes, the genetic acquisition of a specific solution is a liability. On the other hand, if the problem simply vanishes, any added fitness for possessing the solution vanishes with it. The effect can only persist through extended evolutionary time if, somehow, improvements can continually be added to the solution without ever *really* solving the problem. Shielding and goal regression are possible consequences if the Baldwin Effect undermines itself by solving the problem too well.

We can summarize the effects and their relationships this way:

- In an environment that poses a problem for survival, a population arises that survives because it possesses learning ability and an inborn set of goals that happens to be best satisfied when the problem is *anticipated and avoided.*
- *Baldwin Effect:* As generations pass, ways of anticipating and avoiding the problem become incorporated and instinctive.

- A successful inherited ability to avoid a problem—whether a consequence of the Baldwin Effect or not—means that learning to avoid the problem confers little advantage.
- *Shielding:* Genetic information related to learning the ability is less constrained functionally; mutations can accumulate without affecting fitness.
- *Goal regression:* If the inherited ability did arise via the Baldwin Effect, shielding will preferentially affect the original learning ability that the Baldwin Effect relied upon.

In effect, the adaptations selected to avoid a problem tend to flatten the fitness subspace related to learning to solve that problem. Goals can arise whose achievement would actually aggravate the problem, but they can be shielded from a fitness penalty because the inherited ability tends to avoid situations in which the *possibility of achieving* the goals can be discovered.

Given the above descriptions, and scrutinizing Figure 12, a characterization of the significant evolutionary time-scale events in that simulation would include the Baldwin Effect at work in the plant-related genes, and shielding in the carnivore-related genes. Although those seem to be major effects, there are clearly other contributing factors at work in the simulation. Predation as the challenge to survival is one such factor, because the size of the predator population, and thus the severity of the problem, interacts nonlinearly with the ability of the prey to evade. The reduction in the size of the carnivore population, in effect, increases the size of the "flat area" in fitness space, because the base rate of predator-prey interactions is reduced. The effect of shielding is exaggerated, and the agent population risks disastrous "population implosions" when the carnivore population begins to rebound, giving the maladapted learning networks more opportunities for mischief.

Predation can be contrasted with a less reactive problem—such as, say, an ice age—which is largely unaffected by a population's success at keeping warm (ignoring possible long-term issues such as interactions between fire-making and greenhouse gases). In such a case, opportunities for unlearning are hard to avoid completely, so shielding and goal regression tend to incur a fitness penalty more quickly.

It may seem strange that we have claimed such a close coupling between the Baldwin Effect, shielding, and goal regression, given that the simulation of Hinton and Nowlan[16] displays only the first one. In this context, the critical distinction between their approach and ERL is that they have assumed the *a priori* existence of a *criterion of success*. Their organisms are presumed to possess an ability to recognize when the problem has been solved and to therefore stop learning, but they leave open the question of how this ability could be acquired via natural selection. ERL provides an explanation—its evaluation network is an *evolvable criterion of success*—but as we have seen, the price of a mutable evaluation function includes a long-term risk of goal regression.

Goal regression may also give pause to theorists such as Schull[23] who argue that a species as a whole can fruitfully be viewed as an intelligent entity. The Baldwin Effect is a central component of that viewpoint, which likens evolution to species-level learning, and likens individual organism's learning experiences to species-level "hypothetical thoughts." Such viewpoints are not strictly amenable to proof or refutation, but goal regression raises the possibility that sometimes a species may fruitfully be viewed as a fairly stupid entity.[12]

The architectural split between actions and goals in ERL challenges Lloyd's[19] assertion that "organic selection, to the extent that it produces evolutionary change, *eliminates* phenotypic plasticity and replaces it with the genes that produce the local optimal phenotype" (pg. 79, original emphasis). On the one hand, Lloyd's claim is clearly true of the Hinton and Nowlan[16] simulation. In that case, adaptive sites and inherited sites are drawn from the same pool, so more inherited sites necessarily means fewer adaptive sites. On the other hand, the picture is rather different with ERL. One must carefully separate the *source* of individual plasticity—the learning algorithm and the evaluation network—from any *specific goal* that may be represented in the network. Via the Baldwin Effect, a specific goal may cease to be relevant due to shielding and drift away, but some other goal will necessarily replace it. The evaluation network is still there, and the learning algorithm; In ERL, plasticity is not eliminated; at most it is redirected.

6. CONCLUSIONS

We identified two main interaction effects of learning and evolution in our system. The first, known as the *Baldwin Effect*, made it possible for organisms to use learning to stay alive while waiting for successful behaviors to be incorporated directly into the genetic code. We feel this accounts for the superiority of ERL over the systems we studied that used evolution and learning in isolation.

The second interaction effect we encountered had not previously been studied in a computational context. Here we saw that successful inherited behaviors *shielded* (or reduced functional constraints on) the inherited preferences which control learned behavior. This effect appears to account for the long-term instability of the population presented in Ackley and Littman[4] and Section 4.

In addition, the combination of the Baldwin Effect and shielding can lead to the phenomenon of *goal regression*, in which the specific goals that initially facilitated survival are preferentially eroded.

Functional constraints are powerful tools for inferring the relative impacts of learning and evolution. The richness of artificial life datasets allows statistics and fitness models to be analyzed as detailed functions of evolutionary time—rather than as scattered sample points—revealing the dynamical behavior of such systems.

Although in our simulations shielding and goal regression seemed to be liabilities, it is worth noting that they have potential benefits as well. The shielded learning genes might happen upon goals even better the original ones. For example, although it is beyond the capability of the simple agents we simulated, the potential exists in AL for sophisticated agents with shielded plant-learning genes to discover agriculture! A rough calculation shows that world AL could easily support several hundred agents continuously if they controlled the plant population and dispersal with selective feeding. Such a development could follow from just the sort of goal—e.g., to walk away from some meals—that could be deadly at the outset.

There are many obstacles to building and analyzing multiple-scale models. Size and duration acquire multiple interpretations, introducing fundamental ambiguity into such basic concepts as equilibrium and stability. Nonetheless, multiple-scale research efforts such as this one offer hope of uniting the sometimes fractious research groups that are separated only by a scale change. Caughley[10] expressed our sentiments well, while pondering how group selection could possibly occur if natural selection operates solely at the individual level:

> "A resolution to this dilemma must wait for population geneticists to grow weary of their pivotal assumption that a population has no dynamics, and for population dynamicists to abandon the belief that a population has no genetics." (pg. 113)

Computers, like microscopes, are instruments of empirical science. Multiple-scale simulation models offer a way of casting light on elusive phenomena that hide in the cracks between levels due to scale-crossing interaction effects. Between evolutionary theory and population biology, group selection may be such a phenomenon. Between cognitive science and neuroscience, the emergence of mind from brain may be another. The power of the computational microscope is growing by leaps and bounds, and we are just beginning to learn how to use it.

REFERENCES

1. Ackley, D. H. *A Connectionist Machine for Genetic Hillclimbing.* Boston: Kluwer, 1987.
2. Ackley, D. H., and M. L. Littman. "Learning from Natural Selection in an Artificial Environment." In *Proceedings of the International Joint Conference on Neural Networks, IJCNN-90-WASH-DC*, Vol. I, edited by M. Caudhill, 189–193. Hillsdale, NJ: Lawrence Erlbaum, 1990.
3. Ackley, D. H., and M. L. Littman. "Generalization and Scaling in Reinforcement Learning. In *Advances in Neural Information Processing Systems – 2*, edited by D. S. Touretzky. San Mateo, CA: Morgan Kaufmann, 1990.
4. Ackley, D. H., and M. L. Littman. A Video Presentation of "Learning From Natural Selection in an Artificial Environment." Submitted to the *Video Proceedings of the Second Artificial Life Conference.* Redwood City, CA: Addison-Wesley, 1990.
5. Allen, R. B., and M. E. Riecken. "Interacting and Communicating Connectionist Agents." *Proceedings of the International Neural Network Society*, 67, Boston, MA, 1988.
6. Anderson, C. W. "Learning and Problem Solving with Multilayer Connectionist Systems." University of Massachusetts doctoral dissertation in computer science, COINS TR 86–50, 1986.
7. Baldwin, J. M. "A New Factor in Evolution." *Amer. Naturalist* **30** (1896): 441-451, 536–553.
8. Barto, A. G. "Learning by Statistical Cooperation of Self-Interested Neuron-Like Computing Elements." *Human Neurobiology* **4** (1985): 229–256.
9. Belew, R. "Evolution, Learning and Culture: Computational Metaphors for Adaptive Algorithms." CSE Technical Report CS89–156, University of California at San Diego, La Jolla, CA, 1989.
10. Caughley, G. "Plant Herbivore Systems." In *Theoretical Ecology: Principles and Applications*, edited by Robert M. May. Philadelphia, PA: Saunders, 1976.
11. Darwin, C. *On the Origin of Species.* London: John Murray, 1959.
12. Dennett, D. C. "Teaching an Old Dog New Tricks." *Behav. & Brain Sci.* **13** (1990):1, 74–75.
13. Goldberg, D. *Genetic Algorithms in Search, Optimization, and Machine Learning.* Reading, MA: Addison-Wesley, 1989.
14. Gould, S. J. "Through a Lens, Darkly." *Natural History* **(9)89** (1989): 16–24.
15. Hendriks, W., J. Leunissen, E. Nevo, H. Bloemendal, and W. W. de Jong. "The Lens Protein α-Crystallin of the Blind Mole Rat, *Spalax Ehrenbergi*: Evolutionary Change and Functional Constraints." *Proceedings of the National Academy of Sciences* **84** (1987):5320–5324.
16. Hinton, G. E., and S. J. Nowlan. "How Learning Can Guide Evolution." *Complex Systems* **1** (1987): 495–502.

17. Holland, J. H. *Adaptation in Natural and Artificial Systems.* Ann Arbor, MI: University of Michigan Press, 1975.
18. Langton, C. G., ed. *Artificial Life II*, Santa Fe Institute Studies in the Sciences of Complexity, Vol. X, edited by J. D. Farmer, C. G. Langton, C. Taylor, and S. Rasmussen. Redwood City, CA: Addison-Wesley, 1989.
19. Lloyd, E. A. "'Intelligent' Evolution and Neo-Darwinian Straw Men." *Behavioral and Brain Sciences* **13** (1990):1, 79–80.
20. Packard, N. H. "Intrinsic Adaptation in a Simple Model for Evolution." In *Artificial Life*, edited by C. G. Langton, Santa Fe Institute Studies in the Sciences of Complexity, Vol. VI, 141–155. Reading, MA: Addison-Wesley, 1989.
21. Rumelhart, D. E., G. E. Hinton, and J. L. McClelland. "A General Framework for Parallel Distributed Processing." In *Parallel Distributed Processing: Explorations in the Microstructures of Cognition. Volume 1: Foundations.*, edited by D. E. Rumelhart and J. L. McClelland, Ch. 2, 1986.
22. Rumelhart, D. E., G. E. Hinton, and R. J. Williams. "Learning Internal Representations by Error Propagation." In *Parallel Distributed Processing: Explorations in the Microstructures of Cognition. Vol. 1: Foundations.*, edited by D. E. Rumelhart and J. L. McClelland, Ch. 8, 1986.
23. Schull, J. "Are Species Intelligent?" *Behav. & Brain Sci.* **13** (1990): 1, 61–73.
24. Simpson, G. G. "The Baldwin Effect." *Evol.* **7** (1953):110–117.
25. Solbrig, O. T., and D. J. Solbrig. *Introduction to Population Biology and Evolution.* Reading, MA: Addison-Wesley, 1979.
26. Sutton, R. S. "Temporal Credit Assignment in Reinforcement Learning." Ph.D. dissertation, University of Massachusetts, Department of Computer and Information Science, Amherst, MA, 1984. Also University of Massachusetts Technical Report 84-2.
27. Sutton, R. S. "Learning to Predict by the Methods of Temporal Differences." GTE Laboratories technical report TR87–509.1 (Computer and Intelligent Systems Laboratory), Waltham, MA, 1987.
28. Williams, R. J. "Toward a Theory of Reinforcement-Learning Connectionist Systems." Technical Report NU–CCS–88–3, College of Computer Science of Northeastern University. Boston, MA, 1988.
29. Wilson, S. W. "Knowledge Growth in an Artificial Animal." In *Proceedings of an International Conference on Genetic Algorithms and Their Applications*, 16–23, Pittsburgh, PA, 1985.

Richard K. Belew, John McInerney, and Nicol N. Schraudolph
Cognitive Computer Science Research Group, Computer Science & Engineering Department (C-014), University California at San Diego, La Jolla, CA 92093; e-mail: rik@cs.ucsd.edu

Evolving Networks: Using the Genetic Algorithm with Connectionist Learning

It is appealing to consider hybrids of neural-network learning algorithms with evolutionary search procedures, simply because Nature has so successfully done so. In fact, computational models of learning and evolution offer theoretical biology new tools for addressing questions about Nature that have dogged that field since Darwin.[3] The concern of this paper, however, is strictly artificial: Can hybrids of connectionist learning algorithms and genetic algorithms produce more efficient and effective algorithms than either technique applied in isolation? The paper begins with a survey of recent work (by us and others) that combines Holland's Genetic Algorithm (GA) with connectionist techniques and delineates some of the basic design problems these hybrids share. This analysis suggests the dangers of overly literal representations of the network on the genome (e.g., encoding each weight explicitly). A preliminary set of experiments that use the GA to find unusual but successful values for back-propagation (BP) parameters (learning rate, momentum) are also reported. The focus of the report is a series of experiments that use the GA to explore the space of initial weight values, from which two different gradient techniques (conjugate gradient and back propagation) are then allowed to optimize. We find that use of the GA provides much greater confidence in the face of the stochastic variation that can plague gradient techniques, and can also allow training times to be reduced by as much as two orders of magnitude. Computational trade-offs between BP and the GA are considered, including discussion of a software facility that exploits the parallelism inherent in GA/BP hybrids. This evidence leads us to conclude that the GA's *global sampling* characteristics compliment connectionist *local search* techniques well, leading to efficient and reliable hybrids.

1. INTRODUCTION

It is extremely appealing to consider hybrids of neural-network-based learning algorithms with evolutionary search procedures, simply because Nature has so successfully done so. In fact, new computational models of learning and evolution offer theoretical biology new tools for addressing questions about Nature that have dogged that field since Darwin.[3,21] However, these same models have proven interesting enough to computer scientists that they can also be treated as artificial *algorithms*, divorced from the natural phenomena from which the models originally sprung. Considered separately, both connectionist networks and "evolutionary algorithms" have recently drawn a great deal of attention as new forms of adaptive algorithhm. On occasion, the two techniques have been compared.[6] The concern of the current paper is the composition of these two types of algorithms: Can hybrids of connectionist learning algorithms and genetic algorithms produce more efficient and effective algorithms than either technique applied in isolation? This proves to be a very broad question, and the present paper attempts to provide only a survey of results to date. Based on these experiences, we also identify several key areas for further investigation.

Section 2 begins with a brief description of Holland's Genetic Algorithm (GA). While using the GA to guide connectionist learning systems through specification of the networks' *structural* characteristics is perhaps the most natural, there are other hybrids of the two techniques that also seem promising. For example, Section 4 describes experiments that use the GA to find good values for two critical parameters of the BP learning algorithm, learning rate (η) and momentum (α).

Section 5 considers potential hybrids of GA and connectionist algorithms from the perspective of the state spaces they search and their respective methods. In brief, the GA proceeds by *globally sampling* over the space of alternative solutions, while gradient techniques—including BP but also methods like conjugate gradient—proceed by *locally searching* the immediate neighborhood of a current solution. This suggests that using the GA to provide good "seeds" from which BP then continues to search will be effective. We describe several experiments in which the GA is used to select initial values for the vector of weights used by BP and also by conjugate gradient.

Section 6 briefly sketches some of the computational complexity issues arising from GA/BP hybrids. A software facility that exploits the natural parallelism when the GA is used to control multiple instantiations of BP simulation is discussed, and some features of the time and space complexity of the hybrid systems are considered.

2. GENETIC ALGORITHMS

The GA has been investigated by John Holland[19] and students of his for almost twenty years now, with a marked increase in interest within the last few years.[13,14,34] The interested reader is advised to begin a more thorough introduction to these algorithms with the excellent new text by Goldberg.[12]

Attempts to simulate evolutionary search date back as far as the first attempts to simulate neural networks.[11] The basic construction is to consider a population of individuals that each represent a potential solution to a problem. The relative success of each individual on this problem is considered its **fitness**, and used to selectively reproduce the most fit individuals to produce similar but not identical offspring for the next generation. By iterating this process, the population efficiently samples the space of potential individuals and eventually converges on the most fit.

More specifically, consider a population of N individuals x_i, each represented by a **chromosonal** string of L **allele** values. An initial population is constructed at random; call this **generation** g_0. Each individual is evaluated by some arbitrary **environment** function that returns the fitnesses $\mu(x_i) \in \Re$ of each individual in g_0. The evolutionary algorithm then performs two operations. First, its **selection** algorithm uses the population's N fitness measures to determine how many offspring each member of g_0 contributes to g_1. Second, some set of **genetic operators** are applied to these offspring to make them different from their parents. The resulting population is now g_1, these individuals are again evaluated, and the cycle repeats itself. The iteration is terminated by some measure suggesting that the population has converged.

A critical distinction among simulated evolutionary algorithms is with respect to their genetic operators. Often the only genetic operator used is **mutation**: some number of alleles in the parent are arbitrarily changed in the child. This amounts to a *random* search around the most successful individuals of the previous generation, and is therefore not very powerful. The use of a simple mutation operator, coupled with the exponential amplification of good solutions afforded by selective reproduction, produce a powerful adaptive system on their own and some find this sufficient.[11,31]

The central feature of Holland's GA is its use of an additional **cross-over** operator modeled on the biological operation of genetic recombination: during sexual reproduction segments from each of the parents' chromosomes are combined to form the offspring's. One standard version[1] of the cross-over operation picks two points

[1]Because it has proven such an important component of the GA, many other variations of cross-over are under active investigation. For example, one, two or more cross-over points can be selected; these points can be selected non-uniformly over the string, etc.

$1 \leq m, n \leq L$ at random and builds the offspring's bit string by taking all bits between m and n from one parent and the remaining bits from the other parent. For example, if $L = 10, m = 2, n = 6$:

```
Parent(1): 1111111111    Offspring(1): 1100001111
Parent(2): 0000000000    Offspring(2): 0011110000
```

The appeal of the GA is due both to empirical studies that show the cross-over operator works extremely well on real, hard problems, and also to the "schemata" analysis Holland has provided to show why this is the case. Briefly, Holland's Schemata Theorem[19] suggests that the initially random sampling of early generations is concentrated by the GA's search towards those areas of the search space demonstrating better-than-average performance.

At the same time, the crossover operator imposes severe constraints on the genomic representation, as the experiments with the representation of connectionist networks here will demonstrate. Conversely, modern genetics continues to uncover biological mechanisms that are potentially even more powerful operators than crossover.[20] This paper, however, will restrict itself to the GA on the grounds that it currently provides the best balance between empirically demonstrated adaptive power and theoretical understanding.

One key property of the GA is that it works on a population of (binary) bit strings with absolutely no knowledge of the semantics associated with these bits. Its only contact with the environment is the global fitness measure associated with the entire string. This is considered an advantage of the algorithm because it ensures that the GA's success is not related to the semantics of any particular problem. This is not to say that the GA works on all problems equally well, only that these differences can be attributed to the underlying search spaces rather than the semantics of the problem domain.[5]

Having committed ourselves then to the GA and its crossover operator, it is worth noting some of the central "pearls of GA wisdom" that are most salient to the problem of encoding networks onto GA strings. First, our encoding must respect the *schemata*: The representation must allow discovery of small "building blocks" from which larger, complete solutions can be formed. Ideally, this means that we should be able to prove that the Schema Theorem holds with respect to our representation of a network. Second, the well known phenomena of *linkage bias* insists that we do our best to reflect functional interactions with proximity on the string. For example, a great deal of connectionist work has highlighted the role of individual hidden units; localizing the representation of these hidden units on the GA string therefore seems one reasonable strategy..[23,27] Finally, we must worry about the *closure* properties of the GA operators on the network descriptions. It is not strictly necessary that these operators produce valid network descriptions. But unless invalid descriptions are the exception and not the rule, the GA will not get the information about regularities among valid solutions in each new generation it needs to function properly.

3. MAPPING NETWORKS ONTO GA STRINGS

Within these basic guidelines, the ways of representing a weighted graph with a string of bits are limited only by the imagination. One useful dimension along which these alternatives can be organized is what Todd has called "developmental specification": i.e., how complete and literal a representation of the network is encoded on the GA string.[38] At one extreme, it is possible to encode each of the network's weights, in full precision, and then use the GA to solve this as a standard multi-parameter function optimization problem; in this case, there is no role for connectionist learning. At the other extreme, we could emulate biology and use the genetic description as input to a complex *developmental intepretation process* that then constructs the network/phenotype. Between these two extremes are a wide range of mappings in which the GA is used to constrain but not completely determine a network's structure, with connectionist learning processes subsequently embellishing this partial solution.

Surely the most straightforward representation of a connectionist network in a GA string is formed simply by concatenating all of the network's weights.[2] This approach leads to two types of design decision. First, how are each of the real number weights to be represented? Second, in what order are the weights to be concatenated?

3.1 ENCODING REAL NUMBERS

An immediate design issue facing any connectionist/GA hybrid is how the connectionist weights are to be represented on the GA string. Appropriate representation for real values on the basically discrete, binary GA chromosome is a matter of considerable debate within the GA community. Perhaps for this reason, several of the earlier attempts to use the GA with connectionist networks have left real numbers as discrete elements in their representation, thus avoiding this encoding issue.[27,39] The GA is then allowed to search for good combinations of weights, but is not used for finding the value of any one weight. But this *a priori* division of effort is something we hope to avoid; the many successes with which the GA has been used to discover real-valued quantities suggests that it is also unnecessary.[9]

Figure 1 shows the basic features of weight encoding. For each weight, assume first that the real number to be found exists somewhere in the bounded region $[m, M]$. Assume also that B bits have been allocated to represent each weight; these are sufficient to divide the bounded region into 2^B intervals. The B bit index is then Gray-coded to minimize the Hamming distance between indices close in value.[7] Within the specified interval, then, a real number is selected from a random variable uniformly distributed over that interval. Note that this stochastic element

[2] Some tentative results suggest that with this encoding the GA can find weights more quickly than back propagation, but only on fairly deep networks (i.e., with many hidden layers).[28]

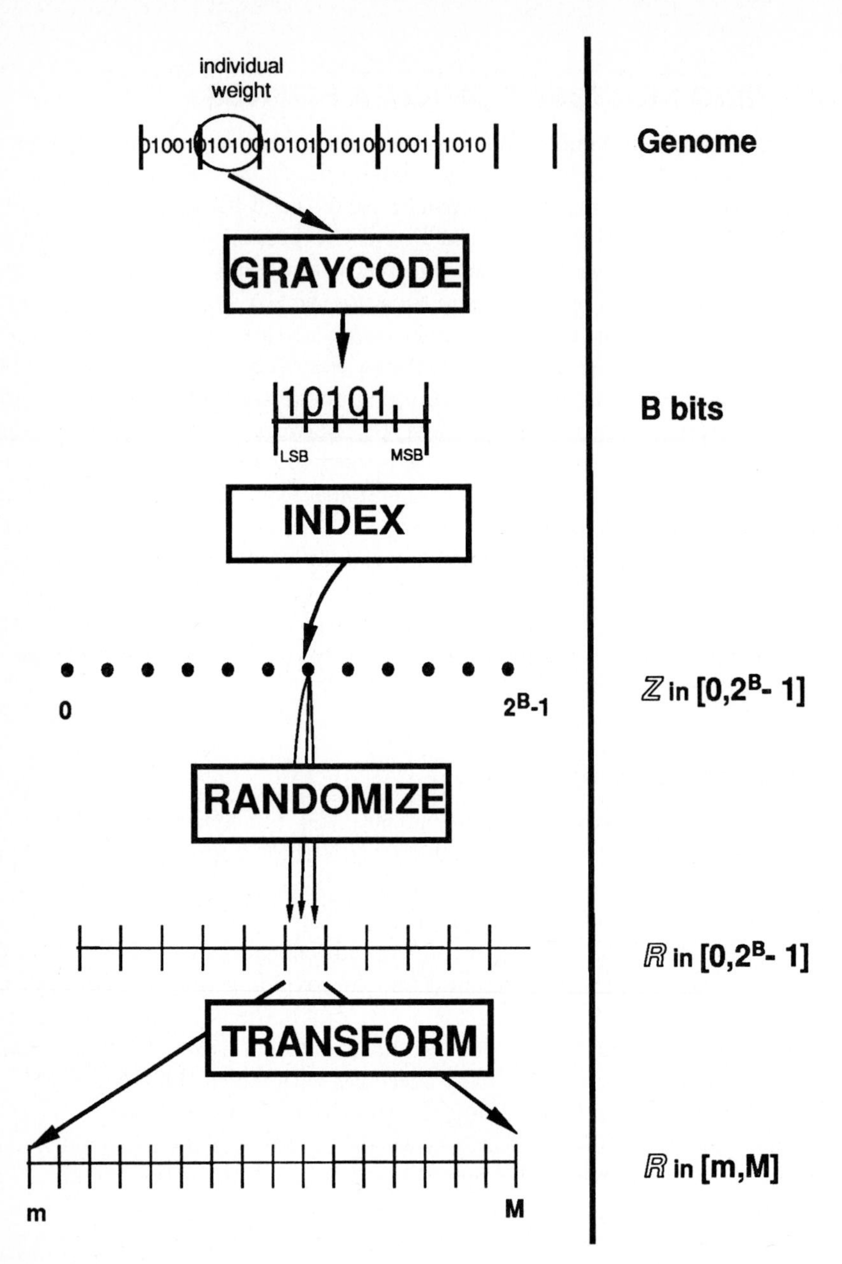

FIGURE 1 Encoding weights.

will often introduce great variability in the resulting networks' performance, as connectionist networks have been shown to be extremely sensitive to small changes in weights.[22] More constructively, it would be desirable if this encoding (the range encoded, the number of bits used) were varied as a consequence of the variability experienced across the population. This kind of *dynamic parameter encoding* for the GA is being explored separately.[35]

3.2 CROSSOVER WITH DISTRIBUTED REPRESENTATIONS

Perhaps the most important feature of representation for the GA is proximity. Because two alleles are much more likely to become separated by a crossover operation if they are far apart on the string than if they are close together, it becomes less and less likely that the GA will be able to discover and exploit nonlinear interactions between any two alleles as they are put farther apart on the string. In our case, this lesson suggests that the best representation will have dependent weights close together on the string just if these weights are functionally dependent on one another.

Consider a standard three-layer, feed-forward network. At least in these networks, the obvious functional units correspond to units in the hidden layer. This suggests that all weights associated with one hidden unit should be placed together on the string. Merrill has performed experiments that substantiate this.[23] Such functional units can be made even more cohesive by introducing "punctuated" crossover operations, which have higher probability of breaking the chromosome at certain punctuated points in the string (e.g., between one hidden unit's weight and another's).[33]

One important property of the solutions learned by networks, however, is that they are generally far from unique. In the context of the GA, this means that crossover among two relatively good parents who have discovered different solutions can lead to abysmal offspring. Consider again the example of a simple three-layer, feed-forward back propagation (BP) network, and consider the solutions it might discover to the "encoder" problem.[3] A typical solution, reported in the PDP volumes,[32] is shown in Figure 2. Note that while this network discovered the same binary encoding scheme a computer scientist might suggest, it also made use of *intermediate* activation values. In general, we can expect such individual variation among the solutions found by connectionist networks, and so in their corresponding genetic descriptions.

Less subtle but just as problematic variations arise because fully isomorphic solutions can be obtained simply by permuting the hidden units. That is, two networks can be identical up to the arbitrarily assigned *indices* of their hidden units. But (at least in the representations considered heretofore) the location of a hidden unit weight on the GA chromosome depends entirely on its (arbitrarily assigned) index!

[3] The encoder problem involves mapping N orthogonal patterns through a hidden layer of $log_2 N$ hidden units onto a set of N orthogonal output patterns.[32]

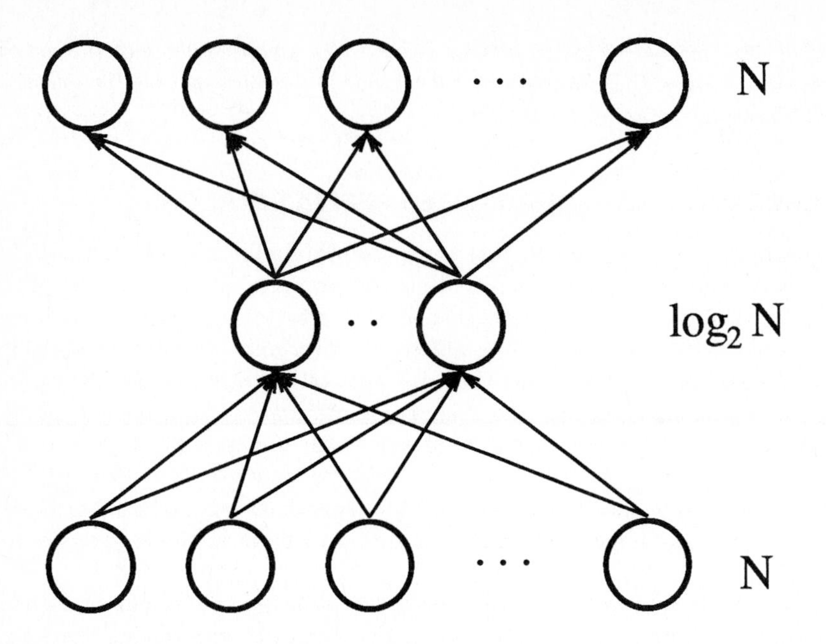

FIGURE 2 Solution to the encoding problem.

The invariance of BP networks under permutation of the hidden units is such a devastating and basic obstacle to the natural mapping of networks onto a GA string that we might consider ways of normalizing network solutions prior to crossover. It seems, for example, that at least in the case of BP networks with a single hidden layer, the differential weighting of the hidden units to the "anchored" (i.e., constant and nonarbitrary) input and output layers might be used to recognize similarly functioning hidden units. Even establishing a correspondence among hidden units of two three-layer networks that have been trained to solve the same problem appears to be computationally intractable, even when we assume that the only difference between the two networks' solutions is a permutation of hidden units. In realistic networks many other different solutions to the same problem can be constructed, for example by reversing the sign of all weights, or taking other "semi-linear combinations"[4] of the weights. We therefore conclude that attempting to normalize networks before combining them is not feasible.

Thus, the specifics of a network's architecture are *underdetermined* by the problem it is trying to solve. Consequently the genetic representations of these varying architectures cannot be expected to share the similarities (schemata) that the GA needs in order to be effective. If very small populations are used with the GA, there

[4]We speak a bit loosely here. Because BP networks depend on nonlinear "squashing" functions, simple linear combinations are not quite adequate.

is not "room" for multiple alternatives to develop. In this case, whichever solution is discovered first comes to dominate the population and resist alternatives. This approach has been used successfully by Whitney.[39] Alternatively, the correspondence between genotype and phenotype can be made less direct; the first step in this direction is discussed in the next section.

3.3 WIRING DIAGRAMS

As we move away from full specification of all network weights on the genetic string, the goal will be to use the GA to specify some *constraints* on a network architecture. Within these constraints, the connectionist learning procedure then does its best to optimize the objective function via weight modification.

One of the most straightforward architectural descriptions for a network is a binary "wiring diagram." The links of a three-layer feedforward network with I input units, H hidden units and O output units are encoded as a binary string of length $H*(I+1+O)+O$, with all links (including the bias) from and to one hidden unit falling contiguously. This wiring specification is given to a simulator that uses BP to do its best to learn the specified task from a series of training examples, and the network's final mean squared error (MSE) becomes the fitness associated with that individual. An entire population of such individuals form a generation. The

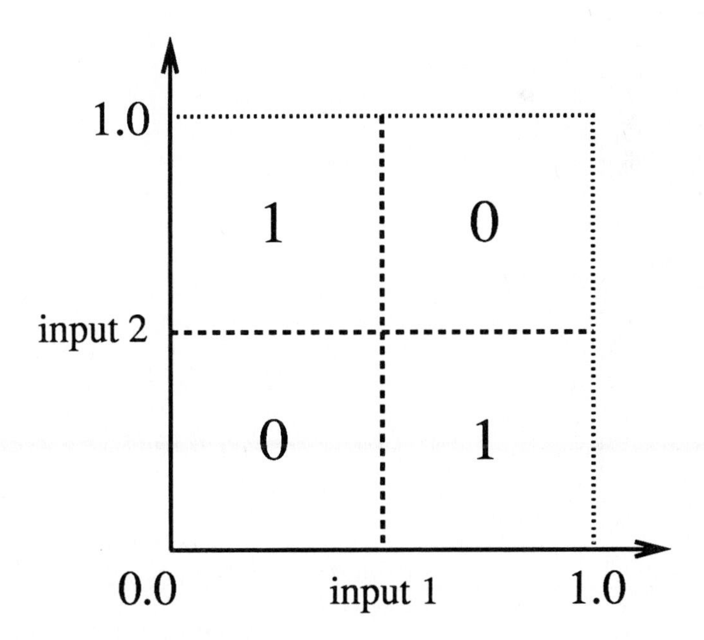

FIGURE 3 Four-quadrant problem.

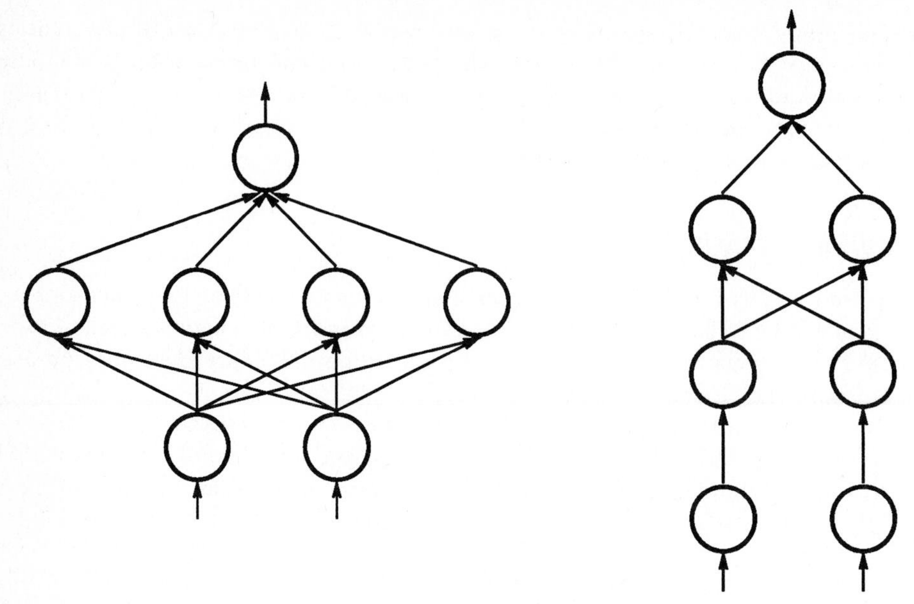

FIGURE 4 Two solutions to the Four-quadrant problem.

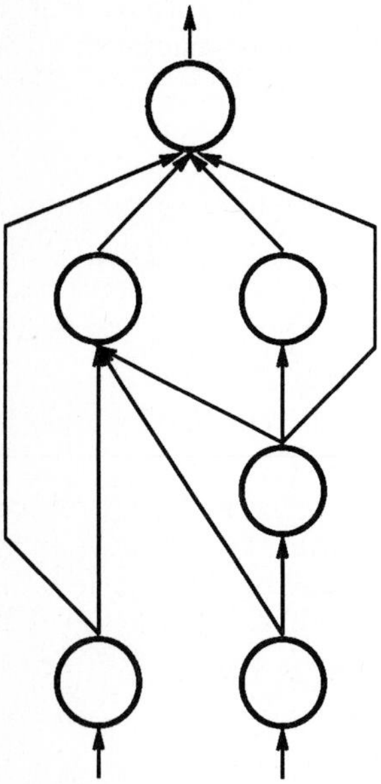

FIGURE 5 GA solutions to the Four-quadrant problem.

GA is used to replicate and alter the binary network descriptions to form a new population, and the cycle is then repeated.

Miller et al. report on experiments using this sort of wiring diagram.[24] Their most striking result was with the Four-quadrant problem, a generalization of the XOR problem to a two-dimensional real interval; see Figure 3. This problem is interesting because it admits at least two types of solution, as shown in Figure 4. The "fat" solution uses only a single layer of hidden units, with the number of hidden units required growing as the desired precision increases. The "skinny" solution makes use of two layers of exactly two hidden units each. Each input unit is connected to only one of the hidden units in the first layer; this layer simply changes the real inputs to binary values. The rest of the network can then solve the problem as a standard XOR network. Since tradeoffs between wide and deep networks are an important and open issue in connectionist learning systems, this problem is particularly valuable because it provides some experience with using the GA to design networks with more than one hidden layer.

Using a network architecture description that allowed either of these solutions to form, the GA in fact consistently found intermediate solutions like that shown in Figure 5. Note that while this network makes use of both hidden layers, it does not solve the Four Quadrant problem in either of the regular fashions described above. Miller et al. also report that the GA consistently created a link directly from the input unit to the output unit, a feature that was allowed by their network architecture description but not anticipated by them. It is important to note, however, that some of these solutions existed in the random initial population with which the GA began, and the GA converged on a population of such individuals in only a few generations. The GA did not therefore play an important role in the discovery of these solutions, and any iterated restart of BP can be expected to have performed similarly.

Several comments should be made about binary wiring diagrams like these. First, experiments such as these impose an unfortunate asymmetry between the adaptation effected by the GA and that done by connectionist learning. Virtually all connectionist learning algorithms allow connections to come to have zero weight, making them act as if the connections were not there. Thus an existing connection can learn to have zero weight, but an absent connection cannot ever become non-zero. We should expect this bias to be exploited by any hybrid adaptive system that combines evolutionary and (connectionist) learning sub-systems.

Second, the binary specification of link presence or absence can easily be generalized to a wider range of constraints on network architectures. For example, Todd suggests that a ternary specification might specify that a link was absent, restricted to positive weights or restricted to negative weights.[38] The range of search allowed the connectionist learning procedure by the genetic network description can be progressively constrained in this fashion until, in the extreme, the GA is specifying each weight exactly. Conversely, the process of dynamic parameter encoding (DPE) can be used to focus the GA's search on those regions with least variability, so that *a priori* divisions of the search between GA and gradient procedures come to be reconsidered.[35]

Third, as soon as the goal of our hybrid algorithm is changed from finding the *weights* for a net that can do the best job, to finding an *architecture and also weights* for that architecture that can do the best job, our search criterion must change correspondingly. More specifically, if the GA is to find good architectures, the function it optimizes must include not only a measure of error (e.g., MSE), but also a measure of the *complexity* of the network. Otherwise, if there is no (fitness) penalty paid for having more links, for example, there will be no adaptive pressure to use more parsimonious representations. The absense of any such complexity term in the Four Quadrant experiments of Miller et al. may account for the fact that the GA found neither the "fat" nor "skinny" solution, but something with (potentially redundant) direct links from input to output; see Section 6.

Finally, wiring diagrams do not avoid the "underdetermined architecture" problems described in the last section. The problem, at least in part, is that the relationship between genotype and phenotype is still inappropriate: features of the network that are inconsequential to its computation (e.g., the indexing of the hidden units) are reflected by radically different genetic descriptions. And so we are pushed another step away from complete network specification and towards the interposition of a developmental process between genotype and phenotype.

3.4 DEVELOPMENTAL PROGRAMS

The GA is generally cast as a function optimizer, with the GA manipulating values of $\underline{x}$ in order to optimize some function $f(\underline{x})$. One critical aspect of biological evolution that is missed in this formulation is that the space of *genotypes* manipulated by genetics is only indirectly related to the space of *phenotypes* which are evaluated by the environment. The process relating genotype to phenotype is, of course, *ontogenetic development.*

The developmental process is an extraordinarily complicated adaptive system in its own right,[29] and attempting to incorporate it within the already complicated hybrids being considered here is problematic.[5] But the incorporation of a developmental interpreter means that the GA can be allowed to search through representations for which it is more well-suited than those derived directly from networks. Just which developmental model will prove most satisfactory in the context of evolution/development/learning hybrids is still an open and important question, but there are a few promising leads.

Harp et al. describe a very interesting model in which the GA searches for a network "blueprint"[17]; see Figure 6. The description of neural networks in terms of "areas" (i.e., sets of units with varying spatial extent), and "projections" (i.e., sets of edges randomly connecting units from one area to another) certainly seems to capture much of the architectural regularity of nervous systems in vertebrates.

These experiments are consistent with only the most basic features of the corresponding biological systems, and we intend to explore more satisfactory models. Purves outlines the basic features of a more elaborate cell adhesion model of neural development[29]; Edelman also desribes an elaborate but idiosyncratic model.[10] The details of retina-to-optic tectum mappings have been described by Cowan.[8] Stork has used a developmental model with the GA to show how evolutionary

[5]For example, the developmental interpreter should, by rights, itself be specified on the genome.

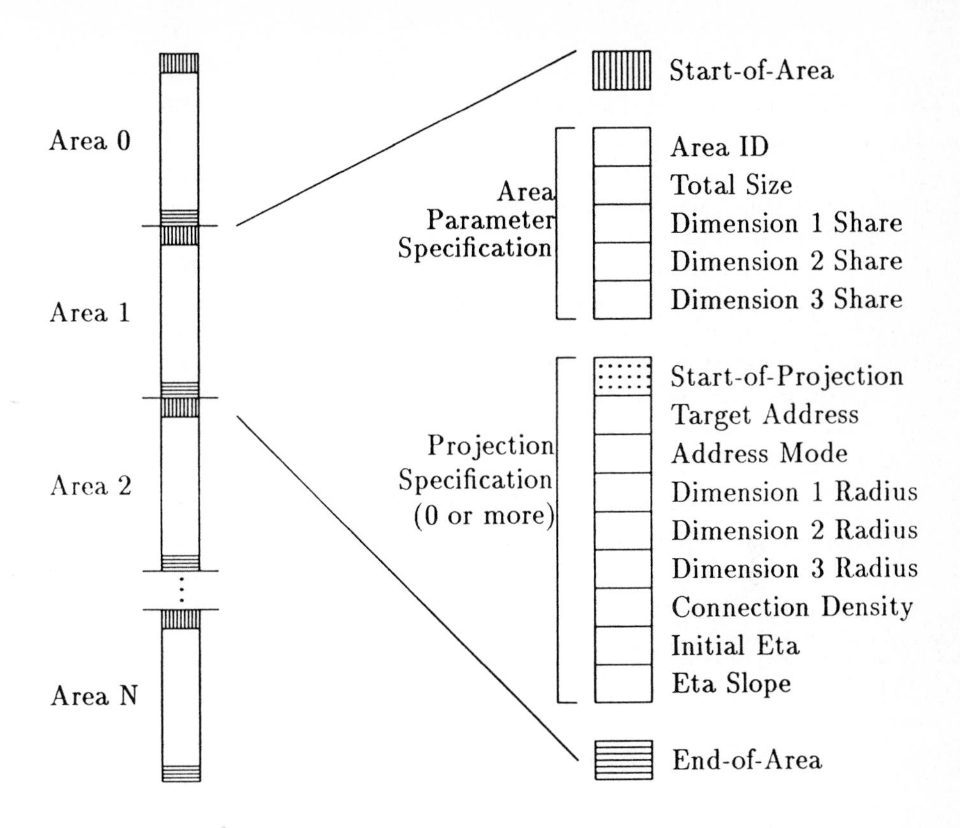

FIGURE 6 Harp blueprint.

"pre-adaptations" may be responsible for certain anomolous neural connections in the modern crayfish.[36]

The range of connectionist network representations for the GA surveyed in this section still leaves much to be explored. Further, the GA can be used with connectionist networks in other ways than specification of network architecture. The next two sections describe experiments that use the GA to control other, non-architectural parameters of connectionist learning systems.

4. TUNING BP PARAMETERS

The central experiments of Section 5 use the GA to find good initial weights from which a gradient descent procedure like back propagation (BP) can reliably converge on a solution. While these investigations are largely orthogonal to the use of the GA for the kind of network architectural definition described in the last section, the well known *symmetry* problem[6] was used because, like Miller et al's Four Quadrant problem (cf. subsection 3.3), it is known to permit two distinctly different network solutions.[25] For the experiments reported here we used the six-bit version of this problem. A three-layer feedforward network with six input units, six hidden units, and one output unit was specified.

Before beginning these experiments, it was necessary to set the learning rate (η) and momentum (α) parameters of BP. These parameters are known to be strongly coupled, dependent on characteristics of the problem being solved, and critical to the successful convergence of the learning procedure. As a result, finding good values for η and α is more art than science and generally a matter for trial-and-error search. Because the GA has often had success at strongly non-linear function optimization problems like this, we began by using the GA to find good values for η and α. The ranges

$$0 \leq \eta \leq 8; \ 0 \leq \alpha \leq 1.0$$

were explored, and each parameter was allocated 10 bits. This unusually large range for η was used because preliminary experimentation with more conservative values (e.g., $\eta \leq 2, \eta \leq 4$) consistently resulted in the GA converging on the maximum value in that range.

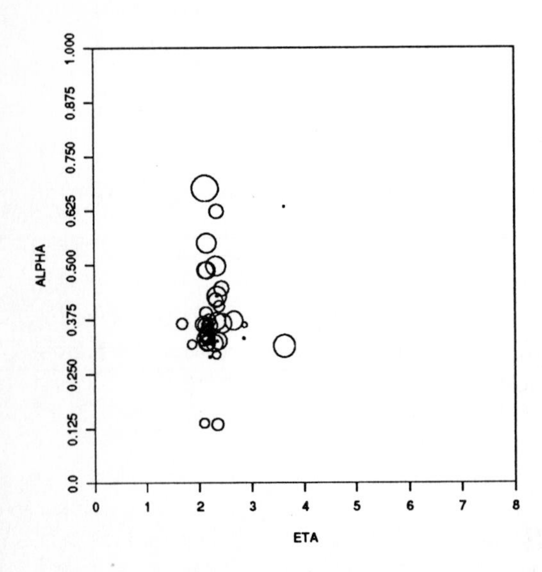

FIGURE 7 Finding η and α with the GA.

[6] Given a binary vector of length $2N$, the net is to produce a value of unity on its single output unit iff, for all input units $I_i = I_{2N-i+1}, i = 1, \ldots, N$.

With the GA selecting values for these two parameters for each individual in a population of size 50, an otherwise standard BP simulation was trained for 200 epochs and its mean squared error (MSE) at the end of this traing was used as the individual's fitness.

Figure 7 shows the results after 200 generations, with the diameter of each circle indicating the reciprocal of the MSE (i.e., larger circles mean better performance) for each individual in the last generation. Conventional wisdom calls for $\eta = 0.1, \alpha = 0.2$, but the GA consistently converged on values of $\eta \approx 2.5$! This means that BP is moving rapidly along the error surface. We conjecture that these values depend on the number of learning epochs ($\mathcal{T}$) we allowed each BP optimization before using the net's MSE value. We used $\mathcal{T} = 200$ epochs, which is an extremely short training period for the six-bit symmetry problem.[7] Networks using more "patient" values for η and α were unable to solve the problem in the time alloted, and only those that "went for broke" were successful. It seems likely that more conservative values for η and α would be found if more trials were allowed, and even more radical values might be found if this number were reduced.[8]

Note also that both large and small diameter circles are sometimes virtually concentric. This means that while the GA consistently converged on fairly narrow ranges for η and α, there is still high variability in the fitnesses of individuals with very similar parameters. This is because the BP simulations have a highly stochastic element, viz., the random initial weights assigned. The values for $\eta = 2.5, \alpha = 0.33$ were robust enough under varying initial weights that they could be exploited by the GA, but the selection of good initial weights is still critical. The identification of these good initial values is the major focus of our next experiments.

5. SAMPLING AND SEARCH

The use of the GA to tune BP parameters has proven practically useful, but there is nothing terribly profound about this type of hybrid. The GA is doing function optimization over a set of parameters in the same way that GAs have been used since DeJong.[9] A more important combination of these technologies arises from the observation that *local search performed by back propagation and other gradient descent procedures is well complemented by the global sampling performed by the*

[7]Randomly restarted BP runs using $\eta = 0.1, \alpha = 0.2$ reliably converge to the solution in approximately 4000 epochs.

[8]It should be noted that the use of these high, quick parameters for BP seem to depend critically, at least in the six-bit symmetry problem, on the order in which training instances are presented. Our simulator allows exemplars to be presented in: sequential order, random order with replacement, random order without replacement, or "batch" training. Because we have had most success on the six-bit symmetry problem using *random with replacement* ordering, these experiments were run with this option.

GA; consider Figure 8. Gradient descent procedures all sample some characteristics of their local neighborhood to determine a direction in which the search is to proceed. Sophisticated techniques for gradient descent (of which BP is only one) efficiently combine characteristics of the local neighborhood to form a good next guess. But, depending on characteristics of the objective function being searched, this local information may be misleading as to the location of the actual optimum. In particular, gradient descent procedures are known to be subject to local minima. Sampling techniques like the GA, on the other hand, are effective because they ensure broad coverage over the entire domain. The GA works by collecting information from the early and virtually uniform sampling of early generations, and then using this information to guide subsequent sampling towards particularly promising regions. The selection of a new element of the domain to evaluate therefore exploits global information from across the domain. Unfortunately, information in the immediate neighborhood of each of these samples plays no role in subsequent GA sampling, meaning the algorithm can come frustratingly close to the solution without actually finding it.

Combining the GA's global sampling with BP's local searching therefore seems an extremely natural and promising form of hybrid. It can be compared to the

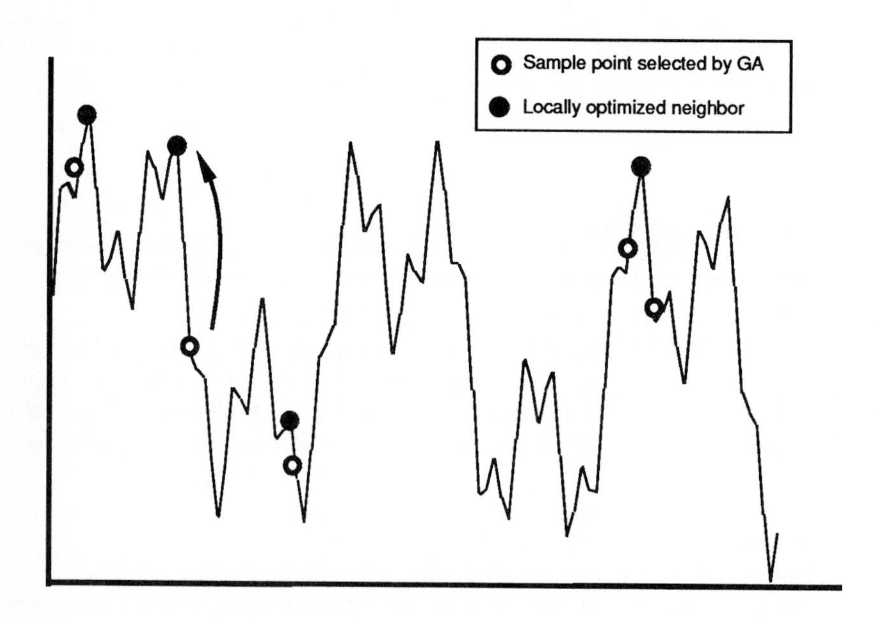

FIGURE 8 Sampling and search.

simple process of hill climbing with restart, with the key advantage of the GA being that it promises to sample in a much better than random fashion.[1,12]

When placed in the context of connectionist networks, the strategy just expressed suggests that the GA be used to create "seeds": starting points from which a connectionist search proceeds. In connectionist learning terms, this corresponds to using the GA to prescribe the *initial weights*, $\underline{\mathbf{W}}(\mathbf{0})$, on a network's links. A schematic view of this hybrid construction is shown in Figure 9. The GA selects an initial weight vector for each individual in a population, each is allowed to learn with BP for some number of trials, and the error rate at which it is performing at this time is considered to be the fitness of that individual.

Thus the GA will sample those regions of weight space from which it is reliably possible to reach good function values via gradient descent. There is a pleasing symmetry to this search, in that the best initial weight vector (found by the GA) is obviously the same as the final weight vector (found by BP); the two algorithms' solutions are interchangable in this respect. It is important to note, however, that the two algorithms are coupled in this symmetric search only if the range of initial weight values being explored by the GA is coextensive with the domain of solution weight vectors ultimately discovered by the gradient descent procedure.

In practice, these two sets are often quite different. For example, the PDP volumes recommend "...small random weights"[32]; Miyata has operationalized this as "...uniformly distributed random numbers between $\pm\frac{1}{2}$."[26] Final weights, on the other hand, can be widely distributed, and often fall outside this initial distribution.[16]

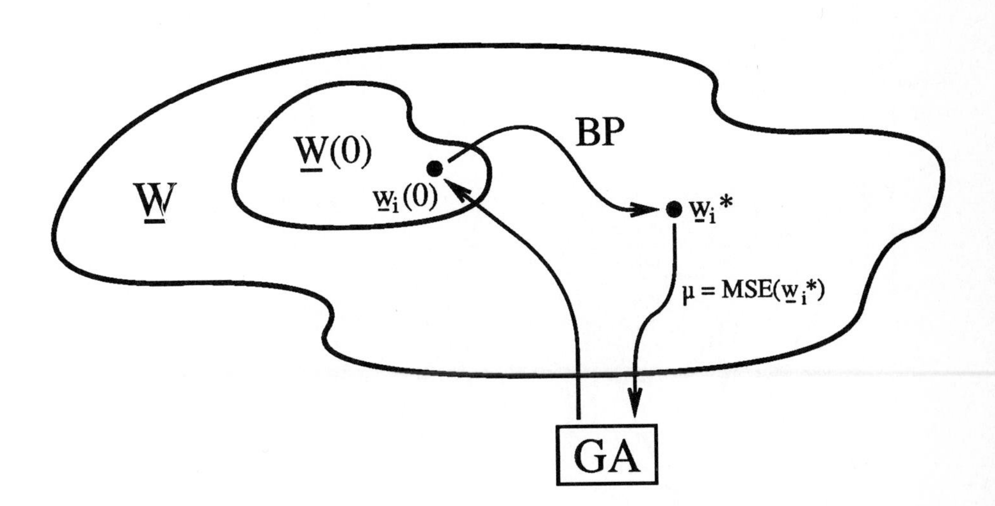

FIGURE 9 Selecting $\underline{\mathbf{W}}(0)$.

The distinction between starting and final points in weight space also complicates our characterization of just what the GA is looking for. While it is true the GA is seeking $\underline{\mathbf{W}}(\mathbf{0})$ that are "close to" the solutions ultimately found by a gradient technique, it is important to note that the relevant measure is not the natural (e.g., Euclidean) distance between initial and final weight vectors. Rather, good $\underline{\mathbf{W}}(\mathbf{0})$ found by the GA are close to good final solutions *with respect to the gradient procedure being used.* Figure 10 caricatures the search regions induced by two different gradient techniques, all begun from the same initial point in weight space. For a particular gradient technique, these regions can be characterized in terms of "isobars" requiring the same computational effort (e.g, BP training epochs). We can imagine error surfaces over which it takes a gradient procedure many iterations to move a short Euclidean distance, and the converse is also true. Further, the range of solutions "reachable" by a gradient procedure varies with the procedure being used; points that are easy to reach via BP may not be reachable via conjugate gradient techniques, and vice versa.

Finally, sampling procedures like the GA require that the gradient procedure *reliably* converge on a good solution. Recent results by Kolen and Pollack demonstrate that "BP is sensitive to initial conditions" (i.e., what we call $\underline{\mathbf{W}}(\mathbf{0})$),[22] and so finding such reliable regions is non-trivial (cf. subsection 5.4). Thus the goal of

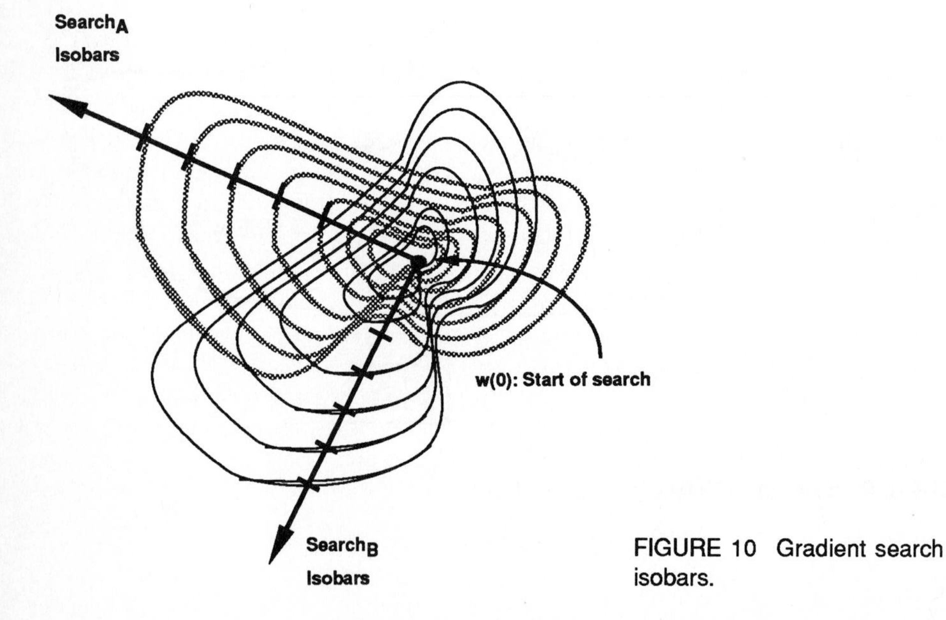

FIGURE 10 Gradient search isobars.

TABLE 1 $\underline{\mathbf{W}}(0)$ Experiments.

Expt.	$\underline{\mathbf{W}}(0)$ *range*	*GA Encoding*	*Gradient proc.*	*Epochs*
1	$\pm 10^5$	Uniform	Conj. grad.	N/A
2	$\pm\frac{1}{2}$	Uniform	Back prop.	200
3	$\pm[10^{-2}, 5]$	Exponential	Back prop.	40

TABLE 2 GA parameters.

Total Trials = 3000
Population Size = 50
Structure Length = 128
Crossover Rate = 0.600
Mutation Rate = 0.001
Generation Gap = 1.000
Scaling Window = 5
Max Gens w/o Eval = 2
Options = acel

our search is somewhat different from most connectionist systems: we are interested in the *distribution* of good solutions rather than simply identifying some of these.

Three sets of experiments were performed to test the feasibility of using the GA as a source of $\underline{\mathbf{W}}(0)$ seeds for gradient techniques that then did their best to optimize further; these are summaraized in Table 1. In the first set a conjugate gradient (CG) method was used, and the range of $\underline{\mathbf{W}}(0)$ explored by the GA was made very large, effectively coextensive with that of the CG procedure. In the second set of simulations, BP was used and the GA was allowed to explore within the more typical range $\pm\frac{1}{2}$. Finally, an expanded range was searched by the GA, but an exponential encoding was used that allowed particularly refined searching of small $\underline{\mathbf{W}}(0)$ values. Basic parameters of the GA were also kept constant across the three experiments; see Table 2. Our group has developed a sophisticated GA simulator, Genesis-UCSD, and it was used for all of the experiments reported here; subsection 6.1 describes a recent extension of this simulator that allows it to run across a distributed network of processors.

5.1 EXPERIMENT 1: CONJUGATE GRADIENT OVER CLOSED DOMAINS

To ensure the closure properties between GA and gradient search procedures mentioned above, the range of initial weights were allowed to run between $\pm 10^5$. Eight bits were allocated to represent each weight. Thus the GA was able to specify an initial weight with extremely poor precision: Each allele value corresponds to a range of approximately 800 (see Section 3 for details). Conjugate gradient (CG) optimization was used in these experiments because it is known to converge to solutions more quickly and reliably than heuristic second-order optimization techniques like BP (with a momentum term).[2][9] The GA was used to create an initial population of $\underline{\mathbf{W}}(\mathbf{0})$ vectors, CG optimized each of these, the MSE of each result was used for the individuals' fitnesses, the GA used these to produce a new population of $\underline{\mathbf{W}}(\mathbf{0})$ vectors, and the cycle repeated itself.

Figure 11 summarizes the basic results of these experiments. The average mean squared error (MSE) of the GA+CG hybrid is plotted on a logarithmic scale, as a function of generation; this curve is labelled **GA+CG(avg)**. To put the results of this hybrid method in perspective, it is appropriate to compare them to use of GA and CG methods used in isolation. For comparison with the CG-alone method, multiple randomly restarted iterations of CG were performed an equivalent number of times and this average baseline is labeled **CG (avg)**. The comparison of averages is appropriate given our interest in *expected* performance, but the simple average does blur information about the underlying distribution. Also, in most applications, the *minimum* of multiple restarts would be used. To facilitate this comparision, multiple randomly restarted iterations of CG were performed an equivalent number of times and the minimum of these is drawn as a baseline labeled **CG (min)**

Alternatively, the hybrid can be compared to search by the GA alone wherein initial $\underline{\mathbf{W}}(\mathbf{0})$ vectors selected by the GA were evaluated without modification by CG. In both cases, the hybrid approach did significantly better than either the gradient technique or the GA used in isolation. Thus, on average, the hybrid of GA+CG can solve the problem more effectively than either a randomly started CG search or GA sampling uninformed by some gradient search.

At the end of 60 generations the GA+CG hybrid had converged significantly.[10] The solutions it found were odd, in that large magnitude weights were used, often greater than $\pm 10^4$. This seems to be due to two factors. First, the GA will sample large $\underline{\mathbf{W}}(\mathbf{0})$ weights more often than small ones, simply because there are so many more of them. Uniformly dividing a large weight range into a finite set of regions results in the low end of the dynamic range being grossly under-represented. (This suggests non-uniform, *exponential* encodings; see subsection 5.3.) Second, the inclusion of even one large weight in the inner product summation performed by

[9]The code implementing the CG method was obtained from the OPT package of Barnard and Cole. This code required extensive modification to allow it to be used for these problems, and to fix a serious bug it contained.

[10]137 out of 384 alleles converged to at least 90%; Bias = 0.888.

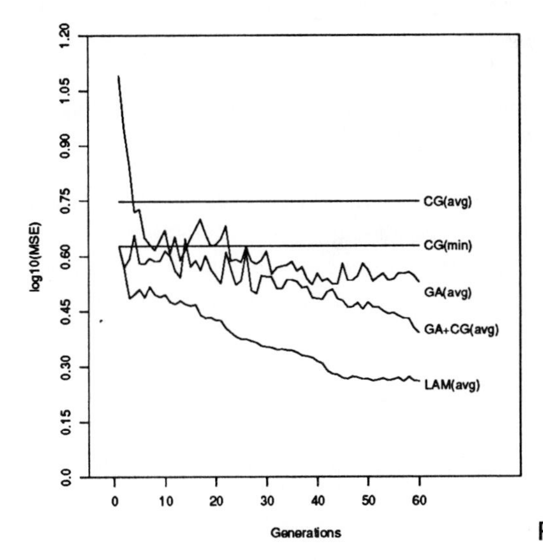

FIGURE 11 GA + CG hybrid.

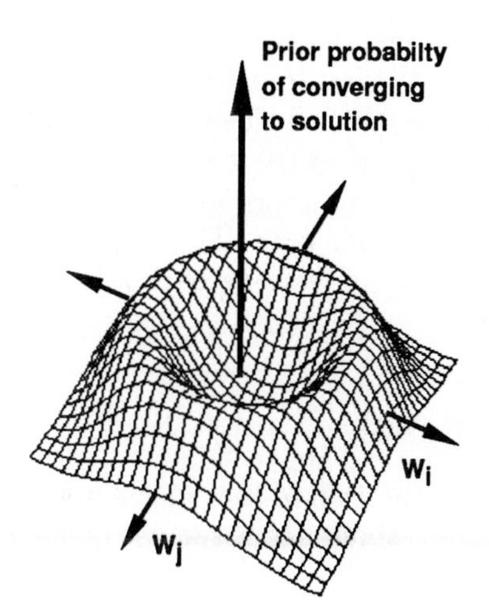

FIGURE 12 "Donut" of good $\underline{\mathbf{W}}(0)$.

connectionist units is enough to push that unit's activity into the asymptotic range of its squashing function, effectively drowning out the contribution of any smaller weights. More precisely, the derivative of the sigmoidal squashing function is near zero for large weights, meaning that almost no error information is available for

BP. The result of these two factors is that the solutions found by these experiments were poor in absolute terms.[11] (BP can find perfect solutions to six-bit symmetry, at least if given enough trials; see Section 6.)

Our picture of the range of good $\underline{\mathbf{W}}(\mathbf{0})$ values now looks like a "donut," with both lower and upper bounds; see Figure 12. If $\underline{\mathbf{W}}(\mathbf{0})$ is too near the origin, the network is unable to break the symmetries of its hidden units so that they all attempt to do the same job, and the network will remain on the "local maximum" of the origin originally described by Rumelhart et al.[32] But if $\underline{\mathbf{W}}(\mathbf{0})$ is too large, the net is drawn to solutions with large weights that grow without bound. Worse, any *one* large weight can push a unit into the asymptotic region of its sigmoidal squashing function, effectively masking any error signal. For these reasons, the remaining two experiments of this report restrict the GA's search for good $\underline{\mathbf{W}}(\mathbf{0})$ values to a limited range.[12]

Before discussing these other experiments, however, another interesting comparison is possible when the GA and gradient methods share coextensive search spaces. The GA+CG hybrid can be compared to a third technique that might be called a "Lamarckian" algorithm: the solution found by CG is remapped into the genetic encoding and this new specification is returned to the GA population in place of the original $\underline{\mathbf{W}}(\mathbf{0})$. That is, instead of giving the GA only the information that some individuals are close (in the gradient search sense) to a good solution, we do our best to "reverse transcribe" the solution actually found back into its genomic correlate, and give this to the GA to act as a new, genetically engineered parent. At the end of the runs shown in Figure 11 this Lamarkian variant appears to do only as well as the standard GA+CG hybrid, but in longer runs the Lamarkian algorithm does significantly better. Note that again the ability to invert from learned solutions back into their genetic correlates depends on these two domains being coextensive. In fact, even in these experiments in which the range of $\underline{\mathbf{W}}(\mathbf{0})$ was extended far beyond normal, the CG often moved to values outside the range of $\pm 10^5$; in this case, the genetic value was given its maximum or minimum value. This is only one, simple example of the difficulty in inverting the results of learning back into their genetic correlates; see Section 7.

5.2 EXPERIMENT 2: BP AND LIMITED $\underline{\mathrm{W}}(0)$ RANGE

With the exception of the folk-wisdom (mentioned in Section 5) that initial weights should be small ($\pm\frac{1}{2}$) and random, little is known about the selection of good

[11]Another possible explanation is that the CG method was used inappropriately. Conjugate gradient techniques work well for moving quickly to the bottom of an attractor basin, but they may not be the best way to find such basins in the first place. This suggests that (yet another!) hybrid method would use BP for a few iterations, to get into a good attractor basin, and then invoke CG to finish mimimization. We are exploring this technique.

[12]In a forthcoming report, we investigate a theoretic characterization of the donut of appropriate $\underline{\mathrm{W}}(0)$ values.[4]

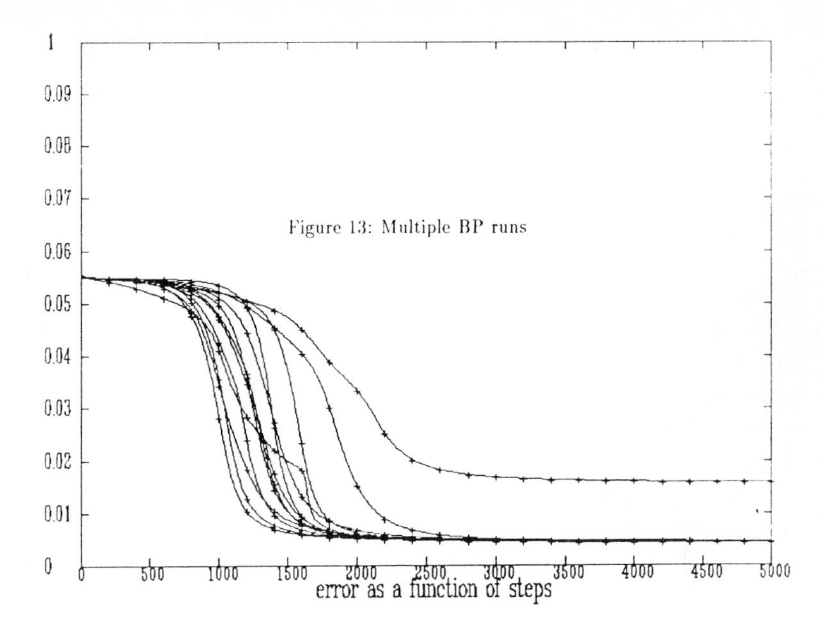

FIGURE 13 Multiple BP runs.

initial weights. Empirically, it has been widely observed that the performance of BP is highly variable with respect to the choice of $\underline{\mathbf{W}}(\mathbf{0})$. Figure 13 is typical. This shows 10 standard[13] runs of BP varying only in their initial values for $\underline{\mathbf{W}}(\mathbf{0})$. All we can say is that sometimes BP works and sometimes it doesn't; this is far from satisfying.

The question to be investigated here is whether the GA can find regions of the $\underline{\mathbf{W}}(\mathbf{0})$ space that reliably lead to good (in the sense of low MSE) networks. The basic construction of the GA+BP hybrid algorithm is the same as the GA+CG hybrid above: The GA was used to create an initial population of $\underline{\mathbf{W}}(\mathbf{0})$ vectors, BP was then used to optimize each of these, the MSE of each result was used for the individual's fitness, the GA used these to produce a new population of $\underline{\mathbf{W}}(\mathbf{0})$ vectors, and the cycle repeats itself. However, for the reasons discussed in the previous section, the range of $\underline{\mathbf{W}}(\mathbf{0})$ explored by the GA was limited to the standard $\pm\frac{1}{2}$. Note also that an unusally high learning rate ($\eta = 2.5$) similar to those discovered in Section 4 was used in these BP simulations.[14] This rapid learning rate made it possible to give each BP network only 200 training epochs before judging its MSE fitness.

[13]Miyata's SunNet simulator[26] was used on the problem of six-bit symmetry with six hidden units; $\eta = 0.1, \alpha = 0.2$.

[14]These experiments were done before the simulations of Section 4 were complete, and hence the values ($\eta = 2.5, \alpha = 0.0$) are less than optimal.

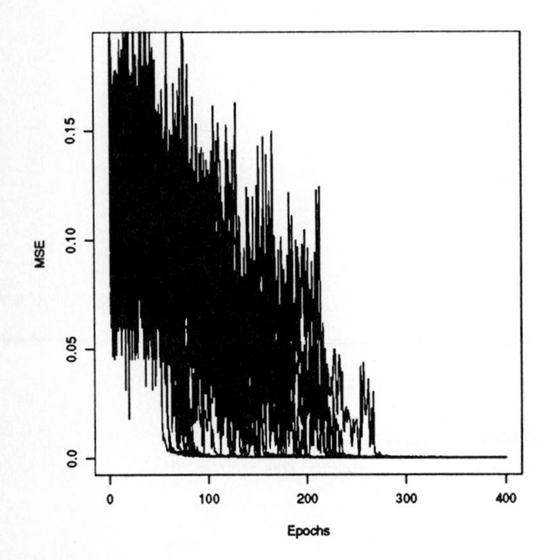

FIGURE 14 GA + BP Initial Population.

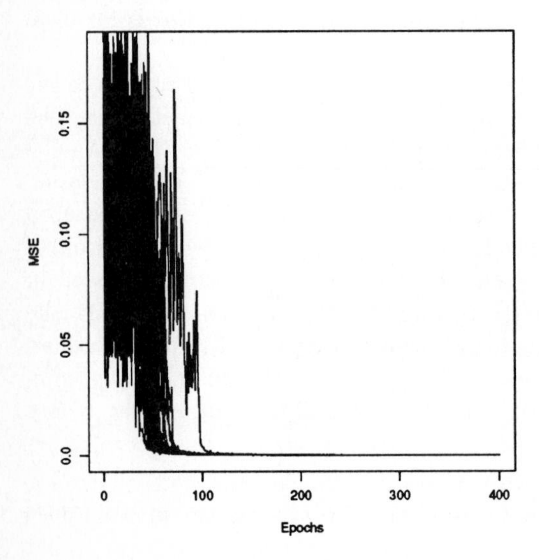

FIGURE 15 GA + BP Final Population.

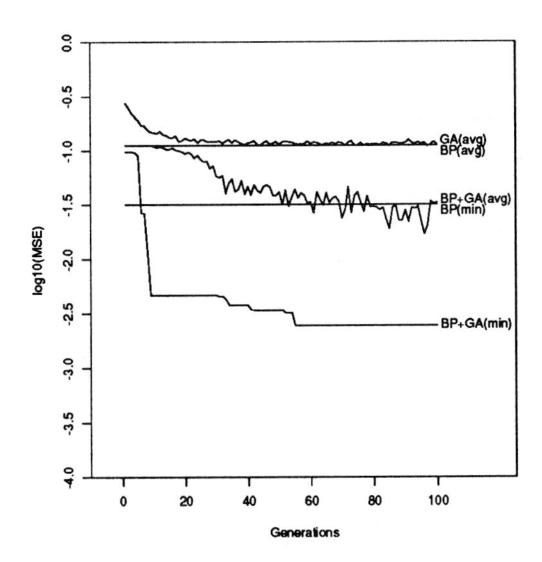

FIGURE 16 GA + BP hybrid.

Figure 14 shows the learning curves for individuals in the initial GA population. Except that the curves are much noiser (due to the high η value used), the same high variability of Figure 13 is exhibited. And this is to be expected since the GA's initial population is also picked in a uniformly random fashion. Figure 15 shows that after 200 generations the GA was able to find a region of $\underline{\mathbf{W}}(\mathbf{0})$ space from which BP can reliably converge on solutions.

As with conjugate gradient, the performance of the BP+GA hybrid can also be compared to the performance of BP and GA used independently; this comparision is shown in Figure 16. Here the generation average mean squared error (MSE) of the BP+GA hybrid is plotted on a logarithmic scale, as a function of generation and labelled **GA+BP(avg)**. The generation average of population in which GA was used to select $\underline{\mathbf{W}}(\mathbf{0})$ and the MSE of these individuals is taken immediately (i.e., without any BP learning) is labeled **GA(avg)**. For comparison with BP, two baselines corresponding to the average and minimum of an equal number of multiple random restarts of a BP simulation are shown as **BP(avg)** and **BP(min)**, resp.

There are several things worth noting in this comparison. First, the GA used by itself fares almost as well as the BP average. This is particularly striking given that the GA is only being allowed to sample in the original $\underline{\mathbf{W}}(\mathbf{0})$ space, $\pm\frac{1}{2}$. Second, use of the BP+GA hybrid creates a population of 50 individuals who, while very different from one another, have MSE performance almost identical with the best found by 5000 random restarts of BP alone. Finally, after only a few generations, the BP+GA hybrid is able to find strictly better individuals than could be found by 5000 independent BP runs, and ultimately finds a much better one. As with the conjugate gradient experiments of subsection 5.1, the hybrid of GA+BP can reliably solve the problem more effectively than either a randomly started BP search or GA sampling without guidance by local information.

5.3 EXPERIMENT 3: BP OVER A CLOSED DOMAIN

The results of the previous two sets of experiments appear to create a paradox: The GA should be allowed to sample over the same "closed" space of weights through which the gradient technique is to search, but if the genome is allowed to encode the entire range of weights, small $\underline{\mathbf{W}}(\mathbf{0})$ weights can be only sparsely represented, and the probability of the GA's sampling the fertile region in which *all* the weights are near zero is very small indeed. This paradox can be resolved, however, if the requirement that the $\underline{\mathbf{W}}(\mathbf{0})$ range be *uniformly* represented on the genome is relaxed. Real numbers are mapped onto the GA chromosome as before, but the final **TRANSFORM** stage (cf. Figure 1) is changed from a linear transformation to an *exponential* one, so as to emphasize sampling of small weights.

Specification of an appropriate exponential sampling transform turns out to be a subtle issue that we are continuing to investigate. For these experiment our initial strategy was simply to select upper and lower bounds for $\underline{\mathbf{W}}(\mathbf{0})$, and the number of bits per weight. Based in part on Hanson and Burr's analysis of two large and well-studied networks,[16] we allowed $\underline{\mathbf{W}}(\mathbf{0})$ to range between approximately:

$$0.01 \leq |\, \underline{\mathbf{W}}(\mathbf{0}) \,| \leq 12$$

and allowed 10 bits/weight.

Using this encoding, the experiments of the last section combining GA and BP were repeated. The first observation was that this encoding was propitious in that it allowed extremely short BP training times. Figure 17 shows the results of combining the GA sampling process with only 40 epochs of BP training; as above, these results are compared to multiple, random BP runs and the use of GA without any BP search. With this short training, the best solution found by the hybrid GA+BP system was still better than that found by an equal number of randomly restarted BP solutions, but the difference in MSE's was less dramatic.[15]

With longer training times (e.g., the 200 epochs used for the previous section's experiments), the problem became too easy for the GA: initial, random generations already contained many good solutions. We conjecture that constructing initial $\underline{\mathbf{W}}(\mathbf{0})$ vectors with many small weights but a few large ones—in effect much like the *final* weight distribution reported by Hanson and Burr—is enough to break hidden unit symmetries and still allow most error information past the sigmoid's derivative. When the training time is then reduced (down to 40 epochs!) to compensate for the facilatory effect of the exponential encoding, the GA is again able to find good regions of $\underline{\mathbf{W}}(\mathbf{0})$.

[15]Still, the hybrid's solution solved the six-bit symmetry problem to criterion while the BP network did not.

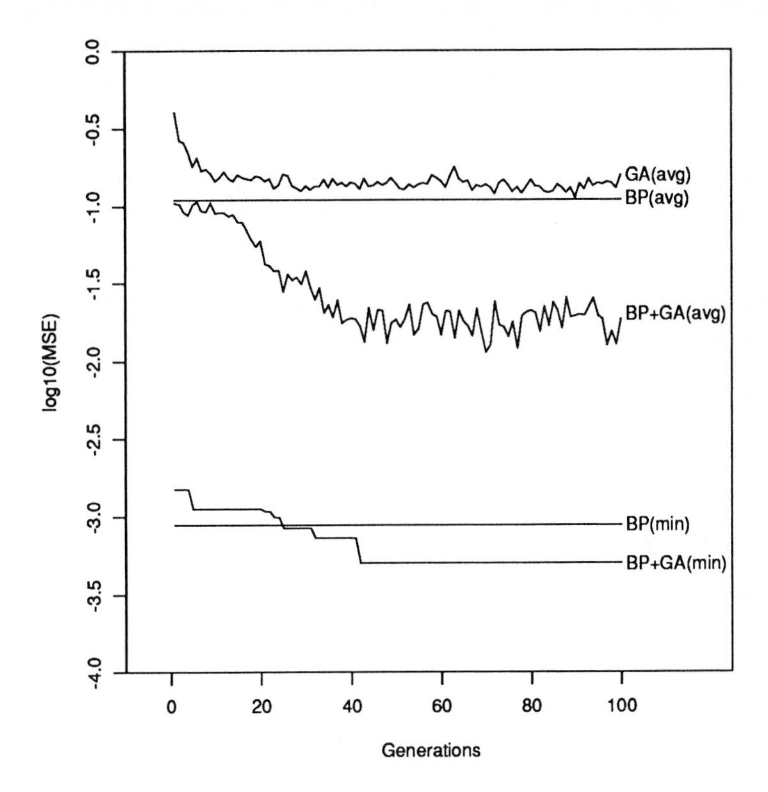

FIGURE 17 Exponential GA + BP hybrid.

5.4 SAMPLING WITH THE GA

The picture we have, then, is of a highly "textured" error surface, with the starting point of a gradient technique's search dramatically influencing its ability to get to satisfactory minima. Some basic bounds on a good $\underline{\mathbf{W}}(\mathbf{0})$ region can limit search to a "donut" around the origin; see Figure 12 and accompanying discussion. But even within this donut there is significant variability.

The experiments of Kolen and Pollack provide an interesting comparison with our own.[22] Among other experiments, they performed a Monte Carlo search through (what we call) $\underline{\mathbf{W}}(\mathbf{0})$ -space for the problem of 2-bit XOR with two hidden units. First, they echo our "donut" observation: "...the magnitude of the initial condition vector (in weight space) is a very significant parameter in convergence time variability." Second, they go further than characterizing the error surface as simply textured, to propse that they are "fractal-like." But the central difference between this work and our own is our use of the GA, effectively in place of their uniformly distributed Monte Carlo iteration. The moral Kolen and Pollack draw from their experiments is that since BP simulations are extremely sensitive to their initial $\underline{\mathbf{W}}(\mathbf{0})$ values, "*...the initial conditions for the network need to be precisely specified or filed in a public scientific database.* [Emphasis in the original]." We believe our conclusion is more optimistic, and certainly less bureacratic: Use of the GA's non-random search allows us to judiciously sample $\underline{\mathbf{W}}(\mathbf{0})$ so as to identify regions from which we can *reliably* converge on good solutions, while simultaneously allowing us to cut training times by as much as two orders of magnitude.

The major finding from these experiments is that the use of the GA to select advantageous initial weights for the BP algorithm is effective. subsection 3.4 outlined our current intuition that further progress with hybrid GA and gradient search procedures will require more sophisticated developmental programs, with the GA specifying broad constraints on patterns of neural connectivity rather than values for any one connection. But it is not that far-fetched to imagine that, at least in some primitive organisms' nervous systems, the genome does specify a coarse pattern of synaptic connections that is then fine-tuned during the organism's lifetime.

6. COMPUTATIONAL COMPLEXITY IN GA/BP HYBRIDS

On first blush, the idea of taking a compute-intensive procedure like BP and duplicating it $\mathcal{O}(100)$ times to form a population, and then using the GA to produce $\mathcal{O}(100)$ such generations seems profligate. And if we are well satisfied with the results of a single BP run, this analysis may be correct. But we have three good reasons to question this analysis. First, BP performance is known to be highly variable under stochastic variation. Consequently, many investigators already use some sort of "outer loop" of multiple, random restarts to improve their confidence in the solutions found. Section 5 argued that the GA's sampling is far superior to such random restarts. Second, there is currently great interest in more elaborate network topologies than the standard single, fully connected hidden layer. However, extending BP and other learning techniques to these new topologies has proven difficult. The recent experiments reported in subsection 3.3, particularly those of Todd et al. and Harp et al., are indications, albeit preliminary ones, that the GA can be a useful tool for exploring these novel architectures.

Finally, most of the experiments reported here have changed the BP algorithm so that its training time is greatly reduced. As algorithm designers, our primary concern must be with the total time taken by the hybrid system. The time complexity of this system is simply the product of the number of generations the GA is run, times the size of the population of each generation times the training time taken by each individual:

$$TotalTime = Generations * PopulationSize * TrainingTime$$

Thus there is a direct tradeoff between the number of generations and the number of trials allocated each individual. Using the GA to produce 100 generations of 50 individuals multiplies the apparatent time complexity by 5000. But subsection 5.2 and subsection 5.3 report on experiments in which the training time was reduced (from 4000 epochs on 6-6-1 symmetry) by factors of 20 and 100, resp. to the same error criterion. Thus the use of faster learning rates and judicious sampling of $\underline{\mathbf{W}}(\mathbf{0})$, the 5000- fold increase in time can be cut to a factor of 50. When the greater assurance in the answers found through the GA's robust, global searching is considered, and compared to the $O(10)$ random restarts often done with BP simulations, our hybrid methods are very competitive.

Much of our current research focuses on a more theoretic basis for the hybridization of GA and gradient techniques. The time complexity issues just mentioned introduce an important new parameter for hybrid connectionist/GA systems like these, viz., how long each connectionist minimization procedure should be allowed to iterate before the value it has found is used to determine that network's fitness. So, are we better off using many GA generations of short-lived individuals, or allowing each individual to search for longer, so as to produce a perhaps more informative value for the GA?[16]

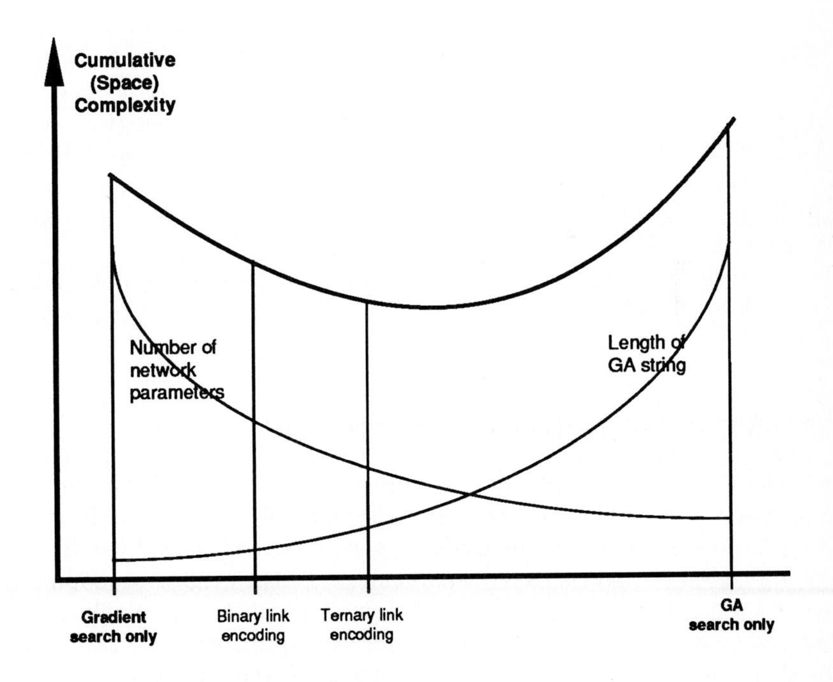

FIGURE 18 Cumulative Complexity.

[16]In other words, whether 'tis nobler to "live fast and die young" or "live long and prosper"?

One view of the space-complexity issues in hybridization is suggested by Figure 18. Subsection 3.3 mentioned a few of the alternative encodings of connectionist networks onto the GA's string that are available. These alternatives can be ordered in terms of the number of free parameters being searched by the GA and gradient procedures. At one extreme we can use only the GA, fully representing in full precision each of the network's weights and leaving BP nothing to do; at the other extreme, BP could be used exclusively. Intermediate between these are hybrids that encode some constraints on the structure of the BP network into the GA's chromosome: a bit specifying a link's presence, several bits specifying allowed weight ranges, etc. When a developmental component is interposed between the GA and BP, complexity issues become even more complicated.

It will take time to understand these tradeoffs completely. As mentioned in subsection 3.3, some measure of a network's complexity must augment the basic performance measure (e.g., MSE), or the GA cannot impose adaptive pressure towards more parsimonious solutions. Rumelhart (personal communication) has attempted to use BP itself to explore new architectures, by including additional terms in the criterion function for number of hidden units, number of links, and even number of "distinct" weights used. Rissanen's "minimum description length" formalism provides a rigorous measure of a model's complexity,[30] and Tenorio and Lee have made an initial attempt to apply this to connectionist network architectures.[37] Characterizing the description length of (binary) GA genomes promises to be more straight-forward, and the stage is then set for measuring the *cumulative* complexity of GA+network solutions in the manner originally suggested by Figure 18. But the time-complexity trade-offs mentioned above must also be incorporated, and interposing developmental programs is even more problematic. This is the core of our future theoretic work.

In the the interim, there is another aspect of our hybrid's computational character that we have already begun to exploit: GA/connectionist hybrids are eminently parallelizable.

6.1 EXPLOITING THE PARALLELISM OF THE GA

The availability of massively parallel computers, let alone large distributed networks of loosely coupled computers, has increased dramatically. Unfortunately, our ability to effectively utilize the vast computational power offered by these parallel systems has not kept pace. Mizell (personal communication) has singled out Monte Carlo-like computations as an example of "embarassingly parallelizable" applications (i.e., so naturally parallel that you get no credit for saying so) that have real utility and do tap the computational power of parallel machines and networks.

The basic "population" model underlying the GA (i.e., each individual in the population independently evaluating the objective function) makes it another candidate for parallelization, embarrased or not. We have exploited this feature of our hybrids in an extension of Grefenstette's GENESIS simulator developed by our

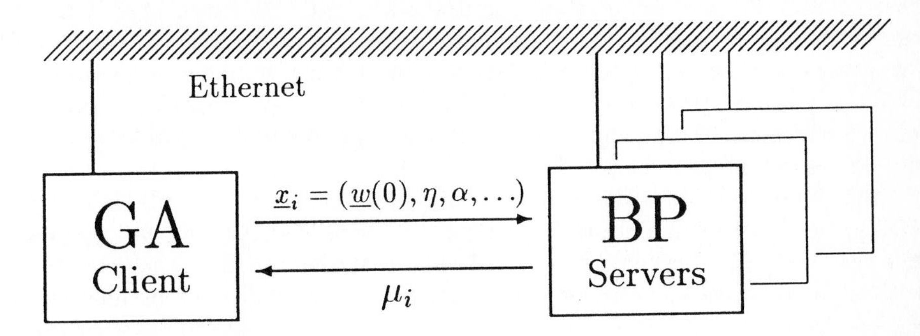

FIGURE 19 Distributed GA Model.

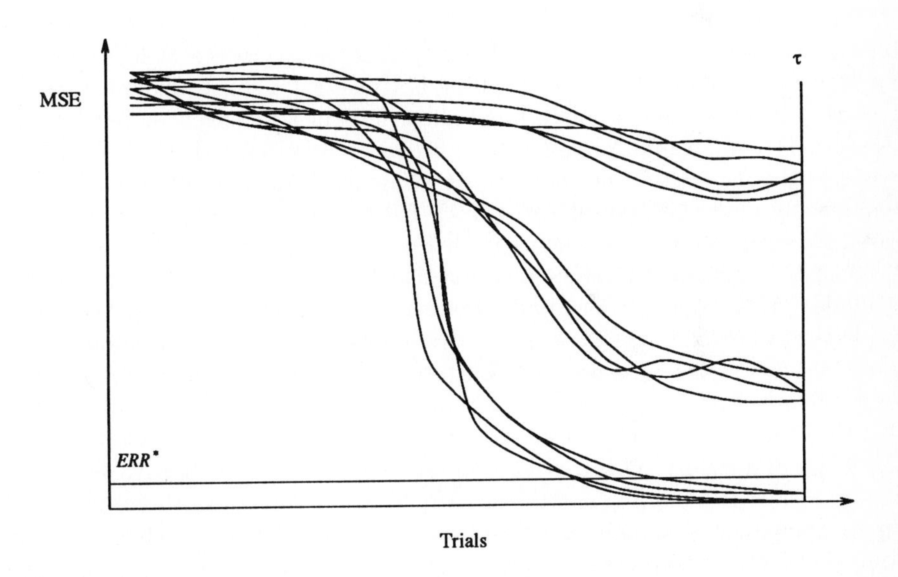

FIGURE 20 Learning curves.

group. The basic architecture of this Distributed GA (DGA) design is shown in Figure 19; Grefenstette has called this a "syncronized master-slave" architecture for the GA.[15] Assume that some number H of hosts are connected via a local area network.[17] A `Host Table` with the name and network addresses of these hosts is

[17]For our simulations, this environment has been a mixture of Sun 3's and 4's and various Vaxen, connected via Ethernet using TCP/IP protocol and NFS.

constructed, a BP Server program is initialized on each. Then, a GA Client program, which is very similar to the standard Genesis package, is begun on one host. This program creates an initial population with each individual corresponding to a particular parameterization of a BP simulation. These may be varied along a number of dimensions,[18] although any one of our experiments has typically explored only one or two of these parameters.

In order to coordinate distribution of BP evaluations to the available servers, a Population Table is maintained, showing the status (Evaluated, Being-Evaluated, or To-Be-Evaluated) of each individual. Initially, a BP parameter packet is "dealt out" to each of the available BP servers. Typically, the population size exceeds the number of BP servers, so as evaluations are completed the Population Table is updated and the idle server is sent a new parameter packet to evaluate. This process repeats until the entire population has been dealt out at least once. After (an adaptively tuned) Timeout Period, any individuals still in the Being-Evaluated status are dealt out again, until all evaluations have been performed. The next generation's evaluations then commence.

This scheme works particularly well with a heterogeneous mix of computers of different processing speeds and user loads because the packets sent to and from the GA server are then randomized, avoiding bottlenecks; a slight amount of randomized delay has been incorporated for similar effect in a homogeneous environment. More recently, the Host Table has been augmented with statistics about how long it is taking each host to perform its BP evaluation. This information is used to dynamically alter the priority with which BP packets are assigned to the various hosts, for example avoiding heavily loaded or completely dead hosts.

The time-complexity issues mentioned above become very significant in the DGA design. For example, one undesirable bottleneck is created by the end of each generation, when all but the last one or two individuals have been evaluated and the rest of the processors must wait; Grefenstette calls this a "semi-syncronized master-slave" architecture.[15] In general, with populations of reasonable size it is unlikely that these last few evaluations are critical, and so it seems reasonable to relax the constraint of a rigidly synchronized generational structure. Similarly, we note that populations of BP learning curves typically look something like those portrayed in Figure 20, with some simulations terminating because they have achieved the error criterion Err^*, and others terminating at a maximum number of training trials $\mathcal{T}$.[19] Ideally, it should be possible for the GA to advantageously manipulate both Err^* and $\mathcal{T}$, for example giving individuals who reach the error criteria quickly a high fitness value, or interupting slow evaluations if the rest of the population is done and has found good values. The interupted processor is then freed to begin work on a potentially more informative evaluation. Finding an appropriate balance

[18]Our simulator allows variation of: Number of hidden units, network wiring, η, α values, initial $\underline{\mathbf{W}}(0)$ weights, squashing functions, training regimes (random, sequential, permuted, batch), and training time.

[19]We are grateful to Peter Todd for suggesting this view.

between the *exploitation* of quickly derived solutions versus the *exploration* of more time-consuming ones is a matter requiring further investigation.

7. CONCLUSION

This paper has reported on a series of experiments combining two popular classes of adaptive algorithms, Genetic Algorithms and connectionist networks. A survey of a wide range of potential mappings of connectionist networks onto the GA genome has convinced us that the most desirable representations for the GA can be expected to be quite removed from the most obvious network representations (Section 3). Our own work is pursuing biologically plausible ontogenic models that create virtual independence between the space of genomes being searched by the GA and the weight space of the connectionist networks (subsection 3.4). We have demonstrated that the GA can be successfully used to tune parameters for the back propagation algorithm, at least for the problems we have investigated (Section 4). A much more substantial class of hybrids uses the GA to *globally sample* the space of initial network weights—$\underline{\mathbf{W}}(\mathbf{0})$ —from which connectionist gradient descent procedures can *locally search* (Section 5). These initial experiments have allowed us to focus several basic theoretic questions arising when GA and connectionist techniques are combined (Section 6), and to develop a distributed version of the GA that exploits the "embarrasingly" parellel nature of GA/connectionist hybrids.

Throughout, our experiments and discussion have remained in the province of computer science, concerned with issues of algorithmic design rather than the natural (genetic, neural, ontogenic) origin of these algorithms. We close with two observations about the computation being performed by these natural systems.

The first observation concerns a fundamental incongruity between the teleologies of biological systems and the artificial ones we run on computers. Much of Section 6 was concerned with refuting objections that wrapping the GA around BP was a waste of (computational) cycles. In our artificial algorithms it is appropriate to worry that a cycle used (for example) to produce a new generation might be better spent allowing an indivudal another training epoch. That is, the total number of cycles is considered a *conserved quantity.* But it is not at all clear that biological populations are burdened by any such constraint. In fact, Evolution seems truly profligate with its "cycles," creating as many redundantly exploring individuals as an environment's "carrying capacity" will allow, and allowing each to live as long as mortality allows. As massively parallel computers become available, using a similarly profligate strategy of redundant search may become more sensible in algorithmic design as well (e.g., Hillis' search for efficient sort routines[18]).

Second, a computational view provides new insight into the exploitation of individuals' learning and the evolution of a species. Despite the intuitive appeal of a theory that would allow individual learning to favorably influence evolution, the biology of genetic reproduction explicitly rules out the *direct*, "Lamarckian" inheritance of acquired (e.g., learned) characteristics. Previous work has demonstrated that at least some of the desirable interactions between learning and evolution can be explained via *indirect* mechanisms, such as the "Baldwin Effect."[3]

Experiments reported here suggest another computational reason why direct Lamarckian inheritance cannot be possible. In particular, Section 5 discussed the complex relationship between the space of initial weights ($\underline{\mathbf{W}}(0)$) searched by the GA and the space of ultimate weights discovered by the gradient techniques of conjugate gradient or BP. But if $\underline{\mathbf{W}}(0)$ is a proper, small subset of the ultimate weight space (as it was in the experiments of subsection 5.2), the solutions found by BP cannot simply be "reverse transcribed" back into the GA's genomic representation! More generally, as the relation between network architecture and genomic specification becomes more and more indirect (for example through the use of the developmental translation programs we advocate) the ability to *invert* this relation diminishes. That is, given a mature, successful individual, it becomes harder and harder to invert the mature *cognitive representation* responsible for the successful performance into his or her original *genetic representation.*

We conjecture further that it is in fact impossible for a system to inherit (at least some) acquired and critical characteristics, not because of biological "implementation details" with reverse transcription (for we know evolution to be terribly inventive), but because it is *computationally* impossible to encode in the *structural* genotype the results of *behavioral* experiments. It is our goal to use our increasing understanding of the computational properties of biologically plausible algorithms like the GA and connectionist networks to cast this conjecture formally. In any case, it appears that features of information processing by natural systems and characteristics of artificial computation are again intimately entwined.

ACKNOWLEDGMENTS

We gratefully acknowledge useful discussions of this work with Peter Todd and Jeffrey Miller. Thanks also to George Wittenberg and Susan Gruber for comments on an earlier draft of this manuscript. Finally we thank the Cognitive Computer Science Research Group of the CSE department at UCSD for generally fomenting this and much other work.

REFERENCES

1. Ackley, D. H. *A Connectionist Machine for Genetic Hillclimbing.* Boston: Kluwer, 1987.
2. Barnard, E., and R. A. Cole. "A Neural-Net Training Program Based on Conjugate-Gradient Optimization." Technical Report, Dept. Computer Science and Engeering, Beaverton, OR: Oregon Graduate Center, 1989.
3. Belew, R. K. "Evolution, Learning and Culture: Computational Metaphors for Adaptive Search." *Complex Systems* **4(1)** 1990.
4. Belew, R. K., N. N. Schraudolph, and W. E. Hart. "Weight Initialization in Neural Networks: An Information Theoretic Approach." In preperation.
5. Bethke, A. "Genetic Algorithms as Function Optimizers." Technical Report, Logic of Computers Group, CCS Dept., University of Michigan, Ann Arbor, MI, 1981.
6. Brady, R. M. "Optimization Strategies Gleaned from Nature." *Nature* **317** (1985): 804–806.
7. Caruana, R., and J. D. Schaffer. "Representation and Hidden Bias: Gray vs. Binary Coding for Genetic Algorithms." In *Proc. Fifth Intl. Conf. on Machine Learning.* Morgan Kaufmann, 1988.
8. Cowan, J. D., and A. E. Friedman. "Development and Regeneration of Eye-Brain Maps: A Computational Model." In *Advances in Neural Info. Proc. Systems 2.* Morgan Kaufmann, 1990.
9. DeJong, K. "Adaptive System Design: A Genetic Approach." *IEEE Transactions on Systems, Man, and Cybernetics*, 1980.
10. Edelman, G. *Neural Darwinism: The Theory of Neuronal Group Selection.* New York: Basic Books, 1987.
11. Fogel, L., A. Owens, and M. Walsh. *Artificial Intelligence Through Simulated Evolution.* New York: Wiley, 1966.
12. Goldberg, D. *Genetic Algorithms in Search, Optimization, and Machine Learning.* Reading, MA: Addison-Wesley, 1989.
13. Grefenstette, J., ed. *Proc. First Intl. Conf. on Genetic Algorithms and their Applications*, 1985.
14. Grefenstette, J., ed. *Proc. 2nd Intl. Conf. on Genetic Algorithms.* Lawrence Erlbaum, 1987.
15. Grefenstette, J. J. (1981). Parallel adaptive algorithms for function optimization. Technical Report CS-81-19, Computer Science Dept., Vanderbilt Univ., Nashville, TN.
16. Hanson, S. J., and D. J. Burr. "What Connectionist Models Learn: Learning and Representation Connectionist Networks." *Behavioral and Brain Sciences* **13(3)** (1990): 471–489.
17. Harp, S., T. Samad, and A. Guha. "Towards the Genetic Synthesis of Neural Networks." In *Proc. Third Intl. Conf. on Genetic Algorithms*, San Mateo, CA: Morgan Kaufmann, 1990.

18. Hillis, W. D. "Co-Evolving Parasites Improve Simulated Evolution as an Optimization Procedure." *Emergent Computation*, edited by S. Forrest. Amsterdam: North Holland, 1990.
19. Holland, J. H. *Adaptation in Natural and Artificial Systems.* Ann Arbor, MI: University of Michigan Press, 1975.
20. Huynen, M. A., and P. Hogweg. "Genetic Algorithms and Information Accumulation During the Evolution of Gene Regulation." In *Proc. Third Intl. Conf. on Genetic Algorithms*, edited by J. D. Schaffer. Washington, DC: Morgan Kaufman, 1989.
21. Kauffman, S., and R. G. Smith. "Adaptive Automata Based on Darwinian Selection." *Physica D* **22** (1986): 68–82.
22. Kolen, J. F., and J. B. Pollack. "Back Propagation is Sensitive to Initial Conditions." *Complex Systems* **4** (1990): 269–280.
23. Merrill, J. W. L., and R. F. Port. "A Stochastic Learning Algorithm for Neural Networks." Technical Report 236, Department of Linguistics and Computer Science, Indiana University, Bloomington, IN, 1988.
24. Miller, G., P. Todd, and S. Hegde. "Designing Neural Networks using Genetic Algorithms." In *Proc. Third Intl. Conf. on Genetic Algorithms*, San Mateo, CA: Morgan Kaufmann, 1989.
25. Minsky, M., and S. Papert. *Perceptrons (expanded edition).* Cambridge, MA: MIT Press, 1988.
26. Miyata, Y. "Sunnet Version 5.2." Technical Report 8708, Inst. for Cognitive Science, UCSD, La Jolla, CA, 1987.
27. Montana, D. J., and L. Davis. "Training Feedforward Networks using Genetic Algorithms." In *Proc. IJCAI.* Morgan Kaufman, 1989.
28. Offutt, D. (1989). "A Reinforcement Learning Algorithm for Training Fully and Bidirectionally-Interconnected Connectionist Networks." (draft), 1989.
29. Purves, D. *Body and Brain: A Trophic Theory of Neural Connections.* Cambridge, MA: Harvard University Press, 1988.
30. Rissanen, J. Stochastic Complexity in Statistical Inquiry." Technical Report RJ-6901, IBM Research Division, Yorktown Heights, NY, 1989.
31. Rizki, M., and M. Conrad. Computing the Theory of Evolution." *Physica D* **42** (1986): 83–99.
32. Rumelhart, D., G. Hinton, and R. Williams. "Learning Internal Representations by Error Propagation." In *Parallel Distributed Processing: Explorations in the Microstructure of Cognition*, edited by J. McClelland and D. Rumelhart, Chapter 8. Cambridge, MA: Bradford Books, 1986.
33. Schaffer, J., and A. Morishima. "An Adaptive Crossover Distribution Mechanism for Genetic Algorithms." In *Proc. Intl. Conf. on Genetic Algorithms and their Applications*, edited by J. Grefenstette, 1985.
34. Schaffer, J. D., ed. *Proc. Third Intl. Conf. on Genetic Algorithms*, Washington, DC: Morgan Kaufman, 1989.
35. Schraudolph, N. N., and R. K. Belew. "Dynamic Parameter Encoding for Genetic Algorithms." *Machine Learning* (1991): to appear.

36. Stork, D. G., S. Walker, M. Burns, and B. Jackson. "Preadaptation in Neural Circuits." In *Proc. IJCNN (Vol. 1)*, New York. IEEE, 1990.
37. Tenorio, M. F., and W. Lee. "Self-Organizing Neural Network for Optimized Supervised Learning." Technical report, School of Electrical Engineering, Purdue University, W. Lafayette, IN, 1989.
38. Todd, P. "Evolutionary Methods for Connectionist Architectures." 1988.
39. Whitley, D., and T. Hanson. "Optimizing Neural Networks Using Faster, More Accurate Genetic Search." In *Proc. Third Intl. Conf. on Genetic Algorithms*, edited by J. D. Schaffer. Washington, DC: Morgan Kaufman, 1989.

David Jefferson, Robert Collins, Claus Cooper, Michael Dyer, Margot Flowers, Richard Korf, Charles Taylor, and Alan Wang
University of California, Los Angeles, CA 90024

Evolution as a Theme in Artificial Life: The Genesys/Tracker System

Direct, fine-grained simulation is a promising way of investigating and modeling natural evolution. In this paper we show how we can model a population of organisms as a population of computer programs, and how the evolutionarily significant activities of organisms (birth, interaction with the environment, migration, sexual reproduction with genetic mutation and recombination, and death) can all be represented by corresponding operations on programs. We illustrate these ideas in a system built for the Connection Machine called Genesys/Tracker, in which artificial "ants" evolve the ability to perform a complex task. In less than 100 generations, a population of 64K "random" ants, represented either as finite-state automata or as artificial neural nets, evolve the ability to traverse a winding broken "trail" in a rectilinear grid environment. Throughout this study, we pay special attention to methodological issues, such as the avoidance of representational artifacts, and to biological verisimilitude.

Artificial Life II, SFI Studies in the Sciences of Complexity, vol. X, edited by C. G. Langton, C. Taylor, J. D. Farmer, & S. Rasmussen, Addison-Wesley, 1991

1. INTRODUCTION

One of the major research themes of the UCLA Artificial Life group is the simulation of biological evolution, especially the exploration of its large-scale behavior (macroevolution). Our simulations do not use differential equations or aggregated models, but are instead direct, fine-grained, "microanalytic" simulations of a large population in a complex environment over many generations. We believe that an organism can be simulated in a natural way by a computer program; just as an organism is born, moves, interacts with its environment, processes information, reproduces with variation, and dies, so too can a program initiate, migrate, interact with its environment, process information, produce modified copies of itself, and terminate. In our simulations, we represent each individual organism separately as a computer program of some kind whose execution represents the sequence of significant events in the organism's life from birth to death. Each organism has a genotype (a bit string) and a phenotype (a program encoded in the bit string), both of which are generally unique in the population. Over many generations of replication, competition, and selection, we expect novel and superior forms to evolve.

Our long-term goal is to study natural evolution through artificial evolution. Today, biologists have only a few ways to study macroevolution: (a) the mathematics of population genetics, (b) laboratory and field experiment, (c) examination of molecular relationships among modern species, and (d) examination of the fossil record. Each method has important limitations. Population genetics today usually cannot give sharp non-equilibrium results for systems with a large number of genes. Experiments are usually confined to small populations for a few generations (except for microorganisms), and usually cannot be perfectly controlled. The fossil record is notoriously biased and incomplete, containing primarily morphological information on those creatures whose life style and body type made them likely to be preserved. In addition, both molecular and fossil studies tell us about how natural evolution proceeded historically, as it was, but not so much about the principles of evolution as it might have been. We hope to add to this list another intellectual tool for the study of evolution, computer simulation. Just as viruses and bacteria can serve as models of the molecular biology of eucaryotic organisms, we believe that evolving populations of computer programs can act as simple, controllable, replicable models for large-scale evolutionary processes. We hope some day that biologists may use simulation to help resolve some of the outstanding foundational problems in evolution, including perhaps questions about modes of speciation, the evolution of cooperation, the unit(s) of selection, and the evolution of sex.

This paper has two parts. The first part is methodological; it sets forth our view of how evolution can profitably be studied by computer in new ways and with new kinds of models. We describe the strengths, and some of the limitations, of our paradigm.

The second part describes a particular system, called the Genesys, that we built to test the computational feasibility of this approach to studying evolution. We show how we have been able to evolve artificial "ants" that are capable of complex behavior, e.g., following a "broken trail" in a grid environment. Genesys is not *per se* an attempt to resolve any biological question. It is rather an ambitious exercise exploring computational and theoretical issues about biologically realistic evolution. Our experience with the most important of these issues is described below.

2. METHODOLOGICAL ISSUES IN ARTIFICIAL EVOLUTION

In attempting to mimic natural evolution many fundamental methodological questions arise. Just what is an organism, and how should it be represented computationally? How do genetic algorithms differ from biologically motivated evolution? How can we be sure, when we purport to create something by evolution, that we have not somehow guided or forced that process by design choices or initial conditions? In this section, we discuss these questions, prior to describing the Genesys system in detail.

CAN ARTIFICIAL EVOLUTION BE OPEN ENDED?

Previously we reported on an earlier evolution simulator, called RAM, in which each organism in the population is represented as a pair containing (a) a parameterized Lisp function, referred to as the organism's "behavior function," and (b) a sequence of parameter values to the behavior function, which act as the organism's genome.[14] Each individual parameter value is like a "gene," and an organism inherits the parameter values of its parent(s), usually modified by mutation (and possibly recombination). All organisms (of the same "species") share a common behavior function, but differ in the genes.

We consider RAM to be a successful experimental system, and still use it today.[3,13,15] It is definitely capable of illustrating evolution of organisms interacting in a common environment. But we are dissatisfied with some of RAM's limitations. First, it can support only small populations (a few hundred organisms) because of the large amount of memory it requires and because it runs on workstation-class machines; and, second, while it is comparatively easy for RAM to simulate asexual reproduction, sexual reproduction and mate selection have to be specially programmed. But one of the most important deficiencies in our view is that the Lisp function that defines the behavior of an organism is not itself subject to mutation or recombination; only the genes (parameters) are. Furthermore, the genes have to be manipulated by mutation operators that are specific to the type and meaning of the gene. An integer gene, for example, might be incremented or decremented, with perhaps a clause preventing it from going negative, or a real-valued gene might be multiplied by a random value near 1.0.

This approach, however, seems ultimately too contrived and too distant from natural genetics. In natural life there are no integers and no reals in the genome.

We do not have separate types of genes, each with its own specific type of mutation operator, and the boundaries between codon regions are not at all respected by the natural "genetic operators" of substitution, deletion, insertion, inversion, and crossover. Because, in RAM, all heritable information is contained in the genes, most of the information determining the structure and behavior of the organism (the behavior function and the Lisp interpreter) is static from generation to generation, and not subject to evolution. Since the modeler specifies in advance both the types of the genes and the mutation operators that apply to them, in effect he specifies a low-dimensional parameter space which is the universe of all genotypes. Evolution is then conducted as a genetic "search" within this space, and most of the structure and behavior of the organism is out of reach of evolution. Mathematically, the confinement of evolution to a small, closed parameter space does not resemble the thousands of degrees of freedom available in the evolution of even the tiniest natural genome. The kind of evolution exemplified by RAM is, unfortunately, not open ended.

In building our new system, Genesys, we wanted to construct a system that corrected these deficiencies. We wanted almost all of the logic of the organism's behavior to be subject to evolution, and we wanted the evolution to be based on genetic algorithms that are biologically defensible, with no knowledge of the environment or the evolutionary "goal" in any way embedded in the choice of genetic operators or any other aspect of the genetic algorithm. These constraints required us to change the representation of organisms and their genomes drastically, away from Lisp functions and typed parameter values, and toward a programming paradigm that would have a *small interpreter* (so that the complexity of an organism's behavior resides more in the program and less in the interpreter) and that would allow *the entire text of the program itself to be subject to evolution.* When representing organisms as programs for evolutionary studies, we believe that an encoded form of the program itself should be the genome.

WHAT IS AN APPROPRIATE REPRESENTATION FOR ORGANISMS?

Given that organisms are to be represented by programs, and that they must be subject to mutation and recombination, what kind of programs should they be? Should they be mathematical automata of some kind? Procedural programs? Rule systems? Logic circuits? Logic programs? Constraint systems? Neural nets? And how should they be encoded into a genome, and what mutation and recombination operators should they be subject to? Should they be encoded as bit (or character) strings, with mutation and recombination at the bit (or character) level? Or should they perhaps be represented a flattened parse trees with mutation and recombination at the lexeme level? Data elements, such as integers, will inevitably be part of such a program. Should they be represented and mutated as atomic objects, with whole-integer operators, or as bit strings with bit flipping as the mutation operator? If the latter, should they be ones-complement, twos-complement, or perhaps gray code? We refer to all of these questions as the problem of the "representation of organisms," a problem we expect to grow in significance as research in artificial evolution continues.

From all these possibilities, we chose two of programming paradigms to work with: finite state automata (FSAs) and recurrent artificial neural nets (ANNs). In both cases, we chose to encode the entire "program" (FSA state transition table, or neural net weights) as a bit string and to subject it to bit-level mutation and recombination operations. We will describe them in more detail in Sections 4 and 5; here we only note that one of the major questions driving us in this research has been: "Which of those two representations, FSAs or ANNs, is 'better' for evolution studies?" Some of us believed that the FSAs were superior, at least for the Tracker task, because that task is essentially finite and sequential, and any deterministic algorithm for it is formally equivalent to some finite automaton. Others of us felt that ANNs were probably superior; ANNs encoded as bit strings would seem to be inherently more robust and well-conditioned under mutation and recombination than the apparently more brittle and less redundant FSAs.

We put a lot of work and debate into trying to answer this question. It turned out that, after going to some length to make the two representations as comparable as possible, empirical trials consistently showed that the FSA representation performed slightly better than the ANN representation on the Tracker task. However, we have not come to a full understanding of why this is so, and we do not necessarily believe the empirical result generalizes in any meaningful way to other tasks, or to larger organisms than the ones we are working with. We never did answer the question fully; probably neither representation dominates the other in all dimensions. In any case, we now believe it to be a false dichotomy, and not such an important question after all, as we will discuss in Section 11.

Koza has described a fascinating system that takes a different point of view from the one we have adopted.[9] He executes genetic algorithms over Lisp programs, but with a different purposes and mechanisms than either RAM or Genesys. Like RAM, he uses Lisp programs as the representation of "organisms" (though he does not refer to them as organisms). But, unlike RAM, where the genome is a list of parameters, he treats Lisp programs as an S-expressions and uses subexpression-oriented mutation and recombination operators on the text of the program itself. His goal is to engineer programs that can solve problems, and he has been able to essentially duplicate our work (among many other things) by evolving Lisp programs that perform the same Genesys/Tracker task that we will describe here. But the problem-solving flavor of his research, which studies essentially goal-directed evolution, is in contrast to our goal, which is the study of natural evolution, which is not fundamentally goal-directed. As a result, there is no need for him to choose genetic or evolutionary mechanisms that closely resemble natural ones, whereas we feel bound to justify our experimental designs biologically. Likewise, with an essentially engineering, nonbiological point of view, there is no need for him to pay the kind of attention to representational artifact that we do here.

HOW CAN WE CONTROL FOR EXPERIMENTER BIAS AND REPRESENTATIONAL ARTIFACTS?

One of the difficult problems that can arise in evolution studies is the "credit assignment" problem. When an evolutionary experiment succeeds in producing a population of organisms that exhibits a certain interesting behavior (e.g. trail-following), how can we decide the extent to which that behavior may have been built into the architecture of the organism or the structure of the initial population? How do we know that there is not a bias built into the mutation operators chosen, or the programming paradigm that the organisms are expressed in, etc? Has evolution really created something novel?

In Genesys, we have taken a number of precautions and used a number of controls to ensure that bias and representational artifact are minimized in our results. First, as we just described, we use two distinct representations, finite automata and artificial neural nets, instead of just one. These two representations are so profoundly different that no artifactual result derived from an experiment with one would be naturally preserved in an experiment with the other. Because we succeed in evolving trail-following organisms in both cases, we feel confident that we are not being fooled by some subtle property of one or the other representation.

We guard against introducing bias through our choice of genetic algorithm and operators by fixing on one genetic algorithm for all experiments, even when we change representation, so that the genetic algorithm cannot be "tuned" to the representation or the experiment. In addition, we always use bit strings to represent the genome of an organism, and we use bit-level genetic operators. Mutation is always simulated by inverting random bits, chosen with equal probability from anywhere in the genome, regardless of where the bit happens to fall with respect to the "meaningful" fields (codons) in the programs. Likewise, we permit crossover with equal probability between any two bits in the genome; we make no effort to guarantee that it occurs between two codons. These decisions, besides preventing bias, have the additional advantage of being biologically realistic, since point mutations and crossover can occur anywhere within any region of DNA, without regard to the role it may play in the resulting organism. And, by choosing the bit string and bit-level genetic operators, we guarantee that the phenotype of an artificial organisms is a very distant and indirect function of the string of bits in its genotype, just as the phenotype of a natural organism is a very indirect function of the string of nucleotides. In both cases, there is no simple way of predicting the change in phenotype that will be caused by a random mutation.

Finally, we further guard against introducing bias through our choice of initial population. We always start with a population of random organisms, i.e., those derived from an initial population of random bit strings, where each bit has an equal probability of being 0 or 1. Of course, all experiments are performed repeatedly with different random seeds. (It has recently been suggested that we might have tried other distributions, i.e., random bit strings that were 99% zeros. That would perhaps have guarded against the possibility that we had somehow "seeded" the population with so much initial variation that evolution proceeded at a faster than normal rate.)

AT WHAT SCALE CAN ARTIFICIAL EVOLUTION OF THIS KIND BE MADE TO WORK?

Even if the idea of evolving ANNs or FSAs encoded in bit strings is right in principle, we were concerned that it might not work on a computational scale that we could afford. In Genesys applied to the Tracker task, which we will describe shortly, we were attempting to run a genetic algorithm with genomes 450 bits long, and with genetic operators that did not reflect anything at all about the task. To our knowledge, very few genetic algorithm experiments had ever been attempted using such long genomes. Since the size of the space of all organisms grows exponentially with the length of the genome, perhaps an astronomical population size, or millions of generations, would be necessary before anything interesting happened. Furthermore, any genetic algorithm must be tuned with several of parameters (e.g., mutation rate, recombination rate, selection fraction, etc.). If our evolutions had failed, it could have been just because we were unsuccessful in finding the "right" combination of parameters for the scale at which we worked.

3. THE TRACKER TASK

With all of these methodological questions and considerations in mind, could we indeed exhibit the evolution of some kind of complex behavior? Our intention was to see if it is possible to produce artificial organisms that exhibit a relatively complex behavior, and to do so entirely by evolutionary methods, with no built-in design bias and with no learning. The evolutionary exercise we now describe was inspired by the behavior of certain species of ants that lay down pheromone trails from a food site to their nest to aid in the process of collective foraging.[8] The trails are "noisy," and fade with time as the pheromone odors disperse. This trail-following ability is quite remarkable, and clearly required the evolution of elaborate nervous system structures that are present in modern ants, but were not present in a sufficiently ancient ancestral population.

We designed a highly simplified task called *Tracker* that resembles ant trail-following at least superficially. The Tracker task requires an ant to follow a crooked, broken trail of black cells in a white toroidal grid. The trail has a series of turns, gaps, and jumps that get more difficult as it progresses. The behavior functions of the ants in the initial population were constructed by a random process, so that while they have built-in "eyes" to sense the environment and built-in "motor apparatus" to move around in it, the population clearly has no built-in bias toward the ability to coordinate sensory input with motor output in such a way as to follow trails. However, after 100 generations of evolution, a large fraction of the population (anywhere from 20% to 80%, depending on the genetic algorithm parameters) is nearly perfect at the task.

Genesys simulates a large population of organisms, each of which is represented as either a finite-state automaton (FSA) or an artificial neural net (ANN). That phenotype (the particular FSA or ANN) is encoded in a genotype represented as a bit string, and the population evolves using a genetic algorithm[7] in which the score of the genotype, which determines the frequency of its representation in the next generation, is a function of how well the organism performs in its environment. Genesys runs on a 16K processor Connection Machine (CM2) with 8K bytes of memory per processor,[6] typically with a population of 64K organisms. Execution takes less than 30 seconds per generation.

Our experiments were conducted using the following task. An "ant" is placed in a two-dimensional grid at the beginning of the broken rectilinear trail shown in Figure1. The grid is toroidal, so that the cells on the left edge are considered to be adjacent to those on the right edge, and the cells on the bottom edge are considered to be adjacent to those at the top. This particular trail, called the John Muir Trail (to distinguish it from other trails we have experimented with) is used throughout this paper. The ant's task is to move from cell to cell, traversing as much of the trail as possible in 200 time steps. Its success is measured by the number of distinct trail cells it traverses. It is free to wander off the trail and pick it up elsewhere, but that is not an efficient strategy; whenever an ant with such a strategy appears during evolution, its lineage presumably dies out.

An ant is considered to erase a cell of the trail as soon as it steps on it. This convention is used because we did not want to give ants "credit" for following the same piece of trail more than once (which would reward backtracking, circling in place, or just no-oping). Instead of incorporating into Tracker, an algorithm to count the number of distinct trail cells the ant steps on, we found it computationally simpler just to erase the trail as it progressed.

The John Muir Trail winds around the environment, easy at first, but presenting ever harder challenges. First it includes three straightaways ending in right turns, followed by a straightaway ending in a left turn. After a while, starting at trail step 42, it takes several turns where the corner cell is missing from the trail. At step 58, there is a straightaway double gap, and then the trail gets progressively more difficult, ending with a series of disconnected knight's moves and long knight's moves. These last ten steps on the trail are extremely difficult compared to the rest of the trail, because each of them requires at least three or four tempos to get from the previous trail cell, and usually several more tempos must be used just searching for the continuation of the trail. The maximum possible score is 89, since that is the number of trail cells.

Each ant is in a sense-and-act loop, receiving one bit of sensory input per unit time about where the trail is, and deciding on two bits of action per unit time. At each time step, it stands on one cell, facing one of the cardinal directions (N, S, E, W), and it can sense whether or not the cell just *ahead* of it is part of the trail (Figure 2). We designed the ant to sense the cell ahead, instead of the cell it is on,

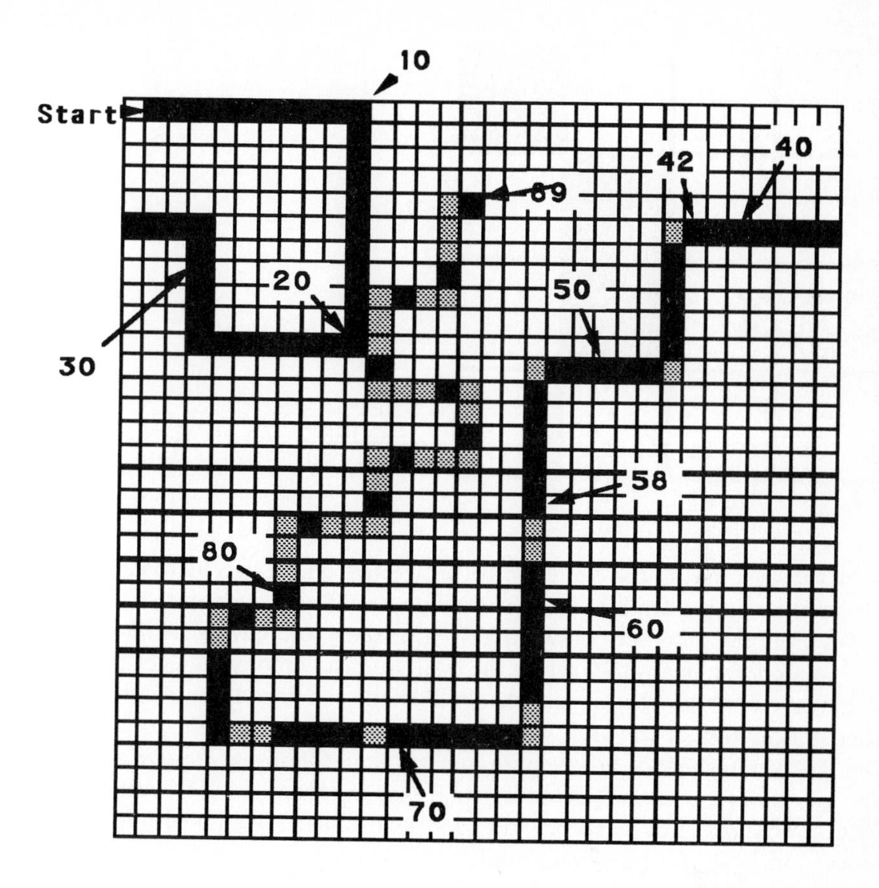

FIGURE 1 The John Muir Trail in 32 × 32 toroidal grid. Scores for reaching various landmarks are indicated. (Gray cells are not part of the trail; they are visual aids to the reader to mark the fastest route of traversal, and appear white to the ants.)

because trail-following strategies with this sensory arrangement are much simpler. After sensing the cell ahead of it, the ant must take one of four actions:

1. move forward one step;
2. turn right (without moving);
3. turn left (without moving); or
4. do nothing.

We chose this menu of moves also for simplicity. The choice to include a no-op (no operation) was in anticipation that algorithms might more easily evolve for trail-following if it were possible for an ant to change its internal state without changing position or orientation on the grid. However, in practice, we found that no-op moves were bred out of the best algorithms. Apparently the tempo that such

FIGURE 2 An ant stands in one cell and sees only the cell ahead of it.

a move consumes is too valuable to be wasted on an action that does not make progress along the trail, because the difference of a single point of the score for an ant is often the difference between having progeny and having its lineage go extinct.

In order for an ant to traverse the trail, it must embody an algorithm that causes it to move forward when it sees the trail on the cell ahead, and to search locally for the continuation of the trail when it does not. It is important to understand that the Tracker task does not require general trail-following ability; it only requires the ants to traverse *this particular trail*, starting from the NW corner facing East. Hence, we should not be surprised to see the evolution of ants with features adapted to the quirks of this particular trail.

4. THE FSA ARCHITECTURE

Figure 3 shows an example of a Finite State Automaton (FSA) that can perform the Tracker task. The circles represent states, and the arcs represent state transitions that the ant undergoes as a function of its current state and its sensory input, each of which takes one time step of the 200 allowed. An arc is labelled with a pair of the form s/a, where s is a 1-bit sensory input indicating whether the ant "sees" a trail cell (1) or a non-trail cell (0); a indicates which of four actions should be taken: move forward (M), turn left without moving (L), turn right without moving (R), and do nothing (N). The FSAs employed in Genesys have up to 32 states.

This example FSA is achieves a score of 42 on the John Muir Trail, and does so with some rather clumsy logic. On the first straightaway of the trail, it walks along

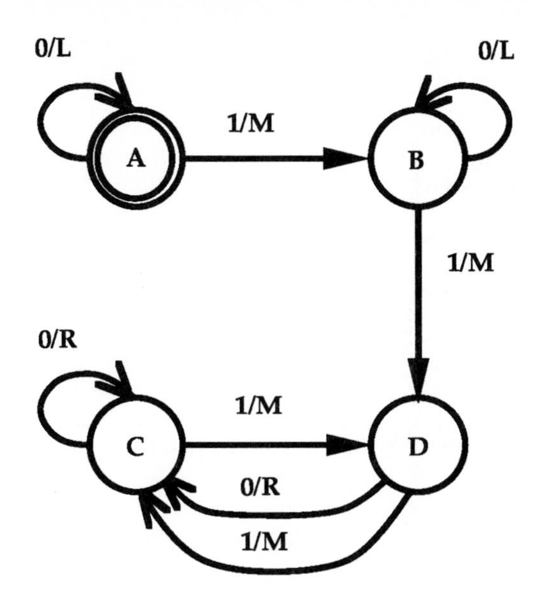

FIGURE 3 The state transition diagram for a 4-state FSA that achieves a score of 42 on the Tracker task. State A is the start state.

in the following sequence of states: ABDCDCDCDC. Thereafter, it never enters states A or B again, but negotiates all turns and straightaways going back and forth from states C to D. Right turns are made in one move; left turns are made using three right turns. At step 42, where there is the first break in the trail, the ant stops, spending the remainder of its life turning right in state C.

In order to be constructed by a genetic algorithm, the FSA must be encoded into a chromosome-like string of bits. The encoding used in Genesys is shown in Figure 4, where the FSA of Figure 3 is arrayed in a table in which each state, input, and move has an arbitrarily assigned binary value. To define the FSA, it suffices to indicate the initial state (state A in this example, coded as 00 in binary), and to enumerate in canonical order the New State and Move columns of the State Transition Table (STT). The genome that encodes the FSA in Figure 3 is shown at the bottom of Figure 4. It is the concatenation of initial state and, in canonical order, the last two columns of the STT. Since Genesys FSAs have up to 32 states (requiring 5 bits to encode), and there are 64 lines in each STT, there are a total of $64 * (5 + 2) + 5 = 453$ bits per genome.

Actual FSAs that arise during evolution are likely, of course, to have fewer than 32 states, since there is no guarantee that all 32, distinct, 5-bit state codes are present in the genome string; even if they are, there is no guarantee that they are reachable from the start state. Some of the rows in the table for a particular FSA may correspond to states it can never enter, and thus represent "unreachable code." An FSA may have even fewer "reachable" states if we consider only those used while traversing the John Muir Trail.

Old State	Input	New State	Action
00	0	00	01
00	1	01	11
01	0	01	01
01	1	11	11
10	0	10	10
10	1	11	11
11	0	10	10
11	1	10	11

FIGURE 4 The binary-coded State Transition Table (STT) for the FSA in Figure 3. The bit string at the bottom is the genome that encodes this organism. Only the initial state and the last two columns (enclosed in the heavy lines) are part of the genome. The first two columns need not be encoded because they are in canonical order.

5. THE ANN ARCHITECTURE

To represent an ant using an ANN, we chose a particular recurrent PDP architecture[2,11,12] with standard sum-and-threshold logic in the connection topology shown in Figure 5. There are two input units with an activation of 1 or 0 according to whether the cell ahead of the ant is on the trail or not. One is activated when there is trail, and the other is activated when there is not. (This choice is historical; one input unit could have sufficed.)

Each input unit is connected to each of five hidden units and to each of four output units. The hidden units are fully connected among themselves, and also to each of the four output units. In each time step, the net receives input from its sensor units, and the activation signals propagate once along every connection in the network. All of the units have Boolean thresholds except the output units. Output units have their inputs summed but not placed at the threshold; instead the ant's next action is determined by which of the output units has the highest activation (with an arbitrary rule for breaking ties). Because this is a recurrent net with five Boolean hidden units, it can behave as though it has up to 5 bits of memory about its past history on the trail. Each ANN is completely deterministic; it does no learning in its lifetime.

To specify an ANN for the trail-following task, we must decide on the values of 63 weights, 5 thresholds, and 5 initial activations. In our model, the nonthreshold activations of the hidden units are 7 bits each; the weights are 6 bits (ones-complement), and the thresholds are 7 bits each (the high-order bits of an implied 9-bit, ones-complement number). The genome representing an ANN ant is

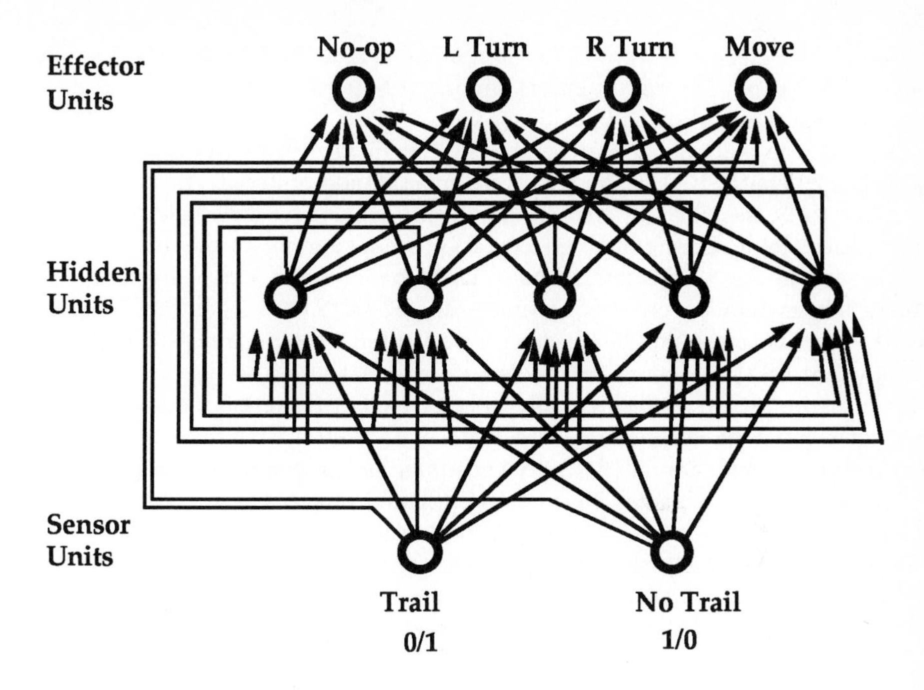

FIGURE 5 Architecture of the recurrent ANN representing an ant.

simply the concatenation of all of the weights, thresholds, and initial activations, which takes $(63 * 6) + (5 * 7) + (5 * 7) = 448$ bits. The concatenation is done in a fixed order that was standardized early in the project.

6. COMPARISON OF FSA AND ANN REPRESENTATIONS

Since, for a long time, we were concerned with deciding which of the two representations, ANNs or FSAs, is "better," we went to some length to assure that the FSA and ANN representations of ants are as similar as possible to make comparisons meaningful. First, we made the two representations approximately equally powerful computationally. We used FSAs that have up to 32 states, the equivalent of up to 5 bits of memory; likewise, we used ANNs with five recurrent hidden units, each of which passes on 1 bit of information, so that they have the equivalent of up to 5 bits of memory as well. Both representations express deterministic algorithms. And, in both cases, an ant's "algorithm" is fixed from birth, so apart from the fact that it can gather a few bits of information about the trail, nothing resembling "learning" takes place during an ant's lifetime.

Second, we made sure that the genomes had similar length: 453 bits for FSAs compared to 448 bits for ANNs. This is important because, other things being equal, we would expect evolution to take longer with a longer genome than with a shorter one, because the space being searched by the genetic algorithm grows exponentially with genome length. If these lengths were substantially different, then the size of the space to be searched by genetic algorithm could have been the dominating performance factor making one representation seem superior to the other.

Because the ANN's are deterministic, with no more than 5 bits of memory, each of them is behaviorally isomorphic to one of the 32-state FSA organisms. (This fact was very useful in our study because, once we created a tool to translate an ANN to an equivalent FSA, all of our other tools for manipulating FSAs, e.g., for animation or state minimization, applied equally well to ANNs.) However, the reverse is not true; not every FSA can be translated into one of our ANNs because there is a great deal of redundancy in the ANN encoding, and in particular FSA ants with the largest numbers of states are underrepresented among the ANN ants. Hence, we must conclude that our ANN ants, as a class, are somewhat narrower and less powerful computationally than our FSA ants (though this is difficult to quantify) despite our attempts to make them comparable. Tasks, defined by a trail and time limit, presumably exist that can be accomplished by some FSA ant, but not by any ANN ant with the same amount of memory.

Many other ways of encodings the FSAs or ANNs into a string could have been used. We could have placed the rows of the FSA table right to left in the string, or placed the bits of the integers in reverse order, or stored the table columnwise, or used gray codes for the integers instead of binary, etc. We could have reordered the weights in the ANN encoding, or chosen different field widths (e.g., 5-bit weights instead of 6-bit). Technically each of these represents a distinct "representation of organisms." The actual encodings we chose were the only ones we tried, though we have no reason to believe it matters very much. Although there are undoubtedly linkage effects in the particular canonical encodings we chose, we have not explored the consequences of other encodings.

7. HOW DIFFICULT IS THE TRACKER TASK, AND HOW DIFFICULT IS IT TO DESIGN AN ANT FOR THE TRACKER TASK?

The John Muir trail was carefully designed so that wherever a straightaway segment of the trail ends, its continuation always begins somewhere in a three-cell-wide extension of the straightaway. The hand-constructed, five-state automaton shown in Figure 6 can traverse the entire trail. Its strategy is to move forward whenever it sees a cell of trail, but when it comes to a point where it does not see the trail in the next cell ahead, it turns right (without moving) and checks for a trail cell there. If it finds one, it moves ahead and continues, but if not, it turns right again. It will turn right a total of four times looking a trail cell. After that, it is facing the original direction, and it will move forward anyway, even though there is no trail cell, and again search the four directions. We chose a trail that could be traversed this way to be certain that at least some small FSA existed that could perform respectably at the task. There is no reason to suppose, however, that this particular FSA, or any one isomorphic to it, ever actually evolved. And, if it did, it could not have survived ultimately because, although it can indeed traverse the entire trail, it requires 314 time steps to do so; it only gets a score of 81 in 200 time steps.

Through evolution we were able to find an FSA with 13 states (the equivalent of less than 4 bits of memory) that can get a perfect score of 89 (see Figure 10).

One can further analyze the difficulty of the task by asking how many bits of information are needed to "memorize" the trail perfectly. If we encode a forward move by a 0, a left turn by the two-bit combination 10, and a right turn by the combination 11, then the entire trail can be unambiguously encoded as a concatenation of "moves" needed to traverse it. Since there are 89 trail cells and 38 non-trail cells also have to be traversed, and since there are 19 turns, then the entire trail can be encoded in $89+38+(2*19)=165$ bits. Our organisms are specified by 450 bits, so it might appear that they could simply evolve to memorize the trail. However, this analysis overlooks that the trail could have be made 100 times longer, and the organisms given 100 times as much time to traverse them. The "information content" of such a trail would be many times larger, but the same 450-bit FSAs and ANNs would be able to successfully traverse it. There is a fundamental difference between evolving programs to traverse a trail and evolving compact data representations of the trail.

We have come to believe that the Tracker task itself is difficult, but not profoundly so, since it can be accomplished by a modest-sized FSA. But just how difficult is the problem of *designing* or *discovering* a high-scoring organism? That

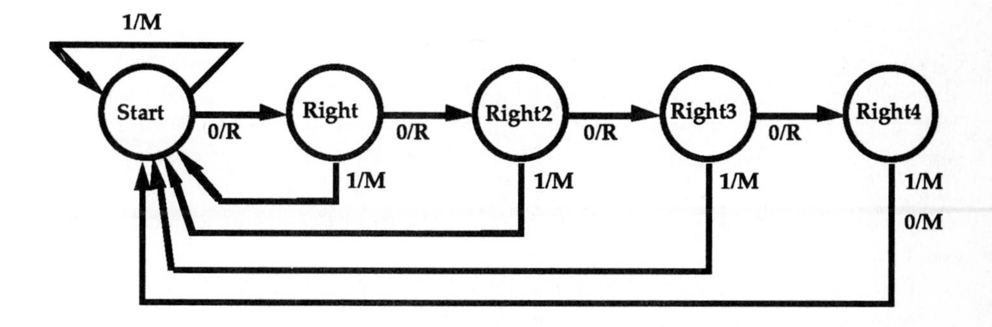

FIGURE 6 Five-state FSA that can traverse the trail in 314 steps, and gets a score of 81 in 200 time steps.

is a completely different issue. Here we are concerned not with searching the environment that the organisms live in, but the searching the space of all organisms. Since the Tracker task is irregular, it does not yield much to analysis. There is no linearity property, for example, to suggest that it is somehow twice as "difficult" to get a score of 80 as a score of 40. Presumably, a skilled human could design an 89-scoring FSA for the Tracker task in an hour or two, and an ANN in somewhat more time, but that is not much of a handle on the difficulty of the design task.

To attempt a quantitative calibration of the design difficulty, we decided to sample the space of all genomes to discover what fraction of them coded for organisms that got each score. We chose a random sample of over 1.3×10^9 organisms, both the FSA and ANN type, and scored each one on the Tracker task to produce the results in Table 1. (In the interest of space we reproduce here only the most interesting regions of the distribution.) The first column shows the score achieved by the animals in the sample. Columns 2–3 show the sample of FSAs. Column 2 shows the exact number of FSAs from the sample that achieved a particular score. Column 3 shows what fraction that represents of the total sample. Columns 4–5 are similar to 2–3, but refer to the ANN representation. The sample was chosen by creating random bit strings of the appropriate length (roughly 450 bits each), translating each string into the appropriate organism, and then running the organism on the Tracker task as usual to tally its score. Since there are $2^{450} \cong 10^{134}$ possible genomes, our sample of $\sim 10^9$ is an extremely small fraction of the total, but is still large enough to reliably sample all but the high end of the distributions.

A glance at Table 1 reveals a number of interesting features. For example, 41% of all FSAs, and 68% of all ANNs receive a score of exactly zero! 90% of both FSAs and ANNs get a score of 10 or less, meaning they cannot make it past the first right turn in the trail. (Of course, a small positive score can also be achieved by striking off in an odd direction and "accidentally" touching a few cells of trail.) By referring to the trail in Figure 1, we can see that a score of more than 32 indicates the ability to make the first left turn; a score of more than 42 shows that the animal can get past the first left turn where the corner cell is missing, and a score of more than 58 shows it can get past the first straightaway double gap (always assuming the organism is following the trail in the obvious order). About 1 in 100,000 FSAs can get past 58, while about 1 in 3,000,000 ANNs can do so. Put another way, in a sample of size 65,536 (the initial generation in most of our evolutionary experiments), there is more than a 55% probability of encountering at least one FSA that can traverse the trail past the double gap, while there is only a 2% probability that there will be such an organism in the initial generation of an ANN run.

The highest FSA score discovered in our sample was 81, of which there were 10 instances. The highest score found among ANNs was 82, of which there was only one. In our entire sample, we never found an FSA that scored higher than the simple one in Figure 6 would score in the 200 time units. We should note that we typically evolve an 81-scoring FSA within 20 generations with a population of

TABLE 1 Interesting Parts of the Score Distribution for FSA and ANN Organisms

Score	Number of FSAs	Fraction of FSAs	Number of ANNs	Fraction of ANNs
0	557,258,091	.413	932,924,367	.683
1	283,207,491	.210	122,460,544	.0896
2	132,410,979	.0981	6,275,613	.00459
3	63,272,402	.0469	944,545	6.91e-04
4	31,819,298	.0236	152,981	1.12e-04
5	18,120,801	.0134	345,287	2.53e-04
6	12,985,610	.00962	20,361	1.49e-05
7	10,063,926	.00746	357,592	2.62e-04
8	9,603,340	.00712	379,791	2.78e-04
9	8,472,800	.00628	38,531	2.82e-05
10	97,907,623	.0726	202,234,068	.1481
⋮	⋮	⋮	⋮	⋮
40	522,627	3.87e-04	119,459	8.74e-05
41	500,301	3.71e-04	896,204	6.56e-04
42	12,110,275	.00897	93,179,890	.0682
43	369,162	2.74e-04	7,127	5.21e-06
44	170,037	1.26e-04	22,279	1.63e-05
45	110,622	8.20e-05	2,436	1.78e-06
⋮	⋮	⋮	⋮	⋮
55	9,505	7.04e-06	365	2.67e-07
56	9,684	7.18e-06	288	2.11e-07
57	10,218	7.57e-06	6,715	4.91e-06
58	125,296	9.28e-05	3,875	2.84e-06
59	5,488	4.07e-06	249	1.82e-07
60	842	6.24e-07	2	1.46e-09
⋮	⋮	⋮	⋮	⋮
75	255	1.89e-07	3	2.19e-09
76	51	3.78e-08	8	5.85e-09
77	45	3.33e-08	0	0.00e+00
78	354	2.62e-07	9	6.58e-09
79	188	1.39e-07	19	1.39e-08
80	64	4.74e-08	19	1.39e-08
81	10	7.41e-09	21	1.54e-08
82	0	0	1	7.32e-10
Totals	1,349,517,312	1.00	1,366,818,816	1.00

65,536, i.e., after having generated at most $20 * 65536 \cong 10^6$ ants, whereas, if we extrapolate from the observed frequency in the sample (which is dangerous since we are extrapolating from the long tail of the distribution), we would expect about one FSA ant in 10^8 to achieve 81. We can only presume that FSA ants that score in the upper 80s are *much* rarer, though there is no feasible way to estimate their frequency by sampling.

In Figure 7 we see the frequency distribution for FSA organisms plotted on a logarithmic scale. There is a clear trend toward exponentially decreasing numbers of organisms as the score increases; but what is most interesting is the pattern of departures from the trend. There are a number of scores, such as 10, 32, 42, 47, 52, 58, 64, and 70, where a sharply larger number of organisms (by a factor of five or more) are clustered than at either neighboring score. A glance back at Figure 1 will show that those scores exactly mark the most challenging features of the John Muir Trail, e.g., the first right turn, the first left turn, the first left-turn-with-gap, the first right-turn-with-gap, the double gap, the first knight's move, etc.

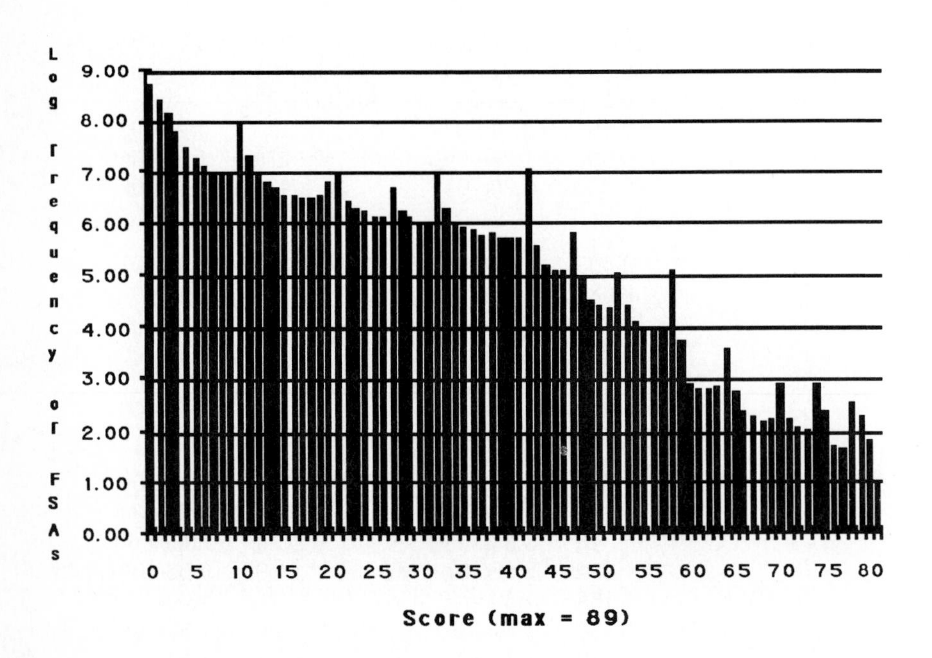

FIGURE 7 Log (base 10) of the number of FSA ants in the sample as a function of their score on the John Muir Trail.

In Figure 8 we see a similar logarithmic plot of the frequency in our sample of the ANN ants. While there is general similarity to the FSA plot, some of the contrasts are striking. From spikes in this distribution, it is still easy to pick out most of the features of the John Muir Trail at scores 10, 27, 32, 42, 47, and 64, but the ratios by which these spikes overshadow the neighboring scores are much larger than for the FSA plot. One curious feature is that there are more ANN organisms that score 57 than 58! We do not know why this should be so, but it is possible that there is a particularly common erroneous path through the trail that happens to achieve that score; we have no other explanation. Another striking difference between the FSAs and the ANNs is that, from scores 60 to 89, the total number of ANN ants in our sample (183) is 61 times smaller than the total number of FSA ants (11,152). This might suggest that high-scoring ANNs are rarer than high-scoring FSAs. Of course, the ANN sample may be unreliable at such high scores, but probably not since it still seems possible to pick out the expected spikes at scores 64, 70, and 74.

From the study of these sample statistics, we have come to view the problem of designing an organism for the Tracker task as being not very difficult up to score

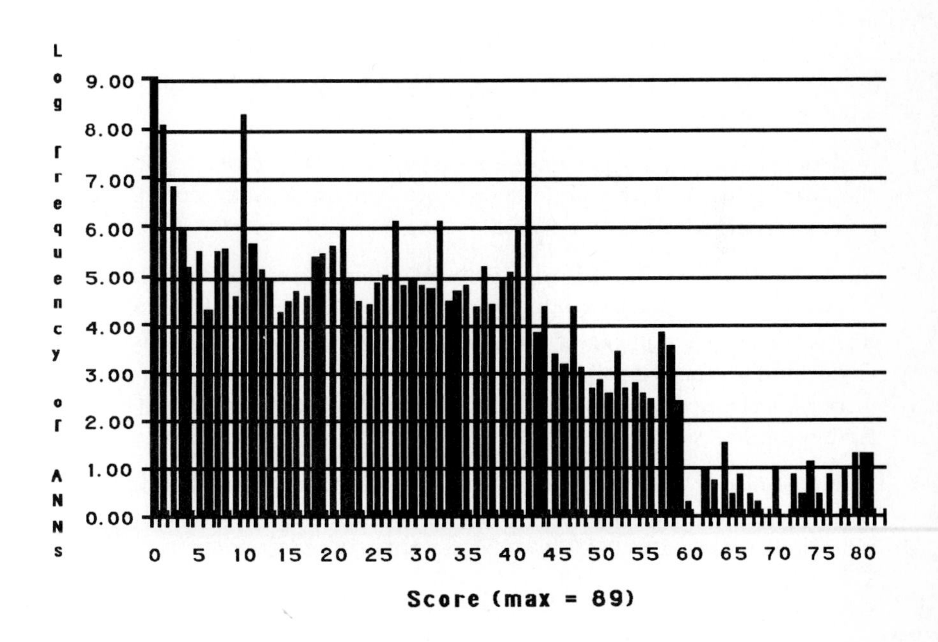

FIGURE 8 Log (base 10) of the number of ANN ants as a function of their score on the John Muir Trail.

81. However, beyond the score of 81, the design task becomes considerably more difficult, not because the John Muir Trail is difficult to traverse *per se*, but because it is difficult to do so in only 200 time steps. It must take over one third of the time to traverse the last eight steps of the trail. Hence, we should expect evolution to proceed in such a way that, at the beginning, there is competition to evolve logic that is basically competent on the trail; in later generations the competition should be for refinements that take advantage of particular features of the trail to save time (steps).

8. THE GENETIC ALGORITHM

The genetic algorithm we used is reasonably standard,[4] though to our knowledge GA's are rarely attempted on bit strings of this length. We begin with a population of 65,536 strings of random bits, each of which is either 448 or 453 bits long (depending on the representation to be used). During a generation, each genome is decoded into either an FSA or an ANN, and all 64K ants are executed for 200 time units (in parallel) on separate copies of the trail. At the end of each generation, all of the ants are scored as to their success on the trail. Those ants scoring highest of their generation (either the top 1% or 10%) are selected for breeding, and all others are discarded. Ties are broken arbitrarily. The new generation is produced by the following procedure.

a. *Mating:* 65,536 pairs of genomes are chosen at random (with replacement) from the selected fraction. No preference is given to those ants scoring higher than others within the selected fraction.

b. *Recombination:* From each pair a single bit string is constructed by random crossover. The crossover probability is typically between 0.5% and 1.0% per bit, so the mean number of crossovers per ant per generation is 2.25 to 4.5. Crossovers are performed without regard to the boundaries of semantic units in the bit string (e.g., state fields in the FSA genome, or weight fields in the ANN genome).

c. *Point mutation:* The recombined string is mutated by random bit-flip operations, again at a rate anywhere from 0.1% to 1% per bit. The mean number of mutations is thus also 0.225 to 2.25 per ant per generation.

Recombination and mutation are illustrated schematically in Figure 9, where two parent bits strings produce one offspring by a two-stage process of random crossover followed by random point mutation.

We performed numerous sensitivity studies to determine that the selection, mutation, and recombination parameters were reasonable and effective. These rates are in line with those reported in Goldberg and Holland[5] for other genetic algorithms.

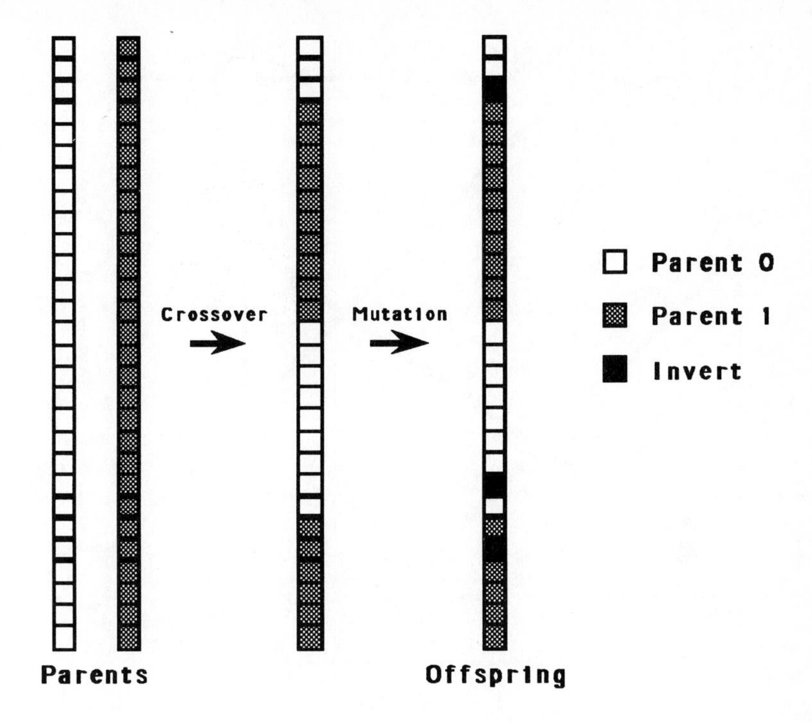

FIGURE 9 Two parents produce one offspring by a two-stage process of random crossover followed by random point mutation.

The qualitative character of the results are extremely robust over wide variations of the evolutionary parameters. Subsequent analysis showed, however, that mutation rates 100-fold smaller would have been just as effective, and in some ways preferable.

9. RESULTS OF THE TRACKER EVOLUTIONS

Figure 10 shows Champ-0, the highest-scoring ant in generation 0 of one run of Genesys with the FSA representation. Since no evolution has taken place, this is just the highest scorer in a particular random sample of 65,536 FSAs. The diagram shown here has been simplified in two ways in order to expose the underlying algorithm: (a) states and transitions that are coded for by the ant's chromosome but are unreachable on the John Muir Trail have been removed; (b) the result was then processed by a state-minimization procedure to remove redundant logic; and (c) the states were renumbered from 0 to 16.

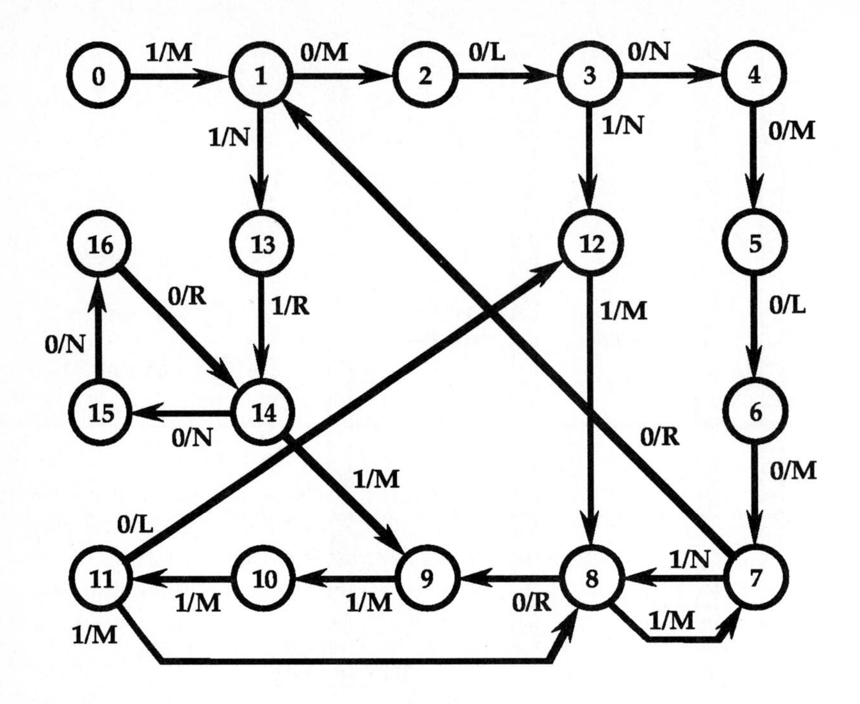

FIGURE 10 Champ-0, the champion FSA in generation 0, which scores 58.

Champ-0 gets a score of 58 on the John Muir Trail. It has great difficulty getting around the first three right turns of the trail, arriving at them in state 7 and taking ten state transitions in each case (1,13,14,15,16,14,15,16,14) before completing the turn in state 9. It has even greater difficulty making the left turn at trail step 32, requiring two executions of the cycle (8,1,2,3,4,5,6,7) before using the right turn sequence (1,13,...) above. On the straight-away segments of the trail, it takes two transitions per step around the 7–8 loop in the diagram, which is a very time-inefficient way to travel. Curiously, Champ-0 makes the left turn, where the corner cell is missing at step 42, more efficiently than it makes the earlier left turn at step 32 where the corner is present! Arriving at the corner in state 8, one trip around the cycle (1,2,3,4,5,6,7,8) makes the turn. Champ-0 runs out of time when it gets to the double gap at trail step 58.

In Figure 11, we show Champ-100, an FSA-ant that is the product of 100 generations of evolution starting from the population of which Champ-0 was the champion. The evolution was conducted with a mutation rate of 0.5% per bit, a recombination rate of 0.5% per bit, and a selection fraction of 1% per generation. Its logic has been simplified in the same way Champ-0's was. Although Champ-100 is descended from the same population that included Champ-0, there is no reason to believe Champ-0 is actually a genetic ancestor of Champ-100, and, in fact, it is rather unlikely.

Champ-100 gets a perfect score of 89, demonstrating that it is indeed possible to traverse this trail in 200 steps. In fact, Champ-100 takes exactly 200 time steps to finish the trail. Since the scoring function does not give any extra credit for reaching the end sooner, there is no selective pressure for any further improvement in performance.

We can observe a number of fascinating features in Champ-100's logic. First, it traverses the straightaway portions of the trail by staying in state 0 or in state 9, thereby taking one unit of time per trail step rather than two. Second, it makes a right turn from state 0 using the (1,12,0) cycle, a far more efficient mechanism than Champ-0 used, though not optimal. Third, the left turn at trail step 32 is accomplished by three right turns. We deliberately placed three right turns in the trail before the first left turn precisely to see if there would be an evolutionary bias toward more efficient right turns. In this case, apparently, there was.

The state sequence (9,10,11) is extremely versatile. It is used in four critical places along the trail: at step 42 (left turn with missing corner), at steps 58 and 74 (straightaway double gap), and at step 64 (forward right knight's jump). Likewise, the sequences (12,7,8,3,4,0) and (6,4,0) are used repeatedly to traverse the "stepping stone" part of the trail from step 78 to 89. Such efficient logic seems exquisitely

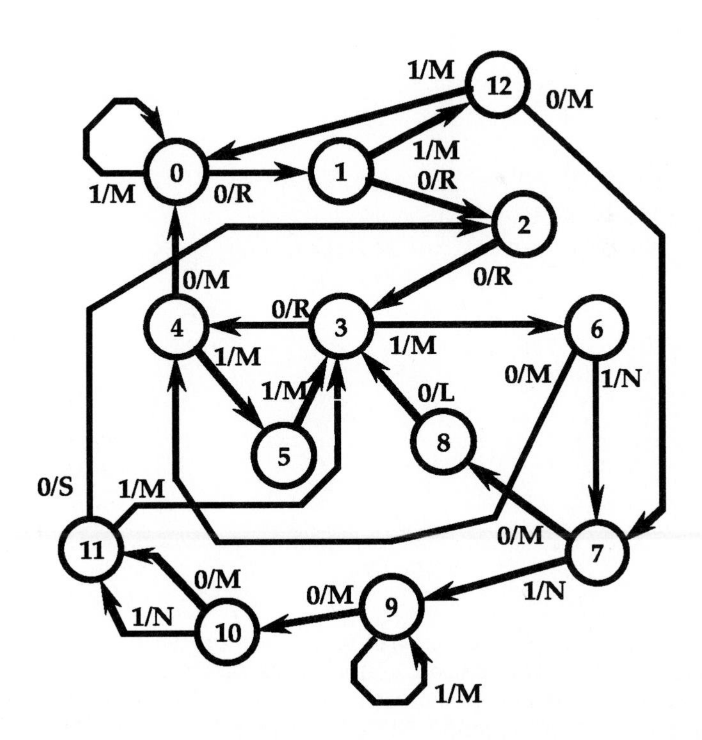

FIGURE 11 Champ-100, the champion FSA in generation 100, which scores 89.

adapted to the features of this particular trail, and suggests that evolution has had the effect of "compiling" knowledge of this environment into the structure of the organism.

In Figures 12 and 13 we chart the evolutionary progress of the entire population in the Tracker task, one chart for FSA ants and one for ANN ants. These are runs

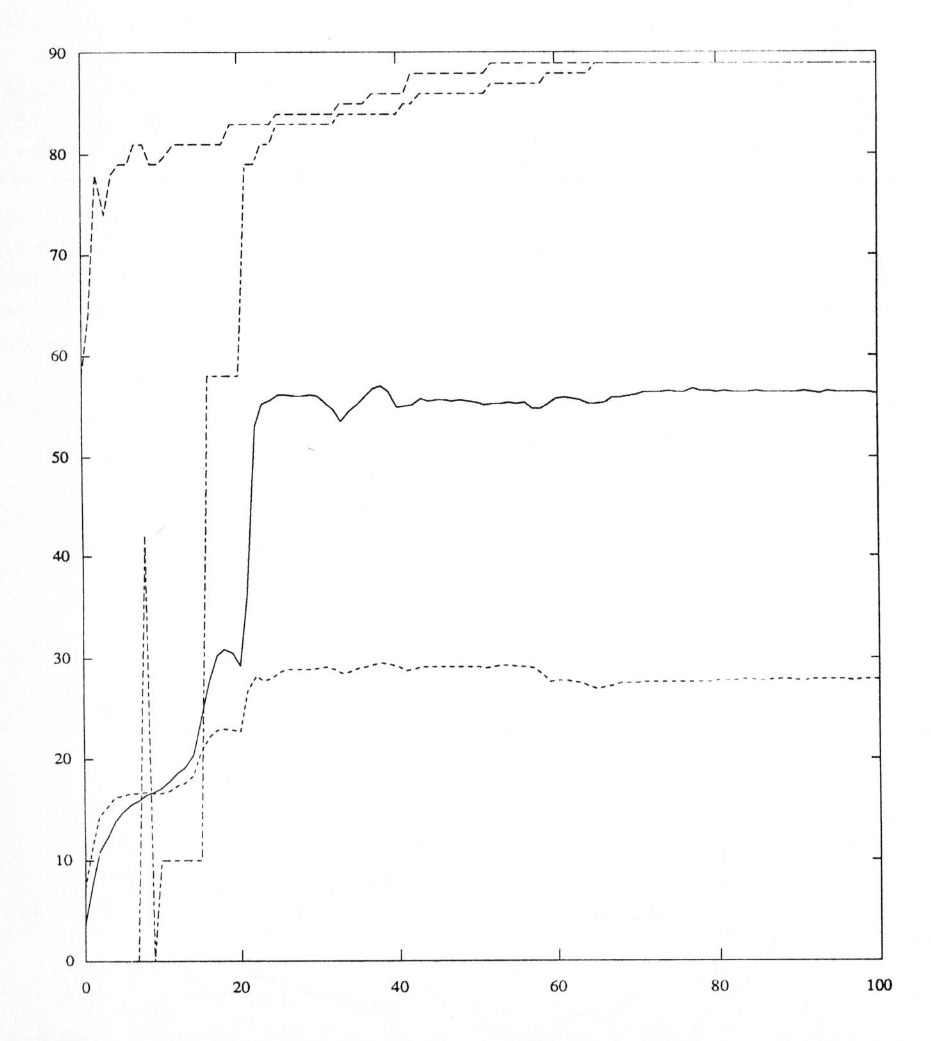

FIGURE 12 Evolution of FSA ants. Mean score shown by solid line (—), standard deviation score by short dashed line (- - - -), maximum score by medium dashed line (— — —), and Mode 1 score by dot and dash line (-—-—-). Execution time: 0:53:11; population: 65536; random seed: 8276342; iterations/generations: 200/environment: 32 × 32; chromosome length: 520; fraction selected: 0.050000; mutation rate: 0.010000; crossover rate: 0.010000; decision maker: fsa; FSA states: 5.

typical of many we have made, and are chosen for presentation here because their parameters are identical. In both cases, the mutation and recombination rates are 1% per bit per generation, and the selection fraction is 5% per generation. Both runs evolve a population of 65,536 for 100 generations. Each took about one hour on a 16K-processor Connection Machine (CM2).

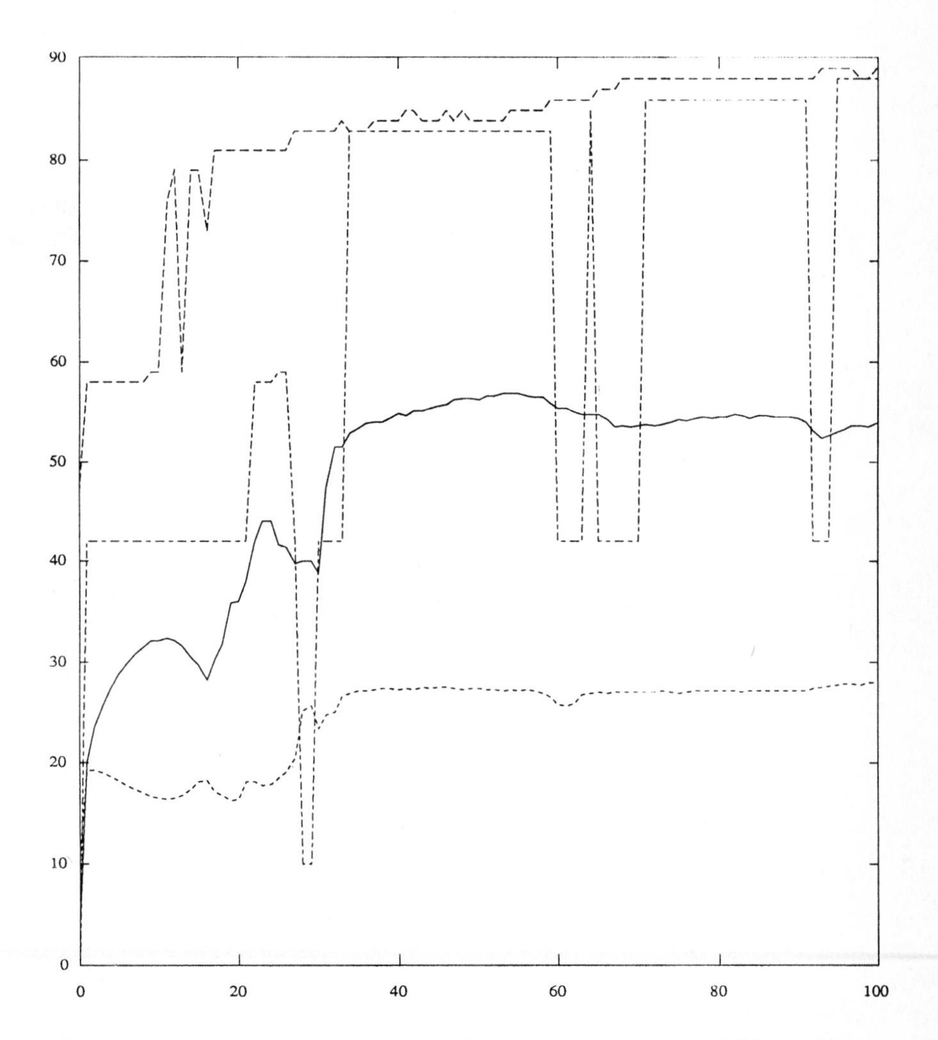

FIGURE 13 Evolution of neural net ants. Mean score shown by solid line (—), standard deviation score by short dashed line (- - - -), maximum score by medium dashed line (– – –), and Mode 1 score by dot and dash line (- – – – -). Execution time: 1:5:22; population: 65536; random seed: 8276342; iterations/generations: 200/environment: 32 x 32; chromosome length: 520; fraction selected: 0.050000; mutation rate: 0.010000; crossover rate: 0.010000; decision maker: net; hidden units: 5.

Figure 12 is a graph of the progress of an evolution with FSA ants. The horizontal axis is the generation number, and the vertical axis is the score on the Tracker task. Four population statistics are plotted as a function of generation:

1. the maximum score achieved by any ant in the generation,
2. the mode (most frequent) score,
3. the mean score, and
4. the standard deviation of the scores.

Initially the mean score was somewhere around 3, but the maximum score was 58, consistent with the statistics given in Section 7. After 15 generations of selection the maximum score is already in the 80s, and, by generation 52, perfect-scoring ants are always present in the population and are the most common score by generation 70. However, the mean score never approaches the maximum score, hovering around 56 for the last 75 generations, and there is always a high standard deviation. The high variance and the great difference between the mean and the max scores is, we have learned, a consequence of the relatively high mutation rate of 1% per bit. We can get the same max score in the same number of generations with a mutation rate two orders of magnitude smaller, and then the mean score reaches the 80s. With a high mutation rate shown here, a large fraction of successful parents have offspring that are destroyed by mutation and hence bring down the average score.

Figure 13 is a similar graph except that the ants are represented as artificial neural nets. All evolution parameters are the same as in Figure 12. We see similar features in Figure 13, i.e., the maximum score starts low (this time at 46) and quickly reaches the 80s (by generation 18). It then takes a long time to reach 89 (generation 94). As in Figure 12, there is a large gap between the maximum score and the mean score, for the same reasons. Comparing Figure 12 to Figure 13 we see that evolution proceeds somewhat more slowly in the ANN representation than in the FSA representation, for reasons unknown. This result was consistent over hundreds of executions with many different parameters.

10. DISCUSSION AND CONCLUSIONS

This research clearly shows, as we had hoped, that it is computationally feasible to produce artificial organisms that can exhibit complex behavior, and to produce them by evolution. And evolution can act on the entire text of the program representing the organisms, at least with populations of 64K organisms and genomes of 450 bits. In that sense, the exercise was a success.

Rather than decide which of the two representations was "better" for the study, we concluded that working with two representations was extremely important for completely different methodological reasons. First, it tends to eliminate the potential problem that we might observe an evolutionary phenomenon which is merely an artifact of the representation, and not a genuine feature of evolution. If we perform each experiment many times, with different representations and different environments, etc., and if we observe the same evolutionary phenomenon in all cases, then we can be more certain that whatever phenomenon we observe is a property of evolution itself, and not an artifact of the representation. Second, it focuses attention on a fundamental issue that all similar studies must face in the future: just what is a good computational representation of living organisms? We expect that this issue will continue to be of fundamental importance in future artificial life studies.

As a result of this study, we have identified a number of important properties that we believe a programming paradigm must have to be suitable as a representation for organisms in biologically motivated studies. Among them are the following:

a. The paradigm must be what we refer to *computationally complete*; i.e., it must be possible for any bounded-memory (finite-state) algorithm to be encoded in it, provided that the size of the representation is large enough. If such organisms are intended for an infinite environment, and have the ability to move around and read and write into it, then the paradigm is Turing-equivalent.

b. It should specify a *simple, uniform model of computation* so the interpreter can be small, and so there is minimal chance to bias the system by choosing a programming model that happens to rich in the very operators or strategies needed to adapt to a particular environment. Furthermore, if the goal is to have a simple model for studying complex natural systems, it is important that organism programs should be manipulable and understandable.

c. It should be *syntactically closed* (or nearly so) under the genetic operators. Mutation and recombination operators must not (usually) transform legal programs into illegal ones. In practice, we have designed our representations so that all bit strings (of the appropriate length) encode for some legal program. With this approach, we get the added benefit that we can start evolution with a population of "random" organisms by just producing a set of random bit strings, a simple and indisputable way of avoiding bias in the initial population from which evolution starts.

d. It should be *well conditioned* under the genetic operators. This requirement is not very formally defined, but essentially requires that "small" mutational changes in the program should (usually) cause "small" changes in its behavior, and that a crossover of two parent programs should usually produce an offspring program whose behavior is in some sense a "mixture" of the parents' behavior. These requirements are probably necessary for evolution to have a chance to succeed in realistic adaptive landscapes. Of course, occasional jumps or discontinuities can be tolerated, and are likely even necessary.

e. It should specify *one time unit* of the organism's life. This basically means that one execution of an organism's program should (a) accept input from its sense organs (both external and internal), (b) possibly change its internal state, and (c) possibly take one or more actions through effectors. An organism's life, then, is the iterated execution of its behavior program.

f. It must *scale well.* This means that the size and time complexity of an organism's program is a modestly growing function of the size of the input to or output from it. (Finite state automata (FSAs), at least as we formulate them in Section 4, fail this test. It takes a transition table with 2^{m+n} rows to specify an FSA with m bits of state and n bits of input, so if the entire transition table is encoded in the genome, then the genome length grows exponentially with the size of the organism's sensory apparatus, which is unacceptable for all but the smallest m and n.)

In RAM, both organisms and the environment are represented as programs.[14] The behavior and evolution of the population in the environment is simulated by the co-execution of all of the organism programs and environment programs. One of the good things about this design choice is that it allows the environment to be just as "active" as the organisms are, and it makes a statement that the line between organism and environment is often arbitrary. It also allows each organism to be part of the environment of the others, so that competitive and cooperative relationships can evolve. Organisms are not of different fundamental stuff than the environment: all are represented as programs and have a symmetric, coequal status in the simulation. But, however attractive this point of view is from a modeling perspective, it does not properly capture the true relationship between organisms and their environment in natural life. For in RAM, organism and environment programs are defined separately, interacting as though each is "outside" the other, whereas in natural life organisms are definitely "inside" the environment, and part of it. This distinction between environment and organism is manifest because, while organisms "move" and "reproduce," these actions are treated as primitive, and they do not occur as the result of any lower-level processes. The organism's behavior function says "move" and, *as if by magic*, the location coordinates of the organism change; the behavior function says "reproduce," and, *as if by magic*, a full copy of the organism (complete with modified genes) appears in no time on a neighboring location. Organisms in RAM do not really have a simulated "physical" body whose natural activity produces motion and reproduction. In natural life, however, both "environment" and "organism" are epiphenomena arising from the same physics below.

Genesys has these limitations of RAM and others as well. While the organisms are represented as programs in Genesys, the environment is almost completely static.

Theoretically it would be extremely fruitful to correct this modeling deficiency, but we know of no way to do that and still retain the principle that an organism is a program. The only really satisfactory way we know to remove the separation between organism and environment is to invent an artificial physics, such as a huge cellular automaton, in which both the organisms and the environmental processes are represented as interacting patterns of activity in the same playing field. For the size of population and the complexity of simulation that we envision today, an artificial physics is computationally out of the question.

As a result, in Genesys, and all succeeding systems that we envision, we will continue to simulate each organism by a program. We will continue to simulate not the organisms themselves, but their *behavior*, in their environments. We may, at a later stage, endow organisms with a "body," e.g., arms and legs, so that the primitive operations are reduced from "move organism" to "move leg." But, for the foreseeable future in our research, there will always be a level at which the organism "interfaces" with the environment, or "communicates" with it, instead of participating in it.

ACKNOWLEDGMENTS

This research was funded by a grant from the W. M. Keck Foundation and by National Science Foundation Biological Facilities grant number BBS 87 14206. We also thank Jon Bentley for extremely detailed correction and comments.

REFERENCES

1. Collins, Robert J., and David R. Jefferson. "AntFarm: Towards Simulated Evolution." This volume.
2. Elman, J. L. "Finding Structure in Time." CRL Tech. Rep. 8801, Center for Research in Language, University of California, San Diego, 1988.
3. Gibson, Robert, Charles E. Taylor, and David R. Jefferson. "Lek Formation by Female Choice: A Simulation Study." *J. Int'l. Soc. Behavioral Ecology* **1(1)** (1990): 36–42.
4. Goldberg, D. E. *Genetic Algorithms in Search, Optimization and Machine Learning.* Reading, MA: Addison-Wesley, 1989.
5. Goldberg, D. E., and J. H. Holland, eds. *Machine Learning,* Vol. 3, Nos. 2-3. Special Issue on Genetic Algorithms. Kluwer Academic Publishers, 1988.
6. Hillis, W. D. *The Connection Machine.* Cambridge: MIT Press, 1985.
7. Holland, John. *Adaptation in Natural and Artificial Systems.* Ann Arbor: The University of Michigan Press, 1975.
8. Holldobler, Bert, and Edward O. Wilson. *The Ants.* Cambridge: Harvard University Press, 1990.
9. Koza, John R. "Genetic Programming: A Paradigm for Genetically Breeding Populations of Computer Programs to Solve Problems." Technical Report No. STAN-CS-90-1314, Computer Science Department, Stanford University, June, 1990.
10. Langton, C. G., ed. *Artificial Life.* Santa Fe Institute Studies in the Sciences of Complexity, proceedings volume VI. Redwood City, CA: Addison-Wesley, 1989.
11. Rumelhart, D. E., and J. L. McClelland, eds. *Parallel Distributed Processing,* Vols. 1 and 2. Cambridge: Bradford Books/MIT Press, 1986.
12. Servan-Schreiber, D., A. Cleeremans, and J. L. McClelland. "Learning Sequential Structure in Simple Recurrent Networks." In *Advances in Neural Information Processing Systems,* edited by D. S. Touretzky, vol. 1, 643–652. 1989.
13. Taylor, Charles E., David R. Jefferson, and Hans Burla. "Habitat-Dependent Dispersal of *Drosophila obscura* and *D. subobscura.*" *Genetica Iberica* **39** (1987): 547.
14. Taylor, Charles E., David R. Jefferson, Scott Turner, and Seth Goldman. "RAM: Artificial Life for the Exploration of Complex Biological Systems." In *Artificial Life,* edited by C. Langton. Santa Fe Institute Studies in the Sciences of Complexity, proceedings volume VI. Redwood City, CA: Addison-Wesley, 1989.
15. Taylor, Charles E., L. Muscatine, and David Jefferson. "Maintenance and Breakdown of the Hydra-Chlorella Symbiosis: A Computer Model." *Proceedings of the Royal Society, London* Series B **B238** (1989): 277–289.

Robert J. Collins† and David R. Jefferson‡
Department of Computer Science, University of California, Los Angeles, Los Angeles, CA 90024; e-mail: †rjc@cs.ucla.edu, ‡jefferso@cs.ucla.edu.

AntFarm: Towards Simulated Evolution

The most easily observed ant behavior is workers foraging for food. Foraging workers do not eat the food, but carry it back to the nest, where it is processed and consumed by all members of the colony. In many species, a high degree of coordination and cooperation between foragers is observed (usually mediated by pheromone communication).

We would like to understand more about the evolution of cooperative foraging. In this paper, we describe a computer program called AntFarm, that simulates the evolution of foraging strategies in colonies of artificial organisms that resemble ants. AntFarm is work in progress, and is being used to investigate issues surrounding simulated evolution of complex behaviors in complex environments, the evolution of cooperation among closely related individuals, and the evolution of chemical communication. We describe our genetic algorithm for simulating evolution. We also discuss the issue of the representation of artificial organisms, and empirically compare several ANN encodings based on their ability to evolve foraging behavior in AntFarm ants.

1. INTRODUCTION

We are attempting to simulate the evolution of complex behavior (rather than physical morphology) in artificial organisms. In this paper, we consider the simulation of organisms that live and reproduce in relatively complex environments, with many sensors (external and internal), and many possible actions at each moment. In addition, the organisms possess internal memory, allowing their behavior to be history sensitive. In the course of its life, each organism is born, makes thousands of decisions (eat, move, etc.), and eventually dies. The reproductive success of each organism is affected by its behavior throughout its lifetime.

We are particularly interested in the evolution of cooperative central-place foraging in ants. AntFarm is a computer program that simulates an evolving population of ant colonies whose reproductive success is a function of the amount of food carried to their nest, producing a selection pressure favoring better foraging strategies. Each colony is made up of a small number of genetically identical ants, whose behavior is specified by an artificial neural network (ANN). In addition to the ability to sense and carry food, the ants can sense and drop pheromones (chemicals used by ants for communication).

AntFarm is work in progress. Eventually we will attempt to determine the conditions that are necessary for the evolution of cooperation (mediated by chemical communication) in central-place foraging. Johnson, Hubbell, and Feener have developed a model of optimal central-place foraging in ant colonies that predicts that the degree of cooperation should be a function primarily of the distribution of food in the environment.[14] While this model predicts when cooperation pays off, under what conditions an optimal strategy will actually evolve is an open question.

So far, we have completed the implementation of AntFarm, and are able to consistently evolve (solitary) foraging behavior. To get to this point, we have had to design both a genetic algorithm that closely resembles natural evolution, and a new ANN representation that is more suitable for evolutionary experiments than those used in previous studies. Our genetic algorithm uses local competition and mating, rather than the usual panmictic (random mating) scheme. In addition, we feel that a clear separation between the genetic algorithm and the simulated world/organisms is necessary to conduct unbiased evolutionary experiments. Hence, our genetic operators are applied to structureless bit-string chromosomes.

The need for an appropriate artificial organism representation has been a major obstacle, which we have recently overcome. A new ANN representation is necessary, because other behavior function representations either are not appropriate for biologically motivated simulated evolution, do not scale well to number of inputs/outputs required for AntFarm, or empirically are not capable of evolving foraging behavior in AntFarm. From this struggle, we have abstracted a number of properties that are necessary for organism representations that are to be used in simulated evolutionary experiments.

2. MICROANALYTIC SIMULATION OF EVOLUTION

The computer simulation of evolving populations is important in the study of ecological, adaptive, and evolutionary systems.[22] Only the simplest genetic systems can be completely solved analytically, and evolutionary experiments in the laboratory or field are usually limited to, at most, a few dozen generations and are difficult to control and repeat. Simulated evolution makes it possible to study evolutionary systems over hundreds or even thousands of generations. By their very nature, computer simulations are easily repeated and varied, with all relevant parameters under the full control of the experimenter.

Most computer simulations in biology (including evolutionary simulations) are based on solving differential equations from mathematical models.[20,21] Due to mathematical limitations, models of evolving systems are usually simple and unrealistic. Complex models that incorporate a large number of both intrinsic factors (e.g., the life history of the organisms) and extrinsic factors (e.g., weather, competitors, etc.) are more accurate and useful. Unfortunately, such complex evolutionary models are difficult or impossible to describe analytically.

Although most biological simulations are equation-based, simulations based on the observation that the execution of a computer program is very similar to the life of an organism have emerged in recent years.[22,23,7,4,24,13] In such simulations, each organism is represented by a program, as are the various environmental processes: the population of executing programs simulates a population of living organisms and the environment. Rather than attempting to capture the complex global dynamics of the population and environment in a set of equations, only the local interactions between the individual organisms and environmental factors are modeled. Based on these relatively simple local interactions, the complex global behavior of the population emerges.

This sort of "life-as-process" simulation is referred to as *microanalytic*, meaning that each individual organism and environmental effect is separately represented, and the biologically significant events in an organism's life are all separately simulated in detail.[2] Each organism in the population is represented as a program, and its life as a process: a detailed sequence of events, including its birth, its interactions with a dynamic environment (potentially including many of the other organisms in the population), its mating and reproduction (if any), and its death.

2.1 BIOLOGICAL ISSUES

While we cannot use simulated evolution to reconstruct an actual situation in the history of natural life, we can explore particular hypotheses, eliminating some and giving credence to others. Such simulations provide the researcher with an artificial world in which to perform evolutionary experiments that can be fully controlled and repeated, and can span a large number of generations. Simulated evolutionary experiments might someday be used to shed light on a number of open evolutionary problems, including:

1. modes of speciation (which of the many hypotheses are most likely, and in which sexual systems and ecological situations),
2. the evolution of mutation and recombination rates,
3. the evolution of information-processing behavior (e.g., sensory-motor integration, communication, etc.),
4. the evolution of sexual reproduction (i.e., why is it maintained in competition with asexual reproduction?),
5. punctuated equilibria (i.e., is it true that most evolution occurs at speciation events, and not within species?),
6. the dynamics of the evolution of predator-prey "arms races,"
7. the influence host-parasite interaction on evolution rates,
8. the stability of ecosystems,
9. the evolution of evolutionarily stable strategies,
10. sexual selection and the evolution of maladaptive characteristics, and
11. the evolution of cooperation (especially among kin).

We have focused our attention on a smaller question: the evolution of central-place foraging strategies in ants. We are exploring the evolution of the use of chemical communication and cooperative foraging within ant colonies.

The dominant insects throughout the world are the ants. All ant species have eusocial societies, characterized by overlapping generations, care of young by adults, and adults divided into reproductive (kings and queens) and nonreproductive (workers) castes. Ants live in colonies ranging in size from a few individuals to more than 20 million, all with a high degree of organization. Nearly all communication between ants is either tactile, visual, or chemical. Large-scale coordination is achieved through the use of pheromones.

Each individual ant is relatively small and simple, typically performing only 20 to 42 distinct behaviors[12]; yet the emergent behavior of the colony as a whole is amazingly complex. In many contexts, myrmecologists treat the whole colony as a single *superorganism.* The unparalleled success of these superorganisms in all parts of the world (perhaps as many as 20,000 species[12]) speaks well for the strength and versatility of the eusocial colony.

Although we are strongly motivated by the example of real ants, we do not feel bound to model them exactly. Our goal is to use AntFarm to verify that approximately optimal cooperative foraging behavior consistent with a model proposed by Johnson et al.[14] can evolve by natural selection, and to explore the conditions favorable to its evolution.

2.1.1 OPTIMAL CENTRAL-PLACE FORAGING Central-place foraging consists of two phases: the search for food and its recovery to a central location.[19] Much of the cost of foraging is associated with search,[6,17] but all of the payoff is from recovery, which consists primarily of transportation of the food to the nest. Foraging strategies that minimize search time will clearly be advantageous.

The Johnson, Hubbell, and Feener model of central place foraging in eusocial insects is fairly complex and the details are beyond the scope of this paper, but it predicts the effect of the size of food patches on the number of foraging workers and the style of foraging that is used. In species that feed on small patches of food, a small number of workers, each foraging alone, is optimal. Recruitment of nestmates to help recover the food does not pay off because the food patches are small. In this model, the search for food (in the absence of recruitment) is assumed to be a random walk beginning at the nest so that the area around the nest is searched many times by different foragers. As the number of foragers increases, the amount of additional area searched decreases. The diminishing returns for additional foragers results in a small foraging force being optimal.

Species that feed on large patches of food should have a large foraging force, with heavy reliance on recruitment. When a patch is too large for the discovering ant to harvest alone, it pays to recruit (rather than rely on rediscovery by other foragers). Recruitment of nestmates to help harvest a known food source can nearly eliminate search costs. With reduced search costs, the diminishing returns for additional workers is not such an important factor, resulting in a large foraging force being optimal.

In real ants, recruitment to harvest food resources takes many different forms.[12] In the simplest case, a second ant is physically led to the food in a process called *tandem running.* More common is *group* recruitment, which uses a short-lived pheromone trail to bring up to a few dozen workers to the food source. The most impressive form is called *mass* recruitment. In mass recruitment, a relatively fixed, long-lived pheromone trail leads hundreds or thousands of workers to the food source. The trail is reinforced by each successful worker. Mass recruitment is used only in species that forage for food that is found in very large clumps.

2.1.2. THE GENETIC ALGORITHM The biological focus of the AntFarm experiments requires us to closely model the process of natural evolution. Genetic algorithms are loosely based on natural evolution, and have been used by computer scientists and engineers as an optimization method for more than 25 years.[11] Unfortunately, traditional genetic algorithms are not well suited for simulated evolution.

Genetic algorithms are typically used to search for good solutions for complex optimization problems,[9] i.e., a string of function parameters that (more or less) optimizes a particular function. A genetic algorithm evolves a population of these strings (chromosomes) by assigning each a score (fitness value), based on the quality of the solution. The likelihood that a particular string will be chosen for mating is a function of its score and the rest of the population. The key to the genetic search is that those chosen to mate reproduce with variation.

The mechanics of genetic algorithms are relatively simple, consisting of four basic parts:

1. assigning fitness scores,
2. selection and mating,
3. recombination, and
4. mutation.

The assignment of fitness scores is wholly dependent on the particular application. New populations are created by the repetition of these steps. Because selection is biased towards strings with higher scores, the populations typically achieve higher and higher scores as generations pass. In this section, we briefly describe our genetic algorithm and informally compare and contrast it with traditional genetic algorithms and natural evolution systems.

In most genetic algorithms, there is only one gender, so that any individual can mate (sexually) with any other individual (although simple extensions allow for two or more genders). The parents of the next generation are selected probabilistically based on their score (defined by the *objective* or *fitness function*) and the scores of all the other members of the population. Let f_i be the fitness score of string i, and N be the number of strings in the population. The probability that string i is chosen to be a parent is usually defined to be something like

$$P(i) = \frac{f_i}{\sum_{j=0}^{N-1} f_j}, \tag{1}$$

and the strings are randomly paired according to this distribution for sexual mating.

Although this panmictic (random global mating) scheme is simple and widely used in genetic algorithms, it is a poor model of real evolution. One of the basic assumptions of Wright's shifting balance theory of evolution is that spatial structure exists in large populations.[5,25,27,28,29,30] The structure is in the form of *demes*,[8] or semi-isolated subpopulations, with thorough gene mixing within a deme, but restricted gene flow between demes. One way that demes can form in a continuous population and environment is *isolation by distance*: the probability that a given parent will produce an offspring at a given location is a function of the geographical distance between the parent and offspring locations.

To simulate isolation by distance in the selection and mating process, we place the artificial organisms on a toroidal, 2-dimensional grid, with one organism per grid location. Selection and mating take place locally on this grid, with each individual competing and mating with its nearby neighbors. In his quantitative analysis of isolation by distance, Wright assumes a normal distribution for parent-offspring distances.

> Normal distributions of parents relative to offspring are to be expected if dispersion occurs by a long succession of random movements...

In our genetic algorithm, the parents are chosen during short random walks that begin at the offspring location, one parent per walk. The highest scoring individual encountered during the random walk is chosen as the parent (breaking ties in favor of those encountered later in the walk). The parents are chosen with replacement, so it possible for the same high-scoring individual to be encountered during both random walks, in which case it would act as both parents for the offspring.

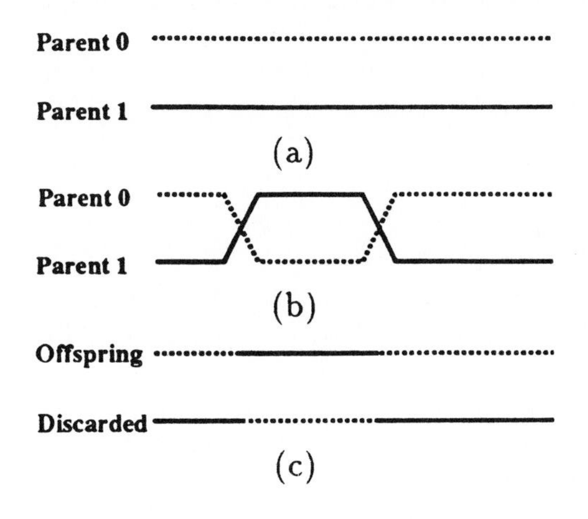

FIGURE 1 A two-point reciprocal recombination. (a) The parent chromosomes are aligned. (b) At a random point, the chromosomes cross. (c) The chromosomes are cut and rejoined at the crossover point, resulting in new gene combinations. One of the chromosomes specifies the offspring, and the other is discarded.

In genetic algorithms used for optimization, it is common to exploit problem-specific or representation-specific information in the implementation of the genetic operators in an effort to speed the search. However, to create unbiased and realistic evolutionary experiments, it is necessary to avoid building the experimenter's preconceptions into the simulation. Therefore, we require a clear separation between the genetic algorithm and the simulated organisms/environment.[2]

The selection phase of the genetic algorithm produces a pair of strings (chromosomes) for each offspring that is to be produced for the next generation. Recombination mixes the genetic information of the parents when producing offspring, so an offspring chromosome contains some of the genetic information from each parent. We only consider *reciprocal recombination*, where equivalent length strings are exchanged (Figure 1). The model of recombination that we use in our genetic algorithm begins with an alignment of the pair of chromosomes (Figure 1(a)). At some random point (or points), the chromosomes cross (Figure 1(b)), then the chromosomes are cut and rejoined at the crossover point(s) (Figure 1(c)).

Our model of crossover is not typical of most genetic algorithms because it operates on the chromosome as a bit string, rather than, for example, a list of parameters. It is defined completely independently of what or how the genetic information is encoded in the chromosome, which is biologically realistic.

Our recombination operation produces two haploid chromosomes. One of these (chosen randomly) is discarded, and the other is retained for use as the offspring chromosome. In practice, we only explicitly generate one of the two chromosomes.

After the process of recombination, the genetic algorithm mutates the new chromosome, producing the final version that describes the offspring. Many classes of mutation appear in natural genetic systems, including base substitution, deletion, frame-shift, insertion, inversion, translocation, duplication, etc. Although most of these types of mutation can make sense in the context of a genetic algorithm, we only consider base substitutions, i.e., the substitution of one nucleotide for another. We simulate a mutation by flipping a bit (change 0 to 1, or 1 to 0) in the bit-string chromosome. Like the recombination operation, this formulation of mutation differs from most genetic algorithm implementations in that the mutations make small changes in a structureless bit string, rather than making small changes to a problem-specific parameter.

2.2. COMPUTATIONAL ISSUES

Microanalytic evolutionary experiments are computationally large in many dimensions (including population size, number of generations, size of the genome, size of the behavior function, size of the sensory/effector/memory apparatus, size of the environment, etc.). Until recently, these experiments were not computationally feasible, and even today parallel computation is required for AntFarm experiments.

The panmictic selection and mating scheme of typical genetic algorithms is not very well suited for a massively parallel implementation, because the survival and mating success of each individual involves global knowledge of the population (Eq. (1)). The local competition/mating scheme that our genetic algorithm is fully distributed, requiring only local information, is both biologically more realistic and well suited for a massively parallel implementation.

3. AntFarm

AntFarm is a microanalytic simulation that evolves group foraging behavior in colonies of ant-like organisms. The AntFarm evolution is driven by the genetic algorithm described above, operating at the level of colonies (superorganisms), not individual ants. The actions (determined by the ant's behavior function) of all of the ants in a colony contribute to its fitness. Each colony has a single chromosome that codes for the behavior functions of all of its ants (all members of a colony are identical, although each ant receives different sensory input, so they behave differently). Fitness is based primarily on the number of pieces of food carried into the nest, so better foraging means a higher score and greater reproductive success, causing selection pressure for better central-place foraging strategies. The initial population consists of randomly generated chromosomes.

AntFarm evolves a population of 16,384 colonies, with 128 ants per colony, for a total of more than two million ants. Each colony lives in its own separate 16×16 grid environment, where each location contains some number of ants along with information about the presence or absence of a nest, the amount of food, and the amount of the pheromone ("odor") at that location (Figure 2).

Color Plates

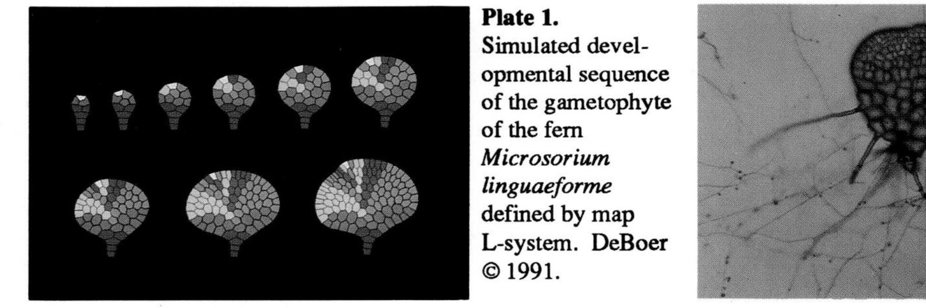

Plate 1. Simulated developmental sequence of the gametophyte of the fern *Microsorium linguaeforme* defined by map L-system. DeBoer © 1991.

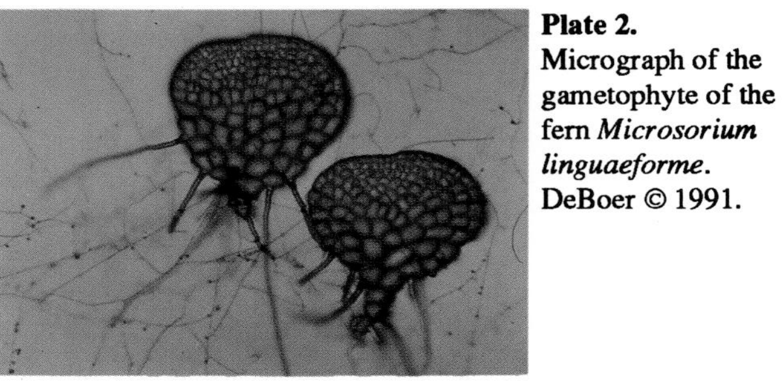

Plate 2. Micrograph of the gametophyte of the fern *Microsorium linguaeforme*. DeBoer © 1991.

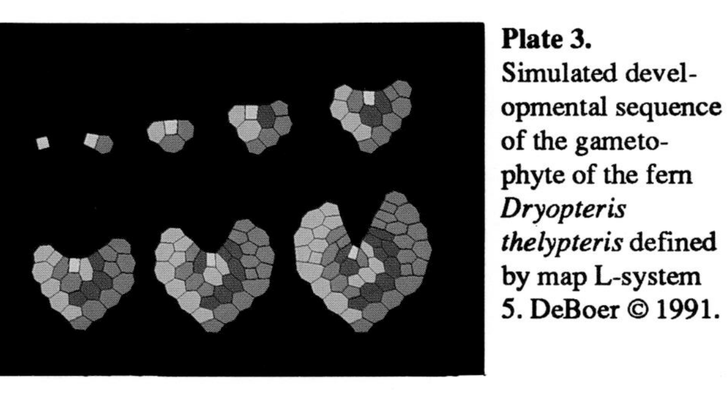

Plate 3. Simulated developmental sequence of the gametophyte of the fern *Dryopteris thelypteris* defined by map L-system 5. DeBoer © 1991.

Plate 4. Developmental sequence of the limpet *Pattela vulgata* defined by map L-system 6 (equatorial view). DeBoer © 1991.

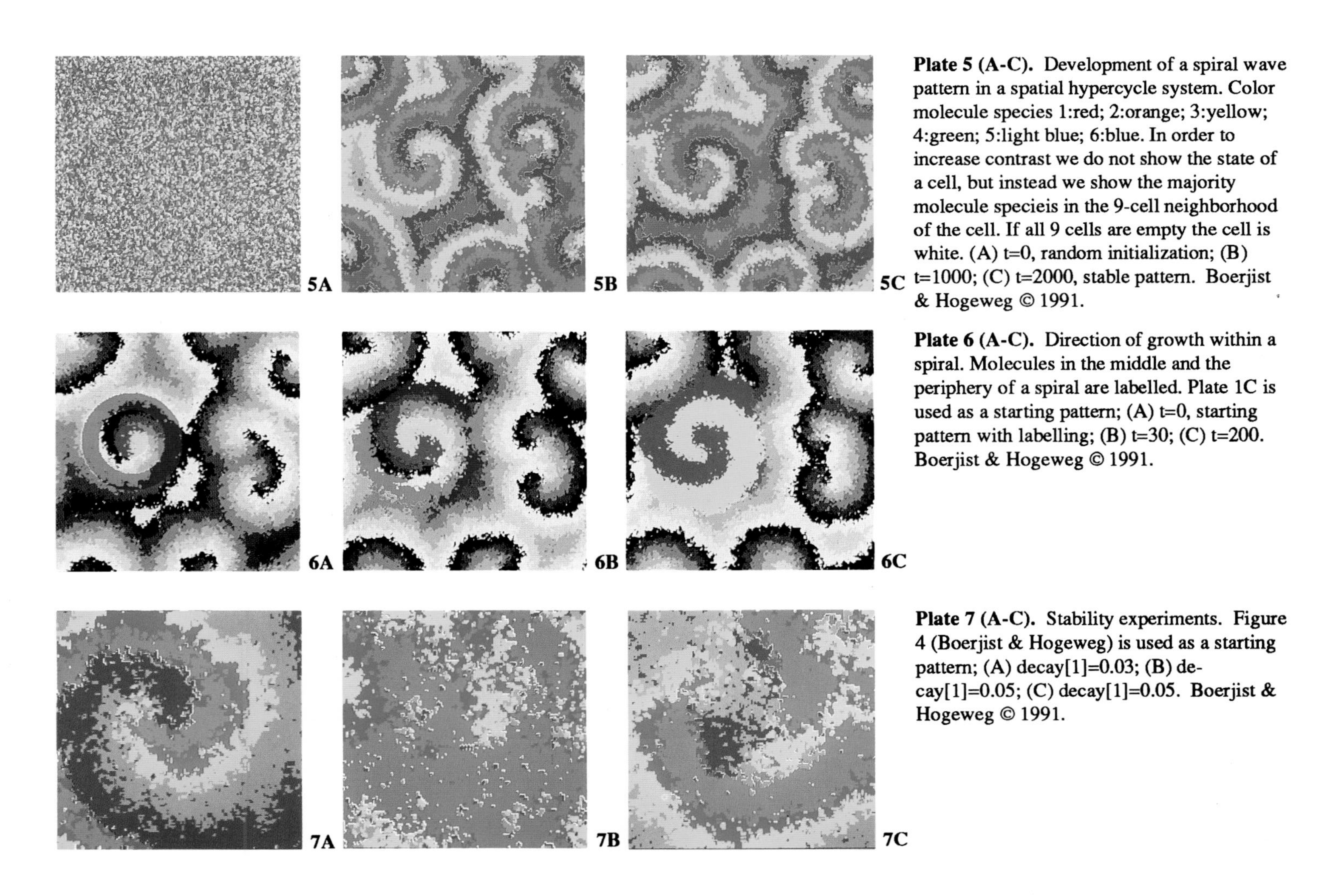

Plate 5 (A-C). Development of a spiral wave pattern in a spatial hypercycle system. Color molecule species 1:red; 2:orange; 3:yellow; 4:green; 5:light blue; 6:blue. In order to increase contrast we do not show the state of a cell, but instead we show the majority molecule specieis in the 9-cell neighborhood of the cell. If all 9 cells are empty the cell is white. (A) t=0, random initialization; (B) t=1000; (C) t=2000, stable pattern. Boerjist & Hogeweg © 1991.

Plate 6 (A-C). Direction of growth within a spiral. Molecules in the middle and the periphery of a spiral are labelled. Plate 1C is used as a starting pattern; (A) t=0, starting pattern with labelling; (B) t=30; (C) t=200. Boerjist & Hogeweg © 1991.

Plate 7 (A-C). Stability experiments. Figure 4 (Boerjist & Hogeweg) is used as a starting pattern; (A) decay[1]=0.03; (B) decay[1]=0.05; (C) decay[1]=0.05. Boerjist & Hogeweg © 1991.

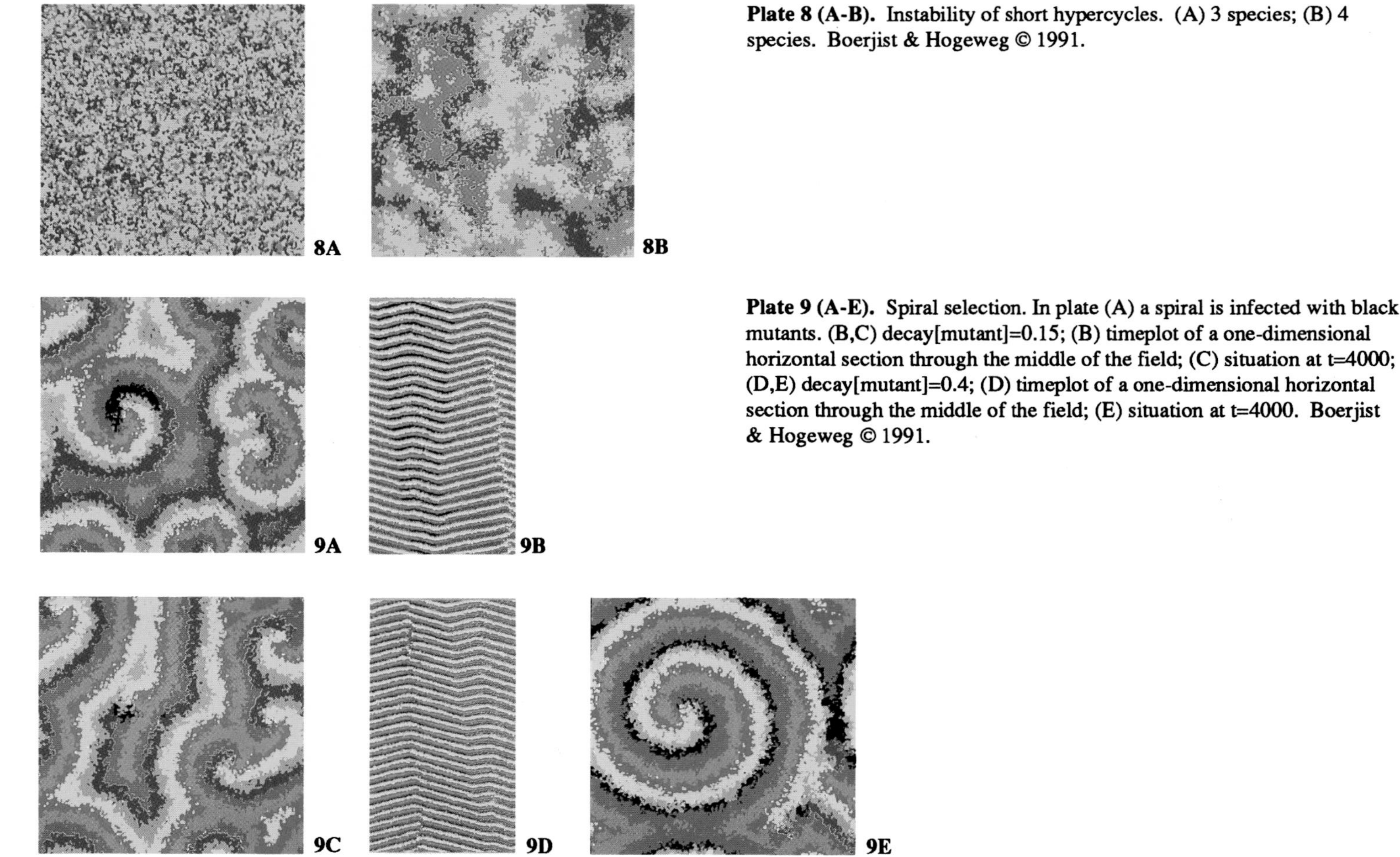

Plate 8 (A-B). Instability of short hypercycles. (A) 3 species; (B) 4 species. Boerjist & Hogeweg © 1991.

Plate 9 (A-E). Spiral selection. In plate (A) a spiral is infected with black mutants. (B,C) decay[mutant]=0.15; (B) timeplot of a one-dimensional horizontal section through the middle of the field; (C) situation at t=4000; (D,E) decay[mutant]=0.4; (D) timeplot of a one-dimensional horizontal section through the middle of the field; (E) situation at t=4000. Boerjist & Hogeweg © 1991.

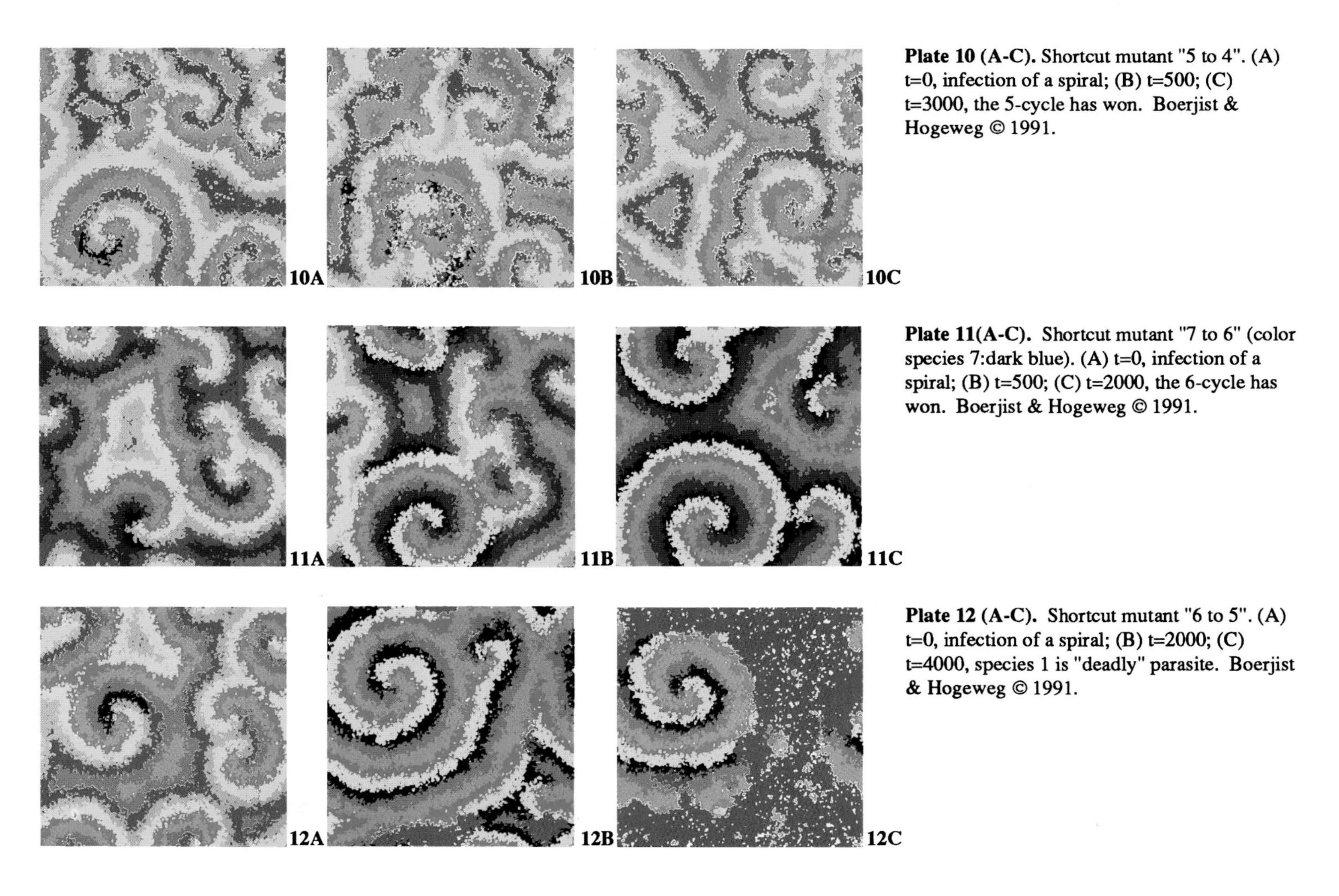

Plate 10 (A-C). Shortcut mutant "5 to 4". (A) t=0, infection of a spiral; (B) t=500; (C) t=3000, the 5-cycle has won. Boerjist & Hogeweg © 1991.

Plate 11(A-C). Shortcut mutant "7 to 6" (color species 7:dark blue). (A) t=0, infection of a spiral; (B) t=500; (C) t=2000, the 6-cycle has won. Boerjist & Hogeweg © 1991.

Plate 12 (A-C). Shortcut mutant "6 to 5". (A) t=0, infection of a spiral; (B) t=2000; (C) t=4000, species 1 is "deadly" parasite. Boerjist & Hogeweg © 1991.

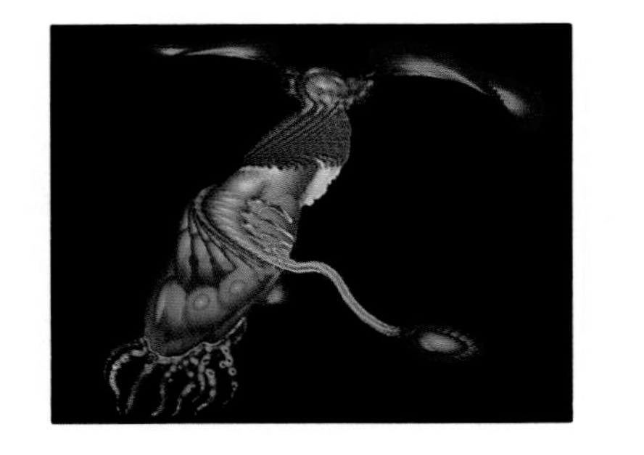

Plate 13. Bec © 1991.

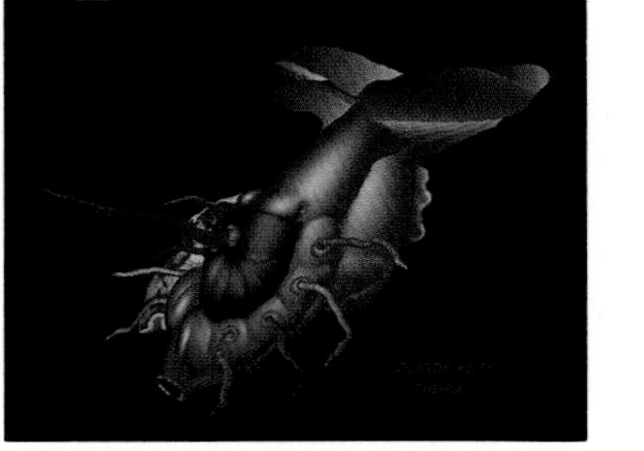

Plate 14. Loligocyste troupa. Bec © 1991.

Plate 15. Ikhnoserpthe. Bec © 1991.

Plate 16. Trupaopalakoule. Bec © 1991.

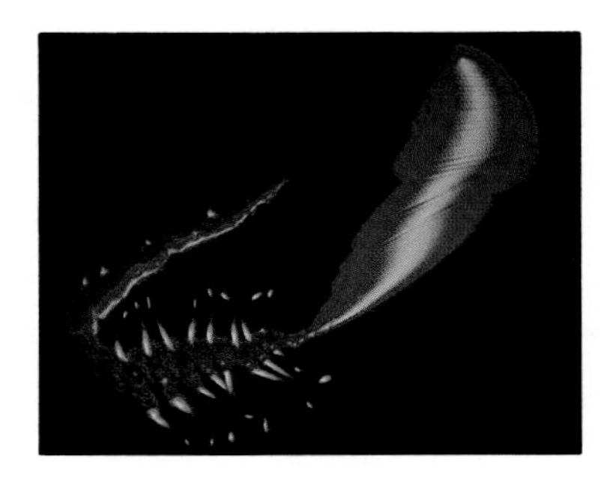

Plate 17. Bec © 1991.

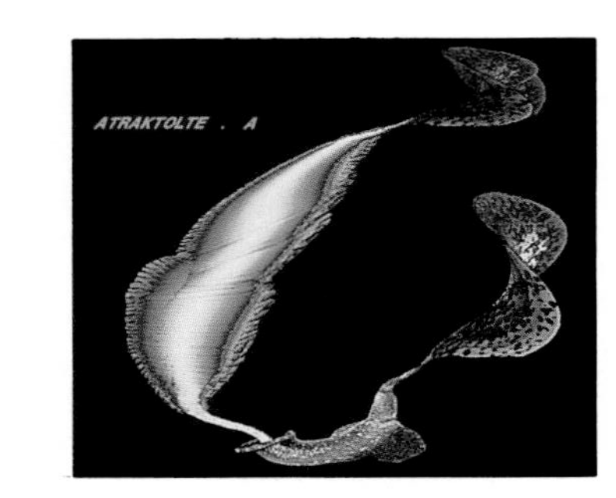

Plate 18. Atraktolte.A. Bec © 1991.

Plate 19. Bec © 1991.

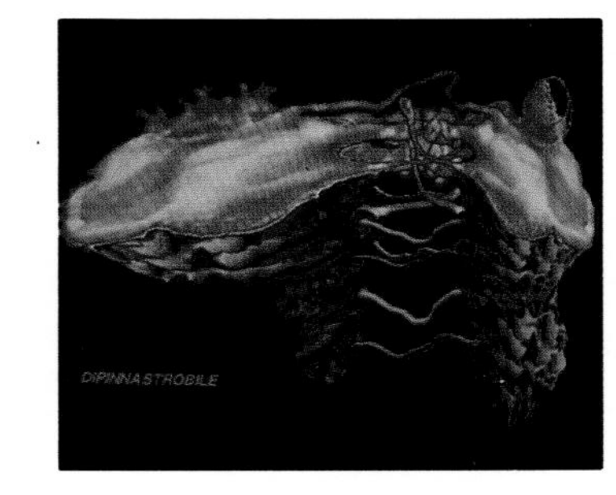

Plate 20. Dipinnastrobilea. Bec © 1991.

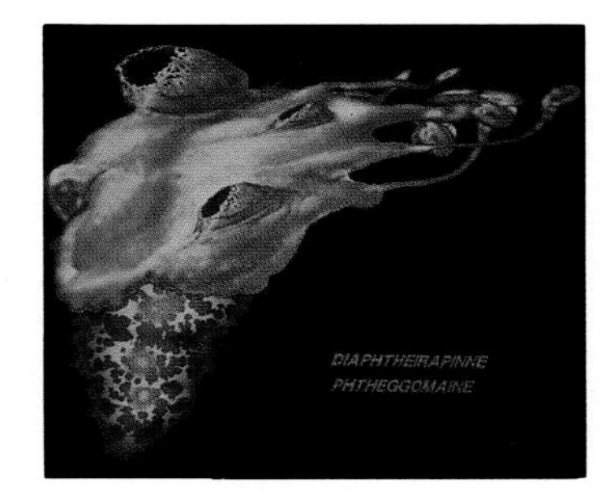

Plate 21. Diaphtherapinne phtheggomaine. Bec © 1991.

Plate 22. Bec © 1991.

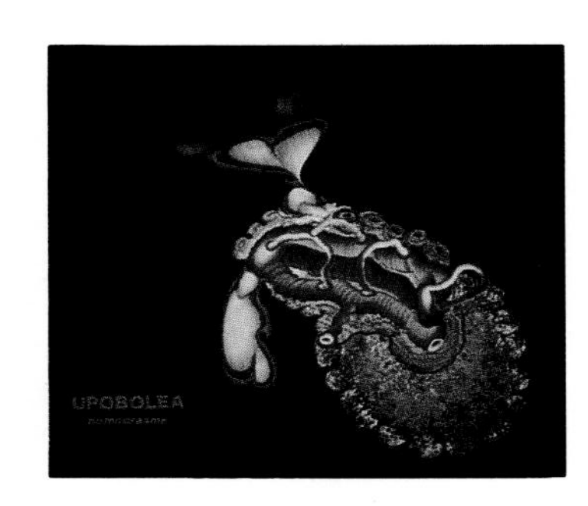

Plate 23. Upobolea nomodrasme. Bec © 1991.

Plate 24. Kremastiktore. Bec © 1991.

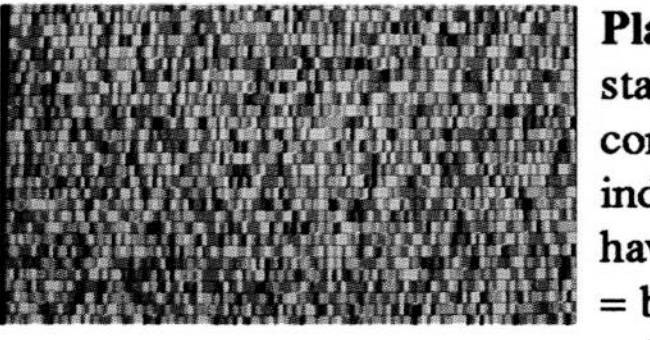

Plate 25. (opcode, pointer)-projection of state space in Venus II; initial correlated core at time 1. The white underscores indicate pointers, and different opcodes have different colors. Color code: DAT = black, MOV = green, ADD = SUB = yellow, JMP = JMZ = JMN = red, DJN = magenta, CMP = blue, and SPL = cyan. The correlations are obtained using the Markov matrices given in Section 4.2 of Rasmussen (this volume). Initially this core is populated with approximately 220 pointers. Note the correlated pair of jump-MOV instructions. Rasmussen © 1991.

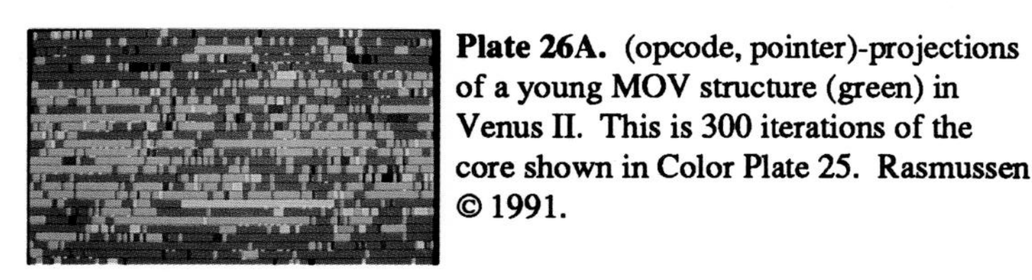

Plate 26A. (opcode, pointer)-projections of a young MOV structure (green) in Venus II. This is 300 iterations of the core shown in Color Plate 25. Rasmussen © 1991.

Plate 26B Part of an (opcode, pointer)-projection of the MOV structure. Address 1000 through 1127 is shown at 140 consecutive iterations (time 141 through 280). Time evolves downwards. Note how the MOV structure emerges around time 250. Rasmussen © 1991.

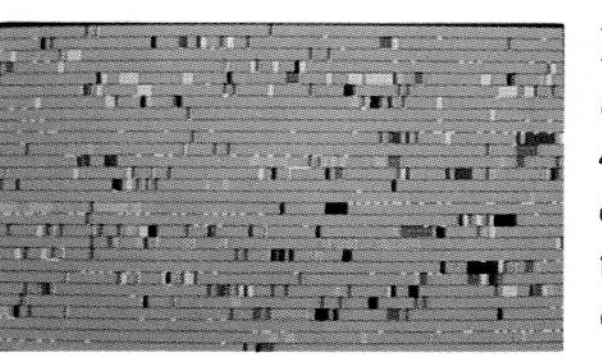

Plate 27. (opcode, pointer)-projection of an SPL-JMP structure (red and cyan) after 40,000 iterations of Venus II. Note the different areas of SPL and JMZ instructions boosted with pointers. Rasmussen © 1991.

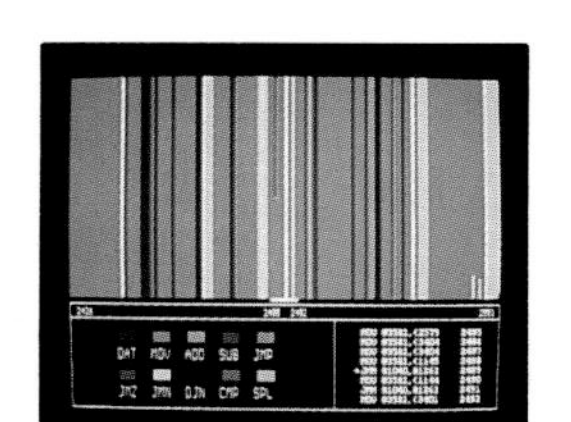

Plate 28. (opcode, pointer)-projection of part of the state space in Venus 1 at time 5500. This is an example of a (noisy) fixed point for the mapping **T**. Most pointers are presently trapped on JMN (·,#·) instructions. Also note the many MOV instructions. The white underscores indicate the pointers, and the different colors indicate differents opcodes. Color code on the figure. Rasmussen © 1991.

Plate 29. (opcode, pointer)-projection of the fully developed MOV-SPL structure after 5000 iterations with Venus I. This is an example of a more complex spatio-temporal structure, which is persistent to computational noise. The white underscores indicate the pointers and the different colors indicate different opcodes. Color code on the figure. Rasmussen © 1991.

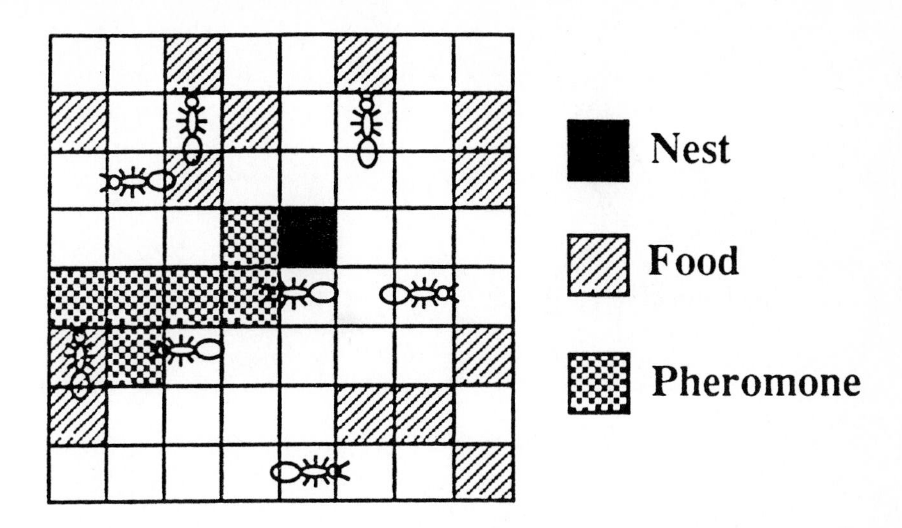

FIGURE 2 The AntFarm environment contains a nest, food, pheromone, and ants. At the beginning of each generation, all the ants are in the nest, food is distributed in the environment, and no pheromones are present. The actual environment is 16×16.

Any pheromones that are dropped by the ants slowly diffuse and eventually disappear. The nest of each colony is located at the center of its environment, and the colony's genetic information is represented by a 25,590-bit haploid chromosome.

Each generation begins with each ant in its nest and its memory initialized to zero. All ants live throughout the entire generation. A score is calculated for each colony based primarily on the amount of food deposited in the colony's nest in 100 time steps, although the "metabolic" costs of ant movement, pheromone production, etc., are also taken into account. Each unit of food is worth 1000 points, each unit of pheromone dropped by an ant costs 0.1, and each other action (move or pickup/drop food) costs 0.1. The inclusion of metabolism in the score results in selection pressure towards more streamlined foraging strategies. During reproduction, both crossovers and mutations occur at a rate of about 0.0001 per bit (about 2.6 mutations and crossovers per colony each generation), which has been shown empirically to be satisfactory.

At the beginning of each generation, the environment is reinitialized so that no pheromone is present and food is placed in a new configuration from a fixed probability distribution. The food pattern seen by each colony in a single generation is identical so that no colony has a chance advantage.

In each of the 100 time steps, the ant's sensory inputs (and 21 bits of internal memory) are processed by its *behavior function* (represented as an ANN), producing a set of actions to perform (Figure 3).

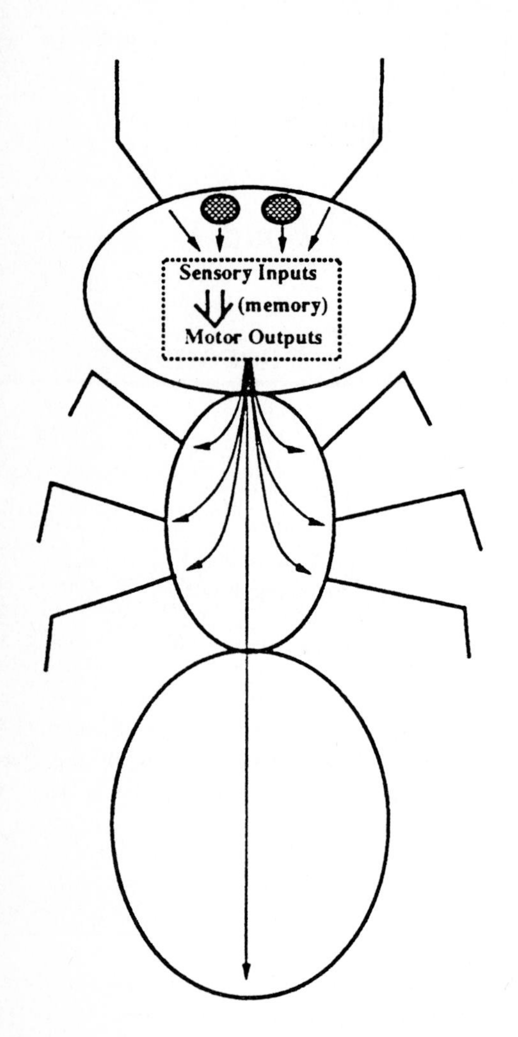

FIGURE 3 The internals of an AntFarm ant. The box represents the behavior function which is computed by an ANN. The behavior function receives sensory inputs and its internal state (memory), and produces motor outputs. Among other things, the motor outputs control locomotion and the production of the pheromone (from the tip of the abdomen, although in AntFarm we do not simulate the morphology of the ant).

An ant has a 3 x 3 sensory array centered on its current location that can sense

1. the presence of food,
2. the presence of a nest, and
3. the amount of pheromone.

In addition, each ant can sense

1. whether or not it is carrying food,
2. the correct direction to its nest (a *compass* sensor), and
3. 4 bits of random input.

In any time step, an ant can decided to do any or all of the following:

1. move to any of the eight neighboring locations,
2. pick up a unit of food (although it can carry a maximum of one unit of food),
3. drop a unit of food, and
4. drop from 0 to 64 units of the pheromone.

We chose not to try to evolve both foraging search strategies and strategies for navigating back to the nest. Real ants typically use elaborate techniques for navigation,[12] often involving memorizing landmarks, calculating average angle of the sun during foraging, etc. We provide the ants with a special sense organ (the "compass") that performs the task of navigation, although the ants still must evolve behavior to interpret and use the compass correctly.

The AntFarm simulation is implemented on a Connection Machine,[10] a massively parallel supercomputer, consisting of up to 65,536 one-bit processing elements. AntFarm is written in C++[18] and uses the CM++[1] interface to the Connection Machine.

3.1. COMPARISON OF AntFarm TO Genesys/Tracker

AntFarm is a direct descendent of the Tracker task studied on the Genesys system.[13] Genesys/Tracker is also a massively parallel, microanalytic-evolutionary simulation, evolving simple organisms that can follow a noisy, broken trail. The behavior of the Genesys organisms is produced by either an FSA or a three-layer, fully connected, recurrent ANN.

The main differences between Genesys/Tracker and AntFarm are a result of the biologically motivated task (central-place foraging) of AntFarm. Since AntFarm is trying to model natural evolution, it is implemented with a more realistic genetic algorithm (local competition and mating, rather than global competition and random mating). In addition, the simulated organisms are more complex in many dimensions (summarized in Table 1).

3.2. REPRESENTATION OF THE BEHAVIOR FUNCTION

The ANN organism representation that was used in Genesys[13] encodes the network as the concatenation of the binary-integer weight (connection strength) values. The strength of each connection is under genetic control, but not the connectivity pattern itself. The connectivity of the network is statically defined, and the weight values are placed in the bit string chromosome in a canonical order.

For reasons described in Section 4, we have departed from the ANN encoding used in Genesys, and we have designed a new way to encode an ANN that places the connectivity pattern of the network under genetic control.[3] Our new encoding consists of K *connection descriptors*; each consists of three parts: the indices of the units that are to be connected (*From* unit, *To* unit) and the weight (strength) of the connection (Figure 4). Certain units are designated as inputs and outputs, and the

TABLE 1 A comparison of AntFarm to Genesys/Tracker. The AntFarm simulation is larger and more complex in many dimensions.

Dimension	AntFarm	Genesys/Tracker
Population	16,384 colonies 2,097,152 ants	65,536 ants
Info/Environment Location	32 bits	1 bit
Level of Selection	Colony	Individual
Sensory Input/Time Step	$\sim$ 200 bits	1 bit
Effector Outputs/Time Step	13 bits	2 bits
Internal Memory (max)	21 bits	5 bits
Genome Size	25,590 bits	450 bits

From	To	Weight
5	1	1
3	4	1
0	6	2
7	4	-1
0	4	2
4	3	2
3	6	-1
0	4	1

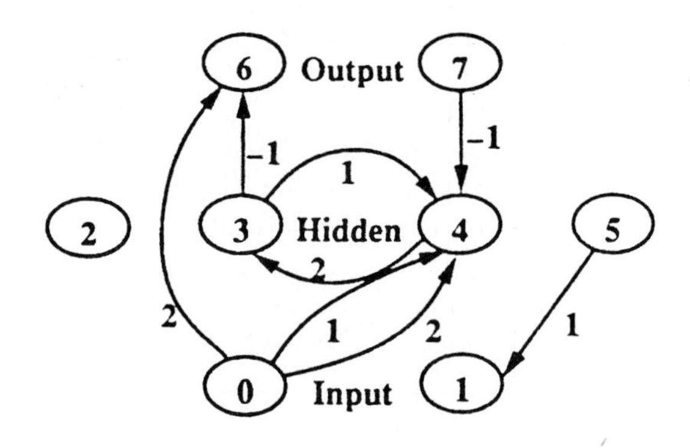

Genotype

101001001011100011000110010111100111000100010100011010011110111000100001

FIGURE 4 The connection descriptors (left), the network (right) and the genotype (bottom) of an ANN encoded with connection descriptors. Each descriptor specifies the pair of units that it connects (*From* and *To* columns), and the strength (*Weight*) of the connection (each of these three fields is 3 bits wide in this example). Note that some units have no connections associated with them, some have no out-going connections, some pairs of units are connected by multiple connections, and recurrent connections are allowed.

rest are hidden units, which can serve as memory for the organism. The genotype is the concatenation of the bit representation of the K connection descriptors (as to complement binary integers).

To convert a set of inputs to a set of outputs (behavior), we transmit one signal across each connection in the network. This consists of adding the product of the *From* unit activation and the weight to the *To* unit accumulator. After all K signals have been transmitted, each accumulator is converted to a Boolean value (positive sums to 1; negative or zero sums to 0) and assigned to the corresponding activation. The output unit activations specify the chosen behavior, and the hidden unit activations describe the memory state of the organism.

All possible connection descriptors are legal, including recurrent connections and multiple connections between pairs of units. Connections leading *From* an output unit or *To* an input unit have no effect on the output of the ANN.

The use of connection descriptors gives this encoding some interesting properties. A mutation might change the value of a particular connection weight, or it can move a connection within the network. A crossover can result in the appearance of a connection that neither parent possesses.

The most important property introduced by this new ANN encoding is the unconstrained and heritable connectivity pattern in the ANN. This freedom is achieved by placing the location and strength of connections under the control of evolution. Another potentially important property of this representation is the position independence of the connection descriptors, which means that a connection descriptor has the same effect no matter where it lies on the chromosome. This allows linkage patterns between functionally related units to evolve. Organisms built with this type of network are competitive with human-designed neural architectures that possess many more connections (see Section 4). Our current encoding is limited in that the number of neurons and connections are not under genetic control.

Here are the details of the AntFarm ANN behavior function. These are features that are available, but particular organisms may "use" (have connected) many fewer:

1. Input Units
   a. 9 units for pheromone density
   b. 9 binary units for presence of food
   c. 9 binary units for presence of a nest
   d. 4 binary units for compass (an optimal path to the nest)
   e. 4 binary units for random noise
   f. 1 binary unit for whether or not it is carrying food
2. Hidden Units
   a. 21 binary units for memory

3. Output Units

   a. 4 binary units for direction to move
   b. 1 binary unit to pick up food
   c. 1 binary unit to drop food
   d. 1 unit to indicate number of units of pheromone to drop

The whole neural network consists of 64 neural units and 1709 connections. The connection weights are encoded in 3 bits and the *From* and *To* each in 6 bits, so the network is specified by 25,590 bits of genome.

4. REPRESENTATIONAL ISSUES

One of the most difficult problems we have encountered thus far has been the search for an appropriate artificial organism representation. Although many organism-based evolutionary simulations have been run, most of the problems and models have been very simple. We encountered serious problems when we attempted to scale the representations to the complexity of the AntFarm organisms.

The representation of an artificial organism in a microanalytic simulation consists of the following parts:

1. genotype: a bit string that encodes the behavior function;
2. development function: the mapping that decodes the genotype to produce the behavior function;
3. behavior function: the program that maps sensory inputs and the memory state into a new memory state and effector outputs; and
4. interpreter: used to execute organism behavior functions.

In AntFarm, the development function and interpreter are fixed for all organisms and for all time; they are not subject to evolution. The genotype, of course, differs from animal to animal, but is static throughout the life of the organism. The behavior function also does not change during the life of an ant; there is no "learning" protocol: the weights and connectivity are static. Complex, history-sensitive behavior can be realized through the use of the 21 bits of internal memory (over 2 million possible memory states), especially in conjunction with feed back connections in the ANN.

We have surveyed a variety of animal representations that have been used in simple evolutionary simulations (e.g., parameterized functions,[22] Lisp S-expressions,[16] finite-state automata,[13] rule systems (e.g., classifier systems),[9] etc.). Unfortunately, none of these representations is appropriate for simulated evolution with the environment/organism complexity of AntFarm.[2] Each of them either scales exponentially in size with the number sensors/effectors (and thus require too much computer memory for use with AntFarm), or inherently requires too much knowledge specific to the artificial world/task to be built in. The inclusion of task-specific information in the organism representation opens the door for systematic biases in our evolutionary experiments, so these representation schemes must be avoided.

The most promising representation that we have examined is based on the ANN programming paradigm: ANNs grow slowly as the number of inputs and outputs increase, their internal computations are simple and fast, they are easily encoded in a bit string, and mutations and crossovers in this bit string representation usually cause little or no change in the function that is computed.

4.1. ARTIFICIAL NEURAL NETWORKS

We began our work with AntFarm using an ANN organism representation with fully connected layers and recurrent connections, like we used in Genesys.[13] With this representation, we were unable to get even simple non-cooperative foraging to evolve. We then tried multi-layer feed-forward ANN networks, and again we failed to evolve foraging behavior.

Our next step was to construct an ANN encoding with as much knowledge of the foraging problem as necessary to get the evolution of foraging behavior. Our aim was to understand why the ANN representations we had used successfully in simpler problems failed in AntFarm.

The foraging task is made up of two separate sub-tasks: searching for food, and returning the food to the nest. The ant can determine which sub-task to perform based on the "carrying food" binary input. Each of these two sub-tasks are separately rather simple. While searching for food, an ant should pay attention to the food sensors, maybe the pheromone sensors (if cooperative foraging is used), maybe the compass and nest sensors (it might want to move away from the nest area), and maybe the random sensors (so a pseudo-random search can be used). While transporting food to the nest, the most important sensor is the compass input; all others can be ignored.

To apply this knowledge of the dual nature of the foraging task, we constructed an ANN behavior function that consists of two fully connected, recurrent networks. One of these networks is invoked when the ant is not carrying food (search), and the other is invoked when the ant is carrying food (transport). We found that ants with behavior functions based on this dual-ANN encoding quickly and consistently evolve (non-cooperative) foraging behavior. This suggests that the problem with the other ANN encodings was that they have difficulty evolving discrete behavior (where a small change in the inputs leads to a large change in behavior). These representations "generalize," so small changes in the inputs are smoothed away, making the evolution of discrete behavior unlikely.

Although we were able to evolve foraging behavior, we still had a serious problem: the dual-ANN representation requires a huge amount of task-specific information. This could bias the evolutionary outcomes of our experiments in subtle (or obvious) ways, which is unacceptable.

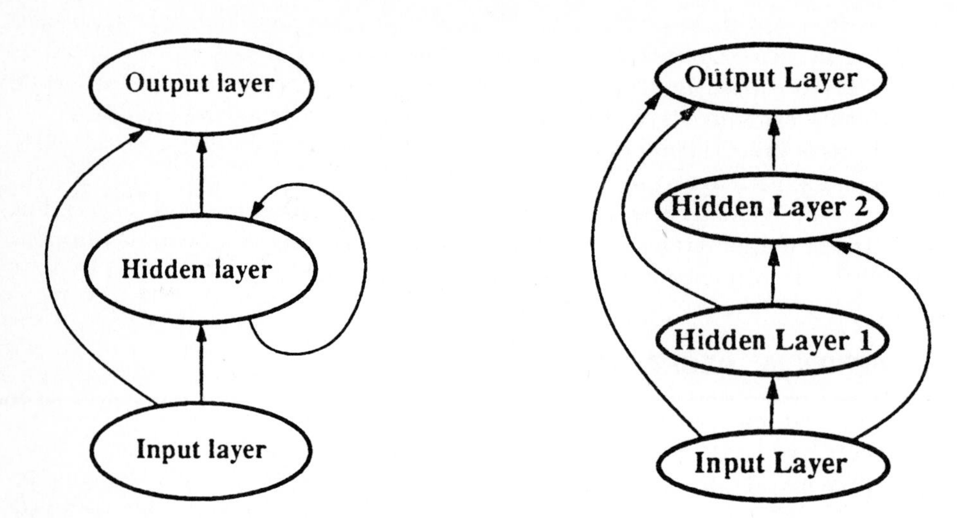

FIGURE 5 The architecture of the networks in ANN2, ANN3, and ANN4. Each arc indicates that the layers are fully connected. ANN2 and ANN4 (left) have a fully recurrent hidden layer. In ANN3, each layer is fully connected to all "forward" layers. ANN3 has five hidden layers (although only two are shown here).

To avoid this problem, we have designed an encoding scheme based on *connection descriptors* (described in Section 4), which we have adopted for use in AntFarm. This decision is based on the fact that the connection descriptor encoding does not allow or require knowledge of the task, and an empirical study (presented below) that shows it is able to evolve foraging behaviors that are as successful as that produced by the human—designed dual—network behavior function.

We have empirically compared four ANN-based behavior functions in AntFarm: a network specified by connection descriptors (ANN1), a three-layer recurrent network (ANN2), a feed-forward network (ANN3), and a behavior function made up of two recurrent networks (ANN4). ANN4 invokes one of the networks when the ant is carrying food and the other when it is not. Comparison with ANN4 allows us to see how well the other representations are able to evolve behavior for two different tasks in one network. In all four behavior functions, all weights are encoded as 3-bit signed integers and all initial activations of the hidden units are initialized to 0 at the beginning of the generation. The connectivity of the ANN2, ANN3, and ANN4 networks is shown in Figure 5. Table 2 summarizes the main parameters of the ANNs. In an attempt to make the comparison fair, we made the different networks approximately the same size (although ANN1 requires more bits to encode it, but it also has far few connections).

TABLE 2 A summary of the ANN behavior functions, including the number of connections, arrangement of hidden units into layers (layers x units/layer), the number of bits of memory, and the number of bits in the chromosome.

ANN	Number of Connections	Hidden Layers x Units/Layer	Bits of Memory	Chromosome Length (bits)
ANN1	682	varies	21	10240
ANN2	2652	1x32	32	7956
ANN3	2612	5x8	0	7836
ANN4	2×1325=2650	2x18	18	7950

TABLE 3 The foraging task is treated as two separate tasks: search for food and transport of the food back to the nest.[1]

BM	Search/Transport	Mean	Max
1	random/random	1.07	6
2	random/compass	15.07	21
3	random+food/compass	20.82	25

[1] Random indicates that only the random inputs are used, compass indicates that the inputs pointing the way to the nest are used, and food indicates that the food sensors are used. Mean and Max refer to the amount of food recovered in the population of 16,384 colonies.

To perform this study, we set the AntFarm parameters as follows. The population consists of 16,384 colonies, each of which contains four identical ants. Each colony forages in its own 16×16 environment. The initial food distribution for each colony in each generation is always the same: one unit of food in each location, except for locations on a straight (horizontal, vertical, or diagonal) line with the nest (for a total of 196 units of food). We chose this food distribution because foraging that only requires walking in a cardinal direction from the nest would involve only a few neurons. Each run is 500 generations long, each lasting 50 time steps. Both the mutation and recombination rates are set at 0.0001 per bit.

How can we tell how well a population is foraging? It is clearly impossible for four ants to carry all 196 units of food in the environment to the nest in only 50 time steps, so how much food can we expect them to recover? We have hand-coded three simple behavior functions (in C++) to serve as foraging benchmarks (Table 3). BM1 forages (both the search phase and recovery phase) using only a random walk. BM2 searches for food with a random walk (ignoring the food sensors) and carries food to the nest by following the compass (the input that always points to the nest). BM3 improves on BM2 by using the the food sensors while searching. These benchmarks provide an absolute measure of foraging efficiency.

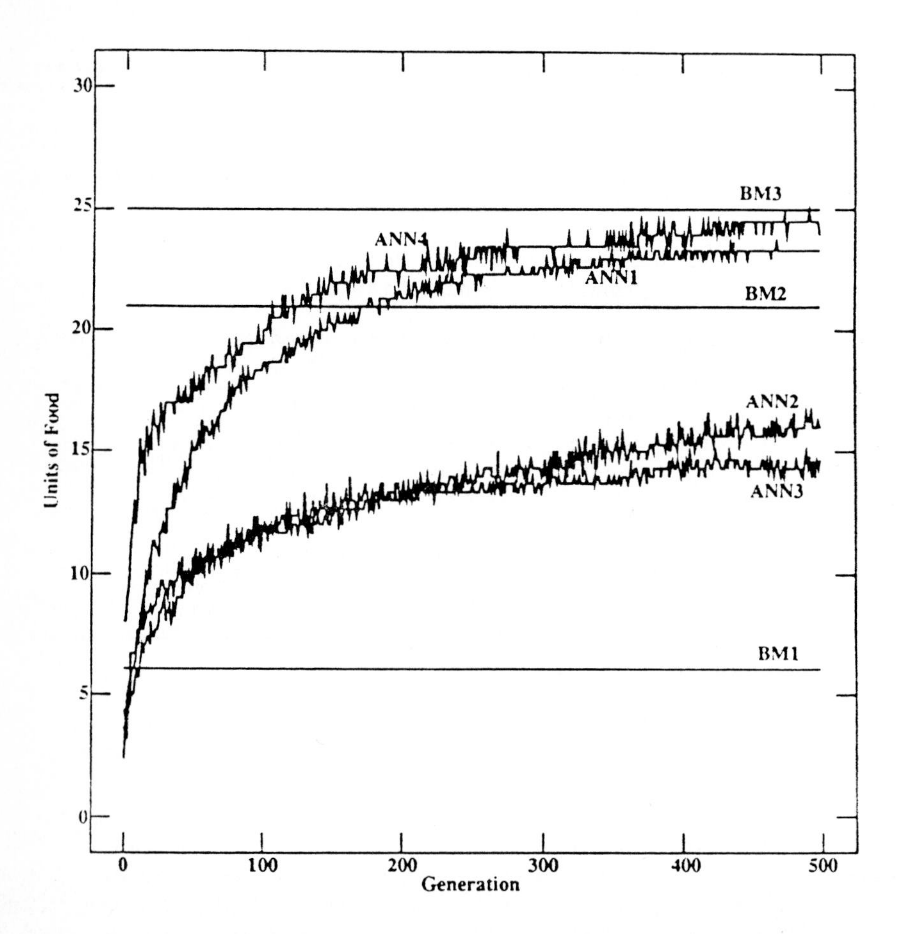

FIGURE 6 The maximum units of food brought back to the nest in a population of 16,384 colonies across 500 generations. Each curve is the average of three runs, differing only in the initial random seed. These simulations required eight days of Connection Machine computation (on 8K processors).

The results are summarized in Figure 6. The dual-network representation (ANN4) out-performs all of the other representations, foraging nearly as effectively as an algorithm that uses the food and compass sensors perfectly. It is not surprising that ANN4 performs the best, since it has a great deal of information about the task built into the representation. The ANN representation based on connection descriptors (ANN1, a scaled down version of the encoding used in AntFarm) is nearly as successful, even though it has been provided with no task-specific information. The other two representations with no built-in knowledge perform rather badly—they are not good at changing their behavior based on whether or not they are carrying food, although the recurrent network (ANN2) does better than the feed forward network (ANN3).

4.2 PROPERTIES OF REPRESENTATIONS

From our exploration of potential artificial organism representations, we have abstracted a number of properties that we believe are necessary for simulated evolution experiments:

1. (approximate) *closure* of the set of legal genotypes under the action genetic operators;
2. *smoothness* of the phenotype under the action genetic operators, i.e., the behavior function should tend to change smoothly as the genotype is changed by mutation and recombination;
3. the ability to *scale* to large behavior functions, i.e., those that can handle large amounts of input and output data without undue increase in genome size;
4. the ability to *evolve* phenotypes that exhibit *both continuous and discrete behaviors* as a function of their inputs; and
5. a *uniform computational model*, i.e., the programming paradigm in which the behavior functions are expressed should not contain features that include any kind of explicit or implicit knowledge of the environment, nor bias toward a particular evolutionary trajectory.

Closure (1) and smoothness (2) are properties of the development function and the genetic operators. Scaling (3), the ability to evolve continuous and discrete behavior (4), and uniformity (5) are all properties of the computational model of the behavior function.

The property of syntactic closure constrains the development function (the encoding of the behavior function into the genotype). To be syntactically closed (approximately), the genetic operators must always (or almost always) produce genomes that translate into syntactically legal behavior functions. An evolutionary system will not work if a single mutation or recombination is likely to transform a genotype that encodes a perfectly good behavior function program into a program that is syntactically illegal.

The smoothness property requires that most changes to the genotype due to the application of genetic operators result in small changes in the behavior function.

For example, a mutation in an ant should usually have a small affect on its foraging algorithm. Of course, it need not *always* cause a small change; some small changes will be fatal, and a few small changes may cause profound but beneficial effects. Still, evolution cannot work if the phenotype space is not relatively smooth as a function of genotype. The encoding and mapping functions should be smooth not only under mutation, but also under recombination, implying that functionally related genes should usually be inherited as a unit (strongly linked). The smoothness property has the effect of requiring the "adaptive landscape"[26,15] to be correlated with respect to the genetic operators. Evolution can successfully searches the space of possible organisms only in correlated adaptive landscapes.

Smoothness is an extension of the closure property. Not only is it required that a legal program be (usually) transformed into another legal program by the genetic operators, but also that it is (usually) transformed into one that is semantically similar to the original.

The scaling properties of a representation are of extreme practical importance because we must be able to store large populations in a reasonable amount of computer memory. Scaling refers to the rate at which the size of the representation grows as a function of the number of inputs, outputs, and bits of internal memory. The size of the representation includes the number of bits in the genotype, the number of bits required to store the decoded behavior function, the amount of time to translate from the genotype to the behavior function, and the amount of time to run a set of inputs through the behavior function to produce the outputs. We are interested in organisms with dozens or hundreds of inputs, outputs, and bits of internal memory.

In AntFarm, foraging requires a combination of both continuous and discrete behaviors (which is probably necessary in any simulation that hopes to evolve complex behavior). Roughly speaking, a behavior function is producing continuous behavior when a small change in the inputs to the function results in a small change in the outputs, and discrete behavior when a small change in the inputs results in a large change in the outputs. In AntFarm, the foraging behaviors that we have evolved consist of two modes: search and transport. Within each mode, the behavior appears to be continuous, but the transition from one mode to the other is determined by the "carrying food" input bit (a very discrete change in behavior). Most or perhaps all complex behaviors will involve the combination of different modes of behavior, and thus require the evolution of discrete behavior.

The final property is that the computational model of the representation should be uniform. In particular, it must be able to describe all desired behavior functions *without designing features of the problem or possible solutions into the representation.* If we are trying to shed light on a biologically motivated hypothesis, the results will be invalid if we bias the organisms toward (or away from) some evolutionary path. The computational model must not have knowledge of the problem or possible solutions embedded in it.

5. FUTURE WORK

The AntFarm project is work in progress. We have designed a genetic algorithm that closely models real evolution, designed an appropriate behavior function, found empirically good values for various parameters (crossover and mutation rates, etc.), and evolved colonies that successfully forage for food.

So far, we have not observed the evolution of cooperative foraging. In the AntFarm model, the use of pheromone trails to lead other workers to a large pile of food is quite complex. Because pheromone trails are nondirectional, following a pheromone trail involves the combined information of the pheromone and compass inputs. Simply walking uphill in the pheromone density does not work, because the pheromone trail will be faintest nearest to the food, due to diffusion. A reasonable trail-following strategy is to move to the location that is furthest from the nest and contains some of the pheromone. This will keep the ant on the trail and moving toward the food source.

Although the model of Johnson et al. indicates under what circumstances cooperation is the optimal strategy, we do not know under what circumstances cooperative strategies will *evolve*. It may be that cooperative foraging is unlikely to evolve in any static environment. We may have to vary the environment, slowly making foraging more difficult.

We plan to perform a systematic study of the effect of food distribution on the evolution of foraging strategies, testing the model of Johnson et al.[14] It will be interesting to see how our artificial evolution differs from biological theory. We are interested in exactly how and when the pheromones are used to communicate information about the food distribution.

We are also interested in the evolution of foraging strategies that are strongly affected by competition. We might find strategies that utilize exploitation or direct interference. A possible strategy for better exploitation of resources under stiff competition would be to forage further from the nest first, beating the neighboring colonies to that food, resulting in a larger foraging area for the colony. Interference strategies might involve disrupting communication of neighboring colonies, either by overwriting existing pheromone trails, or by laying misleading trails. In order to investigate this area, we will run AntFarm in a mode where a single large environment is shared by all colonies during foraging.

ACKNOWLEDGMENTS

We owe thanks to many people who have contributed to ideas and comments to AntFarm, especially Doyne Farmer and Chris Langton. We also acknowledge the contributions of Michael Dyer, Don Feener, Danny Hillis, Andrew Kahng, Adam King, John Lighton, Joe Pemberton, Chuck Taylor, and Greg Werner. This work is supported in part by W. M. Keck Foundation grant number W880615, University of California Los Alamos National Laboratory award number CNLS/89-427, and University of California Los Alamos National Laboratory award number UC-90-4-A-88.

REFERENCES

1. Collins, Robert J. "CM++: A C++ Interface to the Connection Machine." In *Proceedings of the Symposium on Object Oriented Programming Emphasizing Practical Applications*, 1990.
2. Collins, Robert J., and David R. Jefferson. "Representations for Artificial Organisms." *Proceedings of Simulation of Adaptive Behavior*, edited by Jean-Arcady Meyer and Stewart Wilson. Cambridge: MIT Press/Bradford Books, 1991.
3. Collins, Robert J., and David R. Jefferson. "An Artificial Neural Representation for Artificial Organisms." *Proceedings of Parallel Problem Solving from Nature*, edited by Reinhard Männer and David E. Goldberg. New York: Springer–Verlag, 1991.
4. Coulson, Robert N., Joseph Folse, and Douglas K. Loh. "Artificial Intelligence and Natural Resource Management." *Science* **237** (1987): 262–267.
5. Crow, James F. *Basic Concepts in Population, Quantitative, and Evolutionary Genetics.* New York: W. H. Freeman, 1986.
6. Fewell, Jennifer H. "Energetic and Time Costs of Foraging in Harvester Ants, Pogonomyrmex Occidentalis." *Behav. Ecol. Sociobiol.* **22** (1988): 401–408.
7. Fry, John, Charles E. Taylor, and U. Devgan. "An Expert System for Mosquito Control in Orange County California." *Bull. Soc. Vec. Ecol.* **2(14)** (1989): 237–246.
8. Gilmour, J. S. L., and J. W. Gregor. "Demes: A Suggested New Terminology." *Nature* **333** (1939).
9. Goldberg, David E. *Genetic Algorithms in Search, Optimization and Machine Learning.* Reading, MA: Addison–Wesley, 1989.
10. Hillis, W. Daniel. *The Connection Machine.* Cambridge, MA: MIT Press, 1985.
11. Holland, John H. *Adaptation in Natural and Artificial Systems.* Ann Arbor: The University of Michigan Press, 1975.
12. Hölldobler, Bert, and Edward O. Wilson. *The Ants.* Cambridge: Harvard University Press, 1990.
13. Jefferson, David, Robert Collins, Claus Cooper, Michael Dyer, Margot Flowers, Richard Korf, Charles Taylor, and Alan Wang. "The Genesys System: Evolution as a Theme in Artificial Life." This volume.
14. Johnson, Leslie K., Stephen P. Hubbell, and Donald H. Feener, Jr. "Defense of Food Supply by Eusocial Colonies." *Amer. Zool.* **27** (1987): 347–358.

15. Kauffman, Stuart, and Simon Levin. "Towards a General Theory of Adaptive Walks on Rugged Landscapes." *J. Theor. Biol.* **128** (1987): 11–45.
16. Koza, John R. "Genetic Programming: A Paradigm for Genetically Breeding Populations of Computer Programs to Solve Problems." Department of Computer Science, Stanford University, 1990.
17. Lighton, John R. B. "Energetics of Foraging and Recruitment in the Giant Tropical Ant *Paraponera clavata (Hymenoptera: Formicidae)*." Unpublished manuscript.
18. Stroustrup, Bjarne *The C++ Programming Language*. Reading, MA: Addison–Wesley, 1986.
19. Sudd, John H., and Nigel R Franks. *The Behavioural Ecology of Ants.* New York: Chapman and Hall, 1987.
20. Swartzman, Gordon L., and Stephen P. Kaluzny. *Ecological Simulation Primer.* New York: Macmillan, 1987.
21. Taylor, Charles E. "Evolution of Resistance to Insecticides: The Role of Mathematical Models and Computer Simulations." In *Pest Resistance to Pesticides*, edited by George P. Georghiou and Tetsuo Saito. New York: Plenum, 1983.
22. Taylor, Charles E., David R. Jefferson, Scott R. Turner, and Seth R. Goldman. "RAM: Artificial Life for the Exploration of Complex Biological Systems." In *Artificial Life*, edited by C. Langton. Santa Fe Institute Studies in the Sciences of Complexity, Proc. Vol. VI, 275–295. Reading, MA: Addison-Wesley, 1989.
23. Taylor, Charles E., L. Muscatine, and David R. Jefferson. "Maintenance and Breakdown of the *Hydra–Chlorella* Symbiosis: A Computer Model." *Proceedings of the Royal Society of London* **238** (1989): 277–289.
24. Werner, Gregory M., and Michael G. Dyer. "Evolution of Communication in Artificial Organisms." This Volume
25. Wright, Sewall "Evolution in Mendelian Populations." *Genetics* **16** (1931): 97–159.
26. Wright, Sewall. "The Roles of Mutation, Inbreeding, Crossbreeding and Selection in Evolution." *Proceedings of the Sixth International Congress of Genetics* **1** (1932): 356–366.
27. Wright, Sewall. *Evolution and the Genetics of Populations. Volume 1: Genetic and Biometric Foundations.* Chicago: University of Chicago Press, 1968.
28. Wright, Sewall. *Evolution and the Genetics of Populations. Volume 2: The Theory of Gene Frequencies.* Chicago: University of Chicago Press,1969.
29. Wright, Sewall. *Evolution and the Genetics of Populations. Volume 3: Experimental Results and Evolutionary Deductions.* Chicago: University of Chicago Press, 1977.
30. Wright, Sewall. *Evolution and the Genetics of Populations. Volume 4: Variability Within and Among Natural Populations.* Chicago: University of Chicago Press, 1978.

John R. Koza
Computer Science Department, Margaret Jacks Hall, Stanford University, Stanford, CA 94305; email: Koza@Sunburn.Stanford.Edu; phone: 415-941-0336; fax: 415-941-9430

Genetic Evolution and Co-Evolution of Computer Programs

INTRODUCTION AND OVERVIEW

Research in the field of artificial life focuses on computer programs that exhibit some of the properties of biological life (e.g., self-reproducibility, evolutionary adaptation to an environment, etc.). In one area of artificial life research, human programmers write very simple computer programs (often incorporating observed features of actual biological processes) and then study the "emergent" higher-level behavior that may be exhibited by such seemingly simple programs. In this chapter, we consider a different problem, namely, "How can computer programs be automatically written by a computer using only measurements of a given program's performance?" In particular, this chapter describes the recently developed "genetic programming paradigm" which genetically breeds populations of computer programs to solve problems. In the genetic programming paradigm, the individuals in the population are hierarchical compositions of functions and arguments of various sizes and shapes. Increasingly fit hierarchies are then evolved in response to the problem environment using the genetic operations of fitness proportionate reproduction (Darwinian survival and reproduction of the fittest) and crossover (sexual recombination). In the genetic programming paradigm, the size and shape of the hierarchical solution to the problem is not specified in advance. Instead, the size and shape of the hierarchy, as well as the contents of the hierarchy, evolves in response to the Darwinian selective pressure exerted by the problem environment.

This chapter also describes an extension of the genetic programming paradigm to the case where two (or more) populations of computer programs simultaneously co-evolve. In co-evolution, each population acts as the environment for the other population. In particular, each individual of the first population is evaluated for "relative fitness" by testing it against each individual in the second population, and, simultaneously, each individual in the second population is evaluated for relative fitness by testing it against each individual in the first population. Over a period of many generations, individuals with high "absolute fitness" may evolve as the two populations mutually bootstrap each other to increasingly high levels of fitness.

The genetic programming paradigm is illustrated by genetically breeding a population of hierarchical computer programs to allow an "artificial ant" to traverse an irregular trail. In addition, we genetically breed a computer program controlling the behavior of an individual ant in an ant colony which, when repetitively executed by all the ants in the colony, causes the emergence of interesting collective behavior for the colony as a whole. Co-evolution is illustrated with a problem involving finding an optimal strategy for a simple, discrete, two-person competitive game represented by a game tree in extensive form.

BACKGROUND ON GENETIC ALGORITHMS

Genetic algorithms are highly parallel mathematical algorithms that transform populations of individual mathematical objects (typically fixed-length binary character strings) into new populations using operations patterned after (i) natural genetic operations such as sexual recombination (crossover) and (ii) fitness proportionate reproduction (Darwinian survival of the fittest). Genetic algorithms begin with an initial population of individuals (typically randomly generated) and then iteratively (1) evaluate the individuals in the population for fitness with respect to the problem environment and (2) perform genetic operations on various individuals in the population to produce a new population. John Holland of the University of Michigan presented the pioneering formulation of genetic algorithms for fixed-length character strings in 1975.[6] Holland established, among other things, that the genetic algorithm is a mathematically near-optimal approach to adaptation in that it maximizes expected overall average payoff when the adaptive process is viewed as a multi-armed slot machine problem requiring an optimal allocation of future trials in the search space, given currently available information. Recent work in genetic algorithms and genetic classifier systems can be surveyed in Davis,[2] Goldberg,[4] and Schaffer.[14]

BACKGROUND ON GENETIC PROGRAMMING PARADIGM

Representation is a key issue in genetic algorithm work because genetic algorithms directly manipulate the coded representation of the problem and because the representation scheme can severely limit the window by which the system observes its world. Fixed length character strings present difficulties for some problems—particularly problems where the desired solution is hierarchical and where the size and shape of the solution is unknown in advance. The need for more powerful representations has been recognized for some time.[3]

The structure of the individual mathematical objects that are manipulated by the genetic algorithm can be more complex than the fixed length character strings first described by Holland[6] in 1975. Smith[15] departed from the early fixed-length character strings by introducing variable length strings (specifically, strings whose elements were if-then rules, rather than single characters). Holland's introduction of the genetic classifier system[7] continued the trend towards increasing the complexity of the structures undergoing adaptation. The classifier system is a cognitive architecture containing a population of string-based if-then rules (whose condition and action parts are fixed length binary strings) which can be modified by the genetic algorithm.

The recently developed genetic programming paradigm further continues the above trend towards increasing the complexity of the structures undergoing adaptation. In the genetic programming paradigm, the individuals in the population are hierarchical compositions of functions and terminals appropriate to the particular problem domain. The hierarchies are of various sizes and shapes. The set of functions typically includes arithmetic operations, mathematical functions, conditional logical operations, and domain-specific functions. Each function in the function set must be well defined for any combination of elements from the range of every function that it may encounter and every terminal that it may encounter. The set of terminals used typically includes inputs (sensors) appropriate to the problem domain and various constants. The search space is the hyperspace of all possible compositions of functions and terminals that can be recursively composed of the available functions and terminals. The symbolic expressions (S-expressions) of the LISP programming language are an especially convenient way to create and manipulate the compositions of functions and terminals described above. These S-expressions in LISP correspond directly to the "parse tree" that is internally created by most compilers.

The basic genetic operations for the genetic programming paradigm are fitness based reproduction and crossover (recombination).

Fitness proportionate reproduction is the basic engine of Darwinian reproduction and survival of the fittest. It copies individuals with probability proportionate to fitness from one generation of the population into the next generation. In this respect, it operates for the genetic programming paradigm in the same way as it does for conventional genetic algorithms.

The crossover operation for the genetic programming paradigm is a sexual operation that operates on two parental LISP S-expressions (chosen with a probability proportional to fitness) and produces two offspring S-expressions using parts of each parent. Typically the two parents are hierarchical compositions of functions of different size and shape. In particular, the crossover operation starts by selecting a random crossover point in each parent and then creates two new offspring S-expressions by exchanging the sub-trees (i.e., sub-lists) between the two parents. Because entire sub-trees are swapped, this genetic crossover (recombination) operation produces syntactically and semantically valid LISP S-expressions as offspring regardless of which point is selected in either parent.

For example, consider the parental LISP S-expression:

```
(OR (NOT D1) (AND D0 D1))
```

And, consider the second parental S-expression below:

```
(OR (OR D1 (NOT D0))
    (AND (NOT D0) (NOT D1)))
```

These two LISP S-expressions can be depicted graphically as rooted, point-labeled trees with ordered branches. Assume that the points of both trees are numbered in a depth-first way starting at the left. Suppose that the second point (out of six points of the first parent) is randomly selected as the crossover point for the first parent and that the sixth point (out of ten points of the second parent) is randomly selected as the crossover point of the second parent. The crossover points are therefore the NOT in the first parent and the AND in the second parent.

The two parental LISP S-expressions are shown in Figure 1. The two crossover fragments are two sub-trees shown in Figure 2. These two crossover fragments correspond to the bold sub-expressions (sub-lists) in the two parental LISP S-expressions shown in Figure 1. The two offspring resulting from crossover are shown in Figure 3.

Note that the first offspring in Figure 3 is a perfect solution for the exclusive-or function, namely

```
(OR (AND (NOT D0) (NOT D1)) (AND D0 D1)).
```

In addition to the basic genetic operations of fitness proportionate reproduction and crossover, a mutation operation can also be defined to provide a means for

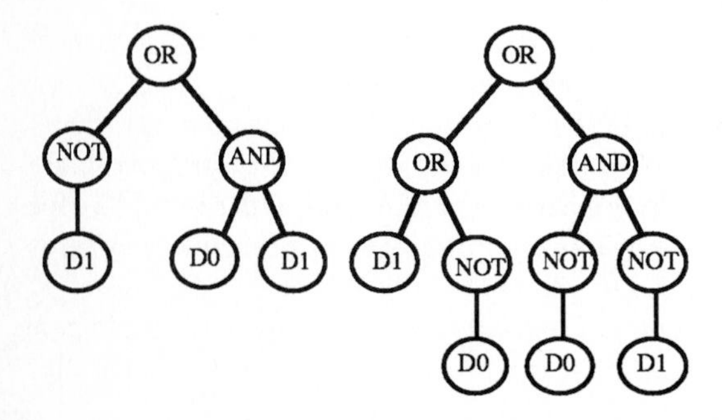

FIGURE 1 Two parental LISP S-expressions.

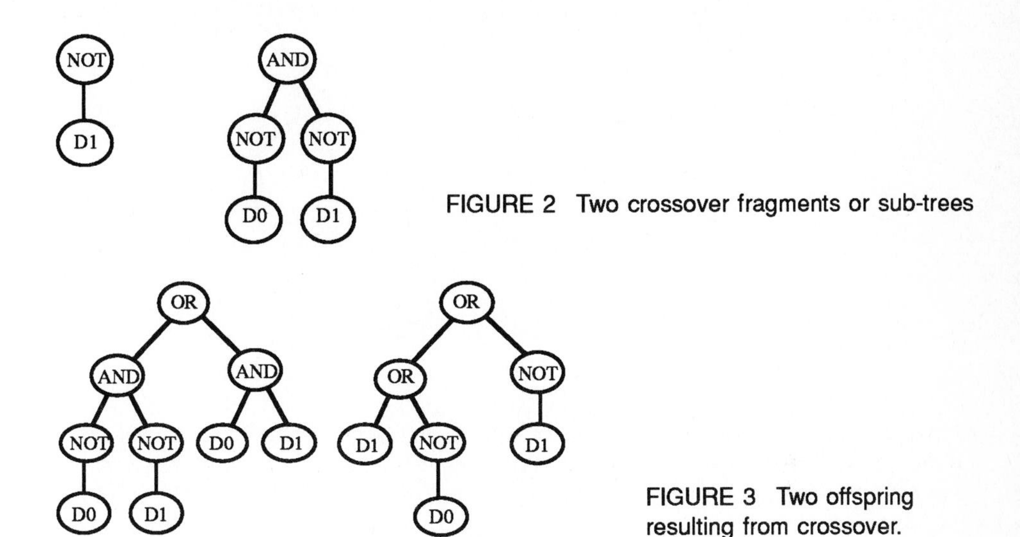

FIGURE 2 Two crossover fragments or sub-trees

FIGURE 3 Two offspring resulting from crossover.

occasionally introducing small random mutations into the population. The mutation operation is an asexual operation in that it operates on only one parental S-expression. The result of this operation is a single offspring S-expression. The mutation operation selects a point of the LISP S-expression at random. The point can be an internal (function) or external (terminal) point of the tree. This operation removes whatever is currently at the selected point and inserts a randomly generated sub-tree at the randomly selected point of a given tree. This randomly generated subtree is created in the same manner as the initial random individuals in the initial random generation. This operation is controlled by a parameter which specifies the maximum depth for the newly created and inserted sub-tree. A special case of this operation involves inserting only a single terminal (i.e., a sub-tree of depth 0) at a randomly selected point of the tree. For example, in the Figure 4, the third point of the S-expression (left) was selected as the mutation point and the sub-expression (NOT D1) was randomly generated and inserted at that point to produce the S-expression (right).

The mutation operation potentially can be beneficial in reintroducing diversity in a population that may be tending to prematurely converge.

Additional details can be found in Koza.[10,11]

We have shown that entire computer programs can be genetically bred to solve problems in a variety of different areas of artificial intelligence, machine learning, and symbolic processing.[10,11] In particular, this new paradigm has been successfully applied to example problems in several different areas, including:

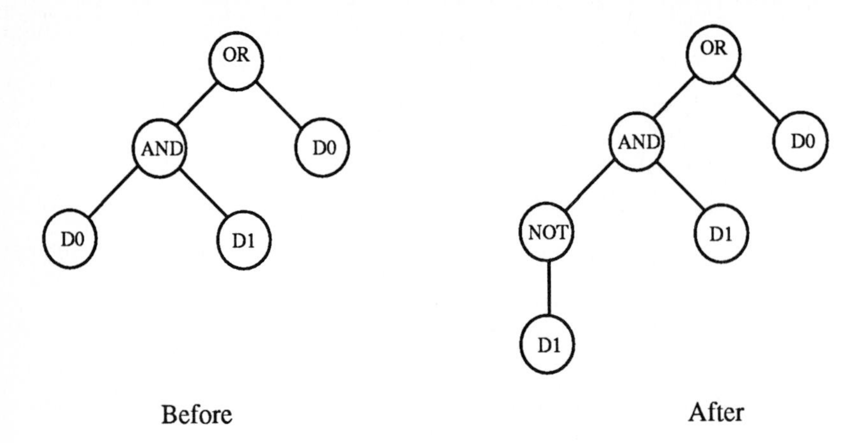

FIGURE 4 Third point of the S-expression (left) was selected as the mutation point and the sub-expression (NOT D1) was randomly generated and inserted at that point to produce the S-expression (right).

- machine learning of functions (e.g., learning the Boolean 11-multiplexer function);
- planning (e.g., developing a robotic action sequence that can stack an arbitrary initial configuration of blocks into a specified order);
- automatic programming (e.g., discovering a computational procedure for solving pairs of linear equations, solving quadratic equations for complex roots, and discovering trigonometric identities);
- sequence induction (e.g., inducing a recursive computational procedure for the Fibonacci and the Hofstadter sequences);
- pattern recognition (e.g., translation-invariant recognition of a simple one-dimensional shape in a linear retina);
- optimal control (e.g., centering a cart and balancing a broom on a moving cart in minimal time by applying a "bang bang" force to the cart);
- symbolic "data-to-function" regression, symbolic "data-to-function" integration, and symbolic "data-to-function" differentiation;
- symbolic solution to functional equations (including differential equations with initial conditions, integral equations, and general functional equations);
- empirical discovery (e.g., rediscovering Kepler's Third Law, rediscovering the well-known econometric "exchange equation" $MV = PQ$ from actual noisy time series data for the money supply, the velocity of money, the price level, and the gross national product of an economy);
- finding the minimax strategy for a differential pursuer-evader game; and
- simultaneous architectural design and training of neural networks.

The genetic programming paradigm permits the evolution of computer programs which can perform alternative computations conditioned on the outcome of intermediate calculations, which can perform computations on variables of many different types, which can perform iterations and recursions to achieve the desired result, which can define and subsequently use computed values and sub-programs, and whose size, shape, and complexity is not specified in advance.

THE "ARTIFICIAL ANT" PROBLEM

In order to illustrate the genetic programming paradigm, we consider the complex planning task devised by Jefferson et al.[9] for an "artificial ant" attempting to traverse a trail.

Jefferson et al. successfully used a string-based genetic algorithm to discover a finite-state automaton enabling the "artificial ant" to traverse the trail.

The setting for the problem is a square 32 x 32 toroidal grid in the plane. The "Santa Fe trail" is a winding trail with food in 89 of the 1024 cells. This trail (designed by Christopher Langton) is considered the more difficult of the two trails tested by Jefferson et al. The trail is irregular and has single gaps, double gaps, single gaps at some corners, double gaps (knight moves) at other corners, and triple gaps (long knight moves) at other corners. The "artificial ant" begins in the cell identified by the coordinates (0,0) and is facing in a particular direction (i.e., east). The artificial ant has a sensor that can see only the single adjacent cell in the direction the ant is currently facing. At each time step, the ant has the capacity to execute any of four operations, namely, to move forward (advance) in the direction it is facing, to turn right (and not move), to turn left (and not move), or to sense the contents of the single adjacent cell in the direction the ant is facing.

The objective of the ant is to traverse the entire trail and collect all of the food. Jefferson et al. limited the ant to a certain number of time steps (200). (See Figure 5.) Jefferson et al. started by assuming that the finite automaton necessary to solve the problem would have 32 or fewer states. They then represented an individual in their population of automata by a 453-bit string representing the state transition diagram (and its initial state) of the individual automaton. The ant's output at each time step was coded as two bits. That is, a total of seven bits specified the action of the automaton for each of the two possible sensory inputs associated with each of the 32 states. The next state of the automaton was coded with five bits. The complete behavior of an automaton was thus specified with a genome consisting of a binary string with 453 bits (five bits representing the initial state of the automaton plus 64 substrings of length seven representing the state transitions). Jefferson et al. then processed a population of 65,536 individual bit strings of length 453 on a Connection Machine™ using a genetic algorithm using crossover and mutation operating on a selected (relatively small) fraction of the population. After 200 generations in a particular run (taking about ten hours on the Connection Machine), they reported that a single individual in the population emerged which attained a perfect score of 89 stones. As it happened, this single individual completed the task in exactly 200 operations.

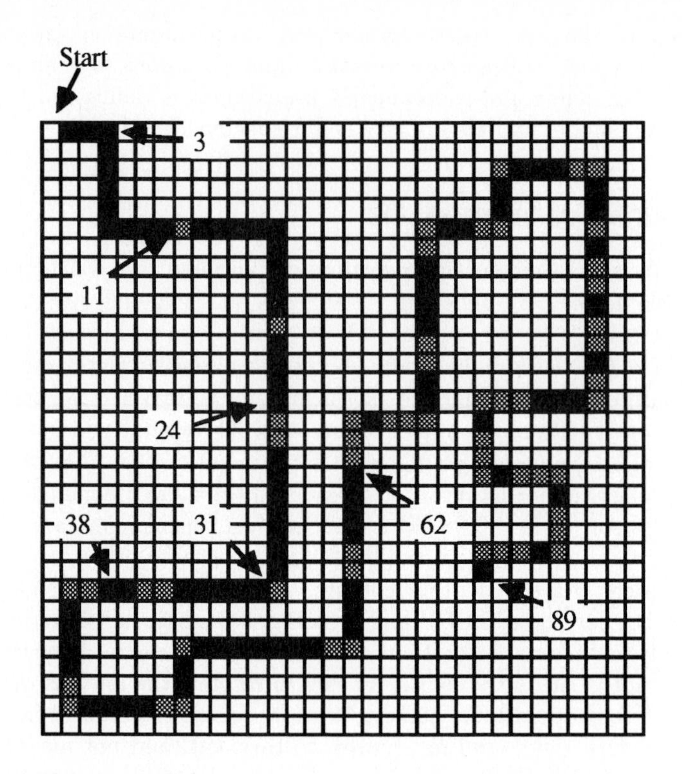

FIGURE 5 Santa Fe Trail where an artificial ant seeks to find all 89 food pellets. (See Jefferson et al.[9])

In our approach to this task using the genetic programming paradigm, we used the function set consisting of the functions F = {IF-SENSOR, PROGN}. The IF-SENSOR function has two arguments and evaluates the first argument if the ant's sensor senses a stone or, otherwise, evaluates the second argument. The PROGN function is the LISP connective (glue) function that sequentially evaluates its arguments as individual steps in a program. The terminal set was T = {ADVANCE, TURN-RIGHT, TURN-LEFT}. These three terminals were actually functions with no arguments. They operate via their side effects on the ant's state (i.e., the ant's position on the grid or the ant's facing direction). Note that IF-SENSOR, ADVANCE, TURN-RIGHT, and TURN-LEFT correspond directly to the operators defined and used by Jefferson et al. We allowed 400 time steps before timing out. Note that we made no assumption about the number of states or complexity of the eventual solution. This problem is such that it cannot be solved in any reasonable amount of time by random search using either Jefferson's approach or our approach.

The initial generation (generation 0) consisted of randomly generated individual S-expressions recursively created using the available functions and available terminals of the problem. Many of these randomly generated individuals did nothing at all. For example, (PROGN (TURN-RIGHT) (TURN-RIGHT)) turns without ever moving the ant anywhere. Other random individuals move without turning [e.g., (ADVANCE)]. Other individuals in the initial random population move forward after sensing food but can only correctly handle a right turn in the trail [e.g., (IF-SENSOR (ADVANCE) (TURN-RIGHT))].

Throughout this chapter (and in virtually all of our experiments), each new generation was created from the preceding generation by applying the fitness proportionate reproduction operation to 10% of the population and by applying the crossover operation to 90% of the population (with reselection allowed). The selection of crossover points in the population was biased 90% towards internal (function) points of the tree and 10% towards external (terminal) points of the tree. For practical reasons (i.e., conservation of computer time), a limit of four was placed on the depth of initial random S-expressions and a limit of 15 was placed on the depth of S-expressions created by crossover. As to mutation, our experience has been that no run using only mutation and fitness proportionate reproduction (i.e., no crossover) ever produced a solution to any problem (although such solutions are theoretically possible given enough time). In other words, "mutating and saving the best" does not work any better for the hierarchical genetic programming paradigm than it does for conventional string-based genetic algorithms. This conclusion as to the relative unimportance of the mutation operation is similar to the conclusions reached by most other research work on string-based genetic algorithms [see, for example, Holland[6] and Goldberg[4]]. Accordingly, mutation was not used here.

In one run, a reasonably parsimonious individual LISP S-expression scoring 89 out of 89 emerged on the seventh generation, namely,

```
(IF-SENSOR (ADVANCE)
     (PROGN (TURN-RIGHT)
          (IF-SENSOR (ADVANCE) (TURN-LEFT))
          (PROGN (TURN-LEFT)
               (IF-SENSOR (ADVANCE)
                    (TURN-RIGHT))
               (ADVANCE)))).
```

This plan is graphically depicted in Figure 6.

This individual LISP S-expression is the solution to the problem. In particular, this plan moves the ant forward if a stone is sensed. Otherwise it turns right and then moves the ant forward if a stone is sensed but turns left (returning to its original orientation) if no stone is sensed. Then it turns left and moves forward if a stone is sensed but turns right (returning to its original orientation) if no stone is sensed. If the ant originally did not sense a stone, the ant moves forward unconditionally as its fifth operation. Note that there is no testing of the backwards directions.

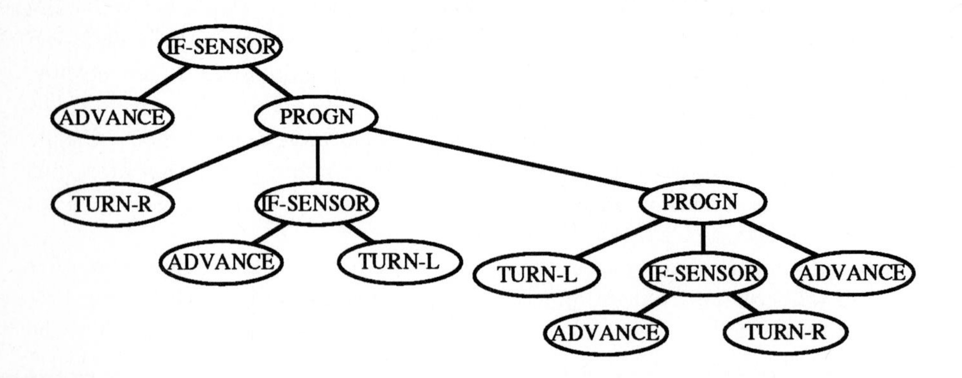

FIGURE 6 Artificial ant solution.

We can measure the performance of a probabilistic algorithm by estimating the expected number of individuals that need to be processed by the algorithm in order to produce a solution to the given problem with a certain probability (say, 99%). Suppose, for example, a particular run of a genetic algorithm produces the desired result with only a probability of success p_s after a specified choice (perhaps non-optimal) of number of generations N_{gen} and population of size N. Suppose also that we are seeking to achieve the desired result with a probability of, say, $z = 1 - \varepsilon = 99\%$. Then, the number K of independent runs required is

$$K = \frac{\log(1-z)}{\log(1-ps)} = \frac{\log \varepsilon}{\log(1-ps)}, \quad \text{where } \varepsilon = 1 - z.$$

For example, we ran 111 runs of the Artificial Ant problem with a population size of 1000 and 51 generations. We found that the probability of success p_s on a particular single run was 43%. With this probability of success, $K = 8$ independent runs are required to assure a 99% probability of solving the problem on at least one of the runs. That is, it is sufficient to process 408,000 individuals to achieve the desired 99% probability. Processing a full 408,000 individuals requires about six hours of computing time on the Texas Instruments Explorer II+™ workstation for this problem. In addition, as Figure 7 shows, the probability of success p_s of a run with a population size of 2000 with 51 generations is 67% so that a population of 2000 requires $K = 4$ independent runs (i.e., 408,000 individuals to be processed) to achieve the desired 99% probability. In contrast, the probability of success of a run with a population of 4000 with 51 generations is 81% so that a population of 4000 requires $K = 3$ independent runs (i.e., 612,000 individuals to be processed) to achieve the desired 99% probability.

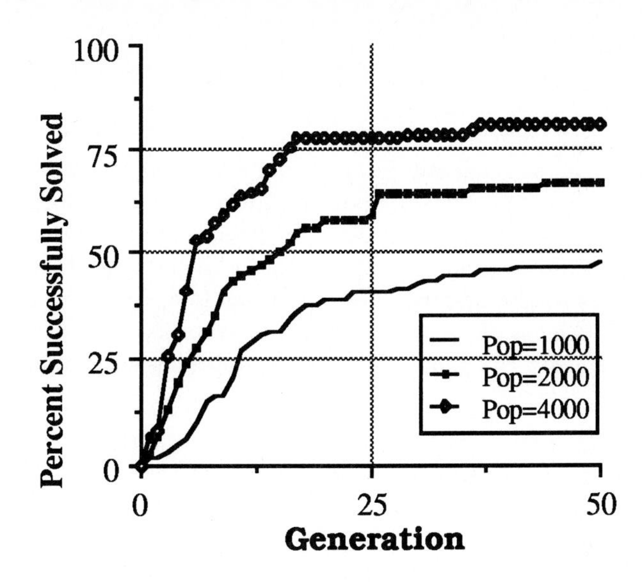

FIGURE 7 Percent of runs of artificial ant on the Santa Fe Trail that are successfully solved in a given number of generations with population size of 1000, 2000, and 4000.

The six hours of computer time required to process the 408,000 individuals required by this problem (either with a population of size 1000 or 2000) on the considerably smaller and slower serial computer (Explorer workstation) represents substantially less computational resources than even a single ten-hour run on the massively parallel Connection Machine with 65,536 processors (even if it were the case that all such ten-hour runs on the Connection Machine were successful in solving the problem). Thus, the genetic programming paradigm is comparatively speedy for this problem.

EMERGENCE OF COLLECTIVE BEHAVIOR IN AN ANT COLONY

Conway's "game of life" and other work in cellular automata, fractals, chaos, and Lindenmayer systems (L-systems) are suggestive of how the repetitive application of seemingly simple rules can lead to complex "emergent" overall behavior. Travers and Resnick[17] have written a program to govern an ant which, when executed by all the ants in a colony, results in overall interesting behavior. In this section, we show how such rules (computer programs) can be evolved using genetic recombination and the Darwinian principle of survival of the fittest as contained in the genetic programming paradigm.

In particular, we show the emergence of interesting collective behavior in a colony of ants by genetically breeding a computer program to govern the behavior of the individual ants in the colony. The goal is to genetically evolve a common computer program governing the behavior of the individual ants in a colony such that the collective behavior of the ants consists of efficient transportation of food to the nest. In nature, when an ant discovers food, it deposits a trail of pheromones as it returns to the nest with the food. The pheromonal cloud (which dissipates over time) aids other ants in efficiently locating and exploiting the food source.

In this example, 144 pellets of food are piled eight deep in two 3-by-3 piles located in a 32-by-32 toroidal area. There are 20 ants. The state of each ant consists of its position and the direction it is facing (out of eight possible directions). Each ant initially starts at the nest and faces in a random direction. Each ant in the colony is governed by a common computer program associated with the colony. The computer program is a composition of the following nine available functions:

- MOVE-RANDOM randomly changes the direction in which an ant is facing and then moves the ant two steps in the random direction.
- MOVE-TO-NEST moves the ant one step in the direction of the nest. This implements the gyroscopic ability of ants to navigate back to their nest.
- PICK-UP picks up food (if any) at the current position of the ant.
- DROP-PHEROMONE drops a pheromone at the current position of the ant (if the ant is carrying food). The pheromone immediately forms a 3-by-3 cloud around the drop point. The cloud decays over a period of time.
- IF-FOOD-HERE is a two-argument function that executes its first argument if there is food at the ant's current position and, otherwise, executes the second (else) argument.
- IF-CARRYING-FOOD is a similar two-argument function that tests whether the ant is currently carrying food.
- MOVE-TO-ADJACENT-FOOD-ELSE is a one-argument function that allows the ant to test for immediately adjacent food and then move one step towards it. If food is present in more than one adjacent position, the ant moves to the position requiring the least change of direction. If no food is adjacent, the "else" clause of this function is executed.
- MOVE-TO-ADJACENT-PHEROMONE-ELSE is a function similar to the above based on adjacent pheromone.
- PROGN is the LISP connective function that executes its arguments in sequence.

Each of the twenty ants in a given colony executes the colony's common computer program. Since the ants initially face in random directions, make random moves, and encounter a changing pattern of food and pheromones created by the activities of other ants, the twenty individual ants almost always have different states and pursue different trajectories.

The fitness of a colony is measured by how many of the 144 food pellets are transported to the nest within the "allotted time" (which limits both the total number of time steps and the total number of operations which any one ant can execute). The goal is to genetically evolve increasingly fit computer programs to govern the colony.

Mere random motion by the 20 ants in a colony will typically bring the ants into contact with only about 56 of the 144 food pellets within the allotted time. Moreover, the ants' task is substantially more complicated than merely coming in contact with food. After randomly stumbling into food, the ants must pick up the food and then move towards the nest. Moreover, while this sequence of behavior is desirable, it is still insufficient to efficiently solve the problem in any reasonable amount of time. It is also necessary that the ants that accidentally stumble into food must also establish a pheromonal trail as they carry the food back to the nest. Moreover, all ants must always be on the lookout for such pheromonal trails established by other ants and must follow such trails to the food when they encounter such trails. In a typical run, 93% of the random computer programs in the initial random generation did not transport even one of the 144 food pellets to the nest within the allotted time. About 3% of these initial random programs transported only one of the 144 pellets. Even the best single computer program of the random computer programs created in the initial generation successfully transported only 41 pellets.

As the genetic programming paradigm is run, the population as a whole and its best single individual both generally improve from generation to generation. In the one specific run which we describe in detail hereinbelow, the best single individual in the population on generation 10 scored 54; the best single individual on generation 20 scored 72; the best single individual on generation 30 scored 110; the best single individual on generation 35 scored 129; and the best single individual on generation 37 scored 142. On generation 38, a program emerged which causes the twenty ants to successfully transport all 144 food pellets to the nest within the allotted time. This 100% fit program is shown below:

```
(PROGN (PICK-UP) (IF-CARRYING-FOOD (PROGN (MOVE-TO-ADJACENT-
      PHEROMONE-ELSE (MOVE-TO-ADJACENT-FOOD-ELSE (MOVE-TO-
      ADJACENT-FOOD-ELSE (MOVE-TO-ADJACENT-FOOD-ELSE
      (PICK-UP))))) (PROGN (PROGN (PROGN (PROGN (MOVE-TO-
      ADJACENT-FOOD-ELSE (PICK-UP)) (PICK-UP)) (PROGN (MOVE-
      TO-NEST) (DROP-PHEROMONE))) (PICK-UP)) (PROGN (MOVE-
      TO-NEST) (DROP-PHEROMONE)))) (MOVE-TO-ADJACENT-FOOD-
      ELSE (IF-CARRYING-FOOD (PROGN (PROGN (DROP-PHEROMONE)
      (MOVE-TO-ADJACENT-PHEROMONE-ELSE (IF-CARRYING-FOOD
      (MOVE-TO-ADJACENT-FOOD-ELSE (PICK-UP)) (MOVE-TO-
      ADJACENT-FOOD-ELSE (PICK-UP))))) (MOVE-TO-NEST)) (IF-
      FOOD-HERE (PICK-UP) (IF-CARRYING-FOOD (PROGN (IF-FOOD-
      HERE (MOVE-RANDOM) (IF-CARRYING-FOOD (MOVE-RANDOM)
      (PICK-UP))) (DROP-PHEROMONE)) (MOVE-TO-ADJACENT-
      PHEROMONE-ELSE (MOVE-RANDOM)))))))
```

The 100% fit program above is equivalent to the simplified program below (except for the special case when an ant is in the nest):

```
1    (PROGN (PICK-UP)
```

```
2        (IF-CARRYING-FOOD
3             (PROGN (MOVE-TO-ADJACENT-PHEROMONE-ELSE
4                      (MOVE-TO-ADJACENT-FOOD-ELSE (PICK-UP)))
5                 (MOVE-TO-ADJACENT-FOOD-ELSE (PICK-UP))
6                 (MOVE-TO-NEST) (DROP-PHEROMONE)
7                 (MOVE-TO-NEST) (DROP-PHEROMONE))
8             (MOVE-TO-ADJACENT-FOOD-ELSE
9                 (IF-FOOD-HERE
10                     (PICK-UP)
11                     (MOVE-TO-ADJACENT-PHEROMONE-ELSE
12                     (MOVE-RANDOM))))))
```

This simplified program can be interpreted as follows: The ant begins by picking-up the food, if any, located at the ant's current position. If the ant is now carrying food (line 2), then the six parts of the PROGN beginning on line 3 and ending on line 7 are executed. Line 3 moves the ant to the adjacent pheromone (if any). If there is no adjacent pheromone, line 4 moves the ant to the adjacent food (if any). In view of the fact that the ant is already carrying food, these two potential moves on lines 3 and 4 generally distract the ant from the most direct return to the nest and therefore merely reduce efficiency. Line 5 is a similar distraction. Note that the PICK-UP operations on lines 4 and 5 are redundant since the ant is already carrying food. These operations reduce efficiency but are otherwise harmless.

Given that the ant is already carrying food, the sequence of MOVE-TO-NEST and DROP-PHEROMONE on lines 6 and 7 is the winning combination that establishes the pheromone trail as the ant moves towards the nest with the food. The establishment of the pheromone trail between the pile of food and the nest is an essential part of efficient collective behavior for exploiting the food source.

The sequence of conditional behavior in lines 8 through 12 efficiently prioritizes the search activities of the ant. If the ant is not carrying food, line 8 moves the ant to adjacent food (if any). If there is no adjacent food but there is food at the ant's current position (line 9), the ant picks up the food (line 10). On the other hand, if there is no food at the ant's current position (line 11), the ant moves towards any adjacent pheromones (if any). If there are no adjacent pheromones, the ant moves randomly (line 12). This sequence of conditional behavior causes the ant to pick up any food it may encounter. Failing that, the second priority established by this conditional sequence causes the ant to follow a previously established pheromonal trail. And, failing that, the third priority of this conditional sequence causes the ant to move at random.

The collective behavior of the ant colony governed by the above 100% fit program above can be visualized as a series of major phases. The first phase occurs when the ants have just emerged from the nest and are randomly searching for food. In Figure 8 (representing time step 3 of the execution of the 100% fit program above), the two 3-by-3 piles of food are shown in black in the western and

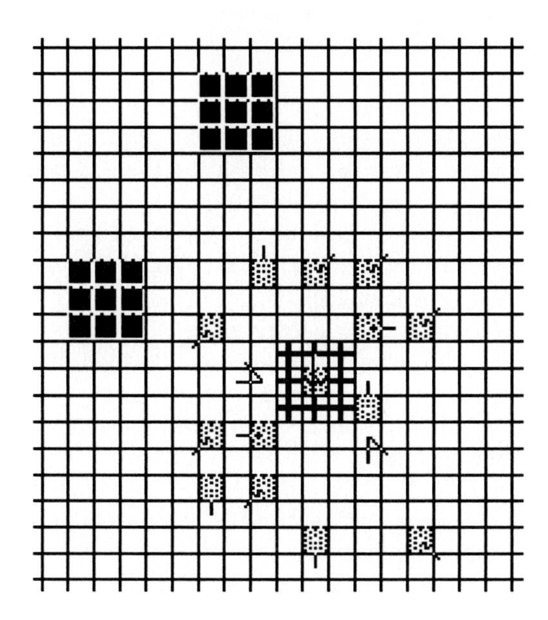

FIGURE 8 Phase 1: The ants have just emerged from nest (near center) and are randomly searching for food (contained in the two 3×3 piles in West and North). Time Step 3 of the execution of the 100% fit program.

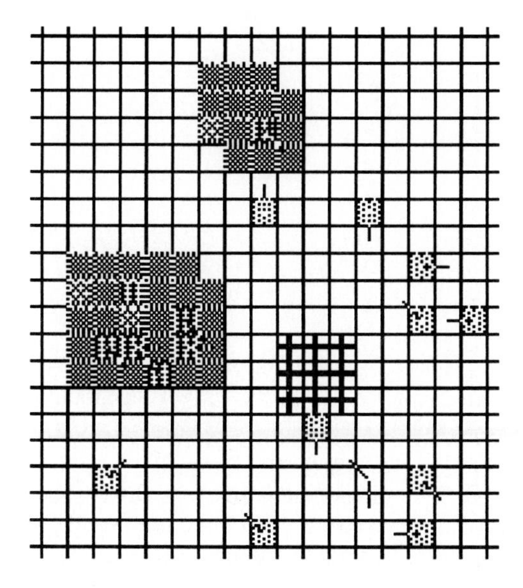

FIGURE 9 Phase 2: Ants have discovered food in both piles, picked food up, and dropped some pheromones. Beginnings of pheromone clouds; time step 12.

northern parts of the grid. The nest is indicated by nine + signs slightly southeast of the center of the grid. The ants are shown in gray with their facing direction indicated.

The second phase occurs when some ants have discovered some food, have picked up the food, and have started back towards the nest dropping pheromones as they go. The beginnings of the pheromone clouds around both the western and northern pile of food are shown in Figure 9 (representing time step 12).

The third phase occurs when pheromonal trails have been established linking both piles of food with the nest. The first two (of the 144) food pellets have just reached the nest in Figure 10 (representing time step 15).

Figure 11 shows the premature (and temporary) disintegration of the pheromonal trail connecting the northern pile of food with the nest while food still remains in the northern pile. The pheromonal trail connecting the western pile of food with the nest is still intact. 118 of the 144 food pellets have been transported to the nest at this point (representing time step 129).

In Figure 12 (representing time step 152), the western pile has been entirely consumed by the ants and the pheromone trail connecting it to the nest has already dissolved. The former location of the western pile is shown as a white area. 136 of the 144 food pellets have been transported to the nest at this point The pheromone trail connecting the nest and the northern pile (with 8 remaining food pellets) has been reestablished.

Shortly thereafter, the run ends with all 144 food pellets in the nest.

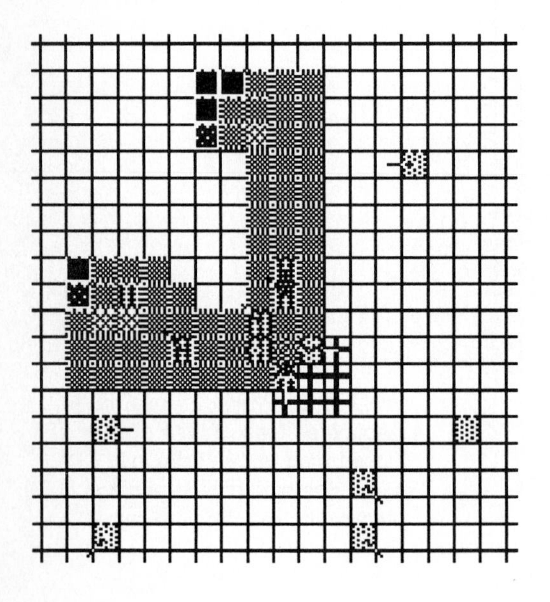

FIGURE 10 Phase 3: pheromonal trails are established between both West and North food pile and nest (near center). First two food pellets reach the nest; time step 15.

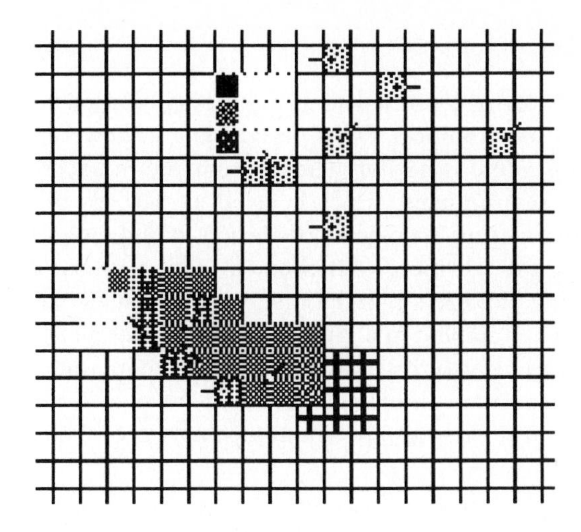

FIGURE 11 Premature (and temporary) disintegration of the pheromonal trail; time step 120.

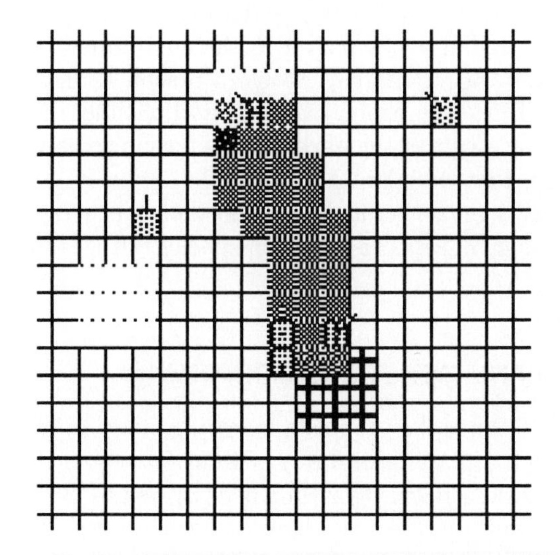

FIGURE 12 The western pile has been entirely consumed by the ants; time step 152.

CO-EVOLUTION IN NATURE

The evolutionary process in nature is often described as if one population of individuals is trying to adapt to a fixed environment. This description is, however, only a first-order approximation to the actual situation. The environment in nature actually consists of both the physical environment (which is usually relatively unchanging) as well as other independently acting biological populations of individuals which are simultaneously trying to adapt to "their" environment. The actions of each of these other independently acting biological populations (species) usually affect all the others. In other words, the environment of a given species includes all the other biological species that contemporaneously occupy the physical environment and which are simultaneously trying to survive. In biology, the term "co-evolution" is sometimes used to reflect the fact that all species are simultaneously co-evolving in a given physical environment.

A biological example presented by Holland illustrates the point.[8] A given species of plant may be faced with an environment containing insects that like to eat it. To defend against its predators (and increase its probability of survival in the environment), the plant may, over a period of time, evolve a tough exterior that makes it difficult for the insect to eat it. But, over a period of time, the insect may retaliate by evolving a stronger jaw so that the insect population can continue to feed on the plant (and increase its probability of survival in the environment). Then, over an additional period of time, the plant may evolve a poison to help defend itself further against the insects. But, then again, over a period of time, the insect may evolve a digestive enzyme that negates the effect of the poison so that the insect population can continue to feed on the plant.

In effect, both the plant and the insects get better and better at their respective defensive and offensive roles in this "biological arms race." Each species changes in response to the actions of the other.

BACKGROUND ON CO-EVOLUTION AND GENETIC ALGORITHMS

In the "genetic algorithm," described by John Holland in his pioneering book *Adaptation in Natural and Artificial Systems*,[6] a population of individuals attempts to adapt to a fixed "environment." In the basic genetic algorithm as described by Holland in 1975, the individuals in the population are fixed-length character strings (typically binary strings) that are encoded to represent some problem in some way. In the basic "genetic algorithm," the performance of the individuals in the population is measured using a fitness measure which is, in effect, the "environment" for the population. Over a period of many generations, the genetic algorithm causes the individuals in the population to adapt in a direction that is dictated by the fitness measure (its environment).

Holland[8] has incorporated co-evolution and genetic algorithms in his ECHO system for exploring the co-evolution of artificial organisms described by fixed-length character strings (chromosomes) in a "miniature world." In ECHO, there is a single population of artificial organisms. The environment of each organism includes all other organisms.

Miller[12,13] has used co-evolution in a genetic algorithm to evolve a finite automaton as the strategy for playing the Repeated Prisoner's Dilemma game. Miller's population consisted of strings (chromosomes) of 148 binary digits to represent finite automata with 16 states. Each string in the population represented a complete strategy by which to play the game. That is, it specified what move the player was to make for any sequence of moves by the other player. Miller then used co-evolution to evolve strategies. Miller's co-evolutionary approach to the repeated prisoner's dilemma using string-based genetic algorithms contrasts with Alexrod's[1] evolutionary approach using genetic algorithms. Axelrod measured performance of a particular strategy by playing it against eight selected superior computer programs submitted in an international programming tournament for the prisoner's dilemma. A best strategy for one player (represented as a 70-bit string with a three-move look-back) was then evolved with a weighted mix of eight opposing fixed computer programs serving as the environment.

Hillis[5] used co-evolution in genetic algorithms to solve optimization problems. Smith[16] discussed co-evolution in connection with discovering strategies for games.

CO-EVOLUTION AND THE GENETIC PROGRAMMING PARADIGM

In the co-evolution version of the genetic programming paradigm there are two (or more) populations of hierarchical individuals. The environment for the first population consists of the second population. And, conversely, the environment for the second population consists of the first population.

The co-evolutionary process typically starts with both populations being highly unfit (when measured by an absolute fitness measure). Then the first population tries to adapt to the "environment" created by the second population. Simultaneously, the second population tries to adapt to the "environment" created by the first population.

This process is carried out by testing the performance of each individual in the first population against each individual (or a sampling of individuals) from the second population. We call this performance the "relative fitness" of an individual because it represents the performance of one individual in one population relative to the environment consisting of the second population. Then, each individual in the second population is tested against each individual (or a sampling of individuals) from the first population.

Note that this measurement of relative fitness for an individual in co-evolution is not an absolute measure of fitness against an optimal opponent, but merely a relative measure when the individual is tested against the current opposing population. If one population contains boxers who only throw left punches, then an individual whose defensive repertoire contains only defenses against left punches will have high relative fitness. But this individual will have only mediocre absolute fitness when tested against an opponent who knows how to throw both left punches and right punches (i.e., an optimal opponent).

Even when both initial populations are initially highly unfit (both relatively and absolutely), the virtually inevitable variation of the initial random population will mean that some individuals have slightly better relative fitness than others. That means that some individuals in each population have somewhat better performance than others in dealing with the current opposing population.

The operation of fitness proportionate reproduction (based on the Darwinian principle of survival and reproduction of the fittest) can then be applied to each population using the relative fitness of each individual currently in each population. In addition, the operation of genetic recombination (crossover) can also be applied to a pair of parents (selected based on its relative fitness).

Over a period of time, both populations of individuals will tend to "co-evolve" and to rise to higher levels of performance as measured in terms of absolute fitness. Both populations do this without the aid of any externally supplied absolute fitness measure serving as the environment. In the limiting case, both populations of individuals can evolve to a level of performance that equals the absolute optimal fitness. Thus, the co-evolution version of the genetic programming paradigm is a self-organizing, mutually bootstrapping process that is driven only by relative fitness (and not by absolute fitness).

Co-evolution is especially important in problems from game theory because one almost never has *a priori* access to a minimax strategy for either player. One therefore encounters a "chicken and egg" situation. In trying to develop a minimax strategy for the first player, one needs a minimax second player against which to test candidate strategies. In checkers or chess, for example, it is difficult for a new player to learn to play well if he does not have the advantage of playing against a reasonably competent player.

CO-EVOLUTION OF A GAME STRATEGY

We now illustrate the co-evolution version of the genetic programming paradigm to discover minimax strategies for both players simultaneously in a simple discrete two-person 32-outcome game represented by a game tree in extensive form.

In the co-evolution, we do not have access to the optimal opponent to train the population. Instead, our objective is to breed two populations simultaneously. Both populations start as random compositions of the available functions and arguments.

Consider the following discrete 32-outcome game whose game tree is presented in extensive form in Figure 13. Each internal point of this tree is labeled with the player who must move. Each line is labeled with the choice (either L or R) made by the moving player. Each endpoint of the tree is labeled with the payoff (to player X).

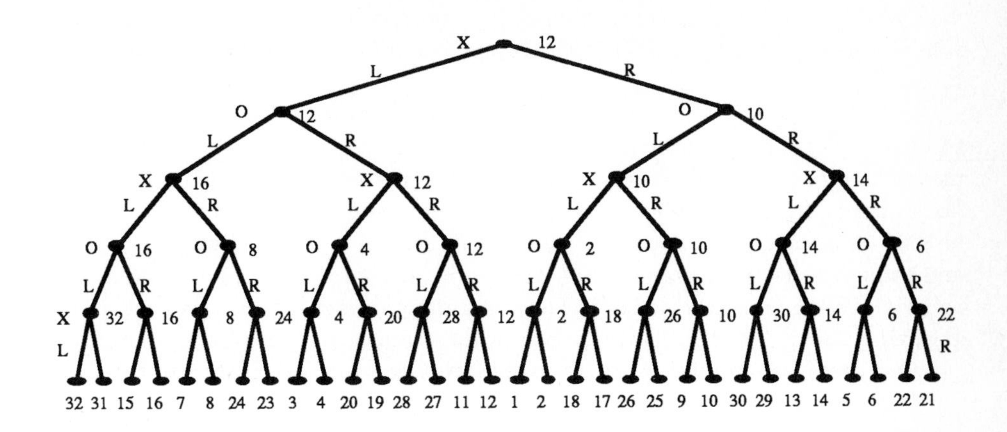

FIGURE 13 32-outcome game tree with payoffs.

This game is a two-person, competitive, zero-sum game in which the players make alternating moves. On each move, a player can choose to go L (left) or R (right). After player X has made three moves and player O has made two moves, player X receives (and player O pays out) the payoff shown at the particular endpoint of the game tree (1 of 32).

Each player has access to complete information about his opponent's previous moves (and his own previous moves). This historical information is contained in five variables XM1 (X's move 1), OM1 (O's move 1), XM2 (X's move 2), OM2 (O's move 2), and XM3 (X's move 3). These five variables each assume one of three possible values: L (left), R (right), or U (undefined). A variable is undefined prior to the time when the move to which it refers has been made. Thus, at the beginning of the game, all five variables are undefined. The particular variables that are defined and undefined indicate the point to which play has progressed during the play of the game. For example, if both players have moved once, XM1 and OM1 are defined (as either L or R) but the other three variables (XM2, OM2, and XM3) are undefined (have the value U).

A strategy for a particular player in a game specifies which choice that player is to make for every possible situation that may arise for that player. In particular, in this game, a strategy for player X must specify his first move if he happens to be at the beginning of the game. A strategy for player X must also specify his second move if player O has already made one move and it must specify his third move if player O has already made two moves. Since Player X moves first, player X's first move is not conditioned on any previous move. But, player X's second move will depend on Player O's first move (i.e., OM1) and, in general, it will also depend on his own first move (XM1). Similarly, player X's third move will depend on player O's first two moves and, in general, his own first two moves. Similarly, a strategy for player O must specify what choice player O is to make for every possible situation that may arise for player O. A strategy here is a computer program (i.e., S-expression) whose inputs are the relevant historical variables and whose output is a move (L or R) for the player involved. Thus, the set of terminals is T = {L, R}.

Four testing functions CXM1, COM1, CXM2, and COM2 provide the facility to test each of the historical variables that are relevant to deciding upon a player's move. Each of these functions is a specialized form of the CASE function in LISP. For example, function CXM1 has three arguments and evaluates its first argument of XM1 (X's move 1) is undefined, evaluates its second argument if XM1 is L (Left), and evaluates its third argument if XM1 is R (Right). Functions CXM2, COM1, and COM2 are similarly defined. Thus, the function set for this problem is F = {CXM1, COM1, CXM2, COM2}. Each of these functions takes three arguments.

Our goal is to simultaneously co-evolve strategies for both players of this game.

In co-evolution, the relative fitness of a particular strategy for a particular player in a game is the average of the payoffs received when that strategy is played against the entire population of opposing strategies.

The absolute fitness of a particular strategy for a particular player in a game is the payoff received when that strategy is played against the minimax strategy for the opponent. Note that when we compute the absolute fitness of an X strategy for our descriptive purposes here, we test the X strategy against four possible combinations of O moves—that is, O's choices of L or R for his moves 1 and 2. When we compute the absolute fitness of an O strategy, we test it against eight possible combinations of X moves—that is, X's choices of L or R for his moves 1, 2, and 3. Note that this testing of four or eight combinations does not occur in the computation for relative fitness. When the two minimax strategies are played against each other, the payoff is 12. This score is known as the value of this game. A minimax strategy takes advantage of non-minimax play by the other player.

As previously mentioned, co-evolution does not use the minimax strategy of the opponent in any way. We use it in this chapter for descriptive purposes only. The co-evolution algorithm uses only relative fitness.

In one run (with population size of 300), the individual strategy for player X in the initial random generation (generation 0) with the best relative fitness was

```
(COM1 L (COM2 (CXM1 (CXM2 R (CXM2 R R R) (CXM2 R L R)) L
      (CXM2 L R (COM2 R R R))) (COM1 R (COM2 (CXM2 L R L)
      (COM2 R L L) R) (COM2 (COM1 R R L) (CXM1 R L R)
      (CXM1 R L L))) (CXM1 (COM2 (CXM1 R L L) (CXM2 R R L)
      R) R (COM2 L R (CXM1 L L L)))) R).
```

This simplifies to

```
      (COM1 L (COM2 L L R) R).
```

This individual has relative fitness of 10.08.

The individual in the initial random population (generation 0) for player O with the best relative fitness was a similarly complex expression. It simplifies to

```
      (CXM2 R (CXM1 # L R) (CXM1 # R L)).
```

Note that, in simplifying this strategy, we inserted the symbol # to indicate that the situation involved can never arise. This individual has relative fitness of 7.57.

Neither the best X individual nor the best O individual from generation 0 reached maximal absolute fitness.

Note that the values of relative fitness for the relative best X individual and the relative best O individual from generation 0 (i.e., 10.08 and 7.57) are each computed by averaging the payoff from the interaction of the individual involved with all 300 individual strategies in the current opposing population.

In generation 1, the individual strategy for player X with the best relative fitness had relative fitness of 11.28. This individual X strategy is still not a minimax strategy. It does not have the maximal absolute fitness.

In generation 1, the best individual O strategy attained relative fitness of 7.18. It is shown below:

```
(CXM2 (CXM1 R R L) (CXM2 L L (CXM2 R L R)) R).
```

Although the co-evolution algorithm does not know it, this best single individual O strategy for generation 1 is, in fact, a minimax strategy for player O. It has maximal absolute fitness in this game. This one O individual was the first such O individual to attain this level of performance during this run. If it were played against the minimax X strategy, it would score 12 (i.e., the value of this game).

This individual O strategy can be graphically depicted as shown Figure 14. This individual O strategy simplifies to

```
(CXM2 (CXM1 # R L) L R).
```

Between generations 2 and 14, the number of individuals in the O population reaching maximal absolute fitness was 2, 7, 17, 28, 35, 40, 50, 64, 73, 83, 93, 98, and 107, respectively. That is, programs equivalent to the minimax O strategy began to dominate the O population.

In generation 14, the individual strategy for player X with the best relative fitness had relative fitness of 18.11. This individual X strategy was

```
(COM2 (COM1 L L (CXM1 R R R)) L (CXM1 (COM1 L L (CXM1
R R R))
     (CXM2 L R R) R)).
```

Although the co-evolution algorithm does not know it, this best single individual X strategy is, in fact, a minimax strategy for player X. This individual X strategy was the first such X individual to attain this level of performance during this run. If it were played against the minimax O strategy, it would score 12 (i.e., the value of this game). This individual X strategy can be graphically depicted as shown in Figure 15. This individual X strategy simplifies to

```
(COM2 (COM1 L L R) L R).
```

Between generations 15 and 29, the number of individuals in the X population reaching maximal absolute fitness was 3, 4, 8, 11, 10, 9, 13, 21, 24, 29, 43, 32, 52, 48, and 50, respectively. That is, programs equivalent to the minimax X strategy began to dominate the X population. Meanwhile, the O population became even more dominated by programs equivalent to the O minimax strategy.

By generation 38, the number of O individuals in the population reaching maximal absolute fitness reached 188 (almost two thirds of the population) and the number of X individuals reaching maximal absolute fitness reached 74 (about a quarter). That is, by generation 38, the minimax strategies for both players were becoming dominant.

Interestingly, these 74 individual X strategies had relative fitness of 19.11 and these 188 individual O strategies had relative fitness of 10.47. Neither of these values equals 12 because the other population is not fully converged to its minimax strategy.

In summary, we genetically bred the minimax strategies for both players of this game without using knowledge of the minimax strategy for either player.

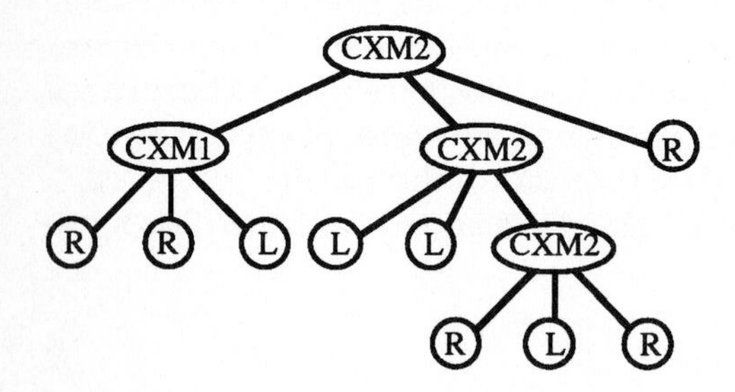

FIGURE 14 Individual O minimax strategy.

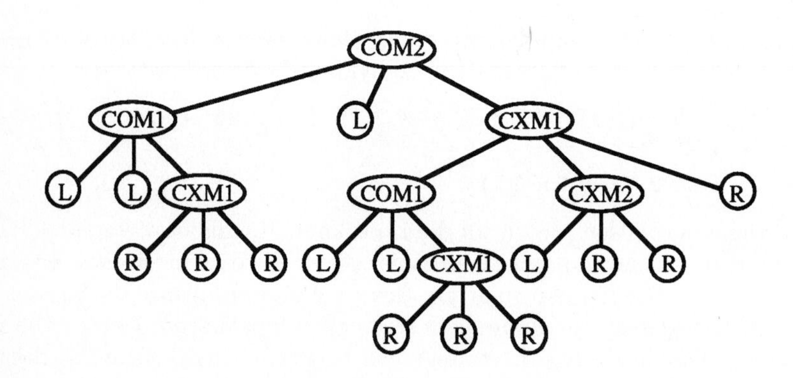

FIGURE 15 Individual X minimax strategy.

CONCLUSION

In this paper, we have demonstrated the use of the newly developed genetic programming paradigm to evolve hierarchical computer programs to solve three illustrative problems. These three illustrative problems are only a small subset of the benchmark problems already successfully solved by the genetic programming paradigm.[10,11] These three illustrative problems highlight some of the features of the genetic programming paradigm as compared to other existing paradigms for machine learning and artificial intelligence (such as neural networks and conventional string-based genetic algorithms) that may commend it for future work in the field of artificial life. These features include the following:

- In the genetic programming paradigm, the size and shape of the solution is not specified in advance, but, instead, evolves as the problem is being solved. For many problems, it is difficult, impossible, or unnatural to try to specify (or restrict) the size and shape of the eventual solution in advance. Moreover, advance specification (or restriction) of the size and shape of the solution to a problem narrows the window by which the system views the world and may well preclude finding the solution to the problem. This dynamic variability of the size and shape of the computer programs in the genetic programming paradigm is in marked contrast to both neural networks and conventional string-based genetic algorithms.
- The genetic programming paradigm evolves solutions that are directly expressed in a natural programming structure that overtly contains the functions and arguments naturally arising from the problem domain itself. Solutions expressed in this way are immediately understandable in the terms of the problem domain and, as a result, are auditable. This is in marked contrast to solutions produced by, for example, neural networks.
- In the genetic programming paradigm, there is no preprocessing of inputs (in contrast to neural networks, conventional string-based genetic algorithms, and most other machine learning paradigms).
- The genetic programming paradigm works with hierarchical structures at each stage of the process. As a result, the solutions are always hierarchical. Hierarchical structures offer the possibility of efficiently and understandably representing solutions to problems and also offer the possibility of scaling up to larger, more significant problems.

In summary, we have shown, by the use the three illustrative problems here (and the wide variety of other seemingly different problems cited from other areas), the power and flexibility of the genetic programming paradigm.

REFERENCES

1. Axelrod, R. "The Evolution of Strategies in the Iterated Prisoner's Dilemma." In *Genetic Algorithms and Simulated Annealing*, edited by L. Davis. London: Pittman, 1987.
2. Davis, L., ed. *Genetic Algorithms and Simulated Annealing.* London: Pittman, 1987.
3. De Jong, Kenneth A. "Genetic Algorithms: A 10 year Perspective." In *Proceedings of an International Conference on Genetic Algorithms and Their Applications*, edited by J. J. Grefenstette. Hillsdale, NJ: Lawrence Erlbaum Associates, 1985.
4. Goldberg, D. E. *Genetic Algorithms in Search, Optimization, and Machine Learning.* Reading, MA: Addison-Wesley, 1989.
5. Hillis, W. Daniel. "Co-Evolving Parasites Improve Simulated Evolution as an Optimization Procedure." In *Emergent Computation: Self-Organizing, Collective, and Cooperative Computing Networks.* edited by S. Forrest. Cambridge, MA: MIT Press, 1990. Also in this volume.
6. Holland, J. H. *Adaptation in Natural and Artificial Systems.* Ann Arbor, MI: University of Michigan Press, 1975.
7. Holland, John H. "Escaping Brittleness: The Possibilities of General-Purpose Learning Algorithms Applied to Parallel Rule-Based Systems." In *Machine Learning: An Artificial Intelligence Approach*, edited by R. S. Michalski, J. G. Carbonell, and T. M. Mitchell, vol. II, 593–623. Los Altos, CA: Morgan Kaufman, 1986.
8. Holland, J. H. "ECHO: Explorations of Evolution in a Miniature World."
9. Jefferson, David, Robert Collins, Claus Cooper, Michael Dyer, Margot Flowers, Richart Korf, Charles Taylor, and Alan Wang. "Evolution as a Theme in Artificial Life: The Genesys System." This volume.
10. Koza, John R. "Hierarchical Genetic Algorithms Operating on Populations of Computer Programs." In *Proceedings of the 11th International Joint Conference on Artificial Intelligence (IJCAI).* San Mateo, CA: Morgan Kaufman, 1989.
11. Koza, John R. *Genetic Programming: A Paradigm for Genetically Breeding Populations of Computer Programs to Solve Problems.* Stanford University Computer Science Department Technical Report STAN-CS-90-1314, June 1990.
12. Miller, J. H. "The Co-Evolution of Automata in the Repeated Prisoner's Dilemma." Santa Fe Institute Report 89-003, 1989.
13. Miller, J. H. "The Evolution of Automata in the Repeated Prisoner's Dilemma." In *Two Essays on the Economics of Imperfect Information.* Ph.D. Dissertation, Department of Economics, University of Michigan, 1988.
14. Schaffer , J. D., ed. *Proceedings of the Third International Conference on Genetic Algorithms.* San Mateo, CA: Morgan Kaufmann, 1989.

15. Smith, Steven F. "A Learning System Based on Genetic Adaptive Algorithms." Ph.D. Dissertation, University of Pittsburgh, 1980.
16. Smith, John Maynard. *Evolutionary Genetics.* Oxford: Oxford University Press, 1989.
17. Travers, Michael, and Mitchel Resnick. "Behavioral Dynamics of an Ant Colony: Views from Three Levels." Videotape from the MIT Media Laboratory. Presented at Conference on Simulation of Adaptive Behavior, Paris, September, 1990.

Bruce MacLennan
Computer Science Department, University of Tennessee, Knoxville, TN 37996-1301; internet address: maclennan@cs.utk.edu

Synthetic Ethology: An Approach to the Study of Communication

A complete understanding of communication, language, intentionality, and related mental phenomena will require a theory integrating mechanistic explanations with ethological phenomena. For the foreseeable future, the complexities of natural life in its natural environment will preclude such an understanding. An approach more conducive to carefully controlled experiments and to the discovery of deep laws of great generality is to study synthetic life forms in a synthetic world to which they have become coupled through evolution. This is the approach of *synthetic ethology*. Some simple synthetic ethology experiments are described in which we have observed the evolution of communication in a population of simple machines. We show that even in these simple worlds we find some of the richness and complexity found in natural communication.

I am an "old bird," ...a Simorg, an "all-knowing Bird of Ages"...

— DeMorgan[10]

1. THE PROBLEM

Language, communication, and other mental phenomena have been studied for many centuries, yet some of the central issues remain unresolved. These include the mechanisms by which language and communication emerge, the physical embodiment of mental states, and the nature of intentionality. I will argue below that answering these questions requires a deep theoretical understanding of communication in terms of the relation between its mechanism and its role in the evolution of the communicators. This is one of the goals of *ethology*, which "is distinguished from other approaches to the study of behavior in seeking to combine functional and causal types of explanation."[24] Our approach differs from traditional ethological methods in that it seeks experimental simplicity and control by studying synthetic organisms in synthetic environments, rather than natural organisms in natural environments; it is thus called *synthetic ethology.*

To explain why we expect synthetic ethology to succeed where other methods have failed, it is necessary to briefly review the previous approaches. In doing this I will focus on a single issue: How can a symbol come to *mean* something?

1.1 PHILOSOPHICAL APPROACHES

Although philosophical methods are quite different from those proposed here, the investigations of several philosophers lend support to synthetic ethology. To see why, consider the denotational theory of meaning, in which the meaning of a word is the thing that it denotes. This theory, which is commonly taken for granted, works well for proper names ("Bertrand Russell" denotes a particular person; "Santa Fe" denotes a particular city), but becomes less satisfactory with increasingly abstract terms. Even for concrete general terms ("dog", "mountain") it is already difficult to say exactly what they denote, as evidenced by 2500 years of debate over the nature of universals. Verbs are even more problematic, and a denotational theory of terms such as "of" and "the" seems hopeless.

In this century denotational theories of meaning came under attack from Wittgenstein and other "ordinary language" philosophers.[34] They pointed out that only a small number of linguistic forms can be understood in terms of their denotation; a more generally applicable theory must ground the meaning of language in its use in a social context. For example, in a simple question such as "Is there water in the refrigerator?", the term "water" cannot be taken to have a simple denotational meaning (such as a certain minimum number of H_2O molecules). Rather, there is a common basis of understanding, grounded in the speaker's and hearer's mutual interests and in the context of the utterance, that governs the quantity, state, purity, spatial configuration, etc., that a substance in the refrigerator should have to elicit a truthful "yes" response. To understand the meaning of "water" we must know the function of the word in its contexts of use. Even scientific terms (e.g., length, mass, energy) acquire their meaning through measurement practices that form a common basis of understanding among scientists.

Heidegger makes very similar points, although with a different purpose.[13,15,16] He shows how our everyday use of language is part of a culturally constituted nexus of needs, concerns, and skillful behavior. In his terms this nexus is a "world," and thus our linguistic behavior both is defined by and contributes to defining the various "worlds" in which we dwell: consider common expressions such as "the world of politics," "the academic world," and "the world of science." Meaning emerges from a shared cultural background of beliefs, practices, expectations, and concerns. (Related ideas are discussed by Preston.[26])

One consequence of these views of language is that the study of language cannot be separated from the study of its cultural matrix. Thus one might despair that we will ever have a scientific theory of meaning. Fortunately, another philosopher, Popper, has shown a possible way out of this difficulty: "The main task of the theory of human knowledge is to understand it as continuous with animal knowledge; and to understand also its discontinuity—if any—from animal knowledge."[25] This is a very unconventional view of epistemology; traditionally philosophers have limited their attention to human knowledge, in particular to its embodiment in human language. Although Heidegger and others have helped to bring non-verbal knowledge into the scope of philosophical investigation, Popper goes a step further, by indicating the importance of animal knowledge.

The importance of Popper's observation for the study of language and the mind is that it encourages us to study these phenomena in the context of simple animals in simple environments. Science usually progresses fastest when it is able to study phenomena in their simplest contexts. We expect this will also be the case with communication and other mental phenomena: we will learn more if we start by studying their simplest manifestations, rather than their most complex (i.e., in humans).

1.2 THE BEHAVIORIST APPROACH

The preceding observations might suggest a *behaviorist* approach, since communication is a behavior and behaviorist experiments often involve simple animals in simple environments. But the behaviorist approach is inadequate for several reasons. First, it suffers from *ecological invalidity.* Animals behave in abnormal ways when put in alien environments, but what could be more alien than a Skinner box? As a result, the behavior of animals in laboratory situations may do little to inform us of their behavior in their natural environments.

Second, behaviorism investigates little snippets of behavior, such as pressing a lever to get some food, an approach that removes these behaviors from the pragmatic context that gives them their meaning. The result is an investigation of meaningless behavior resulting from a *lack of pragmatic context.* An example will illustrate the pitfalls of this approach. On the basis of behavioristic tests it had been thought that honey-bees were color-blind. However, von Frisch showed that in a feeding context they were able to distinguish colors. In the captive, laboratory context the color of lights was not relevant to the bees.[24]

In principle, of course, we could design experimental situations that mimic the natural environment in just the relevant ways and simplify it in ways that don't distort the phenomena. Unfortunately, we don't yet adequately understand the pragmatics of real life, and so we don't know how to design laboratory environments that match the natural environments in just the relevant ways. Therefore, the behaviorist approach is, at very least, premature.

1.3 THE ETHOLOGICAL APPROACH

An alternative approach to the study of communication is found in *ethology*, which is in part a reaction against behaviorism. Ethology recognizes that the behavior of an organism is intimately coupled (through natural selection) with its environment. Therefore, since removing an organism from its environment destroys the context for its behavior, ethology advocates studying animals in their own worlds (or in laboratory situations which closely approximate the natural environment). Unfortunately there are difficulties with this approach.

First, the real world, especially out in the field, is very messy; there are too many variables for clean experimental design. Consider some of the factors that could plausibly affect the behavior of a group of animals: the distribution of other animals and their behavior, the distribution of plants and their growth, the terrain, the weather, ambient sounds and odors, disease agents, etc.[29] Animals are much too sensitive to their environments to permit a cavalier disregard for any of these factors.

Second, there are practical and ethical limits to the experiments we can perform. The ethical limits are most apparent where human behavior is the subject, but the situation differs only in degree where other animals are concerned. Even in the absence of ethical constraints, control of many variables is difficult.[29] Some of the experiments we would most like to perform are completely beyond our capabilities, such as restarting evolution and watching or manipulating its progress.

These two problems—the large number of variables and our inability to control them—make it unlikely that deep ethological laws will be discovered in the field. The history of the other sciences shows that deep, universal laws are most likely to be found when the relevant variables are known and under experimental control. When this is not the case, the best we can hope for is statistical correlation; causal understanding will elude us. Of course, I'm not claiming that empirical ethology is futile, only that it is very hard. Rather I anticipate that synthetic and empirical ethology are complementary approaches to the study of behavior, and that there will be a fruitful exchange between them.

1.4 THE NEUROPSYCHOLOGICAL APPROACH

Behaviorist and ethological investigations of communication are limited in an additional way: they tell us nothing of the *mechanism* by which animals communicate. They are both based on *black-box* descriptions of behavior. On the other hand, deep scientific laws are generally based on a causal understanding of the phenomena. Thus it is important to understand the mechanism underlying meaning and other mental phenomena.[21] Several disciplines investigate the mechanisms of cognition. One is neuropsychology. Unfortunately, the complexity of biological nervous systems is so great that the discovery of deep laws seems unlikely, at least at the current stage of the science. Furthermore, as we've seen, true understanding of communication and other mental phenomena requires them to be understood in their ecological context. Thus a complete theory of communication must unite the neuropsychological and ethological levels. This is far beyond the reach of contemporary science.

1.5 THE ARTIFICIAL INTELLIGENCE APPROACH

Another discipline that investigates cognitive mechanisms is artificial intelligence, but with the goal of creating them, rather than studying their naturally occurring forms. Since AI creates its subject matter, all the variables are in its control, and so it might seem that AI is an ideal vehicle for studying communication, meaning, and the mind. Unfortunately, as is well known, there's much argument about whether AI systems can—even in principle—exhibit genuine understanding. In other words, it is claimed that, since AI systems perform meaningless (syntactic) symbol manipulation, they lack just the properties we want to study: meaningful (semantic and pragmatic) symbol use and genuine intentionality. I will briefly review the key points.

The issue can be put this way: Are AI programs *really* intelligent or do they merely *simulate* real intelligence? Several well-known examples make the difference clear. It has been pointed out that no one gets wet when a meteorologist simulates a hurricane in a computer; there is an obvious difference between a real hurricane and a simulated hurricane.[27,28] Similarly, it is been observed that thinking, like digestion, is tied to its biological context. The same chemical reactions will not be digestion if they take place in a flask, that is, out of the context of a stomach serving its functional role in the life of an organism. By analogy it is claimed that there cannot be any *real* thinking outside of its biological context: just as the flask is not digesting, so the computer is not thinking. It has also been claimed that computers may be able to *simulate* meaningful symbolic activity, but that symbols cannot *really* mean anything to a computer. In particular, any meaning born by machine-processed language is meaning that is derived from *our* use of the language. The rules we put into the machine reflect the meaning of the symbols to us; they have no meaning to the machine. That is, our linguistic behavior has *original intentionality*, whereas machines' linguistic behavior has only *derived intentionality*.[11,12]

1.6 SUMMARY

Here is the problem in a nutshell. If we want to understand what makes symbols meaningful (and related phenomena such as intentionality), then AI—at least as currently pursued—will not do. If we want genuine meaning and original intentionality, then communication must have real relevance to the communicators. Furthermore, if we are to understand the pragmatic context of the communication and preserve ecological validity, then it must occur in the communicators' natural environment, that to which they have become coupled through natural selection. Unfortunately, the natural environments of biological organisms are too complicated for carefully controlled experiments.

2. SYNTHETIC ETHOLOGY AS A SOLUTION

2.1 DEFINITION OF SYNTHETIC ETHOLOGY

The goal of *synthetic ethology* is to integrate mechanistic and ethological accounts of behavior by combining the simplicity and control of behaviorist methods with the ecological and pragmatic validity of empirical ethology. The idea of synthetic ethology is simple: Instead of studying animals in the messy natural world, and instead of ripping animals out of their worlds altogether, we create artificial worlds and simulated organisms (*simorgs*[1]) whose behavior is coupled to those worlds. Since the simulated organisms are simple, we can study mental phenomena in situations in which the mechanism is *transparent*. In brief, instead of *analyzing* the natural world, we *synthesize* an artificial world more amenable to scientific investigation. This is really just the standard method of experimental science.

Synthetic ethology can be considered an extension of Braitenberg's *synthetic psychology*[4] that preserves ecological validity and pragmatic context by requiring that behavior be coupled to the environment. We ensure this coupling by having the simorgs evolve in their artificial world. Synthetic ethology is also related to *computational neuroethology*,[1,2,8] the principal distinction being that that discipline typically studies the interaction of an *individual* organism with its environment, whereas our investigations require the study of groups of organisms.

2.2 REQUIREMENTS OF A SOLUTION

In the following I argue that synthetic ethology does in fact solve the problems discussed above. First observe that, rather than starting with nature in all its glory, as does empirical ethology, or with denatured nature, as does behaviorism, synthetic ethology deals with complete, but simple worlds. Complexity is added only

[1]The simorg (simurg, simurgh), a monstrous bird of Persian legend, was believed to be of great age and capable of rational thought and speech.

as necessary to produce the phenomena of interest, yet the worlds are complete, for they provide the complete environment in which the simorgs "live" or "die" (persist or cease to exist as structures).

Second, observe that because synthetic ethology creates the worlds it studies, every variable is under the control of the investigator. Further, the speed of the computer allows evolution to be observed across thousands of generations; we may create worlds, observe their evolution, and destroy them at will. Also, such use of simorgs is unlikely to be an ethical issue, at least so long as they are structurally simple.

Finally I claim that synthetic ethology investigates real, not simulated, communication. But how can we ensure that linguistic structures really "mean" something, that *communication* is taking place, and not merely the generation and recognition of meaningless symbols? Wittgenstein has shown that we are unlikely to find necessary and sufficient conditions governing our everyday use of words such as "communication," and he has warned us of the pitfalls of removing words from their everyday contexts. Nevertheless, in the very non-everyday context of synthetic ethology, we need a definition that can be applied to novel situations.

As a first approximation, we might say that something is meaningful if it has relevance to the life of the individual. Perhaps we could go so far as to say it must be relevant to its survival—even if only indirectly or potentially. Relevance to the individual cannot be the whole story, however, since there are many examples of communication that do not benefit the communicator (e.g., the prairie dog's warning call, a mother bird's feigning injury). Thus, as a second approximation we can say that something is meaningful if it is relevant to the survival of the language community.

Additional support for this criterion comes from ethology, which has had to grapple with the problem of defining communication.[5,9,29,31] The means that animals use to communicate, both within and between species, are so varied that identifying an act as communication becomes problematic. One animal scratches the bark of a tree; later another animal notes the scratches and goes a different way. Was it a communication act? The first animal might have been marking its territory, which is a form of communication, or it might simply have been sharpening its claws, which is not.

On the one hand, we might say that a communication act has occurred whenever the behavior of one animal influences the behavior of another, but this definition is useless, since it views almost every behavior as communication. On the other hand, we might say that it is not a communication act unless the first animal *intended* to influence the other's behavior, but this criterion requires us to be able to determine the intent of behaviors, which is very problematic. If it is questionable to attribute intent to a bee, it is reckless to attribute it to a simorg: we need a definition of communication that does not appeal to problematic ideas like "intent."

A definition of communication that is very consistent with our approach has been proposed by Burghardt[5,6]:

> Communication is the phenomenon of one organism producing a signal that, when responded to by another organism, confers some advantage (or the statistical probability of it) to the signaler or his group.

This says that communication must be relevant—in an evolutionary sense—to the signaler. In addition it gives us an operational way of determining if a communication act has taken place: we can compare the fitness of a population in the two situations differing only in whether communication is permitted or suppressed. This is the sort of experiment that can be undertaken in synthetic ethology, but that is infeasible for empirical ethology.

2.3 MAKING REAL WORLDS INSIDE THE COMPUTER

The objection may still be made that any communication that might take place is at best simulated. After all, nothing that takes place in the computer is real, the argument goes; no one gets wet from a hurricane in a computer. To counter this objection I would like to suggest a different way of looking at computers. We are accustomed to thinking of computers as abstract symbol-manipulating machines, realizations of universal Turing machines. I want to suggest that we think of computers as programmable mass-energy manipulators. The point is that the state of the computer is embodied in the distribution of real matter and energy, and that this matter and energy is redistributed under the control of the program. In effect, the program defines the laws of nature that hold within the computer. Suppose a program defines laws that permit (real!) mass-energy structures to form, stabilize, reproduce, and evolve in the computer. If these structures satisfy the formal conditions of life, then they are real life, not simulated life, since they are composed of real matter and energy. Thus the computer may be a real niche for real artificial life—not carbon-based, but electron-based.[2] Similarly, if through signaling processes these structures promote their own and their group's persistence, then it is real, not simulated, communication that is occurring.

3. PRELIMINARY EXPERIMENTS

To illustrate the method of synthetic ethology, I will describe several experiments that have been completed. The goal of these experiments was to demonstrate that genuine communication could evolve in an artificial world. A secondary goal was to

[2]There is no claim here, however, that the simorgs used in these experiments are alive.

accomplish this with the simplest procedure possible, so that the phenomena would be most exposed for observation.[3]

3.1 SET-UP

3.1.1 ENVIRONMENT What are the minimum requirements on a world that will lead to the emergence of communication? First, it must permit some simorgs to "see" things that others cannot; otherwise there would be no advantage in communicating. For example, in the natural world the signaler may perceive something which is out of the range of the receiver's senses, or the signaler may be communicating its own internal state, which is not directly accessible to the receiver. Second, the environment must provide a physical basis for communication: something which the signaler can alter and the alteration of which the receiver can detect. Finally, we want the environment to be as simple as possible, so that the phenomena are manifest.

The solution adopted in these experiments is to give each simorg a *local environment* that only it can "see." The states of the local environments, which we call *situations*, are determined by a random process; therefore there is no way they can be predicted. This means that the only way one simorg can reliably predict another's situation is if the second simorg communicates that information to the first. To provide a medium for potential communication there is also a shared *global environment* in which any simorg can make or sense a *symbol*. Any such symbol replaces the previous contents of the global environment; there can be only one symbol in the "air" at a time. See Figure 1 for the topology of the environment.

In these experiments the situations and symbols (local and global environment states) are just natural numbers representing uninterpreted elements of a finite discrete set. Since we are creating an artificial world, there is no need to equip it with familiar environmental features such as temperature, water supply, food supply, etc. We can define the laws of this universe so that the simorgs will survive only if they interact correctly with the uninterpreted states of this artificial environment. Although these states have no interpretable "meaning," they are not simply syntactic, since they are directly relevant to the continued persistence ("survival") of the simorgs.

[3]Our experiment may be contrasted with that of Werner and Dyer, who also observed the evolution of communication, but in a more complicated synthetic world.[32] That such different experimental designs resulted in qualitatively similar observations is evidence that synthetic ethology can reveal general properties of communication.

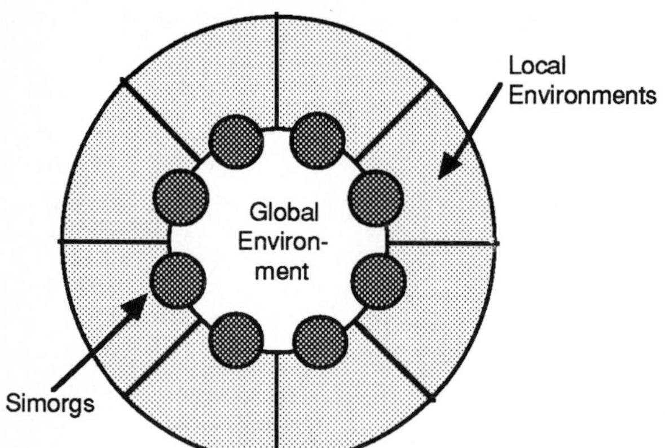

FIGURE 1 Topology of the Environment.

3.1.2 SIMORGS Next consider the simorgs; they should be as simple as possible, yet be capable of evolving or learning complex behaviors. Two simple machine models have the required characteristics, although there are certainly others; they are *finite state machines* (FSMs) and *artificial neural networks* (ANNs). Although ANNs are better models for a variety of reasons,[20] we used FSMs in the experiments described here. (See our progress report for some ANN-based experiments.[23])

Finite-state machines get their name from their internal memory, which at any given time is in one of a finite number of states. In addition, an FSM may have a finite number of sensors and effectors, the states of which are also finite in number. The behavior of an FSM is defined by its *transition table*, which comprises a finite number of discrete rules. For each sensor state s and each internal state i, the table defines an effector state e and a new internal state i'. The machines used in these experiment have only one internal (memory) state. In other words, they have no ability to remember; therefore their response is completely determined by the current stimulus (i.e., their own situation and the shared symbol). In effect, each machine is defined by a table mapping symbol/situation pairs into responses.

There are two kinds of responses, *emissions* and *actions*. The effect of an emission is to change the symbol in the global environment; hence, a response that is an emission must specify the symbol to be emitted. Actions are what must be accomplished effectively for the simorg to survive. Since we are selecting for cooperation we consider a simorg's action effective only if it matches the situation of another simorg. Thus a response that is an action must specify a situation that it is trying to match.

In these experiments we placed an additional requirement on effective action, namely that the action match the situation *of the last emitter*. This increases the selective pressure in favor of communication. Although one may find analogs of this in the natural world (e.g., a predator signaling for appropriate aid in bringing down some prey), the essential point is that we are making an artificial world and so we can define the laws to suit the needs of our experiment.

3.1.3 FITNESS The principal goal of the selective criteria is that they lead to the emergence of communication—without being overly "rigged." In these experiments the environment selects for cooperative activity that requires knowledge of something that cannot be directly perceived, namely another simorg's local environment. Specifically, whenever a simorg acts, its action is compared to the situation of the simorg that most recently emitted. If the two match, then we consider an effective action to have taken place, and both the emitter and actor are given a point of credit. Since several simorgs may respond to a given emitter, a successful emitter can in principle accumulate considerable credit. Each simorg is given an opportunity to respond ten times before all the local environments are changed randomly.[4] This interval is called a *minor cycle*. Credit is accumulated over a *major cycle*, which comprises five minor cycles. The resulting total is considered the simorg's "fitness" for that major cycle, since it measures the number of times the simorg cooperated successfully; it is the criterion by which simorgs are selected to breed or die.

3.1.4 THE BIRTH AND DEATH CYCLE At the end of each major cycle, one simorg is selected to die and two simorgs are selected to breed. This keeps the size of the population constant, which simplifies the simulation and the analysis. Of course, we want the most fit to be most likely to breed and the least fit to be most likely to die.

For reasons discussed later (Section 3.2.3), we use the fitness to determine the *probability* of breeding or dying. In these experiments we made the probability of breeding proportional to the fitness (credit accumulated over one major cycle):

$$p_k = \frac{\phi_k}{P\alpha}$$

where p_k is simorg k's probability of breeding, ϕ_k is its fitness, P is the population size, and $\alpha = P^{-1}\sum_{k=1}^{P}\phi_k$ is the average fitness of the population. (If $\alpha = 0$, we set $p_k = 1/P$.) The probability of dying cannot in general be inversely proportional to fitness. However, we can make it a monotonically decreasing first-degree polynomial of fitness:

$$q_k = \frac{\beta - \phi_k}{P(\beta - \alpha)}$$

where q_k is the probability of dying and β is the fitness of the most fit simorg. (If $\alpha = \beta$, we set $q_k = 1/P$.)

[4]In these experiments the simorgs were serviced in a regular, cyclic fashion. This means that communications with one's nearest neighbors in one direction (say, clockwise) are least likely to be disrupted by other emitters. This may be important in forming "communities" using the same "language" (code).

The offspring is derived from its parents by a simplified genetic process. Each simorg has two transition tables, its genotype and its phenotype. The genotypes of the parents are used to determine the genotype of their offspring by a process described below. In general the genotype defines a developmental process leading to the phenotype, and the phenotype determines the simorg's behavior. In these experiments this process is trivial: the initial phenotype *is* the genotype. Further, if learning is disabled (see Section 3.1.5), then the phenotype remains identical to the genotype.

The genotype of a simorg is a transition table, which defines a response for every symbol/situation pair. Each response is represented, in these experiments, by a pair of numbers, the first of which is 0 or 1, indicating *act* or *emit*, and the second of which is the situation or symbol that goes with the action or emission. The genome itself is just a string containing all these pairs; thus each "gene" defines the response to a given stimulus.

The (unmutated) genotype of the offspring is derived from its parents' genotypes by a process called *crossover*. For purposes of crossover we interpret the genetic string as a closed loop. Two crossover points θ and θ' are selected randomly, and a new genetic string is generated from those of the parents. That is, between θ and θ' the genes will be copied from one parent, and between θ' and θ from the other. Note that our crossover operation never "splits its genes"; it cannot break up a transition table entry. We have found that this leads to faster evolution since the genetic operations respect the structural units of the genetic string. With low probability (0.01 in these experiments), the genetic string may be mutated after crossover. This means that a randomly selected gene is completely replaced by a random allele (i.e., a pair of random numbers in the appropriate ranges).

3.1.5 LEARNING In order to experiment with the effects of learning on the evolution of communication, we have implemented the simplest kind of "single case learning." Specifically, whenever a simorg acts *ineffectively* we change its phenotype so that it *would have acted* effectively. That is, suppose that the global environment state is γ and the local environment state is λ, and that under this stimulus a simorg responds with action λ', but that the situation of the last emitter is $\lambda'' \neq \lambda'$. Then we replace the (γ, λ) entry of the *phenotypic* transition table with the action λ''. (Of course, learning alters the phenotype, not the genotype.) This is a very simple model of learning, and could easily lead to instability; nevertheless it produces interesting results (see Section 3.2).

3.1.6 IMPORTANCE OF OVERLAPPING GENERATIONS Because we are interested in the influence of learning on the evolution of communication, we have done some things differently from typical genetic algorithms.[14,18] GAs typically replace the entire population each generation, with the fitness of the parents determining the frequency with which their offspring are represented in the new generation. In contrast, we replace one individual at a time, with fitness determining the probability of breeding and dying. The difference is significant, because the GA approach prevents the passage of "cultural" information from one generation to the next (through learning). In the current experiment this happens indirectly, since symbol/situation associations are learned through ineffective action. Future experiments may model more direct transmission by having the less successful simorgs imitate the behavior of the more successful. We expect that "cultural" phenomena will be central to understanding the interaction of learning and communication. (See also Belew.[3])

3.1.7 MEASUREMENTS How can we tell if communication is taking place? As noted previously (Section 2.2), Burghardt's definition of communication suggests an operational approach to identifying communication: detect situations in which one simorg produces a signal, another responds to it, and the result is a likely increase in the fitness of the signaler or its group.

In our case, fitness is a direct measure of the number of times that an effective action resulted from a simorg's response to the last emitter. Therefore, the average fitness of the population measures the advantage resulting from actions coincident with apparent communication. But how do we know that the advantage results from communication, and not other adaptations (as it may; see Section 3.2.3)?

I have claimed that synthetic ethology permits a degree of control not possible in natural ethology, and here is a perfect example. We may start two evolutionary simulations with the same population of random simorgs. In one we suppress communication by writing a random symbol into the global environment at every opportunity; in effect this raises the "noise level" to the point where communication is impossible. In the other simulation we do nothing to prevent communication. If true communication—as manifested in selective advantage—is taking place, then the fitness achieved by the two populations should differ. In particular, the *rate* of fitness increase should be significantly greater when communication is not suppressed. This is the effect for which we must watch.

In these experiments we record several fitness parameters. The most important is α, the average fitness of the population (smoothed by a rectangular window of width 50). The second most important is β, the fitness of the most-fit simorg at the end of each major cycle (similarly smoothed). The figures in this chapter show the evolution of α; the evolution of β is qualitatively similar.[22]

I am proposing synthetic ethology as a new way to study communication. Therefore, if by the process just described we find that communication is taking place, then we must see what the simulation can tell us about it. At this stage in the research program we have addressed only the most basic questions: What are the meanings of symbols, and how do they acquire them?

To answer these questions we construct during the simulation a data structure called a *denotation matrix*. This has an entry for each symbol/situation pair, which is incremented whenever there is an apparent communication act involving that pair. If symbols are being used in a haphazard fashion, then all the pairs should occur with approximately the same frequency; the matrix should be quite uniform. On the other hand, if the symbols are being used in a very systematic way, then we should expect there to be one situation for each symbol, and vice versa.[5] Each row and each column of the denotation matrix should have a single nonzero entry, and these should all be about equal; this is a very nonuniform matrix, which we will call the *ideal* denotation matrix. Thus systematic use of symbols can be detected (and quantified) by measuring the *variation* (or *dispersion*) of the denotation matrix.

One of the simplest measures of variation is the standard deviation, which is zero for a uniform distribution, and increases as the distribution spreads around the mean. However, the standard deviation is not convenient for comparing the uniformity of denotation matrices between simulations, since the mean may vary from run to run. Instead, we use the *coefficient of variation* (V), which measures the standard deviation (σ) in units of the mean (μ):

$$V = \sigma / \mu.$$

The coefficient of variation is 0 for a uniform denotation matrix, and for the ideal matrix is $\sqrt{N-1}$ (where N is the number of global or local environment states, which are assumed equal).

Another measure of uniformity is the entropy of a distribution, which is defined[6]:

$$H = -\sum_k p_k \log p_k.$$

This is maximized by the uniform distribution; since there are N^2 equally likely states, its entropy is $2 \log N$. The minimum entropy $H = 0$ is achieved by the "delta distribution" (which makes all the probabilities zero except one). This is not so interesting, however, as the entropy of the ideal matrix, which is easily calculated to be $\log N$. To allow comparisons between simulation runs, we also use a "disorder measure":

$$\eta = \frac{H}{\log N} - 1.$$

This is a scaled and translated entropy, which has the value 1 for a uniform matrix, 0 for the ideal matrix, and -1 for the "overstructured" delta matrix.

There are a variety of other statistical measures that may be used to quantify the structure of the denotation matrix. For example, χ^2 will be 0 for the uniform matrix and maximum for the ideal matrix. Fortunately the results we have observed so far are robust in that they are qualitatively the same no matter what statistics are used.

[5] This is assuming that the number of local environment states equals the number of global environment states, as it does in these experiments. We discuss later (Section 4) the consequences of having unequal numbers of states.

[6] We use logarithms to the base 2, so that our entropy measure is more easily interpreted.

3.2 RESULTS

Unless otherwise specified, the experiments described here used a population size $P = 100$ of finite-state machines with one internal state. Since the number of local and global environment states were the same, $N = 8$, each machine was defined by a transition table containing 64 stimulus/response rules. Simulations were generally run for 5000 major cycles (one birth per major cycle).

3.2.1 EFFECT OF COMMUNICATION ON FITNESS on Figure 2 shows the evolution of the (smoothed) average fitness (α) of a typical random initial population when communication has been suppressed and learning has been disabled. It can be observed to have wandered around the fitness expected for machines that are guessing, $\alpha = 6.25$. (The analysis may be found in an earlier report.[22]) Linear regression detects a slight upward trend ($\dot{\alpha} = 1.55 \times 10^{-5}$). This is a stable phenomenon across simulations, and is explained later (Section 3.2.3).

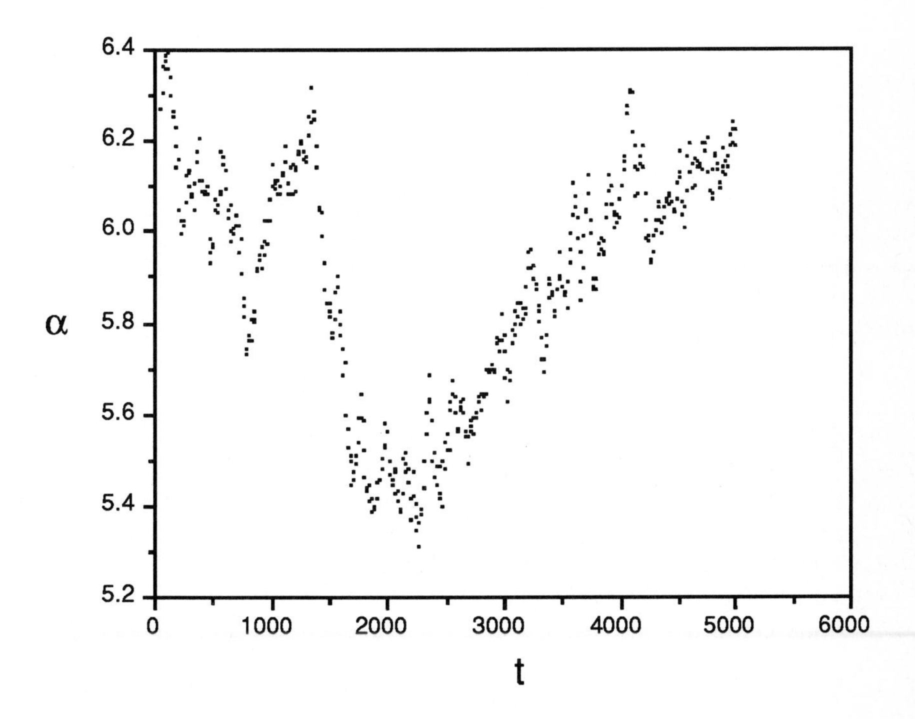

FIGURE 2 Average fitness for 5000 birth periods, with communication suppressed and learning disabled periods. The fitness fluctuates around the value expected for guessing, $\alpha = 6.25$, but has a slight upward trend.

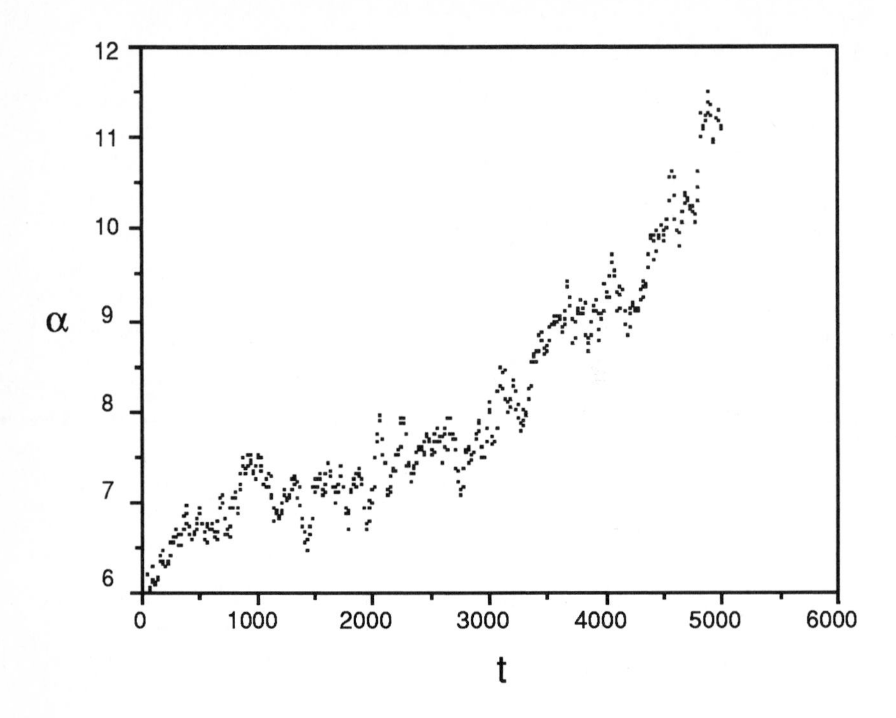

FIGURE 3 Average fitness for 5000 birth periods, with communication permitted and learning disabled. Evolution of communication causes fitness to increase 50 times faster than as when communication was suppressed.

Figure 3 shows the evolution of the average fitness for the same initial population as Figure 2, but with communication permitted (learning still disabled). Within 5000 major cycles the average fitness reaches $\alpha = 11.5$, which is significantly above the guessing level ($\alpha = 6.25$). Furthermore, linear regression shows that the average fitness is increasing over 50 times as fast as when communication was suppressed ($\dot{\alpha} = 8.25 \times 10^{-4}$ vs. $\dot{\alpha} = 1.55 \times 10^{-5}$). We conclude that in this experiment communication has a remarkable selective advantage.

Figure 4 shows the evolution of α for the same initial population, but with communication permitted and learning enabled. First observe that the average fitness begins at a much higher level ($\alpha \approx 45$) than in the previous two experiments. This is because each simorg gets ten opportunities to respond to a given configuration of local environment states. Since learning changes the behavior of a simorg so that its response would have been correct, an incorrect response could be followed by up to nine correct responses (provided no intervening emissions change the global environment). The combination of communication and learning allowed the average fitness to reach 55, which is nearly five times the level reached without learning and nearly nine times that achieved without communication. The rate of fitness increase was $\dot{a} = 2.31 \times 10^{-3}$, which is almost three times as large as that without learning, and nearly 150 times as large as that without communication.

We have observed quantitatively similar results in many experiments. Table 1 (adapted from an earlier report[22]) shows average measurements from several experiments that differ only in initial population.

To better understand the asymptotic behavior of the evolutionary process, we have run several simulations for ten times as long as those previously described. Figure 5 shows the evolution with communication permitted but learning disabled, and Figure 6 shows the evolution of the same initial population, but with communication permitted and learning enabled. In the first case, average fitness reached a level of approximately 20.[7] In the second, (learning permitted) α seems to have reached an equilibrium value ($\overline{\alpha} = 56.6$ in fact); we can also observe an apparent "genetic catastrophe" at about $t = 45,000$.

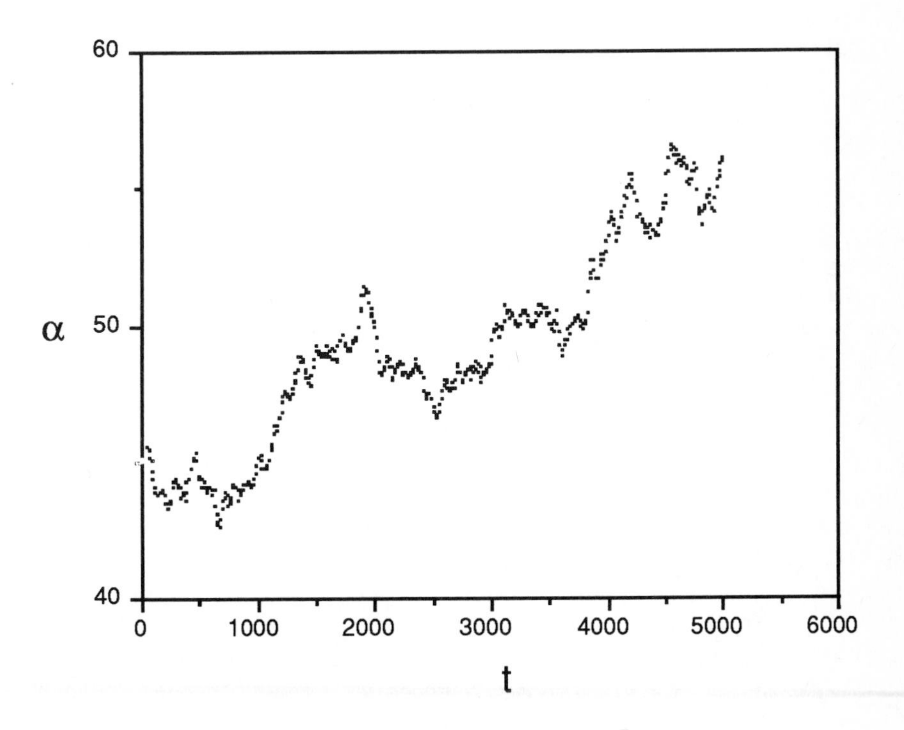

FIGURE 4 Average fitness for 5000 birth periods, with communication permitted and learning enabled. Evolution of communication together with simple learning allows fitness to three times faster than without learning, and 150 times faster than when communication was suppressed.

[7]Under reasonable assumptions, the maximum α achievable without learning by a homogeneous population can be calculated to be 87.5; details are presented elsewhere.[22]

TABLE 1 Average Measurements over Several Random Populations

Measurement	Comm/Learning N/N	Y/N	Y/Y
α	6.31	11.63	59.65
$\dot{a} \times 10^4$	0.36	11.0	28.77
V	0.47	2.58	2.65
H	5.81	3.79	3.87
η	0.94	0.26	0.29

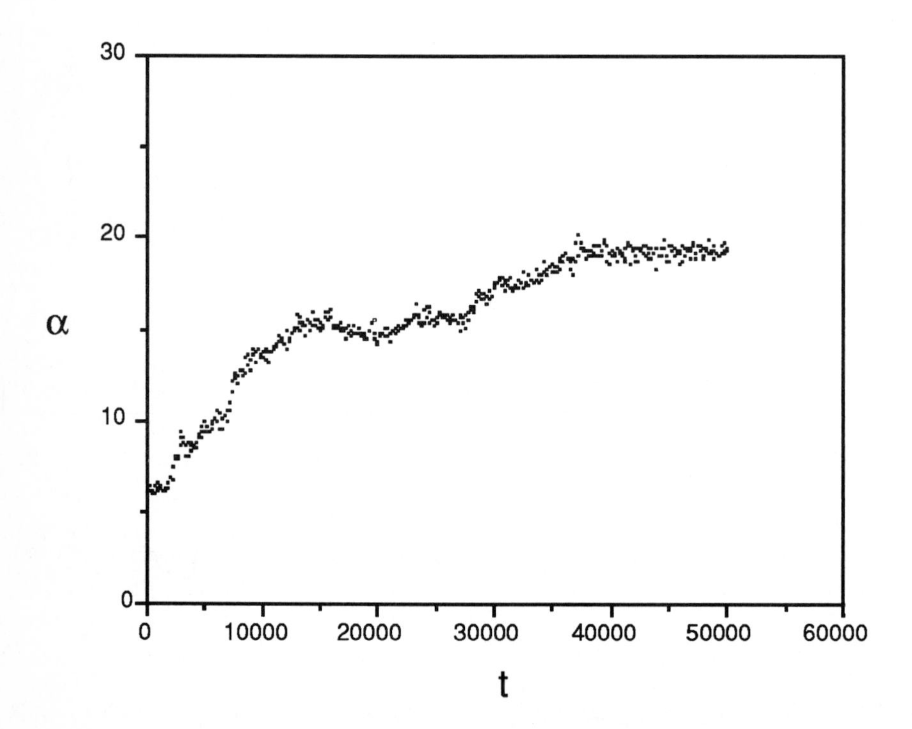

FIGURE 5 Average fitness, with communication permitted and learning disabled over a longer time period, 50000 births.

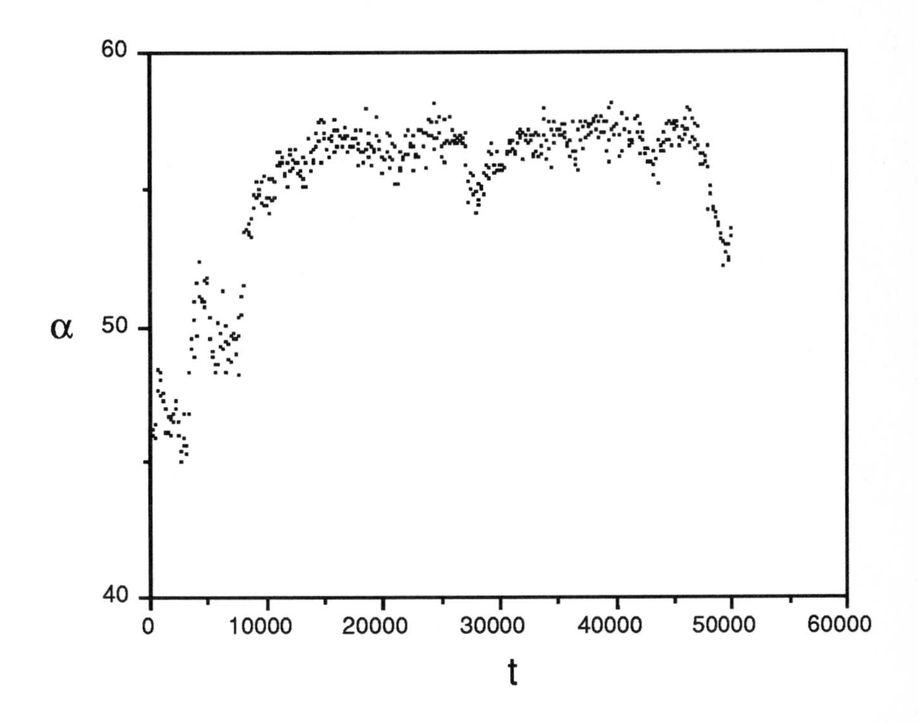

FIGURE 6 Average fitness, with communication permitted and learning enabled over a longer time period, 50000 births. Notice that the fitness has reached a plateau.

The greatly increased fitness that results from not suppressing the signaling process supports the claim that we are observing *genuine* communication. The communication acts have real relevance to the simorgs because they significantly affect the survival of the signaler and its group (cf. Burghardt's definition, Section 2.2).

3.2.2 ANALYSIS OF DENOTATION MATRICES If genuine communication is taking place, then we ought to be able to observe it in more structured use of symbols; therefore we consider the structure of the resulting denotation matrices. First consider Table 2; this is the denotation matrix from the same simulation shown in Figure 2. In the absence of communication and learning we see a very uniform matrix, as measured by its coefficient of variation $V = 0.41$ and entropy $H = 5.87$, which is nearly the maximum possible, 6. This is also reflected in the disorder parameter $\eta = 0.96$; recall that a uniform matrix has $\eta = 1$ and an "ideal" matrix has $\eta = 0$.

TABLE 2 Denotation Matrix, Communication Suppressed and Learning Disabled

symbol	situation 0	1	2	3	4	5	6	7
0	320	138	189	360	266	354	224	89
1	364	130	189	359	261	342	266	75
2	332	126	184	385	252	365	257	82
3	350	125	193	366	257	351	255	98
4	340	119	190	354	254	356	225	78
5	328	145	170	343	244	348	217	86
6	345	119	194	374	214	361	237	78
7	346	149	159	343	242	383	226	83

$V = 0.409451$
$H = 5.868233$
$\eta = 0.9560777$

TABLE 3 Denotation Matrix, Communication Permitted and Learning Disabled

symbol	situation 0	1	2	3	4	5	6	7
0	695	5749	0	1157	0	2054	101	0
1	4242	11	1702	0	0	0	1	0
2	855	0	0	0	0	603	862	20
3	0	0	0	0	1003	430	0	1091
4	0	0	0	0	0	0	2756	464
5	0	0	40	0	548	0	817	0
6	1089	90	1	281	346	268	0	62
7	0	201	0	288	0	0	2	0

$V = 2.272352$
$H = 3.915812$
$\eta = 0.3052707$

TABLE 4 Denotation Matrix, Communication Permitted and Learning Enabled

	situation							
symbol	0	1	2	3	4	5	6	7
0	0	0	2946	0	0	635	4239	3233
1	2084	0	672	1457	0	6701	8517	1284
2	0	0	646	433	0	230	63	879
3	0	1074	446	46	2315	1623	0	1265
4	27850	5504	0	2326	11651	243	3428	20076
5	1301	0	0	854	858	368	0	0
6	13519	2676	0	2223	2391	874	0	644
7	356	226	365	107	1357	27	100	1

$V = 2.165397$
$H = 4.208782$
$\eta = 0.4029273$

Table 3 shows the denotation matrix that results when communication is permitted; even to the eye it is much more structured than Table 2. This is confirmed by our measurements: $V = 2.27$ (cf. $V = 2.65$ for the ideal matrix), $H = 3.92$, $\eta = 0.31$.

Finally, Table 4 is the denotation matrix resulting from both communication and learning. Qualitatively and quantitatively it is very similar to Table 3, but slightly *less* structured. This phenomenon is even more apparent in longer simulations, such as the $t = 50000$ simulations shown in Figures 5 and 6. In these, evolution in the absence of learning produced a denotation matrix having $\eta = -0.2$, indicating an *overstructured* language, whereas evolution with learning produced a less structured language ($\eta = +0.2$) but higher fitness.[8] This phenomenon seems to be consistent with research indicating that there is an optimal degree of structure[19] and that that optimum is more easily achieved with learning.[17]

The "ideal" denotation matrix has one symbol for one situation and vice versa; this is a structure that we might expect to see emerging. For example, in the denotation matrix in Table 3 there is at least one symbol that predominantly denotes a single situation: in 86% of its recent uses, symbol 4 denoted situation 6, in the remainder situation 7. Since these are the only two uses of symbol 4, it seems likely that the denotation matrix reflects two subpopulations (of unequal size) using the

[8]See Tables 17 and 18 in our earlier report[22] for the denotation matrices resulting from these experiments.

same symbol for different situations. More nearly equal subpopulations may be indicated by symbols such as 7, which is used for situations 1 and 3 with nearly equal frequency.

Symbols being used to denote several situations may also result from their being used equivocally by a single population; they could reflect an intermediate stage in the evolution to univocal symbol use. It is difficult to discriminate between these two possibilities on the basis of just the denotation matrix. Doing so requires more detailed analysis of the simorgs in the final population, a process which is straightforward in synthetic ethology, since we have complete access to the structure of the simorgs. (Simple examples of this kind of analysis are presented in our report.[22])

The natural way to interpret the denotation matrix is by rows, which reflects the significance of a symbol to a recipient; ethologists sometimes call this the *meaning* of a signal.[7,29,30,31] We can also look at the denotation matrix by columns, which shows the situation a signaler was expressing by a symbol; ethologists call this the symbol's *message*.[7,29,30,31] Sometimes the two are symmetric. For example, in Table 3 the meaning of symbol 4 is usually (86%) situation 6, and the message 'situation 6' is usually (61%) represented by symbol 4. On the other hand, asymmetries may occur. Symbol 6 usually (51%) means situation 0, but situation 0 is usually (62%) represented by symbol 1. Conversely, situation 2 is usually (98%) represented by symbol 1, but symbol 1 usually (71%) means situation 0.

Even in a synthetic ethology experiment as simple as this, we may begin to observe some of the richness and complexity of real communication. For example, in the actual "language" or code reflected in the evolved denotation matrix—as opposed to the ideal matrix given by theory—we find that there is rarely a one-to-one (univocal) correspondence between symbols and situations. Indeed, it is quite possible that a simorg will attach different significance to a symbol when it is received or when it is emitted; that is, a simorg need not associate the same meaning and message to a given symbol. If this is the case for simorgs, then it would seem foolish to assume that in human languages an utterance has the same pragmatic significance when it is spoken as when it is heard.

The denotation matrix captures the actual use of the code by the entire population over the last 50 major cycles of the simulation. In this sense it is an irreducible description of the message and meaning associated with every symbol. It is irreducible because any attempt to ignore the lesser entries and specify a unique denotational meaning for a symbol will misrepresent the facts of communication. In fact, symbol 4 means situation 7 some (16%) of the time; this is part of the overall meaning of symbol 4 in *that* population at *that* time. To say that symbol 4 *really* means situation 6, and that the rest is noise, is a misrepresentation of the "language."

Given that the denotation matrix is the irreducible description of the code, we see that the evolution of the code is mirrored in the evolution of the denotation matrix. Indeed, in the denotation matrix we may see the code as an emergent nonequilibrium system, which organizes itself by promoting the fitness of simorgs that behave in accord with its emerging structure.[21] This emerging structure is measured by the decreasing entropy of the denotation matrix.

Over time we may observe a changing constellation of meanings associated with a given symbol, and of the symbols representing a given message. We have already seen that these experiments indicate both synonymous and equivocal symbols. The experiments also exhibit both context-sensitive emission and context-sensitive interpretation of symbols. This is because the emission of a symbol by a simorg may depend on the global environment (providing a context) as well as its local environment. Similarly, the response of a simorg to a symbol depends on its situation, which supplies a context. Finally, we observe that the differing use of symbols in various contexts makes it quite possible for every simorg to be using a different dialect of the "language" manifest in the denotation matrix. Even in these simple experiments we can begin to appreciate the complexity of the relation between symbols and their significance.

3.2.3 OTHER OBSERVATIONS In the course of these experiments we have made several observations that provide some insight into the evolution of communication.

All of our experiments in which communication (and learning) is suppressed show a slight upward trend in fitness (see Figure 2 and Table 1). This is surprising, since in the absence of communication it would seem that there is no way to improve on guessing. However, that is not the case, and the way that it can occur is an interesting demonstration of the force of the evolutionary process. To see this, observe that our definition of effective action (Section 3.1.3) permits a kind of "pseudo-cooperation through coadaptation." Specifically, a simorg is credited whenever its action matches the situation of the last emitter, which is also credited. Therefore, if the population contains a group of simorgs that emit only when they are in a fixed subset E of situations, then the possible states of the last emitter will not be equally likely; specifically states in E will be more likely than the other states. Under these conditions a simorg can "beat the odds" by always guessing a situation in E. The coadaptation of such "pseudo-cooperating" groups of simorgs seems to account for the increase of fitness even when communication is suppressed.

We checked this hypothesis in several ways. First, we compared simulations with the usual scoring algorithms to those in which fitness was credited by a match to *any* other simorg (vice just the last emitter); this eliminated the possibility of pseudo-cooperation. As expected, there was no trend in the average fitness. Second, we inspected the denotation matrices; doing so showed that emissions occurred in only a subset of the situations. Third, we calculated the expected average fitness for homogeneous populations and subsets E of the observed size. With the parameters we used, and the observed size 3 for E, we calculated the expected fitness to be $\alpha = 20.83$; in four simulations we observed $\alpha = 20, 29, 21, 23$. Together these are strong evidence in favor of the hypothesis.

Pseudo-cooperation can be eliminated by not favoring a match to the most recent emitter. Unfortunately, this removes much of the selective pressure toward communication (since it makes guessing almost as good a strategy as communicating) and therefore slows the simulations. For this reason we have retained the original scoring rule; in most cases pseudo-cooperation is a low level effect that is unintrusive and can be ignored.

Another observation arose from earlier, unsuccessful experiments. Recall that fitness determines the probability of breeding or dying; there is always a chance that the least fit will breed and that the most fit will die. In earlier experiments we used a simpler approach: breed the two most fit simorgs and replace the least fit. Thus the current algorithm is stochastic, whereas the older one was deterministic (except in the case of fitness ties). The change was made because we never observed the evolution of communication in the deterministic situation.

The reason seems to be as follows. Since only the two most fit simorgs breed, other good, but not great, simorgs are forever excluded from contributing to the gene pool. Since language is hard to get started, it is to be expected that nascent communicators will not be as fit as guessers. Language communities will never evolve, unless they have some chance of breeding, and this seems to be prevented by the brittleness of the deterministic algorithm.

4. CONCLUSIONS

I have argued that a complete understanding of language, communication and the representational capabilities of mental states will require a theory that relates the mechanisms underlying cognition to the evolutionary process. I also argued that the complexity of natural organisms makes it unlikely that such an integrated theory can be found by empirical ethology. Therefore synthetic ethology has been proposed as a complementary research paradigm, since carefully controlled experiments and deep theoretical laws are more likely to be achievable in the comparative simplicity of synthetic worlds.

As an example of synthetic ethology I have described experiments in which we have observed the evolution of communication in a population of simple machines. This was accomplished by constructing a world in which there is selection for cooperation and in which effective cooperation requires communication. The control granted by synthetic ethology permitted us to observe the evolution of *the same population* in two worlds (one in which communication was suppressed, the other permitted), and thus to measure the evolutionary effect of communication. Further, synthetic ethology affords complete access to the structure of the simorgs, thus exposing the mechanisms underlying their communication.

We are hopeful that synthetic ethology will prove a fruitful method for investigating the relation between linguistic and mental structures and the world. The experiments described here are just a beginning, and there are many directions in which to proceed. For example, if the number of situations exceeds the number of symbols, then we would expect the simorgs to string symbols together into "sentences"; this has already been observed, but more experiments are needed to discover the syntax that will emerge and the factors affecting it.

It also seems likely that the complexity of language reflects the complexity of the world. Our experiments to date have used environments that are in one of a finite number of discrete, atomic situations, and the resulting "languages" have been similarly simple. This suggests that we equip our synthetic worlds with environments containing objects in various relationships; we expect this to lead to categories of symbols analogous to the parts of speech (nouns, adjectives, etc.).

To date our experiments have been based on finite, discrete sets of symbols and situations, but much of the natural world is characterized by continuous variation, and both human and animal communication make significant use of continuously variable parameters (loudness, pitch, rate etc.). Ethological studies[33] suggest that discreteness—so called "typical intensity"—will emerge to the extent that communication is noisy, an easy variable to control in synthetic ethology. We hope to address this issue in future experiments and thus identify the principles underlying the emergence of discrete symbolic processes.

ACKNOWLEDGMENTS

Research reported herein was supported in part by Faculty Research Awards (1988, 1989) from the University of Tennessee, Knoxville. I'm also grateful to members of the Cognitive Science Laboratory and the Ethology Program at the University of Tennessee for much helpful advice, and to Beth Preston and the editors for comments on earlier drafts.

REFERENCES

1. Beer, Randall D. *Intelligence as Adaptive Behavior: An Experiment in Computational Neuroethology.* San Diego, CA: Academic Press, 1990.
2. Beer, Randall D., Hillel J. Chiel, and Leon S. Sterling. "A Biological Perspective on Autonomous Agent Design." *Journal of Robotics and Autonomous Systems* **6** (1990): 169–186. Reprinted in: *New Architectures for Autonomous Agents*, edited by Pattie Maes. Cambridge, MA: MIT Press, in press.
3. Belew, Richard K. "Evolution, Learning and Culture: Computational Metaphors for Adaptive Search." *Complex Systems* **4** (1990): 11–49.
4. Braitenberg, Valentino. *Vehicles: Experiments in Synthetic Psychology.* Cambridge, MA: MIT Press, 1984.
5. Burghardt, Gordon M. "Defining 'Communication.'" In *Communication by Chemical Signals*, edited by J. W. Johnston, Jr., D. G. Moulton and A. Turk, 5–18. New York, NY: Appleton-Century-Crofts, 1970.
6. Burghardt, Gordon M. "Ontogeny of Communication." In *How Animals Communicate*, edited by Thomas A. Sebeok, 71–97. Bloomington, IN: Indiana University Press, 1977.
7. Cherry, Colin. *On Human Communication*, 2nd edition, 171, 244. Cambridge, MA: MIT Press, 1966.
8. Cliff, D. T. "Computational Neuroethology: A Provisional Manifesto." University of Sussex, Cognitive Science Research Paper CSRP 162, May 1990.
9. Dawkins, Richard. "Communication." In *The Oxford Companion to Animal Behavior*, edited by David McFarland, 78–91. Oxford, UK: Oxford University Press, 1987.
10. DeMorgan, A. *Budget of Paradoxes*, 329. London, UK: Longmans, Green, 1872.
11. Dennett, Daniel. *The Intentional Stance.* Cambridge, MA: MIT Press, 1987.
12. Dennett, Daniel. "The Intentional Stance." *Behavioral and Brain Sciences* **11** (September 1988): 495–546.
13. Dreyfus, Hubert L. "Introduction." In *Husserl, Intentionality and Cognitive Science*, edited by Hubert L. Dreyfus with Harrison Hall, 1–27. Cambridge, MA: MIT Press, 1982.
14. Goldberg, D. E. *Genetic Algorithms in Search, Optimization, and Machine Learning.* Reading, MA: Addison-Wesley, 1989.
15. Heidegger, Martin. *Being and Time*, translated by J. Macquarrie and E. Robinson. New York, NY: Harper and Row, 1962.
16. Heidegger, Martin. *The Basic Problems of Phenomenology*, translated by Albert Hofstadter. Bloomington, IN: Indiana University Press, 1982.
17. Hinton, G. E., and S. J. Nowlan. "How Learning Can Guide Evolution." *Complex Systems* **1** (1987): 495–501.
18. Holland, J. H. *Adaptation in Natural and Artificial Systems.* Ann Arbor, MI: University of Michigan Press, 1975.

19. Langton, C. G. "Computation at the Edge of Chaos: Phase Transitions and Emergent Computation." *Physica D* **42** (1990): 12–37. Reprinted in: *Emergent Computation*, edited by Stephanie Forrest, 12–37. Amsterdam: North-Holland, 1990.
20. MacLennan, Bruce J. "Logic for the New AI." In *Aspects of Artificial Intelligence*, edited by James H. Fetzer, 163–192. Dordrecht, Holland: Kluwer Academic Publishers, 1988.
21. MacLennan, Bruce J. "Causes and Intentions." *Behavioral and Brain Sciences* **11** (September 1988): 519–520.
22. MacLennan, Bruce J. "Evolution of Communication in a Population of Simple Machines." Technical Report CS-90-99, Computer Science Department, University of Tennessee, Knoxville, January 1990.
23. MacLennan, Bruce, Noel Jerke, Rick Stroud, and Marc VanHeyningen. "Neural Network Models of Cognitive Processes: 1990 Progress Report," Technical Report, Computer Science Department, University of Tennessee, Knoxville, 1990.
24. McFarland, David. *The Oxford Companion to Animal Behavior.* Oxford, UK: Oxford University Press, 1987, 153.
25. Popper, Karl. "Campbell on the Evolutionary Theory of Knowledge." In *Evolutionary Epistemology*, edited by Gerard Radnitzky and W. W. Bartley III, 117. La Salle, IL: Open Court, 1987. Original version: "Replies to My Critics." In *The Philosophy of Karl Popper*, edited by Paul A. Schilpp, 1061. La Salle, IL: Open Court, 1974.
26. Preston, Beth. "Heidegger and Artificial Intelligence." Unpublished, Center for the Philosophy of Science, University of Pittsburgh, 1989.
27. Searle, J. R. "Minds, Brains, and Programs." *Behavioral and Brain Sciences* **3** (1980): 417–424, 450–457.
28. Searle, J. R. "Is the Brain's Mind a Computer Program?" *Scientific American* **262** (January 1990): 26–31.
29. Slater, P. J. B. "The Study of Communication." In *Animal Behavior Volume 2: Communication*, edited by T. R. Halliday and P. J. B. Slater, 9–42. New York, NY: W. H. Freeman, 1983.
30. Smith, W. J. "Message–Meaning Analyses." In *Animal Communication*, edited by T. A. Sebeok, 44–60. Bloomington, IN: Indiana University Press, 1968.
31. Smith, W. J. *The Behavior of Communicating: An Ethological Approach*, Ch. 1. Cambridge, MA: Harvard University Press, 1977. For message/meaning analysis see p. 19.
32. Werner, Gregory M, and Michael G. Dyer. "Evolution of Communication in Artificial Organisms." This volume.
33. Wiley, R. H. "The Evolution of Communication: Information and Manipulation." In *Animal Behavior Volume 2: Communication*, edited by T. R. Halliday and P. J. B. Slater, 156–189. New York, NY: W. H. Freeman, 1983. For typical intensity see pp. 163–164.

34. Wittgenstein, Ludwig. *Philosophical Investigations*, 3rd edition, section 19. New York, NY: Macmillan, 1958.

Gregory M. Werner and Michael G. Dyer
Artificial Intelligence Laboratory, Computer Science Department, UCLA, Los Angeles, CA 90024

Evolution of Communication in Artificial Organisms

A population of artificial organisms evolved simple communication protocols for mate finding. Female animals in our artificial environment had the ability to see males and to emit sounds. Male animals were blind, but could hear signals from females. Thus, the environment was designed to favor organisms that evolved to generate and interpret meaningful signals. Starting with random neural networks, the simulation resulted in a progression of generations that exhibit increasingly effective mate-finding strategies. In addition, a number of distinct subspecies, i.e., groups with different signaling protocols or "dialects," evolve and compete. These protocols become a behavioral barrier to mating that supports the formation of distinct subspecies. Experiments with physical barriers in the environment were also performed. A partially permeable barrier allows a separate subspecies to evolve and survive for indefinite periods of time, in spite of occasional migration and contact from members of other subspecies.

Artificial Life II, SFI Studies in the Sciences of Complexity, vol. X, edited by C. G. Langton, C. Taylor, J. D. Farmer, & S. Rasmussen, Addison-Wesley, 1991

INTRODUCTION

It is our goal to explore the evolution of language from simple genetically controlled signalling to learned patterns of communication that support complex forms of social interaction by simulating environments in which these types of interaction can evolve. As a first step in this direction, we have been exploring the evolution of simple intraspecies signals that are genetically hard coded into the behavior of simulated animals. Such innate signals are commonly found in the animal kingdom.[8]

By simulating environments that exert some pressure to communicate, we believe that we can evolve animal-like communication systems in artificial organisms. As our environments become more complex, we hope to obtain progressively more interesting communication systems. As the animals themselves become more complex, gaining learning ability, for example, we hope to see the evolution of primitive language in our artificial life populations.

Primitive communication is common in many species of real animals. For example, many animals emit signals that communicate their internal states or emotions. Signals representing hunger, fear, anger, or readiness to mate are all common among animal species. Young birds cry for food, lions growl at one another at the site of a kill, frogs issue calls to attract mates, etc. Signals are also used, though less commonly, to communicate something about the state of the external environment to other animals. For example, the location of food or the type of danger present may be signalled. Baboons let the other members of their group know that they have found food so that it can be shared. They also have separate danger signals for the presence of hawks, snakes, and leopards.[8] Such signals seem to have a symbolic component to them. For example, a particular sound literally comes to mean "snake."

This work can also be viewed from another perspective. Agents that work simultaneously and communicate to solve a common problem can be considered a distributed algorithm. We are seeking to find general cases in which distributed algorithms can be evolved to solve non-trivial tasks. Evolving distributed algorithms could help to solve several basic problems in Distributed Artificial Intelligence. (See Bond and Gasser[3] for an overview.) Tasks for which simulated evolution can be useful may be: decomposition of problems into subproblems to be solved by independent agents, selection of communication protocols, and the actual methods of solving each subproblem.

EVOLVING COMMUNICATION

We believe that a number of general principles should be followed when setting up simulations to evolve communication among organisms in artificial environments.

First, there should not be direct pressure on the animals to communicate. Communication should arise as a solution to another problem that has to be solved by the population. It is trivial to set up an evaluation function that directly rewards animals that communicate. This approach, however, does not provide the population with the flexibility to evolve a communication system that the experimenter does not expect. Creative solutions would not be rewarded by an evaluation function that was biased in this way. Animals should not be judged on how well they communicate, but on how well they solve the task at hand. In this way, one can determine how communication aids in tasks that would normally be faced by an evolving population.

Second, it is important to present the populations with natural tasks such as finding food, protecting young, and attracting mates. These are the kinds of tasks that animals and humans faced when creating their communication systems, and such tasks placed important constraints on the development of communication. Since communication plays such an important role in cooperation, tasks that encourage cooperation are good prospects for bringing about communication.

EVOLVING DIRECTIONAL MATING SIGNALS

The specific problem we selected for the population to solve is mate finding. This is a problem that does not require communication, but can be aided by it. A good solution to this problem combines the concurrent evolution of search and signalling strategies.

To put evolutionary pressure on the animals to communicate, we needed to design animals in an environment such that some animals would have information that other animals needed to know but were not capable of finding out for themselves. The animals with this valuable information would have to communicate it to the other animals. The relevant information in this particular simulation is the location of the female animals relative to the male animals. The males do not know the location of female animals, and must listen to directions from females in order to avoid a blind search for mates. To accomplish this, we made the males of our species blind. The females, which we made immobile to avoid having the females simply find the male animals, must produce signals that will guide the males to themselves.

IMPLEMENTATION

The environment for our animals is a simple toroidal grid, 200-by-200 squares. The sides of the grid are wrapped around to avoid having animals getting stuck on the boundaries of the grid. Each of the 40,000 locations in the environment can be empty or occupied by one animal. Typically, we place 800 male animals and 800 female animals into the environment. Therefore, 4% of the locations are occupied at all times.

Each animal in the population has a distinct genome which is interpreted to produce the neural network that controls its actions. Each of the genes of the genome has an 8-bit integer value that corresponds to the connection strength or the bias of a unit in the neural network of the animal.

The neural architecture is a recurrent network in which all hidden units are completely interconnected and have feedback to themselves.[7] Recurrent neural networks differ from feedforward neural nets[12] in that they allow the network to use information about their previous state when producing their current output. Since no learning will take place, the weights and thresholds have been changed to integer values for faster program execution. Weights and biases are simply integers between 127 and -127. All thresholds are zero. The individual locations on the genome encode the weights and biases for the network (Figure 1). Each animal's genome has the encoding for both a male animal and a female animal. The sex of the animal determines which part of the genome is interpreted to create the neural network of the animal.

The female animal is given an "eye" that can sense the location and orientation of animals nearby. Specifically, she can detect any male animal that is within two squares vertically, horizontally, and diagonally of her location. Thus, she sits at the center of a 5-by-5 "visual field." Each of the nearby locations and orientations is associated with an input node in the neural net that is turned on when a male is seen in the particular location and orientation to which the node is sensitive. The female animal has no sensors for seeing or hearing nearby females. Also, unlike males, female animals have no specific orientation.

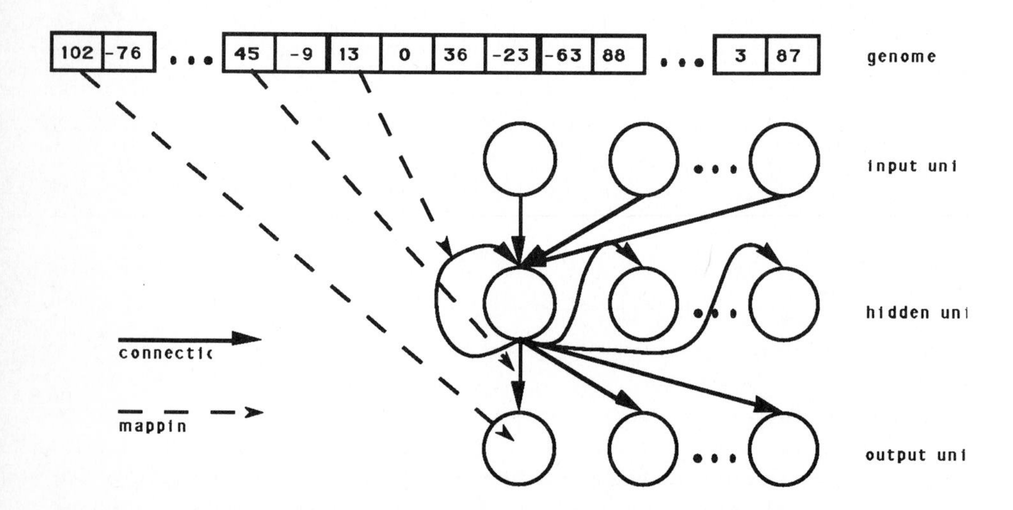

FIGURE 1 Each gene codes a connection weight (or bias) ranging from -127 to 127. Each artificial neuron has a threshold of zero.

The location of a male, within range of the female, produces a pattern of activation on the female's input units ("eye"). The squares around the female are ordered "closest" to "farthest" from her. If more than one male is within her receptive field, only the male that is "closest" is detected. Activation flows through the recurrent net of the female and produces a pattern of activation on her output units. This pattern of activation is interpreted as a "sound" that is transmitted to all of the male animals within the visual field of the female.

The male animals are given an "ear" that can hear these signals produced by nearby females. If more than one female sends a signal to a male, he hears only the sound produced by the female who is closer to him. Ties are broken arbitrarily, but consistently.

The outputs of the male animal are interpreted as moves made by the animal. The four output units of a male correspond to moving forward, standing still, turning left, and turning right. The action corresponding to the output unit with the highest activation is taken.

At each time step, each female's input units are set according to the location of male animals within her visual field. The input values for each of the females are then propagated through the neural net to produce a new output. These outputs, the "sounds" produced by the females, are then fed into the inputs of nearby males. The input values for the males are then propagated to produce the new actions for the male animals. The males are then moved, rotated, or left still, and the time step is complete (Figure 2).

When a male finds a female (moves onto the same grid location that she is on), the animals mate and produce two offspring, a male and a female. The parents' genome, which encode their neural network brains, are combined using the standard genetic operations of crossover and mutation[9] to produce the genome of the offspring. (The mutation rate used was 0.01% per gene. The crossover rate used was 2% per gene.) These offspring replace two old animals in the population and the parents are moved to new locations in the environment so that they can attempt to reproduce again. Simply leaving the animals in the same location would have allowed them to mate repeatedly. The animals removed are selected randomly from the population.

This process, which we call "XGA," is an extension of the typical genetic algorithm (GA) because gene strands are reproduced as soon as they prove their fitness, instead of being compared and reproduced at fixed time intervals. The genetic algorithm[9] traditionally has discrete generations where each of the members of the population is simultaneously judged and possibly paired off with a mate. However, it is important in evolution of language to allow inter-generational communication. For this reason, and for greater realism in simulations, we have made all reproduction asynchronous, thus creating overlapping generations. In addition,

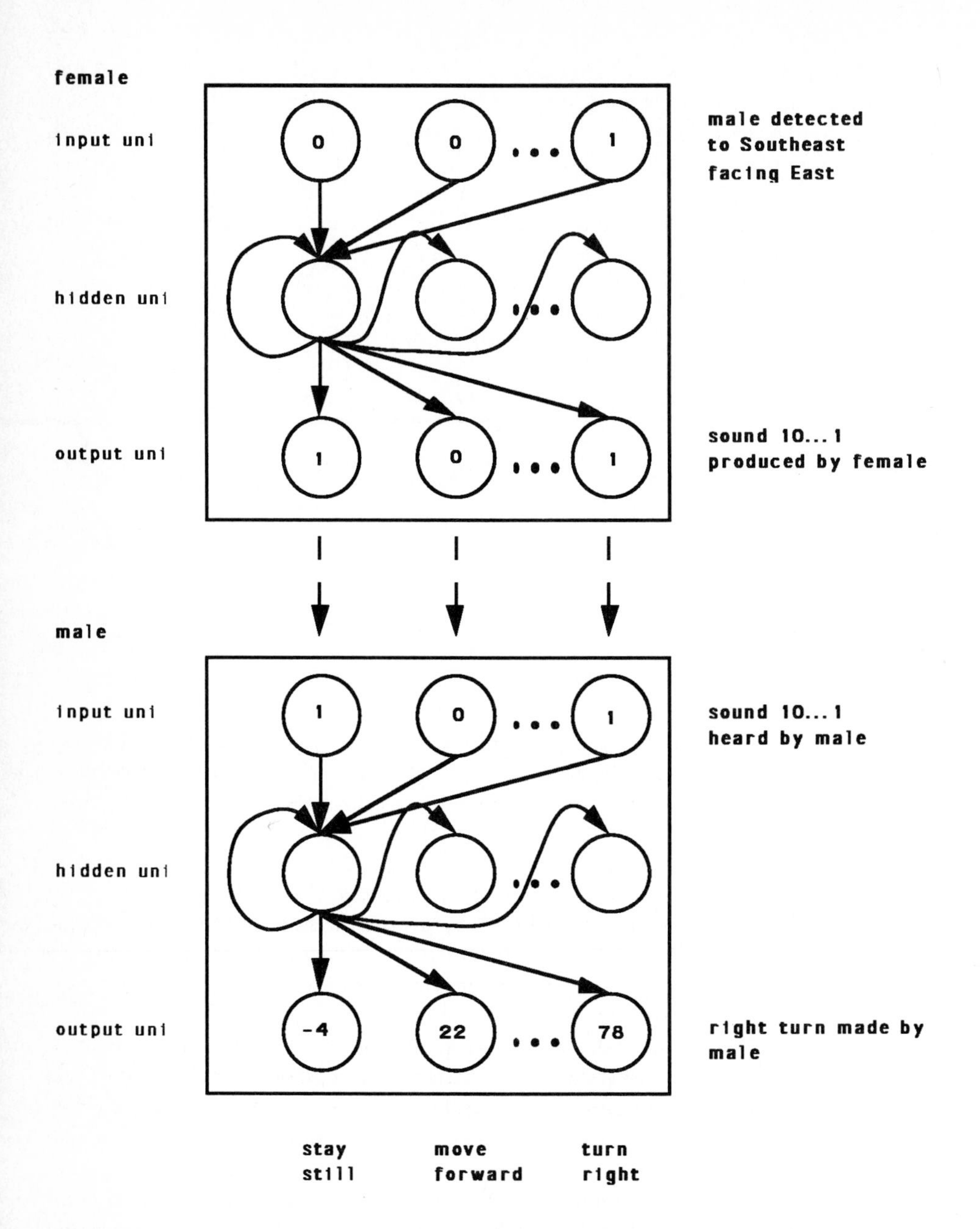

FIGURE 2 Female input units represent positions and orientations of males within female's "visual field." Males interpret female's output to aid in locating female.

in XGA, mates must select one another directly, rather than being mated as a result of some fitness function unrelated to mate selection.

RATIONALE FOR DESIGN

To avoid having males simply learning to home in on a sound, the male was not given the ability to localize the sound source. He only hears the type of sound produced, but cannot determine the direction from which it came. This creates a situation in which the females must produce signals that have a meaning to the males, and the males must correctly interpret this meaning in order for the communication between the animals to be successful.

This formulation of the animals and environment was not meant to be an accurate simulation of any particular species of animal. It was adopted as a simple way of creating selective pressure on animals to communicate.

Within this environment, the best strategy for two animals to find each other is for the female to direct the male to herself. No search strategy employed by males can beat an efficient strategy of the female animals giving directions to males. Any search strategy employed by the males that doesn't include listening to directions from the female will not be as effective as the direction-following strategy. As the males are evolving a search strategy, females will have to evolve a corresponding signalling protocol for the males to follow.

EXPERIMENT

A simulation, with 13-bit wide inputs to the males, was run to determine whether a signalling protocol would evolve that could aid males in finding mates. In addition, the simulation was also run with the sounds produced by the females not copied into the inputs of the male animals. This provided an experimental control to determine whether the communication between males and females was actually used and useful to the males.

With three-bit wide outputs, the females are capable of producing $2^3 = 8$ distinct sounds; similarly, with three-bit wide inputs, the males are capable of recognizing eight different sounds. Since the output of the males consists of movements (left, right, still, and forward), we can interpret the sounds of the females as messages telling the males how to move.

However, the relationship between a specific sound uttered by a female and a move made by a male is arbitrary and depends completely on the genome of each (and its corresponding neural network). For example, one female may produce "011" when a male is one-right-turn-plus-one-forward-move away from her, while another female may produce the distinct sound "101" under identical circumstances. Likewise, upon hearing "011," one male may stand still; another, however, may turn left; yet another may go forward. What each male or female does, given its inputs, is completely determined by the weights on its neural network, which is completely specified by its genome (i.e., no learning occurs in these experiments).

The task is to co-evolve a population of males and females who agree on the same—albeit arbitrary—interpretation of the eight signals. No one "dialect" is *a priori* correct. More than one "sound" can be mapped onto the same motion. For example, a given male may turn right when hearing any seven of the eight possible sounds and then move forward only when hearing the eighth. Although such a male can find a female, it will have a greater chance of getting stuck "spinning in place," than one who interprets more sounds as meaning "move forward." In any case, only one of the large number (8^8 = over 16 million) of all possible mating protocol dialects need co-evolve in order for females to successfully communicate with males.

While we may speak of females as "intending" to communicate, say, the message "turn right" to a male, there really are no intentions as such, since each animal's behavior is completely deterministic. Also, the interpretation of what a female message "means" can itself be problemmatic; for example, it may be the case that females are communicating to males messages more of the sort "hot" (i.e., "you are closer") and "cold" (i.e., "you are farther away"). In the following experiments, we describe the "meanings" of female messages in terms of the motions (right, left, forward, still) taken by the males.

RESULTS

During runs of the simulation, the behavior of the animals changed as improved mate-finding strategies were adopted. These changes in behavior occurred in the following stages.

1. Male animals wandered randomly and female animals signaled randomly. Since the animals started out with a random genome, they had neural networks with random connection strengths and biases. Therefore the population was full of male animals that moved erratically and females that emitted signals while oblivious to their surroundings (Table 1).

TABLE 1 Responses of randomly generated male neural nets to female signals are random (percentage of males making each response at Time = 100).

Signal	Move Forward	Turn Right	Turn Left	Stand Still
000	25	38	9	28
001	19	25	31	25
010	28	22	26	24
011	29	25	29	17
100	25	27	26	22
101	25	25	26	24
110	19	20	31	30
111	22	27	21	30

TABLE 2 Few males interpret a female signal to mean "stand still" (percentage of males making each response at Time = 5000).

Signal	Move Forward	Turn Right	Turn Left	Stand Still
000	74	20	5	1
001	81	12	7	0
010	67	18	14	1
011	79	11	19	1
100	80	7	12	0
101	75	11	14	0
110	56	21	23	0
111	70	14	16	0

2. Males that stood still became extinct. It is never a good strategy in this environment for a male to stand still. He should always move in order to cover the most ground and have the best chance of finding a female. Males that stand still lower their chance of finding a mate and therefore are selected against. Therefore very few animals interpreted any signal as meaning "stay where you are" (Table 2).
3. Males that usually go straight took over the population. Males that spent a large percentage of their time spinning in place were gradually replaced in the population by other males that spent more time covering new ground. Even though males may have evolved so that they would make appropriate moves when near some females, this ability was usually fatal when near a female that used a different signalling protocol. This is because the female's directions would steer him away from her, or very likely, would direct him to spin in place. For this reason, males evolved to simply ignore their inputs (Table 3).
   Moving in a straight line is a good search strategy because by avoiding turns, which cover no new ground, it covers the maximum amount of territory possible. Although this strategy covers only one row or column of the environment, the population is dense enough that on average there will be four females in each row or column of the environment. A male that follows this strategy will find a number of mates directly proportional to the density of the female population.
4. Males appeared that turn when in the same row or column as a female. At this point, females had evolved that produce a signal telling the males how to find them (Table 4).
   Females that did not use these signals, or that gave inappropriate ones, gradually became less common in the population. A male can maximize his chances of reproducing by going straight when not near a female, and by listening to a female's directions when close to one.

TABLE 3 Males ignore their inputs and simply move forward (percentage of males making each response at Time = 7500).

Signal	Move Forward	Turn Right	Turn Left	Stand Still
001	98	2	0	0
001	99	0	0	1
010	98	1	1	0
011	100	0	0	0
100	100	0	0	0
101	99	1	0	0
110	98	1	1	0
111	99	0	1	0

TABLE 4 Males evolve that interpret 101 as "turn left," 110 as "turn right," and the remaining patterns as "move forward" (percent of males making each response at Time = 15,000).

Signal	Move Forward	Turn Right	Turn Left	Stand Still
000	97	2	0	1
001	100	0	0	0
010	98	1	1	0
011	98	0	1	1
100	100	0	0	0
101	22	77	1	0
110	5	2	93	0
111	97	0	3	0

5. Female animals evolve to use the existing signals in more situations. Typically, the first use of a "turn" signal is in the case where a male is adjacent to a female and only needs to turn once to find her. Once males evolve to turn according to this signal, the females use it in more and more situations.
   After about 50,000 time steps, females evolve so that they will signal males who happen to be on the same row or column as the female to turn towards the female. A cross-shaped area (Figure 3) appears in which males will be told to spin until they face the female. All other males sensed by the female will be signalled to go straight. One can see that this strategy is the quickest way for females to guide males to themselves.

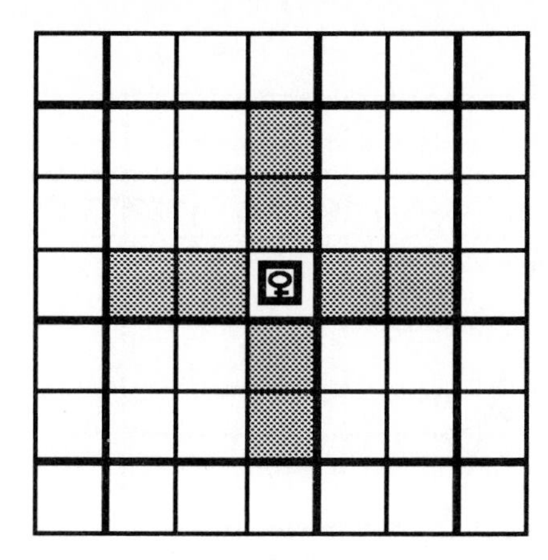

FIGURE 3 Females evolve that tell males in cross-shaped (shaded) area to continue turning until the male is facing the female. Males entering the "receptive field" but outside of the cross-shaped area are told to go straight. Such males will eventually enter the shaded area and from there be guided in to the female.

In about half of the runs, only one signal for turning evolves. Since turning one way can be accomplished by turning the other direction three times, it is possible to do without a signal for one of the turns. Since it is less efficient, however, one would expect that eventually a signal representing a turn in the other direction would evolve.

COMPARISON WITH NON-COMMUNICATING ANIMALS

To test how much the ability to communicate helped these animals, it was compared to the control group in which the males could not hear the signals produced by the females (Figure 4).

Before time = 7500 the population in which the males ignored the females did better than the "listening" males. This is because some of the listening males followed bad directions from a female. These bad directions were the result of incompletely evolved signals. For example, many males got caught because they were repeatedly told by a female to turn in place. They eventually died and were replaced by offspring of a more successful male.

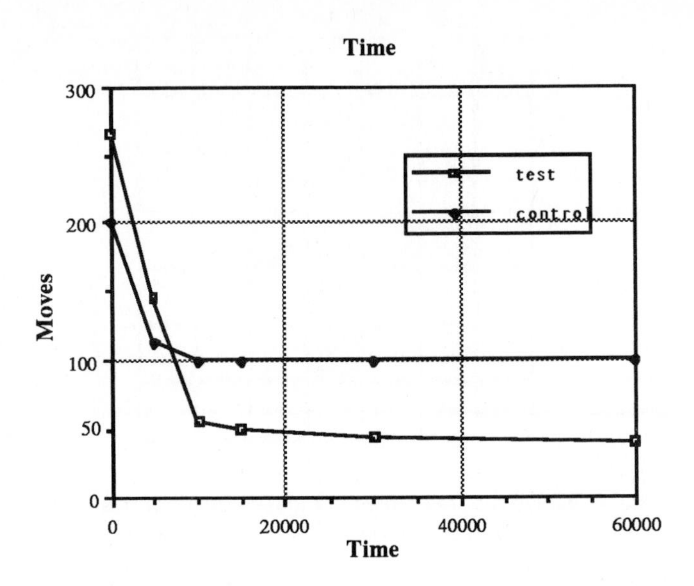

FIGURE 4 Animals with communication evolve to find mates in less than half the time required by animals lacking communication.

Between time = 7500 and time = 15000, the population with "listening" males reproduced more rapidly than the control population, but still below the rate possible using the best strategy that does not involve communication (i.e., moving in a straight line).

By time = 15000, the population with "listening" males reproduced more rapidly than could any population not employing communication.

Note that the control group never reached the maximum possible reproduction rate. Mutations away from the optimal strategy were common enough that about 25% of the animals would behave in non-optimal ways. Since many offspring were produced that were worse than their parents, the average time to find a mate was less than optimal for the population with deaf males. By simply travelling in a straight line, and thereby covering the maximum number of squares possible, the deaf males could bring their average time to find a mate as low as 50 moves. In practice, they never did better than finding a mate every 100 moves. "Listening" males, however, found females in an average time of 40 moves.

EXAMPLES OF ANIMALS USING EVOLVED SIGNALS

EXAMPLE 1

This is an example of the use of the evolved communication protocol from the particular simulation described above.

1. A male is just outside of visual and acoustic range of a female. He does not hear the signal being output by the female, nor does the female see him (Figure 5a).
2. The male has moved into the view of the female animal by moving straight. This was the optimal move given that he couldn't hear anything. At this point the female signals the male to "move forward" (Figure 5b).
3. The female signals the male to continue to "move forward" (Figure 5c).
4. The male has reached a square adjacent to the female, but will continue past her if he does not turn. Appropriately, the female changes her signal to one meaning "turn right" (Figure 5d).
5. Following the signal from the female, the male turns to his right. The female now changes her signal to one meaning "move forward" (Figure 5e).
6. The male then moves onto the square occupied by the female. Notice that the interpretation of a specific signal that evolves will be different in each simulation run. For example, 101 means "move forward" in Figure 5 while it means "turn right" in the run that produced Table 4. Although the specific signals may be different, the overall evolved protocol is the same.

EXAMPLE 2

This is an example of the use of a protocol evolved in a different run of the simulation. In this run, no signal for "turn right" has evolved.

1. A male is just outside of visual and acoustic range of a female (Figure 6a).
2. The male has moved into the view of the female animal by moving straight. The female signals the male to "move forward" (Figure 6b).
3. The female signals the male to continue to "move forward," but uses a different signal with this same meaning (Figure 6c).
4. The male has reached a square in the same column as the female. The female lacks a signal that means "turn right," so she begins signalling for a sequence of left turns that will cause the male to face her (Figure 6d).
5. The female again signals the male to turn left (Figure 6e).

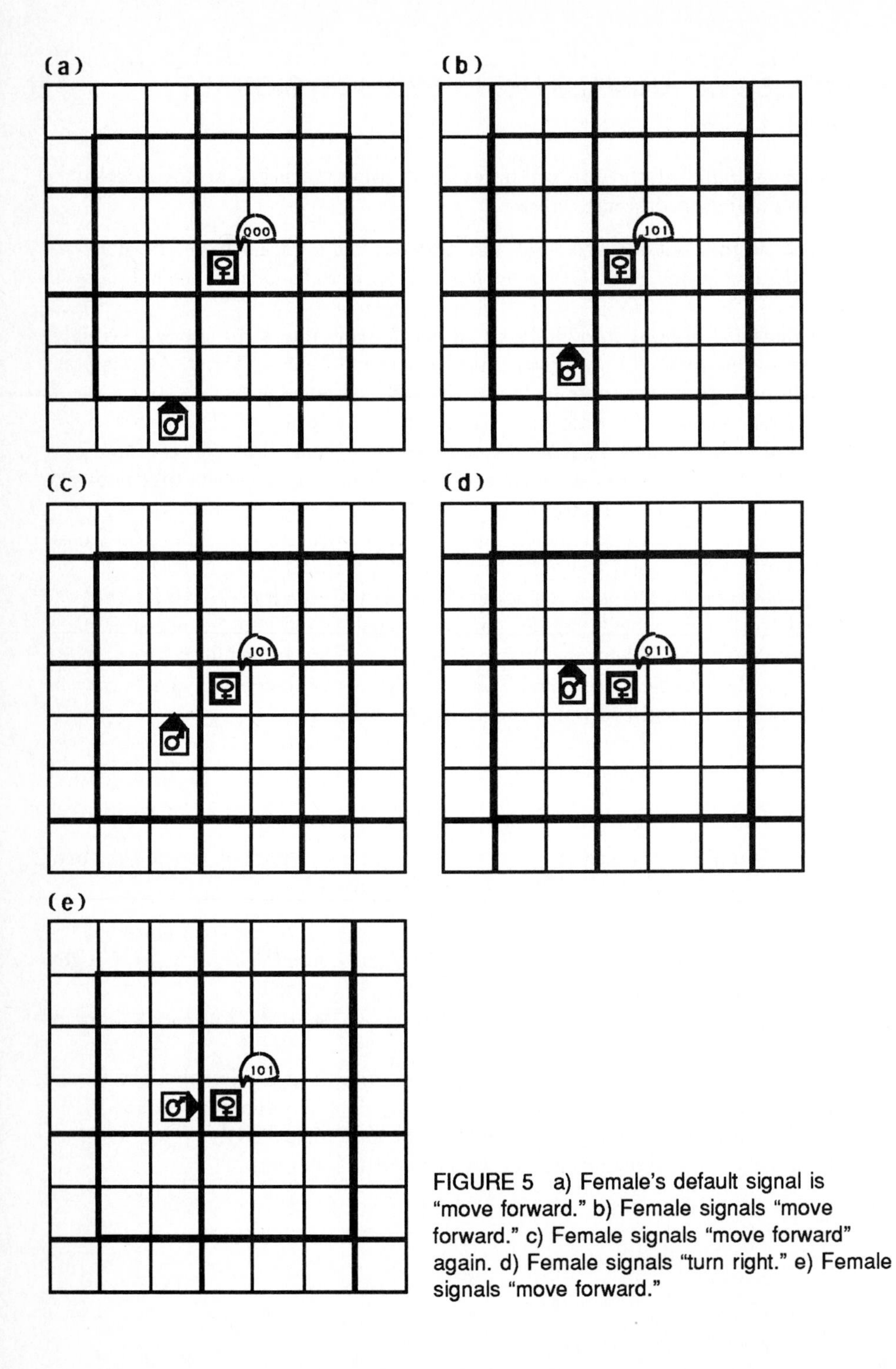

FIGURE 5 a) Female's default signal is "move forward." b) Female signals "move forward." c) Female signals "move forward" again. d) Female signals "turn right." e) Female signals "move forward."

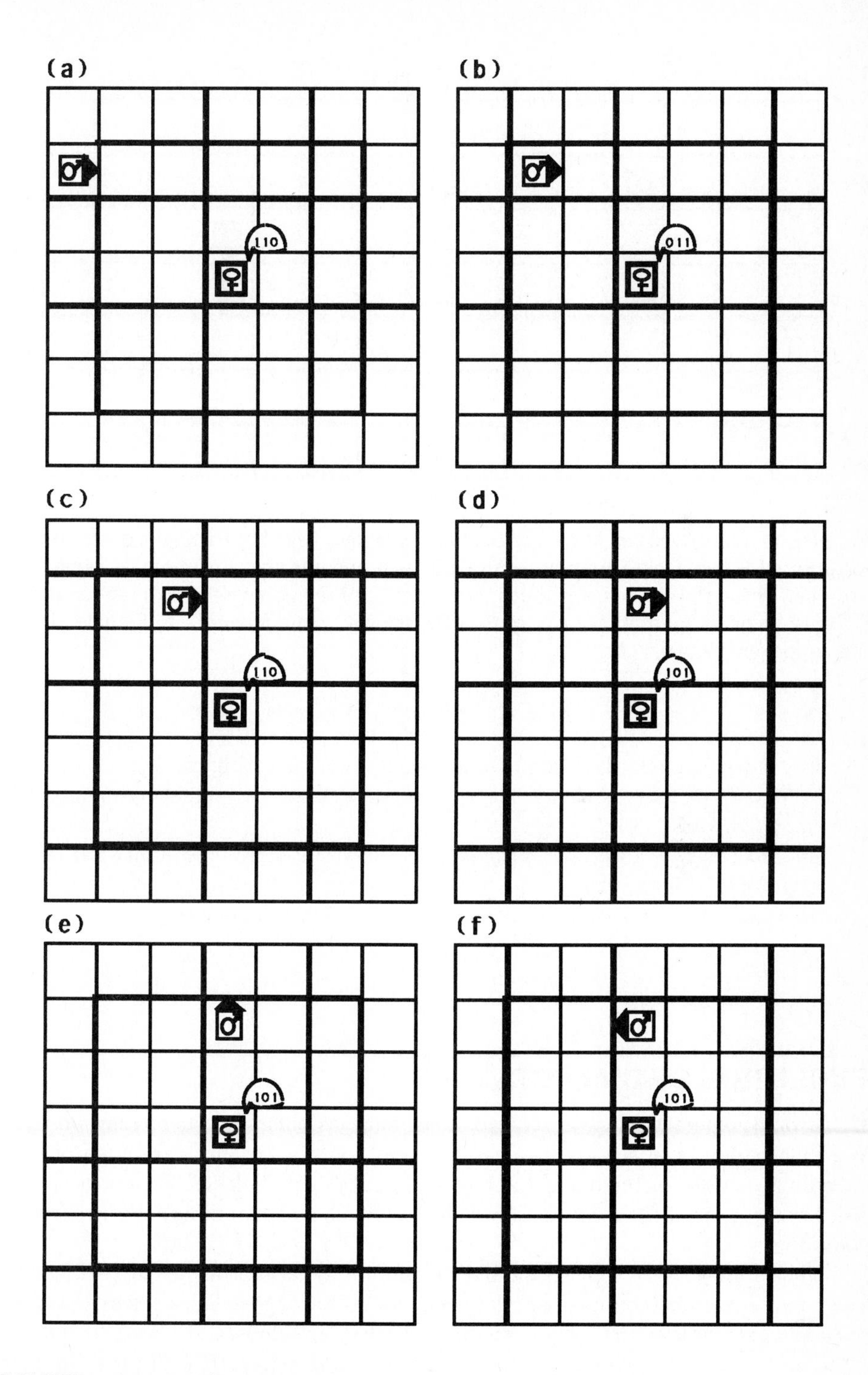

FIGURE 6 (see caption next page)

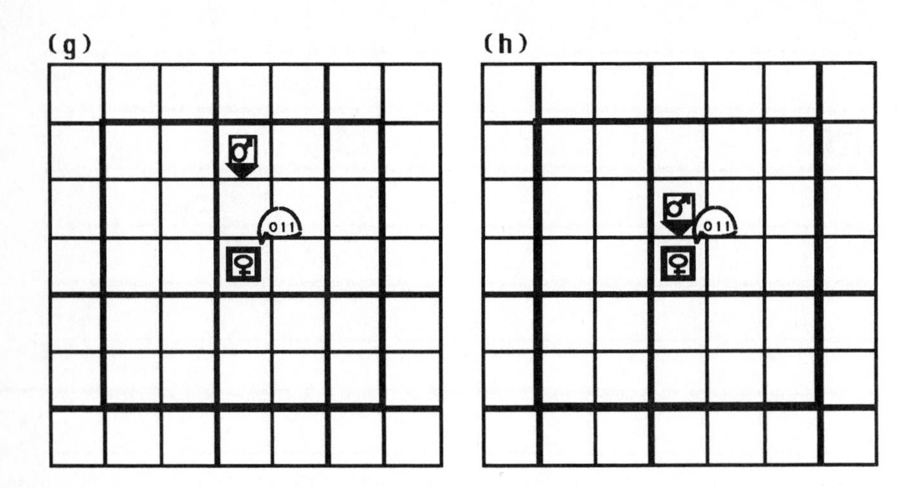

FIGURE 6 (continued) a) Male approaches "receptive field." b) Female signals "move forward." c) Female indicates "move forward" with another signal. d) Female signals "turn left" since she lacks a signal for "turn right." e) Female signals "turn left" again. f) Female signals "turn left" again. g) Female signals "move forward." h) Female signals "move forward" again.

6. The female signals the male to make a final left turn (Figure 6f).
7. Now that the male is facing her, the female changes her signal to one meaning "move forward" (Figure 6g).
8. The male is now adjacent to the female. Only one more "move forward" signal is required (Figure 6h).
9. The male then moves onto the square occupied by the female.

EVOLUTION OF DIALECTS

One would expect that in an environment such as this one, there may be many possible mappings of signals to meanings that could solve the task. Each of these mappings can be considered a different "language" or "dialect." To aid in viewing the evolution of these dialects, we both modified and simplified the experiment somewhat.

The animals were reprogrammed as simple pattern transducers (i.e., hidden layer and recurrent connections were removed). Each gene in the female animal encoded a signal that the female would emit when a male was at a specific location relative to her. Each gene in the male encoded his response to each possible signal from the female. We also reduced to four (i.e., two bits) the number of possible signals that the female could produce. This way, we reduced the total number of types of male animals to $4^4 = 256$ (four possible inputs for each of four possible moves). This representation has the advantages of being much easier to analyze, evolving faster under genetic operations, and running faster.

RESULTS

A series of tables follow that show the evolution of signal responses by the males. Each position in a given table contains two numbers separated by a colon. The first number is the frequency (i.e., number of males) of that particular set of signal responses. The second four-digit number represents the signal responses themselves. The first of these digits represents the response of the male to signal #1 from the female. The second digit represents the response to signal #2, and so on. In these responses, 0 means stay still, 1 means go forward, 2 means turn left, and 3 means turn right. Therefore, the entry 7:0113 means that there are 7 animals that will stand still if they hear signal #1, move forward if they hear #2 or #3, and turn right if they hear #3. Each table contains 256 logically possible dialects, from 0000 (males that always stand still) in the upper left-hand corner to 3333 (males that always turn right) in the lower right-hand corner.

The particular run shown here is interesting because a non-optimal strategy eventually takes over the population. A protocol evolves that only allows "move forward" and "turn right." Apparently, *this dialect dominates because the males that employ it are "bilingual."* They respond correctly to signals from two other common dialects.

Frequency Matrix 1 (Table 5): At time = 0, most of the possible communication protocols are used by at least one male.

Frequency Matrix 2 (Table 6): At time = 8,000, most of the animals that stand still have died out. Several good protocols have become popular including 1311, 1211, 1321, 1112, and 1113. These good strategies allow males to move forward and to turn to find females when necessary. The protocol that eventually wins out, 1313, is used by only one animal.

Frequency Matrix 3 (Table 7): At time = 10,000, protocol 1313 starts to gain in population. We found that females that attracted males using protocol 1311 used only signals #1 and #2. Females that attracted males using 1113 used mainly signals #3 and #4. This allowed males that used protocol 1313 to follow the correct directions whenever it found either of these groups of females. In a sense, *the 1313 males are bilingual.*

TABLE 5 Frequency of Male Signal Responses (time = 0)

4:000	6:0001	5:0002	7:0003	4:0010	6:0011	4:0012	4:0013
12:0020	1:0021	3:0022	4:0023	8:0030	8:0031	2:0032	7:0033
5:0100	7:0101	7:0102	5:0103	7:0110	5:0111	3:0112	8:0113
2:0120	7:0121	6:0122	7:0123	3:0130	9:0131	6:0132	7:0133
8:0200	6:0201	4:0202	3:0203	11:0210	3:0211	4:0212	8:0231
1:0220	7:0221	6:0222	5:0223	8:0230	5:0231	8:0232	2:0233
8:0300	7:0301	2:0302	9:0303	4:0310	9:0311	9:0312	4:0313
10:0320	7:0321	3:0322	5:0323	12:0330	7:0331	6:0332	4:0333
8:1000	6:1001	6:1002	7:1003	7:1010	4:1011	7:1012	2:1013
7:1020	7:1021	0	7:1023	4:1030	8:1031	7:1032	8:1033
11:1100	4:1101	4:1102	6:1103	10:1110	3:1111	3:1112	8:1113
3:1120	5:1121	6:1122	4:1123	3:1130	8:1131	2:1132	7:1133
11:1200	4:1201	6:1202	5:1203	5:1210	6:1211	8:1212	5:1213
7:1220	5:1221	8:1222	8:1223	5:1230	4:1231	7:1232	5:1233
8:1300	4:1301	9:1302	7:1303	8:1310	11:1311	8:1312	6:1313
4:1320	3:1321	10:1322	6:1323	5:1330	2:1331	6:1332	7:1333
7:2000	9:2001	7:2002	11:2003	3:2010	9:2011	5:2012	6:2013
12:2020	8:2021	5:2022	7:2023	7:2030	6:2031	7:2032	10:2033
7:2100	5:2101	8:2102	3:2103	4:2110	7:2111	4:2112	5:2113
11:2120	11:2121	6:2122	6:2123	5:2130	8:2131	4:2132	9:2133
7:2200	4:2201	3:2202	10:2203	7:2210	8:2211	4:2212	7:2213
8:2220	7:2221	9:2222	11:2223	4:2230	5:2231	3:2232	5:2233
6:2300	8:2301	6:2302	7:2303	7:2310	7:2311	5:2312	6:2313
6:2320	4:2321	7:2322	4:2323	3:2330	3:2331	8:2332	3:2333
8:3000	6:3001	8:3002	4:3003	8:3010	4:3011	9:3012	9:3013
4:3020	8:3021	4:3022	9:3023	6:303	10:3031	6:3032	5:3033
9:3100	5:3101	7:3102	3:3103	14:3110	8:3111	5:3112	8:3113
7:3120	5:3121	5:3122	12:3123	3:3130	6:3131	6:3132	6:3133
4:3200	10:3201	6:3202	4:3203	12:3210	6:3211	10:3212	0
14:3220	9:3221	3:3222	12:3223	4:3230	6:3231	7:3232	7:3233
8:3300	4:3301	4:3302	9:3303	8:3310	8:3311	4:3312	3:3313
4:3320	6:3321	11:3322	6:3323	4:3330	5:3331	6:3332	6:3333

Frequency Matrix 4 (Table 8): At time = 12,000, protocol 1213 starts to appear more frequently. This is another "bilingual protocol" combining 1211 and 1113.

Frequency Matrix 5 (Table 9): (18 rows of zeros have been removed.) At time = 14,000, protocol 1313 is now used by a sizeable part of the population. Some good protocols including 1321, 1111, and 1311 are starting to decline.

TABLE 6 Frequency of Male Signal Responses (time = 8000)

0	0	0	0	0	0	0	0
0	0	0	0	0	0	0	0
0	1:0101	0	0	0	0	0	1:0113
0	0	0	0	0	0	0	0
0	0	0	0	0	0	0	0
0	0	1:0222	0	0	0	0	0
0	0	0	0	0	0	1:0312	1:0313
0	0	0	0	0	0	0	1:0333
2:1000	0	0	0	0	73:1011	0	1:1013
1:1020	0	0	0	0	0	2:1032	6:1033
0	1:1101	0	1:1103	1:1110	70:1111	92:1112	64:1113
0	131:1121	14:1122	2:1123	0	3:1131	0	24:1133
1:1200	0	0	0	0	323:1211	5:1212	2:1213
1:1220	4:1221	2:1222	12:1223	0	0	0	2:1233
4:1300	2:1301	1:1302	0	5:1310	547:1311	74:1312	1:1313
0	89:1321	0	1:1323	0	13:1331	0	1:1333
0	0	0	0	0	0	0	0
0	1:2021	0	1:2023	0	0	0	0
0	0	0	0	0	0	0	1:2113
0	1:2121	0	1:2123	0	0	0	0
0	0	0	0	0	0	0	0
0	0	0	0	0	0	0	0
0	0	0	0	0	0	0	0
0	0	0	0	0	0	1:2332	0
0	0	1:3002	0	0	0	0	0
0	0	0	1:3023	0	0	0	0
1:3100	0	0	0	0	0	0	0
1:3120	1:3121	0	0	1:3130	0	0	0
0	0	0	0	0	0	0	0
1:3220	0	0	0	0	0	0	0
0	0	1:3302	0	0	0	0	0
0	0	1:3322	0	0	1:3331	0	0

Frequency Matrix 6 (Table 10): At time = 16,000, protocol 1213 has surged ahead of 1313. The two protocols that it combines are now more common that those that 1313 combines.

Frequency Matrix 7 (Table 11): At time = 20,000, protocol 1313 has become much more common than 1213, even though it is a less successful strategy.

TABLE 7 Frequency of Male Signal Responses (time = 10,000)

0	0	0	0	0	0	0	0
0	0	0	0	0	0	0	0
0	1:0101	0	0	0	0	0	1:0113
0	0	0	0	0	0	0	0
0	0	0	0	0	0	0	0
0	0	0	0	0	0	0	0
0	0	0	0	0	0	0	1:0313
0	0	0	0	0	0	0	0
1:1000	0	0	0	0	45:1011	0	22:1013
1:1020	0	0	0	0	0	1:1032	0
0	0	0	1:1103	0	55:1111	90:1112	127:1113
0	78:1121	5:1122	1:1123	0	0	0	53:1133
0	0	0	0	0	321:1211	2:1212	3:1213
1:1220	2:1221	4:1222	4:1223	0	0	0	1:1233
2:1300	0	1:1302	0	1:1310	613:1311	42:1312	20:1313
0	85:1321	0	0	0	7:1331	0	1:1333
0	0	0	0	0	0	0	0
0	1:2021	0	0	0	0	0	0
0	0	0	0	0	0	0	1:2113
0	0	0	1:2123	0	0	0	0
0	0	0	0	0	0	0	0
0	0	0	0	0	0	0	0
0	0	0	0	0	0	0	0
0	0	0	0	0	0	0	0
0	0	0	0	0	0	0	0
0	0	0	1:3023	0	0	0	0
1:3100	0	0	0	0	0	0	0
0	0	0	0	0	0	0	0
0	0	0	0	0	0	0	0
0	0	0	0	0	0	0	0
0	0	0	0	0	0	0	0
0	0	1:3322	0	0	1:3331	0	0

Frequency Matrix 8 (Table 12): At time = 30,000, the 1313 protocol has reached a point where it can drive the other protocols to extinction.

Frequency Matrix 9 (Table 13): At time = 40,000, only the winning protocol remains.

TABLE 8 Frequency of Male Signal Responses (time = 12,000)

0	0	0	0	0	0	0	0
0	0	0	0	0	0	0	0
0	0	0	0	0	0	0	0
0	0	0	0	0	0	0	0
0	0	0	0	0	0	0	0
0	0	0	0	0	0	0	0
0	0	0	0	0	0	0	1:0313
0	0	0	0	0	0	0	0
1:1000	0	0	0	0	33:1011	0	3:1013
1:1020	0	0	0	0	0	0	1:1033
0	0	0	0	0	40:1111	35:1112	169:1113
0	25:1121	1:1122	0	0	1:1311	0	23:1133
0	0	0	0	0	406:1211	17:1212	20:1213
0	3:1221	0	0	0	0	0	0
0	0	1:1302	0	0	627:1311	9:1312	128:1313
0	53:1321	0	0	0	0	0	0
0	0	0	0	0	0	0	0
0	1:2021	0	0	0	0	0	0
0	0	0	0	0	0	0	0
0	0	0	0	0	0	0	0
0	0	0	0	0	0	0	0
0	0	0	0	0	0	0	0
0	0	0	0	0	0	0	0
0	0	0	0	0	0	0	0
0	0	0	0	0	0	0	0
0	0	0	0	0	0	0	0
0	0	0	0	0	0	0	0
0	0	0	0	0	0	0	0
0	0	0	0	0	0	0	0
0	0	0	0	0	0	0	0
0	0	0	0	0	0	0	0
0	0	1:3322	0	0	0	0	0

TABLE 9 Frequency of Male Signal Responses (time = 14,000)

0	0	0	0	0	0	0	0
⋮	⋮	⋮	⋮	⋮	⋮	⋮	⋮
0	0	0	0	0	0	0	0
1:1000	0	0	0	0	11:1011	0	1:1013
1:1020	0	0	0	0	0	0	0
0	0	0	0	0	27:1111	88:1112	286:1113
0	16:1121	1:1122	0	0	2:1131	0	26:1133
0	0	0	0	1:1210	310:1211	78:1212	20:1213
0	3:1221	0	0	0	0	0	0
0	1:1301	1:1302	0	0	470:1311	20:1312	217:1313
0	18:1321	0	0	0	0	0	0
0	0	0	0	0	0	0	0
0	1:2021	0	0	0	0	0	0
0	0	0	0	0	0	0	0
⋮	⋮	⋮	⋮	⋮	⋮	⋮	⋮
0	0	0	0	0	0	0	0

TABLE 10 Frequency of Male Signal Responses (time = 16,000)

0	0	0	0	0	0	0	0
⋮	⋮	⋮	⋮	⋮	⋮	⋮	⋮
0	0	0	0	0	0	0	0
1:1000	0	0	0	0	7:1011	0	1:1013
1:1020	0	0	0	0	0	0	0
0	0	0	0	0	27:1111	125:1112	192:1113
0	10:1121	1:1122	0	0	1:1131	0	3:1133
0	0	0	0	1:1210	397:1211	102:1212	218:1213
0	0	0	0	0	0	0	0
0	0	1:1302	0	0	326:1311	28:1312	155:1313
0	2:1321	0	0	0	0	0	0
0	0	0	0	0	0	0	0
0	1:2021	0	0	0	0	0	0
0	0	0	0	0	0	0	0
⋮	⋮	⋮	⋮	⋮	⋮	⋮	⋮
0	0	0	0	0	0	0	0

TABLE 11 Frequency of Male Signal Responses (time = 20,000)

0	0	0	0	0	0	0	0
⋮	⋮	⋮	⋮	⋮	⋮	⋮	⋮
0	0	0	0	0	0	0	0
0	0	0	0	0	0	0	0
0	0	0	0	0	0	0	0
0	0	0	0	0	8:1111	72:1112	87:1113
0	5:1121	0	0	0	0	0	0
0	0	0	0	0	222:1211	171:1212	262:1213
0	0	0	0	0	0	0	0
0	0	0	0	0	211:1311	45:1312	517:1313
0	0	0	0	0	0	0	0
0	0	0	0	0	0	0	0
0	0	0	0	0	0	0	0
0	0	0	0	0	0	0	0
⋮	⋮	⋮	⋮	⋮	⋮	⋮	⋮
0	0	0	0	0	0	0	0

TABLE 12 Frequency of Male Signal Responses (time = 30,000)

0	0	0	0	0	0	0	0
⋮	⋮	⋮	⋮	⋮	⋮	⋮	⋮
0	0	0	0	0	0	0	0
0	0	0	0	0	0	0	0
0	0	0	0	0	0	0	0
0	0	0	0	0	0	0	0
0	0	0	0	0	0	0	0
0	0	0	0	0	5:1211	231:1212	183:1213
0	0	0	0	0	0	0	0
0	0	0	0	0	1:1311	40:1312	1140:1313
0	0	0	0	0	0	0	0
0	0	0	0	0	0	0	0
0	0	0	0	0	0	0	0
0	0	0	0	0	0	0	0
⋮	⋮	⋮	⋮	⋮	⋮	⋮	⋮
0	0	0	0	0	0	0	0

TABLE 13 Frequency of Male Signal Responses (time = 40,000)

0	0	0	0	0	0	0	0
⋮	⋮	⋮	⋮	⋮	⋮	⋮	⋮
0	0	0	0	0	0	0	0
0	0	0	0	0	0	0	0
0	0	0	0	0	0	0	0
0	0	0	0	0	0	0	0
0	0	0	0	0	0	0	0
0	0	0	0	0	0	0	0
0	0	0	0	0	0	0	0
0	0	0	0	0	0	0	1600:1313
0	0	0	0	0	0	0	0
0	0	0	0	0	0	0	0
0	0	0	0	0	0	0	0
0	0	0	0	0	0	0	0
⋮	⋮	⋮	⋮	⋮	⋮	⋮	⋮
0	0	0	0	0	0	0	0

ADDING A PHYSICAL BARRIER TO THE ENVIRONMENT

In all of our runs, one particular communication protocol always eventually took over the entire population. To determine whether more than one protocol could survive, we introduced a physical barrier in the environment. This barrier had a permeability that could be modified. A male animal who hit this barrier had a fixed chance of crossing it into another region. If he crossed the barrier, a female was selected from the side from which he came and moved onto the other side. This had to be done since females could not move across the barriers on their own. Males that failed to cross the barrier simply wrapped around in their own toroidal sub-environment. This scheme created an abstract physical barrier that could be modified to provide varied reproductive isolation between sub-populations. This isolation could help sub-populations develop alternate signalling protocols. In addition, we hoped that partial isolation would help a distinct sub-population maintain its own "dialect" in the face of contact from migrating males using foreign dialects.

RESULTS

We found that when the barrier was completely impermeable, different protocols could trivially evolve on each side of the environment. However, we discovered also that even when a great deal of barrier crossing was allowed, the sub-populations could maintain distinct dialects. Once a certain threshold was reached (80% chance to cross when touching the barrier), one of the sub-populations could successfully invade the other and one protocol would end up being used by the entire population. Under that threshold, however, distinct dialects could be maintained indefinitely.

RELATED WORK

MacLennan[10,11] has begun a promising line of research in evolving communication. In his model, animals evolve to produce signals that describe their local environment to other animals. These signals are used by those who hear them to decide on an "action" to take. Animals are rewarded with greater chances to produce offspring when they produce the "action" that corresponds to the local environment of a signalling animal or produce the signal that evokes this action. MacLennan has found that the signals produced by his artificial animals come to represent the state of the animals' local environments. He has also shown that, by including a simple learning process, the speed of this evolution of meaning can be greatly increased.

This work shares our goal of seeking to evolve communication protocols of increasing complexity but differs in several fundamental ways. First, our simulations incorporate a simple, natural task that can be solved using communication as opposed to the abstract environments and actions of the aforementioned model. Tasks similar to those faced by living systems can provide constraints on what types of information are important to communicate. Our signals come to represent tangible things in the simulation (directions) instead of only being symbol associations. MacLennan's abstract formulation, however, is simpler to analyze, and his use of abstract environmental states may make it easier to evolve complex protocols such as those requiring syntax.

Second, our model uses XGA to produce offspring. We believe that use of the XGA, along with the constraint that animals can only communicate with others near them, will be important when attempting to evolve more than one language or dialect within one environment.

FUTURE WORK

In this ongoing series of experiments, we are trying to create tasks that are increasingly difficult so that more complicated information has to be communicated between organisms. We would like to create environments that pressure a population into signalling internal states and intentions among the members, as well as letting one another know about states and events in the environment. We want to keep the tasks for our populations "natural"—similar to the selective pressures that actually brought about the ability to communicate to living systems on earth. We are also examining tasks that require more interesting interactions between animals to accomplish a common goal and that require the individuals to use state information when deciding on what actions and/or signals to produce.

IMPROVING THE MODEL

In order to achieve these research goals we are currently making a number of improvements to our model. First, the physics of the environment is currently very simple. The "sounds" produced by animals are simply copied to other animals nearby. These sounds have no direction or intensity, which would carry a large amount of information. Nor can an animal hear more than one sound at a time. The "vision" possessed by the female animals is also impoverished. The female is only able to detect the presence of a male animal within a very small area. "Vision," in the current simulation also does not take into account factors such as closer objects obscuring those farther away. In later experiments, we will eliminate these flaws. We plan to add sensors onto our animals that can detect the direction of the sound source and the intensity of the sound. Also, the eyes of our animals will no longer be able to see through objects.

The environment is relatively barren in the work described in this paper. The animals have the entire environment to themselves. This will change, as we plan to add a number of new kinds of objects to the environment, such as plants, rocks, pools of water, recognizable offspring, and other species of animals. With these new objects we hope to create a large number of new and realistic tasks for populations to solve.

Metabolism is a key feature which is also lacking in this particular implementation. Including it will force our animals to find food to stay alive (an interesting task by itself), and deaths by starvation will provide a natural way to keep the population at a reasonable size.

Avoiding predators is a very common task for real animals, so we plan to add predatory species to our simulations, hopefully to evolve both predator avoidance and possibly group hunting behavior.

The animals in our current system have a very small number of possible actions. We are adding a few more primitive actions to the animals' repertoire, including the ability to grasp and release objects in the environment, and mating as an action to be selected (versus simply occurring automatically as the result of cohabitating the same square).

The random placement of offspring is another flaw in the current model. By placing offspring near their parents in future models, we will avoid mixing up the animals in the environment. We believe that this, along with the ability to recognize one's offspring, may encourage both speciation and altruism toward kin.

The use of a direct mapping from genome to neural network connection strengths and biases has several major flaws. First, it is clearly biologically implausible. Secondly, it is impossible to use a direct mapping for large neural networks because of the exponentially increasing number of connections contained in them. Finally, this representation does not seem to work well with the genetic algorithm. Experiments with more complex environments and animals have shown that it is extremely difficult to evolve larger neural nets using this simple mapping. We believe that the genome-to-neural net mapping and representation issues pose major research challenges.[5,6]

Lastly, there is a learned component to communication in addition to innate signals. Learning allows the more rapid creation and acceptance of new signals among a population. A new signal can be created through the invention of a new sound. We plan to add a form of unsupervised learning to our organisms which we hope will allow forms of communication to evolve with both innate and learned features.

BIOLOGICAL ISSUES

It is interesting to note that the use of acoustic signals in nature can constitute a behavioral barrier to mating—one that can serve as a basis for later speciation. For example, in nature there are several species of frogs that can produce viable offspring with members of other species. However, the frogs are considered distinct species because the calls made by the frogs insure that such mating almost never happens. The calls made by male frogs to attract females are only pursued by females that are of the same species.[1,2]

We believe that communication protocols could provide a natural way of establishing genetic barriers that spontaneously emerge. This could be useful in exploring a wide variety of biological problems in the origin and maintenance of distinct species in an ecological setting, including kin selection, altruism towards kin, genetic drift, gene flow, mimicry, and parasitism.

CONCLUSIONS

We have shown that it is possible to evolve organisms that communicate to solve a simple mate-finding task. The ability to produce appropriate signals for stimuli co-evolved with the ability to take appropriate actions upon receiving each signal. No explicit pressure on the artificial animals was required to cause this communication to develop. It simply arose as part of a cooperative solution to the mate-finding problem—as it appears to have in nature.

We have also shown that it is possible to vary the genetic algorithm so that the phenotype (neural network) produced by each genome is capable of finding a mate. This variant algorithm, dubbed "XGA," is more realistic than the standard genetic algorithm in which each genome is scored by some evaluation function at fixed intervals and randomly paired off with a similarly scoring animal. The XGA algorithm can be used without losing the ability to evolve useful phenotypes.

Finally, we have shown that subspecies or distinct "dialects" do evolve and compete. Dialects that are "bilingual" (i.e., correctly interpret several signalling protocols) have an increased chance of dominating in the long run. Physical barriers, however, allow distinct dialects to survive indefinitely in different regions, even in cases where partial permeability of a barrier permits some migration across regions and therefore contact between distinct subspecies.

ACKNOWLEDGMENTS

This work was supported in part by a W. M. Keck Foundation grant. Thanks to Rob Collins, David Jefferson, Adam King, Richard Korf, and Charles Taylor for their insightful comments on earlier drafts of this paper. Thanks also to Chris Langton and several anonymous reviewers for their suggestions on an earlier draft of this paper. A special thanks to Ruth Liow for her help in preparing this document.

REFERENCES

1. Blair, W. Frank. "Isolating Mechanisms and Interspecies Interactions in Anuran Amphibians." *Qtr. Rev. of Biol.* **39** (1964): 334–344.
2. Blair, W. Frank. "Mating Call in the Speciation of Anuran Amphibians." *Am. Naturalist* **92** (1958): 27–51.
3. Bond, Alan H., and Les Gasser, eds. *Readings in Distributed Artificial Intelligence.* San Mateo, CA: Morgan Kaufmann, 1988.
4. Bonner, John T. *The Evolution of Culture in Animals.* Princeton, NJ: Princeton University Press, 1980.
5. Collins, Robert J., and David R. Jefferson. "AntFarm: Towards Simulated Evolution." This volume.
6. Collins, Robert J., and David R. Jefferson. "Representations for Artificial Organisms." In *Proceedings of Simulation of Adaptive Behavior*, eds. Jean-Arcady Meyer and Steward Wilson. Cambridge, MA: MIT Press/Bradford Books, in press.
7. Elman, Jeffrey L. "Finding Structure in Time." Technical Report CRL 8801, Center for Research in Language, University of California, San Diego, 1988.
8. Frings, Hubert, and Mable Frings. *Animal Communication.* Norman, OK: University of Oklahoma Press, 1977.
9. Goldberg, David E. *Genetic Algorithms in Search, Optimization, and Machine Learning.* Reading, MA: Addison-Wesley, 1989.
10. MacLennan, Bruce J. "Evolution of Communication in a Population of Simple Machines." Technical Report CS-90-99, Computer Science Department, University of Tennessee, Knoxville, 1990.
11. MacLennan, Bruce J. "Synthetic Ethology: An Approach to the Study of Communication." Technical Report CS-90-104, Computer Science Department, University of Tennessee, Knoxville, 1990.
12. Rumelhart, D. E., G. E. Hinton, and R. J. Williams. Learning Internal Representations by Error Propagation." In *Parallel Distributed Processing*, edited by D. Rumelhart and J. McClelland,Volume I. Cambridge, MA: MIT Press/Bradford Books, 1986.

Edwin Hutchins† and Brian Hazlehurst‡
†Department of Cognitive Science and ‡Department of Anthropology, University of California at San Diego, La Jolla, CA

Learning in the Cultural Process

This paper reports a set of computer simulations that demonstrate a form of adaptation that we believe to be characteristic of human intelligence.

One of the central problems faced by biological and artificial systems is the development and maintenance of coordination between structure inside the system and structure outside the system. That is, the production of useful behavior requires internal structures that respond in appropriate ways to structure in the environment. The processes that give rise to this coordination are generally considered adaptive.

Biological evolution, individual learning, and cultural evolution can all be seen as ways to discover and save solutions to frequently encountered problems; that is, they are processes that generate coordination between internal and external structure.

Creatures that can learn are likely to have a greater range of responses and, as Hinton and Nowlan[4] have shown, learning can actually guide the evolutionary search. Hinton and Nowlan imagine a population of creatures, the behavior of each of which is specified by some number N of alleles. The creatures inhabit a world in which there is a fitness spike associated with just one particular pattern of those

N alleles. If all of the alleles are genetically fixed, the chances of any individual finding the fitness spike is low, and there is no search strategy that beats random search. Thus, for creatures in which behavior is entirely genetically determined, the process of discovering and saving good solutions is blind and relatively slow.[1]

Now suppose that, rather than having all the alleles' settings hardwired, some can be learned (by guessing in Hinton and Nowlan's scheme) during the lifetime of the individual. In this case, any individual whose hardwired alleles correspond to a partial description of the fitness spike has a chance of guessing the rest of the solution. Creatures that are genetically predisposed to learn (guess) the solution to a particular problem in the environment, by virtue of having correct settings on all of the hardwired alleles, are on average more fit than those who cannot guess the solution. These fitter creatures may put more individuals who are also predisposed to learn the solution in the next generation. Hinton and Nowlan show that allowing some of the alleles to be uncommitted in value, and thus learnable, has the effect of putting shoulders on the fitness spike such that the population can "hill climb" to the best genetic solution. This is much more efficient and rapid than a random search for the optimal genome. Still, from the perspective of the population, the learning is slow because the products of individual learning have only very indirect effects (mediated by selection on random variation) on subsequent generations. (See Hinton and Nowlan[4] and Belew[1] for details.)

Culture is a process that permits the learning of prior generations to have more direct effects on the learning of subsequent generations. As predicted by Wilson,[7] the presence of cultural factors may create a selective pressure for the ability to learn itself. Hutchins performed a simple demonstration of this effect by adding a cultural bias factor to the simulation of Hinton and Nowlan. Making the offspring of individuals that have learned the solution to an environmental problem more likely to learn the solution than the offspring of individuals who did not learn has the effect of increasing the relative frequency of alleles that code for learnable responses. That is, adding a cultural effect increases the steady state proportion of uncommitted or learnable alleles in the population. A replication of Hutchins' demonstration and a more complete analysis of this effect was subsequently performed by Belew.[1]

If culture permits the consequences of learning by a prior generation to have direct effects on the learning of a subsequent generation, then could a population, over many generations, be capable of discovering things that no individual could learn in a lifetime? This should be true in spite of the fact that the direct products of individual learning (internal structures) last at most a lifetime. Let us consider this problem in the framework of the coordination of internal and external structure that was presented at the beginning of the paper.

[1]The probability of an individual being genetically predisposed to matching the genome associated with the fitness spike is $1/2^n$, for two-valued alleles. In this case, an evolutionary search for the spike is ineffective since there is no feedback regarding a "close" fit, and therefore no opportunity for co-adapted alleles to retain partial solutions. Furthermore, variation in the genomes of descendents means that even if the solution were found, it would be extremely unlikely to be retained in the population.

In that scheme we had two kinds of structure to be coordinated: external structure—a physical environment, for example—and internal structure—the organization of a nervous system, for example.

Imagine a world in which there is a useful regularity in the environment that is too complex for any individual to learn to predict in a single lifetime. That is, given the rate at which internal structure can be rearranged, it is either not possible or extremely unlikely that any individual will achieve coordination between external and internal structure. How could a useful form of interaction with such a regularity ever be learned? Hinton and Nowlan have demonstrated one method in which parts of the solution are learned genetically so that individuals in future generations are born partly organized and therefore need to do less learning in order to master the regularity. But again, that process is very slow. Hutchins' addition of a cultural bias factor showed that culture can guide the ability to learn which in turn guides evolution. However, any model that reduces all of culture to a single scalar is clearly missing many of the central aspects of cultural phenomena. In particular, culture involves the creation of representations of the world that move within and among individuals. This heavy traffic in representations is one of the most fundamental characteristics of human mental life, yet since it is a phenomenon that is not entirely contained in any individual, it has largely been ignored by cognitive science. If each individual is capable of learning something about the regularity and then *representing* what has been learned in a form that can be used by other individuals to facilitate their learning, knowledge about the regularity could accumulate over time, and across generations.

This introduces a third kind of structure: structure in the environment that is put there by creatures. This is *artifactual* structure. Our inventory now includes *natural structure* in the environment, *internal structure* in the organisms, and *artifactual structure* in the environment. These structures are related to each other through time as shown in Figure 1.

In a cultural world, the internal structure of an organism is shaped by (must achieve coordination with) two kinds of structure in the environment: natural and artifactual structure.

What form should an artifactual representation of the knowledge of the aforementioned natural regularity take?

First, it must itself be a kind of regularity in the environment. Barring mental telepathy, one mind can only influence another by putting some kind of structure in the environment of the other mind. Taking this maxim seriously highlights the importance of the medium and process of transmission of cultural knowledge. Earlier studies of cultural evolution have not directly addressed this issue,[2,3] choosing instead to speak of cultural traits as if they were abstractions without physical form. The danger of that approach is that one ignores the artifactual world and overlooks its capacity as a learning system in its own right.

A second requirement for the artifactual representations of the natural regularity is that the artifacts be strictly symbolic. They must contain no direct (that is,

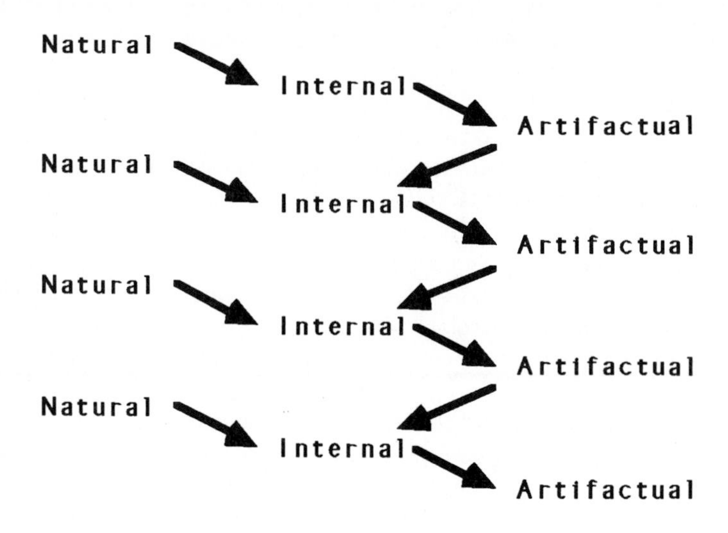

FIGURE 1 The relations of environmental, internal, and artifactual structure. The arrows represent propagation of constraints upon interactions involving these different kinds of structure. Constraints may be propagated by many means.

structurally non-arbitrary) information about the mappings of states of the world onto each other.

It is important to approach this subject with the understanding that culture is not a thing or any collection of things, it is a process. In the human sphere, myths, tools, understandings, beliefs, practices, artifacts, architectures, classification schemes, etc. alone or in combination do not in themselves constitute culture. Each of these structures, whether internal or external, is a *residue* of the cultural process. The residues are, of course, indispensable to the process, but taking them to be culture itself diverts our attention from the nature of the cultural process. In the simulations we present, the artifacts should not be taken to be the culture of the community. Instead they and the internal structures that form in interaction with them are residues of a cultural process.

Consider the scheme diagramed in Figure 1 as a case of intergenerational cultural process. Getting the internal structures into coordination with the natural regularity of the environment requires three kinds of learning: (1) direct learning of the natural regularity in the environment, (2) mediated learning about the natural regularity from the structure of the artifactual descriptions of it, and (3) learning a language that will permit a mapping (in both directions) between the structure of the natural regularity and the structure of artifactual descriptions of it. We return to this in the discussion of the simulation.

Before turning to the details of the simulation we would like to motivate its organization with a "just-so" story about cultural learning.

LEARNING THE RELATION OF MOON PHASE TO TIDE STATE

Up until about two hundred years ago, the hills on which the U.C.S.D. campus is located were inhabited by California Indians. We know from the ethnographic and archaeological record that they hunted deer and rabbits and collected greens in the canyons inland from the present site of the campus. We also know that when the tides were low, they collected shellfish in the many tide pools along the coast. That much is well established. Imagine the sort of problem small groups of hunter-gatherers might have faced. Shellfish is a rich source of protein and easy to get when the tides are low, so when the tides are favorable, it might be worth moving the whole band to the beach. On the other hand, it is a waste of a lot of energy to go all the way to the beach if the tides are not favorable. Furthermore, one can't determine whether the tides are actually going to be low by just looking for a moment. One might be looking at a time when the tide is in an intermediate level. It would therefore by very nice if there was a reliable way to predict the state of the tide without having to go to the cliffs over the beach and watch for many hours to find out what it actually is.

The phase of the moon provides just such a predictor. When the moon is full and when it is new, the gravitational forces of the sun and moon are in phase with each other and they generate large tidal variation. So both very high and very low tides occur on the same day. Figure 2 shows this relationship. This is a regularity we imagine it would have been advantageous for the members of this society to learn. Of course, they already had a language that contained words for the states of the tide and phases of the moon. The problem here is to learn a set of mappings between states of the natural world: to learn an association of phases of the moon to states of the tide.

Since it takes many hours of watching the ocean to see what the state of the tide actually is, and since the sky is not always clear, the opportunities for matching an observed state of the moon with an observed state of the tide are few; possibly too few for any individual alone to learn to predict this regularity. But we imagine that, over time, the community of people could learn something that no individual alone could learn. Since the language is already well developed, the members of the community each learns, as part of growing up, a shared set of mappings between phases of the moon and words for phases of the moon, and between states of the tide and words for states of the tide. We assume that the phenomena of phases of the moon accompanied by labels for phases of the moon and of states of the tide accompanied by labels for states of the tide are frequently available. It is only the

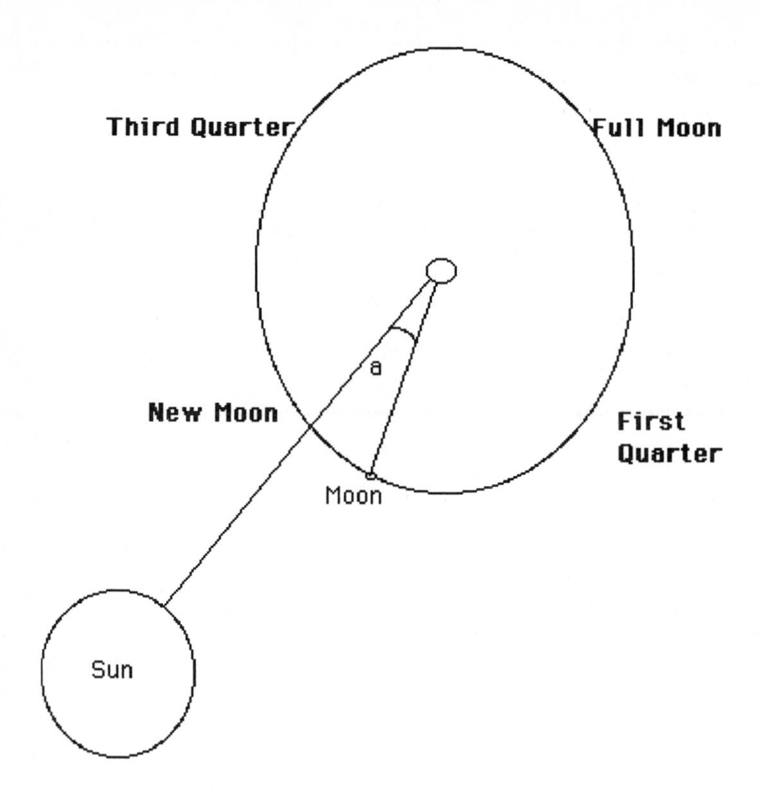

FIGURE 2 The regularity between moon phase and tide state results from a phase relationship between the moon and sun's gravitational effects upon the earth.

conjunction in experience of the phase of the moon and state of the tide that is limited.

THE SIMULATION

The behavior of a community of individual "citizens" through time is simulated. Citizens are composed of connectionist networks that have the ability to learn from both the natural and artifactual structure in the environment. The latter constitutes a type of symbolically mediated task learning. A "generation" is a time step in the simulation during which each citizen in the population has an opportunity to learn the task. Three stochastic factors account for variability in that learning: (a) the quality of the artifact chosen for study, (b) the quality of direct experience with the environment, and (c) the random set of task network connection weights assigned at "birth." After learning from both the natural and artifactual structures, each citizen generates one artifact, gives birth to one novice citizen, and then dies. *All* citizens have the same network architecture and there is no passing of genetic information between generations. Each novice begins life with a random set of connection weights. The only contribution made by an individual to successive generations is a cultural one, a produced artifact.

Each novice of the next generation chooses an artifact from those produced by the previous generation. This choice can be made randomly, or selection can be introduced by having members of the younger generation probabilistically choose an artifact biased by the artifact author's "success." An artifact produced by an author who has learned a lot about the task (and who therefore could be said to "know" a lot about the task solution or environmental regularity), is more likely to be studied by someone in the next generation. The consequences of these two kinds of choice are discussed below.

THE ENVIRONMENT

The representation of the phases of the moon and states of the tide are given in Figure 3. Notice that the representation chosen for this regularity is a continuous version of exclusive or (XOR). There are 28 different moon phase/tide state pairs that constitute direct sensory information about the environment. Each of the 28 pairs was generated by dividing the lunar orbit into 28 segments and encoding the moon phase and tide state for what roughly corresponds to the state of affairs for each day of the lunar month.

Each element of the vector representing the moon phase is a real number between 0 and 1. The first element encodes how much of the left half of the moon is visible, the second value how much of the right half is visible, from an idealized earth. Using this representation, every instance of the two-element, moon phase vector describes a unique point on the unit square. Notice that the four vertices of the square represent the four major moon phases; new, first quarter, full, and third quarter, encoded by 00, 10, 11, and 01 respectively.

The tide state is encoded by a single real number between 0 and 1 that is generated by a transcendental function of the angle between moon and sun with respect to earth (see Figure 3). In particular, each side of the unit square is associated with either a decreasing tide variance (between new and first quarter moons, and between full and third quarter moons) or an increasing tide variance (between first quarter and full moons, and between third quarter and new moons).

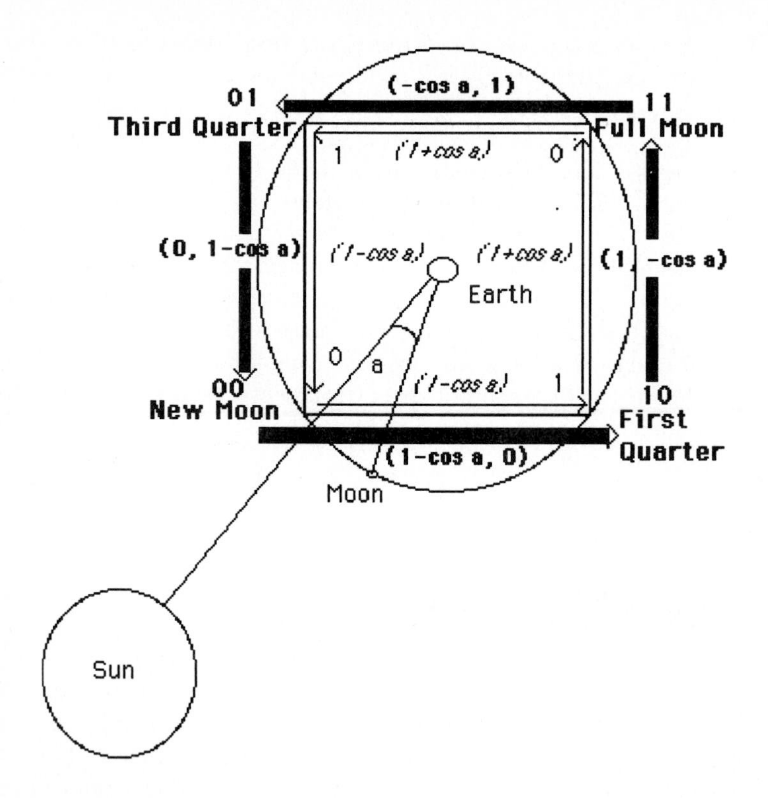

FIGURE 3 Representation of environment. Two-value moon phase shown in **bold**, and one-value tide quality shown in *italics.*

THE LANGUAGE

The simulated citizens must discover how to tell when the tide is "good" (maximum variance) and when the tide is "bad" (minimum variance) using the moon phase as a predictor. The language "spoken" by our citizens is shown in Table 1.

This scheme represents two "lexicons," characterizing two different classes of events: the moon phase and the tide state. Of course each lexicon is not restricted to the prototypical words listed in Table 1. Each placeholder ("bit") in the symbolic descriptors takes on a real value, providing for a theoretically infinite set of such descriptors. The prototypical words represent an externally defined language that is known (by us) to sufficiently characterize the simulated world's behavior.

TABLE 1 Citizen Language

Environment	Symbolic Rep	Physical Rep
	Prototypic Moon Lexicon	Moon Phase
New moon	"1000"	00
First quarter	"0100"	10
Full moon	"0010"	11
Third quarter	"0001"	01
	Prototypic Tide Lexicon	Tide State
Large-variance tide	"01"	0
Small-variance tide	"10"	1

THE ARTIFACTS

An artifact is composed of four pairs of *symbols*. The first element of each pair is a symbol for the phase of the moon and the second is a symbol for the state of the tide. In the artifact creation phase, each citizen symbolically encodes its responses to the moon phases represented by the vertices of the unit square. Figure 4 exemplifies a "perfect artifact," one that describes a perfect association of moon phases to tide states.

This perfect artifact provides us with a method for evaluating artifact quality utilizing a simple distance metric. Artifact quality is defined by the mean squared difference between the corresponding second elements (tide symbols) of a given artifact/perfect artifact comparison. In other words, it is a measure of the difference between the given artifact's tide symbols and those of the perfect artifact, and is thus the extent to which the artifact is a good symbolic representation of the environmental regularity between moon phase and tide state.

CITIZEN ARCHITECTURE

Each citizen is composed of three feed-forward, back-propagation networks: two "language" nets and one "task" net (see Figure 5). Each language net is a standard auto-associating network that is trained to reproduce on its output layer whatever pattern was applied to the input layer. Once trained, the language net provides a mapping between a symbolic description of an event and the event itself. By concatenating instances of these two classes of information (symbolic and physical representations) into one bit string that can be applied to the input units of a language net, the network (after suitable training on this association) can reliably generate: (a) a "symbolic representation" from the experience of an event and (b) an experience of the event from a symbolic representation, via the network's ability to do pattern completion. Using this scheme, each class of symbols (each lexicon,) requires its own language net. Thus each citizen has two language nets for translating the artifacts' encodings of moon phases and tide states, respectively. The task network is a six-unit, one-hidden-layer, XOR network.

	SYMBOLS	
Pair for new moon	1000	01
Pair for first quarter	0100	10
Pair for full moon	0010	01
Pair for third quarter	0001	10
	Moon Phase	Tide State

FIGURE 4 A perfect artifact.

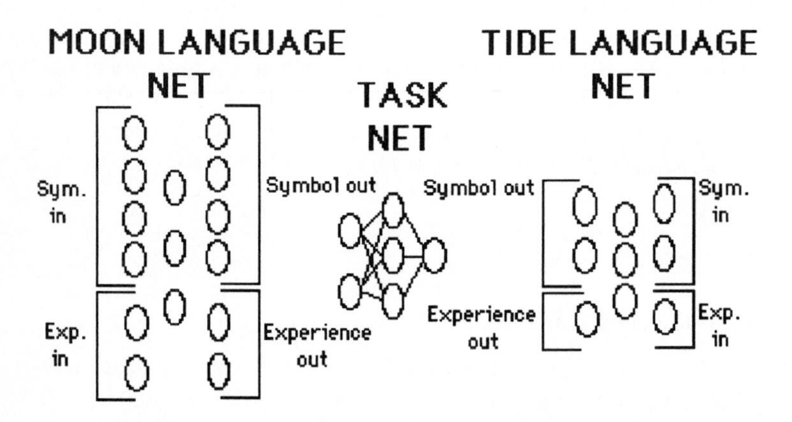

FIGURE 5 Citizen Architecture. Fully connected language nets translate the representational structure of artifacts into "vicarious experience" from which the task network can learn.

THREE KINDS OF LEARNING

There are three kinds of learning that take place in the culture process simulations. First, there is *learning of the language*. In the current implementation, this is simply the process of training the citizens' language nets to associate symbols with events, as described in the last section. For the moment we have left to one side the rather interesting problem of how our citizens might come, by consensus, to utilize a shared lexicon suitable for describing the events in their world. (See Hutchins[5] for a simulation study of this phenomenon.) Here, we take language learning for granted and simply endow citizens with language ability through auto-associative training on the prototypical lexicons.

Second, there is *direct learning from the environment*. This type of learning is the kind that standard network learning employs. Given some representation of the environment, in our case moon phase vectors and tide state scalars, we give the task network a limited amount of simultaneous experience with these two so it may learn to predict one from the other. A random day is chosen from the 28-day lunar cycle and the task network is presented with the moon phase representation for this day on its input layer. The predicted tide state produced on the output unit of this network is then compared with the actual tide state representation for this day. The error is back-propagated to adjust the connection weights in a fashion that will better perform this mapping from moon phase to tide state. This kind of learning does not involve the language faculties and is accomplished by a simple presentation of input and target directly to the task network of the citizen.

Finally, there is *mediated learning*. This learning is characterized by the utilization of language nets to transform externally encoded symbolic descriptions into "vicarious experience" of the events for which they stand. The outputs of the two language nets *produce* the input and target for the task network itself (see Figure 6).

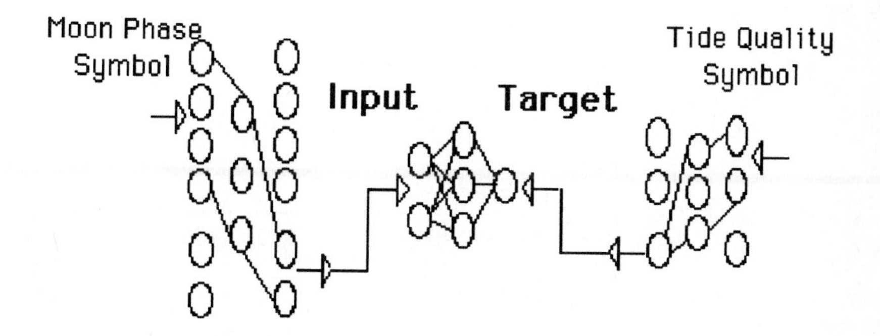

FIGURE 6 Mediated Learning. The language provides interpretations of inputs and targets for task learning.

"Mediation" is taking place at two different levels in this kind of learning. "Inside the skin" of citizens, language faculties mediate between symbolic descriptions and the experience of meaning while "outside the skin" artifacts, structures deposited by other citizens, mediate between events in the world and information about those events.

THE LEARNING PROTOCOL

Figure 7 shows characteristic learning potentials for two of the learning scenarios described in the last section. Each trace on the plot represents the probability of learning two-bit XOR to 0.05 mean squared error criterion (error is averaged over the four cases 00, 01, 11, and 10) as a function of learning trials for the labelled scenario. The two scenarios are: (a) direct learning of the environmental regularity, and (b) mediated learning *from a perfect artifact* utilizing trained language nets to translate the artifact's symbols. Each trace gives estimators for the probability of learning the respective XOR task based on a random sample of 50 starting connection weight configurations. Note that direct learning involves 28 different cases (randomly presented) while mediated learning only involves the four cases on

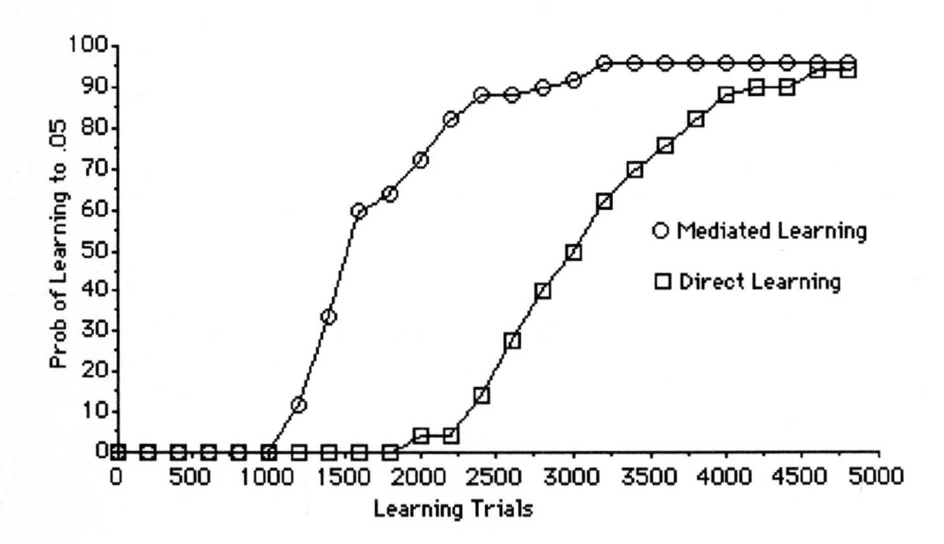

FIGURE 7 Characteristics of two kinds of XOR learning (see main text). Probabilities are based on observed results from random samples of size 50.

which the net is tested. Mediated learning from perfect artifacts is thus an easier task, as is reflected in the more rapid learning rates shown in Figure 7.

How shall we decide the amounts of mediated and direct learning to give each citizen? If all learning was direct, then culture would be irrelevant. If all learning was mediated, then the structure of the world would be irrelevant. Part of the "just so" story was intended to motivate the notion that direct experience of the environmental regularities might be much less available than mediated ones. Thus, we would like to have a total number of trials that permits individuals to learn the task once good artifacts have developed. Simultaneously, we would like the proportion of direct learning to be such that no individual could learn the regularity from direct experience alone. Figure 7 helps us decide what the learning protocol in each citizen's lifetime should be. If we have fewer than 1800 trials of direct learning, the chances of learning the regularity directly are near zero. This sets an upper bound on direct learning. Clearly, even with perfect artifacts, the total number of trials will have to be greater than 1000 to produce any reasonable learning. This sets a lower bound on *total* learning trials.

The actual protocol used called for a citizen to first get 750 epochs of training from one selected artifact. Since each artifact contains four learning instances, this amounts to 3000 trials of mediated learning. Next, the citizen received 260 trials of direct experience learning. Notice that the probability of an individual learning this task to a criterion of 0.05 mean squared error in 260 trials of direct experience is very near zero. Thus, no individual can learn the task alone (see Figure 7). If the culture can generate artifacts that describe the regularity well, the combination of a small amount of direct learning and a large amount of artifact-mediated learning should permit the individual to learn to predict the regularity.

After learning, the citizen produces an artifact by "responding" to a test of its knowledge of the four orthogonal cases: 00, 10, 11, and 01. This process requires a reverse translation of symbol-to-experience; in particular it entails the production of symbols which stand for that citizen's "understanding" of these events (see Figure 8). The production of a symbol for tide state is accomplished via the internal mapping from experience of moon phase to experience of tide state. The artifact thus reflects what the task net has learned about the regularity. We have deliberately excluded internal "propositional" representations that directly link symbols for moon phase to symbols for tide state. The internal models in this simulation are models of the behavior of the natural world, not models of the structure of the artifactual world. Of course, humans do learn the latter sort of representation—they may even be the basis for much of human reasoning—but they introduce unnecessary complexity into this simple world.

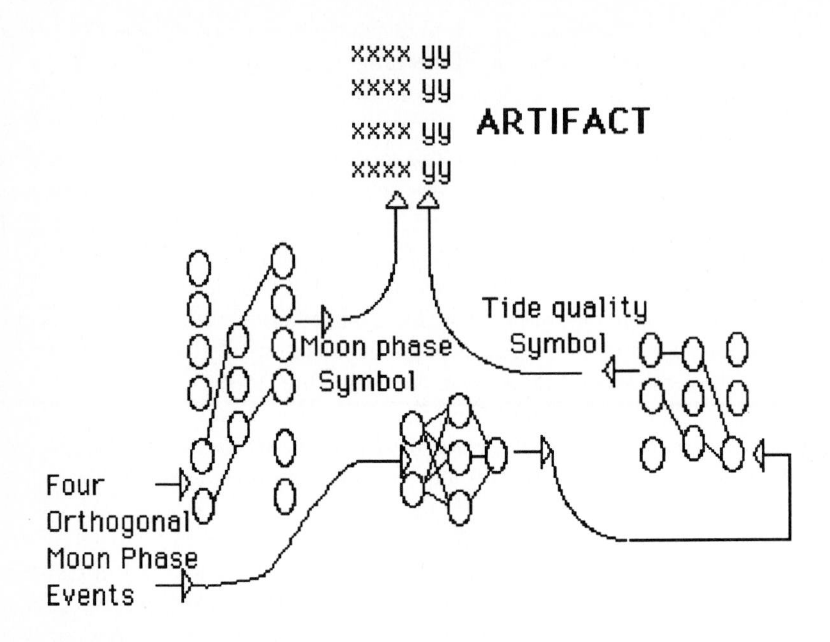

FIGURE 8 Generating an artifact.

RESULTS

Figures 9 and 10 show the results of two simulations run with population sizes of twenty citizens each. There are two traces plotted for each run reporting the generational averages of (1) artifact quality, and (2) the average mean squared error of each citizen's task performance on the four prototypes (00, 01, 11, and 10). As is evident from these figures, the two measures of the culture process track each other quite closely.

The difference between these two simulations is that the one shown in Figure 10 utilized an artifact selection bias. As already mentioned, this amounts to tagging artifacts with a selection probability that is a function of the author's task competence. The probability function utilized simulates a uniform distribution of artifacts based upon the observed task competence of the authors' deviations from that of the most competent author (i.e., the one with the lowest MSE on the prediction task). Selection bias based on author competence seems like a reasonable, though simplified, analog of what takes place in real cultural process.[2]

[2] Boyd and Richerson[2] also utilize this type of biasing in their models of culture and biology as co-evolutionary processes.

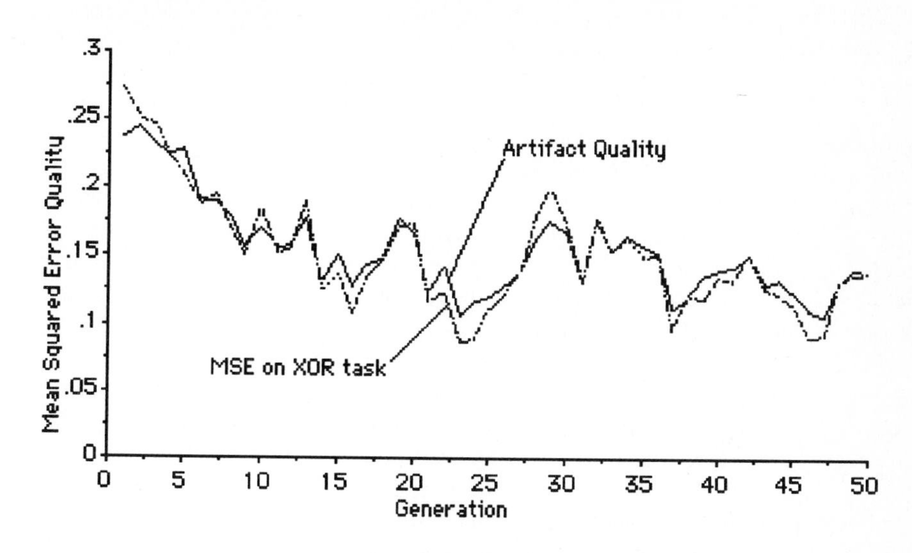

FIGURE 9 Culture process simulation with no artifact selection bias.

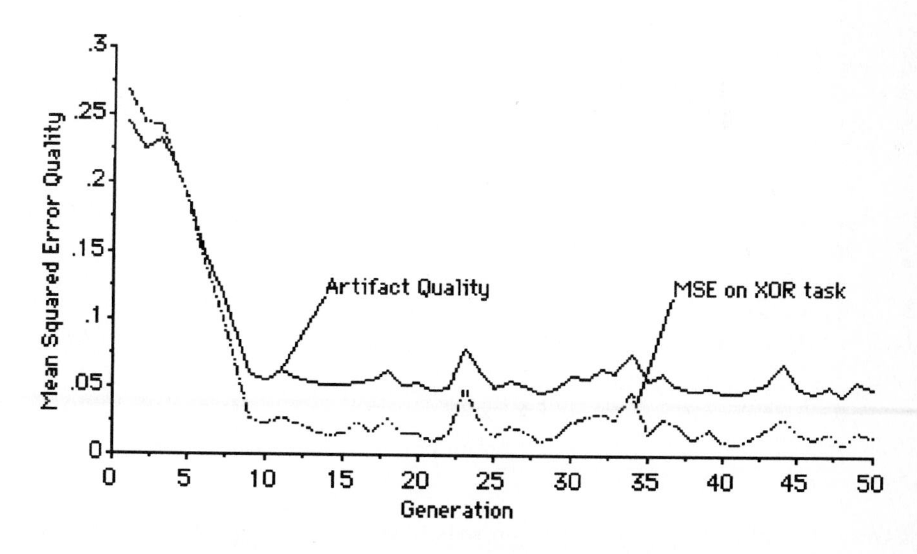

FIGURE 10 Culture process simulation with biased artifact selection.

Without this biasing it appears that while the system is capable of some learning, it is vulnerable to unlucky choices of artifacts to study, resulting in slower and less drammatic learning. Just as artifacts are a bridge from the internal structures of one generation to those of another, so the internal structures of individuals are the bridge between one generation of artifacts and another. If too many individuals in subsequent generations study artifacts created by poor performers, the useful structure that has been built into the good artifacts can be swamped by the noise in the bad ones. This can cause the community to "forget" some of what it knows about the regularity. Nonetheless, Figure 9 shows that, even with random artifact selection, there is an accumulation of knowledge that affords better performance in the task for individuals in later generations.

Finally, Figure 11 shows the effects of participating in a cultural system on the learning abilities of individuals. In the early generations, individuals learn from artifacts with no useful structure, and no member of the community is able to predict the environmental regularity. In later generations using exactly the same learning protocol virtually all of the individuals are able to predict the regularity. This happens even though the individuals in later generations have no greater innate learning abilities than those of the early generations. Clearly this phenomenon results from retention of "successful" knowledge in the artifactual media.

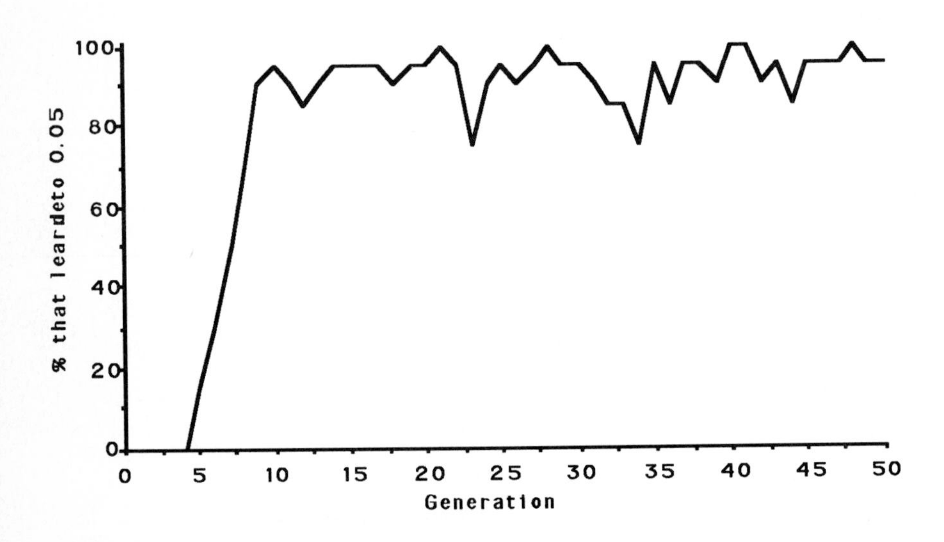

FIGURE 11 Observed population percentages of citizens who learned the task to 0.05 MSE or better during simulation with biased artifact selection (see Figure 10).

Although the simulation presented here is quite primitive, we believe it illustrates, in principle, that the cultural process can be seen, like biological evolution and individual learning, as a way to produce and maintain coordination between internal and external structures. It has been said that culture is the most important invention in the history of life since sex.[6] We hope that in this paper we have been able to show both why the cultural process is so important to human mental life and why in considering the cultural process we must consider the role of artificial as well as natural structure in the environment.

ACKNOWLEDGMENTS

Research support was provided by grant NCC 2-591 to Donald Norman and Edwin Hutchins from the Ames Research Center of the National Aeronautics & Space Administration in the Aviation Safety/Automation Program. Everett Palmer served as technical monitor. Additional support for Edwin Hutchins was provided by a fellowship from the John D. and Catherine T. MacArthur Foundation.

REFERENCES

1. Belew, R. K. "Evolution, Learning, and Culture: Computational Metaphors for Adaptive Algorithms." CSE Technical Report No. CS89-156, University of California, San Diego, 1989.
2. Boyd, R., and P. J. Richerson. *Culture and the Evolutionary Process.* Chicago: The University of Chicago Press, 1985.
3. Cavalli-Sforza, L. L., and M. W. Feldman. *Cultural Transmission and Evolution: A Quantitative Approach.* Princeton: Princeton University Press, 1981.
4. Hinton, G. E., and S. J. Nowlan. "How Learning Can Guide Evolution." *Complex Systems* **1** (1987): 495–502.
5. Hutchins, E. "How to Invent a Lexicon: The Development of Shared Symbols in Interaction." Unpublished manuscript, University of California San Diego, 1990.
6. Sereno, M. "Four Analogies Between Biological and Cultural/Linguistic Evolution." *J. Theor. Biol.* (1990): in press.
7. Wilson, A. C. "The Molecular Basis of Evolution." *Sci. Am.* **253(4)** (October 1985): 164–173.

Computation

Alvy Ray Smith
1001 West Cutting Blvd., Point Richmond, CA 94804

Simple Nontrivial Self-Reproducing Machines

A simple and brief proof of the existence of *nontrivial* self-reproducing machines, as cellular automata (CA) configurations, is presented, which relies only on computation universality. Earlier proofs are book length and rely on "construction universality." Furthermore, simple CA are shown to support nontrivial self-reproduction—hence, simultaneously simple and nontrivial. Nontriviality is guaranteed by the requirement that the machine which reproduces itself is also a universal computer. Biological relevance—or non-relevance—is also briefly discussed, as is trivial self-reproduction, called self-replication.

INTRODUCTION

The purpose of this note is to increase awareness of a cellular automata (CA)[1] result which is over 20 years old, but previously unavailable except in my Stanford

[1]There has been a revival in interest in CA theory since the combined advent of chaos, fractals, and interactive computer graphics in the 1980s, inspired principally by Wolfram and the other authors.[39] There is a surprising ignorance by much of the post-revival literature of the hundreds of papers written during the preceding 20 years or so. Part of the problem is the lack of knowledge that CA had many other names then: iterative arrays, tessellation automata, cellular spaces, modular

Ph.D. dissertation,[25] abbreviated in an old conference proceedings,[24] and hinted at in exercises in Arbib.[5] This is a *brief* proof of the existence of self-reproducing machines which are nontrivial in the sense that they are capable of computing any computable function. In fact, several different proofs are given.

This is to be compared to numerous lengthy or complex proofs. John von Neumann introduced the concept in a book[35] which also created the field of CA theory (from a suggestion by Stanislaw Ulam). We will use the term "machine" in the same sense as von Neumann—a finite collection of non-quiescent cells in a CA. Both his CA (29-state cells) and his self-reproducing configuration (40,000 or so nonquiescent cells) were complex and inspired others[3,6,10,33] to simplify. Von Neumann's lengthy proof—book length—is constructive. The other proofs just mentioned are also constructive and book length, except for Arbib[3] which is short but uses very complex cells. The proof exhibited here is an existence proof only, but it is just two pages long. The constructions of the constructive proofs are very tedious. This plus the fact that the principal result in all cases is that self-reproduction is logically possible makes an existence proof sufficient. And, as we shall see, the existence proof herein is actually stronger than that term might suggest, it being theoretically possible to compute a construction from the proof.

More importantly the proof here reduces the problem of self-construction to a computation problem, which means that no machinery beyond ordinary computation theory is required for self-reproduction. This is the real reason for the brevity of the proof. It will be seen to turn on invocation of the famous Recursion Theorem of recursive function theory (see Rogers,[23] for example).

The result and its relation with the literature is brought up-to-date.

RECURSIVE FUNCTION THEORY

The branch of mathematics which deals with those computations that digital computers can perform is called recursive function theory. This theory tells us that there exist "effective enumerations" of the partial recursive functions; this simply means that all computer programs can be listed by some one computer program (which itself must be in the list, of course). Let's call this list "the List." Then the ith program (partial recursive function) in the List is denoted List(i).[2]

Two well-known theorems used herein are these:

arrays, polyautomata, etc. The references herein can be used as pointers into this literature (e.g., see Smith[30]).

[2] As usual a *total* function is one defined for every element of its (input) domain, whereas a *partial* function may be undefined for some elements of its domain. A total recursive function corresponds to a computer program which halts on every input, and a partial recursive function to one that may not halt on some inputs (for example, it might go into an infinite loop).

THE UNIVERSAL TURING MACHINE THEOREM In the List of all programs is a special program, U = List(j), and a total recursive function e of two variables such that, for an arbitrary program P = List(i) and an arbitrary input x to that program, $U(\ e(P,\ x)\ ) = P(x)$ if $P(x)$ is defined, and is undefined if $P(x)$ is undefined. We will call U in this case a *universal Turing machine function* U and e the *encoder function* associated with U. In other words, one of the programs in the List of programs can simulate all others, given an encoding of each program and its input. U is the mathematical representation of today's digital computer. It was Alan Turing who made this wonderful discovery (in 1937) of a universal computer and initiated the digital computer revolution, so we name the function after him.

We assume that an encoder function is simple—that it is not complex enough to hide a universal computer, for example. We also assume that an encoder function is total recursive on (program, input) pairs. In fact, we assume it is *birecursive*, meaning that it can be uniquely decoded in a computable way.

In practice, a generalization of the above is allowed. Thus, an encoding of an arbitrary program and its input is given to a universal computer. The universal computer simulates the given program on its given input and generates the output of the simulated program. In the Universal Turing Machine Theorem statement above, the output of the universal computer is exactly the output of the simulated program. In practice, a machine is still considered universal if it outputs an encoding of the simulated output rather than the simulated output itself. As for the program/input encoding function, the output encoding function is assumed to be simple. We shall call universal computers *decodeless* if they do not require an output encoding function.

It is worth pointing out here that any universal computer which is not decodeless can be made so by appending a "termination subroutine" to its program which decodes the encoded output before halting. This is possible because we assume decoding is computable. Clearly, this larger program is still a program and, hence, in the List. So there are lots of universal computer programs (or functions) in the List. The next important theorem says this more formally.

THE RECURSION THEOREM If h is a total recursive function from programs into programs, then there is a fixed-point program P such that $P = h(P)$. This is just a way of saying that any computation shows up in the List many times—that the List is redundant, certainly in any computable way (represented by h)—or that there are many different programs that ultimately compute the same thing.

It's not hard to understand why. Consider a program which performs the computation of adding 1 to its input. There is another program which does the same thing but (stupidly) executes a subroutine which counts to a million before exiting. Since any number can be substituted for "million" in the preceding sentence and each resulting program is different, then the List must contain them all or a countable infinity of programs for adding 1. In this case, h is the function that appends to any program the subroutine which does nothing but count to a million. The h introduced in this paper is far more interesting and the resulting fixed-point program of the Recursion Theorem far more profound—it is crucial to our desired result.

Since P in the Recursion Theorem above can be computed (e.g., see Arbib[5]), our existence proof will be constructive, in this theoretical sense.

Alan Turing invented, in the course of his studies leading to the universal computer theorem, a theoretical computer now called a *Turing machine*. This very simple device is instrumental in our results here, and details will be provided in a later section. It is mentioned here because each program in the List can be thought of as a Turing machine. In other words, a list of all Turing machines could be the List we refer to above, and both of the important theorems above apply to it. In fact, this is why the first theorem is called the Universal Turing Machine Theorem. The universal Turing machine is that Turing machine in the List which simulates all other Turing machines in the List. Several leading mathematicians in the 1930s and 1940s spent considerable effort proving that several apparently different ways of describing computation were all equivalent. We therefore feel free today to use "program," "Turing machine," "effective computation," and "partial recursive function" interchangeably for the same concept although the formalisms are often different.

CONSTRUCTION VS. COMPUTATION

Usually it is only the computational abilities of CA which are investigated. It is also of interest to study the "constructional" abilities—the construction of one configuration by another. This was one of the original motivations for the study of CA. As mentioned above, von Neumann[35] introduced the CA, conceived as an infinite chessboard, each square, or "cell," of which represents a copy of a single finite automaton (computer), as an environment in which to study the logical intricacies of biological reproduction. This paper is also along these lines.

It is not at all clear what "construction" should be defined to be. In fact, Holland[12] has given arguments which show that Moore-type CA are inherently incapable of supporting construction in the full generality that term might support—namely, construction of, or simulation of, arbitrary networks of finite automata.[3] This is no surprise in view of the existence of Garden-of-Eden configurations (patterns which cannot arise from computation, but can only exist at time zero by external programming of the CA) in even trivial CA.[25] Lieblein[15] has characterized the sequences of configurations which can be attained in Moore-type CA. We shall be interested in sequences which are readily realized in a Moore-type CA and will not even find it necessary to define construction—computation will be sufficient.

The type of construction process which is treated in this paper is called *self-reproduction*[20] where the definition of this term is assumed: Let c_0 be an initial

[3] A *Moore-type*, or sequential, CA is one with memory at each cell so that there is a delay between any input and its associated output. This is distinguished from *Mealy-type* CA which allow instantaneous (combinational) output from inputs at each cell. The von Neumann CA and all others considered in this paper are Moore-type.

configuration or subconfiguration in some (infinite) CA. Then c_0 self-reproduces in the (Moore) sense if, at some time $t > 0$, there exist at least two disjoint copies of c_0; at time $t' > t$, there exist at least three disjoint copies of c_0; etc. To say there exists two disjoint copies of a configuration c in configuration c' means, of course, that there is a translation δ such that $c' = c \cup \delta c$. Two configurations are *disjoint*, of course, if their non-quiescent cells are disjoint.

As Moore points out, this definition permits trivial CA phenomena to be interpreted as self-reproduction. For example, the configuration consisting of a single non-quiescent cell (a single 1 on a background of 0's) self-reproduces in the one-dimensional CA with K_1 template (a cell and one of its nearest neighbors—see Smith[27] for details) and transition function f which maps the current state of a cell, given the current state of its neighbor, to next-state 1 (that is, $f(q_0, q_1) = 1$ except for $f(0, 0)$ which by definition of quiescent background must be 0). Thus, care will be exercised to ensure the nontriviality of any self-reproducing configuration introduced in this paper. In particular, we will insist, as is the historical precedent, that a self-reproducing configuration be a universal computer. This definition of nontriviality is perhaps overly restrictive for biological purposes, but we as yet do not have a more lifelike complexity requirement.

TURING MACHINE SIMULATION

The environment of the self-reproducing machines to be introduced below will be CA of the simple Turing-machine simulation type introduced in Smith.[24,26] We first review Turing machines and then outline the simple simulations of them by CA we require.

A Turing machine is a conceptually simple device consisting of a one-dimensional *tape* divided lengthwise into squares and a *head* which scans along the tape in either direction. The tape is assumed infinite in extent and each square is initially blank, but for a finite input section. The head has a finite amount of memory, expressed as a finite number of states the memory can be in. The head can be thought of as containing the program to be executed by the Turing machine. At any one step, the Turing machine head can read or write the square currently being scanned and can then move either one square left or right. It can read or write a finite set of symbols, of which one is the blank, on its tape. We will refer to an (m, n) Turing machine M and mean an m-symbol, n-state Turing machine. It is surprising that this simple device can compute any computable function but this is what the Universal Turing Machine Theorem says. Minsky has found the simplest known universal Turing machine to be a (4, 7) Turing machine; he has also found a (6, 6) universal Turing machine.[19] These are not decodeless universal computers; they assume the simulated output tape is encoded.

The notion of simulating a Turing machine by a CA is straightforward (and rediscovered numerous times): A row of CA cells simulates the squares on a Turing machine tape while at the same time simulating the head of the Turing machine. The state set of the CA cell is chosen to have two coordinates: one simulates the symbol on the Turing machine tape; the other simulates the state of the Turing machine head or the absence of the head at that cell. Appropriate choice of transition function operating on the nearest-neighbor neighborhood makes it easy to directly simulate the Turing machine. It is easy to show that an (m, n) Turing machine can be simulated by a one-dimensional CA with $m \times (n+1)$ states per cell. Without working too much harder, this can be improved to $m + 2n$ states per cell.[26] Thus, the (4, 7) universal computer (or the (6, 6)) can be simulated by an 18-state, 3-neighbor CA.[4] We will say that Turing machine M is *wired into* CA Z_M or that Z_M has M *wired in*.

For notational convenience, the following devices will be employed:

$$x \rightarrow_P y \tag{1}$$

means that Turing machine program P acting on initial tape x halts with y (and nothing else but blank squares) on the tape as its final result.

$$x \rightarrow_P y \rightarrow_P y' \rightarrow_P \ldots \tag{2}$$

means that Turing machine program P acting on initial tape x alters the tape until y is on the tape at some time $t > 0$. P proceeds to compute on this tape until y' appears on it at some time $t' > t$; and so on. This is a way of representing the temporal sequence of tapes generated by a Turing machine treated as generator of recursively enumerable sets instead of as a computer of partial recursive functions.

$$\uparrow x \tag{3}$$

indicates that a Turing machine control head is scanning the leftmost (nonblank) symbol in the string x written on its tape.

Exploiting the ease of simulation of Turing machines by CA leads directly to self-reproducing CA in the next section. Then, in the following section, very simple self-reproducing CA are derived, also using the simulation of Turing machines.

SELF-REPRODUCTION: SIMPLE COMPUTATION, COMPLEX CA

In this section we will present a self-reproducing CA—a CA with a self-reproducing configuration—which is not necessarily simple, but is easy to describe. A Turing machine computation is described which does the desired thing. This Turing machine is wired in a CA which is then self-reproducing. The CA is, thus, as complex as required by the Turing machine that is wired in. In other words, the wired-in Turing machine "does everything." But first, just for exercise, we use the technique to prove self-reproduction where we relax the nontriviality requirement. We call this trivial self-reproduction *self-replication*.

[4]See the last section and last footnote for improvements to this.

SELF-REPLICATION, OR TRIVIAL SELF-REPRODUCTION

Wire in machine M which duplicates its input tape, repositions its head, duplicates all tape to the right of this position, and then repeats these steps forever. This can be represented as follows:

$$\uparrow x \rightarrow_M (x, \uparrow x) \rightarrow_M (x, x, \uparrow x) \rightarrow_M \cdots \tag{4}$$

where the vertical arrow indicates the position of the head after the computation indicated by the horizontal arrow immediately to its left.

Note that any one-dimensional configuration can be made to self-replicate in this scheme. Just embed a "head" cell at the left end of a given "tape" configuration; program the head cell to its initial state. Thus, we have proved the following theorem:

THEOREM 1 Let S be the set of finite one-dimensional configurations on finite state set Q. Then there exists CA Z_M in which s is a self-reproducing configuration, for all $s \in S$.

Since this result can be readily generalized to the d-dimensional case, we have answered in the affirmative a question raised by Lieblein[15]: Can any configuration be made to self-reproduce? Closely related to the theorem is the work of Waksman.[37] He has shown that a one-dimensional string can be made to self-replicate in a one-dimensional CA specially designed for the given string. Self-replication in the Waksman scheme proceeds in real time except for an initial "setting-up" time. "Realtime" here means that a string of length n is reproduced in n time steps. Note that our self-replication scheme is slower than real time (although linear time in n), but one CA serves for all strings.

NONTRIVIAL SELF-REPRODUCTION

Wire in universal Turing machine U' such that

$$\uparrow e'(P,x) \rightarrow_{U'} (e'(P,x), y, \uparrow e'(P,x)) \rightarrow_{U'} (e'(P,x), y, e'(P,x), y, \uparrow e'(P,x)) \\ \rightarrow_{U'} \cdots$$

where e' is the encoding of programs and tapes required by U' and $x \rightarrow_P y$. Thus, U' first makes a copy of its input string. This can be done in such a way that the head of the universal machine never moves left of its initial position (at the leftmost symbol of its initial input string). Then, it simulates the computation of program P on its input x. If this computation halts with output y, then U' writes y on its output to the right of the copy of the input string $e(P,x)$. It is well known to Turing machine programmers that this simulation can be made to happen in the half-infinite space to the right of the input string copy so as not to destroy it during the simulation. Then another copy of the input string $e(P,x)$ is written to the right of y. Finally, the machine repositions its head to the leftmost symbol of the rightmost copy of string $e(P, x)$ and reenters its start state. Since U' is designed to ignore everything to the left of its start state position, the infinite computation represented above is achieved. Hence:

THEOREM 2 For any Turing machine P, there exists a CA $Z_{U'}$, and configuration c such that c self-reproduces and simulates P.

Note that P could be universal if desired. Hence:

Corollary 2.1. There exists a CA $Z_{U'}$, and configuration c such that c self-reproduces and is universal.

This corollary establishes the existence of universal (meaning computation universal) self-reproducing machines, but we are now also interested in obtaining the simplest such machines (in terms of state count per cell). Notice that U' above has to do a lot more than just simulation of an arbitrary computation. In particular, its restriction to half-infinite tapes will increase its complexity.

SELF-REPRODUCTION: COMPLEX COMPUTATION, SIMPLE CA

In this section we derive the main result of this paper. As opposed to the section above, the self-reproducing configuration "does everything." That is, for simplicity of the simulating CA, the wired-in computer U does as little as possible—i.e., nothing more is required of U than it be universal. The (4, 7) universal machine suggests to us just how simple (in terms of state and symbol count) U can be. In fact, since the simplest (decodeless) universal program U does just the following:

$$e(P',\ x') \rightarrow_U y'$$

where e is the encoding function required by U and

$$x' \rightarrow_{P'} y'\ ,$$

then a program P is desired such that

$$\uparrow x \rightarrow_P (e(P,x), y, \uparrow e(P,x)) \rightarrow_P (e(P,x), y, e(P,x), y, \uparrow e(P,x)) \rightarrow_P \ldots$$

where y is the result of some computation on x. For example, y might be a string of d background (blank or quiescent) symbols, b^d, so that $e(P, x)$ is "separated" from the second $e(P, x)$.

Note that if such a program P exists, then self-reproduction is straightforward. Simply take CA Z_U with U wired in and embed initial subconfiguration $e(P, x)$. Then the following situation holds:

$$\uparrow e(P,x) \rightarrow_U (e(P,x), y, \uparrow e(P,x)) \rightarrow_U (e(P,x), y, e(P,x), y, \uparrow e(P,x)) \rightarrow_U \ldots .$$

That is, the configuration representing $e(P, x)$ is self-reproducing. We now show that the desired program P exists and, furthermore, that it can be chosen to be

universal. This result should be compared to a similar, but "one-shot," result on self-describing machines obtained by Lee.[14] He proved that there is a program P such that

$$x \rightarrow {}_P e(P, x)$$

with x the blank tape. Thatcher[32] created an actual example of such a machine.

LEMMA 3 For an arbitrary birecursive encoding function e from (program, tape) pairs to tapes and for an arbitrary partial recursive function g, with $g(x) = y$, there exists a self-describing machine with program P such that

$$\uparrow x \rightarrow {}_P(e(P, x), y, e(P, x)) \rightarrow {}_P(e(P, x), y, \uparrow e(P, x)) \rightarrow {}_P \ldots .$$

PROOF Define function h from programs into programs such that

$$\uparrow x \rightarrow {}_{h(Q)}(e(Q, x), y, \uparrow e(Q, x)) \qquad (a)$$

and

$$\uparrow e(Q, x) \rightarrow {}_{h(Q)}(e(Q, x), y, \uparrow e(Q, x)) . \qquad (b)$$

That is, $h(Q)$, for arbitrary program Q, is a program which first checks the input tape for a string of form $e(Q, x)$, for arbitrary x. This it can do because both e and Q are given it. If the input tape is not of form $e(Q, x)$, then $h(Q)$ executes the action indicated in (a) and described in temporal order below. As above, all computation is carried out in the singly infinite tape to the right of the initial head position which is the leftmost symbol of the initial tape.

1. $h(Q)$ first encodes the given program Q and input x with the given encoding function e to obtain $e(Q, x)$ and writes this on the tape while destroying the initial tape x.

2. It computes the given function g on argument x and writes the result y, if it is defined, to the right of the code $e(Q, x)$.

3. It writes the code $e(Q, x)$ again to the right of string y.

4. It repositions its head to the leftmost symbol of the rightmost code string $e(Q, x)$ and halts.

If the initial tape is of form $e(Q, x)$, then $h(Q)$ skips step (1) above and proceeds immediately to steps (2)–(4).

h is a total recursive function. Q and $h(Q)$ may not halt (see step 2 above), but $h(Q)$ is defined for every Q. Hence, by the Recursion Theorem, there exists a fixed-point program P for h—i.e., $h(P) = P$. For this program, (a) and (b) above become respectively

$$\uparrow x \rightarrow {}_{h(P)=P}(e(P, x), y, \uparrow e(P, x))$$

and

$$\uparrow e(P,x) \rightarrow_P (e(P,x), y, \uparrow e(P,x))$$

where $y = g(x)$ is defined. The lemma follows immediately if the final (halt) state of P is identified with its initial state. Clearly, if the halting version of P exists in the list of all programs, then the nonhalting version must also though, of course, with a different name or index.

THEOREM 4 Let Z_U be a universal CA with decodeless universal Turing machine U wired in. Then there exists a configuration c in Z_U which is self-reproducing and computes an arbitrary given partial recursive function g.

COROLLARY 4.1 There exists a CA Z_U and configuration c such that c self-reproduces and is universal.

PROOF Choose g in the theorem to be a universal Turing machine function.

I had hoped to prove the stronger result with the word "decodeless" omitted from the theorem above (and in fact claimed to have done so until my error was pointed out by Kristian Lindgren and Mats Nordahl). Then the universal CA of Smith[26] would have immediately implied the existence of one-dimensional self-reproducing universal machines with 2, 3, 4, 5, and 7 states per cell, for example.[5] These universal CA, however, are derived from Minsky's simplest known universal Turing machines which are not decodeless. I claim without proof that there are extremely simple decodeless universal Turing machines, perhaps adapted from Minsky's with the addition of one or two states, but leave it to some reader to find such a machine. It is only necessary that a universal machine be extended to decode its own output. This should be far simpler than the extensions beyond universality required by U' in the preceding section. Then the simplest known non-trivial self-reproducing machine immediately follows from the theorem and corollary above.

In comparison, von Neumann exhibited a 29-state universal CA capable (in two-dimensional) of supporting self-reproduction. His neighorhood—the so-called von Neumann neighborhood—consists of the five nearest-neighbor cells, including the cell itself. His construction is very lengthy and complex. Codd[10] was able to reduce the state count to 8 states per cell but his construction is also long and even more complicated. Codd's construction uses the so-called Moore neighborhood consisting of the 9 nearest-neighbor cells in two dimensions. Then Banks[6] created

[5] The complete selection of m-state × n-neighbor one-dimensional universal CA is 2×21, 3×13, 4×9, 5×8, 7×6, 11×4, 18×3, and 40×2.

the simplest known two-dimensional self-reproducing CA with 4 states and the von Neumann neighborhood.[6] By greatly increasing cell complexity,[3] Arbib was able to describe the processes simply. Here we have demonstrated both simple CA and simple descriptions by deriving CA with low-state count and using recursive function theory for compact constructions in them. Whereas, the earlier work in this area was confined to two dimensions, the results here are most elegant in one dimension although applicable to two or more dimensions. And finally, it is also striking that nowhere have we had to define "construction" in a CA to obtain self-reproduction—in fact, computation universality has been shown sufficient. The notion of "construction universality" used in earlier proofs is not required here.

Several comments about the class of self-reproducing machines developed here are in order. First, reproducing schemes based on the Recursion Theorem have been mentioned in several other places.[21,23,33] However, in all these cases the reproduction is one-shot—the process ceases after production of one offspring. The scheme of Theorem 4 produces offspring ad infinitum.

Second, if a self-reproducing configuration is interpreted to be "alive" only if it can reproduce again and again, then the machines of Theorem 4 are not alive. Once one of these machines computes and reproduces a single time, it becomes inactive. Of course, there is "no room" for another reproduction, but we will ignore this (degenerate) Malthusian dilemma and present a method for keeping a parent alive.

What we want is an augmentation of the scheme in the proof of Theorem 4 such that the head cell (active site) is also reproduced. This is how to accomplish the desired task:

1. A special symbol # is written after each string y, if y is defined.
2. When this situation occurs

$$\ldots, \#, e(P, x), y, \#, \uparrow e(P, x),$$

the cell representing # has the head cell in its neighborhood. When the head cell goes into its initial state H, the transition function f of the CA takes note of the juncture of # and H and creates a new head cell (indicated by an arrow pointing down):

$$\ldots, \#, e(P, x), y, \#, \uparrow e(P, x) \rightarrow \ldots, \#, e(P, x), y, \downarrow \#, \uparrow e(P, x).$$

[6]Banks[6] also demonstrated a 2-state, Moore neighborhood (two-dimensional) universal CA and a 3-state, von Neumann neighborhood universal CA. More attention has been paid to another 2-state, Moore neighborhood (two-dimensional) CA which is universal.[8,36] It is the so-called game of Life popularized in the *Scientific American* in 1970 and 1971.[11] Codd[10] proves that no 2-state, von Neumann neighborhood universal CA can exist (hence, no $k = 2$, $r = 1$, Class 4 CA, in the terminology of Wolfram,[39] can be universal without changing the definition of universality (as done in Lindgren,[16] see last footnote)). In any event, none of these CA are known to support self-reproduction.

3. The new head cell propagates to the left until it encounters the next # to the left at which time it is in a position to reactivate the parent, and it does so by going into state H.

This scheme has the attraction of being a fully functional CA, with lots of computations proceeding in parallel, and not just a simulated Turing machine. (See Hurd[13] and Smith[29] for demonstrations of how CA exceed Turing machines.)

If other dimensions are available, then in step (3) the new head can, while propagating left, move the entire string between consecutive # states up one unit in another dimension. This, however, still leads to a Malthusian overcrowding (see Moore[20]—his "population" theorem goes through for our generalized cellular automata). Löfgren[17] gives a partial solution to this problem in terms of "birth" and "death" rates and "complexity," or number of cells per self-reproducing configuration. He does not, however, give the mechanics of self-reproduction assumed in his theory—each machine is known only to reproduce and be erased at certain rates.

The third and most important comment on our self-reproduction scheme is concerned with its biological relevance. Have we designed machines which actually *construct* other machines, or are the procedures introduced here just fancy copying routines somehow distinct from construction? Even with the modifications suggested in the second comment above, the model is still very non-biological (as are all the other tessellation models) in one obvious way—it is not parallel but serial in the extreme. Also the concept of growth is ignored in the sense that in our model a full-grown multi-cell offspring is constructed by the parent—not an undeveloped single-cell "egg" which grows under its own control into a replica of its parent.

Arbib[4] has exhibited an example of morphallactic regeneration of an elementary "worm" which proceeds in a highly parallel fashion. The worm is a one-dimensional string of identical cells divided into equal sections of head, body, and tail states. If a length of the worm is destroyed, the remaining cells reapportion themselves into equal thirds of head, body, and tail. This resembles the morphallaxis of hydra in the biological world. (A simple exercise for the interested reader is to embed a worm like that of Arbib in a one-dimensional CA designed so that the worm grows its missing parts in the correct proportion.[29]) Similarly, Lieblein[15] has detailed the CA realization of one of the highly parallel tessellation-like models of biological phenomena[7] such as reproduction, evolution, mutation, crossing-over, and even creation. Vitányi[34] added sexuality. Case[9] pursues self-describing computations which build distortions of themselves. My own closest approach to biological uses of the theory is Smith.[31]

TOTALISTIC SELF-REPLICATION AND SELF-REPRODUCTION

The recent revival of interest in CA has concentrated on the *totalistic* case, where the transition function is required to be an arithmetic function of the neighborhood states. A result that has been rediscovered numerous times is that trivial self-reproduction—that is, self-replication—is possible in totalistic CA.[2,18,22,38]

For nontrivial self-reproduction, the relevant result is due to Albert and Culik[1] who have discovered a one-dimensional universal CA that uses only the nearest neighbors for a neighborhood and 14 states per cell.[7] Inspection of their proof reveals that their CA has a background wave of activity which converts the quiescent state of the CA to a preparatory state. So there is a constantly expanding wavefront of multiple state changes to effect this preparatory state. Their universal computation operates in the midst of this "big bang" expanding wave. The proofs herein do not go through for this background activity. Nevertheless, with appropriate changes to the definitions—to allow an expanding background rather than a quiescent background—I believe the arguments could be made to go through. I have not attempted to do so.

[7] And shattered my record of 19 years for the simplest known one-dimensional universal CA with the 3 nearest-neighbor neighborhood! Mine had 18 states,[26] but the authors of Lindgren[16] show that a careful selection of state subset of my solution suffices for a 13-state solution. In fact, they do better than this: They show a new 9-state solution. (That is for an apples-to-apples comparison, for universal computation against an initially quiescent background. By relaxing the quiescent background requirement to permit periodic initial backgrounds, they are able to show a 7-state solution.) The most surprising result to me of the Albert-Culik work is that *totalistic* CA suffice for universality. Combining two of their results yields a 56-state, 3-neighbor, one-dimensional totalistic universal CA.

REFERENCES

1. Albert, Jürgen, and Karel Culik, II. "A Simple Universal Cellular Automaton and Its One-Way and Totalistic Version." *Complex Systems* **1** (1987): 1–16. Their CA simulates any other CA, a new form of CA universality. Their Theorem 1 is the one-dimensional case of Theorem 3.3 in Smith.[27]
2. Amoroso, Serafino, and Gerald Cooper. "Tessellation Structures for Reproduction of Arbitrary Patterns." *J. Comput. Sys. Sci.* **5** (1970): 455–464. There is an entire literature on tessellation automata, which are CA with programs—that is, the local transition function can change, in an SIMD way, at every step.
3. Arbib, Michael A. "Automata Theory and Development: Part I." *J. Theoret. Biol.* **14** (1967):131–156.
4. Arbib, Michael A. "Self-Reproducing Automata—Some Implications for Theoretical Biology." *Towards a Theoretical Biology, 2: Sketches*, edited by C. H. Waddington. Edinburgh University Press, 1969.
5. Arbib, Michael A. *Theoreis of Abstract Automata.* Prentice-Hall, 1969. The jacket of this textbook is my first cover art. It represents a self-reproducing machine!
6. Banks, E. Roger. "Cellular Automata." AI Memo No. 198, MIT Artificial Intelligence Lab, Cambridge, MA, 1970. His 2-state-9-neighbor, two-dimensional universal CA predates the universality proof of Life.
7. Barricelli, Nils Aall. "Numerical Testing of Evolution Theories. Part I: Theoretical Introduction and Basic Tests." *Acta Biotheoretica* (parts I/II) **16** (1962). This contains a large number of computer graphics of the type prominent in Wolfram,[39] which is surprising considering the crude computers available at the time.
8. Berlekamp, Elwyn R., John H. Conway, and Richard K. Guy. *Winning Ways for Your Mathematical Plays*, Vol. 2. New York: Academic Press, 1982. I have not read this reference, but it apparently contains Conway's proof of the universality of Life.
9. Case, John A. "Note on Degrees of Self-Describing Turing Machines." *J. ACM* **18** (1971): 329–338.
10. Codd, Edgar Frank. "Propagation, Computation and Construction in 2-Dimensional Cellular Spaces." Technical Report 06921-1-T, ORA, The University of Michigan, 1965. Also *Cellular Automata*, ACM Monograph Series. New York: Academic Press, 1968.
11. Gardner, Martin. "On Cellular Automata, Self-Reproduction, the Garden of Eden and the Game 'Life.'" *Sci. Amer.* **224** (1971): 112–117. I did the cover of this issue of the magazine. It is a finite CA recognizing the palindrome

TOOHOTTOHOOT in real time (see Smith[29]). See also Vol. 223, pp. 120–123, for Gardner's first Life article.

12. Holland, John H. "Universal Embedding Spaces for Automata." In *Cybernetics of the Nervous System*, edited by N. Wiener and J. P. Schade. Progress in Brain Research Series, Vol. 17. New York: Elsevier, 1965.
13. Hurd, Lyman P. "Formal Language Characterizations of Cellular Automata Limit Sets." *Complex Systems* **1** (1987): 69–80. Proves that one-dimensional CA can compute more than Turing machines! Presumably this is because of the CA ability to compute infinitely. See also Smith.[29]
14. Lee, Chester Y. "A Turing Machine Which Prints Its Own Code Script." *Mathematical Theory of Automata.* Microwave Research Institute Symposia Series, Vol. 12. Brooklyn, NY: Polytechnic Press, 1963.
15. Lieblein, Edward. "A Theory of Patterns in Two-Dimensional Tessellation Space." Ph.D. Dissertation, University of Pennsylvania, 1968. An underappreciated work, the first to show the regular set characterization of cellular automata patterns.
16. Lindgren, Kristian, and Mats G. Nordahl. "Universal Computation in Simple One-Dimensional Cellular Automata." *Complex Systems* **4** (1990): 299–318.
17. Löfgren, Lars. "Self-Repair as a Computability Concept in the Theory of Automata." *Proc. of the Symp. on Math. Theory of Automata.* Brooklyn, NY: Polytechnic Institute of Brooklyn, 1962.
18. Merzinich, Wolfgang. "Cellular Automata With Additive Local Transition." *Proceedings of the First International Symposium on Category Theory Applied to Computation and Control*, 186–194. Amherst, MA, Department of Computer and Information Sciences, University of Massachusetts, 1974.
19. Minsky, M. *Computation: Finite and Infinite Machines.* Englewood Cliffs, NJ: Prentice-Hall, 1969.
20. Moore, Edward F. "Machine Models of Self-Reproduction." *Proceedings of Symposia in Applied Mathematics, American Mathematical Society* **14** (1962):17–34.
21. Myhill, John. "The Abstract Theory of Self-Reproduction." In *Views on General Systems Theory*, edited by M. D. Mesarovic. John Wiley: New York, 1964.
22. Ostand, Thomas J. "Pattern Reproduction in Tessellation Automata of Arbitrary Dimension." *J. Comput. System Sci.* **5** (1971): 623–628.
23. Rogers, Hartley, Jr. *Theory of Recursive Functions and Effective Computability.* San Francisco: McGraw-Hill, 1967.
24. Smith, Alvy Ray. "Simple Computation-Universal Cellular Spaces and Self-Reproduction." *Proceedings of the 9th SWAT* (1968): 269–277. SWAT = IEEE Symposium on Switching and Automata Theory. At the 16th conference the

name changed to FOCS = IEEE Symposium on the Foundations of Computer Science. Although I have published several papers in this conference series (see also Smith[28]), I am most proud of my illustration which has been used by FOCS as its proceedings cover for the last 17 years, starting with the 14th conference.

25. Smith, Alvy Ray. "Cellular Automata Theory." Technical Report No. 2, Digital Systems Laboratory, Stanford University, Stanford, California, 1969. My Ph.D. dissertation. There are several bugs in this report, fixed in the publications below.
26. Smith, Alvy Ray. "Simple Computation-Universal Cellular Spaces." *J. ACM* **18** (1971): 339–353.
27. Smith, Alvy Ray. "Cellular Automata Complexity Trade-Offs." *Info. & Control* **18** (1971): 466–482. General CA theory.
28. Smith, Alvy Ray. "Two-Dimensional Formal Languages and Pattern Recognition by Cellular Automata." *Proceedings of the 12th SWAT* (1971): 144–152. A paper *Pattern Recognition by Finite Cellular Automata*, based on this work was accepted for publication by the *J. Comput. System Sci.* in 1975, but I never completed the submission since I changed to computer graphics about this time. Since then most of the results of this paper were published in *Picture Languages: Formal Models for Picture Recognition*, edited by Azriel Rosenfeld, Chapters 3 and 5. Academic Press, New York, 1979.
29. Smith, Alvy Ray. "Real-Time Language Recognition by One-Dimensional Cellular Automata." *J. Comput. System Sci.* **6** (1972): 233–253. Relationship of formal languages and one-dimensional CA.
30. Smith, Alvy Ray. "Introduction to and Survey of Polyautomata Theory." In *Automata, Languages, Development*, edited by Aristid Lindenmayer and Grzegorz Rozenberg, 405–422. Amsterdam and New York: North-Holland, 1976. This paper contains an extensive bibliography on CA. The book is the conference proceedings for what I propose be called the Artificial Life 0 conference, held in Noordwijkerhout, The Netherlands, April, 1975. Other participants were Karel Culik, Pauline Hogeweg, John Holland, Aristid Lindenmayer, and Stanislaw Ulam.
31. Smith, Alvy Ray. "Plants, Fractals, and Formal Languages." *Computer Graphics* **18** (1984): 1–10. I have called the L-systems or Lindenmayer systems used here dynamic cellular graph automata in the taxonomy of Smith.[30]
32. Thatcher, James W. "The Construction of a Self-Describing Turing Machine." In *Proceedings on the Symposium of the Mathamatical Theory of Automata*, 165–171. New York: Polytechnic Press, 1963.
33. Thatcher, James W. "Universality in the von Neumann Cellular Model." Technical Report 03105-30-T, ORA, The University of Michigan, 1964. Also

in *Essays on Cellular Automata*, edited by Arthur W. Burks, University of Illinois Press, 1970, a compendium of early papers.

34. Vitányi, Paul M. B. "Sexually Reproducing Cellular Automata." *Mathematical Biosciences* **18** (1973): 23–54.
35. von Neumann, John. *The Theory of Self-Reproducing Automata*, edited by Arthur W. Burks. Urbana: University of Illinois Press,1966. Von Neumann performed the work, completed here, in 1952-53.
36. Wainwright, Robert. "Life Is Universal." Proceedings of the Winter Simulation Conference, ACM, Washington, DC, 1974. I have not read this article but assume it is a presentation of the argument by R. William Gosper, circulated among Life cognoscenti who were the primary readers of Wainwright's Lifeline newsletter.
37. Waksman, Abraham. "A Model of Replication." *J. ACM* **16(1)** (1969).
38. Winograd, Terry A. "Simple Algorithm for Self-Replication." AI Memo No. 197, MIT Artificial Intelligence Lab, Cambridge, Massachusetts, 1970.
39. Wolfram, Stephen. *Theory and Applications of Cellular Automata*. Singapore: World Scientific, 1986. Represents well the combination of dynamic systems theory with CA, which revitalized the field in the 1980s.

Eugene H. Spafford
Software Engineering Research Center, Department of Computer Sciences, Purdue University, West Lafayette, Indiana 47907-2004 ; email: spaf@cs.purdue.edu

Computer Viruses—A Form of Artificial Life?

INTRODUCTION

There has been considerable interest of late in computer viruses. One aspect of this interest has been to ask if computer viruses are a form of artificial life, and what that might imply.

This paper is a condensed, high-level description of computer viruses—their history, structure, and how they relate to some properties that might define artificial life. It provides a general introduction to the topic without requiring an extensive background in computer science.

The interested reader might pursue Spafford et al.,[12] Cohen,[1] Denning,[3] and Hoffman[7] for more detail about computer viruses and their properties. The description in this paper of the origins of computer viruses and their structure is taken from Spafford et al.[12]

WHAT IS A COMPUTER VIRUS?

The term *computer virus* is derived from and analogous to a biological virus. The word *virus* itself is Latin for *poison*. Viral infections are spread by the virus (a small shell containing genetic material) injecting its contents into a far larger body cell. The cell then is infected and converted into a biological factory producing replicants of the virus.

Similarly, a computer virus is a segment of machine code (typically 200–4000 bytes) that will copy its code into one or more larger "host" programs when it is activated. When these infected programs are run, the viral code is executed and the virus spreads further. Viruses cannot spread by infecting pure data; pure data is not executed. However, some data, such as files with spreadsheet input or text files for editing, may be interpreted by application programs. For instance, text files may contain special sequences of characters that are executed as editor commands when the file is first read into the editor. Under these circumstances, the data is "executed" and may spread a virus. Data files may also contain "hidden" code that is executed when the data is used by an application, and this too may be infected. Technically speaking, however, pure data itself cannot be infected.

WORMS

Worms are another form of software that is often referred to by the uninformed as a computer virus. The Internet Worm of November 1988 is an example of one of these programs.

Unlike viruses, worms are programs that can run independently and travel from machine to machine across network connections; worms may have portions of themselves running on many different machines. Worms do not change other programs, although they may carry other code that does, such as a true virus.

In 1982, John Shoch and Jon Hupp of Xerox PARC (Palo Alto Research Center) described the first computer worms.[9] They were working with an experimental, networked environment using one of the first local area networks. While searching for something that would use their networked environment, one of them remembered reading *The Shockwave Rider* by John Brunner, written in 1975. This science fiction novel described programs that traversed networks, carrying information with them. Those programs were called *tapeworms* in the novel. Shoch and Hupp named their own programs *worms*, because in a similar fashion they would travel from workstation to workstation, reclaiming file space, shutting off idle workstations, delivering mail, and doing other useful tasks.

Few computer worms have been written in the time since then, especially worms that have caused damage, because they are not easy to write. Worms require a network environment and an author who is familiar not only with the network services and facilities, but also with the operating facilities required to support them once they have reached the machine. The Internet worm incident of November, 1988 clogged machines and networks as it spread, and is an example of a worm.[11,10]

Worms have also appeared in other science fiction literature. Recent "cyberpunk" novels such as *Neuromancer* by William Gibson[6] refer to worms by the term "virus." The media has also often referred incorrectly to worms as viruses. This paper focuses only on viruses as defined here. Many of the comments about viruses and artificial life may also be applied to worm programs.

OTHER THREATS

There are many other kinds of *vandalware* that are often referred to as viruses, including *bacteria, trojan horses, logic bombs,* and *trapdoors.* These will not be described here. The interested reader can find explanations in Spafford et al.,[12] Hoffman,[7] and Denning.[3]

NAMES

As the authors of viruses generally do not name their work formally and do not come forward to claim credit for their efforts, it is usually up to the community that discovers a virus to name it. A virus name may be based on where it is first discovered or where a major infection occurred, e.g., the *Lehigh* and *Alameda* viruses. Other times, the virus is named after some definitive string or value used by the program, e.g., the *Brain* and *Den Zuk* viruses. Sometimes, viruses are named after the number of bytes by which they extend infected programs, such as the *1704* and *1280* viruses. Still others may be named after software for which the virus shows an affinity, e.g., the *dBase* virus. In the remainder of this paper, viruses are referred to by commonly accepted names. Refer to Spafford et al.[12] or Stang[13] for detailed lists of virus names and characteristics.

A HISTORY LESSON

The first use of the term *virus* to refer to unwanted computer code occurred in 1972 in a science fiction novel, *When Harley Was One*, by David Gerrold.[1] The description of *virus* in that book does not fit the currently accepted definition of computer virus—a program that alters other programs to include a copy of itself. Fred Cohen formally defined the term *computer virus* in 1983.[1] At that time, Cohen was a graduate student at the University of Southern California attending a security seminar. The idea of writing a computer virus occurred to him, and in a week's time he put together a simple virus that he demonstrated to the class. His advisor, Professor Len Adelman, suggested that he call his creation a computer virus. Dr. Cohen's thesis and later research were devoted to computer viruses.

It appears, however, that computer viruses were being written by other individuals, although not named such, as early as 1981 on Apple II computers.[4] Some

[1] The recent reissue of Gerrold's book has this subplot omitted.

early Apple II viruses included the notorious "Festering Hate," "Cyberaids," and "Elk Cloner" strains. Sometimes virus infections were mistaken for trojan horses, as in the "Zlink virus" [sic] which was a case of the Zlink communication program infected by "Festering Hate." The "Elk Cloner" virus was first reported in mid-1981.

It is only within the last three years that the problem of viruses has grown to significant proportions. Since the first infection by the *Brain* virus in January 1986, up to April 1, 1991, the number of known viruses has grown to over 200 distinctly different IBM PC viruses. The problem is not restricted to the IBM PC, and now affects all popular personal computers. Mainframe viruses do exist for a variety of operating systems and machines, but all reported to date have been experimental in nature, written by serious academic researchers in controlled environments.

Where viruses have flourished is in the weak security environment of the personal computer. Personal computers were originally designed for a single dedicated user—little, if any, thought was given to the difficulties that might arise should others have even indirect access to the machine. The systems contained no security facilities beyond an optional key switch, and there was a minimal amount of security-related software available to safeguard data. Today, however, personal computers are being used for tasks far different from those originally envisioned, including managing company databases and participating in networks of computer systems. Unfortunately, their hardware and operating systems are still based on the assumption of single trusted user access, and this allows computer viruses to flourish on those machines.

FORMAL STRUCTURE

True viruses have two major components: one that handles the spread of the virus, and a manipulation task. The manipulation task may not be present (has null effect), or it may act like a logic bomb, awaiting a set of predetermined circumstances before triggering. These two virus components will be described in general terms, and then more specific examples will be presented as they relate to the most common personal computer: the IBM PC. Viruses on other machines behave in a similar fashion.

A NOTE ABOUT MAINFRAME VIRUSES As already noted, viruses can infect minicomputers and mainframes as well as personal computers. Laboratory experiments conducted by various researchers have shown that any machine with almost any operating system can support computer viruses. However, there have been no documented cases of true viruses on large multi-user computers other than as experiments. This is caused, in part, by the greater restrictions built into the software and hardware of those machines, and by the way they are usually used. My further comments will therefore be directed towards PC viruses, with the understanding that analogous statements could be made about mainframe viruses.

STRUCTURE For a computer virus to work, it somehow must add itself to other executable code. The viral code must be executed before the code of its infected host (if the host code is ever executed again). One form of classification of computer viruses is based on the three ways a virus may add itself to host code: as a shell, as an add-on, and as intrusive code.

Shell Viruses. A shell virus is one that forms a "shell" (as in "eggshell" rather than "Unix shell") around the original code. In effect, the virus becomes the program, and the original host program becomes an internal subroutine of the viral code. An extreme example of this would be a case where the virus moves the original code to a new location and takes on its identity. When the virus is finished executing, it retrieves the host program code and begins its execution.

Add-On Viruses. Most viruses are add-on viruses. They function by appending their code to the end of the host code, or by relocating the host code and adding their own code to the beginning. The add-on virus then alters the start-up information of the program, executing the viral code before the code for the main program. The host code is left almost completely untouched; the only visible indication that a virus is present is that the file grows larger.

Intrusive Viruses. Intrusive viruses operate by replacing some or all of the original host code with viral code. The replacement might be selective, as in replacing

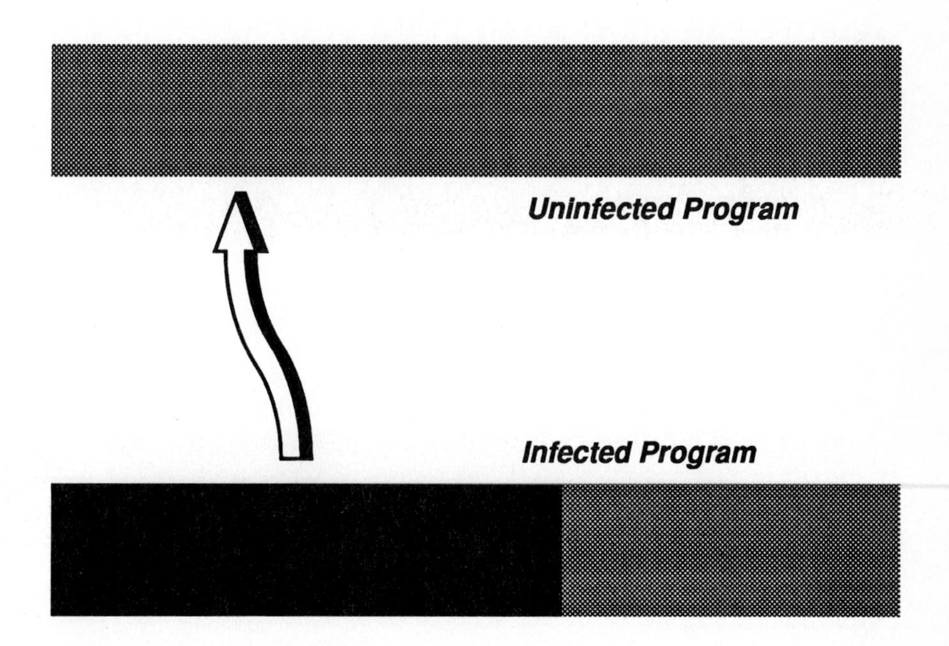

FIGURE 1 Shell Virus Infection.

a subroutine with the virus, or inserting a new interrupt vector and routine. The replacement may also be extensive, as when large portions of the host program are completely replaced by the viral code. In the latter case, the original program can no longer function.

TRIGGERS Once a virus has infected a program, it seeks to spread itself to other programs, and eventually to other systems. Simple viruses do no more than this, but most viruses are not simple viruses. Common viruses wait for a specific triggering condition, and then perform some activity. The activity can be as simple as printing a message to the user, or as complex as seeking particular data items in a specific file and changing their values. Often, viruses are destructive, removing files or reformatting entire disks.

The conditions that trigger viruses can be arbitrarily complex. If it is possible to write a program to determine a set of conditions, then those same conditions can be used to trigger a virus. This includes waiting for a specific date or time, determining the presence or absence of a specific set of files (or their contents), examining user keystrokes for a sequence of input, examining display memory for a specific pattern, or checking file attributes for modification and permission information. Viruses also may be triggered based on some random event. One common trigger component is a

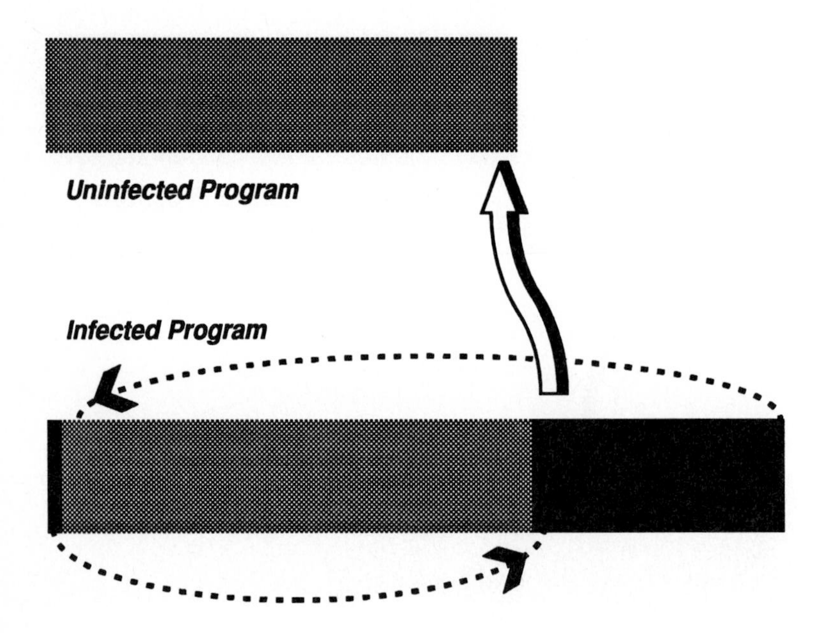

FIGURE 2 Add-On Virus Infection.

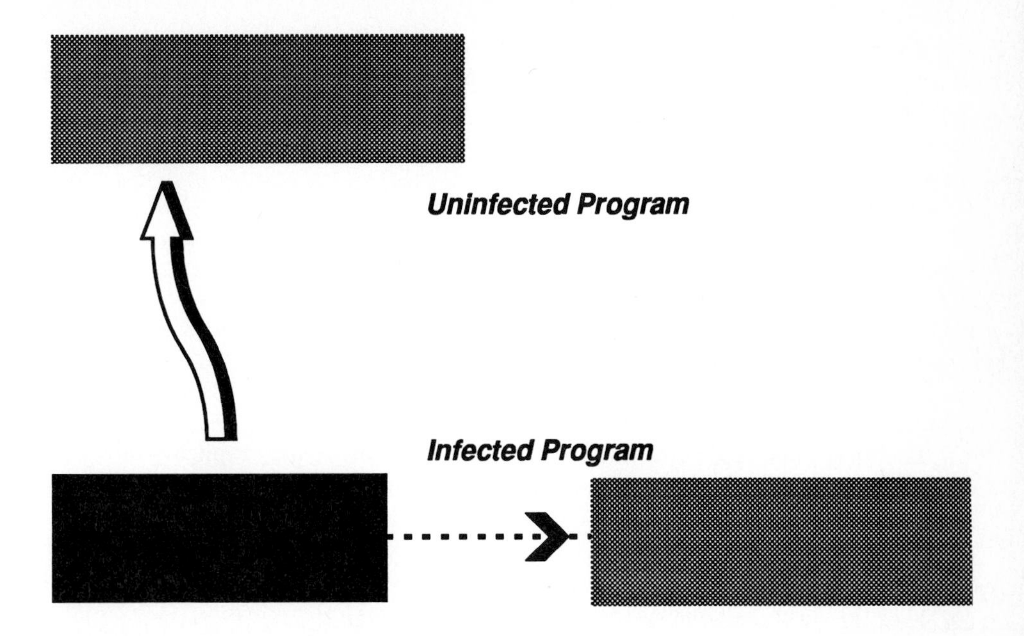

FIGURE 3 Intrusive Virus Infection.

counter used to determine how many additional programs the virus has succeeded in infecting—the virus does not trigger until it has propagated itself a certain minimum number of times. Of course, the trigger can be any combination of these conditions, too.

HOW DO VIRUSES SPREAD?

Computer viruses can infect any form of writable storage, including hard disk, floppy disk, tape, optical media, or memory. Infections can spread when a computer is booted from an infected disk, or when an infected program is run. It is important to realize that often the chain of infection can be complex and convoluted. A possible infection might spread in the following way:

1. A client brings in a diskette with a program that is malfunctioning (because of a viral infection).
2. A consultant runs the program to discover the cause of the bug—the virus spreads into the memory of the consultant's computer.
3. The consultant copies the program to another disk for later investigation—the virus infects the copy utility on the hard disk.
4. The consultant moves on to other work preparing a letter—the virus infects the screen editor on the hard disk.

5. The system is switched off and rebooted the next day—the virus is cleared from memory, only to be reinstalled when either the screen editor or copy utility is used next.
6. Someone invokes the infected screen editor across a network link, thus infecting his or her own system.

THE THREE STAGES OF A VIRUS' LIFE

For a virus to spread, its code must be executed. This can occur either as the direct result of a user invoking an infected program, or indirectly through the system executing the code as part of the system boot sequence or a background administration task.

The virus then replicates, infecting other programs. It may replicate into only one program at a time, it may infect some randomly chosen set of programs, or it may infect every program on the system. Sometimes a virus will replicate based on some random event or on the current value of the clock. The different methods will not be presented in detail because the result is the same: there are additional copies of the virus on your system.

Finally, most viruses incorporate a manipulation task that can consist of a variety of effects (some odd, some malevolent) indicating the presence of the virus. Typical manipulations might include amusing screen displays, unusual sound effects, system reboots, or the reformatting of the user's hard disk.

ACTIVATING A VIRUS The IBM PC can be used as an example to illustrate how a virus is activated. Viruses in other types of computer systems behave in similar manners.

The IBM PC Boot Sequence. This section gives a detailed description of the various points in the IBM PC boot sequence that can be infected by a virus. We will not go into extensive detail about the operations at each of these stages; the interested reader may consult the operations manuals of these systems, or any of the many "how-to" books available.

The IBM PC boot sequence has six components:

1. ROM BIOS routines
2. Partition record code execution
3. Boot sector code execution
4. IO.SYS and MSDOS.SYS code execution
5. COMMAND.COM command shell execution
6. AUTOEXEC.BAT batch file execution

ROM BIOS. When an IBM PC, or compatible PC, is booted, the machine executes a set of routines in ROM (read-only memory). These routines initialize the hardware and provide a basic set of input/output routines that can be used to access the disks, screen, and keyboard of the system. These routines constitute the basic input/output system (BIOS).

ROM routines cannot be infected by viral code (except at the manufacturing stage), as they are present in read-only memory that cannot be modified by software. Some manufacturers now provide extended ROMs containing further components of the boot sequence (e.g., partition record and boot sector code). This trend reduces the opportunities for viral infection, but also may reduce the flexibility and configurability of the final system.

PARTITION RECORD. The ROM code executes a block of code stored at a well-known location on the hard disk (head 0, track 0, sector 1). The IBM PC disk operating system (DOS) allows a hard disk unit to be divided into up to four logical partitions. Thus, a 100Mb hard disk could be divided into one 60Mb and two 20Mb partitions. These partitions are seen by DOS as separate drives: "C," "D," and so on. The size of each partition is stored in the partition record, as is a block of code responsible for locating a boot block on one of the logical partitions.

The partition record code can be infected by a virus, but the code block is only 446 bytes in length. Thus, a common approach is to hide the original partition record at a known location on the disk, and then to chain to this sector from the viral code in the partition record. This is the technique used by the New Zealand virus, discovered in 1988. (See Figures 4 and 5.)

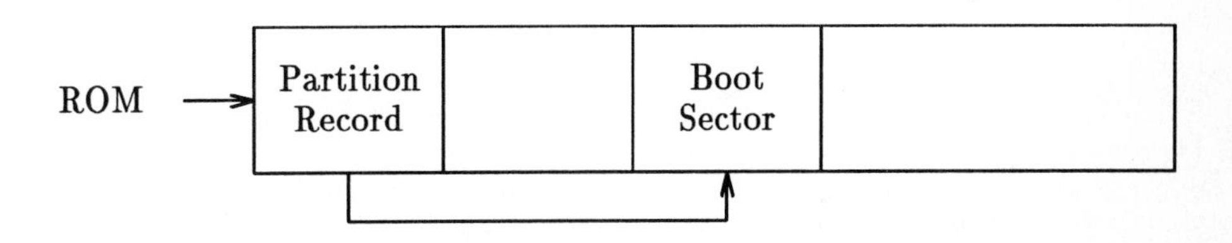

FIGURE 4 Hard disk before infection.

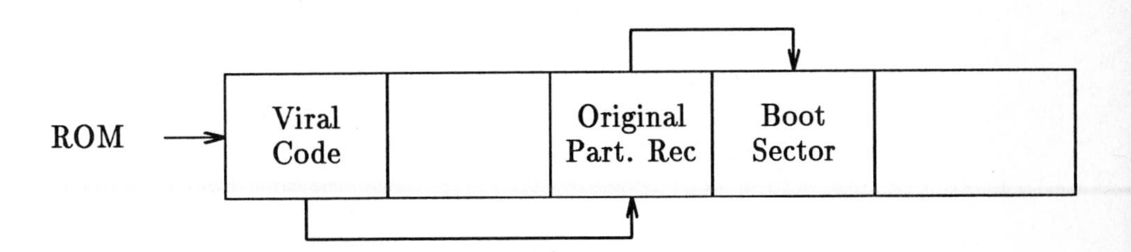

FIGURE 5 Hard disk after infection by New Zealand Virus.

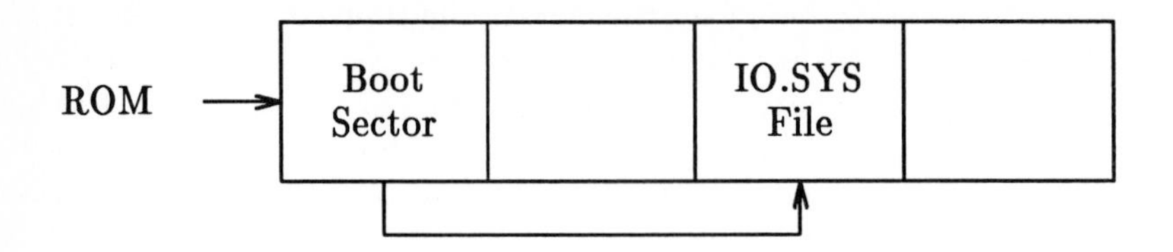

FIGURE 6 Floppy disk before infection.

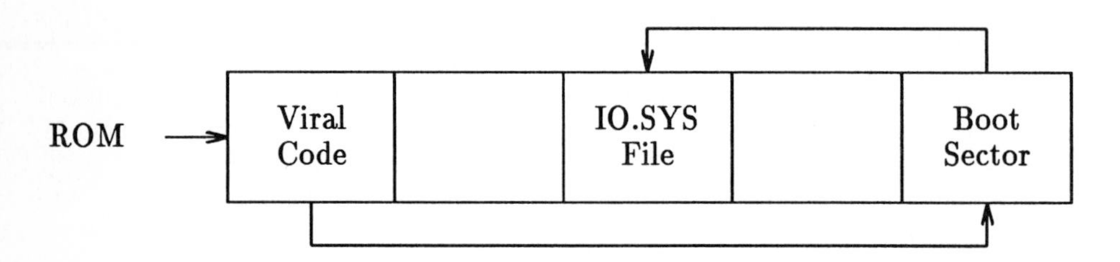

FIGURE 7 After Alameda Virus Infection.

BOOT SECTORS. The partition record code locates the first sector on the logical partition, known as the boot sector. (If a floppy disk is inserted, the ROM will execute the code in its boot sector, head 0, track 0, sector 1.) The boot sector contains the BIOS parameter block (BPB). The BPB contains detailed information on the layout of the filing system on disk, as well as code to locate the file IO.SYS. That file contains the next stage in the boot sequence. (See Figure 6.)

A common use of the boot sector is to execute an application program, such as a game, automatically; unfortunately, this can include automatic initiation of a virus. Thus, the boot sector is a common target for infection.

Available space in the boot sector is limited, too (a little over 460 bytes is available). Hence, the technique of relocating the original boot sector while filling the first sector with viral code is also used here.

A typical example of such a "boot sector" virus is the *Alameda* virus. This virus relocates the original boot sector to track 39, sector 8, and replaces it with its own viral code. (See Figure 7.) Other well-known boot sector viruses include the *New Zealand* (on floppy only), *Brain, Search*, and *Italian* viruses. Boot sector viruses are particularly dangerous because they capture control of the computer system early in the boot sequence, before any anti-viral utility becomes active.

MSDOS.SYS, IO.SYS. The boot sector next loads the IO.SYS file, which carries out further system initialization, then loads the DOS system contained in the MSDOS.SYS file. Both these files could be subject to viral infection, although no known viruses target them.

COMMAND SHELL. The MSDOS.SYS code next executes the command shell program (COMMAND.COM). This program provides the interface with the user, allowing execution of commands from the keyboard. The COMMAND.COM program can be infected, as can any other .COM or .EXE executable binary file.

The COMMAND.COM file is the specific target of the *Lehigh* virus that struck Lehigh University in November 1987. This virus caused corruption of hard disks after it had spread to four additional COMMAND.COM files.

AUTOEXEC BATCH FILES. The COMMAND.COM program is next in the boot sequence. It executes a list of commands stored in the AUTOEXEC.BAT file. This is simply a text file full of commands to be executed by the command interpreter. A virus could modify this file to include execution of itself. Ralf Burger has described how to do exactly that in his book *Computer Viruses—A High Tech Disease.* His virus uses line editor commands to edit its code into batch files. Although a curiosity, such a virus would be slow to replicate and easy to spot. This technique is not used by any known viruses "in the wild."

Infection of a User Program. A second major group of viruses spreads by infecting program code files. To infect a code file, the virus must insert its code in such a way that it is executed before its infected host program. These viruses come in two forms:

1. *Overwriting.* The virus writes its code directly over the host program, destroying part or all of its code. The host program will no longer execute correctly after infection.
2. *Non-overwriting.* The virus relocates the host code, so that the code is intact and the host program can execute normally.

A common approach used for .COM files is to exploit the fact that many of them contain a jump to the start of the executable code. The virus may infect the programs by storing this jump, and then replacing it with a jump to its own code. When the infected program is run, the virus code is executed. When the virus finishes, it jumps to the start of the program's original code using the stored jump address. (See Figure 8.)

Notice that in the case of the overwriting virus, the more complex infection strategy often means that all but a small block of the original program is intact. This means that the original program can be started, although often it will exhibit sporadic errors or abnormal behavior.

Memory-Resident Viruses The most "successful" viruses to date exploit a variety of techniques to remain resident in memory once their code has been executed and their host program has terminated. This implies that, once a single infected program has been run, the virus potentially can spread to any or all programs in the system. This spreading occurs during the entire work session (until the system is rebooted to clear the virus from memory), rather than during a small period of time when the infected program is executing viral code.

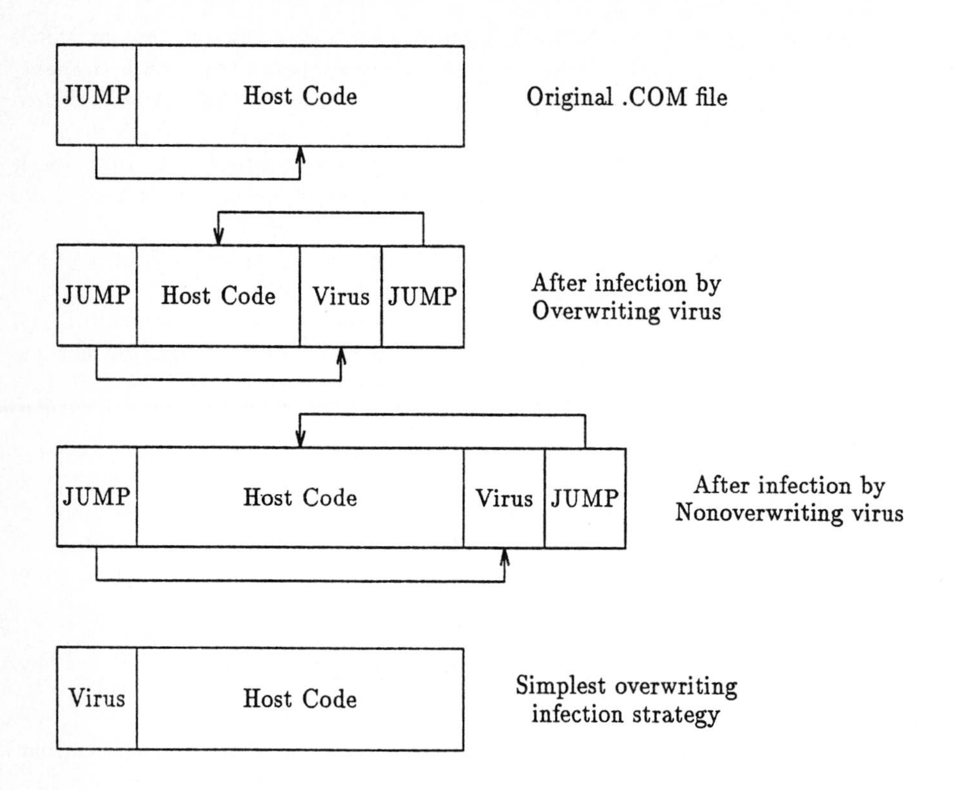

FIGURE 8 Infection of user applications.

Thus, the two categories of virus are:

1. *Transient.* The viral code is active only when the infected portion of the host program is being executed.
2. *Resident.* The virus copies itself into a block of memory and arranges to remain active after the host program has terminated. The viruses are also known as TSR (**T**erminate and **S**tay **R**esident) viruses.

Examples of memory-resident viruses are all known boot sector viruses, the *Israeli*, *Cascade*, and *Traceback* viruses.

If a virus is present in memory after an application exits, how does it remain active? That is, how does the virus continue to infect other programs? The answer is that it also infects the standard interrupts used by DOS and the BIOS so that it is invoked by other applications when they make service requests.

The IBM PC uses many interrupts (both hardware and software) to deal with asynchronous events and to invoke system functions. All services provided by the BIOS and DOS are invoked by the user storing parameters in machine registers, then causing a software interrupt.

When an interrupt is raised, the operating system calls the routine whose address it finds in a special table known as the *vector* or *interrupt* table. Normally, this table contains pointers to handler routines in the ROM or in memory-resident portions of the DOS (see Figure 9). A virus can modify this table so that the interrupt causes viral code (resident in memory) to be executed.

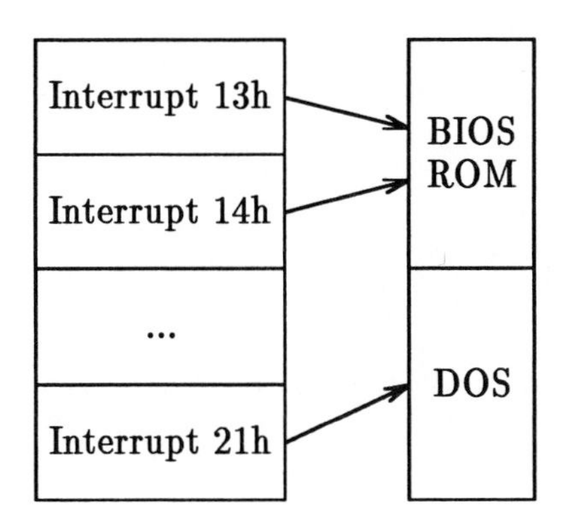

FIGURE 9 Normal interrupt usage.

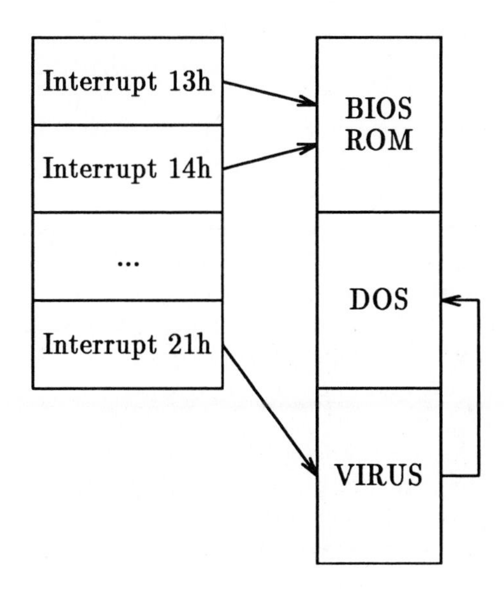

FIGURE 10 Interrupt vectors with TSR virus.

By trapping the keyboard interrupt, a virus can arrange to intercept the CTRL-ALT-DEL soft reboot command, modify user keystrokes, or be invoked on each keystroke. By trapping the BIOS disk interrupt, a virus can intercept all BIOS disk activity, including reads of boot sectors, or disguise disk accesses to infect as part of a user's disk request. By trapping the DOS service interrupt, a virus can intercept all DOS service requests including program execution, DOS disk access, and memory allocation requests.

A typical virus might trap the DOS service interrupt, causing its code to be executed before calling the real DOS handler to process the request. (See Figure 10.)

REPLICATION STRATEGIES

Types. Viruses can also be grouped into four categories, based on the type of files they infect:

1. Boot sector viruses that only infect boot sectors (or rarely, partition records)
2. System viruses that are targeted against particular system files, such as the DOS command shell
3. Direct viruses that scan through the DOS directory structure on disk looking for suitable files to infect
4. Indirect viruses that wait until the user carries out an activity on a file (e.g., execution of a program) before infecting it

Transient viruses are always direct in that they attempt to infect one or more files (usually in the same directory or home directory) before terminating. Resident viruses can be either direct or indirect (or worse, both). The recently reported *Traceback* virus infects any file executed (indirect), while also incrementally scanning the directory structure (direct).

In general, indirect viruses are slower to spread, but often pass unnoticed as their infection activities are disguised among other disk access requests.

Signatures to Prevent Reinfection. One problem encountered by viruses is that of repeated infection of the host, leading to depleted memory and early detection. In the case of boot sector viruses, this could (depending on strategy) cause a long chain of linked sectors. In the case of a program-infecting virus (or link virus), repeated infection may result in continual extension of the host program each time it is reinfected. There are indeed some viruses that exhibit this behavior (e.g., the Israeli virus extends .EXE files 1808 bytes each time they are infected).

To prevent this unnecessary growth of infected files, many viruses implant a unique *signature* that signals that the file or sector is infected. The virus will check for this signature before attempting infection, and will place it when infection has taken place; if the signature is present, the virus will not reinfect the host.

A virus signature can be a characteristic sequence of bytes at a known offset on disk or in memory, a specific feature of the directory entry (e.g., alteration time or file length), or a special system call available only when the virus is active in memory.

The signature is a mixed blessing. The virus would be easier to spot if reinfections caused disk space to be exhausted or showed obvious disk activity, but the signature does provide a method of detection and protection. Virus sweep programs are available that scan files on disk for the signatures of known viruses, as are "inoculation" routines that fake the viral signature in clean systems to prevent the virus from attempting infection.

VIRUSES AS ARTIFICIAL LIFE

Now that we know what computer viruses are, and how they spread, we can ask if they represent a form of artificial life. The first, and obvious, question is "What is life?" Without an answer to this question, we will be unable to say if a computer virus is "alive."

One list of properties associated with life was presented in Farmer and Belin.[5] That list included:

1. *Life is a pattern in space-time* rather than a specific material object.
2. *Self-reproduction*, in itself or in a related organism.
3. *Information storage of a self-representation.*
4. *A metabolism* that converts matter/energy.
5. *Functional interactions with the environment.*
6. *Interdependence of parts.*
7. *Stability under perturbations* of the environment.
8. *The ability to evolve.*
9. *Growth or expansion.*

Let us examine each of these characteristics in relation to computer viruses.

VIRUSES AS PATTERNS IN SPACE-TIME

There is an obvious match to this characteristic. Viruses are represented by patterns of computer instructions that exist over time on many computer systems. Viruses are not associated with the physical hardware, but with the instructions executed (sometimes) by that hardware.

SELF-REPRODUCTION OF VIRUSES

One of the primary characteristics of computer viruses is their ability to reproduce themselves (or an altered version of themselves). Thus, this characteristic is met.

INFORMATION STORAGE OF A SELF-REPRESENTATION

This, too, is an obvious match for computer viruses. The code that defines the virus is a template that is used by the virus to replicate itself. This is similar to the DNA molecules of what we recognize as organic life.

VIRUS METABOLISM

This property involves the organism taking in energy or matter from the environment and using it for its own activity. Computer viruses use the energy of computation expended by the system to execute. They do not convert matter, but make use of the electrical energy present in the computer to traverse their patterns of instructions and infect other programs. In this sense, they have a metabolism.

FUNCTIONAL INTERACTIONS WITH THE VIRUS'S ENVIRONMENT

Viruses perform examinations of their host environments as part of their activities. They alter interrupts, examine memory and disk architectures, and alter addresses to hide themselves and spread to other hosts. They very obviously alter their environment to support their existence. Many viruses accidentally alter their environment because of bugs or unforeseen interactions. The major portion of damage from all computer viruses is a result of these interactions.

INTERDEPENDENCE OF VIRUS PARTS

Living organisms cannot be arbitrarily divided without destroying them. The same is true of computer viruses. Should a computer virus have a portion of its "anatomy" excised, the virus would probably cease to function normally, if at all. Few viruses are written with superfluous code, and even so, the working code cannot be divided without destroying the virus.

VIRUS STABILITY UNDER PERTURBATIONS

Computer viruses run on a variety of machines under different operating systems. Many of them are able to compromise (and defeat) anti-virus and copy protection mechanisms. They may adjust on-the-fly to insufficient storage, disk errors, and other exceptional events. Some are capable of running on most variants of popular personal computers under almost any software configuration—a stability and robustness seen in few commercial applications.

VIRUS EVOLUTION

It is here that viruses display a difference from systems we traditionally view as "alive." No computer viruses evolve as we commonly use the term, although it is conceivable that a very complex virus could be programmed to evolve and change. However, such a virus would be so large and complex as to be many orders of magnitude larger than most host programs, and probably bigger than the host operating systems. Thus, there is some doubt that such a virus could run on enough hosts to allow it to evolve.

Mutations of viruses do exist, however. There are variants of many known viruses, with as many as 15 known for some IBM PC viruses. The variations involved can be very small, on the order of two or three instructions difference, to major changes involving differences in messages, activation, and replication. The source of these variations appears to be programmers (the original virus authors or otherwise) who alter the viruses to avoid anti-viral mechanisms, or to cause different kinds of damage.

There is also one case where two different strains of a Macintosh virus are known to interact to form infections unlike the "parents," although these interactions usually produce "sterile" offspring that are unable to reproduce further.[8]

GROWTH

Viruses certainly do exhibit growth. Some transient viruses will infect every file on a system after only a few activations. The spread of viruses through commercial software and public bulletin boards is another indication of their wide-spread replication. One reasonable set of estimates had the number of computer virus infections in 1989 at a level 50% above the 1988 rate.[2] The number of new virus "species" reported in the first four months of 1990 underwent a 15-fold increase over the same period in 1989. Clearly, computer viruses are exhibiting major growth.

OTHER BEHAVIOR

As already noted, computers viruses exhibit "species" with well-defined ecological niches based on host machine type, and variations within these species. These species are adapted to specific environments and will not survive if moved to a different environment.

Some viruses also exhibit predatory behavior. For instance, the DenZuk virus will seek out and overwrite instances of the Brain virus if both are present on the same system. Other viruses exhibit territorial behavior—marking their infected domain so that others of the same type will not enter and compete with the original infection.

SUMMARY AND COMMENTS

Our examination of computer viruses leads us to the conclusion that they are very close to what we might define as "artificial life." Rather than representing a scientific achievement, this probably represents a flaw in our definition. To suggest that computer viruses are alive also implies to me that some part of their environment—the computers, programs, or operating systems—also represents artificial life. Can life exist in an otherwise barren and empty ecosystem? A definition of "life" should probably include something about the environment in which that life exists.

I would also be disappointed if computer viruses were considered as the first form of artificial life, because their origin is one of unethical practice. Viruses created for malicious purposes are obviously bad; viruses constructed as experiments and released into the public domain are likewise unethical, and poor science besides: experiments without controls, strong hypotheses, and the consent of the subjects. Facetiously, I suggest that if computer viruses evolve into something with artificial consciousness, this might provide a doctrine of "original sin" for their theology.

More seriously, I would suggest that there is something to be learned from the study of computer viruses: the importance of the realization that experimentation with systems in some way (almost) alive can be dangerous. Computer viruses have caused millions of dollars of damage and untold aggravation. Some of them have been written as harmless experiments, and others as malicious mischief. All have firmly rooted themselves in the pool of available computers and storage media, and they are likely to be frustrating users and harming systems for years to come. Similar but considerably more tragic results could occur from careless experimentation with organic forms of artificial life. We must never lose sight of the fact that "real life" is of much more importance than "artificial life," and we should not allow our experiments to threaten our experimenters.

REFERENCES

1. Cohen, Fred. "Computer Viruses." Ph.D. thesis, University of Southern California, 1985.
2. Cook, William, J., Assistant U.S. Attorney, Chicago, quoting Bell Labs' estimates.
3. Denning, Peter J. *Computers Under Attack: Intruders, Worms and Viruses.* Reading, MA: ACM Press (Addison-Wesley), 1990.
4. Dellinger, Joe. Private communication.
5. Farmer, J. Doyne, and Alletta d'A. Belin. "Artificial Life: The Coming Evolution." In *Proceedings in Celebration of Murray Gell-Man's 60th Birthday.* Cambridge: Cambridge University Press, 1990. Reprinted in this volume.
6. Gibson, William. *Neuromancer.* New York: Ace/The Berkeley Publishing Group, 1984.
7. Hoffman, Lance J. *Rogue Programs: Viruses, Worms, and Trojan Horses.* New York: Van Nostrand Reinhold, 1990.
8. Norstad, J. "Disinfectant On-Line Documentation," Edition 1.8. Manual, Northwestern University, June 1990.
9. Shoch, John F., and Jon A. Hupp. "The 'Worm' Programs—Early Experiments with a Distributed Computation." *CACM* **25(3)** (1982): 172–180.
10. Spafford, Eugene H. "An Analysis of the Internet Worm." In *Proceedings of the 2nd European Software Engineering Conference*, edited by C. Ghezzi and J. A. McDermid, 446–468. Berlin: Springer-Verlag, 1989.
11. Spafford, Eugene H. "The Internet Worm: Crisis and Aftermath." *CACM* **32(6)** (1989): 678–687.
12. Spafford, Eugene H., Kathleen A. Heaphy, and David J. Ferbrache. *Computer Viruses: Dealing with Electronic Vandalism and Programmed Threats.* Arlington, VA: ADAPSO, 1989.
13. Stang, David J. *Computer Viruses*, 2nd edition. Washington, DC: National Computer Security Association, 1990.

Philosophy/Emergence

Elliott Sober
Philosophy Department, University of Wisconsin, Madison, Wisconsin 53706

Learning from Functionalism—Prospects for Strong Artificial Life

I want to explore an analogy: artificial intelligence (AI) is to psychology as artificial life (AL) is to biology. Since I am interested in philosophical issues concerning both sides of this equation, I will jump back and forth between the philosophy of psychology and the philosophy of biology.

1. TWO USES FOR COMPUTERS

There are two quite different roles that computers might play in biological theorizing. Mathematical models of biological processes are often analytically intractable. When this is so, computers can be used to get a feel for the model's dynamics. You plug in a variety of initial condition values and allow the rules of transition to apply themselves (often iteratively); then you see what the outputs are.

Computers are used here as aids to the theorist. They are like pencil and paper or a sliderule. They help you think. The models being investigated are about life. But there is no need to view the computers that help you investigate these models as alive themselves. Computers can be applied to calculate what will happen when a bridge is stressed, but the computer is not itself a bridge.

Population geneticists have used computers in this way since the 1960's. Many participants in the Artificial Life research program are doing the same thing. I see nothing controversial about this use of computers. By their fruits shall ye know them. This part of the AL research program will stand or fall with the interest of the models investigated. When it is obvious beforehand what the model's dynamics will be, the results provided by computer simulation will be somewhat uninteresting. When the model is very unrealistic, computer investigation of its properties may also fail to be interesting. However, when a model is realistic enough and the results of computer simulation are surprising enough, no one can deny the pay-off.

I just mentioned that computers may help us understand bridges, even though a computer is not a bridge. However, the second part of the artificial life research program is interested in the idea that computers are instances of biological processes. Here the computer is said to be alive, or to exemplify various properties that we think of as characteristic of life.

This second aspect of the Artificial Life research program needs to be clearly separated from the first. It is relatively uncontroversial that computers can be tools for investigating life; in contrast, it is rather controversial to suggest that computers are or can be alive. Neither of these ideas entails the other; they are distinct.

Shifting now from AL to AI, again we can discern two possible uses of computers. First, there is the use of computers as tools for investigating psychological models that are too mathematically complicated to be analytically solved. This is the idea of computers as tools for understanding the mind. Second and separate, there is the idea that computers are minds. This latter idea, roughly, is what usually goes by the name of strong AI.

Strong AI has attracted a great deal of attention, both in the form of advocacy and in the form of attack. The idea that computers, like paper and pencil, are tools for understanding psychological models has not been criticized much, nor should it have been. Here, as in the case of AL, by their fruits shall ye know them.

So in both AI and AL, the idea that computers are tools for investigating a theory is quite different from the idea that computers are part of the subject-matter of the theory. I'll have little more to say about the tool idea, although I'll occasionally harp on the importance of not confusing it with the subject-matter idea. The following table depicts the parallelism between AI and AL; it also should help keep the tool idea and the subject matter idea properly separated:
Of course, "strong" does not mean plausible or well defended; "strong" means daring and "weak" means modest.

Where did the idea that computers could be part of the subject matter of psychological and biological theories come from? Recent philosophy has discussed the psychological issues a great deal, the biological problem almost not at all. Is it possible to build a case for living computers that parallels the arguments for thinking computers? How far the analogy can be pushed is what I want to determine.

TABLE 1 Parallelism between AI and AL

	Psychology	Biology
Computers are tools for investigating	weak AI	weak AL
Computers are part of the subject matter of	strong AI	strong AL

2. THE PROBLEM OF MIND AND THE PROBLEM OF LIFE

The mind/body problem has witnessed a succession of vanguard theories. In the early 1950's, Ryle[12] and Wittgenstein[18] advocated forms of logical behaviorism. This is the idea that the meaning of mentalistic terms can be specified purely in terms of behavior. Ryle attacked the Myth of the Ghost in the Machine, which included the idea that mental states are inner causes of outward behavior.

In the mid-1950's to mid-1960's, behaviorism itself came under attack from mind/brain identity theorists. Australian materialists like Place[10] and Smart[14] maintained that mental states are inner causes. But more than that, they argued that mental properties would turn out to be identical with physical properties. Whereas logical behaviorists usually argued for their view in an *a priori* fashion, identity theorists said they were formulating an empirical thesis that would be borne out by the future development of science.

The identity theory can be divided into two claims. They claimed that each mental object is a physical thing. They also claimed that each mental property is a physical property. In the first category fall such items as minds, beliefs, memory traces, and afterimages. The second category includes believing that snow is white and feeling pain. This division may seem a bit artificial—why bother to make separate claims about my belief that snow is white and the property of believing that snow is white? In a moment, the point of this division will become plain.

Beginning in the mid-1960s the identity theory was challenged by a view that philosophers called functionalism. Hilary Putnam,[11] Jerry Fodor,[6] and Daniel Dennett[3] argued that psychological properties are multiply realizable. If this is correct, then the identity theory must be rejected.

To understand what multiple realizability means, it is useful to consider an analogy. Consider mousetraps. Each of them is a physical object. Some are made of wire, wood, and cheese. Others are made of plastic and poison. Still others are constituted by bunches of philosophers scurrying around the room armed with inverted wastepaper baskets.

What do all these mousetraps have in common? Well, they are all made of matter. But more specifically, what properties do they share that are unique to them? If mousetraps are multiply realizable, then there is no physical property that all mousetraps, and only mousetraps, possess.

Each mousetrap is a physical thing, but the property of being a mousetrap is not a physical property. Here I am putting to work the distinction between object and property that I mentioned before.

Just as there are many physical ways to build a mousetrap, so, functionalists claimed, there are many physical ways to build a mind. Ours happens to be made of DNA and neurons. But perhaps computers could have minds. And perhaps there could be organisms in other species or in other galaxies that have minds, but whose physical organization is quite different from the one we exemplify. Each mind is a physical thing, but the property of having a mind is not a physical property.

Dualism is a theory that I have not mentioned. It claims that minds are made of an immaterial substance. Identity theorists reject dualism. So do functionalists. The relationship between these three theories can be represented by saying how each theory answers a pair of questions (see Table 2).

I mentioned earlier that identity theorists thought of themselves as advancing an empirical thesis about the nature of the mind. What about functionalism? Is it an empirical claim? Here one finds a division between two styles of argument. Sometimes functionalists appear to think that the meaning of mentalistic terminology guarantees that the identity theory must be false. This *a priori* tendency within functionalism notwithstanding, I prefer the version of that theory that advances an empirical thesis. It is an empirical question how many different physical ways a thinking thing can be built. If the number is enormously large, then functionalism's critique of the identity theory will turn out to be right. If the number is very small (or even one), then the identity theory will be correct. Perhaps the design constraints dictated by psychology are satisfiable only within a very narrow range of physical systems; perhaps the constraints are so demanding that this range is

TABLE 2

	Are mental ___ physical?		
	Dualism	Functionalism	Identity Theory
Objects	NO	YES	YES
Properties	NO	NO	YES

reduced to a single type of physical system. This idea cannot be dismissed out of hand.

If philosophers during the same period of time had been as interested in the problem of life as they were in the problem of mind, they might have formulated biological analogs of the identity theory and functionalism. However, a biological analog of mind/body dualism does not have to be invented—it existed in the form of vitalism. Dualists claim that beings with minds possess an immaterial ingredient. Vitalists claim that living things differ from inanimate objects because the former contain an immaterial substance—an *élan vital*.

Had the problem of life recapitulated the problem of mind, the triumphs of molecular biology might then have been interpreted as evidence for an identity theory, according to which each biological property is identical with some physical property. Finally, the progression might have been completed with an analog of functionalism. Although each living thing is a material object, biological properties cannot be identified with physical ones.

Actually, this functionalist idea has been espoused by biologists, although not in the context of trying to recapitulate the structure of the mind/body problem. Thus, Fisher[5] says that "fitness, although measured by a uniform method, is qualitatively different for every different organism." Recent philosophers of biology have made the same point by arguing that an organism's fitness is the upshot of its physical properties even though fitness is not itself a physical property. What do a fit cockroach and a fit zebra have in common? Not any physical property, any more than a wood and wire mousetrap must have something physical in common with a human mouse catcher. Fitness is multiply realizable.

There are many other biologically interesting properties and processes that appear to have the same characteristic. Many of them involve abstracting away from physical details. For example, consider Lewontin's[7] characterization of what it takes for a set of objects to evolve by natural selection. A necessary and sufficient condition is heritable variation in fitness. The objects must vary in their capacity to stay alive and to have offspring. If an object and its offspring resemble each other, the system will evolve, with fitter characteristics increasing in frequency and less-fit traits declining.

This abstract skeleton leaves open what the objects are that participate in a selection process. Darwin thought of them as organisms within a single population. Group selectionists have thought of the objects as groups or species or communities. The objects may also be gametes or strands of DNA, as in the phenomena of meiotic drive and junk DNA.

Outside the biological hierarchy, it is quite possible that cultural objects should change in frequency because they display heritable variation in fitness. If some ideas are more contagious than others, they may spread through the population of thinkers. Evolutionary models of science exploit this idea. Another example is the economic theory of the firm; this describes businesses as prospering or going bankrupt according to their efficiency.

Other examples of biological properties that are multiply realizable are not far to seek. Predatory/prey theories, for example, abstract away from the physical details that distinguish lions and antelopes from spiders and flies.

I mentioned before that functionalism in the philosophy of mind is best seen as an empirical thesis about the degree to which the psychological characteristics of a system constrain the system's physical realization. The same holds for the analog of functionalism as a thesis about biological properties. It may be somewhat obvious that some biological properties—like the property of being a predator—place relatively few constraints on the physical characteristics a system must possess. But for others, it may be much less obvious.

Consider, for example, the fact that DNA and RNA are structures by which organisms transmit characteristics from parents to offspring. Let us call them hereditary mechanisms. It is a substantive question of biology and chemistry whether other molecules could play the role of a hereditary mechanism. Perhaps other physical mechanisms could easily do the trick; perhaps not. This cannot be judged *a priori*, but requires a substantive scientific argument.

3. WHAT ARE PSYCHOLOGICAL PROPERTIES, IF THEY ARE NOT PHYSICAL?

In discussing functionalism's criticism of the mind/brain identity theory, I mainly emphasized functionalism's negative thesis. This is a claim about what psychological states are not; they aren't physical. But this leaves functionalism's positive proposal unstated. If psychological properties aren't physical, what are they, then?

Functionalists have constructed a variety of answers to this question. One prominent idea is that psychological states are computational and representational. Of course, the interest and plausibility of this thesis depends on what "computational" and "representational" are said to mean. If a functionalist theory entails that desk calculators and photoelectric eyes have beliefs about the world, it presumably has given too permissive an interpretation of these concepts. On the other hand, if functionalism's critique of the identity theory is right, then we must not demand that a system be physically just like us for it to have a psychology. In other words, the problem has been to construct a positive theory that avoids, as Ned Block[1] once put it, both chauvinism and liberalism. Chauvinistic theories are too narrow, while liberal theories are too broad, in their proposals for how the domain of psychology is to be characterized.

4. BEHAVIORISM AND THE TURING TEST

Functionalists claim that psychological theories can be formulated by abstracting away from the physical details that distinguish one thinking system from another. The question is: How much abstracting should one indulge in?

One extreme proposal is that a system has a mind, no matter what is going on inside it, if its behavior is indistinguishable from some other system that obviously has a mind. This is basically the idea behind the Turing Test. Human beings have minds, so a machine does too, if its behavior is indistinguishable from human behavior. In elaborating this idea, Turing was careful that "irrelevant" cues not provide a tip off. Computers don't look like people, but Turing judged this fact to be irrelevant to the question of whether they think. To control for this distracting detail, Turing demands that the machine be placed behind a screen and its behavior standardized. The behavior is to take the form of printed messages on a tape.

Besides intentionally ignoring the fact that computers don't look like people, this procedure also assumes that thinking is quite separate from doing. If intelligence requires manipulating physical objects in the environment, then the notion of behavior deployed in the Turing Test will be too meager. Turing's idea is that intelligence is a property of pure cogitation, so to speak. Behavior limited to verbal communication is enough.

So the Turing Test represents one possible solution to the functionalist's problem. Not only does thought not require a physical structure like the one our brains possess. In addition, there is no independently specifiable internal constraint of any kind. The only requirement is an external one, specified by the imitation game.

Most functionalists regard this test as too crude. Unfortunately, it seems vulnerable to both type-1 and type-2 errors. That is, a thing that doesn't think can be mistakenly judged to have a mind and a thinking thing can be judged to lack a mind by this procedure (see Table 3).

In discussions of AI, there has been little attention to type-2 errors. Yet, it seems clear that human beings with minds can imitate the behavior of mindless computers and so fool the interrogator into thinking that they lack minds. Of perhaps more serious concern is the possibility that a machine might have a mind, but have

TABLE 3 Errors in the Turing Test

	The subject thinks	The subject does not think
The subject passes	ok	type-1
The subject does not pass	type-2	ok

beliefs and desires so different from those of any human being that interrogators would quickly realize that they were not talking to a human being. The machine would flunk the Turing Test, because it cannot imitate human response patterns; it is another matter to conclude that the machine, therefore, does not have a mind at all.

The possibility of type-2 error, though real, has not been the focus of attention. Rather, in order to overcome the presumption that machines can't think, researchers in AI have been concerned to construct devices that pass the Turing Test. The question this raises is whether the test is vulnerable to type-1 errors.

One example that displays the possibility of type-1 error is due to Ned Block.[2] Suppose we could write down a tree structure in which every possible conversation that is five hours or less in duration is mapped out. We might trim this tree by only recording what an "intelligent" respondent might say to an interrogator, leaving open whether the interrogator is "intelligent" or not. This structure would be enormously large, larger than any current computer would be able to store. But let's ignore that limitation and suppose we put this tree structure into a computer.

By following this tree structure, the machine would interact with its interrogator in a way indistinguishable from the way in which an intelligent human being would do so. Yet the fact that the machine simply makes its way through this simple tree structure strongly suggests that the machine has no mental states at all.

One might object that the machine could be made to fail the Turing Test if the conversation pressed on beyond five hours. This is right, but now suppose that the tree is augmented in size, so that it encompasses all sensible conversations that are ten hours or less in duration. In principle, the time limit on the tree might be set at any finite size—four score and ten years if you like.

Block draws the moral that thinking is not fully captured by the Turing Test. What is wrong with this branching structure is not the behaviors it produces but how it produces them. Intelligence is not just the ability to answer questions in a way indistinguishable from that of an intelligent person. To call a behavior "intelligent" is to comment on how it is produced. Block concludes, rightly I think, that the Turing Test is overly behavioristic.

In saying this, I am not denying that the Turing Test is useful. Obviously, behavioral evidence can be telling. If one wants to know where the weaknesses are in a simulation, one might try to discover where the outputs mimic and where they do not. But it is one thing for the Turing Test to provide fallible evidence about intelligence, something quite different for the test to define what it is to have a mind.

A large measure of what is right about Searle's[13] much-discussed paper "Minds, Brains, and Programs" reduces to this very point. In Searle's Chinese room example, someone who speaks no Chinese is placed in a room, equipped with a manual, each of whose entries maps a story in Chinese (S) and a question in Chinese about that story (Q) onto an answer in Chinese to that story (A). That is, the man in the room has a set of rules, each with the form

$$S + Q \rightarrow A\,.$$

Chinese stories are sent into the room along with questions about those stories, also written in Chinese. The person in the room finds the $S+Q$ pairing on his list, writes down the answer onto which the pair is mapped, and sends that message out of the room. The input/output behavior of the room is precisely what one would expect of someone who understands Chinese stories and wishes to provide intelligent answers (in Chinese) to questions about them. Although the system will pass the Turing Test, Searle concludes that the system that executes this behavior understands nothing of Chinese.

Searle does not specify exactly how the person in the room takes an input and produces an output. The details of the answer manual are left rather vague. But Block's example suggests that this makes all the difference. If the program is just a brute-force pairing of stories and questions on the one hand and answers on the other, there is little inclination to think that executing the program has anything to do with understanding Chinese. But if the manual more closely approximates what Chinese speakers do when they answer questions about stories, our verdict might change. Understanding isn't definable in terms of the ability to answer questions; in addition, how one obtains the answers must be taken into account.

Searle considers one elaboration of this suggestion in the section of his paper called "The Brain Simulator Reply (Berkeley and MIT)."[13] Suppose we simulate "the actual sequence of neuron firings at the synapses of the brain of a native Chinese speaker when he understands stories in Chinese and gives answers to them." A system constructed in this way would not just duplicate the stimulus/response pairings exemplified by someone who understands Chinese; in addition, such a system would closely replicate the internal processes mediating the Chinese speaker's input/output connections.

I think that Searle's reply to this objection is question-begging. He says "the problem with the brain simulator is that it is simulating the wrong things about the brain. As long as it simulates only the formal structure of the sequence of neuron firings at the synapses, it won't have simulated what matters about the brain, namely its causal properties, its ability to produce intentional states."

Although Searle does not have much of an argument here, it is important to recognize that the denial of his thesis is far from trivial. It is a conjecture that might or might not be right that the on/off states of a neuron, plus its network of connections with other neurons, exhaust what is relevant about neurons that allows them to form intentional systems. This is a meager list of neuronal properties, and so the conjecture that it suffices for some psychological characteristic is a very strong one. Neurons have plenty of other characteristics; the claim that they are all irrelevant as far as psychology goes may or may not be true.

There is another way in which Searle's argument recognizes something true and important, but, I think, misinterprets it. Intentionality crucially involves the relationship of "aboutness." Beliefs and desires are about things in the world outside the mind. How does the state of an organism end up being about one object, rather than about another? What explains why some states have intentionality whereas others do not? One plausible philosophical proposal is that intentionality involves a causal connection between the world and the organism. Crudely put, the reason my term "cat" refers to cats is that real cats are related to my use of that term in some specific causal way. Working out what the causal path must be has been a difficult task for the causal theory of reference. But leaving that issue aside, the claim that some sort of causal relation is necessary, though perhaps not sufficient, for at least some of the concepts we possess, has some plausibility.

If this world/mind relationship is crucial for intentionality, then it is clear why the formal manipulation of symbols can't, by itself, suffice for intentionality. Such formal manipulation is purely internal to the system, but part of what makes a system have intentionality involves how the system and its states are related to the world external to the system.

Although conceding this point may conflict with some pronouncements by exponents of strong AI, it does not show that a thinking thing must be made of neuronal material. Consistent with the idea that intentionality requires a specific causal connection with the external world is the possibility that a silicon-chip computer could be placed in an environment and acquire intentional states by way of its interactions with the environment.

It is unclear what the acquisition process must be like, if the system is to end up with intentionality. But this lack of clarity is not specific to the question of whether computers could think; it also applies to the nature of human intentionality. Suppose that human beings acquire concepts like *cat* and *house*; that is, suppose these concepts are not innate. Suppose further that people normally acquire these concepts by causally interacting with real cats and houses. The question I wish to raise is whether human beings could acquire these concepts by having a neural implant performed at birth that suitably rewired their brains. A person with an implant would grow up and feel about the world the way any of us do who acquired the concepts by more normal means. If artificial interventions in human brains can endow various states with intentionality, it is hard to see why artificial interventions into silicon computers can't also do the trick.

I have tried to extract two lessons from Searle's argument. First, there is the idea that the Turing Test is overly behavioristic. The ability to mimic intelligent behavior is not sufficient for having intelligence. Second, there is the idea that intentionality—aboutness—may involve a world/mind relationship of some specifiable sort. If this is right, then the fact that a machine (or brain) executes some particular program is not sufficient for it to have intentionality. Neither of these conclusions shows that silicon computers couldn't have minds.

What significance do these points have for the artificial life research program? Can the biological properties of an organism be defined purely in terms of environment/behavior pairings? If a biological property has this characteristic, the Turing Test will work better for it than the test works for the psychological property that Turing intended to describe. Let's consider photosynthesis as an example of a biological process. Arguably, a system engages in photosynthesis if it can harness the sun's energy to convert water and CO_2 into simple organic compounds (principally CH_2O). Plants (usually) do this in their chloroplasts, but this is just one way to do the trick. If this is right, either for photosynthesis or for other biological properties, then the first lesson I drew about the Turing Test in the context of philosophy of mind may actually provide a disanalogy between AI and AL. Behaviorism is a mistake in psychology, but it may be the right view to take about many biological properties.

The second lesson I extracted from Searle's argument concerned intentionality as a relationship between a mental state and something outside itself. It is the relation of aboutness. I argued that the execution of a program cannot be the whole answer to the question of what intentionality is.

Many biological properties and processes involve relationships between an organism (or part of an organism) and something outside itself. An organism reproduces when it makes a baby. A plant photosynthesizes when it is related to a light source in an appropriate way. A predator eats other organisms. Although a computer might replicate aspects of such processes that occur inside the system of interest, computers will not actually reproduce or photosynthesize or eat unless they are related to things outside themselves in the right ways. These processes involve actions—interactions with the environment; the computations that go on inside the skin are only part of the story. Here, then, is an analogy between AI and AL.

5. THE DANGER OF GOING TOO FAR

Functionalism says that psychological and biological properties can be abstracted from the physical details concerning how those properties are realized. The main problem for functionalism is to say how much abstraction is permissible. A persistent danger for functionalist theories is that they err on the side of being too liberal. This danger is especially pressing when the mathematical structure of a process is confused with its empirical content. This confusion can lead one to say that a system has a mind or is alive (or has some more specific constellation of psychological or biological properties) when it does not.

A simple example of how this fallacy proceeds may be instructive. Consider the Hardy-Weinberg Law in population genetics. It says what frequencies the diploid genotypes at a locus will exhibit, when there is random mating, equal numbers of males and females, and no selection or mutation. It is, so to speak, a "zero-force law"—it describes what happens in a population if no evolutionary forces are at work (see Sober[16]). If p is the frequency of the A allele and q is the frequency of a (where $p + q = 1$), then in the circumstances just described, the frequencies of AA, Aa, and aa are p^2, $2pq$, and q^2, respectively.

Consider another physical realization of the simple mathematical idea involved in the Hardy-Weinberg Law. A shoe manufacturer produces brown shoes and black shoes. By accident, the assembly line has not kept the shoes together in pairs, but has dumped all the left shoes into one pile and all the right shoes into another. The shoe manufacturer wants to know what the result will be if a machine randomly samples from the two piles and assembles pairs of shoes. If p is the frequency of black shoes and q is the frequency of brown ones in each pile, then the expected frequency of the three possible pairs will be $p^2, 2pq$, and q^2. Many other examples of this mathematical sort could be described.

Suppose we applied the Hardy-Weinberg Law to a population of *Drosophila*. These fruit flies are biological objects; they are alive and the Hardy-Weinberg Law describes an important fact about how they reproduce. As just noted, the same mathematical structure can be applied to shoes. But shoes are not alive; the process by which the machine forms pairs by random sampling is not a biological one.

I wish to introduce a piece of terminology: the Shoe/Fly Fallacy is the mistaken piece of reasoning embodied in the following argument:

Flies are alive.
Flies are described by law L.
Shoes are described by law L.

Hence, shoes are alive.

A variant of this argument focuses on a specific biological property—like reproduction—rather than on the generic property of "being alive."

Functionalist theories abstract away from physical details. They go too far—confusing mathematical form with biological (or psychological) subject matter—when they commit the Shoe/Fly Fallacy. The result is an overly liberal conception of life (or mind).

The idea of the Shoe/Fly Fallacy is a useful corrective against overhasty claims that a particular artificial system is alive or exhibits some range of biological characteristics. If one is tempted to make such claims, one should try to describe a system that has the relevant formal characteristics but is clearly not alive. The Popperian attitude of attempting to falsify is a useful one.

Consider, for example, the recent phenomenon of computer viruses. Are these alive? These cybernetic entities can make their way into a host computer and take over some of the computer's memory. They also can "reproduce" themselves and undergo "mutation." Are these mere metaphors, or should we conclude that computer viruses are alive?

To use the idea of the Shoe/Fly Fallacy to help answer this question, let's consider another, similar system that is not alive. Consider a successful chain letter. Because of its characteristics, it is attractive to its "hosts" (i.e., to the individuals who receive them). These hosts then make copies of the letters and send them to others. Copying errors occur, so the letters mutate.

I don't see any reason to say that the letters are alive. Rather, they are related to host individuals in such a way that more and more copies of the letters are produced. If computer viruses do no more than chain letters do, then computer viruses are not alive either.

Note the "if" in the last sentence. I do not claim that computer viruses, or something like them, cannot be alive. Rather, I say that the idea of the Shoe/Fly

Fallacy provides a convenient format for approaching such questions in a suitably skeptical and Popperian manner.

6. FROM HUMAN COGNITION TO AI, FROM BIOLOGY TO AL

One of the very attractive features of AI is that it capitalizes on an independently plausible thesis about human psychology. A fruitful research program about human cognition is based on the idea that cognition involves computational manipulations of representations. Perhaps some representations obtain their intentionality via a connection between mind and world. Once in place, these representations give rise to other representations by way of processes that exploit the formal (internal) properties of the representations. To the degree that computers can be built that form and manipulate representations, to that degree will they possess cognitive states.

Of course, very simple mechanical devices form and manipulate representations. Gas gauges in cars and thermostats are examples, but it seems entirely wrong to say that they think. Perhaps the computational view can explain why this is true by further specifying which kinds of representations and which kinds of computational manipulations are needed for cognition.

Can the analogous case be made for the artificial-life research program? Can an independent case be made for the idea that biological processes in naturally occurring organisms involve the formation and manipulation of representations? If this point can be defended for naturally occurring organisms, then to the degree that computers can form and manipulate representations in the right ways, to that degree will the AL research program appear plausible.

Of course, human and animal psychology is part of human and animal biology. So it is trivially true that some biological properties involve the formation and manipulation of representations. But this allows AL to be no more than AI. The question, therefore, should be whether life processes other than psychological ones involve computations on representations.

For some biological processes, the idea that they essentially involve the formation and manipulation of representations appears plausible. Thanks to our understanding of DNA, we can see ontogenesis and reproduction as processes that involve representations. It is natural to view an organism's genome as a set of instructions for constructing the organism's phenotype. This idea becomes most plausible when the phenotypic traits of interest are relatively invariant over changes in the environment. The idea that genome represents phenotype must not run afoul of the fact that phenotypes are the result of a gene/environment interaction. By the same token, blueprints of buildings don't determine the character of a building in every detail. The building materials available in the environment and the skills of workers also play a role. But this does not stop us from thinking of the blueprint as a set of instructions.

In conceding that computers could exemplify biological characteristics like development and reproduction, I am not saying that any computers now do. In particular, computers that merely manipulate theories about growth and reproduction are not themselves participants in those processes. Again, the point is that a description of a bridge is not a bridge.

What of other biological properties and processes? Are they essentially computational? What of digestion? Does it involve the formation and manipulation of representations? Arguably not. Digestion operates on food particles; its function is to extract energy from the environment that the system can use. The digestive process, *per se*, is not computational.

This is not to deny that in some systems digestion may be influenced by computational processes. In human beings, if you are in a bad mood, this can cause indigestion. If the mental state is understood computationally, then digestion in this instance is influenced by computational processes. But this effect comes from without. This example does not undermine the claim that digestion is not itself a computational process.

Although I admit that this claim about digestion may be wrong, it is important that one not refute it by trivializing the concepts of representation and computation. Digestion works by breaking down food particles into various constituents. The process can be described in terms of a set of procedures that the digestive system follows. Isn't it a short step, then, to describing digestion as the execution of a program? Since the program is a representation, won't one thereby have provided a computational theory of digestion?

To see what is wrong with this argument, we need to use a distinction that Kant once drew in a quite different connection. It is the distinction between following a rule and acting in accordance with a rule. When a system follows a rule, it consults a representation; the character of this representation guides the system's behavior. On the other hand, when a system acts in accordance with a rule, it does not consult a representation; rather, it merely behaves as if it had consulted the rule.

The planets move in ellipses around the sun. Are they following a rule or are they just acting in accordance with a rule? Surely only the latter. No representation guides their behavior. This is why it isn't possible to provide a computational theory of planetary motion without trivializing the ideas of computation and representation. I conjecture that the same may be true of many biological processes; perhaps digestion is a plausible example.

In saying that digestion is not a computational process, I am not saying that computers can't digest. Planetary motion is not a computational process, but this does not mean that computers can't move in elliptical orbits. Computers can do lots of things that have nothing to do with the fact that they are computers. They can be doorstops. Maybe some of them can digest food. But this has nothing to do with whether a computational model of digestion will be correct.

I have focused on various biological processes and asked whether computers can instantiate or participate in them. But what about the umbrella question? Can computers be alive? If Turing's question was whether computers can think, shouldn't the parallel question focus on what it is to be alive, rather than on more fine-grained concepts like reproduction, growth, selection, and digestion?

I have left this question for last because it is the fuzziest. The problem is that biology seems to have little to tell us about what it is to be alive. This is not to deny that lots of detailed knowledge is available concerning various living systems. But it is hard to see which biological theories really tell us about the nature of life. Don't be misled by the fact that biology has lots to say about the characteristics of terrestrial life. The point is that there is little in the way of a principaled answer to the question of which features of terrestrial life are required for being alive and which are accidental.

Actually, the situation is not so different in cognitive science. Psychologists and others have lots to tell us about this or that psychological process. They can provide information about the psychologies of particular systems. But psychologists do not seem to take up the question "What is the nature of mind?" where this question is understood in a suitable, nonchauvinistic way.

Perhaps you are thinking that these are the very questions a philosopher should be able to answer. If the sciences ignore questions of such generality, then it is up to philosophy to answer them. I am skeptical about this. Although philosophers may help clarify the implications of various scientific theories, I really doubt that a purely philosophical answer to these questions is possible. So if the sciences in question do not address them, we are pretty much out of luck.

On the other hand, I can't see that it matters much. If a machine can be built that exemplifies various biological processes and properties, why should it still be interesting to say whether it is alive? This question should not preoccupy AL any more than the parallel question should be a hang up for AI. If a machine can perceive, remember, desire, and believe, what remains of the question of whether it has a mind? If a machine can extract energy from its environment, grow, repair damage to its body, and reproduce, what remains of the issue of whether it is "really" alive?

Again, it is important to not lose sight of the if's in the previous two sentences. I am not saying that it is unimportant to ask what the nature of mind or the nature of life is. Rather, I am suggesting that these general questions be approached by focusing on more specific psychological and biological properties. I believe that this strategy makes the general questions more tractable; in addition, I cannot see that the general questions retain much interest after the more specific ones are answered.

7. CONCLUDING REMARKS

Functionalism, both in the study of mind and the study of life, is a liberating doctrine. It leads us to view human cognition and terrestrial organisms as examples of mind and life. To understand mind and life, we must abstract away from physical details. The problem is to do this without going too far.

One advantage that AL has over AI is that terrestrial life is in many ways far better understood than the human mind. The AL theorist can often exploit rather detailed knowledge of the way life processes are implemented in naturally occurring organisms; even though the goal is to generalize away from these examples, real knowledge of the base cases can provide a great theoretical advantage. Theorists in AI are usually not so lucky. Human cognition is not at all well understood, so the goal of providing a (more) general theory of intelligence cannot exploit a detailed knowledge of the base cases.

An immediate corollary of the functionalist thesis of multiple realizability is that biological and psychological problems are not to be solved by considering physical theories. Existing quantum mechanics is not the answer, nor do biological and psychological phenomena show that some present physical theory is inadequate. Functionalism decouples physics and the special sciences. This does not mean that functionalism is the correct view to take for each and every biological problem; perhaps some biological problems are physical problems in disguise. The point is that, if one is a functionalist about some biological process, one should not look to physics for much theoretical help.

The Turing Test embodies a behavioristic criterion of adequacy. It is not plausible for psychological characteristics, though it may be correct for a number of biological ones. However, for virtually all biological processes, the behaviors required must be rather different from outputs printed on a tape. A desktop computer that is running a question/answer program is not reproducing, developing, evolving, or digesting. It does none of these things, even when the program describes the processes of reproduction, development, digestion, or evolution.

It is sometimes suggested that, when a computer simulation is detailed enough, it then becomes plausible to say that the computer is an instance of the objects and processes that it simulates. A computer simulation of a bridge can be treated as a bridge, when there are simulated people on it and a simulated river flowing underneath. By now I hope it is obvious why I regard this suggestion as mistaken. The problem with computer simulations is not that they are simplified representations, but that they are representations. Even a complete description of a bridge—one faithful in every detail—would still be a very different object from a real bridge.

Perhaps any subject matter can be provided with a computer model. This merely means that a description of the dynamics can be encoded in some computer language. It does not follow from this that all processes are computational. Reproduction is a computational process because it involves the transformation of representations. Digestion does not seem to have this characteristic. The AL research program has plenty going for it; there is no need for overstatement.

REFERENCES

1. Block, N. "Troubles with Functionalism." In *Perception and Cognition: Issues in the Foundations of Psychology*, edited by W. Savage. Minnesota Studies in the Philosophy of Science, Vol. IX. University of Minnesota Press, 1975.
2. Block, N. "Psychologism and Behaviorism." *Phil. Rev.* **90** (1981): 59-43.
3. Dennett, D. *Brainstorms* Cambridge: MIT Press, 1978.
4. Dretske, F. "Machines and the Mental." *Proceedings and Addresses of the American Philosophical Association* **59** (1985): 23–33.
5. Fisher, R. *The Genetical Theory of Natural Selection.* New York: Dover Books, 1930 (1958).
6. Fodor, J. *Psychological Explanation.* New York: Random House, 1968.
7. Lewontin, R. "The Units of Selection." *Annual Review of Ecology and Systematics* **1** (1970): 1–14.
8. Pattee, H. "Simulations, Realizations, and Theories of Life." In *Artificial Life*, edited by C. Langton, 63–78. Santa Fe Institute Studies in the Sciences of Complexity, Proc. Vol. VI. Redwood City, CA: Addison-Wesley, 1989.
9. Penrose, R. *The Emperor's New Mind.* Oxford: Oxford University Press, 1990.
10. Place, U. T. "Is Consciousness a Brain Process?" *Brit. J. Psy.* **47** (1956): 44–50.
11. Putnam, H. "The Nature of Mental States." *Mind, Language, and Reality.* Cambridge: Cambridge University Press, 1967 (1975).
12. Ryle, G. *The Concept of Mind.* New York: Barnes and Noble, 1949.
13. Searle, J. "Minds, Brains, and Programs." *Behavior and Brain Sciences* **3** (1980): 417–457.
14. Smart, J. J. C. "Sensations and Brain Processes." *Phil. Rev.* **68** (1959): 141–156.
15. Sober, E. "Methodological Behaviorism, Evolution, and Game Theory." *Sociobiology and Epistemology*, edited by J. Fetzer, 181–200. Dordrecht: Reidel, 1985.
16. Sober, E. *The Nature of Selection.* Cambridge: MIT Press, 1984.
17. Turing, A. "Computing Machinery and Intelligence." *Mind* **59** (1950): 433–460.
18. Wittgenstein, L. *Philosophical Investigations.* Oxford: Basil Blackwell, 1953.

Steen Rasmussen
Complex Systems Group, Theoretical Division (T-13), and Center for Nonlinear Studies, MS-B258, Los Alamos National Laboratory, Los Alamos, New Mexico 87545 USA; e-mail: steen@t13.lanl.gov (arpa-net)

Aspects of Information, Life, Reality, and Physics

PROLOGUE

In this paper, I shall comment on some fundamental questions which inevitably arise in our attempts to understand *life*. It is surprising to me that these attempts confront us with both foundation problems in physics and with philosophical questions about the definition of reality. At some level, a part of the following conception seems similar to Wheeler's ideas[15] of the "Meaning Circuit" and to Sakharov's[13] and Wheeler's ideas[9] of pre-geometry. These similarities exist despite the fact that both the central problem and the current context are very different, and the following may, therefore, give a new angle for discussing some of the foundation problems in physics. The emerging ontological questions are, of course, reformulations of classical questions. They involve aspects of: the solopsism problem, the mind-body problem, and the problem of other minds. However, these ontological questions are rooted in a new setting which allows us not only to ask these questions in a new way, but also to draw some new conclusions. This new angle to the foundation problems in physics as well as to the philosophical problems are prompted by the development and the use of computers. It is the emergence of this new tool that has allowed us to view these problems in a new perspective. I believe that our attempts to understand life force us to develop a new conception of the relations between information, life, reality, and physics.

The intended format is as short as possible so that the ideas do not vanish in moderations and discussion. This means that much of the relevant discussion, a lot of important questions, and many references are excluded. I shall pose postulates, comment on these postulates, raise questions, and draw conclusions. I lay out arguments to see what follows from the postulates and, thereby, provide reasons for a closer investigation of the truth of these postulates. Note that I am *not* claiming that all the following postulates are true; I am merely exploring the logical consequences of assuming them true.

1. INFORMATION AND LIFE

POSTULATE 1. *A universal computer at the Turing machine level can simulate any physical process (Physical Church-Turing thesis).*

Since we can construct universal computers, we know that our physics supports universal computation. Since we, up until now, have been able to simulate the information transformation rules of any physical mechanism for which these rules are known (to know and formulate the rules is the difficult part of this), it is tempting to claim that physical processes themselves can be viewed as computations. This idea is clearly discussed by Fredkin[3] and may in fact be a good working hypothesis. However, we do not know whether Nature is able to support a more powerful and general form of information processing than that supported by a universal computer of the Turing class.

POSTULATE 2. *Life is a physical process.*

This idea was originally due to John von Neumann,[11] and has recently been clearly reformulated by Chris Langton.[4] Life is associated with the functional organization of the different parts of an organism and does *not* depend on the properties of the hardware in which it is implemented. Details of such a functional organization will, of course, be determined by the actual hardware, but these details are not central for life *per se*. Note that not any kind of hardware can support or implement living processes.

COROLLARY 1. Hence from Postulates 1 and 2 it should be possible to simulate living processes on a universal computer. Still, a simulation of a living process is not the same as the real thing.

POSTULATE 3. *There exist criteria by which we are able to distinguish living from non-living objects.*

It is very difficult to set up objective criteria by which we can *define* life, and I do not think that we can do it in a satisfactory manner at present. We have not yet been able to synthesize life, and, therefore, we do not really know what life is. No matter how precisely we try to define life, there always seems to be a "grey" zone between the non-living and the living, in which we will have difficulties in deciding whether a given object is alive or not. There is a vast literature about what properties life should have. See, for instance, Schrödinger,[14] Monod,[10] Mayr,[7] or Farmer and Belin.[2] Although there is no formal agreement on this question, we all have an intuitive notion of what is alive and what is not. A vague definition of a living system should include a notion of *a metabolism, adaptive organism-environment responses, reproduction,* and *evolvability.* For sake of argument we claim that postulate 3 is possible in principle.

COROLLARY 2. From postulate 3 it should, for instance, be possible to determine if some potential, future computer process is alive or not.

2. LIFE AND REALITY

POSTULATE 4. *An artificial organism must perceive a reality R_2, which, for it, is just as real as our "real" reality, R_1, is for us (R_1 and R_2 may be the same).*

One of the criteria for a process to be alive involves adaptive organism-environment responses. This implies that even the simplest living object, for example, a hypothetical process implemented on a computer, must have a primitive notion of itself and its surrounding environment. Such responses imply the existence of an internal model of the world. The living object perceives a reality. We assume that a *sufficient* condition for the existence of a reality is the existence of life. There is a vast philosophical literature on what reality is which I will not go into. Here, I shall refer to John Wheeler's[15,16] "Meaning Circuit." The basic idea behind this concept is that the world is a self-synthesized system of existence. On one hand, physics provides the means for communication (light, sound, etc). Reality can, thereby, acquire its meaning through a conscious conception of the world, via an organization of the information we get from our senses. On the other hand, physics also gives rise to chemistry and biology, and through them, an observer participation, namely the emergence of life and later the evolution of man.

POSTULATE 5. *R_1 and R_2 have the same ontological status.*

Assuming R_2 exists in a computer, its properties may be very different from the properties of R_1. From a logical point of view it is possible to create interactions in

R_2 which do not in any direct way obey the physics in R_1. We can, thereby, create a more general physics in our universal machines than the physics we know. Such an independence exists although R_2 in a material way is embedded in R_1. Due to the physical embedding, we can effect R_2 from R_1 by "pulling the plug" or through a re-programming of the code that supports R_2. However, R_2 should, in principle, also be able to effect R_1, since self-programming abilities[12] are likely to exist in R_2, and, thereby, is a re-programming of interface systems also possible from "the inside."

In postulate 4 we argued that a reality obtains its meaning through the existence of an observer. Since R_2 is being perceived, just as R_1 is being perceived, R_2 becomes a reality with equal ontological status as R_1, whenever R_2 has a living observer. Why should one have priority over the other?

Note that all R_2 processses have a *physical* instantiation in R_1. The physical instantiation of a modern computer process is electron relocations in semiconductors in R_1.

COROLLARY 3. We can now use postulate 5 to rephrase corollary 1 and say that the ontological status of a living process is *independent* of the hardware that carries it. This perspective removes the problem with the term "simulate," indicating that processes are occuring in "some" hardware. Since the two processes have the same ontological status the one cannot be more real than the other. Real life in a digital computer should, thereby, be possible.

3. REALITY AND PHYSICS

POSTULATE 6. *It is possible to learn something about the fundamental properties of realities in general, and of R_1 in particular, by studing the details of different R_2's. An example of such a property is the physics of a reality.*

We may be able to make some interesting conclusions, if we, as a guiding example for our thoughts, use current low-level computational attempts, such as those presented in this volume, to set up environments to develop living processes (e.g., the systems programmed at the level of the "local physics" of the system). For the following arguments it is important to note that the structure of the relation between a living process, implemented on a computer, and its computational environment and the relation between a living process outside the computer, implemented in carbon chemistry, and its environment, is the same. This follows from postulates 4 and 5.

The common sense view of our physical world is that there is an absolute physical reality independent of us, where the geometry of this reality is a "three-dimensional box" in which all the events occur. This is also refered to as the Newtonian view. We know, however, from modern physics (geometrodynamics) that matter and geometry are coupled, and that it is impossible to define one without the other.[1] The computational approach we are undertaking indicates that one can go a step deeper: Matter, as well as geometry, can be derived from information processing. It is not possible to separate the topological (geometrical) and the functional (the matter related: particles, forces, reactions) properties from each other.[1] No single piece of code defines the topology nor the functional entities. Both the geometry and the forces in these universes are created through the *logical interactions* carried by the software and the hardware on the virtual and physical universal machine. We, thereby, end up with a conception very close to John Wheeler's and Andre Sakharov's idea of pre-geometry: *The calculus of propositions as the basis for everything.*[9,16] According to Wheeler, pregeometry precedes physics, just as logical rules for functional interaction precedes whatever properties our computational systems have. How far this analogy between our physics and our computational systems will hold is at present unknown.

EPILOGUE

Once again, let me stress that I am not claiming the truth of each of the above postulates, but exploring some of the consequences of assuming them to be true. I am making explicit the assumptions that lead to conclusions, which are held by many people in the field, in particular for the conclusions in section 1 (*Information and Life*). It is my hope that an uncovering of the underlying assumptions will make it clearer what kinds of questions to ask and where a discussion on these controversial issues should occur.

Along the current effort to formulate physics in terms of information theory (see Zurek[17]), Lloyd's[5] and Miller's[8] work on an information metric is a step towards the definition of a geometry of interactions. It would be an interesting and worthwhile project to continue along these lines and try to define geometry in terms of functional interactions in computational systems.

With respect to life and reality, I am not sure what I believe myself. Something instinctively tells me that whatever can emerge "in there" is of another nature and cannot really be alive. Anyway, these problems will probably clear up in the near future, when "things in there" become more obvious—some day we may even be able to ask them....

[1]Viewing geometry, and hence space, in the way I do here, has some similarities with Mach's[6] view on space, e.g. as the totality of instantaneously distances between all material points.

ACKNOWLEDGMENTS

Thanks to almost everybody I met during the fall of 1989 and the spring of 1990 both at Los Alamos and at the Santa Fe Institute. I am in particular grateful to William Wootters who initially gave direct pointers to relevant literature on pregeometry and took the time to discuss these issues with me. I am also grateful to Mark Bedau, Jeffery Davitz, Doyne Farmer, Rasmus Feldberg, Stuart Hameroff, Carsten Knudsen, Chris Langton, and Warner Miller who have critically reviewed earlier versions of the paper.

REFERENCES

1. Einstein, A. "Relativity and Problems on Space." In *Relativity*, Appendix V. New York: Bonanza Books, 1952 (first ed. 1916).
2. Farmer, D., and A. Belin. "Artificial Life, The Coming Evolution." This volume.
3. Fredkin, E. "Digital Mechanics." *Physica D* **45** (1990): 254–270.
4. Langton, C. "Artificial Life." In *Artificial Life*, edited by C. Langton. Santa Fe Institute Studies in the Sciences of Complexity, Vol. VI, 1-47. Redwood City, CA: Addison-Wesley, 1989.
5. Lloyd, S. Private comunication, 1990.
6. Mach, E. *The Science of Mechanics, A Chritical and Historical Account of its Development*, (English translation by T. McCormack), La Salle, ILL: Open Court Publications, 1960. (Original: Die Mechanic in Ihrer Entwicklung Historiche Kritich Dragestellt, Brockhaus, Lipzig, 1912).
7. Mayr, E. *The Growth of Biological Thought.* Cambridge: Harvard University Press, 1982.
8. Miller, W. Private communication and the talk: Towards Information Geometry, given at the Santa Fe Institute conference on physics, information, and complexity, held April, 1990, Santa Fe, New Mexico, USA.
9. Misner, C., K. Thorne, and J. Wheeler. "Beyond the End of Time." In *Gravitation*, Ch 44. Freeman, 1973.
10. Monod, J. *Chance and Necessity.* New York: Knopf, 1972.
11. von Neuman, J. *Theory of Self-Reproducing Automata*, edited and completed by A. W. Burks. Urbana: University of Illinois Press, 1966
12. Rasmussen, S., C. Knudsen, and R. Feldberg. "Dynamics of Programmable Matter." This volume.
13. Sakharov, A. "Vacuum Quantum Fluctuations in Curved Space and the Theory of Gravitation." *Sov. Phys. Doklady* **12(11)** (1968): 1040–1041.
14. Schrödinger, E. *What is Life.* Cambridge, 1943.

15. Wheeler, J. A. "World as System Self-Synthesized by Quantum Networking." *IBM Journal of Research and Development* **32(1)** (1988): 4–16.
16. Wheeler, J. A. "Information, Physics, Quantum: The Search for Links." In *Complexity, Entropy, and the Physics of Information*, edited by W. H. Zurek. Santa Fe Institute Studies in the Sciences of Complexity, Vol. VIII, 3–28. Redwood City, CA: Addison-Wesley, 1990.
17. Zurek, W. H., ed. *Complexity, Entropy, and the Physics of Information*, Santa Fe Institute Studies in the Sciences of Complexity, Vol. VIII. Redwood City, CA: Addison-Wesley, 1990.

Peter Cariani
Department of Systems Science, State University of Binghamton; Present affiliation: Eaton-Peabody Labs, Massachusetts Eye and Ear Infirmary, 243 Charles St, Boston, MA 02114; email: eplunix!peter@eddie.mit.edu

Emergence and Artificial Life

"Anyone who looks at living organisms knows perfectly well that they can produce other organisms like themselves.... Furthermore, it's equally evident that what goes on is actually one degree better than self-reproduction, for organisms appear to have gotten more elaborate in the course of time. Today's organisms are phylogenetically descended from others which were vastly simpler than they are, so much simpler, in fact, that it's inconceivable how any kind of description of the later, complex organism could have existed in the earlier one."
—John von Neumann[63]

THE PROBLEM OF EMERGENCE

The problem of emergence classically involved the origins of qualitatively new structures and functions which were not reducible to those already in existence,[5,20,32,33,46,64] In its most general form it encompasses all questions of the appearance of fundamental novelty in the world, the origins of order, and the increase in hierarchical complexity.[25,39,40,47,52,59]

The pragmatic relevance of emergence is intimately related to Descartes Dictum: how can a designer build a device which outperforms the designer's specifications?[2] If our devices follow our specifications too closely, they will fail to improve on those specifications. If, on the other hand, they are not in any way constrained by our purposes, they may cease to be of any use to us at all. Thus, the problem of emergence is the problem of specification vs. creativity, of closure and replicability vs. open-endedness and surprise. Its solution entails finding ways of building devices which circumvent Descartes Dictum. Emergent devices to be useful in amplifying our own creativity must have both a degree of structural autonomy relative to us as well as richness of potential structure. There has been a long-standing debate—from Leibnitz to Lady Lovelace to the present—over whether purely computational devices are capable of fundamentally-creative, truly emergent behavior. This paper will discuss various kinds of devices capable of emergent behaviors and take up the question of whether we can by purely computational means amplify our capacities as observers and actors in the physical world.

EMERGENCE: COMPUTATIONAL, THERMODYNAMIC, AND RELATIVE TO A MODEL

Currently there are three largely separate discourses on the problem of emergence, each with its own interpretation of the problem and distinct view of the world (Table 1). These three major conceptions of emergence might be called *computational emergence, thermodynamic emergence, and emergence relative to a model.*

Computational emergence is the view that emergent, complex global forms can arise from local computational interactions.[15,21,22] The result is a "bottom up" computational approach which is highly compatible with connectionist ideas now in vogue. Because its ontology only admits of micro-deterministic computational interactions (of which chaotic processes can generate apparently nondeterministic macro-interactions), there is a strong platonic component to the world view: the ideal forms of computational behaviors can be abstracted completely from their material substrates, and the material world can be left completely for a virtual one. In assuming rule-governed, bottom-up organization rather than semi-autonomous levels of organization, computational emergence tacitly incorporates the older reductionist assumption that micro-orders determine macro-orders but not vice versa.[20] There is a small literature which has critically addressed some of the specific problems with generating emergent behaviors in computer simulations in particular and micro-deterministic processes in general.[7,8,12,23,37,39,40,44,45,52,55,56,57,58] While computational emergence has been slow to provide adequate, nonarbitrary definitions of exactly what distinguishes emergent computations from nonemergent ones, some efforts are currently being made to sharpen the criteria.[14] *If we randomly come across a computer simulation and we have no clue as to its purpose, can we tell if its computations are emergent?*

TABLE 1 A survey of three contemporary discourses on emergence

theory	account of the origins of order	ontology	exemplars	research program
Computational Emergence (formally based)	order-from-order Macro-order from micro-determinism Macro-indeterminism through mathematical chaos	Discrete Universe Monism Microscopic Computational Rules Platonist	phase changes simulated turbulence Kauffmann nets chaos & fractals in nature	realize emergent behaviors through cellular automata and evolutionary computer simulations. (Langton, Toffoli)
Thermodynamic Emergence (physically based)	Order-from-noise discrete macro-structures (symbols=attractors) from continuous micro-processes New structures emerge through fluctuations	Continuous Universe Monism Continuous Physical Laws Materialist	origins of life condensation of DNA code catalytic networks formation of dissipative structures	Develop a thermodynamic theory to describe how stable, complex structures can arise far from equilibrium. Apply this to biological & social systems & upwards. (Prigogine, Iberall)
Emergence Relative to a Model (functionally based)	Form-from-formlessness Order-from-chaos Definition-from-ambiguity Processes of linking symbols to the world result in new functions and extend the realm of symbolic activity	Symbol-Matter Hylomorphism Complementarity between Different Points of View Pragmatist	evolution of new sensory & effector organs construction of new tools, languages, & models	Realize emergent functions through the construction of semantically adaptive devices. Augment the capabilities of the observer. (Pattee, Pask, Cariani)

Thermodynamic emergence finds its way into the Artificial Life discourse through realizations (osmotic growths), simulations (reaction-diffusion-systems, autocatalytic cycles), and dynamical theory (attractors and dissipative substrates of symbolic activity), although the necessary thermodynamic underpinnings of a theory of life have yet to be considered by the community at large. Active efforts are underway to connect the emergence of thermodynamic structures with theories of living organization,[12,48,60,65] perspectives which should be highly relevant to Artificial Life as a whole in the future. A major question is how to connect thermodynamic theories of structural stability with the appearance of new *functions*. Given a thermodynamic description of a material system, can we deduce which (kinds of) functions can or cannot be implemented?

The emergence-relative-to-a-model perspective arises out of the epistemological issues surrounding the successive enlargement of perceptual, cognitive, and behavioral organismic repertoires over biological evolution and their counterparts in the evolution of scientific models. These issues necessarily involve the evolutionary origins of symbols and coding relations in biological systems as well as the appearance of novel functions over time which are not reducible to previously existing functions. Hence while computational emergence is concerned with what kinds of global behaviors can be built up through the operation of local rules without altering the local rules, emergence relative to a model is concerned with the formation of global structures which then constrain and alter local interactions.[37,38,39,40] An example is the formation of new sensory organs, whose global interaction patterns vis-à-vis the world lead to metabolic changes contingent upon the sensed state of the world. While computational emergence sees macroscopic order as a consequence of microscopic order, theoretical biology sees the symbols and coding relations as arising through nonsymbolic, rate-dependent dynamics. Thus, it is always important to bear in mind that the discrete symbols in biological systems have nonsymbolic material substrates, and the origins of new symbolic primitives must be organized out of this nonsymbolic, analog realm.

While computational emergence is primarily a formal, mathematically based conception, and thermodynamic emergence is a physically based materialism, emergence relative to a model can be seen as a functionally based hylomorphism.[29] Here matter has as many properties as one can measure; a given object can support radically different types of behavior, depending upon how one has chosen to observe it. In addition, many of the functions and organizational properties of interest cannot be reduced to purely structural descriptions; we are confronted directly with the problem of defining and recognizing living organization and biological functionalities in the structure-oriented realm of our simulations.[17,16,37,39,40,51,52,53] How do we come to recognize the "amoeba in the pot" if we are only given the local concentrations of the chemical species? Clearly, the concept of an amoeba is not reducible to a set of points in a multidimensional chemical concentration space. While logical and structural descriptions can generally be ordered into consistent formal relations, functional measures usually confound complete translation from other measures (e.g., swimming reduced to floating and kicking and not drowning and...). To explain the origins of new functionalities, in our models as well as in the biological organisms themselves, one needs new observables to appear which are not simply logical combinations of pre-existing ones. This is the central epistemological problem of creating new observational primitives, and the key to constructing devices with the means to create their own functional categories.[36] If we simply specify the observational primitives outright, as we do for the primitive features of a connectionist device, then the device will not have the structural autonomy needed to construct new primitives, and its usefulness will be limited to those tasks it can perform within the categories we have given it. To build devices which find new observational primitives for us, they must be made epistemically autonomous relative to us, capable of searching realms for which we have no inkling. But to embark on this very ambitious enterprise, we must be first be very clear exactly how we would recognize the emergence of a new observational primitive in an organism or device we are observing. We need a precise definition of emergence.

EMERGENCE RELATIVE TO A MODEL

The emergence-relative-to-a-model view[8,54,55,56] sees emergence as the deviation of the behavior of a physical system from an observer's model of it. Emergence then involves a change in the relationship between the observer and the physical system under observation. If we are observing a device which changes its internal structure and consequently its behavior, we as observers will need to change our model to "track" the device's behavior in order to successfully continue to predict its actions. This perspective is close to the conception of "self-organizing systems" proposed by Ashby, Pask, Beer, and von Foerster thirty years ago.[4,36] While Howard Pattee and Robert Rosen have explicated similar concepts using the language of dynamical systems, this paper will present a complementary account in terms of systems-theoretic distinctions and cybernetic mechanisms.

THE MODELLING RELATION

The basic modelling methodology utilized here was first explicated by von Helmholtz and Hertz a century ago, and elaborated by the physicists Bohr and Heisenberg as they dealt with the foundational problems raised by quantum mechanics (Figure 1). The framework was further extended by Ashby[3] to form the basis of systems theory and by Rosen[51,54,55,56] to form the basis for a theoretical biology.[16,18,19]

In this perspective the modelling framework consists of two kinds of processes: measurements, which relate the symbols in the formal part of the model to the world at large, and computations , which relate the symbols in the formal part to other symbols via rule-governed operations. In semiotic terms, computations are logically

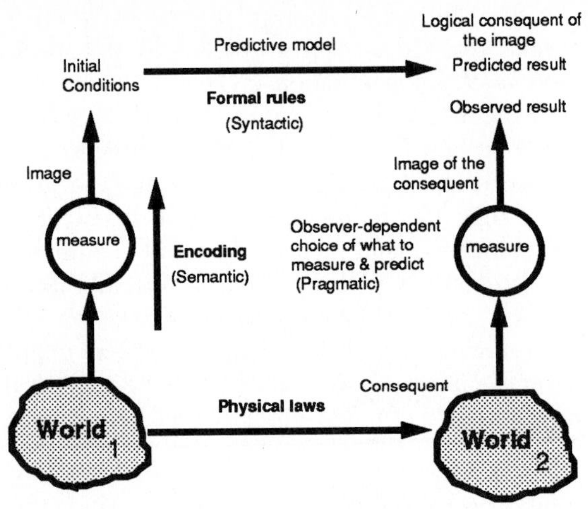

FIGURE 1 Semiotic relations in scientific models. The Hertzian commutation relation between symbols and the world.

necessary, convention-governed *syntactic* operations, while measurements are empirically contingent, materially governed *semantic* operations. The choice of appropriate measurements and computations which determine how well the model satisfies the purposes of the observer constitute intentional, *pragmatic* operations. These three kinds of operations, syntactic, semantic, and pragmatic (necessary, contingent, and intentional, respectively), form the irreducible orthogonal functional axes of the modelling relation. Thus a model consists of (1) a set of observables, implemented through contingent connections to the world (measuring devices, sensors), (2) a set of formal relations on the states of the observables implemented through physical devices with reliable state-transition behavior, (3) a set of state variables of interest which the observer seeks to predict, (4) evaluative criteria for that framework, and (5) means of changing either the measurements or the computations made (by building or recalibrating measuring devices or by changing computational rules, reprogramming). All of these aspects of the modeling relation can be made accessible to a public realm of a community of observers (Helmholtz's *locus observandi*) through calibration processes. The much-discussed tacit parts of scientific theories, the images and conceptual contexts through which the observers interpret these relations, affect the choices of observables and computations, but not their results once they have been selected.

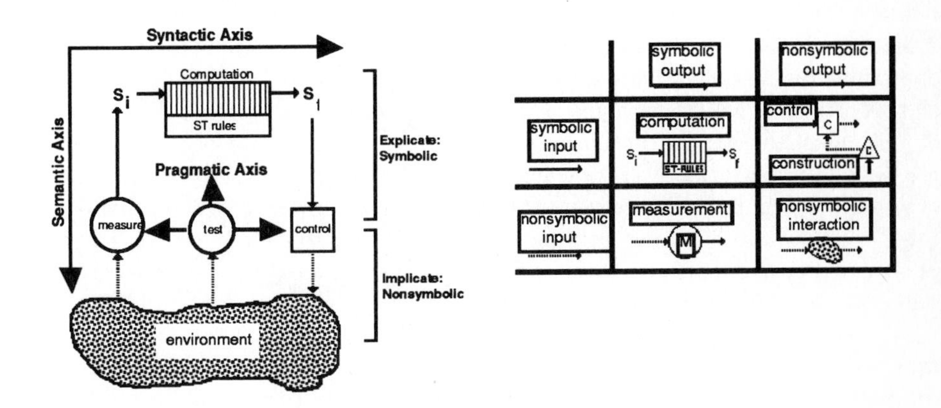

FIGURE 2 (a) Basic semiotic relations in organisms and devices. (b) Basic semiotic operations.

This basic modeling framework can be enlarged to describe devices and organisms by adding a set of effectors through which the device or organism influences the world beyond its boundaries.[8,24,39,40,55,56,58] *Thus the epistemological framework used to describe the operations of the modeling relations can also be used to describe modeling relations as they appear in organisms, devices, and other observers.* If we consider a robot with sensing elements, computational elements, and effector elements, then the corresponding basic operations will be those of *measurement, computation*, and *control.* Since effectors take the symbolic outputs of the computational part and act on the nonsymbolic world, they, along with measurements, determine the relation of the symbols in the computational part of the device to the world at large. Hence both measurements and controls are semantic operations.

Because a physical system will always contain multitudes of variables which by necessity will lie unrepresented in our finite models and because the choice of the subset of the relations that we choose to include in our models is up to the observer, it is imperative that we specify how we are viewing the system (what are the observables that constitute our *observational frame*) when we construct our definitions or make statements about the physical world. Depending upon how we look at the electronic devices on our desks (e.g., even in terms of volts, millivolts, or microvolt descriptions), we can see the material system as either completely rule governed or highly stochastic. By specifying the observational frame, we can avoid unproductive arguments over whether the device is "really" deterministic or nondeterministic—we can subject the question to an unambiguous empirical test and come to an agreement, on the outcome.

OBSERVED SYSTEMS AND THEIR BEHAVIORS

It is helpful to adopt Ashby's concepts of "system" and "state" in these discussions. A *state* is a joint property of an observer and the physical system being observed, an observable distinction: "By a state of a system is meant any well-defined condition or property that will be recognized if it occurs again."[3] A *system* consists of a set of distinguishable states chosen by the observer. The behavior of this system will consist of the patterns of state transitions that it undergoes over some period of observation (Figure 3). These state transitions can be apparently determinate (e.g., $4 \rightarrow 5$) or they can be apparently indeterminate (e.g., $16 \rightarrow 4$, $16 \rightarrow 18$), depending upon whether a given predecessor state is always observed to transit to the same successor state. Within this frame, those transitions which are determinate can be described in terms of a rule relating the states (4 and 5); such that one can duplicate exactly the apparent behavior of the system by using this rule. When a such rule can be written, the state can be treated as a symbolic type, abstracted from its material substrate. In this situation we lose no information about the trajectory of the system if we drop the distinction between the two states (4 and 5) by merging them. We can label such state-determined "apparently necessary" transitions as computations and nondeterministic, contingent transitions as measurements, controls, and nonsymbolic transductions, depending upon how they are connected to computational transitions.[8] The method gives us a clear means of deciding whether we are seeing a computation, measurement, control, or nonsymbolic interaction taking place in a material system under observation. Immediately we can decide whether a given material system observed via a particular frame is implementing "computations." This a distinct advantage over purely formal definitions of computation, whose mapping to the physical world is left unspecified (e.g., Is the amoeba in our test tube a Turing machine? On what basis do we decide?). Likewise, these criteria offer a possible solution to "the measurement problem," the question of when a measurement is judged to have taken place.[44,45]

Those states that participate in state-determined, computational transitions can be considered "symbolic" states, and the rest can be considered "nonsymbolic" states. To simplify the discussion, we will talk in terms of computational, syntactic transitions (those which can be described in terms of rule-governed transitions between symbolic states) and noncomputational, semantic transitions (those contingent measurement and control transitions which connect symbolic and nonsymbolic states). Those transitions in which no symbolic states participate can be identified as nonsymbolic interactions, and these transitions are neither syntactic nor semantic, since no symbolic states are involved. In effect we have systems-theoretic, operational definitions for the basic operations of the observer-actor, be it a device, organism, or some other appropriately organized physical system. This also gives us the means of distinguishing between syntactic, semantic, and pragmatic relations.

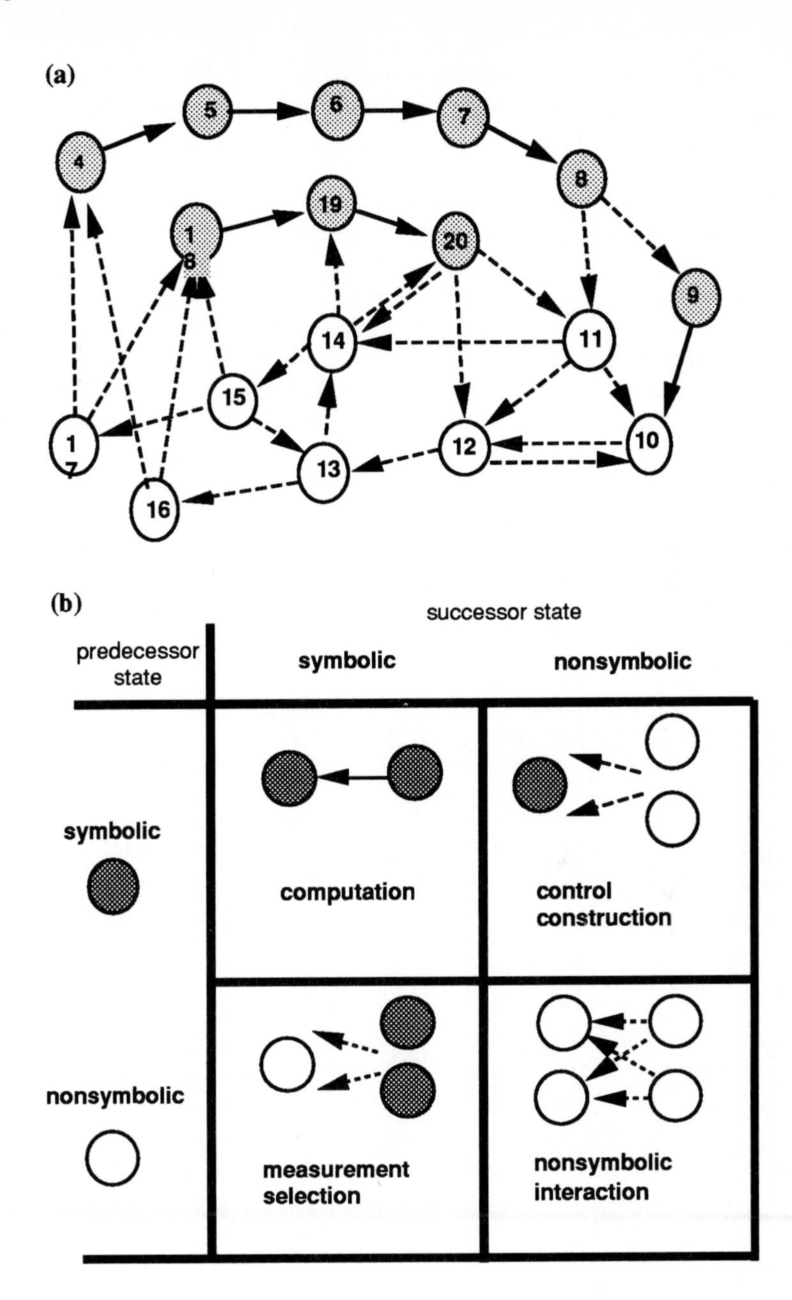

FIGURE 3 (a) An observed state diagram. (b) Observed-state correlates of the basic semiotic operations.

The result is a phenomenally grounded *semiotic functionalism* incorporating syntactic, semantic, and pragmatic relations, rather than a logically grounded computational functionalism which only considers syntactic operations as its constituent relations.

These distinctions are also relevant for an operational definition of living organization, a central theoretical problem for Artificial Life. If we lack such an empirically decidable definition, we have no means of deciding whether our simulated organisms are living or not, and no means of evaluating our research efforts.[13,30,31,37,52,53] Thus far, the most thoughtful attempts at a definition have come from theoretical biology and cybernetics and have as their basis a network which is constantly reproducing its parts through a recurrent set of production relations: Rashevsky's "relational biology,"[50] metabolism-repair (M, R) systems,[28,55,56] semantically closed systems,[37,42] autopoietic systems,[13,26,30,31,62] self-modifying systems,[11,16,18,19] self-reproducing automata,[63] and recurrent neural networks.[27] The semiotic functionalism outlined above may provide a means for grounding our recognition of the organism/environment boundary: regions of computations (rule-governed order) bounded by contingent measurement and control transitions.

TYPES OF EMERGENT BEHAVIOR

> "The main point to realize is that all knowledge presents itself within a conceptual framework adapted to account for previous experience and that any such frame may prove too narrow to comprehend new experiences."
>
> —Niels Bohr[6]

Once we have fixed our observational frame we can talk precisely about emergence: whether the behavior of the physical system in question has changed with respect to the frame and in what ways it has changed. If we observe the system for a period of time, we can build up a model including all of the syntactic and semantic state transitions we have previously observed and we can see if new state-transition patterns appear over time as the system subsequently transits through its states. The model thus constitutes the observer's expectations of how the system will behave in the future. If we are confident that our measuring apparatuses are stable (via calibration tests), then any observed deviations from the observer's model can be reasonably interpreted as structural changes in the organism or device under observation.

If no new state-transitions arise, then the device or organism is *nonemergent* with respect to the observer's frame. The observer would need to make no change in his/her model in order to include all the observed state transitions within the system. We would infer from this nonemergent behavior that the structure of the organism or device was stable relative to our observational frame. Whatever other physical processes are going on in the material system, they do not affect the relation between the observer and the material system.

If the pattern of syntactic state transitions changes such that new computational transitions are formed, then the device or organism will appear to be *syntactically emergent*. From such behavior, we could infer that the device or organism was implementing a new computation within the same set of symbolic states. We could interpret this behavior as signifying the switching of the computations performed by the device or organism, contingent upon some other process or mechanism unrepresented in the model.

If the pattern of semantic state-transitions changes such that new measurement or control transitions are formed, then the device or organism will appear to be *semantically emergent*. This would signify changes in the structure of the device which make its semantic operations contingent upon other factors outside the observational frame. In the case of an emergent measurement, the organism or device would appear to have acquired the ability to sense changes in the world which were unrepresented in the observer's model. The observer could then have access to this new organism/device observable by watching the state-transition behavior of the organism/device. An observer trying to track the device would find it necessary to add an observable to the model whenever a new sensor was constructed by the device. The apparent dimensionality of the device's state space would thus increase by one.

Syntactic and semantic emergence thus are not mutually exclusive possibilities: both processes can be happening simultaneously in the same system.

EMERGENCE AND ADAPTIVE DEVICES

What sorts of devices exhibit the various types of emergent behavior? A comprehensive taxonomy of adaptive devices incorporating combinations of fixed and performance-dependent sensing, computing, and effecting elements can be developed corresponding to the types of emergent behaviors listed above.[8,36,66] According to the plasticities of the respective functionalities, the capacities and limitations of each type of adaptivity can be outlined (Table 2). Nonemergent devices (Figure 4) are those with fixed syntactic parts (nonadaptive computational devices) and fixed semantic parts (nonadaptive robotic devices). Syntactically emergent devices (Figure 5(a)) are those devices having performance-contingent syntactic parts (adaptive devices). Semantically emergent devices (Figure 5(b)) are those with performance-contingent semantic parts (evolutionary devices). Devices which are both syntactically and semantically adaptive (general evolutionary devices, Figure 6(b)) would appear both syntactically and semantically emergent.

TABLE 2 Summary of capacities and limitations of some of the device types

device type	plasticity	capacities	limitations
fixed-computational	fixed syntax	reliable execution of pre-specified rules	limited to pre-specified rules and states
fixed-robotic	fixed syntax fixed semantics	reliable execution of fixed percept-action combinations	no feedback or learning from environment
adaptive	adaptive syntax fixed semantics	performance-dependent optimization of percept-action coordination	limited to precept & action categories fixed by the sensors & effectors
general evolutionary	adaptive syntax adaptive semantics	creation of new percept & action categories: performance-dependent optimization within these categories	time to construct & test new sensors & effectors may be very long

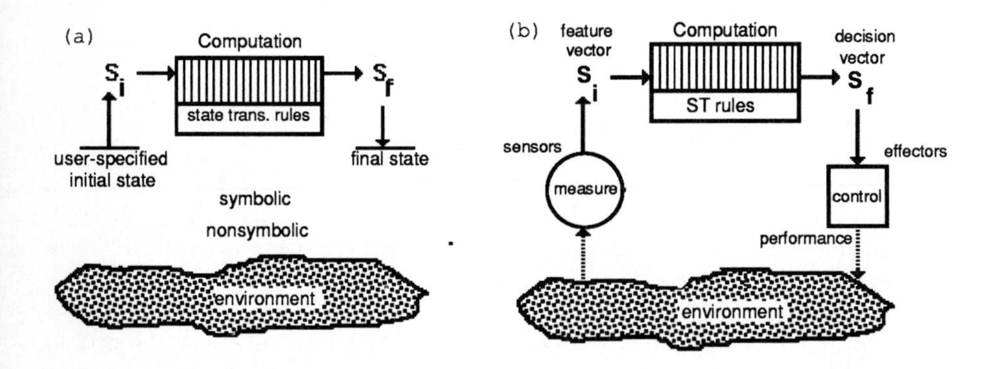

FIGURE 4 Two types of nonadaptive devices: (a) fixed computational and (b) fixed robotic devices.

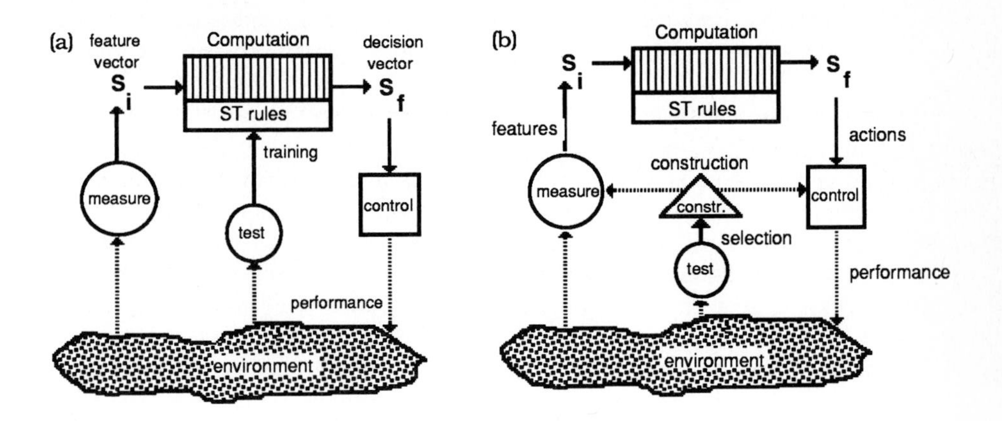

FIGURE 5 Two types of adaptive robotic devices: (a) syntactically adaptive and (b) semantically adaptive.

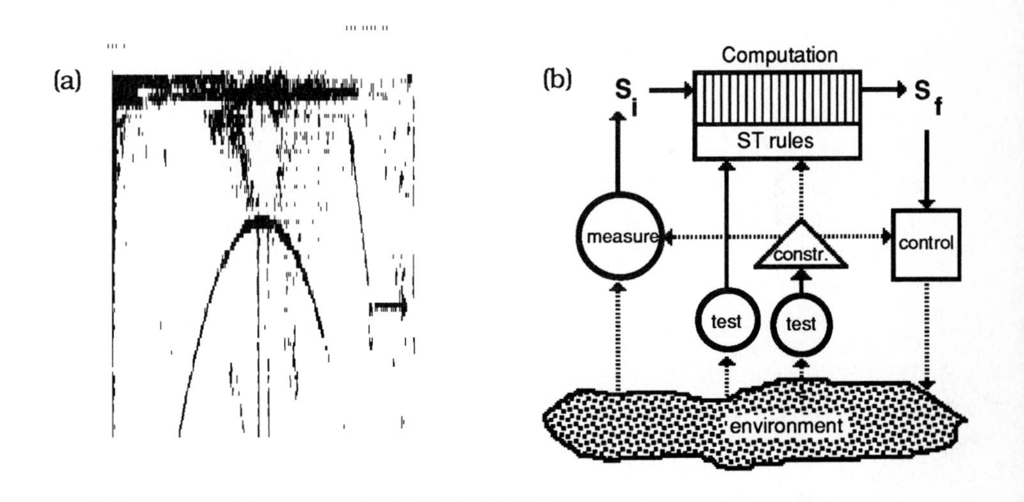

FIGURE 6 (a) Pask's electrochemical device (left) and (b) a general evolutionary device (right).

Nonadaptive computational devices. If we examine a computer with fixed programs, and we take as our observational frame the total machine states of the device, then the device will appear to us as a completely state-determined system.

The device will yield consistent, replicable results—we can implement a formal system (in Hilbert's sense) using the states of the device as our tokens. As long as there are no hidden inputs upon which the computational outcome depends, then the device will continue to behave in the same way given the same initial states. No new state transitions will be created when we run the device, since it has been designed to perform independently of fluctuations in its environment. Hence the device will appear to be nonemergent relative to the observational frame of total machine states. There will be other physical processes going on within the device (heat flows, humidity changes, magnetic fields), but they will not make a difference on the behavior as seen through our observational frame of all its discrete-voltage gate states.

The same is true for computer programs and simulations, which have some subset of the total machine states as their states. As long as we are observing all of the program and simulation states (state-variable values, pseudo-random number generator states) and all the input values, then the program/simulation will give us the exact same behavior over time, and will appear nonemergent. We will discuss the implications for artificial life simulations below.

Nonadaptive robotic devices. Likewise fixed robotic devices will have nonemergent behavior, owing to the stability of their sensing, computing, and effecting elements. The sensing and control elements will have contingent transitions, but no new ones will arise if these elements are stable and enough time has been taken to exhaust the repertoire of possibilities.

Syntactically-adaptive devices (or adaptive computational devices) alter their computational parts contingent upon their performance in an external environment. Typically the device incorporates some performance measure and training rules which alter its computational part to improve performance. Consequently, if the observer chooses his/her frame to coincide with the computational states of the device, the pattern of computational transitions will change with every training iteration. Between training iterations the pattern of computational transitions will be stable and the device will appear nonemergent. Trainable classifiers and controllers, neural nets, and genetic algorithms will appear to be syntactically emergent as long as performance measure on which the computational changes are based appears as a contingent event to the observer. Typically this is the case. The user of an adaptive device does not know beforehand whether a given computational change will improve or degrade the device's performance, so the testing of the change appears contingent.

Semantically adaptive devices construct and select their sensing and effecting elements contingent upon their performance. The evolution of sensory and effector organs, the immune system, and activity-dependent, sensory-motor specializations (the musician's ear, the athlete's reflexes) are natural examples. The development of sensory prostheses and scientific measuring instruments (telescopes, radiometers, molecular assays, etc.) and effector prostheses (hand tools, motor vehicles, weaving looms, chemical plants, etc.) are artificial examples if their human designer-constructors are included.

Apparently there is but one device in the engineering literature with the explicit goal of adaptively creating new sensors *de novo*.[34,35,36] In the late 1950's the cyberneticist Gordon Pask constructed an electrochemical device along the lines of an adaptive coherer (Figure 8(a)). Although a stable analog, connectionist network could have been implemented with his apparatus, Pask designed his device to be conditionally sensitive to nonelectrical perturbations in its environment, such that it could be tuned with the appropriate rewards. Given a set of electrodes in an aqueous ferrous sulfate/sulphuric acid solution, ferrous threads could be adaptively grown to become sensitive to sound. In about half a day the assemblage could be trained to discriminate between two frequencies of sound: "the evolution of an ear."[35]

A device of this type could be combined with an adaptive computed neural network to form a general evolutionary device (Figure 8(b)): the electrochemical assemblage would evolve the feature primitives; hence the semantics of the feature space that the neural network would adaptively partition. We could thereby achieve adaptation on both syntactic and semantic levels: the resulting assemblage would have the means of both creating its own feature primitives and adapting its computations within those sets of primitives. Such a device would be *epistemically autonomous*, capable of choosing its own semantic categories as well as its syntactic operations on the alternatives.

A network of such devices would be capable of generating new, independent inter-element signalling channels in an apparently open-ended manner, thus increasing the dimensionality of its signal space. Such considerations may be relevant to multiplexing and creation of new signalling possibilities in real neurons.[10] A still greater degree of autonomy would entail devices which not only determined their own syntactic and semantic relations, but their pragmatic relations as well. Such *motivationally autonomous* devices would have mechanisms for constructing their own performance-measuring apparatuses, hence the criteria which control the selection of their syntactic and semantic functions. Such devices would not be useful for accomplishing *our* purposes as their evaluatory criteria might well diverge from our own over time, but this is the situation we face with other autonomous human beings, with desires other than our own, and the dilemma faced by all human parents at some point during the development of their children.

IMPLICATIONS FOR THE INTERPRETATION OF ARTIFICIAL LIFE SIMULATIONS

The emergence-relative-to-a-model view has deep implications for the interpretation of artificial life simulations. All computer simulations can be described in terms of finite-state automata, as networks of computational state transitions, as formal symbol manipulation systems. As observer-programmers we can always find a frame which will make our simulation appear nonemergent. If we choose our observables to coincide with the stable computational states of the finite state automaton being implemented by the simulation, then we will always see it as a nonemergent state-determined system. Here the states of our frame will generally correspond to the discrete values that the state variables of the simulation can take on and the state transitions will correspond to all of the simulation rules which govern the values of the state variables. Every time the simulation is run with the same initial conditions, the simulation will transit through the same trajectory of variable values. The computer simulation will be completely replicable; there will be no deviation of the simulation's behavior from the model of possible trajectories built up by the observer. Thus, from this perspective many of the breakout strategies that have been proposed to make artificial life simulations "open-ended" and "emergent" will simply not improve the situation because they do not change the formal, completely replicable nature of the process. Increasing the size of the simulation, adding new layers of simulation rules, simulating random or chaotic processes, or representing genotypes and phenotypes will not in any way change the replicability of the simulation; hence these changes will be ineffective at transforming a previously nonemergent simulation into an emergent one.

Likewise, the inclusion of global patterns, such as cycles, waves, gliders, or other complex forms into the consideration of cellular automata behavior also fails to change the replicability of these simulations. Typically, the appearance of these patterns is thought to be an emergent event in the device itself: previously the pattern was absent, now it is present. For the purposes of judging whether an emergent event has occurred, we need to be careful not to shift frames of reference in these situations, from talking in terms of microstates and pixel states before and "higher level" features afterwards. If we start to observe the device in terms of individual pixels, we must continue to do so in those terms throughout. If we wish to include complex pixel patterns (e.g., cycles, waves, moving patterns which look to us like a horse galloping), they need to be in our state descriptions from the start, or they will remain in the realm of tacit, private observation, unrecognized by our public model. These other distinctions can then added to the observational frame, but as long as the mechanisms for recognizing the complex pixel patterns are reliable and this recognition is solely a function of the pixel pattern presented, the replicability of the observed behavior of the simulation will be unaffected; hence the simulation will still appear nonemergent. That the behavior of the cellular automata by itself is nonemergent should not obscure its potentially emergent effects on the people intently watching the screen. They are busy recognizing new patterns and forming new concepts. *The interesting emergent events that involve artificial life simulations reside not in the simulations themselves, but in the ways that they change the way we think and interact with the world.* Rather than emergent devices in their own right, these computer simulations are catalysts for emergent processes in our minds; they help us create new ways of seeing the world.

MEASUREMENTS AND CONTROLS IN SIMULATED ORGANISMS

One of the hopes for artificial life is that we can gain useful knowledge of the material world around us by the construction of computer models and simulations. We know that various computational functions can be carried out by higher levels of computational organization—the subcognition as computation approach. It has been a matter of debate within the community, however, whether artificial life simulations are capable of realizing measurements, as opposed to merely representing them.[44,45] At the outset, there should be no doubt in anyone's mind that we cannot make measurements on the material world outside of our computers solely by programming computations within our computers. *Operationally we cannot carry out measurements on the world about us if we completely control the actions of the measuring devices.* The measurements become a consequence of our actions rather than a process contingent upon factors outside our control. An example would be if we immersed a thermometer in a heat bath whose temperature we controlled completely—the readings we would get would cease to tell us anything about the world outside the heat bath. If we remove the human controller and close the loop by connecting the output of the thermometer to the heating element of the bath, then the thermometer-bath system will appear as a circular series of state-determined computations. To the extent that the system is not contingent upon fluctuations from outside the bath, it is informationally closed with respect to that external world.

Our simulated organisms are likewise part of informationally closed computational cycles, purposely insulated from the nonsymbolic world outside the simulations. This is exactly why we cannot make measurements on the physical world through programs and simulations that we specify completely. From the vantage point of total simulation states, simulated organisms appear as completely specified syntactic objects, having completely rule-governed behavior, and any contingent operations we have attempted to implement (e.g., perceptions) will lose their contingency once they are embedded in the simulation. Our simulated organisms are like the heat bath we control: they reflect our actions perfectly; hence we only learn about the world of artificial conventions we have constructed. *Whatever information or knowledge we gain from performing computational operations, it cannot be the same kind of knowledge or information we gain by making measurements.* We face the analytic-synthetic distinction in yet another context, the general question of exactly what it is we learn when we do mathematics.[49]

Perhaps *we* cannot make measurements through our simulations, but what about our simulated organisms on *their* simulated environments? The important question here is not whether the organism or robot is "simulated" or "material," since as limited observers we see the system only through a series of state transitions (we don't know what is on the other side of our measuring devices that give us our observable states). The critical question becomes how do we distinguish a measurement process in our simulation from all the computations? If the *form* of the behavior is all that matters, we should be able to do this with an unlabelled, uninterpreted simulation. But if we have access to all of the simulation states, then our simulation will appear to us as a large state-determined system. There is no way to make the distinction, since all state transitions will appear to be necessary ones.

It has been suggested that a way out of this dilemma is to "take the perspective of the artificial organism" by discarding our knowledge of the simulated environment. As external observers, if we restrict our states to those of the organism and leave the simulated environment out of the system, then the organisms will indeed appear to be implementing contingent operations and there will be apparent informational exchange. We as external observers who no longer have access to the total simulation states can now gain empirical information about the simulated environment, since we no longer control it. Once we do this, however, we cannot vacillate between ignorance and omniscience. Once we have disavowed all knowledge of the environment, we can no longer interpret our simulated organisms as purely formal entities, since their internal states will be contingent on "hidden variables" unrepresented in our truncated frame. Either we view our simulated organisms through the lens of total simulation states and see them as purely computational, nonemergent formal objects, or we view them through a restricted lens and see them as informationally open, potentially emergent systems implementing measurements and controls as well as computations. *We cannot see both pictures at the same time.*

Finally, because all of our simulations are discrete representations of limited detail, even if we adopt partial ignorance, the simulated organisms will only evolve their sensory distinctions up to the granularity with which the environment is represented in the simulation program. This may be very coarse, or it may be very fine, but it will be finite. While we will be able to see our artificial organisms as making new measurements, computations, and controls over time, eventually the set of simulated possibilities will be exhausted and there will be no more emergent behavior.

In the end the functional closure of computer simulations is due to the stability properties of the physical substrates on whose motions they depend. Digital computers are *closed-state* devices because of the finite number and stable nature of the discrete attractors they utilize for their state spaces. As functional objects, they are *defined* by this discrete, stable organization. Biological organisms and mixed digital-analog devices, in contrast, as *open-state* devices can continue to proliferate new discrete attractors (new symbol primitives) because of the continuous, contingently stable nature of their energy landscapes.[7] New distinctions can then arise out of the continuum, in a constant interplay between the differentiated and the undifferentiated, the symbolic and the indefinite.

ACKNOWLEDGMENTS

The author would like to acknowledge vibrant discussions with Julio Fernandez, Gail Fleischaker, Michael Hudak, Cliff Joslyn, George Kampis, Chris Langton, Eric Minch, Gordon Pask, Howard Pattee, James Pustejovsky, Robert Rosen, and Rod Swenson. I am indebted to Paul Pangaro, curator of the Pask archives, for his assistance researching the early literature on self-organizing systems and Pask devices.

REFERENCES

1. Ashby, W. R. *Design for a Brain*, 281, 291. London: Chapman & Hall,1952.
2. Ashby, W. R. *An Introduction to Cybernetics.* London: Chapman & Hall, 1956 .
3. Ashby, W. R. "Principles of the Self-Organizing System." In *Modern Systems Research for the Social Scientist: A Sourcebook*, edited by W. Buckley. Chicago: Aldine, 1968. Reprinted from *Principles of Self-Organization*, edited by H. von Foerster and G. Zopf. New York: Pergamon Press, 1962.
4. Bergson, Henri. *Creative Evolution*, 1946. New York: Random House, 1911.
5. Bohr, Niels. "Unity of Knowledge." In *Atomic Physics and Human Knowledge.* New York: John Wiley, 1958. Reprinted by Ox Bow Press, Woodbridge, CT, 1987.
6. Carello, C., M. T. Turvey, P. N. Kugler, and R. E. Shaw. "Inadequacies of the Computer Metaphor." *Handbook of Cognitive Neuroscience*, edited by M. Gazzaniga. New York: Plenum Press, 1984.
7. Cariani, Peter. "On the Design of Devices with Emergent Semantic Functions." Ph.D. dissertation, Department of Systems Science, State University of New York at Binghamton, Ann Arbor: University Microfilms, 1989.
8. Cariani, P. "Implications from Structural Evolution: Semantic Adaptation." In *Proceedings of the International Joint Conference on Neural Networks, Washington, D.C., January, 1990*, vol. I, 47–50. Hillsdale, NJ: Lawrence Eirlbaum Associates.
9. Cariani, P. "Adaptivity and Emergence in Biological Organisms and Artificial Devices." *World Futures: The Journal of General Evolution.* London: Gordon & Breach, in press.
10. Csanyi, V. *Evolutionary Systems and Society: A General Theory.* Durham, NC: Duke University Press, 1990.
11. Denbigh, K. G. *An Inventive Universe.* New York: George Braziller, 1975.
12. Fleischaker, G. R. "Origins of Life: An Operational Definition." In *Origins of Life and Evolution of the Biosphere*, July, 1990.

13. Forrest, S. "Emergent Computation: Self-Organizing, Collective, and Cooperative Phenomena in Natural and Artificial Computing Networks." *Proceedings of the Ninth Annual Center for Nonlinear Studies and Computing Division Conference*, Technical Report LA-UR-89-4087, Los Alamos National Laboratory, May, 1989.
14. Hillis, Daniel. "Intelligence as an Emergent Behavior; or the Songs of Eden." *Daedalus* (Winter, 1988): 175–189.
15. Kampis, G. *Self-Modifying Systems in Biology and Cognitive Science: A New Framework for Dynamics, Information, and Complexity.* New York: Pergamon, in press.
16. Kampis, G. "Some Problems of System Descriptions I: Function, II: Information." *Int. J. General Systems* **13** (1987): 143–171.
17. Kampis, G. "The Idea of Self-Modification." *Int'l. J. Systems Research & Info. Science* **3** (1988): 15–19.
18. Kampis, G. "On the Modelling Relation." *Systems Research* **5(2)** (1988): 131–144.
19. Klee, R. L. "Micro-Determinism and Concepts of Emergence." *Philosophy of Science* **51** (1984): 44–63.
20. Langton, C. "Studying Artificial Life with Cellular Automata." *Physica D* **22** (1986): 120–149.
21. Langton, C. "Artificial Life." In *Artificial Life*, edited by C. Langton. SFI Studies in the Sciences of Complexity, Proc. Vol. VI. Redwood City, CA: Addison-Wesley.
22. Lilienfeld, R. *The Rise of Systems Theory: An Ideological Analysis.* New York: Wiley-Interscience, 1978. See pp. 86–88 regarding emergence and machine creativity.
23. Lotka, A. J. "Physical Aspects of Organic Evolution." *Bull. of Math. Biophys.* **10** (1948): 103–115.
24. Maruyama, M. "Heterogenistics: An Epistemological Restructuring of Biological and Social Sciences." *Cybernetica* **20** (1977): 69–86.
25. Maturana, H. "Autopoiesis." In *Autopoiesis: A Theory of Living Organization*, edited by M. Zeleny. New York: North-Holland, 1981.
26. McCulloch, W. Embodiments of Mind, rev. 1988 edition. Cambridge, MA: MIT Press, 1965.
27. Minch, E. "The Representation of Hierarchical Structure in Evolving Networks." Ph.D. dissertation, Department of Systems Science, State University of New York at Binghamton, Ann Arbor: University Microfilms, 1987.
28. Modrak, D. K. *Aristotle: The Power of Perception.* Chicago: University of Chicago, 1987.
29. Moreno, A., J. Fernandez, and A. Etxeberria. "The Necessity of a Definition for Life." Poster, Second Workshop on Artificial Life, Santa Fe, February, 1990.

30. Moreno, A., J. Fernandez, and A. Etxeberria. "Cybernetics, Autopoiesis, and Definitions of Life." In *Proceedings European Meeting on Cybernetics and Systems Research (E.M.C.S.R. 1990), Gordon Pask Symposium, Vienna, April, 1990*. [publisher?]
31. Morgan, C. L. *Emergent Evolution*. London: Northgate and Williams, 1927.
32. Nagel, E. *The Structure of Science*. New York: Harcourt, Brace & World, 1961.
33. Pask, G. "Physical Analogues to the Growth of a Concept." In *Mechanization of Thought Processes: Proceedings of a Symposium , National Physical Laboratories, November 1958*. London: H.M.S.O., 1958.
34. Pask, G. "The Natural History of Networks." In *Self-Organizing Systems*, edited by M. C. Yovits and S. Cameron. New York: Pergamon Press, 1960.
35. Pask, G. *An Approach to Cybernetics*. New York: Harper & Brothers, 1961. In his preface Warren McCulloch writes of Pask's device, "With this ability to make or select proper filters on its inputs, such a device explains the central problem of experimental epistemology. The riddles of stimulus equivalence or of local circuit action in the brain remain only as parochial problems."
36. Pattee, H. H. "The Physical Basis of Coding and Reliability in Biological Evolution." In *Towards a Theoretical Biology. 1.Prolegmena*, edited by C. H. Waddington. Chicago: Aldine, 1968.
37. Pattee, H. H. "The Nature of Hierarchical Controls in Living Matter." In *Foundations of Mathematical Biology*, edited by R. Rosen, Vol. I. New York: Academic Press, 1972.
38. Pattee, H. H. "Physical Problems in the Origin of Natural Controls." In *Biogenesis, Evolution, Homeostasis*, edited by A. Locker. New York: Pergamon Press, 1973.
39. Pattee, H. H. "The Physical Basis of the Origin of Hierarchical Control." In *Hierarchy Theory: The Challenge of Complex Systems*, edited by H. H. Pattee. New York: George Braziller, 1973.
40. Pattee, H. H. "The Complementarity Principle and the Origin of Macromolecular Information." *Biosystems* **11** (1979): 217–226.
41. Pattee, H. H. "Cell Psychology: An Evolutionary View of the Symbol-Matter Problem." *Cognition and Brain Theory* **5** (1982): 325–341.
42. Pattee, H. H. "Instabilities and Information in Biological Self-Organization." In *Self-Organizing Systems: The Emergence of Order*, edited by F. E. Yates. New York: Plenum Press, 1988.
43. Pattee, H. H. "The Measurement Problem in Artificial World Models." *Biosystems* **23** (1989): 281–290.
44. Pattee, H. H. "Simulations, Realizations, and Theories of Life." In *Artificial Life*, edited by C. Langton. SFI Studies in the Sciences of Complexity, Proc. Vol. VI. Redwood City, CA: Addison-Wesley, 1989.
45. Pepper, Stephen. "Emergence." *J. Philosophy* **23** (1926): 241–245.
46. Piatelli-Palmarini, Massimo. "How Hard is the Hard Core of a Scientific Paradigm?" In *Language and Learning. The Debate between Jean Piaget and*

Noam Chomsky, edited by M. Piatelli-Palmarini. Cambridge, MA: Harvard University Press, 1980.
47. Prigogine, I. *From Being to Becoming.* San Francisco: W. H. Freeman, 1980.
48. Proust, J. *Questions of Form: Logic and the Analytic Proposition from Kant to Carnap.* Minneapolis: University of Minnesota, 1989. In interpreting artificial life simulations, we are immediately confronted with the analytic-synthetic distinction, and all of the age-old debates between Platonic idealism and Aristotelian hylomorphism.
49. Rashevsky, N. *Mathematical Biophysics*, Vols I & II. New York: Dover, 1960.
50. Rosen, R. *Life Itself. A Comprehensive Inquiry into the Nature, Origin, and Fabrication of Life.* New York: Columbia University Press, in press.
51. Rosen, R. "On the Generation of Metabolic Novelties in Evolution." In *Biogenesis, Evolution, Homeostasis*, edited by Alfred Locker. New York: Pergamon Press, 1973. This paper directly addresses the central questions of mathematical models and computer simulations of emergence, the closed nature of formal notational systems, and the difficulties inherent in generating functional descriptions from structural ones.
52. Rosen, R. "Biological Systems as Organizational Paradigms." *Int. J. General Systems* **1** (1974): 165–174.
53. Rosen, R. *Fundamentals of Measurement and Representation of Natural Systems.* New York: North Holland, 1978.
54. Rosen, R. *Anticipatory Systems.* New York: Pergamon Press, 1985.
55. Rosen, R. "Organisms as Causal Systems which are not Mechanisms: An Essay into the Nature of Complexity." In *Theoretical Biology and Complexity. Three Essays on the Natural Philosophy of Complex Systems*, edited by R. Rosen. Orlando, FL: Academic Press, 1985.
56. Rosen, R. "Causal Structures in Brains and Machines." *Int. J. General Systems* **12** (1986): 107–126.
57. Rosen, R. "On the Scope of Syntactics in Mathematics and Science: The Machine Metaphor." In *Real Brains Artificial Minds*, edited by J. Casti and A. Karlqvist. New York: North-Holland, 1987.
58. Salthe, S. *Evolving Hierarchical Systems.* New York: Columbia University Press, 1985.
59. Swenson, Rod. "Emergent Attractors and the Law of Maximum Entropy Production: Foundations of a Theory of General Evolution." *Systems Research* **6** (1989): 187–197.
60. Toffoli, Tomaso. "Physics and Computation." *Int. J. Theor. Phys.* **21(3/4)** (1982): 165–175.
61. Varela, F. *Principles of Biological Autonomy.* New York: North Holland, 1980.
62. von Neumann, J. "Theory and Organization of Complicated Automata." In *Papers of John von Neumann on Computing and Computer Theory*, edited by W. Aspray and A. Burks. Cambridge, MA: MIT Press, 1987.
63. von Uexküll, J. *Theoretical Biology.* New York: Harcourt, Brace & Co., 1926.

64. Weber, B., D. Depew, and J. D. Smith, eds. *Entropy, Information, and Evolution: New Perspectives on Physical and Biological Evolution*. Cambridge, MA: MIT Press, 1988.

Louis Bec
Institut Scientifique de Recherche Paranaturaliste, 14, Avenue du Griffon, 84700 Sorgues, France

Eléments d'Epistemologie Fabulatoire

En complément des contributions scientifiques, la communication du zoosystémicien doit être considerée comme une tentative visant à montrer que tout acte de modélisation concernant "l'Artificial Life" construit inevitablement un artefact épistemologico-esthétique du vivant.

A travers cette communication, le zoosystémicien s'attachera donc à developper des points de vue differents qui pourraient contribuer à éclairer et à enrichir l'AL.

Il considerera toute recherche en AL comme une modélisation permanente de systèmes complexes, composés de multiples sous-systèmes (biographiques, scientifiques, artistiques, technologiques, épistemologiques, esthétiques, éthiques, idéologiques, sociologiques ....).

C'est pourquoi le zoosystémicien determiné postulera que toute entreprise de réduction mécaniste, que tout cloisonnement verouillant les diverses disciplines, sont des obstacles majeurs au complet développement de l'AL, de ses théories et de ses pratiques.

Il est convaincu que la radicalité des interogations qu'elle impose, revisite certains fondements de la culture occidentale et propose en consequence, une nouvelle redistribution des savoirs.

Aussi le zoosystémicien consciencieux s'interessera-t-il particulièrement à certains "operateurs" qui travaillent à la transformation des connaissances d'une façon sous-jacente et participent efficacement au développpement de l'AL.

Artificial Life II, SFI Studies in the Sciences of Complexity, vol. X, edited by C. G. Langton, C. Taylor, J. D. Farmer, & S. Rasmussen, Addison-Wesley, 1991

PREMIER OPERATEUR: LA TRANSVERSALITÉ

Les pratiques modélisatrices de "l'Artificial Life" tissent à travers les domaines scientifiques, artistiques, et technologiques, et plus particulièrement au sein des sciences du vivant, des sciences de l'artificiel, et des systèmes de représentations artistiques, un reseau specifique de rélations.

Ces relations sont en constantes mutations travaillées qu'elles sont par:

a. La pression de pulsions "demiurgiques" dans lesquelles sommeille depuis toujours, au sein des systèmes de représentation artistiques et mécaniques, une fascination pour toute création du vivant (sculptures, golem, fetiches, cires anatomiques, automates, ...).

b. Les poussées des techno-sciences qui tendent à simuler, et plus encore, à vouloir "fabriquer" du vivant. Elles trouvent actuellement leurs lieux d'expressions dans la génétique, le génie génétique, les biotechnologies, la bio-informatique, les automates, la robotique, ...

c. Les conséquences determinantes des inter-actions entre differents domaines qui, par l'emprunt des expériences, des methodes, des modèles, ne s'embarassent pas des catégorisations dogmatiques et constituent un continuum de recherche sur le vivant qu'il apparait arbitraire de subdiviser. Cela a pour effet de deplacer la césure qui existait entre la pensée directe et rationnelle (scientifique et technologique) et la pensée indirecte et symbolique (artistique).

Le zoosystémicien tentera de montrer que ce déplacement ouvre un vaste champ de "création" aux modes de connaissances et d'expressions, et plus particulièrement encore, aux sciences du vivant et aux sciences de l'artificiel.

SECOND OPERATEUR: LES PROCEDURES DE "SAISIE" DU VIVANT

L'étude d'une typologie de gestes modélisateurs du vivant met en evidence le deplacement de cette cesure ainsi que ses consequences modélisatrices.

Le zoosystémicien zélé se livrera à une rapide description assortie de commentaires d'une serie de geste qui, selon lui, sont à l'origine de la modélisation du vivant et constituent les fondement de l'AL.

a. Représentation et correspondance: modélisation analytique
    i. mise à plat du vivant
    ii. mise en volume du vivant

b. Cohérence: modélisation systémique

i. mise en mouvement du vivant

ii. mise en transformation du vivant

TROISIEME OPERATEUR: L'UPOKRINOMENOLOGIE

Le zoosystémicien conçoit l'upokrinomenologie comme une activité operatoire et artificialisante qui interroge de toute part, l'incapacité du vivant à saisir le vivant.

A travers les differentes modélisations de l'animal, de l'animal-machine à l'animal-auto-organisé en passant par l'animal-programmé, le zoosystémicien propose une metazoologie artificielle et biaisée en construisant des leurres heuristiques, les *upokrinomenes*. Il propose une strategie hypocrisique pour élaborer les bases d'une épistemologie fabulatoire, l'*upokrinomenologie*.

TRANSVERSALITE ET "ARTIFICIAL LIFE"

1. Il faut insister sur le fait que le zoosystémicien opiniâtre est fortement convaincu que tout acte de modélisation concernant l'AL est une modélisation systémique qui met non seulement en jeu des paramètres scientifiques et technologiques (biologiques, génétiques, physiologiques, morphogénétiques, ethologiques, informatiques, etc.), mais aussi des "fonctions artistiques et cognitives" (représentations, imaginaires, visualisations, ...).

C'est pourquoi il avance la thèse suivante: *Les modèles de correspondance qui sont le plus souvent actuellement utilisés, sont inadaptés.*

Il ne sera probablement plus question à l'avenir de représentation ou de simulation tendant à imiter les apparences d'un objet ou d'un comportement existant, mais bien de "façonner un processus endogène," un système visant à une certaine autonomie et produisant sa propre représentation.

Les modélisations les plus pertinentes seraient donc dans ce cas, des modélisations visant à une "cohérence," capable de lier, sans déperdition, les données quantitatives et syntaxiques aux données qualitatives et semantiques, et de produire par chimerisation un "méta-modèle" ou par modélisation d'autres "potentialités métaboliques."

2. La deuxième thèse avancée par le zoosystémicien: *La césure courrement admise, entre les arts d'une part et les sciences et les technologies d'autre part, n'est qu'un archaïsme tenace, un masquage qui perdure pour des raisons en grande partie idéologiques.*

Il affirme que cette cesure s'est déplacée et qu'elle est d'une autre nature.

Elle separe maintenant deux substrats épistemologiques qui nourrissent des projets et des objectifs profondement opposés.

Ce déplacement propose une nouvelle situation épistemologique qui libére les activités théoriques et pratiques de l'AL des contingences étroitement scientifiques et l'engage à faire émerger un imaginaire specifique, un "*bio-techno-imaginaire*" en mixant totalement les domaines scientifiques, artistiques, et technologiques.

Tant que le lieu de comparaison entre les pratiques scientifiques, techniques, et entre les pratiques artistiques, se situait au sein d'une épistemologie de la modélisation analytique ou expressive, la représentation du vivant (substitut ou correspondance entre le vivant et un modèle de ce vivant) ne pouvaient que produire des zônes d'interactions peu prospectives.

Mais l'émergence des sciences de l'artificiel, des sciences de la communication, l'explosion des technosciences et des sciences du vivant, ainsi que la transformation sous toutes ses formes, des domaines artistiques, proposent à nouveau un champ fusionel.

Ces deux substrats épistemologiques s'expriment et s'opposent souvent violement et sourdement au sein du débat social et culturel actuel.

Ils focalisent chacun de leur côtés et à leur manière, des positions existentielles et idéologiques contraires. Ils produisent des activités artistiques, scientifiques, et technologiques differentes. Ils attirent des artistes, des chercheurs, et des techniciens dont les objectifs et les statuts divergent.

Dans le premier substrat, les rélations arts-sciences-technologies developpent des zônes d'interactions, de convergence, de divergence, et d'impregnation.

Tout en conservant leurs specificités, chacunes des activités artistiques, scientifiques et technologiques élaborent des strategies de déchiffrement du monde, tentant de décrire ou modéliser des phénomènes de la nature pour mieux en comprendre les mécanismes et mieux en exprimer les multiples apparences.

Les pôles scientifiques et technologiques s'ils continuent de s'appuyer sur une épistemologie positiviste et analytique, n'en demeurent pas moins ébranlés par de recentes découvertes et s'interrogent de plus en plus sur les fondements idéologiques des avançées scientifiques, sur les aspects culturels et intellectuels de leur devenir. Ils ouvrent des lieux de contact avec des disciplines diverses dont les disciplines artistiques.

Celles-ci, de leurs côtés, s'emploient pour une part, à se dégager des contraintes de leurs specificités et déploient une vaste activité exploratoire qui va de la déconstruction des langages et du brouillage des codes jusqu'à des formes multimedia, associant ainsi des modes d'expressions et des techniques les plus diversifiés.

Il en résulte un paysage composé de zones relationnelles, dans les quelles ou à partir des quelles s'échangent des fragments d'expérimentations artistiques, des aspects conceptuels, des "recettes" méthodologiques, des modèles scientifiques, et des savoir-faire technologiques.

Tout cela provoquant par un effet de serre philosophique et culturel, un phénomène d'impregnation.

Dans de nombreuses réalisations artistiques, ces concepts scientifiques et ces pratiques techniques peuvent agir en tant que tremplin à l'imaginaire, comme des décentrements provocateurs.

Ils peuvent alimenter des analogies, des métaphores, des coïncidences, des utilisations instrumentales et provoquer, par un dépaysement defocalisant, la surection d'une idée, d'une image mentale, d'une émotion et entrainner des productions dans un supplement d'étrangeté ou de fiction logique.

Mais, malgre cela, le schéma des divisions classiques entre les pratiques artistiques et scientifiques est conservé. Il n'en continue pas moins à perpetuer des dogmatismes lourds et réactionnaires.

Le second substrat évolue autour d'une épistemologie des sciences de l'artificiel et des techno-sciences. Il met en jeu des activités dont une des finalités est de construire un univers. Il favorise une culture qui semble valoriser les procédures technologiques, systémiques et méthodologiques et entrainne une transformation radicale de la nature des collaborations entre des chercheurs, des techniciens, et des créateurs.

Si pour le premier substrat la connaissance n'est connaissance que si elle refléte le monde tel qu'il est, avec ses caractères d'objectivité, le second s'exprime dans le projet qui fonde les sciences de l'artificiel: la construction d'un monde pour le connaître

Il s'édifie autour du comment de la conception, du comment de la construction des représentations symboliques par lesquels s'élaborent les artifices.

Les pratiques artistiques, scientifique, et technologiques au paravant dissociées, ne sont plus face à une théorie dans la quelle et la connaissance et l'expression doivent réfleter une réalité ontologique et objective, mais se trouvent enrolées conjointement dans l'élaboration d'un projet, dans sa mise en ordre, et dans l'organisation d'un monde constitué par les expériences du modélisateur ou des modélisateurs.

L'association intime et concrete des connaissances et des expressions aux procédures technologiques devient une toute première necessité.

Dans ce substrat, ces technologies ne peuvent plus être considerées comme de simples outils d'executions, neutralisés, mais comme des sous-systèmes determinants, qui temoignent eux aussi d'un univers symbolique et fantasmatique.

Ce dialogue, entre "l'intentionalité créatrice" du modélisateur, la variabilité des connaissances et la charge de l'univers symbolique et fantasmatique des technologies, fait naître une forme d'imagination prospectante et inventive, proposant des hybridations bio-techno-imaginaires.

Ces formes d'hydridations induisent un *principe programatique de "délégation,"* propre à toute vie artificielle.

Dans cette situation, il n'est plus possible d'opposer ou de comparer bord à bord les domaines artistiques, scientifiques, et techniques dans leurs specificités, à travers leurs méthodes et leurs productions.

Il faut s'exercer à des écarts épistemologique pour tenter de repérer les lieux ou la pensée s'origine et se developpe dans l'acte de recherche et de conception.

Il faut mettre à l'exercice de nouvelles approches pour tenter de cheminer dans les regions souterraines là où s'élaborent les hypothèses confuses, là où il est possible de repérer et de saisir les imbrications entre les activités fantasmatiques, symboliques, logiques, et rationnelles.

Il faut utiliser de nouvelles méthodes comme la modélisation des systèmes complexes pour faire émerger ces potentialités virtueles.

Le zoosystémicien prophétique, quant à lui, considere qu'une autre épistemologie devient necessaire, une épistemologie élargie à l'esthétique, à l'éthique, à l'idéologique et au technologique, une épistemologie très fortement traversée par une activité imaginaire.

UNE EPISTEMOLOGIE FABULATOIRE...

Le zoosystémicien arrive à se convaincre que c'est dans ces lieux que se fondent réellement les pratiques de "l'Artificial Life."

GESTES MODELISATEURS

Le vivant deploie une activité incessante dans la "saisie" du vivant.

Du predateur qui tend des pièges au généticien qui manipule les processus biologiques, du berger qui se domestique au dessinateur qui métaphorise ou codifie la représentation, le vivant, en s'adonnant à de bien curieuses attitudes comportementales, fabrique un "TAS" de productions artificielles, dans les quelles regnent à la fois l'ingéniosité, la rigueur, et la fantaisie. Si la discipline de "l'Artificial Life" est, comme la définit Chris Langton dans un premier temps, "the study of man-made systems that exhibit behaviors characteristic of natural systems," il est possible en étudiant certains gestes producteurs d'objet, de faire apparaître une "gestuelle modélistique" specifique dans la quelle les pratiques de l'AL sont inscrites.

Cette "gestuelle modélistique" part de la saisie carnassiere du vivant, sur et par le vivant, passe par des simulations de systèmes naturels, pour parvenir à la construction de systèmes autonomes qui tendent dans leurs positions les plus extrêmes à ne plus obeir aux biologies et comportements connus.

Elle trace par le même, les étapes successives des modélisations du vivant qui, d'une implication bestiale du modélisateur, passe par les phases d'une modélisation analytique en procédant par des substituts (sujet-modèle-objet) pour proposer présentement, une modélisation qui tend à l'auto-organisation.

Elle montre que c'est donc au coeur de ce dispositif, là, où se construisent les strategies de connaissances les plus inventives, les strategèmes et les ruses les plus tortueuses que se sont installées les composantes de "l'Artificial Life."

Le cadre reduit de cet intervention ne permet d'en développer tous les aspects. Quatre gestes peuvent pourtant rendre plus explicite cette thèse.

Les gestes de mettre à plat et en volume le vivant, peuvent être considerés comme constituant le dispositif central des systèmes de représentation et de correspondance.

Les gestes de mise en mouvement et de mise en transformation autonome du vivant, peuvent être considerés comme générant à des niveaux divers, les modèles de cohérence.

LE GESTE DE METTRE A PLAT LE VIVANT

Le geste de mettre à plat le vivant consiste à en dérouler la peau, à l'étaler et à lire sur les traces inscrites, la confluence de deux univers: *le révèlé et l'enfoui.*

Ce geste permet de révéler aussi le milieu qui s'exprime sur cette peau et l'organique interne qui trouve, à travers elle, les voies de communications avec ce milieu.

La mise à plat du vivant consiste à établir un cadre d'intervention, permet de determiner un support, de cadastrer une partie du monde et de construire un enclos: celui de la représentation.

Sur des supports divers, le modélisateur, vivant, memorise des émotions, des observations qui s'effectuent durant le passage plus ou moins bref du vivant dans le cadre.

Ce geste de mettre à plat le vivant, facilite la fixation du décrit, de l'écrit. Il permet la codification, la désignation, l'ennoncé, la classification, et une certaine modélisation par un arsenal de techniques.

Il élimine la troisième et la quatrième dimension qui cachent certaines des apparences et compliquent les caractéristiques du vivant.

Chacune de ses parties seront ainsi restituées sur le "plat" par des artifices divers:

- multiplication des points de vue,
- perspectives et dessins anatomiques éclatés,
- peintures, dessins, et études documentaires,
- schéma anatomiques, physiologiques, etc., et
- photographies, radiographies, etc.

Ce geste permet de differencier les limites organiques avec l'exterieur, tout comme de cadastrer les cloisonnements qui structurent le dedans organique.

La hantise et l'observation du compliqué, amène le grossissement et le changement d'echelle, par une microscopie de plus en plus performante.

L'imagerie médicale procede, elle aussi, par un découpage tomographique, par des lamellisations du scanner, par un pélliculage radiographique et par un compartimentage scintigraphique ou thermographique.

Le vivant est ainsi envisagé comme un feutilletage embryologique artificialisé.

Enfin, le support proposé par la mise à plat du vivant est le lieu de la mise à l'exercice de la gestualité du vivant, de la griffe concrète des éléments disponibles dans sa memoire.

C'est donc le lieu de l'inscription sismographique de ce vivant par matérialisation graphique du système nerveux et musculaire. C'est un geste qui s'affirme solidaire de l'organique, qui exprime le temperament, la psychologie, et les capacités conceptuelles du mod elisateur.

LE GESTE DE METTRE EN VOLUME LE VIVANT

Le geste de mettre en volume le vivant consiste à le fixer, pour mieux l'étudier, dans une matière solide, permanente et lui conserver ses apparences et ses particularités d'objet saisi.

Il consiste donc à le bloquer dans un espace euclidien, à l'enfermer dans un bocal ou dans une cage à grille de coordonnées.

Il conjugue l'observation et la connaissance des comportements pour construire une tridimensionnalité basée sur un principe de correspondance, à même de restituer une identité.

Chaque partie de la surface renvoie à un organique interne.

Dans le cas de la taxidermie, l'organique interne est remplaçé par une matière non perissable, celle-çi sera par la suite recouverte de la peau même du vivant, donc de son apparence réelle.

Le geste de mettre en volume le vivant ne peut se satisfaire des seuls aspects formels exterieurs.

Il s'engage inevitablement dans le voyage endogène et endoscopique, dans le compliqué organique, dans la diversité des structures, et des matières molles et chaudes (cires anatomiques).

Il opère des dissections precises, et tout en tentant de ne pas detruire le "global," il conserve et étudie les organes dans leurs situations topologiques et spatiales.

Il propose une vision réaliste, mécaniste, et théatralisé du dedans.

Le geste qui construit ce type de modèles appartient à ce geste très specialisé, qui s'inscrit dans un acte general d'humanisation par la fabrication des outils et des instruments technologiques pour des tâches précises.

Une autre phase s'amorce avec le geste de mettre en mouvement le vivant.

Ce geste repond à une vision mécanique et cybernetique qui privilegie la définition du vivant par l'animé. Il anticipe le geste de mise en transformation où les principes de délégation et d'autonomie s'inscrivent comme les bases de "l'Artificial Life."

LE GESTE DE METTRE EN MOUVEMENT LE VIVANT

Le geste qui tente de mettre en mouvement des modèles du vivant consiste à introduire un mécanisme artificiel dans un modèle. Ce qui donne à ce modèle l'une des apparences du vivant la plus caractéristique: la mobilité.

Il correspond aux notions de l'animal/machine et de l'animal/programmé. Ce geste vise donc à animer des objets construits dans un univers, par du vivant et pour comprendre ce vivant. Il espère par simulation s'en rendre maître sur les plans sensibles, perceptifs, symboliques, comportementaux, logiques, et techniques.

L'objet mis en mouvement peut-être une endo ou une exo-mécanique.

La marionnette, par exemple, est une exo-mécanique, une sorte de prolongement du vivant, une prothèse que le vivant implante dans l'univers de l'imaginaire. En elle, se rejoint la réalité mouvante qu'elle figure et la simulation qu'elle propose.

Les automates simulent des comportements complexes rappellant certains aspects du métabolisme biologique par l'utilisation de nouveaux matériaux, par la miniaturisation croissante des parties mécaniques et par l'apport de l'électronique et de l'informatique.

Ils sont en décalage constant vis-à-vis d'une réalité conventionnelle et de la représentation qu'ils en donnent.

Cette situation les rend disponibles et donc mobiles.

Ils deviennent doublement opératoire tant au plan d'une animation matérielle convaincante que sur le plan d'explorations scientifiques inédites et fantasmatiques.

C'est aussi le même geste qui opère dans la construction des organes artificiels. Les progrès réalisés dans l'élaboration des biomateriaux et de la micro-électronique permettent:

- La construction de protèses de remplacement mécanique d'organes disparus et l'assistance à des organes défaillants.
- L'inscription au sein du système du vivant, de prothèses mixtes remplaçant les aspects sensoriels et moteurs par des "artifices" en liaisons avec le système nerveux.

LE GESTE DE METTRE EN TRANSFORMATION LE VIVANT

Le geste de mettre en transformation le vivant consiste à abandonner l'instrumentation mécanique pour passer au control differé, par simulation ou strategie heuristique des processus biologiques. Il correspond aux notions de l'animal auto-organisé.

Le geste de mettre en transformation le vivant permettant peut-être, pour la première fois à une éspèce d'accéder à son patrimoine génétique, écrit une histoire qui passe par la domestication et la zootechnie.

Il touche à la multiplication et à l'amélioration du vivant dans ses formes végétales et animales, par le biais de l'alimentation, de la réproduction, de l'évolution dirigée et la création de milieux artificiels.

Il s'illustre surtout par des manipulations génétiques qui grâce aux developpements de la biologie moleculaire et de la génétique, peuvent modifier l'aspect, la viabilité et le comportement du vivant.

Cette gestuelle modélistique met en evidence deus voies paralelles qui se developpent actuellement:

- La première est celle de "l'Artificial Life" ou d'une "techno-mimétique" qui developpe par la bio-informatique une instrumentologie heuristique sous forme d'automates conceptuels et technologiques.
  Elle éspère, en élargissant la connaissance du vivant, mener, au-delà des limites restreintes de la biologie paroissiale terrestre, une exploration plus extensive pour imaginer ou rencontrer d'autres formes possibles de vie.

- La seconde est celle de la génétique ou d'un "métamorphisme biologique" qui projète de procéder à des mutations à partir de la matière et de son métabolisme.
  Elle éspère la multiplication et l'amelioration des conditions du vivant dans ses formes végétales et animales, par le biais de l'alimentation, de la réproduction, par la création de milieux artificiels et surtout par la manipulation génétique.

Ces deux voies trouvent des points de jonction. Elles se completent et se compénétrent en échangeant des aspects théoriques, des expérimentations, des modélisations, ...

Elles proposent, par l'émergence de forme de vie artificialisée à des niveaux divers, le plus boulversant prolongement de la théorie de l'évolution, en plaçant le phénomène de la recherche scientifique au coeur des problèmes de la création et inversement.

Elles s'inscrivent dans un cadre phylogénétique de conception et de production artificielle proliferante du vivant.

- Au delà d'une expérience créatrice essentiellement psychologique et sociale, produisant des objets culturels et artistiques.
- Hors d'une expèrience positiviste de la connaissance scientifique produisant du savoir objectif.
- Mais dans un prolongement plus fondamental, inscrit au sein des caractéristiques du vivant: *Une proliferation et une complexification par des strategies adaptatives les plus diversifiées.*

Si cette activité que l'homme a deployé, jusqu'à ce jour, pour saisir, maîtriser et simuler le vivant n'était qu'une ébauche, la plus visible d'une "pulsion" irresistible, d'une entreprise plus vaste et plus definitive qui serait de "créer du vivant."

Si l'accès au patrimoine génétique en même temps que l'approche du vivant par des strategies et des objets technologiques levaient peu à peu cet interdit fabricatoire éthique et esthétique.

L'impact de telles actions, dont la portée n'est pas mesurable, au plan éthique, épistemologique, esthétique, economique, et idéologique serait tel qu'il risquerait de renvoyer certains des postulats et des activités qui ont construit nos civilisations au rang d'une archéologie nécéssaire mais caduque.

L'exemple de la représentation du vivant par la sculpture, considerée comme une constante dans toute forme de societé, illustre bien ces propos.

Que devient le geste du sculpteur qui en taillant en polissant, gravant, moulant une matière inerte, rêve de golem?

Ne cède-t-il pas la place à un autre geste qui greffe, sélectionne, congèle, implante, chimerise?

Ne s'abandonne-t-il pas à un autre geste qui simule, programme, construit des automates, des robots, un geste qui élabore des programes techno-morphogenetiques et les imprime dans le processus évolutif d'une "sculpture vivante"?

Dans ce cas les cloneurs ou les bio-informaticiens, sont-ils les vrais sculpteurs, les ultimes réalisateurs d'un projet qui a cheminé par les voies de l'art et de la science pendant de longs siecles?

L'UPOKRINOMENOLOGIE, OU LA PRATIQUE D'UNE ÉPISTEMOLOGIE FABULATOIRE

INTERVENTION DU ZOOSYSTÈMICIEN À SANTA FE, 8/2/90

Je ne suis pas un bon zoologue ni un bon biologiste.

Je ne suis pas non plus un bon informaticien et je crois qu'en tant qu'artiste cela ne vaut guère mieux.

Je ne suis qu'un modeste zoosystémicien qui n'a fait aucun effort pour le devenir, car de l'âge de 4 ans, j'ai su que je n'allais être qu'un Artefact et j'ai bien l'impression que cela a empiré depuis.

J'ai obtenu avec éclat mon diplôme de zoosystémicien.

Ce diplôme m'a été décerné par l'Institut Scientifique de Recherche Paranaturaliste, institute que j'avais pris soin de fonder quelles que années plus tot et dont je suis le seul diplôme et apparemment le seul president.

Comme tout zoosystémicien qui se respecte, je me suis mis, depuis 1972, avec l'obstination industrieuse, propre à mon corps professionnel, à élaborer des zoosystèmes.

Ce qui, comme chacun le sait, consiste à tenter de faire émerger du dessous des apparences de ce qui est appellé la zoologie, une *hypozoologie*, dans un espace zoosystémique.

Ce qui revient à concretiser des morphogeneses sous-jacentes à partir d'activités fabricatoires, imaginaires, fantasmatiques, symboliques, rationnelles, logiques, méthodologiques, et technologiques.

Cette activité m'amène à faire de l'*upokrinomenologie*, c'est à dire à reflechir sur les concepts d'une zoologie de l'hypocrisie et à modéliser des organismes: les *upokrinomenes.*

Pour comprendre cette curieuse manie, il faut dire que tout zoosystémicien est inhibé, dès son plus jeune âge, par une pédagogie de methodologie hypocrisique.

Cette pédagogie lui propose comme exercice la construction de leurres pour leurrer les autres tout en se leurrant lui-même.

Ce qui le condamne, sa vie durant, à produire dans les zoosystèmes qu'il élabore un continuum d'organismes bestialisés, intermitants, periodiques, bifurquants et dont certains, dans des cas bien particuliers peuvent avoir la malchance d'appartenir à une classification zoologique conventionnelle.

En d'autres termes, cette pratique consiste à développer un stratagème évolutif, en créant autour du phénomène central une accumulation hétérogene d'operateurs-bestialisés.

Il s'agit, pour le zoosystémicien ambitieux, d'élargir ou de faire émerger, le champ "hypocrisique" du vivant, c'est à dire *la complexité du dialogue rusé qu'entretient le vivant avec le vivant* et d'en produire un *expansé heuristique.*

Il s'agit pour lui, de faire devier toutes lectures univoques et instrumentalistes du vivant, vers des multiplicateurs équivoques, algorithmiques, et inconsiderés.

Il peut, à tout moment, se déclarer d'une manière naïve: *peupleur, artefacteur* ou pire *bestialeur.* Dans ses moments de grande lucidité, le zoosystémicien propose quatre axes principaux de travail:

1. Développer des prolongements complementaires, prospectifs ou aberrants à partir du schéma classificatoire actuel de la zoologie (bovideologie, vampyroteuthis infernalis, loligo vulgaris, ...).
2. Implanter des peuplements inventifs dans les lacunes zoologiques réperées au sein de la biomasse dans les quelles les zoologistes craintifs ne s'aventurent pas (élitoniens, théreutes, tarakhophères, ...).
3. Révisiter des virtualités avortées lors des phases originelles ou inédites de l'évolution (sulfanogrades, ...).
4. Elaborer sans contrainte, toute modélisation zoosytémique comme écart heuristique irreverenciux (astatères, molunothes, épistrephomes, ...).

Pour lui l'upokrinomenologie est un laboratoire de méthodologies vrillées dans lequel une memoire culturelle interrogeant une memoire biologique et inversement, peut produire des chimères polymorphes.

Il semble évoluer au sein d'un stratagème en déséquilibre d'une parodie pour un exode continuel et fictif, d'une strategie du canular affublée de pirouettes et de ricochets, d'une pantomine ironique.

Sa détermination est grande à interroger les pratiques scientifiques et technologiques au même titre que les activités artistiques, comme des productions symptomatiques.

Il élabore des systèmes zoologiques arbitraires et imaginaires dans lesquels des zoomorphies singulières, des biologies curieuses, des zoosemiotiques aberrantes se développent. Ce qui lui permet par des approches plurielles, de réagir contre l'ultra-analytique et de proposer l'opportunité de concevoir, de construire et de maîtriser de vastes ensembles zoologisés qui dependent du modélisateur, de ses intentions claires ou diffuses, de son environnement culturel, économique, politique, et social.

Cette modélisation fait apparaître des réseaux inedits, des noeuds communicatoires indétectables habituellement.

Elle élabore une strategie qui au lieu de déchiffrer un phénomène de manière analytique ou sensible, conçoit et construit un projet signifiant, un agencement de signes.

Pour un zoosystémicien consequent et "branché," modéliser voudrait donc dire: *élaborer un système zoologique par une construction artificielle, qu'il est possible de doter de propriétés variables en accordant un rôle discretionnaire au modélisateur.*

Le paradigme de l'animalité est le lieu privilegié de l'intervention du zoosystémicien diplomé.

Le zoosystémicien, en élaborant cette upokrinomenologie, travaille sur l'animalité du vivant (biologique, zoologique, symbolique, fantasmatique, ...) avec la ferme conviction que le paradigme de l'animalité est bien plus dessiné et signifié par une semantique plurielle que par les éléments d'une zoologie objective.

Là, il a l'intime conviction que ce sont les formes toutes entières d'une représentation incluant l'observateur qui sont condensées et inscrites dans l'animalité.

Le rapport de l'homme à l'animal n'a été finalement tout au long de son histoire, qu'une modélisation, dans tous les sens du terme, qu'une simulation en creux d'une forme dramatique d'incommunicabilité.

Le travail du zoosystémicien est d'affirmer par une outrance systèmique et fabulatoire les facettes de cette "cohabitation" probablement une des sources les plus riches et les plus fondamentales de l'imaginaire et de la connaissance humaine.

C'est pourquoi la tâche du zoosystémicien est de faire apparaître une metazoologie qui s'enfoncerait d'une toute autre manière dans le terrain des significations et qui émergerait du dessous des apparences.

En élaborant des zoosystèmes paralelles ou complementaires, il chimerise des éléments de sa pensée directe et rationnelle à des éléments de sa pensée indirecte et symbolique.

En produisant une biologie systémique et des upokrinomenes, il implante au sein d'une zoologie "scientifique," une zoologie en expansion, une zoologie de la mutabilité, une zoologie "hypophanique," une animalisation potentielle qui interroge d'une manière biaisée, les vacances semantiques de la représentation du vivant.

The Future

J. Doyne Farmer† and Alletta d'A. Belin‡
†Complex Systems Group, Theoretical Division, and Center for Nonlinear Studies, Los Alamos National Laboratory, Los Alamos, NM 87545 and Santa Fe Institute, 1120 Canyon Road, Santa Fe, NM 87501 and ‡Shute, Mihaly, and Weinberger, P.O. Box 2768, Santa Fe, New Mexico 87504

Artificial Life: The Coming Evolution

Within fifty to a hundred years, a new class of organisms is likely to emerge. These organisms will be artificial in the sense that they will originally be designed by humans. However, they will reproduce, and will evolve into something other than their initial form; they will be "alive" under any reasonable definition of the word. These organisms will evolve in a fundamentally different manner than contemporary biological organisms, since their reproduction will be under at least partial conscious control, giving it a Lamarckian component. The pace of evolutionary change consequently will be extremely rapid. The advent of artificial life will be the most significant historical event since the emergence of human beings. The impact on humanity and the biosphere could be enormous, larger than the industrial revolution, nuclear weapons, or environmental pollution. We must take steps now to shape the emergence of artificial organisms; they have potential to be either the ugliest terrestrial disaster, or the most beautiful creation of humanity.

PREFACE

This paper was originally prepared for a conference held in honor of Murray Gell-Mann's 60th birthday, whose theme was "Where Are Our Efforts Leading?" The speakers were presented with a list of sixteen questions, which Gell-Mann had picked out as great challenges in science and human affairs. We were asked to pick one or more of these challenges and to examine the efforts of society to address them.

The challenges were very broad; a few examples from the list are:

6. To understand the common features of complex adaptive systems, such as the brain and the mind, the immune system, biological evolution, prehistoric chemical evolution, the generation of new strategies in computers, the evolution of human language, and the rise and fall of human cultures.
9. To find ways to cope with human "tribalism" in its many manifestations, including national, ethnic, and religious rivalries.
11. To build worldwide institutions, formal and informal, that may permit mankind to continue to avert large-scale catastrophe through management of conflict and management of the biosphere.

We chose artificial life as our subject for this symposium, since we felt that it was a topic that was pertinent to many of these challenges, but which had not received much popular attention. This paper was written for a popular, non-scientific audience; readers that are already familiar with artificial life may wish to skip the latter part of section 2. It was submitted for publication in the proceedings of that conference entitled "Proceedings in Celebration of Murray Gell-Mann's 60th Birthday" (Cambridge University Press ©1991) and is reprinted by permission of the publisher and the authors.

1. INTRODUCTION

Murray Gell-Mann posed some difficult questions for this symposium. Among them are: Where will our efforts lead in 50 to 100 years? What are the most important challenges that we face, for both science and society? What should people be thinking about that they are not properly aware of?

One answer to each of these questions concerns the advent of "artificial life." Within the next century we will likely witness the introduction on earth of living organisms originally designed in large part by humans, but with the capability to reproduce and evolve just as natural organisms do. This promises to be a singular and profound historical event—probably the most significant since the emergence of human beings.

The study of artificial life is currently a novel scientific pursuit—a quest to understand some of the most fundamental questions in physics and biology. This field is in its infancy. There are very few researchers actively engaged in the study of artificial life, and as yet there are far more problems than solutions. Studying artificial life has the potential to put the theory of evolution in a broader context and to help provide it with a firmer mathematical basis.

The advent of artificial life also has deep philosophical implications. It prompts us to reexamine our anthropocentric views and raises numerous questions about the nature and meaning of life. In addition, the study of artificial life may help us to understand, guide, and control the emergence of artificial life on earth, thereby averting a potential disaster and perhaps helping to create beautiful and beneficial new life-forms instead.

2. WHAT IS ARTIFICIAL LIFE?

In a recent book on the subject,[14] the discipline of artificial life was defined by Chris Langton as "the study of man-made systems that exhibit behaviors characteristic of natural living systems." A primary goal of this field is to create and study artificial organisms that mimic natural organisms.

We are used to thinking of evolution as a phenomenon specific to life on earth. Biology as it is commonly practiced is, in this sense, a parochial subject. The only example of life at hand is carbon-based life on earth. All life-forms on earth involve the same basic mechanisms. They all reproduce and develop under the control of the protein and DNA-templating machinery. However, it is not at all clear that this is the *only* possible basis for life. It is easy to conceive of other forms of life, in different media, with a variety of different reproductive and developmental mechanisms.

One motivation for thinking about life at this level of generality is the question, "If we ever make contact with life from other planets, will our science of biology help us understand it?" The answer depends very much on how universal the characteristics of life on earth are to all life-forms. Since we know nothing about life on other planets, it is a difficult question to answer. It seems probable, however, that much of our biology will simply be inapplicable to other life-forms. A central motivation for the study of artificial life is to extend biology to a broader class of life-forms than those currently present on the earth, and to couch the principles of biology in the broadest possible terms.

2.1 WHAT IS LIFE?

In order to state how something artificial might also be alive, we must first address the question of what life is, as generally as we can. To see why this is a difficult question, consider a related question to the one above: If we voyage to another planet, how will we know whether or not life is present? If we admit the possibility that life could be based on very different materials than life on earth, then this becomes a difficult task. Obviously, we cannot answer this question unless we have a general definition of what it means to be "alive." At present we do not have a good answer to this question.[1]

Nonetheless, we will make an attempt to state some of the criteria that seem to bear on the nature of life. There seems to be no single property that characterizes life. Any property that we assign to life is either too broad, so that it characterizes many nonliving systems as well, or too specific, so that we can find counter-examples that we intuitively feel to be alive, but that do not satisfy it. Albeit incomplete and imprecise, the following is a list of properties that we associate with life:

- *Life is a pattern in spacetime*, rather than a specific material object. For example, most of our cells are replaced many times during our lifetime. It is the pattern and set of relationships that are important, rather than the specific identity of the atoms.
- *Self-reproduction*, if not in the organism itself, at least in some related organisms. (Mules are alive, but cannot reproduce.)
- *Information storage of a self-representation.* For example, contemporary natural organisms store a description of themselves in DNA molecules, which is interpreted in the context of the protein/RNA machinery.
- *A metabolism* which converts matter and energy from the environment into the pattern and activities of the organism. Note that some organisms, such as viruses, do not have a metabolism of their own, but make use of the metabolisms of other organisms.
- *Functional interactions with the environment.* A living organism can respond to or anticipate changes in its environment. Organisms create and control their own local (internal) environments.
- *Interdependence of parts.* The components of living systems depend on one another to preserve the identity of the organism. One manifestation of this is the ability to die. If we break a rock in two, we are left with two smaller rocks; if we break an organism in two, we often kill it.
- *Stability under perturbations* and insensitivity to small changes, allowing the organism to preserve its form and continue to function in a noisy environment.
- *The ability to evolve.* This is not a property of an individual organism, but rather of its lineage. Indeed, the possession of a lineage is an important feature of living systems.

Another property that might be included in this list is growth. Growth is not a very specific property, however; there are many inanimate structures such as mountains, crystals, clouds, rust, or garbage dumps that have the ability to grow.

[1] One attempt has been made by Schrödinger in his book, "What is Life?" However, the discussion is heavily based on life as we know it rather than life as it might be, and as Schrödinger himself admits, the description is highly incomplete. Perhaps the best discussion of this issue is that of Monod.[18] He defines life in terms of three qualities: (1) teleonomic or "purposeful" behavior; (2) autonomous morphogenesis; and (3) invariance of information. The latter two are similar to some of the criteria we present here, but the first criterion seems as difficult to define as life itself.

Many mature organisms do not grow. Once they replicate, viruses do not usually grow.

It is not clear that life should be an either/or property. Organisms such as viruses are in many respects midway between what we normally think of as living and non-living systems. It is easy to conceive of other forms, for example the "proto-organisms" in some origin of life models,[1,7,8] that are "partially alive." In a certain sense societies and ecosystems may be regarded as living things. It seems more appropriate to consider life as a continuum property of organizational patterns, with some more or less alive than others.

This list is far from adequate—an illustration of the poverty of our understanding. We hope that as the field of artificial life develops, one of its accomplishments will be to give a sharper definition of what it means to be alive.

2.2 EXAMPLES OF OTHER LIFE-FORMS

The creation of new life-forms will almost certainly broaden our understanding of life, for several reasons:

- The act of construction is instructive about the nature of function.
- Artificial life-forms provide a broader palette, making it easier to separate the universal from the parochial aspects of life.
- Dissection and data gathering are potentially much simpler, particularly for life-forms that exist only inside a computer.

In the latter sense artificial life is to biology as physics is to astronomy: In astronomy we can only observe, but in physics we can perform experiments to test our hypotheses, altering the universe to enhance our understanding of it. Life, however, is a collective phenomenon, the essence of which is the interaction of the parts—too large an alteration results in death. Our ability to dissect or alter the form of natural organisms is limited. In contrast, we have complete knowledge of artificial organisms inside a computer, and furthermore we have the ability to alter their structure as well as that of the artificial universe in which they reside.[2] Similarly, by recreating new forms of life inside a test tube, we may understand these underlying principles more thoroughly.

There are many possible media for artificial organisms. They might be made of carbon-based materials in an aqueous environment, similar to natural organisms; they might be robots, made of metal and silicon; or they might be abstract mathematical forms, represented as patterns of electrons existing only inside a computer.

[2] For a provocative and entertaining discourse on the potential ethical problems involved in the study of artificial organisms, see "The Experiment" by Stanislaw Lem.[15]

2.2.1 COMPUTER VIRUSES Much of current research in artificial life focuses on computer programs or elements of computer programs that might be considered living organisms. It may be difficult to understand how this may be life, so we will begin by discussing the notorious example of computer viruses. Although computer viruses are not fully alive, they embody many of the characteristics of life, and it is not hard to imagine computer viruses of the future that will be just as alive as biological viruses.

These viruses are computer programs that reproduce themselves, typically designed as practical jokes by computer hackers. They are a diverse lot, and can live in many different media. For example, many viruses spend most of their life on floppy disks. Suppose a friend gives you a floppy disk that is infected with a virus. When you put the disk into your personal computer, the virus attempts to copy itself into the machine; when you insert another floppy disk, the virus attempts to copy itself onto the new disk. If the virus is effective, you may discover that, perhaps without your knowledge, it has infected all your floppy disks. If it is virulent, you may find that it takes up a great deal of space on your floppy disks, or that when it enters your machine, it causes the machine to spend much of its time executing the virus program rather than the task that you *want* the machine to perform. If the virus is really malevolent, it may destroy other programs that you have stored on your floppy disks.

A computer virus is certainly not life as we know it. It is just a pattern, a particular magnetic configuration on a floppy disk, or a particular set of electronic states inside a computer. Is the computer virus alive?

Note that a computer virus satisfies most, and potentially all, of the criteria that we have stated:

- A computer virus is a pattern on a computer memory storage device.
- A computer virus can copy itself to other computers, thereby reproducing itself.
- A computer virus stores a representation of itself.
- Like a real virus, a computer virus makes use of the metabolism of its host (the computer) to modify the available storage medium. The computer virus can direct the conversion of electrical energy into heat to change the composition of a material medium—it uses energy to preserve its form and to respond to stimuli from other parts of the computer (its environment).
- A computer virus senses changes in the computer and responds to them in order to procreate.
- The parts of a computer virus are highly interdependent; a computer virus can be killed by erasing one or more of the instructions of its program.
- Although many viruses are *not* stable under large electrical perturbations, by the nature of the digital computer environment, they are stable to small noise fluctuations. A truly robust virus might also be stable under some alterations of its program.
- Computer viruses evolve, although primarily through the intermediary of human programmers; an examination of the structure of computer viruses naturally places them in a taxonomic tree with well-defined lineages. For current

computer viruses random variation in computer virus programs is almost always destructive, although some more clever viruses contain primitive built-in self-alteration mechanisms that allow them to adapt to new environments, or that make them difficult to detect and eliminate. Thus, contemporary viruses do not evolve naturally.

Although computer viruses live in an artificial medium that we cannot directly see, they nonetheless possess most of the properties we have listed as characteristic of life, except possibly the last two. Computer viruses are already more than just a curiosity, and software infected by viruses is becoming increasingly common. During the fall of 1988, a computer virus propagated across the ARPA network (a fast communication link built by the defense department for interconnecting geographically separated computers), and brought computer operations at many major universities and national laboratories to a standstill.

Computer viruses are just one of many possible artificial life-forms, selected for discussion because they have already emerged, and because they illustrate how artificial life-forms can appear to be fundamentally different from more familiar contemporary biological life-forms. Because of their instability and their dependence on human intervention in order to evolve, they not as fully "alive" as their biological counterparts. However, as computers become more prevalent, more complex, and more highly interconnected, we suspect that so will computer viruses. Eventually it is likely that a computer virus will be created with a robust capacity to evolve, that will progress far beyond its initial form.

One example of computer organisms that evolve within a restricted environment is already provided by the VENUS simulation of Rasmussen et al.[20] Their work was inspired by a computer game called "Core Wars," in which hackers create computer programs that battle for control of a computer's "core" memory.[6] Since computer programs are just patterns of information, a successful program core wars is one that replicates its pattern within the memory, so that eventually most of the memory contains its pattern rather than that of the competing program.

VENUS is a modification of Core Wars in which the computer programs can mutate. Furthermore, each memory location is endowed with "resources," which, like sunshine, are added at a steady rate. A program must have sufficient resources in the regions of memory it occupies in order to execute. The input of resources determines whether the VENUS ecosystem is a "jungle" or a "desert." In jungle environments Rasmussen et al. observe the spontaneous emergence of of primitive "copy/split" organisms starting from (structured) random initial conditions. Note that since these "organisms" are contained by a highly specialized computer environment, there is no possibility of escape into the computer operating system. Such a protocol for containment is followed by all responsible researchers in artificial life.

2.2.2 MACHINES AND AUTOMATA A machine may be defined as "an apparatus consisting of interrelated parts with separate functions." Like an organism, a machine can break or die. One of the main features that distinguishes machines from organisms is the ability for self-reproduction. However, as demonstrated by John von Neumann in the late 1940's, it is possible, at least in principle, to build self-reproducing machines. Von Neumann imagined an "environment" filled with spare parts. The hypothetical machines in this environment had descriptions of themselves, and "construction arms" for acquiring and assembling the spare parts, all under the control of a computer. He sketched out the basic principles that such self-reproducing machines might follow, and laid out a blueprint for how they might operate.

Such a mechanical world is too complicated for simple mathematical analysis. Von Neumann, like contemporary researchers in artificial life, wanted to study the emergence and functioning of life in order to discover the basic principles that distinguish life from non-life. He was searching for an abstract environment to facilitate the study of these questions, in which simple patterns can be created that have lifelike properties. His hope was that by creating environments that give rise to pseudo-organisms he could gain an understanding of the fundamental properties of life itself.

Toward this end he turned to an abstract mathematical world, whose inhabitants are mathematical patterns. Following a suggestion of Stan Ulam's, he postulated a world consisting of a two-dimensional latticework of abstract "states," that change at discrete times according to a deterministic rule that depends only on the value of the neighboring states. This interaction rule may be thought of as defining the "physics" of a toy universe. Such a set of discrete states, together with a rule that changes them based of the states of their neighbors, is called a *cellular automaton*. In this world he demonstrated that there was a particular configuration of states with the capability to reproduce itself. The resulting construction is complicated to describe in detail. Roughly speaking, he constructed an initial pattern that contained a description of itself. Because of the particular rules he chose for the toy universe, the information from this description could flow out through a "constructing arm" (also consisting entirely of abstract states) so that the organism could "build" a copy of itself.

A simpler example of a cellular automaton is the *game of life*.[10] Imagine a checkerboard. Each square is either "alive" (has a piece on it) or "dead" (empty). Each square has a neighborhood, defined as the eight adjoining squares. To make a "move" each square examines its neighbors in order to decide whether it will be alive or dead when the move is completed. If it is dead, and two or three of the squares in its neighborhood are alive, then after the move is completed, it is alive. If it is alive, and three of the squares in its neighborhood are alive, then after the move is completed, it is alive. Otherwise it is dead. This procedure is followed for each position with the pieces fixed in place, and then the positions are updated simultaneously.

This game is so simple that, unless you have seen it before, you may find it hard to believe that it can give rise to very complex structures. For example, there are "gliders," simple oscillating patterns that propagate across the game board; "glider guns," which periodically emit gliders; and "self-reproducing glider guns," which make glider guns.

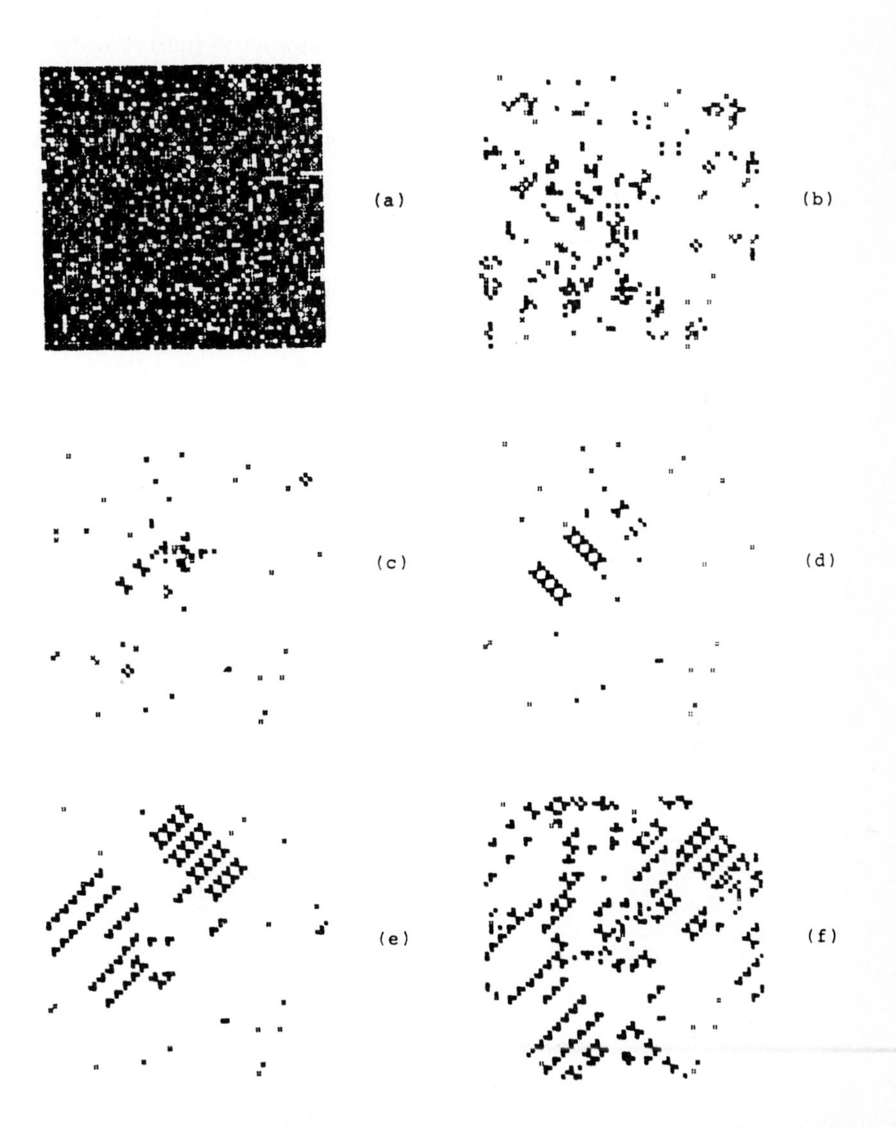

FIGURE 1 A cellular automaton with robust self-reproducing patterns, discovered by Chris Langton. Figure (a) shows a random initial condition on a square lattice; the eight possible states are represented by distinct patterns of dots. As time evolves the density of blank states increases, as shown in (b) and (c). In (c) we already see the seeds of self-reproducing patterns; as time progresses these patterns grow by replicating themselves. There are several reproducing patterns, which compete with each other for space, as shown in (d), (e), and (f).

Like the contemporary computer viruses, the self-reproducing objects in the game of life are not very stable. A small perturbation in their patterns typically destroys the replicating structures. Furthermore, if the game of life is run from a random initial condition, it typically settles down into static or simple periodic configurations. There are, however, other cellular automaton rules that are similar to the game of life, for which self-replicating structures seem to be quite robust. One set of examples, recently discovered by Chris Langton, is shown in Figure 1.

Langton has also made models for the formation of colonies,[13] as shown in Figure 2.

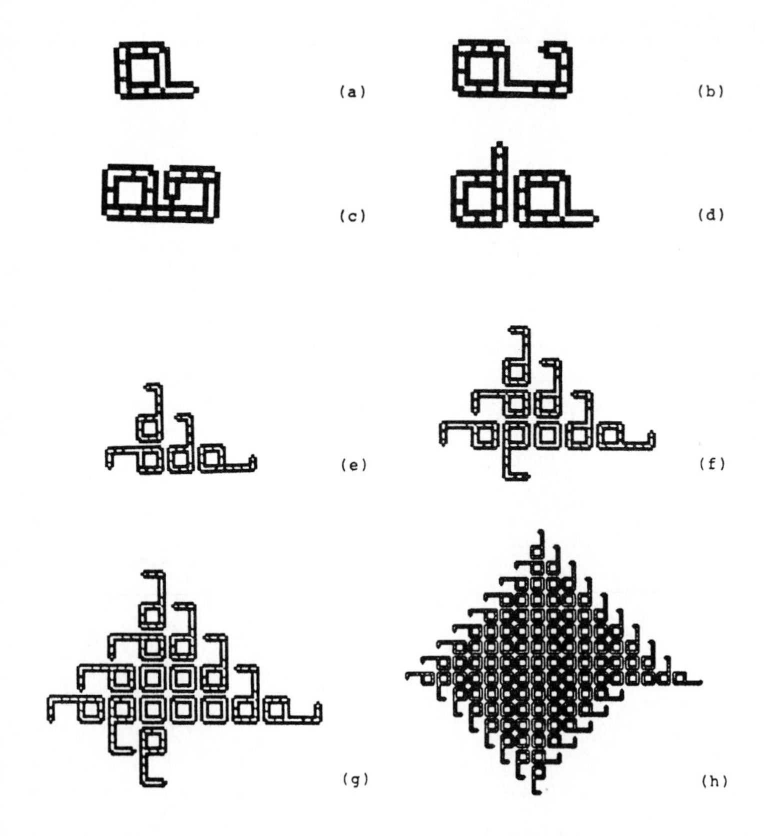

FIGURE 2 A cellular automaton model of self-reproduction, due to Chris Langton. Signals propagating around the "Adam" loop (a) cause the short arm to grow and curl back on itself (b,c,d), producing an offspring loop (e). Each loop then goes on to produce further offspring, which also reproduce (f). This process continues indefinitely, resulting in an expanding colony of loops (g,h), consisting of a "living" reproductive fringe surrounding a growing "dead" core, as in the growth of a coral.

An initial pattern reproduces itself on adjacent squares, in a manner reminiscent of the growth of a coral reef.

Along a somewhat different line, Richard Dawkins has created simple forms called "biomorphs" that evolve under artificial selection.[5] The "breeder" begins with a random pattern of lines connected to form a "tree." The geometric pattern of these lines is specified by simple rules, the details of which are given by simple abstract "genes." These rules are recursive, i.e., their form is the same at every level of the tree, and they can be applied to themselves. The biomorphs reproduce by making copies of themselves which differ from each other due to random mutation of their genes. The breeder selects the biomorphs he or she finds pleasing, and lets them breed again. In only a few steps, it is possible to create forms that are reminiscent of many different organisms. A few examples are shown in Figure 3.

The ease with which specific biomorphs are created illustrates the importance of recursive operations in generating evolvable biological forms.

Do these worlds give rise to life? So far, with the possible exception of the copy/split organisms of VENUS, or the robust self-reproducing automata of Langton, we would have to say that the answer is probably no. The key problem is finding the right combination of stability and variability. In most of the examples above, the self-reproducing patterns are destroyed by the slightest change. They are so fragile that they have difficulty evolving beyond their initial form. The robust replicating structures in both VENUS and Langton's automata are robust, but so far they have not been able to evolve beyond a fairly simple level of complexity. Discovering how to make such self-reproducing patterns more robust so that they evolve to increasingly more complex states is probably the central problem in the study of artificial life.

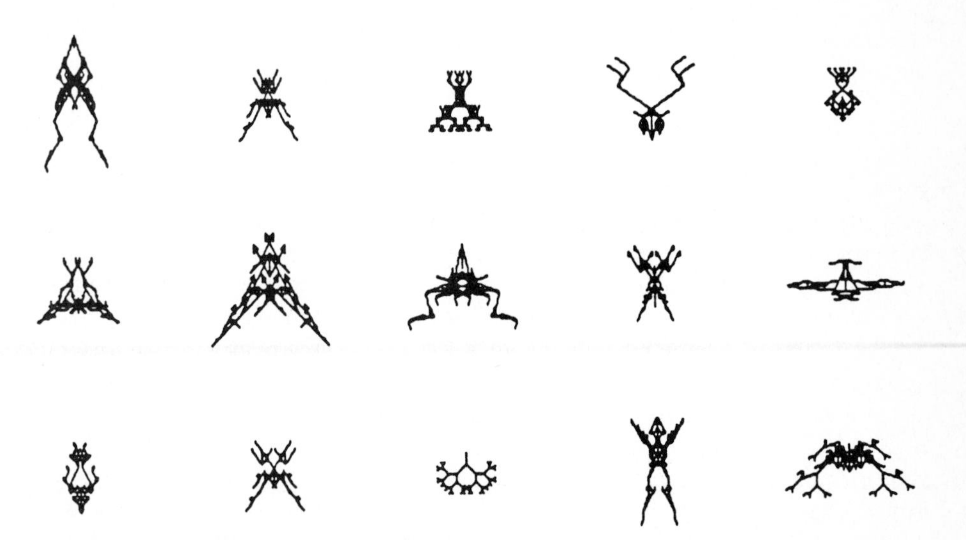

FIGURE 3 "Biomorphs," created through random variation of a simple "genome" and artificial selection of desirable features.

2.2.3 GENETIC ENGINEERING AND ARTIFICIAL WETWARE To highlight the contrast with carbon-based naturally occurring life-forms, in the discussion so far we have mainly addressed silicon-based artificial life-forms. However, artificial life can also occur in the wet, carbon-based medium of contemporary organisms.

The preponderance of examples of contemporary life-forms are bags of mostly water, built out of proteins, nucleic acids, lipids, and other organic compounds. The genome containing a self-representation is a DNA molecule. It is essentially a book, an instruction manual for the construction and operation of an organism. The message is written in an alphabet consisting of four letters, corresponding to four nucleotide molecules. The detailed message that distinguishes one organism from another is contained in the sequence of nucleotides along the DNA chain. The machinery of the cell, consisting of proteins, lipids, etc., reads this message and constructs replicas of the cell, much as does von Neumann's automaton.

In a certain sense carbon-based artificial life-forms have been with us since the advent of animal husbandry. By circumventing natural selection and replacing it with artificial selection, we alter the genome, creating varieties and hybrids that would never exist in the natural world. Nonetheless, artificial breeding is a comparatively weak tool, and the plant and animal forms it has produced are all relatively close to naturally occurring forms. The techniques of modern biochemistry and genetic engineering promise to take us far beyond this, giving us much more control over the genome, and the potential ability to create artificial life-forms that are radically different from natural life-forms.

There are two paths for the emergence of artificial life in the organic medium. The first path, which has already produced a variety of artificial life-forms, is genetic engineering. By directly manipulating the genome, we can modify existing life-forms. The second path, "artificial wetware," returns to the most primitive level, attempting to recreate the origin of life, or perhaps to generate whole new roots of the evolutionary tree.

Genetic Engineering. Under normal circumstances the message contained in a DNA molecule is invisible to us, and can only be read by the cell itself. In recent years, however, we have acquired the ability to peer into the cell and translate the sequence of nucleotides into sequences of human-readable symbols, A, G, C, and T. It is quite possible that we will be able to sequence the entire human genome, which consists of more than a billion nucleotides, by the year 2000. In other words, we will be able to read off the entire sequence of letters comprising the message that defines a particular human being. There is still a large step before we can *understand* what this information actually means and anticipate the effect of making changes in the sequence; at the moment the language is much more foreign than any human language, with semantics that lie in an entirely different realm. When we acquire the ability to interpret the messages of the genome, we will be able to "design" living things, change their form, cure them of hereditary diseases, make them bigger or smaller, or more or less intelligent. We will be able to create new species with properties radically different from those of natural organisms.

There are many potential commercial applications for genetic engineering. Bacteria have been genetically engineered to perform a variety of useful tasks. For example, the "*Ice*$^-$" bacterium protects plants against damage from freezing. Other bacteria have been designed so that they increase nitrogen fixation in plants or help clean up hazardous waste sites. Genetic engineering of fungi promises to improve industrial production of antibiotics and other useful chemicals. Plants have been genetically engineered so that they are resistant to infectious agents, or produce more and better food. Pigs have been genetically engineered to produce better meat, by elevating their level of bovine growth hormone, making them more like cattle. This is an example of how delicate success in genetic engineering can be—these artificial pigs also acquired a variety of unacceptable health problems, making them unviable freaks.

Some applications to humans are already in place. A particularly promising one is "gene therapy," in which defective or mutant genes are fixed by genetically engineered viruses that either replace or supplement the defective genetic information. This gives us the potential to cure many disorders of the bone marrow, liver, central nervous system, some kinds of cancer, and hormone imbalances. Gene therapy only involves the reproductive machinery of the cell, and not that of the whole organism, so that changes are not transmitted to the offspring. Another technique, called homologous recombination, makes it possible to replace a defective gene in the reproductive cells, forever altering future generations.

Through techniques such as homologous recombination, we have the capacity to change the human species by eliminating deleterious genes. The evolution of human beings thus comes under conscious human control. Initially these changes will be minor adjustments, such as the elimination of diabetes and other genetic diseases. As we acquire more knowledge of the function and interpretation of the code, we will also acquire more capability to *add* new features, as we already do for bacteria, plants, and even some mammals. Should we choose to exercise the option of making such changes, we may give rise to "human" beings that are quite different from current homo sapiens.

Making alternative organic life-forms from scratch. The quest to discover the origins of life has led to a great deal of speculation about the simplest possible life-forms. Since our record of the earliest life-forms is extremely poor, it is generally agreed that the only experimental test is to recreate life "from scratch" in the laboratory. This forces us to seriously consider the issue of what it means to be alive, and also raises that possibility that, rather than recreating the origin of contemporary life, we might create a whole new evolutionary tree, whose material basis is quite different from that of contemporary life.

The basic building blocks of contemporary life are amino acids, which form proteins, and nucleic acids, which form DNA. As demonstrated by Miller and Orgel,[17] amino acids are easily synthesized under artificial conditions. It is more difficult to make proteins, although Sidney Fox,[9] has demonstrated that it is possible to make similar molecules called "protenoids," which form bacterium-sized protenoid spheres with some suggestively lifelike properties. Nucleotides can also be formed in the laboratory, providing the proper protein enzymes are present.

However, there are still several crucial problems that remain to be solved before we will be able to directly recreate lifelike behavior from non-living material.

3. HOW AND WHEN?

Whether or not we study it as a scientific pursuit, we suspect artificial life will emerge in one form or another for economic reasons. We feel that this is unavoidable because of the economic incentives. The timetable and detailed mechanisms are are still uncertain, but the imperative is quite clear. In any case the implications for our civilization and ecosystem are dramatic.

3.1 HOW WILL IT HAPPEN?

We feel that artificial life will emerge gradually, slowly becoming a part of our day-to-day lives. There are many possible avenues for this; probably many of them will be explored simultaneously. True artificial life will be preceded by a series of stages in which we come closer and closer to the real thing.

It is often the case that technological developments are anticipated, at least in spirit, in speculative fiction. Just as Jules Verne anticipated much of the technology of the twentieth century, many aspects of artificial life that may appear in the twenty-first century and beyond have been anticipated in the fiction and nonfiction of this century. The possibility of self-reproducing machines was anticipated as early as 1929 by J. D. Bernal.[2] More recently the poet laureate of artificial life, Stanislas Lem, has written many books that are populated by a variety of artificial life-forms, and that suggest how and why they might arise. For example, in his recent book *Fiasco*,[16] a group of space explorers searching for life travels to what seems from a distance to be a planet with a ring around it, similar to Saturn. However, on closer inspection the ring turns out to be composed of attack satellites and anti-missile weapons. It originally began as a "star wars" defense shield against land-based nuclear attack. As each side learned to jam the operations of the others' technology, more and more autonomous control was given to the satellites. Since material was difficult to transport into space, they made them self-reproducing. The ring evolved and developed into an ecology of hostile, autonomous organisms, beyond the control of the parental planet. Unfortunately, in view of modern developments, this scenario is all too believable.

More peaceful applications are already beyond the realm of science fiction. For example, NASA sponsored a summer study group to investigate the feasibility of making self-reproducing aluminum mining modules on the moon.[12] The purpose was to design aluminum mining machines capable of mining aluminum, making copies of themselves, and catapulting aluminum into a near-zero gravity orbit between the earth and the moon where it can be used to build a space station. The machines use the aluminum they mine to manufacture replacement parts. Although the initial investment would be large, once the seed machinery is in place, because of the ability to reproduce, the amplification of the initial investment is almost unlimited. The NASA study concluded that this could be accomplished by placing only 100 metric tons of material on the surface of the moon.

Outer space provides a favorable medium for artificial life. Although the conditions in space are hostile to biological organisms, machines do not breathe oxygen, do not require water, are naturally powered by solar energy, and elegantly driven by "solar sails," which employ the solar wind as a motive force. Machines thrive where humans perish. If we ever wish to explore the solar system and make use of the tremendous natural resources that exist outside of earth, self-reproducing machines provide the natural way to accomplish this task. Because of the enormous potential economic returns, self-reproducing machines are likely to emerge as the natural tool for space exploration.

The emergence of artificial life will probably have antecedents on earth that are not as dramatic as self-reproducing aluminum mining modules in outer space. Indeed, we are already coming close to such possibilities. The Macintosh computer, for example, is produced in factories with virtually no human intervention, machines producing other machines. Microchip fabrication is under increasing levels of computer control, from the layout of printed circuit boards to etching of the actual chips. As computers become more sophisticated and more integrated into our lives, and as we become more dependent on them, they will exert more control on us and on themselves. We have already discussed how computer networks form an "agar," fostering the formation of computer viruses. It seems that whenever there is a medium capable of supporting large amounts of specific information, organizational patterns emerge that propagate themselves by taking over the resources of this medium. As our society becomes increasingly information intensive, it automatically acquires increasing potential to support artificial life-forms.

Carbon-based artificial organisms are already a reality. At this point they do not play a major role in our lives, but then genetic engineering is a new technology whose potential has only begun to be explored. As the power of this technology develops, we will inevitably come to rely more and more heavily on genetic engineering to face the problems caused by overpopulation and the limits of our resources. It is only a question of time before we begin to apply genetic engineering to human beings. Elimination of genetic-related diseases will probably occur without a great deal of controversy. But once this is accepted, more controversial measures will begin to be considered. Some changes, while potentially desirable for society, may be very difficult to bring about. For example, we could use genetic engineering to make human beings smaller. Small people take up less space and consume fewer resources, and if we were all significantly smaller, we could support the same number of people and place far less strain on our planet. Nonetheless, who would be the first to volunteer?

A critical point will occur when we acquire the ability to modify the intelligence of our offspring. If this can be done simply and reliably, there will probably be many volunteers. Although the political and social difficulties may be substantial, as our society becomes increasingly complex, the demand for increased intelligence will grow. In a relatively short amount of time, we may find "human" beings that are quite different from current homo sapiens, new generations of men and women as anticipated by Stapleton.[23]

3.2 WHEN WILL IT HAPPEN?

The easy answer to this question is that it has already happened. Computer viruses and genetic engineering are a reality, a tangible demonstration that artificial life is not only a subject for science fiction. However, neither of these are self-sufficient life-forms; both computer viruses and genetically engineered life-forms require human beings to create them. This does not say that they are not alive—there are many natural organisms that cannot exist without other organisms. It merely says that their evolutionary development depends on symbiotic relationships with other parts of the ecosphere.

Before artificial life is achieved on a broader scale, so that it contains all the rich possibilities of natural life, there are still technological developments that need to occur. These developments are significantly different in detail for carbon-based and silicon-based organisms, although the general problems are related.

For carbon-based artificial life-forms, we need a much more comprehensive and efficient capability to read and alter the the genome. Sequencing or "reading" a genome is currently a very labor-intensive task. We have complete sequences for only a few of the most primitive organisms. Nonetheless, technological developments in this area are relatively easy to anticipate, and it seems likely that in the twenty-first century we will be able to read large genomes relatively easily. Similarly, techniques for manipulating the genome, i.e., making specific alterations in the sequence of nucleotides, are developing at a rapid pace, and we can expect that in the twenty-first century this will be a relatively easy matter.

The real limiting factor to the development of carbon-based artificial life is *understanding* the language of the genome, so that we can anticipate the effect of making a given change. This is complicated by the fact that genes do not act independently—their actions are highly dependent on those of other genes. This interdependence makes it very difficult to anticipate the effect of a given change. Solving this problem requires a much more complete understanding of how a living organism functions.

For computer-based life-forms, the needed developments are naturally divided into two areas: hardware and software. Of these, the development of hardware, the raw computational machinery, is much easier to predict. The development of software is analogous to understanding the language of the genome—we need fundamental breakthroughs and a comprehensive understanding, and its development is much more difficult to predict.

We will first examine the development of hardware: Since the advent of computers, our ability to compute has increased at a steady exponential rate. Up until now computational power has increased by a factor of roughly 1000 every 20 years. This implies that by about the year 2030, if we follow the same growth curve, we will have computer hardware roughly a million times as powerful as that we possess now.[19] At this point, we will have computers whose power is roughly comparable to that of the human brain.

It is, of course, difficult to compare the power of the human brain to the power of a computer. Their capabilities are quite different. Roughly speaking, though, the raw hardware power of the human brain can be estimated in terms of the number of neurons and their speed as computational elements. These figures are not known with any precision, but a ball park figure places the number of neurons at 10^{10}, the switching speed of a neuron at 100 bits per second, and the storage capacity at 100 bits per neuron. Using this estimate, and extrapolating the rate of growth of computer technology, we can expect that by about the year 2025, we will have computers with roughly the computational power of a human brain. Our estimate may easily be wrong by a factor of 1000, but as long as the available computational power grows exponentially, this makes only a very small difference in the time for the hardware potential of artificial computers to reach equivalence with the human brain. Even if the estimates of the power of the human brain are off by a factor of 1000, the crossover point still shifts by only 20 years.[3]

In any case, the complexity of the human brain is probably more than that needed for life. The "hardware" that makes up a simple bacterium is certainly far less complex than that of the human brain. Its true complexity is difficult to estimate, but it is quite possible that contemporary computers already have enough hardware power to simulate the essential information-processing functions of a bacterium.

The time for the emergence of software is more difficult to assess, and places a more severe limit on the emergence of artificial life than the development of sufficient hardware. Conventional computer languages and computer programs follow very different principles than those of the brain or of the machinery that controls the cell. The underlying principles behind biological organisms are robust and adaptable. New approaches evolve spontaneously, without conscious intervention. In contrast, conventional computer programs are not robust; they are easily broken by small changes. Spontaneous evolution is difficult.

To create artificial computer-based life that is robust, which can survive fluctuations in its environment and evolve as freely as biological life, we must solve several fundamental problems in the design of computer software. We must make software that is *adaptable*, with learning algorithms that allow computer programs to profit from experience. Ultimately we need computer programs capable of writing other computer programs, with "goal-seeking" behavior that allows programs to function in ill-specified environments. We need computer software that can innovate, and

[3] See Hans Moravec[19] for more detailed treatments of these issues.

add onto itself in response to its "needs." Solving these problems is one of the fundamental goals in the study of artificial life. These are also central problems in the related field of artificial intelligence.

New approaches to artificial intelligence include computer programs that mimic aspects of real biological neurons,[21] and computer programs that alter themselves through "genetic" manipulations very much like those employed by our reproductive machinery.[11] However, we are still lacking several principles needed to build living systems. It is unclear at this stage whether all that is needed are a few broad fundamental theoretical breakthroughs, or whether we still face a long trail of piecemeal and highly specialized discoveries. In the latter case, the timetable for the broad emergence of robust artificial life-forms might be extended significantly.

The advent of computer viruses illustrates the immediacy of artificial computer-based life. Although contemporary computer viruses are not very robust in the face of changes in their programs, they can nonetheless be quite long lived. We believe that the ability to make stable, self-reproducing artificial life-forms only awaits a few conceptual breakthroughs. In this case, artificial life might fully emerge by the middle of the next century.

Note that the development of carbon-based and computer-based life-forms are highly complementary processes. The technology for sequencing and manipulating the genome is highly dependent on computers and developments in computer-based artificial intelligence. Developing an understanding of the language of the genome is likely to be highly dependent on increasingly more sophisticated computer simulations of the functioning of organisms. In turn, this understanding is likely to guide us in developing the principles for computer-based artificial life. And eventually, genetic engineering of more intelligent humans is likely to have an impact on all of these problems.

4. THE BIG PICTURE

4.1 EVOLUTION AND SELF-ORGANIZATION

We are accustomed to thinking of evolution as an explicitly Darwinian phenomenon, specific to biological organisms, involving competing processes of random mutation and natural selection. However, it is possible to take the broader view that biological evolution is just one example of the tendency of matter to organize itself as long as the proper conditions prevail.

This concept of evolution was originally introduced by Herbert Spencer in the mid-nineteenth century.[22] He defined evolution as "a change from an incoherent homogeneity to a coherent heterogeneity." According to Spencer, evolution is a process giving rise to increasing differentiation (specialization of functions) and integration (mutual interdependence and coordination of function of the structurally differentiated parts). He viewed evolution as the dominant force driving the spontaneous formation of structure in the universe, including the formation of matter, stars, geological formations, biological species, and social organizations. Thus, Darwinian evolution is just a special case of a broader principle.

In Spencer's view, evolution is the antagonist of dissolution. His notion of dissolution is essentially what physicists call the second law of thermodynamics. According to the second law, disorder, or *entropy*, tends to increase in the absence of an input of energy. This is an embodiment of the familiar principle that it is easier to make a mess than to clean it up. In nature organized forms of energy such as light or the bulk motion of matter tend to turn into disorganized energy (heat), i.e., disordered atomic motion.

When organized energy streams down onto earth, much of it simply turns into disorder, in the form of heat. However, something else also happens, which seems to be quite the opposite: Processes of differentiation cause oceans, clouds, wind, rain, and geologic formations. These are processes of *organization* rather than *disorganization*. It is not that they disobey the second law of thermodynamics, but rather that the second law does not tell the full story. While there is an overall net increase of disorganization at the molecular level, at higher levels, under favorable circumstances there is an inexorable tendency for an increase of order. Life is, of course, the primary example.

The theory of organization is much less developed than the theory of disorganization. We have a precise formulation of the second law, but at this point, there are no good general theories for self-organization. In its broadest sense, the study of artificial life is an avenue that can help us make a broader theory of evolution more precise. By producing tangible examples of self-organization in simple mathematical models, we hope to understand why nature has an inexorable tendency to organize itself, and to discover the laws under which this process operates.

4.2 LAMARCKIAN *VS.* DARWINIAN EVOLUTION

Viewed in a broad context, the advent of artificial life is significant because it signals the possibility of a major change in the manner in which evolution as a whole takes place. The first such change probably occurred with the creation of the first self-reproducing organisms. Before this the spontaneous formation of structure relied on more indirect processes of self-organization. With self-reproduction it became possible to directly transmit information and patterns from the past to the future. It also made it possible to incrementally change this structure through Darwinian evolution, a process of random mutation and natural selection. Under that process, small changes take place during the process of reproduction, producing random variations in the offspring. If these changes are not favorable then the offspring may die out. If these changes are favorable, however, then the offspring reproduce more frequently, passing these changes on so that they propagate. *Only* genetic information is transmitted. Acquired characteristics, such as good muscles developed through exercise or the wisdom acquired in one's lifetime, cannot be transmitted to subsequent generations directly. Darwinian evolution is the fundamental mechanism that has designed the flora and fauna of earth. Viewed in the broad sense of Spencer, self-reproduction provided a new mechanism for evolution, signaling a major speedup in the rate at which evolution as a whole took place.

An alternate mechanism of biological evolution was postulated by Lamarck. He believed in the transmission of acquired characteristics to subsequent generations. He believed, for example, that if a giraffe stretched its neck and made it longer, then its offspring would have longer necks. We now know that this is not true for biological organisms. However, there is an important context in which it is true: the evolution of culture.

A culture can be viewed as a kind of organism built out of individuals and social units. New ideas in a culture compete for prominence within the culture. These ideas propagate through our modes of communication, largely language and writing. Ideas and their concomitant modes of behavior are selected according to their usefulness to the society, and cultures evolve through the course of time. More successful ideas supplant other ideas, mimicking the survival of the fittest that we associate with biological evolution. In contrast to biological evolution, in social evolution acquired characteristics are passed on to subsequent generations. Cultural evolution is essentially a Lamarckian process.[4]

The capability of cultural evolution for bringing about effective change on a timescale far faster than biological evolution demonstrates the power of Lamarckian evolution. Although cultural evolution occurs on a limited basis in other organisms, such as monkeys and birds, its true potential has been manifested only in humans. The emergence of language, with the attendant amplification in the ability to transmit cultural information, wrought an enormous change in collective human behavior. In a very short period of time, perhaps only fifty thousand years, human culture has given us the ability to send people to the moon, to destroy life on our planet, and to create life. The pace of cultural evolution is strikingly fast when compared to the much slower pace of biological evolution. This is not surprising; change happens much more efficiently when acquired characteristics can be transmitted directly, and when innovation comes as a result of conscious design rather than random guessing.

Viewed in the broad terms of Spencer, the introduction of culture, with its more rapid mechanism of Lamarckian evolution, can be viewed as a watershed event in the history of evolution as a whole—a "phase change" accelerating the global evolutionary process. However, in the absence of biological change the scope and possibilities of cultural change are limited. The human brain is limited in its ability to assimilate the vast quantities of information generated by our culture. Increasingly we turn to tools made specially to help us in these tasks, computer memories that are capable of storing information much more efficiently than we can ourselves. These tools are gradually becoming much more than passive memories, actively performing many of the functions that we would otherwise perform.

Artificial life provides the possibility for Lamarckian evolution to act on *the material composition of the organisms themselves.* Once we can manipulate the

[4] In analogy with genes, Dawkins has characterized the fundamental units of cultural evolution as *memes.*[4]

genome directly, once we understand how the genome is built and can anticipate the effects of changing it, we can modify our offspring according to our perception of their needs. This is true for both the silicon-based genomes of computing machines and the carbon-based genomes of genetically engineered biological organisms. Unlike the original concept of Lamarck, this does not happen automatically, but rather through the intermediary of consciousness. The giraffe's longer neck is not automatically passed on as a result of stretching. Instead, the giraffe realizes that it would be nice if its offspring could have longer necks, and does appropriate genetic engineering to make this happen.

In artificial computer-based life-forms, the genetic material will almost certainly be under direct control, in computer readable and easily modifiable form. Initially, of course, such organisms may not be very smart. The most likely event is that the genomes will be modified by humans, to effect a good design for some commercial purpose. We then have a *symbiotic* Lamarckian evolution, in which one species modifies the genome of another, genetically engineering it for the mutual advantage of both. In a sense, this is what we have done all along with our technology—automobiles, for example, "evolve" as we manipulate their genomes (blueprints). With artificial life there is the potential for the control of the genome to be given to the products of our technology, thus creating self-modifying, autonomous tools. As artificial life-forms achieve higher levels of intelligence, the ability to modify their own genomes will become increasingly more feasible.

Assuming that artificial life-forms become dominant in the far future, this transition to Lamarckian evolution of hardware will enact another major change in the global rate of evolution, comparable to the enormous acceleration that occurred with the advent of culture. The distinction between artificial and natural will disappear. This will be a landmark event in the history of the earth, and possibly the entire universe.

5. THE CONSEQUENCES FOR HUMANITY

The study of artificial life may potentially answer some very important questions in biology and the theory of evolution. It also provides a tool to address some of the most fundamental philosophical questions, such as: What does it mean to be alive? What are the underlying physical and mathematical processes that give rise to life? How does nature spontaneously create order from chaos? What are the mechanisms of creativity and self-organization?

Artificial life-forms will probably emerge whether or not we choose to study artificial life as a scientific discipline. Artificial life-forms have the capacity to evolve beyond contemporary life. At first, they will be quite unsophisticated, simple tools that we have built to satisfy our needs. Ultimately, however, economic and political pressures will drive artificial life-forms to greater degrees of sophistication, until their complexity and information-processing capabilities are comparable or superior to those of humans. This may engender competition with humans.

What should our attitude be? It is natural to fear the unknown, particularly when it involves a possible threat to our species. It is easy to imagine nightmare scenarios in which cold, malevolent machines or vicious genetically engineered creatures overwhelm humanity. Viewed in this way, artificial life becomes a threat to our survival to which we must respond, something that must be eliminated so that human beings can continue to prosper without competition.

We should, however, use care before automatically taking such a view. In the challenges issued for this symposium, Murray Gell-Mann has asked us to address the dangers of "human tribalism." Humanity has traditionally been self-centered, eager to exalt itself and to regard itself as the sublime creation of God, squarely in the center of the universe for the rest of time. We have now evolved somewhat away from this narcissistic view. We now know that we are the inhabitants of an average planet orbiting an average star in an average galaxy. We may also surmise that this moment in cosmic history was arrived at through an evolutionary process of change which will replace us at the next moment.

The natural order of evolution is change. No species has persisted forever. Individual species are altered and replaced through an evolutionary process of modification and succession that continually alters the composition of the flora and fauna of earth. There is no reason to believe that we are immune to this. It seems quite natural that we, too, will evolve and change with the passage of time, giving rise to new species in the genus homo. With artificial life this evolutionary change may not follow such a continuous path; although we give rise to new species, they may be our own direct conscious creations and radically different in form from ourselves.

Another topic that Murray asked us to address in this symposium is the "preservation of cultural and biological diversity." We now have the possibility to *create* cultural and biological diversity. With the advent of artificial life, *we may be the first species to create its own successors.* What will these successors be like? If we fail in our task as creators, they may indeed be cold and malevolent. However, if we succeed, they may be glorious, enlightened creatures that far surpass us in their intelligence and wisdom. It is quite possible that, when the conscious beings of the future look back on this era, we will be most noteworthy not in and of ourselves but rather for what we gave rise to. Artificial life is potentially the most beautiful creation of humanity. To shun artificial life without deeper consideration reflects a shallow anthropocentrism.

But the path is fraught with danger. Short-sighted fear and hatred all too often dominate the activities of human beings. At the outset at least, we will shape the form and innate drives of artificial organisms. A particularly frightening scenario comes from the potential military uses of artificial life. There are many military applications for which artificial life-forms would be extremely useful, from battlefield robots to satellite warfare. We can only hope that we have the collective wisdom to make treaties and suppress such applications before they occur. As we have seen with nuclear weapons, political forces make the consensus necessary to dismantle existing weapons systems extremely difficult to achieve. Once self-reproducing war machines are in place, even if we should change our mind and establish a consensus, dismantling them may become impossible—they may be literally out of our control. An escalated technological war involving the construction of artificial armies would likely end by destroying the participants themselves, and would give rise to a generation of life-forms that might be even more hostile and destructive than their human ancestors.

Artificial life-forms will be shaped by the forces that create them. If, instead of building war machines, we use our technology for productive purposes, they may bring us a wealth of resources that will greatly enhance our well-being. Ultimately, they may evolve to be far more intelligent than humans, and capable of intellectual feats that we cannot even dream of. Such intelligence might result in enlightened behavior that is inconceivable to lower forms of life such as us.

If we can shape artificial life in a positive direction, the bittersweet consequences to humanity can be visualized by analogy to Arthur C. Clarke's book, *Childhood's End.*[3] In this story he imagines that the children on earth acquire the ability of mental telepathy. This ability makes them into an enlightened race whose collective powers are far greater than those of ordinary humans. However, as a result, they are so beyond their parents that they become strangers to them. Their parents are left with feelings of glory for the harmony and greatness of what they see their children will accomplish, but simultaneously they feel sadness that they cannot participate.

In discussing artificial life here we have been intentionally provocative. Our vision of the future may not be accurate. We hope that, whether or not you agree with us, you will be stimulated to address this issue. If we are right the advent of artificial life is the greatest challenge facing humanity, an inevitability that we must shape and set in motion in the proper direction. If the future is to do justice to the nobler attributes of humanity, then we must take positive action.

ACKNOWLEDGMENTS

We appreciate valuable comments and criticism from Christian Burks, Walter Fontana, John Holland, Chris Langton, Norman Packard, and Steen Rasmussen.

The "Bizarro" cartoon by Dan Piraro is reprinted by permission of Chronicle Features, San Francisco, CA.

REFERENCES

1. Bagley, R. J., J. D. Farmer, S. A. Kauffman, N. H. Packard, A. S. Perelson, and I. M. Stadnyk. *Modeling Adaptive Biological Systems.* Technical Report LA-UR-89-571, Los Alamos National Laboratory, 1989. To appear in *Biosystems.*
2. Bernal, J. D. "title." In *The World, The Flesh, and The Devil,* edited by E. P. Dutton. 1929.
3. Clarke, A. C. *Childhood's End.* San Diego: Harcourt Brace Jovanovich, 1963.
4. Dawkins, R. *The Selfish Gene.* Oxford: Oxford University Press, 1976.
5. Dawkins, R. "The Evolution of Evolvability." In *Artificial Life,* edited by C. Langton, 201. Santa Fe Institute Studies in the Sciences of Complexity Proc. Vol. VI. Redwood City, CA: Addison-Wesley, 1989.
6. Dewdney, A. K. "Computer Recreations: In the Game Called Core War Hostile Programs Engage in a Battle of Bits." *Sci. Amer.* **250(5)** (1984): 14–22.
7. Farmer, J. D., S. A. Kauffman, and N. H. Packard. "Autocatalytic Replication of Polymers." *Physica D* **22** (1986): 50–67.
8. Farmer, J. D. "A Rosetta Stone for Connectionism." *Physica D* **42** (1990): 153–187.
9. Fox, S. W., K. Harada, and J. Kendrick. "Production of Spherules from Protenoids and Hot Water." *Science* (1959): 129.
10. Gardner, M. "Mathematical Games: The Fantastic Combinations of John Conway's New Soliaire Game 'Life.'" *Sci. Amer.* **223(4)** (1970): 120–123.
11. Holland, J. H. "Escaping Brittleness: The Possibilities of General Purpose Learning Algorithms Applied to Parallel Rule-Based Systems." In *Machine Learning II,* edited by R. S. Mishalski, J. G. Carbonell, and T. M. Mitchell, 593–623. Morgan-Kaufman, 1986.
12. Laing, R. "Replicating Systems Concepts: Self-Replicating Lunar Factory and Demonstration." In *Advanced Automation for Space Missions,* edited by R. Freitas and W. P. Gilbreath, 189–335. NASA Conference Publication 2255, 1982.
13. Langton, C. G. "Studying Artificial Life with Cellular Automata." *Physica D* **22** (1986): 120–149.
14. Langton, C. G. ed. *Artificial Life.* Santa Fe Institute Studies in the Sciences of Complexity Proc. Vol. VI. Redwood City, CA: Addison-Wesley, 1989.
15. Lem, S. "The Experiment." *The New Yorker,* 1978.
16. Lem, S. "title." *Fiasco.* San Diego: Harcourt Brace Jovanovich, 1986.
17. Miller, S. L., and L. E. Orgel. *Science* **30** (1959).
18. Monod, J. *Chance and Necessity.* New York: Knopf, 1971.
19. Moravec, H. *Mind Children: The Future of Robot and Human Intelligence.* Cambridge: Harvard University Press, 1988.
20. Rasmussen, S., R. Feldberg, M. Hindsholm, and C. Knudsen. *Core Evolution: Development of Assembler Automata in the Computer Memory.* Technical report, Los Alamos National Laboratory, 1989. To appear in *Physica D.*

21. Rummelhart, D., and J. McClelland. *Parallel Distributed Processing*, vol. I. Cambridge: MIT Press, 1986.
22. Spencer, H. *First Principles*. 1962.
23. Stapleton, O. *Last and First Men*. 1931.

Index

A

B

D

E

F

G

O

P

Q

R

S

W

X

B

Z

The Addison-Wesley **Advanced Book Program** and the SANTA FE INSTITUTE would like to offer you the opportunity to learn about our new "Studies In the Sciences of Complexity" titles and workshops in advance. To be placed on our mailing list and receive pre-publication notices and special offers, just **fill out this card completely** and return to us.

Title, Author, and Code # of this book: **Date purchased:**

______________________________ ______________

Name ______________________________

Title ______________________________

School/Company ______________________________

Department ______________________________

Street Address ______________________________

City ____________________ State __________ Zip __________

Telephone (   ) ______________________________

Where did you buy this book?

☐ Bookstore
☐ Mail Order
☐ School (Required for Class)
☐ Campus Bookstore (individual Study)
☐ Toll Free # to Publisher
☐ Professional Meeting
☐ Publisher's Representative
☐ Other ______________________________

Please define your primary professional involvement:

☐ Academic: Professor
☐ Academic: Student
☐ Academic: Researcher
☐ Industry: Administrator
☐ Industry: Researcher
☐ Industry: Technician
☐ Government: Administrator
☐ Government: Researcher
☐ Government: Technician

Check your areas of interest.

200 ☑ SFI

201 ☐ Agriculture
202 ☐ Anthropology
203 ☐ Artificial Intelligence
204 ☐ Astronomy
205 ☐ Atmospheric Sciences
206 ☐ Biological Sciences
207 ☐ Chemistry
208 ☐ Computer Sciences
209 ☐ Communication Sciences
210 ☐ Dentistry
211 ☐ Economics
212 ☐ Education
213 ☐ Engineering
214 ☐ Geology/Geography
215 ☐ History/Philosophy Science
216 ☐ Industrial Science
217 ☐ Information Sciences
218 ☐ Mathematics
219 ☐ Medical Sciences
220 ☐ Pharmaceutical Sciences
221 ☐ Physics
222 ☐ Political Sciences
223 ☐ Psychology
224 ☐ Social Sciences
225 ☐ Statistics
226 ☐ OTHER ______________________________
(please specify)

Of which professional scientific associations are you an active member?

________ ________ ________ ________ ________ ________

________ ________ ________ ________ ________ ________

Would you like to be sent information about the SANTA FE INSTITUTE and its workshops?

☐ Yes ☐ No

fold and staple

No Postage
Necessary
if Mailed in the
United States

BUSINESS REPLY MAIL
FIRST CLASS PERMIT NO. 828 REDWOOD CITY, CA 94065

Postage will be paid by Addressee:

ADDISON-WESLEY PUBLISHING COMPANY, INC.®

Advanced Book Program
350 Bridge Parkway
Redwood City, CA 94065-1522